The Biophysical Chemistry of
Nucleic Acids & Proteins

Thomas E. Creighton

Helvetian Press

© Thomas E. Creighton 2010

Published by Helvetian Press 2010

www.HelvetianPress.com

HelvetianPress@gmail.com

All rights reserved. Reproduction of this book by photocopying or electronic means for non-commercial purposes is permitted. Otherwise, no part of this book may be reproduced, adapted, stored in a retrieval system or transmitted by any means, electronic, mechanical, photocopying, or otherwise without the prior written permission of the author.

ISBN 978-0-9564781-1-5

CONTENTS

Preface xix

Common Abbreviations xxi

Glossary xxv

Section I: Macromolecules

1. **Configurations and conformations** 1
   1.1. Stereochemistry 2
      1.1.A. Chirality 3
         1. Enantiomers 5
         2. Racemic mixtures 6
         3. Diastereomers 7
         4. Epimers and epimerization 8
         5. *Cis* and *trans* isomers 9
      1.1.B. Prochiral 10
      1.1.C. Tautomers 12
   1.2. Conformations 13
      1.2.A. Torsion angle 15
      1.2.B. Dihedral angle 16
   1.3. Conformations of idealized polymers 17
      1.3.A. Random coils 18
         1. End-to-end distances 19
         2. Radius of gyration 21
         3. Characteristic ratio 21
      1.3.B. Excluded volume effects and theta solvents 22
         1. Covalent cross-links 23
   1.4. Structure databases: structures on the WEB 24

Section II: Nucleic acids

2. **DNA structure** 25
   2.1. Polynucleotides 26

2.1.A. The deoxyribose group   29
2.1.B. Properties of the bases   31
2.1.C. Modifications of the bases   37
2.2. DNA three-dimensional structures   40
    2.2.A. Base pairing and stacking   42
    2.2.B. Double helices   46
        1. B-DNA   52
        2. A-DNA   53
        3. Z-DNA   54
    2.2.C. Other DNA structures   56
        1. Hoogsteen base pairs   56
        2. Triple helices   57
        3. H-DNA: intramolecular triple helices   60
        4. Four-stranded structures: guanine quartet   60
        5. i- motif   63
        6. Inverted repeat sequences and palindromes   64
        7. Helical junctions: cruciforms, Holliday junctions   66
        8. Parallel-stranded DNA duplexes   66
2.3. DNA as a polyelectrolyte: hydration and counterions   68
2.4. DNA flexibility and dynamics: curving, twisting, stretching   71
    2.4.A. Local flexibility   74
    2.4.B. Hydrogen exchange   75
2.5. Binding of small molecules   77
    2.5.A. Binding to the minor groove   77
    2.5.B. Intercalation   78
        1. Ethidium bromide   78
        2. Psoralen photo cross-linking   79
2.6. Chemical modification as a probe of structure   80
    2.6.A. Cross-linking   85

3. DNA topology   86

3.1. Supercoiling and superhelices: topoisomers   90
3.2. Linking number, Lk   92
    3.2.A. Linking difference, ΔLk 92
        1. Relaxed duplex DNA   93
    3.2.B. Superhelix density, σ 94
3.3. Topoisomerases   95
3.4. Twist and writhe   96
    3.4.A. Twist, Tw   96
    3.4.B. Writhe, Wr   97
3.5. DNA topology and geometry   98
    3.5.A. Experimental characterization of DNA topology   99
        1. Electron microscopy   99
        2. Gel electrophoresis to separate topoisomers   101

3. Intercalation by ethidium bromide	101
4. Two-dimensional gel electrophoresis	101
3.6. Energetics of supercoiling	103
3.6. B. Energy distribution of topoisomers	104
3.6. C. Topology-dependent binding of ligands	105
3.7. DNA wrapped around the nucleosome	106

4. RNA structure — **108**
- 4.1. Secondary structure of RNA — 112
  - 4.1.A. Hairpin loops — 116
  - 4.1.B. Tetraloops — 117
  - 4.1.C. Bulges and internal loops — 118
- 4.2. Tertiary structure of RNA — 118
  - 4.2.A. Common structural motifs — 120
    1. Pseudoknots — 120
    2. Coaxial helices: interhelical stacking — 121
    3. A-minor motif — 122
    4. Dinucleotide platform — 123
    5. "Kissing" hairpin loops — 123
    6. Ribose zipper — 124
    7. Uridine turn — 125
    8. Tetraloop/receptor interactions — 126
    9. Roles of ions — 126
  - 4.2.B. Transfer RNA structures — 127
  - 4.2.C. Ribozyme structures — 129
- 4.3. Quaternary structure of RNA — 133
- 4.4. RNA structure prediction — 134
  - 4.4.A. Prediction of secondary structure — 134
    1. Thermodynamic approach — 134
    2. Phylogenetic approach — 135
  - 4.4.B. Prediction of tertiary structure — 136

5. Denaturation, renaturation, and hybridization of nucleic acids — **139**
- 5.1. Denaturation of double-stranded nucleic acids — 139
  - 5.1.A. Methods for monitoring denaturation — 140
  - 5.1.B. Double-stranded DNA — 142
    1. Thermal melting — 143
    2. Denaturants — 144
    3. pH — 145
    4. Salt effects — 145
    5. Prediction of the T_m — 146
  - 5.1.C. Double-stranded RNA — 148
  - 5.1.D. DNA • RNA heteroduplexes — 149
  - 5.1.E. Single-stranded nucleic acids — 149
    1. Physical stretching — 150
- 5.2. Unfolding and refolding of single-stranded RNA molecules — 150

5.2.A. Transfer RNA unfolding/refolding	153
5.2.B. Ribozyme unfolding/refolding	154
5.2.C. Unfolding using mechanical force	155
5.3. Renaturation, annealing, and hybridization	157
5.3.A. Competing intramolecular structures in individual single strands	162
5.3.B. C_0t and R_0t curves	163
5.3.C. Probe hybridization	165
1. Stringency	167
2. Analyzing the extent of complementarity	168
3. *In situ* hybridization	170
5.4. DNA mimics: peptide nucleic acids	170
5.4.A. Chemistry and synthesis	172
5.4.B. Hybridization properties	172
1. PNA•DNA and PNA•RNA duplexes	173
2. $(PNA)_2$•DNA triplexes	173
3. Strand invasion: binding to double-stranded DNA	173
4. PNA duplexes and triplexes	175
5.4.C. Structures of PNA complexes	175
6. Manipulating nucleic acids	**177**
6.1. Replicating DNA	177
6.1.A. DNA polymerase	178
6.1.B. DNA ligase	181
6.1.C. Polymerase chain reaction (PCR)	182
6.2. Producing RNA	185
6.2.A. RNA replication: RNA replicases	185
6.2.B. Transcription: DNA-dependent RNA polymerases	186
1. Single-subunit phage DNA-dependent RNA polymerases	188
6.2.C. Reverse transcription: RNA into DNA	189
6.2.D. Antisense oligonucleotides	190
6.3. Cloning	191
6.3.A. Expression vectors	193
6.3.B. cDNA libraries	194
6.3.C. Restriction enzymes	195
6.3.D. Restriction maps	196
6.4. Sequencing DNA	198
6.4.A. Isolating the DNA fragments to be sequenced	199
6.4.B. Chain-termination, Sanger method	199
6.4.C. Separating the DNA fragments by size	203
6.4.D. Alternative approaches	204
6.5. Sequencing RNA	205
6.5.A. Direct sequencing of oligoribonucleotides	205
6.5.B. Identifying modified nucleotides	207
6.6. Chemical synthesis of DNA	208
6.6.A. Protecting groups for 2´-deoxynucleosides	211
6.6.B. Coupling methods	211

1. Phosphotriester procedure	211
2. Phosphoramidite procedure	212
3. *H*-Phosphonate procedure	213
6.6.C. Solution-phase DNA synthesis	215
6.6.D. Solid-phase DNA synthesis	216
6.6.E. Site-directed mutagenesis	218
6.7. Chemical synthesis of RNA	220
6.7.A. Protecting the 2′-hydroxyl group	220
6.7.B. RNA synthesis in solution	222
6.7.C. Solid-phase RNA synthesis	223

Section III: Proteins

7. Polypeptide structure   **227**

7.1. Polypeptide chains	227
7.2. Amino acid residues	229
7.2.A. Glycine (Gly)	230
7.2.B. Nonpolar amino acid residues (Ala, Leu, Ile, Val)	232
7.2.C. Hydroxyl residues (Ser, Thr)	232
7.2.D. Arginine (Arg)	233
7.2.E. Lysine (Lys)	234
1. Acetylation by anhydrides	236
2. Amidination	236
3. Guanidination	237
4. Schiff base formation	238
5. Carbamylation	239
7.2.F. Histidine (His)	239
7.2.G. Acidic residues (Asp, Glu)	241
7.2.H. Amide residues (Asn, Gln)	242
1. Deamidation	243
7.2.I. Cysteine (Cys)	245
1. Alkylation of thiol groups	245
2. Thiol addition across double bonds	246
3. Binding of metal ions	247
4. Oxidation of thiol groups	247
5. Disulfide bonds	248
6. Thiol-disulfide exchange	249
7. Dithiothreitol, dithioerythritol	252
8. Ellman's reagent	253
7.2.J. Methionine (Met)	254
7.2.K. Phenylalanine (Phe)	255
7.2.L. Tyrosine (Tyr)	256
7.2.M. Tryptophan (Trp)	257
7.2.N. Imino acid (Pro)	258
7.2.O. Selenocysteine (Sec)	259
7.2.P. Physical properties and hydrobicities of amino acid residues	260

1. Hydrophilicities	261
2. Hydrophobicities	261
7.3. Protein detection	267
7.3.A. Biuret reaction	267
7.3.B. Lowry assay	268
7.3.C. Ninhydrin	268
7.3.D. Fluorescamine	269
7.3.E. Coomassie brilliant blue	270
7.3.F. Ponceau S	271
7.4. Peptide synthesis	271
7.4.A. Chemistry of polypeptide chain assembly	272
1. Chemical ligation of peptide fragments	275
7.4.B. Solution or solid phase?	276
7.4.C. Peptide libraries	278
7.5. Peptide and protein sequencing	280
7.5.A. Amino acid analysis	280
1. Peptide bond hydrolysis	281
2. Quantifying amino acids	282
3. Counting residues	283
7.5.B. Fragmentation of a protein into peptides	285
1. Proteolytic enzymes	285
2. Chemical methods of cleavage	288
7.5.C. Peptide mapping	289
7.5.D. Diagonal maps	291
1. Isolating peptides containing certain amino acids	291
2. Identifying disulfide bonds	292
7.5.E. Sequencing	293
1. Amino-terminal and carboxyl-terminal residues	294
2. Sequencing from the N-terminus: the Edman degradation	295
3. Sequencing from the C-terminus	298
4. Sequencing by mass spectrometry	299
7.5.F. Protein sequences from gene sequences	301
1. Post-translational modifications	302
7.6. Primary structures of natural proteins: evolution at the molecular level	306
7.6.A. Homologous genes and proteins.	307
1. Detecting sequence homology	310
2. Aligning homologous sequences	313
3. Orthologous / paralogous genes and proteins	314
4. Nature of amino acid sequence differences	315
5. Rates of divergence	318
6. Roles of selection.	321
a. Neutral mutations and negative selection	322
b. Positive selection for functional mutations	323
7.6.B. Gene rearrangements and the evolution of protein complexity.	324
1. Gene duplications	324

2. Protein elongation by intra-gene duplication	325
3. Gene fusion and division	326
7.6.C. Protein engineering	326

8. Polypeptide conformation — 329
    8.1. Local flexibility of the polypeptide backbone: the Ramachandran plot — 329
    8.2. Random-coil polypeptide chains — 334
        8.2.A. Statistical properties — 334
        8.2.B. Rates of conformational change — 336
            1. X-Pro peptide bond *cis/trans* isomerization — 337
    8.3. Regular structures — 338
        8.3.A. α-Helix — 339
            1. 3_{10}- and Π-helices — 342
        8.3.B. β-Sheet — 342
        8.3.C. Polyglycine — 344
        8.3.D. Polyproline — 344
    8.4 α-Helix formation from a random coil — 346
        8.4.A. Factors stabilizing the α-helix — 347
        8.4.B. Helix-coil transitions — 348
        8.4.C. Helix-coil models — 351
            1. Zimm-Bragg model — 351
            2. Lifson-Roig model — 352
        8.4.D. Trifluoroethanol — 353
    8.5. Fibrous proteins — 353
        8.5.A α-Fibrous structures: coiled coils — 354
        8.5.B. β-Fibrous structures — 358
        8.5.C. Collagens — 359

9. Protein structure — 362
    9.1. Three-dimensional structures of globular proteins: molecular complexity — 363
        9.1.A. Tertiary structure: the overall fold — 366
        9.1.B. Secondary structure: regular local structures — 369
            1. Helices — 371
            2. β-Structure — 372
            3. β-Bulge — 374
        9.1.C. Reverse turns: changing direction — 375
            1. β-Turns. — 376
            2. γ-Turns. — 377
            3. Omega loops — 378
        9.1.D. Supersecondary structures: common motifs — 378
            1. β-Hairpin and β-meander — 379
            2. β-helix and β-roll: β-solenoids — 380
            3. Cystine knot — 381
            4. Greek key motif — 381
            5. Jelly roll motif — 382
            6. Four-helix bundle — 383

7. Epidermal growth factor (EGF) motif	384
8. Interleukin-1 motif: trefoils	384
9. Kringle domain	385
9.1.E. Contact Maps	386
9.1.F. Interiors and exteriors.	388
9.1.G. The solvent: interactions with water	393
9.1.H. Quaternary structure: initiating macromolecular assembly	394
1. Symmetry	398
2. Asymmetry and approximate symmetry	400
3. Interfaces	402
4. Oligomerization and domain swapping	403
5. Filamentous arrays	405
9.1.I. Classification of tertiary structures: order out of diversity	405
1. α Structures	405
2. β Structures	406
3. αβ Structures	407
a. TIM, $(\alpha/\beta)_8$ barrels	407
4. α + β and other structures	408
5. Protein structure classification databases	408
9.2. Membrane proteins: avoiding water	409
9.2.A. Helical superfolds	411
9.2.B. β-Barrel membrane proteins	413
9.2.C. Monotopic and bitopic membrane proteins	414
9.2.D. Interactions with the membrane	414
9.3. Proteins with similar folded conformations: evolution in 3-D	415
9.3.A. Homologous proteins: protein families	415
1. Structural homology within a polypeptide chain	421
9.3.B. Structural similarity without apparent sequence homology: surprises	421
9.3.C. Sequence similarity without structural homology: new folds?	422
9.4. Protein structure prediction	424
9.4.A. *Ab initio* predictions: the ultimate goal	425
9.4.B. Secondary structure prediction: a one-dimensional problem	426
1. Identifying transmembrane helices: hydropathy	427
9.4.C. Homology modeling	429
1. Inverse folding problem	430
2. Threading protein sequences	430
9.4.D. *De novo* protein design	432
10. Physical properties of folded proteins	**434**
10.1. Solubilities and volumes of proteins in water	436
10.1.A. Hydration layer	437
10.1.B. Partial volumes	438
10.2. Chemical reactivities	441
10.2.A. Ionization: electrostatic effects	445
10.3. Isotope (hydrogen) exchange	447

10.3.A. Exchange in macromolecules	447
10.3.B. Solvent penetration model	449
10.3.C. Local unfolding mechanism	449
1. EX1 mechanism	450
2. EX2 mechanism	450
10.4. Flexibility detected crystallographically	451
10.4.A. Effects of different crystal lattices	451
10.4.B. The temperature factor: mobility or disorder?	452
10.5. Flexibility detected by NMR	453
10.5.A. NMR time scales	453
10.5.B. Aromatic ring flipping	454
10.6. Varying the temperature	456
10.7. Effects of high pressure	457
10.7.A. Adiabatic compressibility	457
10.7.B. Isothermal compressibility	458
10.7.C. Structural effects of high pressure	458
10.8. Spectral properties	459
10.9. Integral membrane proteins	460
11. Protein denaturation: unfolding and refolding	**462**
11.1. Reversible unfolding at equilibrium	463
11.1.A. Reversibility of denaturation	463
11.1.B. Cooperativity of unfolding	465
11.1.C. Denaturants	468
11.1.D. Heat denaturation	472
11.1.E. Cold denaturation	474
11.1.F. pH denaturation	476
11.1.G. Denaturation by high pressure	478
11.1.H. Breakage of disulfide bonds	480
11.2. Unfolded proteins	480
11.2.A. Molten globule	484
11.2.B. Conformational equilibria in polypeptide fragments	486
11.3. Protein stability	488
11.3.A. Physical basis of protein stability	491
1. Water and co-solvents	496
a. Stabilizers	496
b. Destabilizers	497
c. The role of water	497
11.3.B. Effects of varying the primary structure	498
1. Natural proteins of exceptional stability	498
2. Mutagenic studies	499
11.3.C. Structural stability of membrane proteins	503
11.4. Protein refolding *in vitro*	504
11.4.A. Refolding of single-domain proteins	505
1. Characterizing the transition state for folding	507
2. Kinetic schemes for folding	509

11.4.B. Kinetic determination of folding	510
1. Bacterial proteinases	511
2. Serpin proteinase inhibitors	512
11.4.C. Folding coupled to disulfide formation	513
11.4.D. Proteins with multiple domains	516
11.4.E. Proteins with multiple subunits	516
11.4.F. Competition with aggregation and precipitation	517
11.4.G. Differences with folding *in vivo*	518

Section IV: Functions

12. Ligand binding by proteins — 519

12.1. General properties of protein-ligand interactions	520
12.2. Metalloproteins	525
12.2.A. Chelation: synergy between ligands	527
12.2.B. Zinc-binding proteins	529
12.2.C. Metallothioneins	531
12.2.D. Iron-transport and storage proteins	532
12.2.E. Blue-copper proteins	534
12.3. Calcium-binding proteins	535
12.3.A. EF-hand calcium-binding proteins	536
1. Calmodulin and troponin C	538
12.3.B. Carboxylation and hydroxylation of Asp, Asn and Glu residues	539
12.4. NAD- and nucleotide-binding proteins	540
12.4.A. Dinucleotide binding motif	541
12.4.B. Mononucleotide-binding motif	543
12.5. Allostery: interactions between different binding sites	543
12.5.A. Structural models	544
1. Sequential model: direct interactions	544
2. Concerted model: quaternary structure changes	545
3. Comparison of the sequential and concerted models	546
12.5.B. Hemoglobin and myoglobin	547
1. Structure	548
2. Oxygen binding	549
3. Cooperativity of oxygen binding	552
4. Heterotropic interactions	554
5. Bohr effect	555
6. Allosteric mechanism of hemoglobin	557
12.5.C. Negative cooperativity	560
1. Negative cooperativity or heterogeneity of sites?	562

13. Nucleic acid/protein interactions — 564

13.1. Techniques for measuring protein-DNA interactions	566
13.1.A. Filter-binding assays	567
13.1.B. Gel retardation assay	568
13.1.C. Footprinting	570
13.2. Principles of protein-DNA recognition	572

13.2.A. Specificity of DNA-protein binding	573
1. Specific interactions	576
2. Nonspecific complexes	578
3. Water-mediated contacts	579
4. Dehydration effects	579
5. Release of condensed counterions	581
13.2.B. Changes in the protein conformation	582
13.2.C. Changes in the DNA conformation	582
13.3. DNA-binding structural motifs	585
13.3.A. Helix-turn-helix motif	587
1. Lac repressor	590
2. Lambda CI and Cro repressors	591
3. Homeodomains	592
4. POU domains	593
5. Trp repressor	595
6. Cyclic AMP receptor protein (CRP) / Catabolite gene activator protein (CAP)	596
13.3.B. TATA-binding protein (TBP)	598
13.3.C. Zinc-containing DNA-binding motifs	599
1. Zinc fingers	599
2. Steroid hormone receptors	602
3. GAL4 type	603
13.3.D. bZip and helix-loop-helix domains	604
13.3.E. β–Sheets: Methionine repressor	605
13.3.F. Histone fold	607
13.3.G. Bacterial type-II DNA-binding proteins: heat-unstable (HU) and integration host factor (IHF)	608
13.3.H. Single-strand DNA-binding proteins	609
1. Prokaryotic singe-strand DNA binding proteins	609
2. Eukaryotic replication protein A	610
3. OB (oligonucleotide/oligosaccharide binding) fold	611
13.4. RNA-binding proteins	613
13.4.A. Ribonucleoprotein (RNP) domain	616
13.4.B. Double-stranded RNA-binding domain	616
13.4.C. KH domain	618
13.4.D. MS2 bacteriophage coat protein	619
13.4.E. Recognizing transfer RNAs	620
1. Class-I glutaminyl-tRNA synthetase	621
2. Class-II aspartyl-tRNA synthetase	622
3. Class-II seryl-tRNA synthetase	623
4. Elongation factor EF-Tu	624
13.4.F. The ribosome	625
14. Catalysis	**628**
14.1. Chemical catalysis	630

14.2. Enzyme kinetics: Michaelis-Menten	630
14.2.A. The Michaelis-Menten equation	633
14.2.B. K_m (Michaelis constant)	635
14.2.C. Turnover number (k_{cat})	635
14.2.D. Lineweaver-Burke plot	635
1. Eadie-Hofstee plot	636
2. Lineweaver-Burke *versus* Eadie-Hofstee	637
14.2.E. Kinetics of individual enzyme molecules	637
14.3. Enzyme kinetic mechanisms with multiple substrates	638
14.3.A. Sequential mechanisms	639
1. Ordered mechanisms	640
2. Random mechanisms	640
14.3.B. Non-sequential mechanisms: Ping-Pong	640
14.3.C. Initial rate equations	641
1. Steady-state ordered and rapid-equilibrium mechanisms	641
2. Equilibrium ordered mechanism	643
3. Ping-pong mechanism	643
14.3.D. Dead-end inhibitors	645
14.3.E. Competitive inhibition	646
1. Linear competitive inhibition	647
2. Hyperbolic competitive inhibition	649
14.3.F. Noncompetitive inhibition	649
1. Linear noncompetitive inhibition	649
2. Hyperbolic noncompetitive inhibition	650
14.3.G. Uncompetitive inhibition	651
14.3.H. Substrate inhibition	652
14.3.I. Product inhibition	654
14.3.J. Haldane relationship	655
14.3.K. Isotope exchange at equilibrium	656
1. Ping-pong mechanism	657
2. Sequential reactions	658
14.3.L. Slow- and tight-binding enzyme inhibitors	659
14.4. Mechanisms of enzyme catalysis	662
14.4A. Reactions on the enzyme	666
14.4.B. Stabilizing the transition state	668
1. Transition state analogues	669
14.4.C. Entropic contributions	672
14.4.D. Bisubstrate analogues	675
1. Ap_5A	675
2. *N*-phosphonacetyl-L-aspartate (PALA)	676
14.4.E. Induced fit	677
14.4.F. Covalent catalysis	680
14.4.G. Cofactors, coenzymes, and prosthetic groups	681
1. Pyridoxal phosphate	681
14.4.H. Suicide substrates	683

14.4.I. Cryoenzymology	685
14.4.J. Time-resolved crystallography	685
14.4.K. Polymeric substrates: processivity	686
14.4.L. Enzyme function *in vivo*: toward "perfection"	688
14.4.M. One example: tyrosyl tRNA synthetase	691
1. Editing of amino acid activation	695
14.5. Catalytic antibodies	696
14.6. Catalytic nucleic acids: ribozymes and deoxyribozymes (DNAzymes).	698
14.6.A. Natural reactions	698
14.6.B. Ribozyme structure and catalysis	701
14.6.C. Selection for novel ribozymes and deoxyribozymes	702
14.6.D. Ligand-binding nucleic acids: aptamers	704
15. Enzyme regulation	**706**
15.1. Allosteric enzymes	707
15.1.A. Allosteric models	708
15.1.B. Structural aspects	709
15.1.C. Aspartate transcarbamoylase	710
15.1.D. Phosphofructokinase	717
1. Mechanism of phosphoryl transfer	718
2. Allosteric properties of *E. coli* PFK	719
15.1.E. Threonine synthase	720
15.2. Covalent regulation	721
15.2.A. Phosphorylation	721
1. Phosphorylation in eukaryotes	722
2. Phosphorylation in prokaryotes	724
3. Protein phosphorylation in signal transduction networks	725
4. Specificity of protein phosphorylation	725
5. Effects of phosphorylation on the properties of proteins	726
6. Methods to characterize protein phosphorylation	727
15.2.B. Glycogen phosphorylase	728
1. Regulation by phosphorylation	731
2. Allosteric properties of glycogen phosphorylase	733
15.2.C. Adenylylation	735
1. Glutamine synthetase	736
15.2.D. Proteolysis: turning zymogens into proteinases	738
1. Trypsin family of serine proteinases	738
2. Carboxyl proteinases	740
3. Metalloproteinases	742

~ PREFACE ~

The field of molecular biology continues to be the most exciting and dynamic area of science and is predicted to dominate the 21st century. Only by investigating biological phenomena at the molecular level is it possible to understand them in detail. Such understanding is vital for advances in medicine, and the pharmaceutical industry that produces new drugs and cures is greatly dependent upon molecular biology. But molecular biology also contributes to the understanding of what human beings are and how they fit into this universe.

This volume builds on its companion volume, *The Physical and Chemical Basis of Molecular Biology*. It will be most intelligible and useful if the reader is aware of the information in that volume.

Proteins and nucleic acids are the primary subjects of molecular biology. They carry, transmit, and express the genetic information that defines each living organism. It is vital to understand how these molecules function.

The first chapter is an introduction to the covalent structures and conformations of macromolecules. The next four chapters deal with the nucleic acids. The structural and chemical properties of DNA are the basis of its central role in storing and transmitting the genetic information (Chapter 2). DNA molecules tend to be immensely long, equivalent to a rope that is many kilometers long, which gives them special topological properties that must be accommodated (Chapter 3). The structure of RNA differs from DNA only very slightly, but this gives it remarkably different properties and functions (Chapter 4). The abilities of individual strands of DNA and RNA to base-pair with other strands with complementary nucleotide sequences are central to many techniques of molecular biology and increasingly to molecular medicine (Chapter 5). The ability to manipulate nucleic acids is central to molecular biology and described in Chapter 6.

The next six chapters deal with proteins, starting with the chemical properties of polypeptide chains and the implications of their covalent structures (Chapter 7). The conformational properties of polypeptides determine the structures that proteins can adopt (Chapter 8), to produce three-dimensional structures of incredible diversity and amazing functional properties (Chapter 9). Proteins in solution have very important dynamic properties that are crucial for their biological activities (Chapter 10). They also have a propensity to lose their folded structures and unfold, and how proteins do this and how they manage to fold to their native three-dimensional structure remains a major question (Chapter 11).

The final four chapters describe the most fundamental functional properties of proteins and nucleic acids. Central to the functions of proteins is their interactions with other molecules (Chapter 12).

Some of the physiologically most important interactions are those between proteins and nucleic acids (Chapter 13). The most impressive and important property of proteins and nucleic acids is their ability of catalyze the rates of chemical reactions by many orders of magnitude, and usually incredibly specifically (Chapter 14). Such potent chemical capabilities must be controlled very closely (Chapter 15).

The references listed were chosen to be those that would best provide the interested reader with entry to the literature. They should not be assumed to be those most important for the subject.

No one person can be expert in all the areas of molecular biology, so I have made ample use of the work of many others more expert than me, but too numerous to specify. Very special thanks are due to Eric Martz of the University of Massachusetts for making available the program Firstglance in Jmol (http://firstglance.jmol.org). It is incredibly useful for examining protein structures and, at least as important, is very easy to use.

Of course, shortcomings and errors in this volume are totally my responsibility, for which I apologize in advance. Criticisms and suggestions would be welcome and can be sent to me at HelvetianPress@gmail.com.

Thomas E. Creighton

~ COMMON ABBREVIATIONS ~

a:	atto (10^{-18})
Å:	Ångstrom (= 0.1 nm)
A:	adenine
ac:	alternating current
ADP:	adenosine diphosphate
Ala:	alanine residue of a protein
AMP:	adenosine monophosphate
Arg:	arginine residue of a protein
Asn:	asparagine residue of a protein
Asp:	aspartic acid residue of a protein
ATP:	adenosine triphosphate
bp:	base pairs
BPTI:	bovine pancreatic trypsin inhbitor
C_p:	heat capacity at constant pressure
C:	cytidine
cal:	calorie (= 4.184 joules)
CD:	circular dichroism
ccDNA:	closed circular DNA
cDNA:	complementary DNA
CDP:	cytidine diphosphate
cmc:	critical micelle concentration
CMP:	cytidine monophosphate
CTP:	cytidine triphosphate
Cys:	cysteine residue of a protein
Da:	Dalton
dc:	direct current
ddNTP:	dideoxynucleoside triphosphate
DNA:	deoxyribonucleic acid
dNTP:	deoxynucleoside triphosphate

e:	mathematical number (2.718) that is the base of natural logarithms
e:	unit of atomic charge (1.602×10^{-19} C)
EDTA:	ethylenediamine-N,N,N',N'-tetracetic acid
EGTA:	ethyleneglycol bis(β-aminoethyl ether) N,N,N',N'-tetracetic acid
EPR:	electron paramagnetic resonance
ESR:	electron spin resonance
Et:	ethyl group (-$CH_2 - CH_3$)
f:	femto (10^{-15})
FPLC:	fast-protein liquid chromatography
g:	gram
g:	gravitational constant (9.81 m s^{-2} or 6.673×10^{-11} N m^2/kg^2)
G:	Gibbs free energy
G:	Guanine
G:	Giga (10^9)
Gdm:	guanidinium
GDP:	guanosine diphosphate
Gln:	glutamine residue of a protein
Glu:	glutamic acid residue of a protein
Gly:	glycine residue of a protein
GMP:	guanosine monophosphate
GSH:	glutathione, thiol form
GSSG:	disulfide form of glutathione
GTP:	guanosine triphosphate
H:	enthalpy
h:	Plank's constant (1.584×10^{-34} cal s; 6.626×10^{-34} J s)
His:	histidine residue of a protein
HPLC:	high-performance liquid chromatography
I:	inosine base
Ile:	isoleucine residue of a protein
IPG:	immobilized pH gradient
IR:	infrared
J:	joule
K:	absolute temperature
kB:	Boltzmann's constant (3.298×10^{-24} cal K^{-1}; 1.381×10^{-23} J K^{-1})
k:	rate constant
k:	kilo (10^3)
K_{eq}:	equilibrium constant
K_M:	Michaelis constant
l:	liter (10^{-3} m^3)

Leu:	leucine residue of a protein
Lys:	lysine residue of a protein
μ:	micro (10^{-6})
m:	milli (10^{-3})
m:	meter
M:	molar (moles/liter)
M;	mega (10^6)
Me:	methyl group ($-CH_3$)
Met:	methionine residue of a protein
mRNA:	messenger RNA
n:	nano (10^{-9})
N:	Newton (1 kg m s^{-2} = 1 J m^{-3})
N_A:	Avogadro's number (6.022 x 10^{23} mol^{-1})
NAD:	nicotinamide adenine dinucleotide
NADH:	reduced form of NAD
NADP:	nicotinamide adenine dinucleotide phosphate
NADPH:	reduced form of NADP
n_H:	Hill coefficient
NMR:	nuclear magnetic resonance
NTP:	nucleoside triphosphate
p:	pico (10^{-12})
p.p.m.:	parts per million
p.s.i.:	pounds per sqare inch (= 4.88 kg per square meter)
PAGE:	polyacrylamide gel electrophoresis
PCR:	polymerase chain reaction
PDB:	Protein Data Bank
Phe:	phenylalanine residue of a protein
pI:	isoelectric point
P_i:	inorganic phosphate
PP_i:	inorganic pyrophosphate
Pro:	proline residue of a protein
Pu:	purine
Py:	pyrimidine
r.m.s.:	root-mean-square
R:	gas constant (1.987 cal mol^{-1} K^{-1}; 8.315 J mol^{-1} K^{-1}, = $N_A k_B$)
redox:	reduction/oxidation
RF:	radio-frequency
RNA:	ribonucleic acid
r.p.m.:	revolutions per minute

S:	entropy
S:	Svedberg unit of sedimentation (10^{-13} s)
SDS:	sodium dodecyl sulfate
Sec.:	section
Ser:	serine residue of a protein
T:	temperature
T:	tera (10^{12})
T:	Tesla
T:	Thymine
TFA:	trifluoroacetic acid
Thr:	threonine residue of a protein
TLC:	thin-layer chromatography
tRNA:	transfer RNA
Trp:	tryptophan residue of a protein
TTP:	thymidine triphosphate
Tyr:	tyrosine residue of a protein
U:	uracil
UDP:	uridine diphosphate
UMP:	uridine monophosphate
UTP:	uridine triphosphate
UV:	ultraviolet
Val:	valine residue of a protein
V_{max}:	maximum velocity of an enzyme-catalyzed reaction
z:	zepto (10^{-21})

~ GLOSSARY ~

Ab initio: starting from the beginning or first principles.

Abscissa: the horizontal axis of a graph.

Achiral: having no chirality, with a plane of symmetry, so that the mirror image is identical.

Acid: any compound that can supply a proton.

Activation energy: the free energy barrier ($\Delta G^\ddagger$) that must be overcome for a chemical reaction to occur.

Active site: the region of an enzyme where the substrate binds and the chemical reaction occurs.

Activity, chemical: concentration corrected for nonideality.

Activity coefficient: constant multiplied by the concentration to give the chemical activity.

Adduct: a chemical group added to another molecule.

Adiabatic: occurring without loss or gain of heat.

Aerobic: in the presence of air or oxygen.

Agonist: substance that produces the same response as a hormone.

Algorithm: set of instructions that define a method, usually a computer program.

Allostery: binding of a ligand at one site on a macromolecule affects another site on the same molecule.

Amphiphile: molecule having both a hydrocarbon part and a polar part, so that it localizes at interfaces between hydrocarbons and water.

Amphoteric: containing both acidic and basic groups.

Anaerobic: in the absence of air or oxygen.

Analyte: molecule being analyzed by some technique.

Anion: negatively-charged molecule.

Anisotropic: exhibiting properties with different values when measured in different directions.

Annealing: association of oligonucleotides by forming base pairs between them.

Anode: positively-charged electrophoresis terminal toward which negatively charged molecules (anions) migrate.

Anomers: the two isomers that result when sugar molecules are linked together, due to the C-1 atom becoming asymmetric, usually designated as alpha or beta.

Antagonist: substance that prevents the response of a hormone.

Antigen: any molecule recognized specifically by an antibody.

Antisense: oligonucleotide complementary in nucleotide sequence to an original strand, designated the "sense" strand.

Apoenzyme: enzyme without its coenzyme.

Aprotic: incapable of donating a proton.

Aptamer: nucleic acid that was selected to bind a specific ligand.

Artifact: something created by humans.

Autoradiography: detection of radioactivity by its effect on photographic emulsions.

Bacteriophage: a virus that multiplies in bacteria.

Base: any compound that can accept a proton.

Bohr effect: influence of pH on the oxygen affinity of hemoglobin.

Boltzmann distribution: the population of a species in equilibrium with others is proportional to the negative exponential of its energy.

Buffer: a mixture of acidic and basic forms of a reagent that tends to keep the pH constant.

Calorie: the amount of heat necessary to raise the temperature of 1 g of water from 15°C to 16°C.

Canonical: conforming to a general rule.

Catalyst: any substance that increases the rate of a chemical reaction without being consumed in that reaction.

Catenane: two molecules linked together topologically.

Cathode: negatively-charged electrophoresis terminal toward which positive charges migrate.

Cation: positively-charged molecule.

Chaotropic: biologically disruptive.

Chelate: multiple interactions between several groups of a molecule and a metal ion.

Chemical potential: the partial molar Gibbs free energy.

Chemiluminescence: light emission during chemical reactions that results from the decay of excited species.

Chiral: consisting of nonsuperimposable isomers that are mirror images of each other.

Chromatin: genomic DNA and associated proteins, as found in chromosomes.

Chromophore: a molecule or moiety that absorbs light and appears colored.

Chromosome: self-replicating structure of DNA and proteins that contains the genetic information.

Cis-acting: acting on the same molecule.

Clone: replicas of all or part of a macromolecule, or a cell, produced by replication.

Coding region: nucleic acid segment that contains the linear arrangement of codons specifying (by the genetic code) the order of amino acid residues in a protein.

Codon: a three-nucleotide unit in a gene or messenger RNA used by the genetic code.

Coenzyme: a molecule required by a number of different enzymes and used as a substrate, alternating between two forms, such as NAD and NADH.

Cofactor: a molecule required by a number of enzymes for their catalytic activity, but not changed in the reaction.

Cognate: recognized specifically.

Coherent: all the waves have the same phase.

Coherent scattering: the scattered waves interfere to produce a single resultant wave in a given direction.

Colligative: depending upon the number of molecules, not on their identities.

Complexome: all the protein complexes of the cell.

Configuration: three-dimensional arrangement of atoms at the chiral center of a molecule.

Conformation: the three-dimensional structure of a large molecule defined by rotations about covalent bonds.

Cooperativity: phenomenon by which one event on a molecule increases or decreases the probability of further such events.

Corepressor: small molecule that increases the affinity of a repressor for its operator.

Co-solute, co-solvent: additional compounds, such as salts or denaturants, that are added to aqueous solutions of macromolecules.

Covalent: involving the sharing of one or more pairs of electrons.

Covariation: correlated variation of two or more variables.

Cryo-: pertaining to very low temperatures.

Cryoprotectant: substance that protects against low temperatures, especially freezing.

Cryosolvent: solvent that remains liquid at very low temperatures.

Denaturant: reagent that causes proteins or nucleic acid molecules to unfold.

Dialysis: adding or removing small molecules from a solution by their diffusion across a semipermeable membrane.

Diamagnetic: having only paired electrons and a negative magnetic susceptibility; diamagnetic substances move out of magnetic fields.

Diastereomers: molecules with different chirality.

Diffraction: scattering of radiation from atoms or molecules organized in an ordered array.

Diffusion: the spontaneous movement of molecules due to their kinetic energy.

Dipole: separation of charge within a molecule.

Diprotic: having two acid groups.

Disulfide: two sulfur atoms linked by a single covalent bond.

Eclipsed: being behind another.

Elastic scattering: the scattered beam has the same energy as the incident beam.

Electrolyte: consisting of ions.

Electronegative: attracting electrons.

Electrophile: molecule or group that is electron-deficient and reacts with nucleophiles.

Electrophilic catalysis: increase in rate of a reaction by stabilization of a negative charge that develops in its transition state.

Electrophoresis: movement of molecules or particles under the influence of an electric field.

Ellipticity: difference in absorbance of left- and right-circularly polarized components of plane-polarized radiation, measured in circular dichroism.

Empirical: based on experiment and observation, rather than theoretical.

Enantiomer: one of two isomers of a chiral compound.

Endo-: acting on the interior residues of a polymer.

Enhancer: transcription factor required for the expression of a gene.

Enzyme: protein that catalyzes a chemical reaction.

Epitope: Sites on a molecule recognized directly by an antibody.

Equilibrium: state of a chemical reaction in which the forward and backward rates are equal, so there is no net change in the concentrations of the reactants and products.

Eukaryote: organism whose cells contain a true nucleus; all organisms other than viruses, bacteria, and blue-green algae.

Exciton: high-energy excited state of a molecule or array of molecules.

Exo-: acting on the terminal residues of a polymer.

Exocytosis: release from the interior of a cell.

Exon: segment of a gene that is present in the mature messenger RNA and used in translation.

Extensive property: one that depends upon the size of the system (e.g. mass, volume).

Fatty acid: long-chain aliphatic carboxylic acid normally found esterified to glycerol.

Fluorography: detection of radioactivity by the fluorescence emitted by a scintillator in close contact.

Fluorescence: emission of light from molecules in excited electronic states.

Fluorescence lifetime: average amount of time between absorption of light and emission as fluorescence.

Fluorophore: fluorescent molecule or moiety.

Free radical: molecule containing one or more unpaired electrons.

Gene: the basic unit of genetic information, usually a segment of DNA whose expression results in the production of a messenger RNA that most often is translated into a protein.

Genetic code: the way in which the codons of a messenger RNA are read during translation and formation of a polypeptide chain.

Genetic: information present in genes.

Genetics: mechanisms by which the genetic information is transferred from one generation to the next.

Genome: the DNA and genes contained in the whole set of chromosomes present in a cell.

Genotype: the genetic constitution of an organism.

Globular: having a compact folded molecular structure.

Glycolipid: sugar linked to one or more fatty acyl groups.

Glycoprotein: protein with carbohydrate units attached covalently.

Glycosidase: enzyme that hydrolyzes glycosidic links in carbohydrate polymers.

Half-life: time taken for radioactivity or a reactant to decrease to half its original value.

Hapten: a small molecule that mimics part or all of the antigenic site of a larger molecule and interferes with its binding to an antibody.

Hard ion: one that is small, compact and not readily polarized and tends to interact with other hard ions.

Heat capacity: that quantity of heat required to increase the temperature of a system of substance by one degree centigrade or Kelvin.

Heteroduplex: a double-stranded nucleic acid molecule in which the two strands are not identical.

Heterotropic: involving different molecules.

Heuristic: involving observation and trial-and-error methods.

Histone: small basic protein that, with DNA, forms the nucleosome.

Holoenzyme: enzyme including its coenzyme.

Homeostasis: relatively stable system of interdependent elements.

Homologous: proteins or nucleic acids that arose from a common evolutionary ancestor and consequently have related sequences.

Homotropic: involving the same type of molecules.

Hormone: a chemical, nonnutrient, intercellular messenger that is effective at very low concentrations.

Hybridization: annealing complementary strands of nucleic acid.

Hydration: association with water.

Hydrogen bond: a noncovalent bond between a hydrogen donor and acceptor.

Hydrolysis: breaking a covalent bond by reaction with water.

Hydrophilic: attracted to water.

Hydrophobic: not attracted to water, but to nonpolar environments.

Hypertonic: having a high osmotic pressure.

Hypervariable region: a segment of genomic DNA characterized by considerable variation in the number of tandem repeats or a high degree of polymorphism due to point mutations.

Hyphenate: to extend one technique by combining it with another, such as LC-MS (liquid chromatography combined with mass spectrometry).

Hypotonic: having a low osmotic pressure.

Hysteresis: when the forward and reverse processes follow different paths, due to slow equilibration of the system.

Immunogen: substance that elicits an antibody response.

In silico: performed with a computer.

In situ: in position.

In vacuo: in a vacuum.

In vitro: in the test-tube.

In vivo: in the living organism.

Incoherent scattering: scattered radiation that is the sum of the individual scattered waves, with no interactions between them.

Inducer: small molecule that decreases the affinity of a repressor for its operator.

Inelastic scattering: the scattered beam has either greater or lesser energy than the scattered beam, having exchanged energy with the scatterer.

Intensive property: one that is independent of the size of the system (e.g. temperature).

Intron: segment of a gene that is removed from the messenger RNA before it is translated.

Ionic: having a net charge.

Ionizing radiation: photons or sub-atomic particles with sufficient energy to produce ionization events while passing through matter.

Isoelectric point: pH at which a molecule exhibits no net charge and does not migrate in an electric field.

Isoionic point: the pH of a solution containing only the macromolecule of interest from which all other ions, except for H^+ and OH^-, have been removed.

Isomers: compounds with the same molecular formula but differing in the nature or sequence of bonding of their atoms, or in their spatial arrangement.

Isothermal: at constant temperature.

Isotonic: having the same osmotic pressure.

Isotopes: atoms with the same atomic number (protons) but different mass numbers (protons plus neutrons)

Isotropic: exhibiting properties with the same values when measured along axes in all directions.

Isozyme, isoenzyme: enzyme that is closely related in sequence, structure and activity to another enzyme, usually from the same source.

Kinase: enzyme that transfers a phosphate group to a protein, usually from ATP.

Le Chatelier principle: to every action there is an equal and opposite reaction.

Lectin: protein other than an antibody that recognizes specific polysaccharides.

Lewis acid: atom having empty d electron orbitals that act as electron sinks.

Ligase: enzyme that covalently join the ends of nucleic acids through a phosphodiester bond.

Lipid: molecule soluble in nonpolar organic solvents.

Lipoprotein: complex of proteins (apolipoproteins) and lipids.

Lone pair: pair of valence electrons that are not involved in covalent bond formation.

Lyophilization: removal of volatile solvent by subjecting a frozen solution to a vacuum, so that the solvent sublimes but any nonvolatile materials are left behind.

Lysogeny: integration of a viral genome into that of the cell, rather than lytic multiplication.

Lytic: leading to lysis of the cell.

Macrostate: state of a system defined by its macroscopic properties.

Melting temperature: temperature at mid-point of a thermally-induced transition.

Mesophile: organism that grows optimally at normal physiological temperatures.

Messenger RNA: the RNA copy of a gene that is produced by transcription and used for translation into a polypeptide chain.

Microstate: state of a system defined by its individual molecules.

Minisatellites: regions of tandem repeats in the genome.

Mismatch: any base pair other than the normal A•T (or U) and G•C.

Molality: moles of substance per 1000 g of solvent

Molarity: moles of substance in 1000 ml of solution

Mole: mass in grams of the molecular weight of a molecule, containing N_A molecules.

Molecular weight: the sum of the atomic weights of all the atoms in a molecule.

Molecule: the smallest unit of matter that can exist by itself and retain all the properties of the original substance.

Monochromatic: composed of a single wavelength.

Monoclonal antibodies: homogeneous antibodies synthesized by a population of identical antibody-producing cells.

Monodisperse: homogeneous population of molecules

Morphology: external shape adopted by a solid.

Mutation: a change in the structure of the genome DNA, usually of the nucleotide sequence, that is passed on to future generations.

Nascent: newly-synthesized.

Negative stain: visualizing a structure by observing the shell that it leaves in an amorphous solid medium.

Nonpolar: having no functional or reactive groups, only inert hydrocarbons.

Nuclease: enzyme that hydrolyzes phosphodiester bonds of nucleic acids.

Nucleophile: molecule or group that is electron-rich and reacts with electrophiles.

Nucleosome: fundamental unit of chromatin, consisting of histones and about 200 base pairs of double-stranded DNA.

Nucleotide: monomer that upon polymerization generates a nucleic acid, either DNA or RNA.

Oligonucleotide: a short polynucleotide, usually 2 to 20 nucleotides in length.

Oligosaccharide: linear or branched carbohydrate consisting of 2 to 20 monosaccharides, linked by glycoside bonds.

Operator: region on a DNA molecule upstream of a gene at which a repressor binds and blocks transcription.

Operon: several linked genes subject to the same control.

Ordinate: the vertical axis of a graph.

Osmolyte: molecule that contributes significantly to the osmotic pressure.

Osmosis: net movement of molecules through a semipermeable membrane.

Osmotic pressure: pressure required to stop osmosis.

Oxidase: enzyme that catalyzes an oxidation using O_2 as the electron acceptor, without incorporating the O atoms into the product.

Oxidation: a chemical reaction that removes electrons, often by transferring them to O_2 to produce water.

Oxidoreductase: enzyme that catalyzes a reaction involving electron transfer to or from an external electron carrier, usually a redox protein.

Oxygenase: enzyme that catalyzes the reaction between O_2 and an organic substrate, adding O atoms to the substrate.

Palindrome: nucleotide sequence from the 5′- end to the 3′-end that is the same as its complement on the other strand.

Paramagnetic: having unpaired electron spins; a paramagnetic substance tends to move into a magnetic field.

Peptidase: enzyme that cleaves the peptide bonds of peptides or small proteins.

Peptide: a short linear segment of amino acids linked by peptide bonds.

Peptide bond: covalent bond between the α-amino and α-carboxyl groups of two amino acids.

pH: negative logarithm of the hydrogen ion activity or concentration.

Phage, bacteriophage: a virus that replicates in bacteria.

Phenotype: the observable properties of an organism, resulting from the interaction of the genotype and environment.

Phospholipid: any lipid containing phosphate, usually referring to lipids based on 1,2-diacylglycerol-3-phosphate.

Phylogeny: pathway by which genes, nucleic acids, proteins, individuals, species, or populations arose and diverged during evolution.

Piezoelectric: changing its shape in response to an electric field.

pK_a: the pH at which a polar chemical group is half ionized.

Plasmid: DNA molecule that is stably inherited genetically without becoming part of a chromosome.

Plectonemic: interwound.

Polar: having functional, reactive, and ionizable groups.

Polarized light: light that exhibits different properties in different directions at right angles to its direction.

Polarizer: instrument that polarizes light.

Polyamide: repeated amide (-CO-NH-) units, usually formed by polymerization of amino and carboxyl groups.

Polychromatic: comprised of many colors.

Polymer: compound formed by polymerization and consisting essentially of repeating structural units.

Polymerase: enzyme that catalyzes polymerization.

Polymorphism: ability to assume different forms.

Polynucleotide: a linear polymer produced by condensation of nucleotides.

Post-transcriptional: occurring after biosynthesis of an RNA molecule.

Post-translational: occurring after biosynthesis of a polypeptide chain.

Precess: to undergo a relatively slow gyration of the rotation axis of a spinning body about another line intersecting it so as to describe a cone.

Primary structure: sequence of amino acid or nucleotide residues in a protein or nucleic acid.

Probe: a labelled oligonucleotide used to detect complementary sequences in DNA or RNA.

Prokaryote: organism that lacks a true nucleus; a bacterium, virus, or blue-green algae.

Promoter: region on a DNA molecule upstream of a gene at which an RNA polymerase binds and initiates transcription.

Prosthetic group: any chemical group of a protein that was not part of the primary structure but acquired by binding another molecule.

Proteinase (protease): enzyme that hydrolyzes peptide bonds.

Proteome: all the proteins produced by the cell.

Pseudoknot: highly structured RNA secondary structural motif.

Quaternary structure: involving aggregation of two or more individual protein or nucleic acid molecules.

Radian: the angle subtended by an arc of a circle equal to its radius (57°).

Radical ion: free radical with a positive or negative charge.

Radioactivity: emission of ionizing radiation by atoms.

Radioisotopes: isotopes that have unstable nuclei and decay to a stable state by the emission of ionizing radiation.

Random coil: a flexible polymer in which the conformational properties of each residue are independent of those of other residues not close in the covalent structure.

Receptor: structure on a cell that binds other molecules specifically and produces a response.

Redox protein: protein that can exist reversibly in more than one oxidation state; usually it has a cofactor that handles the electrons.

Reduction: a chemical reaction that adds electrons, often by transferring H atoms.

Relaxation time: time required for a change to reach $1/e$ (0.368) the final state; the reciprocal of the rate constant for the reaction.

Replication: copying of DNA or RNA molecules to make multiple identical copies.

Repressor: a protein that binds to an operator to prevent transcription of a gene.

Reptating: a linear polymer moving in a snake-like manner.

Residue: individual amino acid unit of a polypeptide chain.

Resonance: two or more alternative electronic structures are required to describe a molecule.

Restriction enzyme: enzyme that recognizes a specific sequence of DNA and cleaves the backbone at or near this sequence.

Reverse transcriptase: enzyme that uses a single-stranded RNA molecule as template to synthesize a complementary strand of DNA.

Reverse transcription: the synthesis of a complementary DNA molecule from an RNA template, catalyzed by a reverse transcriptase.

Ribozyme: RNA molecule with catalytic ability.

Root-mean-square: the square root of the average value of the squares of the individual values, weighted by the probability of that value occurring.

Secondary structure: local conformation adopted by interactions only between residues close in the sequence of a protein or nucleic acid.

Sedimentation: movement of molecules or particles in a gravitational field.

Semipermeable: permeable to some molecules but not others.

Singlet state: having zero electronic spin.

Soft ion: one that is large and relatively polarizable and tends to interact with other soft ions.

Solute: that constituent of a solution that is considered to be dissolved in the other, the solvent.

Spin: intrinsic angular momentum of a nucleus or an unpaired electron that induces magnetic momentum.

Splicing: removal of introns from a messenger RNA precursor.

Stereoisomers: isomers differing only in the spatial arrangement of their atoms.

Sticky end: single-stranded nucleotides protruding from the end of a double-stranded nucleic acid, which may hybridize with another single-stranded nucleic acid with a complementary sequence.

Stochastic: random.

Substrate: the specific compound on which an enzyme acts.

Tandem repeat: end-to-end duplication of a series of identical or almost identical segments of DNA (usually of 2 to 80 base pairs).

Tautomers: isomers that are readily interconverted spontaneously and normally exist together in equilibrium.

Tertiary structure: the folded conformation adopted by a substantial segment of polypeptide or oligonucleotide chain involving interactions between groups distant in the covalent structure.

Tetrahedral: in the shape of a regular tetrahedron with four identical faces.

Torus: a shape like a donut, with a hole in the middle.

Trans-acting: acting on other molecules.

Transcription: expression of the nucleotide sequence of a gene into a messenger RNA with a complementary base sequence.

Transcription factor: regulatory protein that binds to a promoter or to a nearby sequence of DNA to facilitate or prevent initiation of transcription.

Transition metal: element with incompletely filled d subshell of electrons or that gives rise to cations with incompletely filled d subshells.

Transition state: the least stable species that occurs during a chemical reaction; its free energy determines the rate of the reaction.

Translation: expression of the genetic information of a messenger RNA into a polypeptide chain.

Triplet state: an atom or molecule with total spin quantum number of one.

Tunneling: quantum-mechanical phenomenon by which a particle can move from one energy state to another by penetrating, rather than traversing, an energy barrier as a wave.

Valence electron: outer electrons of an atom that are involved in forming chemical bonds.

Vicinal: adjacent sites in a molecule.

Virus: a structure comprised of proteins and nucleic acids that can infect a host cell and replicate to produce many more such structures.

Zwitterion: molecule with both positively- and negatively-charged groups.

~ CHAPTER 1 ~

CONFIGURATIONS AND CONFORMATIONS

Molecules are generated by the formation of covalent bonds between pairs of atoms, in which the two atoms share electrons. A covalent bond forms when atoms individually do not have enough electrons for a complete octet: if two atoms can complete their octets by sharing electrons, they can do so by forming a covalent bond. Covalent bonds can be explained only by quantum mechanics, but here it is necessary simply to recognize that covalent bonds are generally not broken in isolation under most conditions experienced in molecular biology. When a covalent bond is broken, as in a chemical or enzymatic reaction, it is generally exchanged with another covalent bond to a different atom. Consequently, covalent bonds define the structures and properties of small molecules, and those of large molecules are determined by the covalent structures of the smaller substituents from which they are made.

The most important molecules in biology are proteins and the nucleic acids deoxyribonucleic acid (DNA) and ribonucleic acid (RNA); all are macromolecules characterized by their very large sizes and high molecular weights. These giant molecules can contain many thousands, millions, even billions, of atoms. Fortunately, these macromolecules are **polymers**, produced by linking together in a linear fashion only a few relatively simple monomers: four nucleotides in the case of DNA and RNA, and 20 amino acids in the case of proteins. In each of these cases, each residue, i, of the chain consists of two parts: group X comprises the backbone and is the constant, repeating part of the polymer, while the side-chains (A, B, C, ...) attached to the backbone are variable:

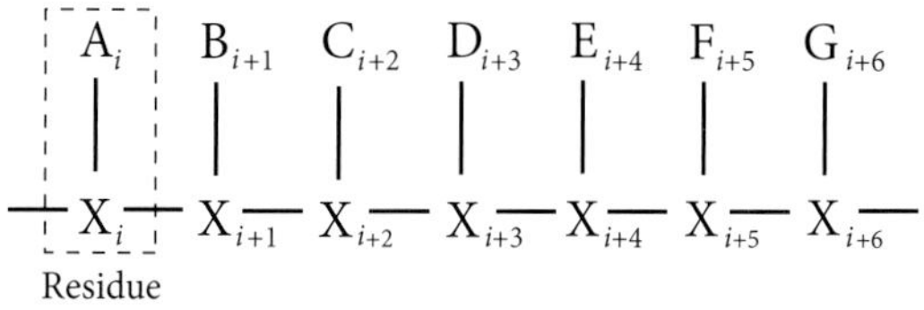

$$\begin{array}{ccccccc} A_i & B_{i+1} & C_{i+2} & D_{i+3} & E_{i+4} & F_{i+5} & G_{i+6} \\ | & | & | & | & | & | & | \\ -X_i- & X_{i+1}- & X_{i+2}- & X_{i+3}- & X_{i+4}- & X_{i+5}- & X_{i+6}- \end{array} \tag{1.1}$$

Residue

The side-chains connected to the backbone are all the same in **homopolymers**, as in carbohydrates or polymers made chemically, but they are variable in **copolymers**; the natural proteins and nucleic acids are extreme examples with several different types. Normally the individual residues are indexed from 1 to n, starting from one end of the polymer chain and finishing at the other. The bonds between the residues are numbered similarly, with bond i joining residues i and $i + 1$; there are then $n - 1$

bonds linking the *n* residues. Normally the backbone primarily has a structural role, while the sidechains contain the functional groups. In spite of the enormous sizes of proteins and nucleic acids, it is possible to determine their detailed covalent structures because, knowing the detailed structures of all the possible monomers, it is necessary only to determine their linear sequence in the polymer.

The detailed structures of the monomers are extremely important, because they determine the global properties of the macromolecule. They occur many times in the polymer, and their structures are multiplied many times over. Many of the monomers occur in only one of several possible isomers; for example natural proteins are composed solely of L-amino acids and nucleic acids of D-ribose or D-deoxyribose. While these details of the structure might seem very minor and mundane, they have extremely important consequences for the three-dimensional (3-D) structures of biopolymers and their functions. These consequences even extend to the macroscopic level; for example, the left/right asymmetry of all but the simplest microorganisms is believed to result from asymmetry at the atomic level of certain molecules.

Biopolymers. A. G. Walton & J. Blackwell (1973) Academic Press, NY.

An Introduction to Macromolecules. L. Mandelkern (1983) Springer-Verlag, NY.

Introduction to Macromolecular Science. P. Munk (1989) Wiley-Interscience, NY.

Advanced Organic Chemistry, 2nd edn. J. March (2000) Wiley-Interscience, NY.

Virtual exploration of the chemical universe up to 11 atoms of C, N, O, F: assembly of 26.4 million structures (110.9 million stereoisomers) and analysis for new ring systems, stereochemistry, physicochemical properties, compound classes, and drug discovery. T. Fink & J. L. Reymond (2007) *J. Chem. Inf. Model.* **47**, 342–353.

1.1. STEREOCHEMISTRY

Isomers are two molecules that share the same elemental formula but have different structures. A simple example is ethanol and dimethyl ether:

Dimethyl ether   Ethanol

(1.2)

They have the same elemental formula C_2H_6O but different covalent structures. Their 3-D structures are indicated here by solid tapered bonds that project above the plane of the paper, while open tapered bonds project below the plane. **Structural isomers like these cannot be interconverted without breaking chemical bonds.** They can differ dramatically in their 3-D structures and in their chemical, physical and biological properties. The potential for different isomers increases dramatically with the size of the molecule and is especially acute with macromolecules.

Isomers can differ in various ways. Ethanol and dimethyl ether (Equation 1.2) are **structural isomers** because the number and type of bonds linking the atoms are different. **Geometric isomers** differ in their geometrical arrangement of bonds:

$$\underset{\text{Maleate}}{\overset{-O_2C}{\underset{H}{\diagdown}}C=C\overset{CO_2^-}{\underset{H}{\diagup}}} \qquad \underset{\text{Fumarate}}{\overset{H}{\underset{-O_2C}{\diagdown}}C=C\overset{CO_2^-}{\underset{H}{\diagup}}} \qquad (1.3)$$

Fumarate and maleate are geometric isomers because they differ only in the rotation about the double bond. Double and triple bonds are not readily rotated, so such geometric isomers are not readily interconverted. The double bond makes these molecules planar, with all the C and H atoms in the plane of the paper. The two carboxyl groups are highlighted to emphasize that they are on the same side in maleate but on opposite sides in fumarate; they can be said to be *cis* and *trans*, respectively (Section 1.1.A.5).

Two molecules are **stereoisomers** if they differ only in the spatial orientation of those atoms that cannot be rapidly interconverted by rotation about single bonds. Stereoisomers contain the same number and type of bonds and have the same chemical name, except for a prefix (e.g. D or L) that is sometimes used to discriminate between them. Stereoisomers are divided into **enantiomers**, molecules with nonsuperimposable mirror images (Section 1.1.A.1), and **diastereomers**, which comprise all other types of stereoisomers (Section 1.1.A.3). **Tautomers** are a specialized class of isomers that are distinguished by their ability to equilibrate rapidly (Section 1.1.C).

Stereoisomers, diastereomers and enantiomers are specialized types of isomers, in which the differences in the molecules are due solely to the **configuration**, which identifies the spatial arrangement of atoms within the structure of a molecule. Configurations are interconverted only by altering the chemical bonds between atoms, and they should be distinguished from **conformations** (Section 1.2), which maintain all covalent bonds and differ only in rotations about single bonds. The two terms configuration and conformation should not be confused.

Complete relative stereochemistry of multiple stereocenters using only residual dipolar couplings. J. Yan *et al.* (2004) *J. Am. Chem. Soc.* **126**, 5008–5017.

Isolation of isomers based on hydrogen/deuterium exchange in the gas phase. U. Mazurek *et al.* (2004) *Eur. J. Mass Spectrom.* **10**, 755–758.

The evolution of stereochemistry. H. D. Arndt (2006) *Angew. Chem. Int. Ed. Engl.* **45**, 4542–4543.

Mechanistic inferences from stereochemistry. I. A. Rose (2006) *J. Biol. Chem.* **281**, 6117–6119.

1.1.A. Chirality

A structure is **chiral** if it cannot be superimposed on its mirror image; examples are our left and right hands. The nonsuperimposable mirror-image isomers are **enantiomers**. The most common structural feature of chiral molecules is the presence of a tetrahedral atom, such as C, with four different substituents, which is by definition a **chiral center**. There are two different ways to arrange four different substituents around a tetrahedral atom, and they are mirror images. Chiral centers include the hydroxymethylene (–HCOH–) carbons of carbohydrates (Equation 1.6) and the C^α atoms of the α-amino acids:

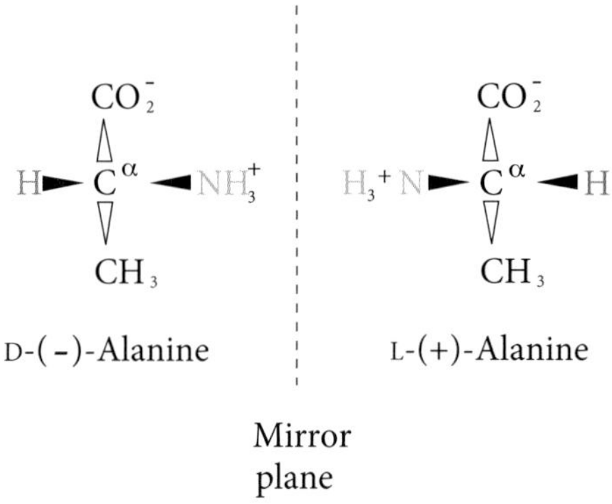

$$\text{D-(-)-Alanine} \qquad | \qquad \text{L-(+)-Alanine}$$

Mirror plane

(1.4)

The exceptional amino acid is glycine, which has two H atoms bonded to the C^α atom (Section 7.2.A). The L and D designate the absolute configuration of the chiral center, while the + and − indicate in which direction an aqueous solution of the enantiomer will rotate the plane of polarized light. Enantiomers can be distinguished because they interact differently with polarized light; this is the basis of optical rotation and circular dichroism.

A tetrahedral chiral center is not required for chirality, neither does the presence of a chiral center require that the molecule be chiral. A molecule with an internal mirror plane may have chiral centers but not be chiral; these are **meso** compounds. Because of their internal symmetry, these molecules can be superimposed on their mirror image. For example, the meso form of tartaric acid, HO_2C–CHOH–HOCH–CO_2H, has two chiral centers, at each of the two middle C atoms, but there is a mirror plane between them:

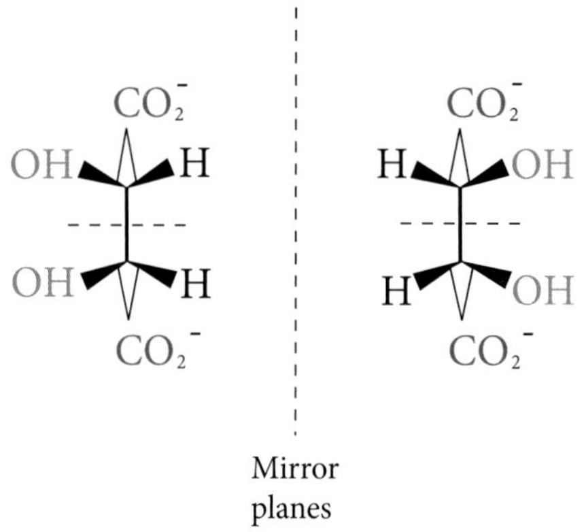

Mirror planes

(1.5)

Consequently, these mirror images can be superimposed by a rotation of 180°, unlike the two isomers of alanine (Equation 1.4).

Supramolecular chirality of self-assembled systems in solution. M. A. Mateos-Timoneda et al. (2004) *Chem. Soc. Rev.* **33**, 363–372.

Stereolabile chiral compounds: analysis by dynamic chromatography and stopped-flow methods. C. Wolf (2005) *Chem. Soc. Rev.* **34**, 595–608.

Nonlinear optical spectroscopy of chiral molecules. P. Fischer & F. Hache (2005) *Chirality* **17**, 421–437.

A novel spectroscopic probe for molecular chirality. N. Ji & Y. R. Shen (2006) *Chirality* **18**, 146–158.

Absolute configuration of chirally deuterated neopentane. J. Haesler *et al.* (2007) *Nature* **446**, 526–529.

1. Enantiomers

Enantiomers are recognized most readily by determining whether each chiral center could have an opposite configuration (Equation 1.4). Enantiomers interact identically with achiral compounds, but they can interact differently with other chiral objects. The physical property that differentiates enantiomers is the direction in which they rotate plane-polarized light, as in optical rotatory dispersion and circular dichroism. Thus, two enantiomers are differentiated as either dextrorotatory (+) or levorotatory (−), depending on whether the rotation of the polarized light is clockwise or counterclockwise, respectively (Equation 1.4). Solutions containing an excess of either of the enantiomers rotate polarized light and are said to be **optically active**.

Fischer introduced a general procedure that designated enantiomers as either D- or L- based on whether the nonhydrogen substituent was on the right or left when the molecule was drawn as a **Fischer projection**:

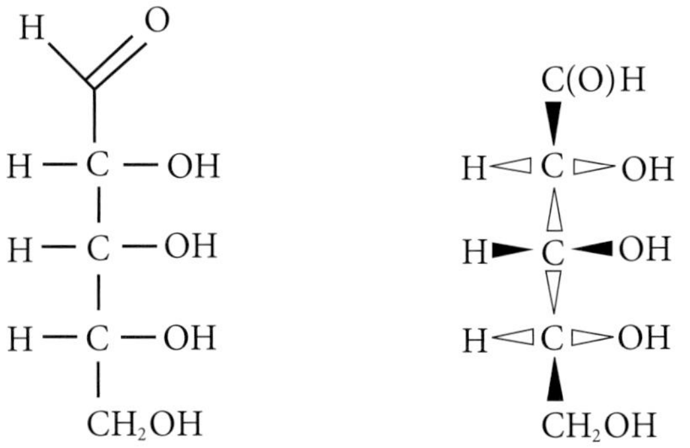

$$\begin{array}{c}
\text{H}\diagdown\!\!\!\diagup\text{O} \\
\text{C} \\
\text{H}-\text{C}-\text{OH} \\
| \\
\text{H}-\text{C}-\text{OH} \\
| \\
\text{H}-\text{C}-\text{OH} \\
| \\
\text{CH}_2\text{OH} \\
\text{Fischer} \\
\text{projection}
\end{array}
\qquad
\begin{array}{c}
\text{C(O)H} \\
\blacktriangledown \\
\text{H}\triangleleft\text{C}\triangleright\text{OH} \\
\triangle \\
\text{H}\blacktriangleright\text{C}\blacktriangleleft\text{OH} \\
\triangledown \\
\text{H}\triangleleft\text{C}\triangleright\text{OH} \\
\blacktriangle \\
\text{CH}_2\text{OH} \\
\text{Stereochemical} \\
\text{drawing}
\end{array}
\qquad (1.6)$$

By convention the Fischer projection has the vertical bonds directed away from the viewer and the horizontal bonds directed out towards the viewer. In the stereochemical drawing, the solid tapered bonds project out from the plane of the paper, whereas those that are open project below. The example of Equation 1.6 is D-ribose. The configuration of each chiral carbon in D-ribose is D because the nonhydrogen substituent is drawn to the right in the Fischer projection. For sugars, the enantiomer is defined by the bottom chiral carbon when the carbon chain is oriented vertically with the carbonyl carbon at the top; thus the ribose depicted in Equation 1.6 is D.

The Fischer nomenclature leads, however, to ambiguities. A more rigorous and unambiguous method of identifying configuration for specifying absolute configuration of chiral tetrahedral centers as either R or S (from the Latin 'rectus' and 'sinister', respectively) is generally accepted. The procedure requires the assignment of priority to the four substituents that generate a **chiral center**, followed by a procedure to identify the arrangement as either R or S. There are four rules for assigning the priorities of substituents bonded to the same atom.

(1) The priority is assigned in order of decreasing atomic number of the four atoms directly bonded to the chiral center.

(2) When two or more atoms cannot be distinguished by step 1, the atoms bonded to each of the atoms with equal priority are given their own priorities. If the atoms of greatest priority do not differentiate the two groups, the second atoms are compared, then the third.

(3) Heavier isotopes are given priority over lighter isotopes; for example ^2H is given precedence over ^1H and ^{14}C over ^{12}C.

(4) Double bonds count as two bonds to the same atom.

The relative priorities of the four substituents are assigned in step 1 and labeled 1 to 4:

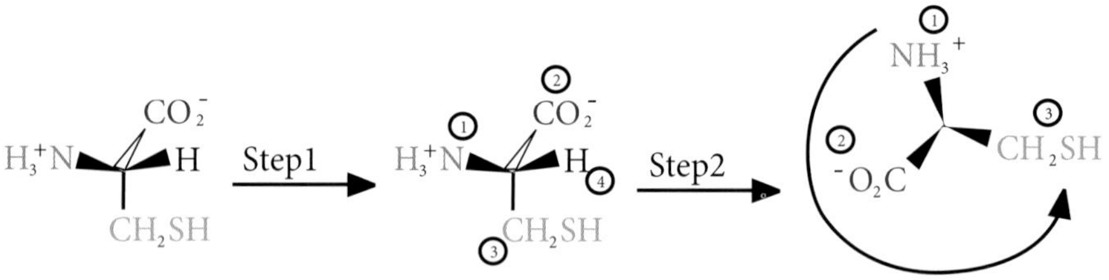

$$\text{L-Cysteine} \xrightarrow{\text{Step 1}} \quad \xrightarrow{\text{Step 2}} \quad (S)\text{-Cysteine} \tag{1.7}$$

The carboxylate carbon here takes precedence over that with the thiol group because it has three bonds to oxygen. In step 2, the bond to the lowest priority group is oriented directly away from the observer, and the remaining three groups are viewed from above. If the arrangement of highest to lowest priority of the three groups is clockwise, the chiral center is assigned the R configuration; if counterclockwise it has the S configuration. The assignment of the S configuration in Equation 1.7 results from the counterclockwise arc that connects the substituents in order of precedence.

Determination of the interconversion energy barrier of enantiomers by separation methods. J. Krupcik et al. (2003) *J. Chromatogr. A* **1000**, 779–800.

A simple method to determine concentration of enantiomers in enzyme-catalyzed kinetic resolution. R. C. Zheng et al. (2007) *Biotechnol. Lett.* **29**, 1087–1091.

Molecular quantum similarity and chirality: enantiomers with two asymmetric centra. S. Janssens et al. (2007) *J. Phys. Chem.* **111**, 3143–3151.

---

2. Racemic mixtures

A mixture with equal amounts of two enantiomers of a compound is known as a **racemic mixture** or **racemate**. Two enantiomers have identical free energies when in a homogeneous environment, such as in solution, so when a chiral compound is formed by the chemical reaction of two nonchiral reactants, the product will be a racemic mixture. The exception is when the reaction involves an asymmetric catalyst, such as an enzyme (Chapter 14). Any process that catalyzes the interconversion of enantiomers (i.e. a **racemization**) will necessarily result in a racemic mixture being formed. In the absence of any other chiral compound, the free energies of formation of two enantiomers must be identical.

Racemic macromolecules for use in X-ray crystallography. J. M. Berg & L. E. Zawadzke (1994) *Curr. Opinion Biotechnol.* **5**, 343–345.

Enantiomer-selective activation of racemic catalysts. K. Mikami *et al.* (2000) *Acc. Chem. Res.* **33**, 391–401.

Advances in chiral separation using capillary electromigration techniques. G. Gubitz & M. G. Schmid (2007) *Electrophoresis* **28**, 114–126.

---

3. Diastereomers

Diastereomers are stereoisomers that are not enantiomers. There are several common types. Molecules containing carbon–carbon and carbon–nitrogen double bonds will exist as two different geometric diastereomers if both of the double-bonded atoms have two different substituents (Equation 1.3). These diastereomers are differentiated with the designations *cis* and *trans* (Section 1.1.A.5).

The other common form of diastereomer occurs when a molecule contains more than one chiral center, with one having the same configuration in the two molecules but not the other:

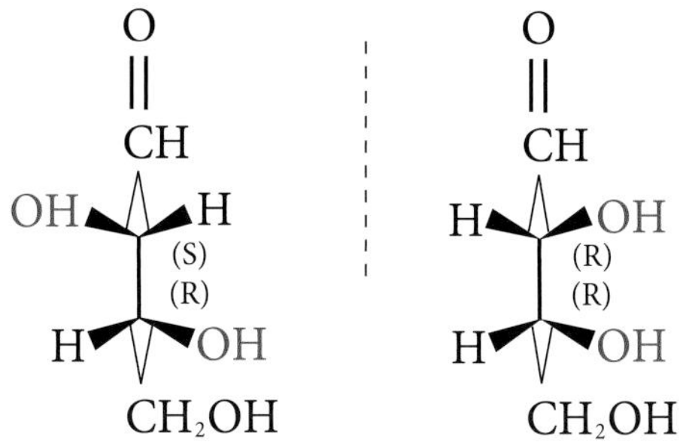

$$\begin{array}{cc} \text{D-threose} & \text{D-erythrose} \end{array} \tag{1.8}$$

The two diastereomers are not mirror images. **Diastereomers have different chemical properties**.

Two atoms are considered to be **diastereotopic** if they would generate different diastereomers upon substitution with a different isotope (Section 1.1.B). Diastereotopic atoms are chemically different and can be distinguished by nuclear magnetic resonance (NMR).

Diastereoisomerism, contact points, and chiral selectivity: a four-site saga. R. Bentley (2003) *Arch. Biochem. Biophys.* **414**, 1–12.

Evaluation of experimental strategies for the development of chiral chromatographic methods based on diastereomer formation. N. R. Srinivas (2004) *Biomed. Chromatogr.* **18**, 207–233.

Total chemical synthesis and X-ray crystal structure of a protein diastereomer: [D-Gln 35]ubiquitin. D. Bang *et al.* (2005) *Angew. Chem. Int. Ed. Engl.* **44**, 3852–3856.

Crystallization-induced diastereomer transformations. K. M. Brands & A. J. Davies (2006) *Chem. Rev.* **106**, 2711–2733.

4. Epimers and Epimerization

Diastereomers related by the inversion of configuration at a *single* chiral center are known as epimers. This definition excludes enantiomers such as D- and L-alanine (Equation 1.4) because they are not diastereomers. It also excludes diastereomers that are related by inversion of more than a single chiral center. For example, D-glucose and D-galactose are epimers, as are D-glucose and D-mannose, but galactose and mannose are not:

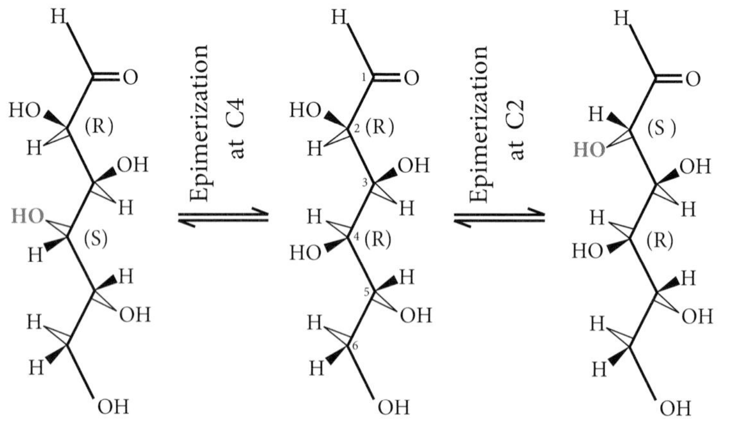

D-Galactose ⇌ (Epimerization at C4) ⇌ D-Glucose ⇌ (Epimerization at C2) ⇌ D-Mannose

(1.9)

The configurations at C2 and C4 are labeled and distinguish these three sugars. Glucose is an epimer of both mannose and galactose because it differs from each by the configuration of a single chiral center. Mannose and galactose have different configurations at both C2 and C4 and therefore are not epimers.

The chemical conversion of one epimer to another is called **epimerization**; it contrasts with the racemization of enantiomers.

Mechanistic aspects of enzymatic carbohydrate epimerization. J. Samuel & M. E. Tanner (2002) *Nat. Prod. Rep.* **19**, 261–277.

Understanding nature's strategies for enzyme-catalyzed racemization and epimerization. M. E. Tanner (2002) *Acc. Chem. Res.* **35**, 237–246.

5. Cis *and* Trans *Isomers*

Molecules containing carbon–carbon and carbon–nitrogen double bonds can exist as two different geometric diastereomers if both of the double-bonded atoms have two different substituents. These diastereomers are differentiated with the designation *cis* and *trans*, or more formally as *Z* and *E* (from the German 'zusammen' for together and 'entgegen' for opposite). When similar substituents, or those with the highest priority, are on the same side of the ring or double bond, the configuration is referred to as *cis* or *Z*; when they are on opposite sides, they are referred to as *trans* or *E*. For example, maleate is the *cis* form of fumarate (Equation 1.3). In potentially ambiguous cases, the priority of the substituents of each double-bonded atom is determined by the rules for deciding priorities (Section 1.1.A.1).

The conformation about a single bond may also be noted as *cis* or *trans*, particularly when all of the substituents lie in a plane. For example, the highlighted C2 and C3 hydroxyl groups of the furanose ring form of ribose are *cis*:

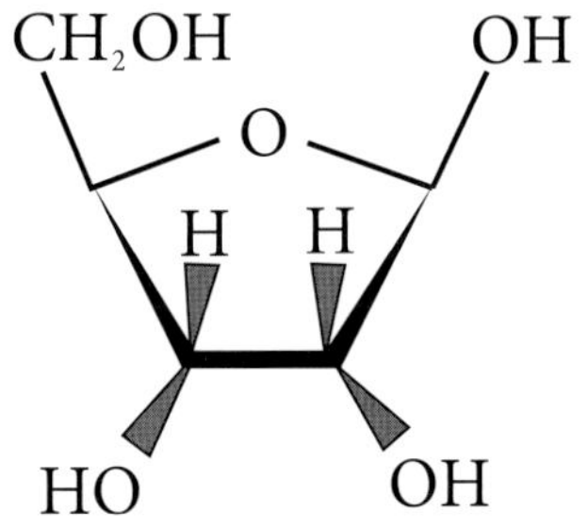

$$(1.10)$$

The *cis* conformation of a peptide bond has both C^α on the same side of the C–N amide bond:

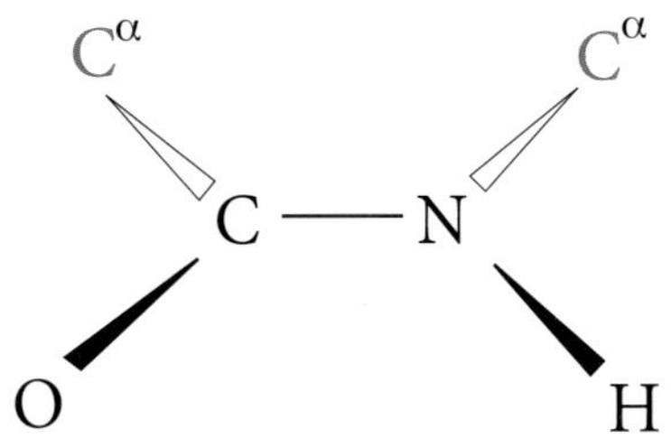

$$(1.11)$$

The peptide bond has partial double-bond character (Section 8.1), which tends to keep the four atoms planar. The notation *s-cis* is used to emphasize that the conformation about a single bond is being described, as in a *s-1-cis* long-chain aldehyde:

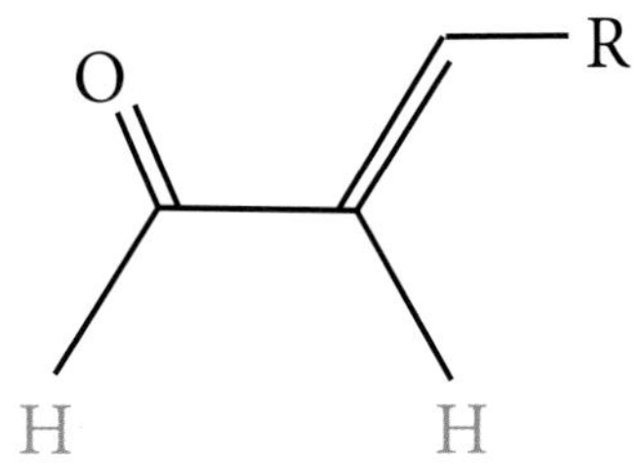

$$(1.12)$$

An example would be retinal.

Cis and *trans* isomers can often be interconverted by rotation about the central linkage, but at widely varying rates. The bond order of the central linkage correlates with the magnitude of the energy barrier to rotation. It ranges from high values (slow rotations) for double bonds, via an intermediate range for linkages having a partial double-bond character, down to low values (rapid rotation) for C–C single bonds. The magnitude of the rotational barrier determines whether or not the individual isomers are readily interconverted; if readily interconverted, they are considered different conformations; otherwise, they are different configurations. There is the potential for confusion in such instances unless the time scale is specified.

In molecular biology, the terms *cis* and *trans* are also often used to refer to genetic elements on the same or different nucleic acid molecules.

1.1.B. Prochiral

The designation prochiral indicates that substitution of an atom with a different isotope will alter the chirality of the molecule. A **prochiral center** is an atom with two identical substituents where substitution of either one with a different isotope would make that atom a chiral center. If this would generate enantiomers, the groups are **enantiotopic**, while if the substitution would produce diastereomers, the groups are **diastereotopic** (Section 1.1.A.3).

For example, the C^α atom in the amino acid glycine:

$$\text{pro-}R \quad \begin{array}{c} H \\ \diagdown \\ C^\alpha \\ \diagup \quad | \\ H \quad NH_3^+ \end{array} \diagup CO_2^- \quad \text{pro-}S$$

(1.13)

is a prochiral center because substitution of either of the two nearly identical H^α atoms with a heavier isotope would generate a chiral molecule. The two H^α atoms are enantiotopic because the replacement of one would generate different enantiomers of glycine. Enantiotopic groups are distinguished by using the prefix 'pro-' with the specification of the configuration generated. In chemical structures, the prochirality of an atom can be identified by R or S. The forward H atom of glycine is pro-S because substitution with ^2H there would generate (S)-[2-^{2}H]glycine, as indicated by the counterclockwise arc connecting the atoms in order of priority. In the peptide glycyl-L-serine, where the serine residue has a chiral center, the glycine H^α atoms are diastereotopic because substitution of either glycine H would generate diastereomers.

In the case of citrate:

$$\begin{array}{c}
\text{Pro-(2S,3R)} \quad H \diagdown \quad \diagup CO_2^- \\
\phantom{\text{Pro-(2S,3R)}} \quad C\text{-}2 \\
\text{Pro-(2R,3R)} \quad H \diagup \quad | \\
\phantom{\text{Pro-(2R,3R)}} \quad {}^-O_2C \diagdown C\text{-}3 \\
\phantom{\text{Pro-(2R,3R)}} \quad HO \diagup \quad \diagdown \\
\phantom{\text{Pro-(2R,3R)}} \quad H \diagup C\text{-}4 \diagdown CO_2^- \\
\phantom{\text{Pro-(2R,3R)}} \quad H
\end{array}$$

(1.14)

the H_2C groups 2 and 4 are enantiotopic; C3 is a prochiral center because substitution of any of the H atoms or carboxyl group attached to either C2 or C4 will generate chirality there. The upper (C2) $-CH_2-COO^-$ is the pro-R group because substituting an isotope anywhere in the functional group would generate (3R)-citrate. C2 and C4 are also prochiral centers, because substituting either methylene H atom would generate a chiral center. The two H atoms on C2 are diastereotopic, because substituting deuterium at either of these positions would generate diastereomers; two new chiral centers would be generated at C2 and C3. The chirality at C3 would be the same for both H atoms, but different at C2.

Double-bonded C and N atoms that are planar have three substituents. They will be prochiral if the three substituents are different, because the two faces of the planar molecule are not equivalent. The faces of the molecule are differentiated by the designation of Re or si based on the priorities of the three substituents according to the rules of priority for substituents (Section 1.1.A.1). The two faces of pyruvate are not equivalent:

$$\text{L-lactate} \quad \xleftarrow{2H} \quad \text{Re-face} \quad \begin{array}{c} O \\ \| \\ C \\ / \diagdown \\ CH_3 \quad CO_2^- \end{array} \quad \text{Si-face} \quad \xrightarrow{2H} \quad \text{D-lactate}$$

(1.15)

The circle connecting the three different substituents in their priority order describes a clockwise motion when viewed from the left, but a counterclockwise motion when viewed from the right. The two faces of pyruvate are enantiotopic because addition of chemical groups to the opposite sides of the central C atom would generate enantiomers. As shown in Equation 1.15, addition of ^2H to the Re face would generate L-lactate, whereas addition to the opposite face would generate D-lactate. The faces would be diastereotopic if addition to the opposite faces would generate diastereomers.

The concept of enantiotopic and diastereotopic groups is important in molecular biology because these groups interact differently with chiral molecules, such as proteins (Chapter 12).

Prochirality revisited. An approach for restructuring stereochemistry by novel terminology. S. Fujita (2002) *J. Org. Chem.* **67**, 6055–6063.

Analysis of the ¹³C NMR spectra of molecules, chiral by isotopic substitution, dissolved in a chiral oriented environment: towards the absolute assignment of the pro-R/pro-S character of enantiotopic ligands in prochiral molecules. P. Lesot *et al.* (2004) *Chemistry* **10**, 3741–3746.

1.1.C. Tautomers

Tautomers are structural isomers that interconvert rapidly, so all the isomers will exist in a solution of the compound at equilibrium. In most cases, tautomers are generated by structural and electronic rearrangements caused by moving a single proton. This can occur via the solvent, with one water molecule or hydroxide ion removing the proton and another water molecule donating a proton. For example, glyceraldehyde-3-phosphate and dihydroxyacetone-phosphate are both tautomers of the enediol intermediate in their interconversion:

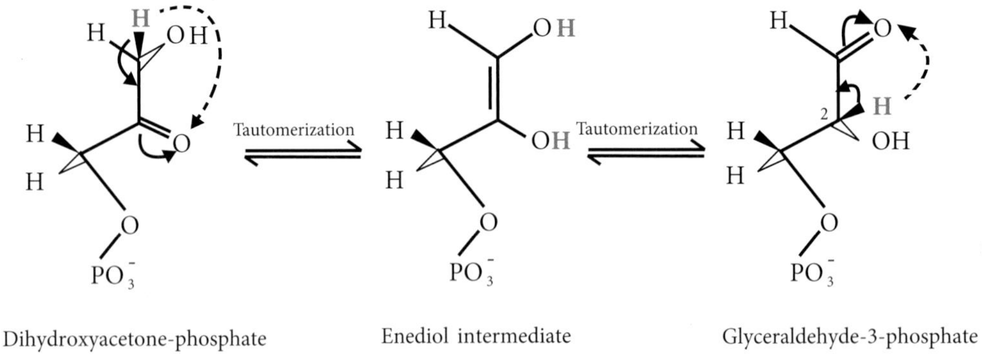

$$\text{Dihydroxyacetone-phosphate} \rightleftharpoons \text{Enediol intermediate} \rightleftharpoons \text{Glyceraldehyde-3-phosphate} \tag{1.16}$$

Glyceraldehyde-3-phosphate and the enediol are structural isomers that result when a proton is transferred from C2 to the aldehyde oxygen, as indicated by the dashed arrow; transfer of the proton is accompanied by a rearrangement of the electrons, as indicated by the solid curved arrows. Similarly, dihydroxyacetone-phosphate and the enediol intermediate are tautomers, as a single proton transfer can also cause their interconversion. The two H atoms that move in each tautomerization of Equation 1.16 are indicated in bold.

The phrase 'rapidly' is not precise, so two molecules that are each tautomers of a third molecule may not be tautomers. For example, in Equation 1.16 glyceraldehyde-3-phosphate and dihydroxyacetone-phosphate are not interconverted rapidly and are not considered to be tautomers, even though both are considered tautomers of the enediol intermediate.

Another form of tautomerism that is important in molecular biology is the existence of both open-chain and ring forms of monosaccharides. Thus the three common forms of glucose α–gluco-pyranose ⇌ Aldose ⇌ β-gluco-pyranose

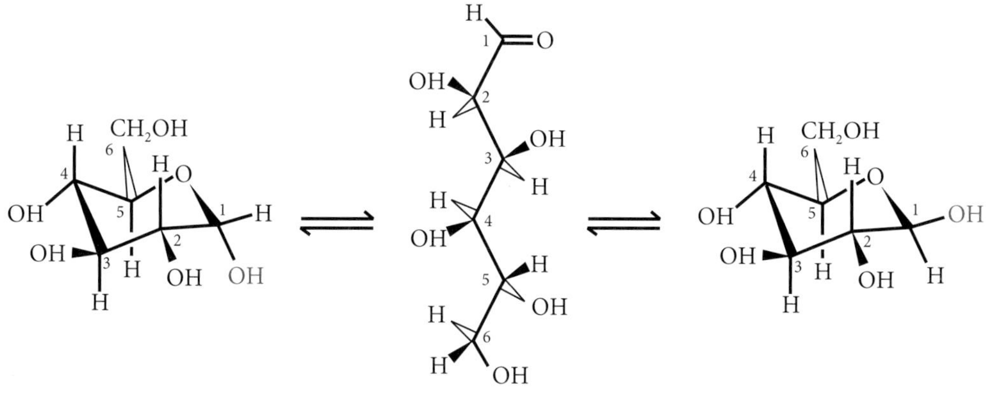

(1.17)

are all tautomers because they interconvert rapidly in aqueous solution. The two cyclic pyranose forms are also epimers and diastereomers (Section 1.1.A).

The existence of tautomers can be detected spectroscopically if each tautomeric form gives rise to different spectral features. For example, resonances from all three tautomeric forms of glucose (Equation 1.17) are apparent simultaneously by proton NMR (^{1}H-NMR). One tautomer can predominate if it is stabilized in some way, such as by incorporating it into a crystal lattice or into a larger molecule. The equilibration of the tautomers in Equation 1.17 in aqueous solution can be monitored using the change in optical rotation after dissolving a crystal of either of the pure pyranose forms in water.

Often one tautomeric form is more stable than the other in solution, but the less-favored tautomer might be the biologically active form. Identifying the correct tautomers of the nucleobases of nucleic acids was a crucial step in discovering the double-helical structure of DNA, as it governs the pairing of bases by hydrogen bonding (Section 2.2.A). Keto-enol tautomerism, comparable to that at C1 in Equation 1.17, can change a carbonyl group, which is a hydrogen-bond acceptor, to a hydroxyl, which can be a hydrogen-bond donor. Tautomeric forms of all five nucleic acid bases exist in solution but the rare tautomer occurs $<10^{-4}$ of the time.

Tautomerism of sterically hindered Schiff bases. Deuterium isotope effects on ^{13}C chemical shifts. A. Filarowski et al. (2005) *J. Phys. Chem. A* **109**, 4464–4473.

Differential solvation and tautomer stability of a model base pair within the minor and major grooves of DNA. F. Y. Dupradeau et al. (2005) *J. Am. Chem. Soc.* **127**, 15612–15617.

Hydrogen-bonded nucleic acid base pairs containing unusual base tautomers: complete basis set calculations at the MP2 and CCSD(T) levels. J. Rejnek & P. Hobza (2007) *J. Phys. Chem. B* **111**, 641–645.

1.2. CONFORMATIONS

Different conformations are nonidentical spatial arrangements of the atoms of a molecule achieved solely by rotations about single covalent bonds. Molecules with identical covalent structures but different conformations are known as **conformers**. The ability of two conformers to assume the same geometry spontaneously and become identical differentiates them from stereoisomers. The terms 'conformers' and 'stereoisomers' should not be confused, nor should 'configuration' and 'conformation'.

Conformers are usually interconverted rapidly because single bonds can usually rotate rapidly. Unless certain conformers are stabilized specifically, it is difficult, if not impossible, to isolate individual conformers. Any slow bond rotations will provide exceptions to this rule; one is rotation about the C–N peptide bond of polypeptides (Equation 1.11). It has partial double-bond character, so it is planar and the two possible *cis* and *trans* conformations are only interconverted slowly (Section 8.2.B.1).

Any conformation of a molecule of known covalent structure can be specified by the rotations about its single bonds, generally measured by either the **torsion angle** (Section 1.2.A) or the **dihedral angle** (Section 1.2.B). In general, interactions between neighboring atoms, usually steric repulsions, mean that not all bond rotations have the same free energy and are equally probable. Ethylene glycol can serve as a simple model:

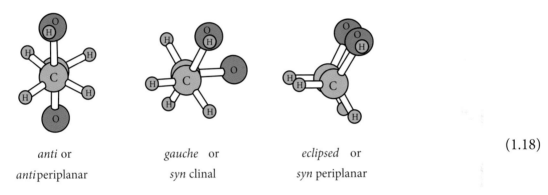

anti or
*anti*periplanar

gauche or
syn clinal

eclipsed or
syn periplanar

(1.18)

Individual torsion angles may be described qualitatively as ***anti*, *gauche*** or ***eclipsed*** (Section 1.2.A.1), which are more formally named ***anti*periplanar, *syn*clinal** and ***syn*periplanar**. The term *gauche* generally means not lying in a plane, while *anti* is used to describe the relative orientation of two substituents on adjacent atoms of a molecule when their torsion angle is about ±180°. When the large substituents (hydroxyl groups in this case) are superimposed, the conformation is described as *eclipsed*; steric clashes between the two hydroxyl groups mean that this conformation is the least stable. When they are staggered by 60°, the conformation is *gauche*. When they are opposite, staggered by 180°, it is *trans* or *anti*; this conformation would have the fewest clashes between the hydroxyl groups and be the most stable. With constrained systems, such as cyclic molecules, certain common combinations of torsion angles can lead to descriptive names for groups of atoms; for example, the 2'-*endo* conformation of ribose describes all five of the torsion angles of the furanose ring (Figure 2-5).

A large biological macromolecule can have a stable, fixed 3-D structure, referred to as its **native conformation**, if it has sufficient stabilizing interactions between its various atoms. Although most of the conformation is fixed and it is considered a single conformation, parts of the molecule may still be flexible and able to undergo rotations about certain bonds. The average overall conformation can be considered a **macro-conformation**, whereas the variations resulting from flexibility define various **micro-conformations**. Interconverting different macro-conformations requires a cooperative change of a number of bond rotations simultaneously, whereas micro-conformations are interconverted by changes of just one or a few bond rotations. A cooperative change of a number of bond rotation angles, such as occurs in protein unfolding (Chapter 11) or the melting of double-stranded DNA (Chapter 5), will usually occur only slowly or infrequently under physiological conditions, although it can be speeded up dramatically under denaturing conditions. This relatively slow interconversion of the two macro-conformations is described as a **conformational change**, in which the macromolecule has been converted from one family of micro-conformations to an experimentally distinguishable family of other micro-conformations.

Conformational analysis. E. L. Eliel *et al.* (1967) Wiley-Interscience, NY.

1.2.A. Torsion Angle

Torsion angles within a molecule refer to the rotations about individual covalent bonds linking a pair of atoms; they are defined using two further atoms bonded to the first pair of atoms. Therefore, four atoms connected by three consecutive covalent bonds are used to define the torsion angle of the middle bond:

(1.19)

The torsion angle of the bond connecting atoms B and C is determined by looking down this bond and measuring the angle that the bond between atoms A and B must be rotated through to eclipse the bond between atoms C and D. Torsion angles are usually expressed as having values between −180° and +180°; the value is zero when the flanking bonds are eclipsed, and positive if the front bond is rotated in a clockwise direction. For biological macromolecules, the backbone torsion angles are defined by four contiguous backbone atoms.

Torsion angles about single bonds may also be described qualitatively. They are distinguished as being (a) positive or negative, (b) syn or anti and (c) periplanar or clinal:

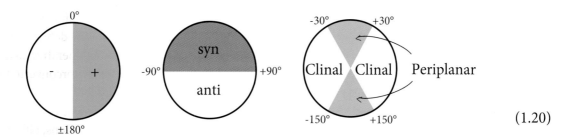

(1.20)

These define eight different rotamers that are indicated with their two-letter abbreviations:

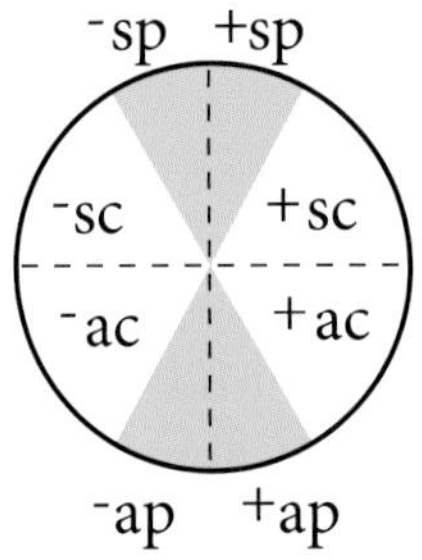

(1.21)

Torsion angles are important because they are commonly used to define the conformations of molecules and any differences between them.

Application of torsion angle molecular dynamics for efficient sampling of protein conformations. J. Chen *et al.* (2005) *J. Comput. Chem.* **26**, 1565–1578.

On updating torsion angles of molecular conformations. V. Choi (2006) *J. Chem. Inf. Model.* **46**, 438–444.

1.2.B. Dihedral Angle

A dihedral angle is the angle formed by intersecting lines normal to two planes:

(1.22)

In this example, one plane is defined by atoms A, B and C, the other by atoms B, C and D. The dihedral angle θ is defined as the angle between the two lines perpendicular to each plane. The preferred nomenclature for rotations about single bonds in a molecule uses the conventions described for torsion angles. Unlike torsion angles, however, the dihedral angle can be measured between the two planes defined by any group of three atoms, whether or not they are bonded together. When four atoms are specified in order, such as atoms A, B, C and D, the dihedral angle is measured between the plane containing atoms A, B and C and the plane containing atoms B, C and D. When the four atoms are not linked by consecutive covalent bonds, and a torsion angle can not be defined, this dihedral angle is known as an **improper dihedral angle** or **improper torsion angle**.

Straightening out the dihedral angles. J. Clauwaert & J. Z. Xia (1993) *Trends Biochem. Sci.* **18**, 317–318.

Pairwise NMR experiments for the determination of protein backbone dihedral angle ϕ based on cross-correlated spin relaxation. H. Takahashi & I. Shimada (2007) *J. Biomol. NMR* **37**, 179–185.

1.3. CONFORMATIONS OF IDEALIZED POLYMERS

The conformations adopted by a polymer will depend on the structures and conformational properties of the monomers, the way that they are covalently linked together, and the environment, especially the relative interactions of the polymer with the solvent and with itself. In the simplest case, **where each monomeric unit is the same and each adopts only a single conformation, the linear polymer will adopt a helical conformation** (Figure 1-1). A helix is defined as a point that rotates at a given distance around an axis z while moving parallel to that axis. In a helical macromolecule, the helix will be characterized by the angle α and the translation p along axis z between adjacent monomers. It can be specified by the **helical repeat**, the number of monomers per turn (which need not be an integer), and the **pitch** of the helix, the vertical distance between adjacent turns. Whether the helix is left- or right-handed is determined by the sense of the rotation needed to advance along z: the helix should be considered a screw that needs to be advanced by turning it with a screwdriver. If the direction of rotation is that indicated by the fingers of the right hand, the helix is right-handed. Otherwise it is left-handed. Note that a helical repeat of exactly 2 does not produce a helical molecule but one that takes a zig-zag path (Figure 1-1C).

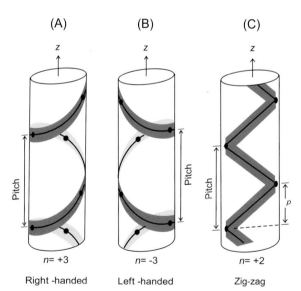

Figure 1-1. Description of helical conformations generated when each monomer of a polymer adopts the same conformation. The two helices in (A) and (B) are identical except for their handedness. Both have three monomers per turn, n, so $\alpha = 120°$ and pitch = $3\,p$. If $n = 2$, as in (C), the polymer does not adopt a helical path but is straight and zig-zag, as will be found in the β-strands of proteins (Figure 8-9); $\alpha = 180°$ and pitch = $2\,p$. In each case, the z-axis is vertical.

If, on the other hand, each monomer can adopt more than one conformation, the polymeric chain is likely to adopt a wide variety of micro-conformations, unless it is subject to interactions that favor one or more individual conformations over all the others. The 3-D structure will be specified by the torsion angles (Section 1.2.A) adopted by each monomer.

The ability to adopt a number of conformations is an entropic factor that stabilizes the flexible state, known as the **conformational entropy**. A single conformation will be adopted only if the interactions stabilizing that particular conformation are sufficiently strong to overcome the conformational entropy tending to keep the polymer unfolded. The number of micro-conformations and the conformational entropies of polymer chains can be very large. For example, if each residue can adopt an average of j conformations, and there are N residues in the polymer chain, the total number of conformations possible will be approximately j^N. If the reasonable assumption is made that j is 8 and N is 500, there will be 10^{452} ($= 8^{500}$) conformations possible, a truly astronomical number. Of course, some of these conformations will not be feasible because they would have atoms of the polymer overlapping in space, the **excluded volume effect**. There is still, however, much scope to be conservative and predict an astronomical number of conformations. For example, even a short polymer of 100 residues in which each residue could adopt only two different conformations could adopt more than 10^{30} different polymer conformations. If all these conformations have similar free energies, each would have only a very small probability of occurring in a molecule. At 25°C, the free energy contribution of the conformational entropy for a polymer in which each residue can adopt 10 conformations will be 1.36 N kcal/mol. Consequently, for any one conformation to predominate will require, on average, stabilizing interactions >1.36 kcal/mol/residue. In the absence of such stabilizing interactions, a polymer will tend to exist in many different conformations. Yet proteins and nucleic acids, as will be shown in Chapters 2, 4 and 9, are able to adopt single folded conformations that predominate. Such stable conformations can be considered macro-conformations, in contrast to the micro-conformations that are adopted only transiently (Section 1.2).

A disordered polymer will usually have so many possible micro-conformations that not all can possibly be present within a population of molecules at any instant of time. For example, a reasonable sample consisting of 1 μmol of a polymer will contain only 6×10^{17} molecules. Moreover, micro-conformations can be converted no more rapidly than rotations can occur within the polymer backbone, which requires at least 10^{-10} s, so a single molecule can sample no more than 10^{+10} conformations per second, which is likely to be only a small fraction of the conformations possible. Although it may not be possible for all conformations to be present or sampled, very many micro-conformations will be present within a sample of polymer molecules, and only statistical averages of the properties of the population of molecules can be given.

With so many conformations possible, the conformational properties of random polymers are best calculated statistically, using the mathematical procedures developed for synthetic polymers. Such calculations require detailed knowledge of the conformational properties of the monomeric unit of the polymer and the relative energies of all of its possible micro-conformations. Note that with N monomer units, there are only $N - 1$ linkages between them, and the torsion angles of only $N - 2$ such linkages specify the conformation (specifying the other linkage will merely fix the orientation of the molecule in space).

Conformations of Macromolecules. T. M. Birshtein & O. B. Ptitsyn (1964) Wiley-Interscience, NY.

Statistical Mechanics of Chain Molecules. P. J. Flory (1969) John Wiley, NY.

1.3.A. RANDOM COILS

Polymers in which the conformational properties of each residue are independent of the conformations of all other residues, except for those adjacent in the polymer chain, are known as **random coils**.

Frequently the statistical properties are calculated of an ideal **unperturbed random coil**, in which the 3-D covalent structure of the polymer is considered, along with the conformational properties of each monomer unit, but not any interactions between distant parts of the polymer. In the 'unperturbed' state, no account is taken of the excluded volume effect, so impossible conformations in which nonbonded atoms occupy the same space are included. This is unrealistic but makes the calculations more feasible.

The average properties of such polymers are often compared with those of the hypothetical **random-flight chain** or **freely jointed chain**. This is not a realistic model either, but simply a mathematical string of vectors of fixed lengths representing the bonds between adjacent atoms; the atoms are not included, the chain has no volume, all bond angles have equal probability, and all rotations about the bonds are equally likely. A somewhat more realistic model is the **freely rotating chain**, in which a constraint of fixed-bond angles between monomers is introduced. When the actual conformational preferences of the monomer unit are taken into account, by permitting only the most favorable possible torsion angles, the model is known as the **rotational isomeric state** model. In this case, however, the rotations about adjacent bonds are not independent, as the corresponding atoms would interact, so the allowed torsion angles must be those for pairs of neighboring bonds rather than single bonds. Calculations on polymers where each monomer unit contributes more than one covalent bond to the polymer backbone often simplify the architecture by using a single **virtual bond** for each monomer, a vector joining the comparable atoms of adjacent monomeric units. For example, the torsion angles of the three covalent bonds that make up the backbone of one amino acid residue (Figure 8-1) can be replaced by one rotation about a virtual bond linking adjacent C^α atoms. In spite of its simplification, the rotational isomeric state model can simulate experimental data reasonably well. Such computations are complex, however, and outside the scope of this volume.

Random-coil behavior and the dimensions of chemically unfolded proteins. J. E. Kohn *et al.* (2004) *Proc. Natl. Acad. Sci. USA* **101**, 12491–12496.

Secondary structures in long compact polymers. R. Oberdorf *et al.* (2006) *Phys. Rev. E* **74**, 051801.

---

1. End-to-end Distances

Of greatest interest with random polymers are the averages and the variation of their physical dimensions. The **root mean square** (r.m.s.) value of the distance, r, between two atoms of a hypothetical random-flight chain is given by:

$$<r^2>_0^{1/2} = n^{1/2} \, l \qquad (1.23)$$

where n is the number of linkages between monomers ($= N - 1$, where N is the number of monomers in the chain) and l is the distance between monomers in the polymer backbone. The angle brackets in Equation 1.23 indicate that it is the average over all conformations, and the subscript zero refers to the unperturbed state. Note that the dimensions of such a random coil increase only with the square root of the number of residues in the polymer chain.

The calculated distribution of end-to-end distances is usually expressed as either the **Gaussian distribution function** or the **radial distribution function**, which are illustrated in Figure 1-2 for a hypothetical random-flight chain. The Gaussian distribution function, W(x, y, z) dx dy dz, gives the probability that the end of the polymer chain is within the volume dx dy dz at coordinates (x, y, z); the origin is taken as the other end of the chain. This distribution is spherically symmetrical, so it is usually expressed as the radial distribution function, W(r) dr, which is the probability that the two ends of the chain are within a distance r and r + dr of each other. For unperturbed random-coil chains, the scale for r in random-flight chains is simply increased by the factor $C_n^{1/2}$ (Section 1.3.A.3).

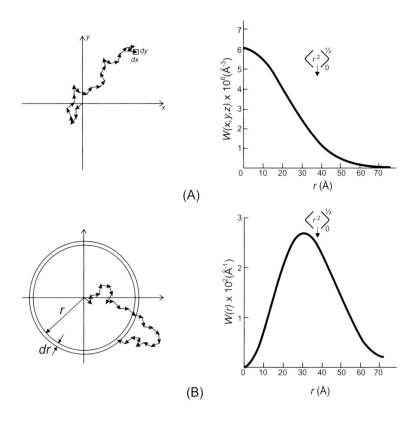

Figure 1-2. Illustration of Gaussian (A) and radial (B) distribution functions for the end-to-end distance of a freely jointed chain. On the *left* of each is a two-dimensional representation of how each distribution function is defined, giving the probability that the other end of the chain will lie within the enclosed area. The distribution functions are given by:

$$W(x,y,z)\, dx\, dy\, dz = \left(\frac{\beta}{\sqrt{\pi}}\right)^3 \exp(-\beta^2 r^2)\, dx\, dy\, dz$$

$$W(r)\, dr = \left(\frac{\beta}{\sqrt{\pi}}\right)^3 \exp(-\beta^2 r^2)\, 4\pi\, r^2\, dr$$

where $r^2 = x^2 + y^2 + z^2$, $\beta = (3/2nl^2)^{1/2}$ and n is the number of freely jointed bonds of length l. On the *right* of each is the calculated distribution for a freely jointed polypeptide chain of 100 residues and a virtual bond length of 3.8 Å. The r.m.s. distance, $<r^2>_0^{1/2} = 38$ Å is indicated. The probability of the Gaussian distribution function reaches a maximum near the origin, whereas that of the radial distribution function approaches zero. The latter is simply a mathematical consequence of the decreasing volume of the spherical shell between r and $r + dr$ as r decreases. From T. E. Creighton (1993) *Proteins: structures and molecular properties*, 2nd edn, W. H. Freeman, NY, p. 178.

One of the most direct methods of measuring distances between residues in a random coil-like polymer is to attach a fluorescent donor group to one and an acceptor to the other and measure the efficiency of the fluorescence energy transfer between them. It ideally varies inversely with the sixth power of the distance between them, but only if the donor and acceptor groups introduced have random orientations and do not interact with each other or modify the properties of the polymer; unfortunately fluorescent groups are usually large hydrophobic moieties that almost certainly interact with each other.

End-to-end distribution function of stiff polymers for all persistence lengths. B. Hamprecht & H. Kleinert (2005) *Phys. Rev. E* **71**, 031803.

Scaling exponents and probability distributions of DNA end-to-end distance. F. Valle *et al.* (2005) *Phys. Rev. Lett.* **95**, 158105.

End-to-end distance distributions and intrachain diffusion constants in unfolded polypeptide chains indicate intramolecular hydrogen bond formation. A. Moglich *et al.* (2006) *Proc. Natl. Acad. Sci. USA* **103**, 12394–12399.

---

2. Radius of Gyration

Another statistical measure often used with random coils is the average **radius of gyration (R_g)**, which is defined as the r.m.s. distance of the collection of atoms from their common center of gravity. For the random-flight chain:

$$<R_g>_0^2 = \frac{nl^2}{6} \frac{n+2}{n+1} \tag{1.24}$$

For large values of n, this becomes:

$$<R_g>_0^2 = \frac{nl^2}{6} = \frac{<r^2>_0}{6} \tag{1.25}$$

This relationship holds for the unperturbed states of all very long polymers, so the radius of gyration is simply 0.408 (= $6^{-1/2}$) times the average end-to-end distance.

3. Characteristic Ratio

The flexibility of an actual random coil is more limited than that of the hypothetical random-flight chain, and its dimensions tend to be considerably greater than those predicted by Equations 1.24 and 1.25. This increased stiffness is usually expressed as the **characteristic ratio**, C_n:

$$C_n = \frac{<r^2>_0}{nl^2} \tag{1.26}$$

where $<r^2>_0$ is the square of the observed average end-to-end distance for the actual random coil, and nl^2 is this value for the random-flight chain (Equation 1.23). The value of C_n increases with increasing length of the polymer, n, reflecting the stiffness of the polymer backbone, so that segments distant in the polymer still do not have random orientations. In very long chains, however, C_n approaches a limit designated as C_∞; in this case, the distant segments are behaving as a truly random-flight chain.

Polypeptide chains of L-amino acids have values of C_∞ of approximately 9 (Section 8.2), while single-stranded polynucleotides generally have values of 10–15. Calculations in which all sterically allowed conformations are given equal weight predict values of C_∞ of 2.0–4.5, so energy differences between sterically allowed conformations must bias the chain towards greater extension.

A real polymer chain may be approximated as a polymer of freely jointed **statistical segments**, chosen to be long enough so that each is randomly oriented with respect to all other such segments. The stiffer the polymer chain, the longer the statistical segment. The statistical segment of a polypeptide chain is about 10 residues or 36 Å.

Another useful parameter is the **persistence length**, a. This is defined as the average projection of the end-to-end distance vector on the first bond of the chain, in the limit of infinite chain length. It can be considered a measure of the length over which the chain persists in the same direction as the first bond. For polymer chains composed of identical bonds of length l, the persistence length a is closely related to the limiting characteristic ratio, C_∞:

$$a = (C_\infty + 1)\, l/2 \tag{1.27}$$

For random polypeptide chains, the value of the persistence length a is about 20 Å, or nearly six residues, while for double-stranded DNA it is approximately 500 Å, roughly 150 base pairs, depending upon the conditions. Double-stranded DNA is clearly much less flexible than an individual polypeptide chain.

Calculations of the φ–ψ conformational contour maps for *N*-acetyl alanine *N'*-methyl amide and of the characteristic ratios of poly-L-alanine using various molecular mechanics force fields. C. H. Lee & S. S. Zimmerman (1995) *J. Biomol. Struct. Dyn.* **13**, 201–218.

1.3.B. Excluded Volume Effects and Theta Solvents

Within an ensemble of many micro-conformations of an unfolded polymer, some will have all atoms in contact with the solvent, whereas others will have various parts of the polymer in contact with each other. Conformations where atoms would overlap in space are excluded, of course, which is the **excluded volume effect**. The relative energies of the individual micro-conformations will vary, depending upon the relative energetics of the different interactions. In a poor solvent, which does not interact favorably with a polymer, those conformations with contacts within the chain will be favored over those interacting only with the solvent. Consequently, the average dimensions of the population will be reduced. In contrast, a solvent that interacts favorably with the polymer will favor those conformations that are more extended, and the average dimensions of the population will be expanded. In a Θ **solvent**, known as **theta conditions**, the interactions of the polymer with the solvent are balanced to be the same energetically as those within the polymer. Polymers usually have relatively poor solubilities in Θ solvents.

In a Θ solvent, the solvent-induced preference of the chain for relatively compact conformations counterbalances exactly the effects of excluded volume. A real polymer then has dimensions like those calculated for the unperturbed state, in which excluded volume effects are ignored. Such ideal solvents are, however, generally not practical with biological polymers, such as polypeptide and polynucleotide chains, where there are a variety of side-chains on the polymer.

Contributions of short-range and excluded-volume interactions to unperturbed polymer chain dimensions. H. Yamakawa & T. Yoshizaki (2004) *J. Chem. Phys.* **121**, 3295–3298.

The optimized Rouse–Zimm theory of excluded volume effects on chain dynamics. J. H. Kim & S. Lee (2004) *J. Chem. Phys.* **121**, 12640–12649.

Corrections to scaling and crossover from good- to theta-solvent regimes of interacting polymers. A. Pelissetto & J. P. Hansen (2005) *J. Chem. Phys.* **122**, 134904.

---

1. Covalent Cross-links

The probability that the ends of a chain are spatially near each other in the polymer gives the probability that two functional groups on the polymer that are separated by the same number of residues will interact by reacting chemically with each other or forming a covalent cross-link. This would be an intramolecular reaction, and its probability can be expressed as the **effective concentration** of the two residues relative to each other within the polymer. **The ends of relatively flexible polymers with 30–100 residues are usually measured to have effective concentrations in the region of 10^{-3} M.** In contrast, the values measured are usually in the range of only 10^{-7} to 10^{-9} M for double-stranded DNA molecules with up to 10,000 base pairs, because DNA is relatively inflexible. The values can vary dramatically when residues are close in the polymer and will depend on the particular stereochemistry of the polymer. With longer chains, the effective concentration decreases with increasing length in proportion to $n^{-3/2}$.

Cross-linking two parts of a polymer chain with a covalent bond decreases the number of conformations that are possible and the conformational entropy of the random coil. The magnitude of this effect can be calculated by considering the probability, in the absence of the cross-link, that the ends of a random coil corresponding to the points of the cross-linkage would lie simultaneously within a small volume element, V (Figure 1-2). The smaller this probability, the greater the effect on the conformational entropy of constraining the ends to lie within this volume by a covalent cross-link. **The further apart in the covalent structure the residues that are cross-linked, the greater the decrease in conformational entropy.** Unfortunately, it is not certain what volume element V is appropriate for any particular example, and the energetic consequences of cross-links cannot be calculated with any degree of confidence.

Most importantly, **covalent cross-links stabilize any folded conformations with which they are compatible, because they destabilize the disordered state by decreasing its conformational entropy.**

Dissecting the roles of individual interactions in protein stability: lessons for a circularized protein. D. P. Goldenberg (1985) *J. Cell Biochem.* **29**, 321–335.

Loops, linkages, rings, catenanes, cages, and crowders: entropy-based strategies for stabilizing proteins. H. X. Zhou (2004) *Acc. Chem. Res.* **37**, 123–130.

Loop entropy and cytochrome c stability. L. Wang *et al.* (2005) *J. Mol. Biol.* **353**, 719–729.

1.4. STRUCTURE DATABASES: STRUCTURES ON THE WEB

Structure databases contain the 3-D structures of biological macromolecules determined by X-ray crystallography and NMR. The primary structure database is the Protein Data Bank (PDB), an archive of all publicly available 3-D structures of proteins, nucleic acids, carbohydrates, viruses and biomolecular complexes. The PDB is basically a collection of text files, each containing a structure entry deposited by those who determined the structure. An entry file consists of (1) text information of definition, source, references and comments, (2) the sequence information, (3) the secondary structure information and (4) the 3-D coordinates of all the atoms, as well as crystallographic structure factors and NMR experimental data. There is an efficient search engine for finding files of the desired structures. Other useful structure databases include the Nucleic acid Database (NDB) for nucleic acids and the Cambridge Structural Database (CSD) for organic and metal–organic compounds. The World Wide Web addresses for these databases are given in Table 1-1. The majority of databases in molecular biology are publicly available and freely accessible via the Web.

Table 1-1. Web addresses for the structure databases

Database	Address
PDB	www.rcsb.org
NDB	ndbserver.rutgers.edu
CSD	www.ccdc.cam.ac.uk

To view the known structures of macromolecules, it is recommended that you find the name of the data file in the PDB, then view the molecule using the program Jmol at http://molvis.sdsc.edu/fgij/index.htm.

Protein structure database search and evolutionary classification. J. M. Yang & C. H. Tung (2006) *Nucleic Acids Res.* **34**, 3646–3659.

Conformational specificity of non-canonical base pairs and higher order structures in nucleic acids: crystal structure database analysis. S. Mukherjee *et al.* (2006) *J. Comp. Aided Mol. Des.* **20**, 629–645.

The CATH domain structure database: new protocols and classification levels give a more comprehensive resource for exploring evolution. L. H. Greene *et al.* (2007) *Nucleic Acids Res.* **35**, D291–297.

~ CHAPTER 2 ~

DNA STRUCTURE

The discovery of the well-known double-helix structure of deoxyribonucleic acid (DNA) (Figure 2-1) was the seminal event in the origin of the field of molecular biology. This structure demonstrated graphically how just the three-dimensional (3-D) structures of biological macromolecules could be sufficient to illuminate the mechanisms of their biological functions. The DNA double-stranded helix held together by complementary base pairs immediately revealed how DNA could store and transmit all the genetic information in the genome of a living organism. It is now clear that the canonical Watson–Crick DNA double helix is not a rigid structure but, depending upon the conditions and the local nucleotide sequence, can adopt variant conformations that are both subtle and important. Even markedly different structural forms of DNA have been discovered. The structure of DNA is still central to molecular biology.

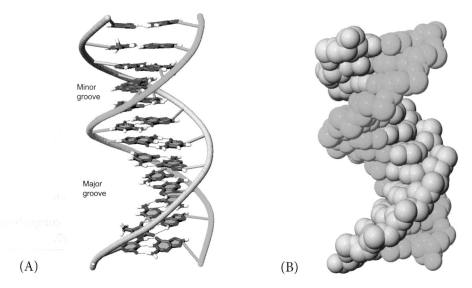

Figure 2-1. The double-stranded helical structure of B-DNA. This structure was determined by NMR with the dodecamer in which one chain has the sequence 5′–d(CTCGGCGCCATC)–3′ and the other is the complementary 5′–d(GATGGCGCCGAG)–3′. (A). The detailed structures of the nucleobases are presented, but the deoxyribose-phosphate backbone is depicted as a *simple cylinder*. The major and minor grooves are indicated, and some hydrogen bonds between bases are indicated as *dashed lines*. (B). Space-filling representation of the same structure in which all the atoms are depicted as spheres with their van der Waals radii. Figure generated from PDB file *2hkb* using the program Jmol.

DNA occurs naturally almost exclusively as the genetic material of organisms, the chromosomes containing the genetic information. Generally there is one DNA molecule per chromosome, and its detailed structure contains all the genetic information.

Principles of Nucleic Acid Structure. W. Saenger (1984) Springer-Verlag, NY.

Understanding DNA. C. R. Calladine & H. R. Drew (1992) Academic Press Inc., CA.

DNA Structure and Function. R. R. Sinden (1994) Academic Press Inc., CA.

Nucleic Acids in Chemistry and Biology. G. M. Blackburn & M. J. Gait (eds) (1996) Oxford University Press, NY.

Bioorganic Chemistry: nucleic acids. S. M. Hecht (ed.) (1996) Oxford University Press, NY.

Nucleic Acids: structures, properties and functions. V. A. Bloomfield *et al.* (2000) University Science Books, CA.

2.1. POLYNUCLEOTIDES

The nucleic acids DNA and ribonucleic acid (RNA) (Chapter 4) are polynucleotides that are produced by polymerizing nucleotides, either deoxynucleotides or ribonucleotides. Each **nucleotide** building block consists of three components: a sugar (**deoxyribose** or **ribose**), a phosphate group and a nucleobase. Two of the bases are derivatives of **purine** (Pu), **adenine** (A) and **guanine** (G), and two of **pyrimidine** (Py), **cytosine** (C) with **thymine** (T) in the case of DNA and **uracil** (U) in the case of RNA. When the bases are attached covalently to only the sugar moiety, they are known as the **nucleosides**: **deoxyadenosine, deoxyguanosine, deoxycytidine** and **thymidine** in the case of DNA (Table 2-1). In each nucleoside, the N9 atom of a purine or N1 of a pyrimidine is bonded in a **glycosidic bond** to the C1′ atom of the sugar (by convention, the numbers of atoms of the sugar are given primes). A **nucleotide** is a nucleoside joined to one or more phosphate groups by an ester linkage to the ribose or deoxyribose sugar (Figure 2-2 and Table 2-1).

Table 2-1. Names and abbreviations of the nucleotide constituents of nucleic acids and cyclic nucleotides

Name	Abbreviation
Ribonucleotides	
Uridine 5′-monophosphate, 5′-uridylic acid	UMP, pU
Guanosine 5′-monophosphate, 5′-guanylic acid	GMP, pG
Cytidine 5′-monophosphate, 5′-cytidylic acid	CMP, pC
Adenosine 5′-monophosphate, 5′-adenylic acid	AMP, pA
Deoxyribonucleotides	
Deoxythymidine 5′-monophosphate, 5′-deoxythymidylic, acid, thymidine 5′-monophosphate, 5′-thymidylic acid	dTMP, pdT
Deoxyguanosine 5′-monophosphate, 5′-deoxyguanylic acid	dGMP, pdG
Deoxycytidine 5′-monophosphate, 5′-deoxycytidylic acid	dCMP, pdC
Deoxyadenosine 5′-monophosphate, 5′-deoxyadenylic acid	dAMP, pdA

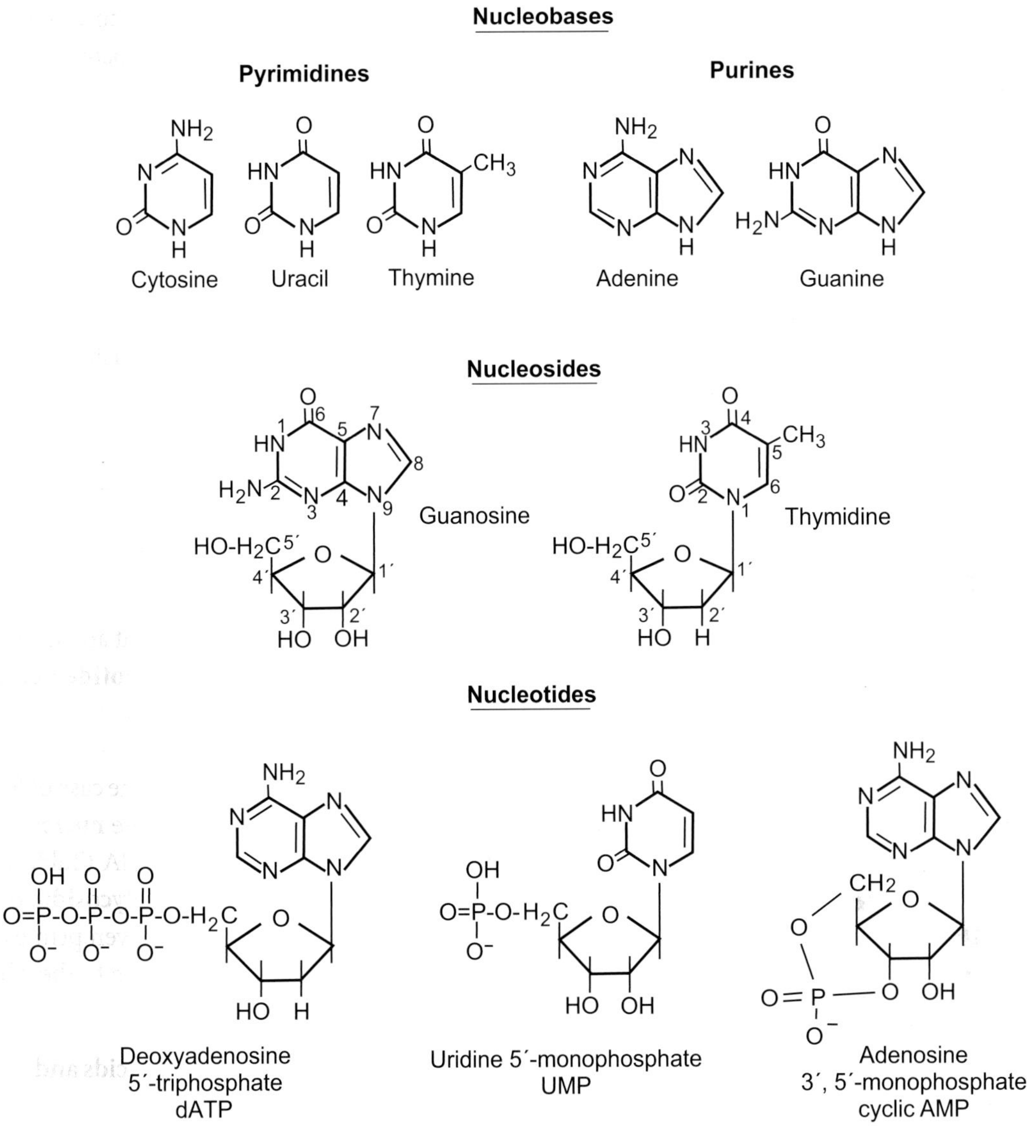

Figure 2-2. Structures of the five bases used in DNA and RNA and of representative nucleosides and nucleotides. The *syn* conformation (Equation 2.1) is shown for the nucleosides and nucleotides for reasons of space, but the *anti* conformation predominates in RNA and DNA.

Polynucleotides like DNA and RNA are produced by polymerizing the nucleotides, linking them by phosphodiester bridges, in which the 3'-hydroxyl group of the sugar moiety of one nucleotide is esterified to a phosphate group, which is, in turn, joined to the 5'-hydroxyl group of the adjacent sugar (Figure 2-3). **The chain of sugars linked by phosphodiester bridges comprises the backbone of the nucleic acid molecule**, which is invariant. The chemical structure of a short strand of DNA, containing one each of the four nucleotides, is illustrated in Figure 2-3. **The genetic information in the polynucleotide is made up of the sequence of the four bases that protrude from the backbone.**

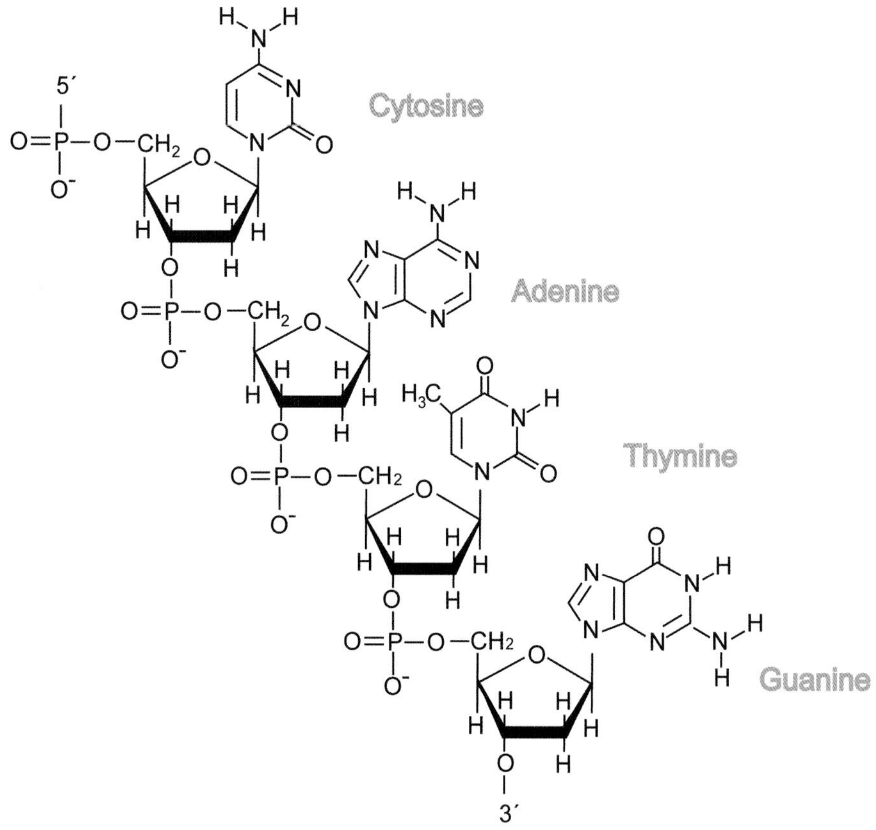

Figure 2-3. The chemical structure of a single strand of DNA.

In polynucleotides, the nucleotide building blocks are phosphorylated on both the C5′ and C3′ sugar atoms. Whether an isolated nucleotide is phosphorylated on the 5′- or 3′-sugar atom depends upon its origin. Most commonly, the phosphate group is attached to the C5′ atom, to produce a 5′-nucleotide. In certain nucleotides used in signal transduction in cells, such as cyclic AMP and cyclic GMP, a single phosphate group is esterified simultaneously to C3′ and C5′.

A linear polynucleotide has a 5′-end and a 3′-end, either or both of which can be phosphorylated. By convention, the sequence of a polynucleotide chain is described in the 5′- to 3′-direction. The general abbreviation for 5′-nucleotides is **pN**; for 3′-nucleotides it is **Np**. For convenience, DNA oligonucleotides are denoted with their sequence in a form like d(CGCGAATTCGCG). For short sequences, a more specific designation would indicate the phosphate groups ('p'), such as pCGCGAATTCGCG or pCpGpCpGpApApTpTpCpGpCpG.

The phosphate groups cause nucleotides and polynucleotides to be highly acidic and to exist as anions at neutral pH. The primary and secondary pK_a values for the 5′-phosphates of nucleotides are about 1 and 6, respectively, so they have two negative charges at neutral pH. Each phosphodiester group of DNA has an intrinsic pK_a of about 1 and will bear one negative charge at all except extremely acidic pH values. Its ionization is complicated in DNA, however, because DNA has many such groups in close proximity that affect each other and bind counterions tightly, so such polyelectrolytes have complicated ionization behavior (Section 2.3). Ionization of the nucleotides and polynucleotides prevents them from crossing cell membranes.

RNA (Chapter 4) differs from DNA in having ribose as the sugar and the base uracil instead of thymine. Ribose differs from deoxyribose only in having an additional O atom, while uracil differs from thymine only in lacking the methyl group. Yet these two small differences cause DNA and RNA to have different structures, functions and properties. The absence of the 2′-hydroxyl group on the ribose ring causes DNA to be about 100-fold less susceptible to alkaline hydrolysis than RNA (Figure 4-2). Thymine is believed to be used in DNA instead of uracil because it can be distinguished from the uracil that arises spontaneously in DNA by any deamination of cytosine nucleotides (Figure 2-33); if the cell detects uracil in DNA, an enzyme converts it back to cytosine. Both of these changes seem appropriate in view of the need for DNA to be extremely stable as the carrier of genetic information. In contrast, RNA is a much more dynamic molecule that need not be long-lived in most cases, but when necessary can be protected by binding proteins (Section 13.4). It is important to use the 'deoxy' prefix when there might be confusion with the ribonucleosides of RNA. It is usually omitted for thymidine, however, because this base is not normally encountered in RNA molecules; it is, however, found in some transfer RNA (Section 4.2.B), when it is referred to as ribosylthymidine or 5-methyluridine.

A single-stranded polynucleotide is very flexible, because rotations can take place about six bonds of each nucleotide of the backbone (Figure 2-4). There are very significant steric restrictions on what combinations of torsion angles can occur, but to depict this as with a Ramachandran plot for proteins (Figure 8-3) would be impractical, as it would require six dimensions. The number of distinct conformations for each of the bond rotations is believed to be about one for torsion angle α, two for δ, ζ and χ, and three for β, γ and ε.

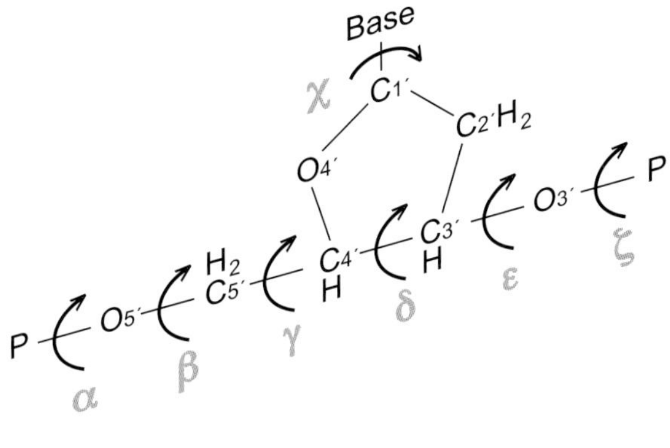

Figure 2-4. Definition of the torsion angles α, β, γ, δ, ε and ζ of a polynucleotide sugar-phosphate backbone and of the glycosyl angle χ.

2.1.A. The Deoxyribose Group

Sugars such as ribose and deoxyribose normally exist in solution as three different tautomeric forms in equilibrium, two cyclic forms and one linear (Equation 1.17). Replacing the H atom normally on C1′ by a base, to generate a nucleoside, however, locks the sugar into the cyclic form. In DNA and RNA, the base lies on the same side of the plane of the sugar as the 5′-carbon, and the configuration of the *N*-glycosidic link is β (Figure 2-2). This stereospecificity is possible because the nucleosides and nucleotides are synthesized by asymmetric enzymes that are usually stereospecific.

The deoxyribose ring does not have all five atoms lying in the same plane. The five torsion angles of the ribose ring are designated υ_0 to υ_4, starting with the bond between the O atom and C1′ and ending with that from C4′ to the O. All five torsion angles would be zero in the planar conformation, but this would be very strained. Such strain is relieved by having at least one of the five atoms lie out of the plane, by up to 0.5 Å, introducing a pucker to the ring. The carbons most likely to be out of the plane are the 2′ and 3′. If the nonplanar atom is on the same side as the base and the 5′-carbon, the conformation is said to be *endo*; if on the other side, the conformation is *exo*. These conformations can interchange rapidly, and they do so on the 10^{-9} s time scale in single stranded nucleic acids, where one conformation is not greatly favored over another. The ribose ring conformations observed most frequently in nucleic acids are illustrated in Figure 2-5. The ribose pucker is important for nucleic acids because it governs the relative orientations of the phosphate substituents (Figure 2-6). In the most common helical structures (Section 2.2.B), the C2′-*endo* conformation occurs in B-DNA, C3′-*endo* in A-DNA. The purine nucleotides are C3′-*endo* in Z-DNA and the pyrimidine residues C2′-*endo*.

Nucleotides and nucleosides can exist with either of two orientations about the glycosidic bond (as defined by the torsion angle χ about the C1′–N glycosyl bond; Figure 2-4), commonly designated as *syn* and *anti*:

$$\text{(structures of syn and anti adenosine nucleotides)} \tag{2.1}$$

Rotation about this bond is hindered, so these two most stable conformations are not readily interconverted. For pyrimidines, only the *anti* conformation is stable, because otherwise the sugar residue would sterically interfere with the pyrimidine's C2 O atom. The *anti* conformation is found in all DNA structures, except for Z-DNA, where the purine residues are *syn*. In spite of that, most of the structures shown here are represented as *syn* because it is more compact to draw.

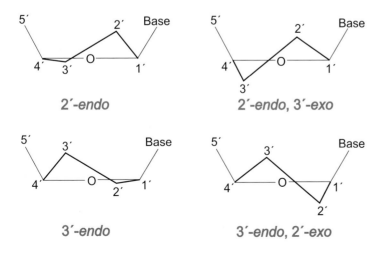

Figure 2-5. The conformations of the ribose and deoxyribose rings observed most frequently in nucleic acids. The plane of the ring is observed edge on, with the bond between C2′ and C3′ in front and the O atom at the rear.

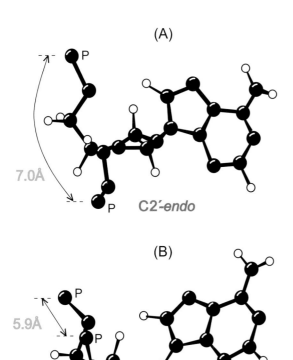

Figure 2-6. Effects on the backbone of nucleotides in the (A) C2′-*endo* conformation and (B) the C3′-*endo* conformation. These conformations occur in A-DNA and B-DNA, respectively. The distances between adjacent P atoms in the sugar-phosphate backbone are indicated.

2.1.B. Properties of the Bases

Each base is essentially planar and rigid, although its bonds are subject to the usual bending and stretching vibrations. The methyl group of thymine in isolation rotates on the 10^{-10} s time scale. The peripheral amino groups rotate more slowly, probably due to hydrogen-bond interactions with the aqueous solvent; that of cytosine rotates at an exceptionally slow rate of 35 revolutions per second (35 s^{-1}) at 0°C.

The bases are not ionized at neutral pH, but they can lose and accept protons at extremes of pH (Figure 2-7). The adenine, guanine and cytosine bases have amino groups that are very weak bases, with pK_a values of 2–4.5, so they are ionized only at acidic pH values. The bases guanine, uracil and thymine can lose a proton above pH 9. The presence of the ionized phosphate group on mononucleotides raises these pK_a values by 0.2–0.6 pH units (Table 2-2), and those of deoxyribonucleotides are 0.1–0.3 pH units greater than those of ribonucleosides. The variations in the pK_a values, plus the absence of an amino group on uracil and thymine, can be used to adjust the pH so that the various nucleotides have different net negative charges and can be separated readily by ion-exchange chromatography or electrophoresis.

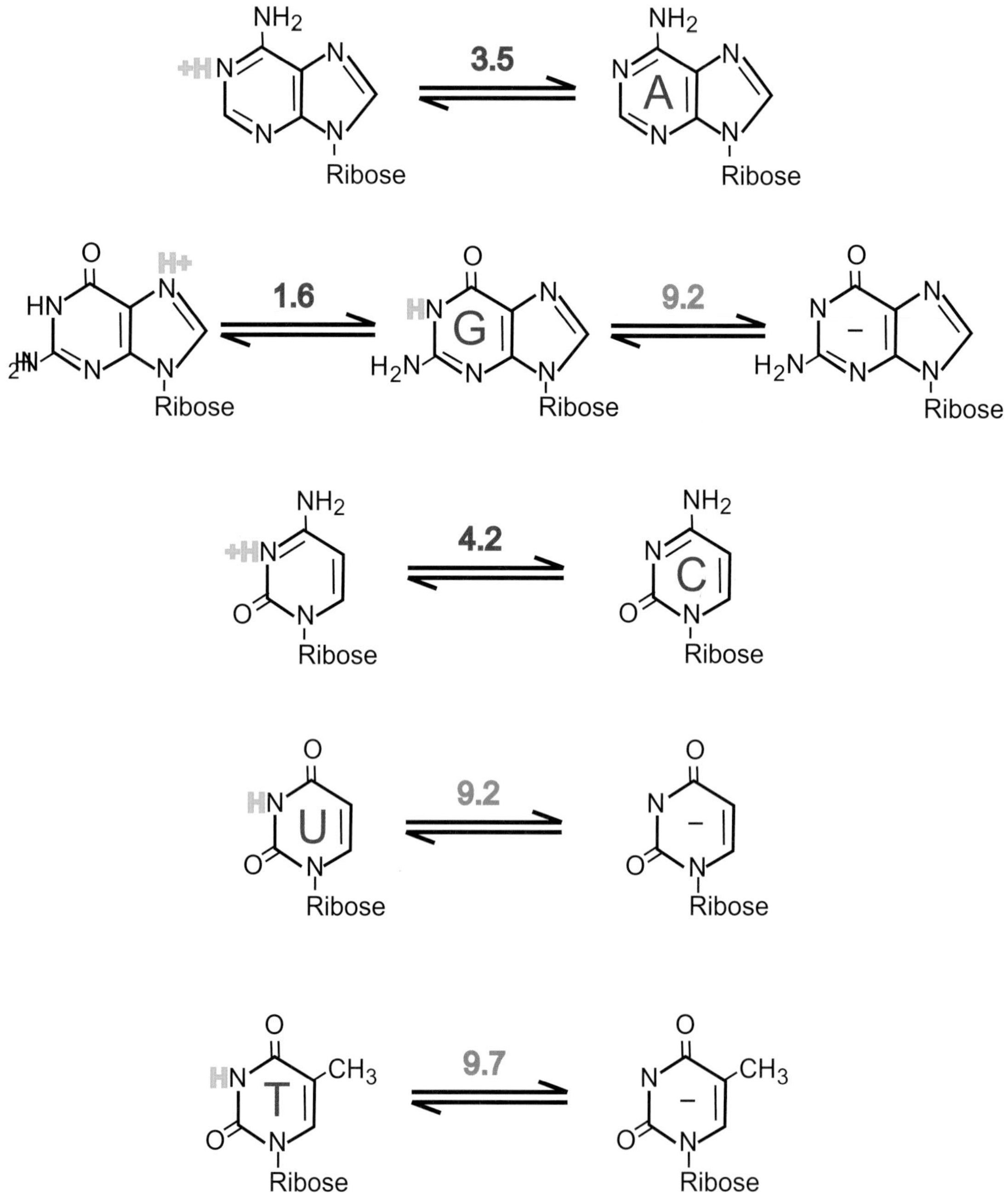

Figure 2-7. Ionization of the bases in nucleosides. The pK_a value is given above the *arrows*. The proton removed or added is indicated in *green*.

Table 2-2. Ionization and UV Absorbance Properties of Nucleoside Monophosphates.

Nucleotide	pK$_a$ of ring –NH$_2$	UV absorbance[a]	
		λ_{max} (nm)	Molar absorbance (10^3 M^{-1} cm^{-1})
AMP	3.8	257	15.0
CMP	4.5	280	13.2
GMP	2.4	256	12.2
UMP		262	10.0
dAMP	4.4	258	14.3
dCMP	4.6	280	13.5
dGMP	2.9	255	11.8
dTMP		267	10.2
cAMP		256	14.5
cGMP		256.5	11.4

[a] Measured at pH 1 to 2.
Data from C.K. Mathews (1999) In *Encyclopedia of Molecular Biology* (T.E. Creighton, ed.) Wiley-Interscience, New York, p. 1676.

The peripheral groups on the bases are extremely important, as they can serve as donors or acceptors in hydrogen bonds (Figure 2-8). Each base has both donor and acceptor sites, so each has the potential to hydrogen bond with any of the bases. These determine their crucial tendency to pair with each other (Section 2.2.A). The peripheral groups are not fixed, however, as each base can occur in several tautomers (Section 1.1.C), which tends to reverse their hydrogen-bonding potentials (Figure 2-9). These other tautomeric forms are rare, however, and those described in Figure 2-8 predominate under normal conditions.

The aromatic nature of the bases causes them to absorb ultraviolet (UV) light with characteristic absorbance spectra (Table 2-2), which allows them to be detected, identified and quantified. Excitations of the electrons in the bases are classified as being either perpendicular to the plane of the base or in the plane. The in-plane transitions involve primarily the π electrons of the base (those on the planar surface of the base) and thus are designated as π→π* transitions. The out-of-plane transitions involve excitation of nonbonding electrons of the N or O atoms into the π electron system and are designated as n→π*. The π→π* transitions predominate in the absorbance of light, but the n→π* transitions can be important for circular dichroism. The spectral properties of adenine, guanine and cytosine are dependent on the pH, due to the ionization of their amino groups, which also helps in their identification. The close proximity of bases in helical polynucleotides alters their absorbance properties, which is very useful for monitoring the breakdown of the helical structure (Section 5.1).

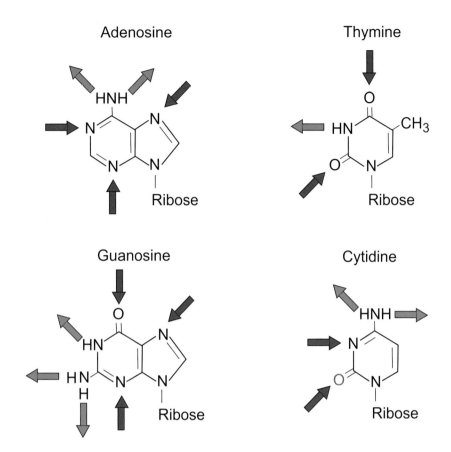

Figure 2-8. Hydrogen-bonding capabilities of the bases of DNA. H atoms that may be donated in hydrogen bonds are indicated by *blue arrows*, whereas potential acceptors of H atoms are indicated by *red arrows*.

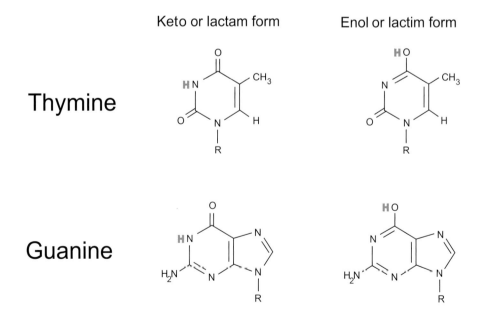

Figure 2-9. Possible tautomeric conversions for thymine and guanine bases. The hydrogen that changes position is indicated in *blue*. Cytosine and adenine residues can undergo similar proton shifts, in some cases involving the amino group attached to the ring.

The nucleobases are only very slightly fluorescent at room temperature in aqueous solution. Their quantum yields vary between 3×10^{-5} and 1.2×10^{-4}, while their fluorescent lifetimes are in the picosecond range. These weak fluorescent properties are not very useful, as they can be overwhelmed by impurities.

The hydrophobicities of the bases are important for their interactions; they were measured by partitioning the bases between water and cyclohexane, the organic solvent considered most pertinent to hydrophobicity, as it contains a minimum of water. The ribose moiety is so polar that nucleosides are almost totally excluded from cyclohexane. Consequently, the ribose moiety was substituted by methyl, butyl and tetrahydrofuryl moieties:

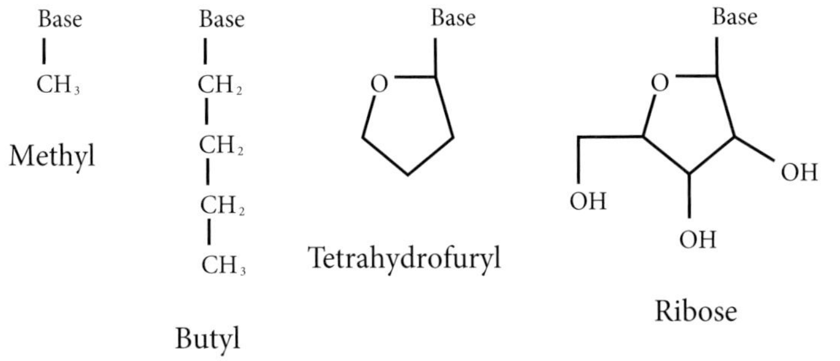

(2.2)

The free energies of transfer measured in these systems are tabulated in Table 2-3. The relative hydrophobicities of the bases are compared in Figure 2-10.

Table 2-3. Relative hydrophobicities of the nucleic acid bases as measured by the free energy of transfer from water to cyclohexane of their butyl and tetrahydrofuryl derivatives

	Free energy of transfer (kcal/mol)	
Base	**Butyl derivative**	**Tetrahydrofuryl derivative**
Purine	1.1	2.6
Thymine	2.2	4.1
Adenine	2.8	4.3
Uracil	3.7	5.0
Pyrimidin-2-one	4.1	5.6
Hypoxanthine	5.7	7.1
Cytosine	6.0	7.3
Guanine	6.7	7.5

Data from P. Shih *et al.* (1998) *J. Mol. Biol.* **280**, 421–430.

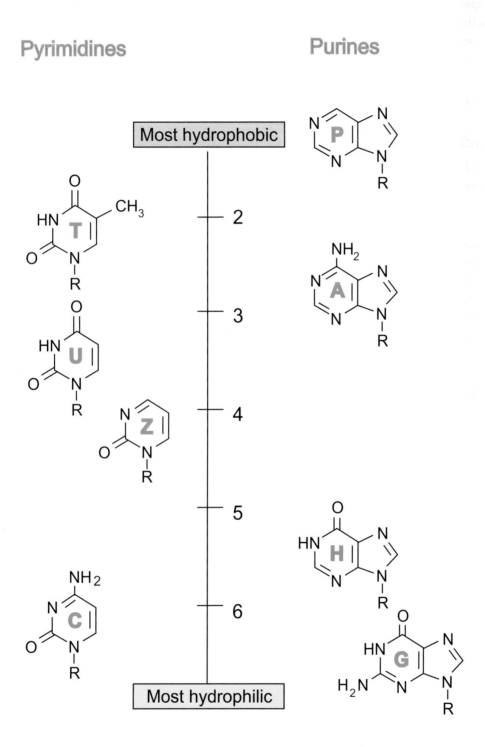

Figure 2-10. The relative hydrophobicities of the nucleic acid bases. The hydrophobicity scale is in units of kcal/mol. The structures of the bases are given: P, purine; T, thymine; A, adenine; U, uracil; Z, pyrimidin-2-one; H, hypoxanthine; C, cytosine; G, guanine. Data from P. Shih *et al.* (1998) *J. Mol. Biol.* **280**, 421–430.

The heat capacities of the bases are significantly more positive than would be predicted on the basis of just their surface areas accessible to water. N, O and P atoms of small molecules and amino acids make negative contributions to ΔC_p, probably because they increase the distortion of water hydrogen bonds. The anomalous properties of the nucleobases may indicate that many of their polar atoms are hydrated by water molecules whose hydrogen bonds are less distorted than in bulk water, which would produce a more positive ΔC_p.

2.1.C. Modifications of the Bases

The nucleotides of DNA can be modified by a number of chemical processes (Section 2.6). Some of these are very important *in vivo*, as any change of the base of a nucleotide is likely to produce a mutation in the organism, a change in the genetic information that is passed on to the next generation. The bases in DNA are sensitive to the action of numerous chemicals, such as nitrous acid (HNO_2) and hydroxylamine (NH_2-OH), and various alkylating reagents, such as dimethyl sulfate (Section 2.6). Alkylating agents are strong electrophiles that can attack chemically the approximately 20 nucleophilic sites in DNA. Bases that are alkylated become positively charged and have altered chemical properties, including their hydrogen-bonding patterns. DNA also can undergo numerous spontaneous changes, most commonly deamination, especially of cytosine to uracil, and the loss of purine bases (Figure 2-33-C) by direct attack of oxygen radicals. Oxidation and radiation can cause many types of damage to DNA; UV light damages DNA bases by many different mechanisms, including the production of oxygen radicals that lead to oxidative damage and the formation of dimers between adjacent pyrimidine bases by addition across their C5–C6 double bonds:

$$\text{(2.3)}$$

Thymine residues are particularly susceptible to this reaction, although cytosine dimers and thymine–cytosine pairs are also produced.

Each human genome of 3 billion base pairs is believed to suffer between 100 and 5000 of these modifications each day! Fortunately, enzymes are present to reverse most of these unavoidable modifications.

Most natural nucleic acids contain some bases that have been modified specifically; the most common are illustrated in Figure 2-11. Most widespread is **5-methylcytosine**, which is present in up to 5% of the cytosine residues in most eukaryotic DNAs. These cytosine bases are modified by **DNA methyltransferases**, enzymes that catalyze the transfer and covalent attachment of a methyl group from the cofactor *S*-adenosylmethionine to DNA, with the release of *S*-adenosylhomocysteine. Methylation occurs at cytosine bases to form C^5-methylcytosine or N^4-methylcytosine, or at adenine bases to generate N^6-methyladenine (Figure 2-11). Such DNA methylation is involved in biological phenomena ranging from replication of prokaryotic DNA and protection of host DNA to gene regulation in eukaryotes and embryonic development. Other methylated bases are found in RNA (Chapter 4).

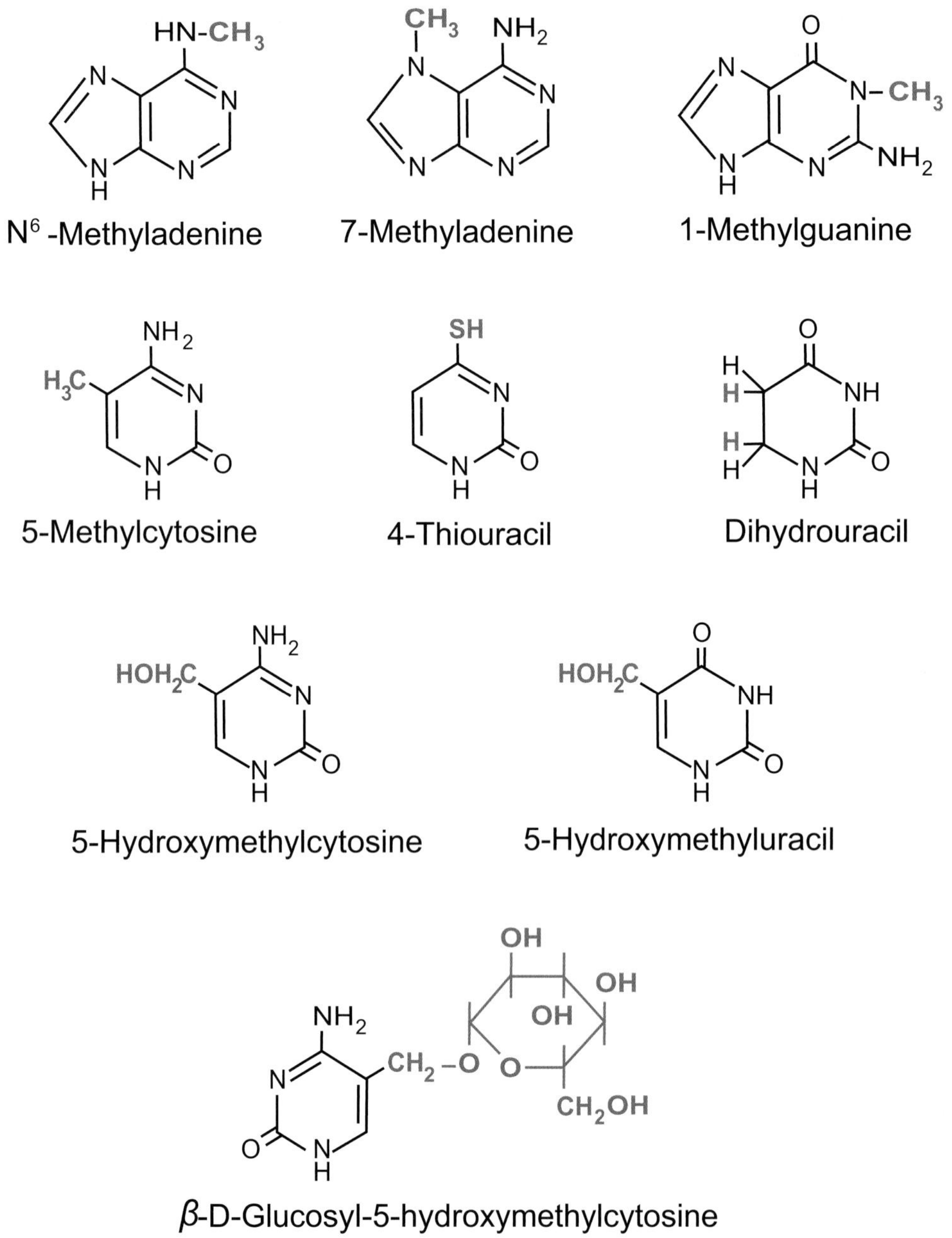

N⁶-Methyladenine

7-Methyladenine

1-Methylguanine

5-Methylcytosine

4-Thiouracil

Dihydrouracil

5-Hydroxymethylcytosine

5-Hydroxymethyluracil

β-D-Glucosyl-5-hydroxymethylcytosine

Figure 2-11. Structures of the most common modified nucleobases found in natural nucleic acids. The modifications are in *red*.

Certain organisms, such as the T-even (T2, T4 and T6) bacteriophages, contain four forms of 5-methylcytosine in place of all the cytosine nucleotides: the unmodified 5-methylcytosine, when glycosylated at the hydroxymethyl position with glucose in either the *alpha* or *beta* configuration, or when glycosylated with cellobiose (glucosyl-1,4-β-D-glucose). Other bacteriophages contain 5-hydroxymethyluracil in place of all thymine bases. This modification can also occur by oxidation of thymine residues.

A large number of nucleoside and nucleobase analogs (Figure 2-12) have been synthesized and developed as drugs useful in treating cancer and infections caused by viruses, bacteria and other parasites. These analogs usually act after being converted enzymatically to analogs of the deoxyribonucleotides. This must be done within the cell, because nucleotides generally cannot cross cell membranes, due to their ionization. The modified nucleotides usually function by inhibiting DNA replication in cells, because deoxyribonucleotides normally function only as building blocks for DNA synthesis.

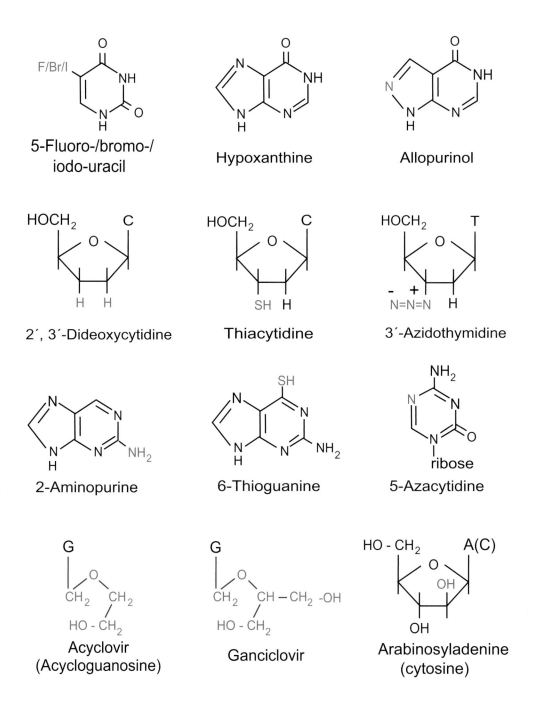

Figure 2-12. Structures of some important and useful analogs of nucleobases and nucleosides.

The fluorinated derivatives 5-fluorouracil and 5-fluorodeoxyuridine, after conversion to the nucleotides, inhibit the synthesis of dTTP and consequently the replication of DNA. The pyrimidines modified by bromine are useful experimentally because of the greater density imparted by the Br atom, making it possible to separate brominated DNA from normal DNA by density-gradient centrifugation. The Br atom is similar in size to a methyl group, so 5-bromouracil can be mistaken as thymine by the cell. Other analogs have the ribose moiety replaced by arabinose, which is the 2-epimer of ribose (Section 1.1.A.4).

Nucleobases in molecular recognition: molecular adducts of adenine and cytosine with COOH functional groups. S. R. Perumalla *et al.* (2005) *Angew. Chem. Int. Ed.* **44**, 7752–7757.

Efforts towards expansion of the genetic alphabet: pyridone and methyl pyridone nucleobases. A. M. Leconte *et al.* (2006) *Angew. Chem. Int. Ed.* **45**, 4326–4329.

Hydration of mononucleotides. D. Liu *et al.* (2006) *J. Am. Chem. Soc.* **128**, 15155–15163.

Syn- and *anti-*conformations of 5′-deoxy- and 5′-O-methyl-uridine 2′,3′-cyclic monophosphate. T. Grabarkiewicz & M. Hoffmann (2006) *J. Mol. Model.* **12**, 205–212.

Thermodynamic properties of enzyme-catalyzed reactions involving cytosine, uracil, thymine, and their nucleosides and nucleotides. R. A. Alberty (2007) *Biophys. Chem.* **127**, 91–96.

2.2. DNA THREE-DIMENSIONAL STRUCTURES

DNA molecules normally exist as two base-paired strands, associated in opposite orientations. An adenine residue in one strand is paired with a thymine in the other strand (A•T or T•A) **and each guanine residue is paired with a cytosine** (G•C or C•G; Figure 2-13). Consequently, the nucleotide sequence of each strand is sufficient to dictate the sequence of the other strand. This is vital to the replication of DNA molecules, in that the two strands are separated and the sequence of each is used as a template to synthesize the other strand with the complementary sequence (Section 6.1). Single-stranded DNA molecules occur only rarely in nature, primarily in a few bacterial viruses where the single-stranded DNA is packaged and protected within the virus. Under most circumstances *in vivo*, a single-stranded DNA molecule will be used as a template to synthesize the second strand, catalyzed by a DNA polymerase enzyme (Section 6.1). Even though the two strands are not linked by any covalent bonds and can be dissociated in various ways (Section 5.1), together they are commonly referred to as a single molecule.

How sequence defines structure: a crystallographic map of DNA structure and conformation. F. A. Hays *et al.* (2005) *Proc. Natl. Acad. Sci. USA* **102**, 7157–7162.

NMR structures of damaged DNA. M. Lukin & C. de Los Santos (2006) *Chem. Rev.* **106**, 607–686.

Exploring DNA structure with Cn3D. S. G. Porter *et al.* (2007) *CBE Life Sci. Educ.* **6**, 65–73.

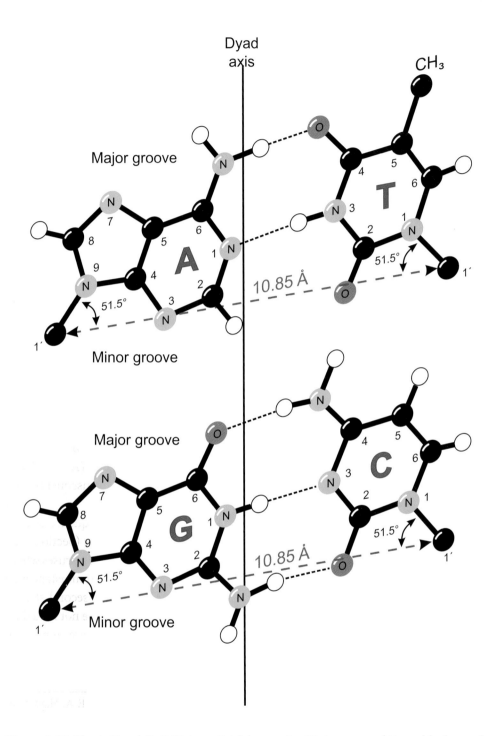

Figure 2-13. The A•T and G•C Watson–Crick base-pairs. H atoms are *white*, and hydrogen bonds are shown as *dashed lines*. The *black spheres* are C atoms, the *green spheres* are N atoms, and the *red spheres* are O atoms. The distance between the C1′ atoms of the two ribose moieties is the same in both base pairs and makes equal angles to the glycosidic bonds to the bases. This gives DNA a series of pseudo-two-fold symmetry axes (often referred to as dyad axes) that pass through the center of each base pair; this symmetry axis is perpendicular to the axis of the double helix. The edges of the base pair that contact the major and minor grooves of the double helix are indicated.

2.2.A. Base Pairing and Stacking

The base pairs normally found in DNA are known as the **Watson–Crick base pairs**, after the discoverers of the DNA double helix. **The specificity of base pairing is provided by both hydrogen bonds and the shapes of the pairs.** The most obvious determinant of base pairing is the hydrogen bonds between the two pairs. An equally important consideration, however, is their identical geometries: the line joining the C1′ atoms of the complementary ribose moieties has the same length in both base pairs and makes equal angles to the glycosidic bonds to the bases; consequently, **both base pairs are perfectly compatible with the same structure of the backbone**. This gives DNA a series of pseudo-two-fold symmetry axes (often referred to as **dyad axes**) that pass through the center of each base pair perpendicular to the axis of the double helix. Note that there are three hydrogen bonds between cytosine and guanine, but only two between adenine and thymine, and C•G base pairs correspondingly tend to be the more stable (Table 5-1). It should also be noted that these hydrogen-bonding patterns require the correct tautomeric state (Section 1.1.C) of each base. Other tautomers are possible (Figure 2-9 and Equation 2.8) and many other types of base pairs are possible and known to occur (Figure 2-14) in unusual DNA structures (Section 2.2.C). The Watson–Crick base pairings are among those that are intrinsically most stable, even in the absence of the double helix, but the double helix further minimizes the other possibilities.

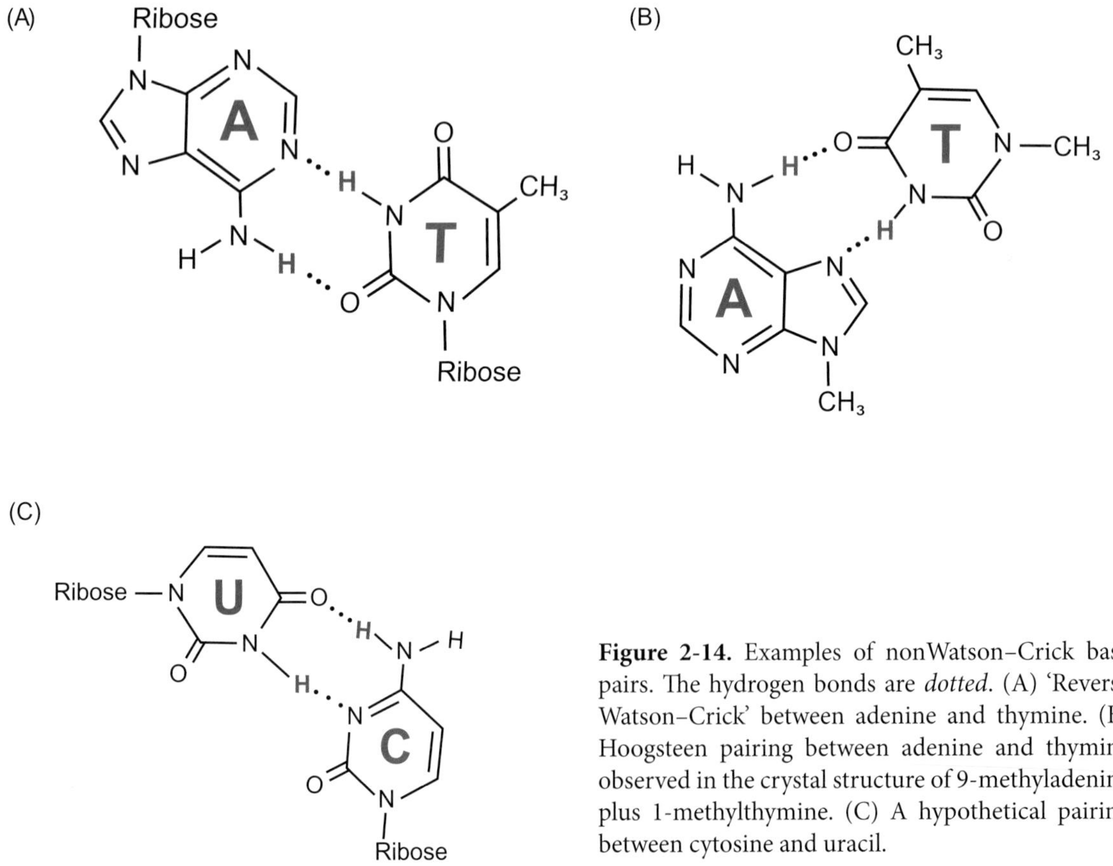

Figure 2-14. Examples of nonWatson–Crick base pairs. The hydrogen bonds are *dotted*. (A) 'Reverse Watson–Crick' between adenine and thymine. (B) Hoogsteen pairing between adenine and thymine observed in the crystal structure of 9-methyladenine plus 1-methylthymine. (C) A hypothetical pairing between cytosine and uracil.

Bases in a DNA double helix are stacked on top of each other and in close contact, although they usually do not maximize their contact area but are offset somewhat (Figure 2-17); the resultant structure is probably a compromise between the optimum geometry of base pairing and that of the deoxyribose-phosphate backbone. The base pairs within a DNA duplex often have significant departures from a strict coplanar geometry. The conformational parameters associated with the DNA base pairs are defined in Figure 2-15.

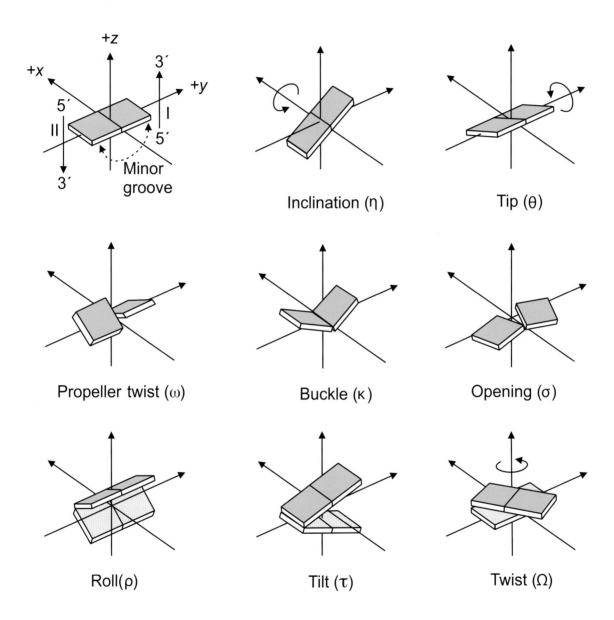

Figure 2-15. Definitions of various base-pair conformations (*upper two rows*) and two successive base pairs (*bottom row*). The motions of the bases in the *top row* are coordinated, whereas they are opposed in the *middle row*.

The stacking interaction is believed to be a complex phenomenon that is stabilized by charge transfer and van der Waals interactions between the π electrons of the aromatic rings; it also minimizes the exposure of the bases to the aqueous solvent. The bases are relatively polarizable and not remarkably hydrophobic (Figure 2-10), due to their polar N and O atoms, but they have large interfaces.

Stacking is an intrinsically favorable interaction for the bases in water, in that individual purines and pyrimidines free in aqueous solution tend to form aggregates of at least five bases, depending upon the concentration. The equilibrium constant for stacking one base on another is in the range of 1–10 M^{-1}. Base stacking is energetically more favorable than pairing because the individual hydrogen bonds formed between bases are not much more stable than those that the bases form with water, so there is little to be gained energetically by them forming hydrogen bonds with each other. On the other hand, base stacking is favored in water because it minimizes the contact of nonpolar surfaces with water. In the absence of water, the bases do interact by hydrogen bonding rather than base stacking.

The stacking interaction has been measured in several model systems (Tables 2-4 and 2-5). One is the stacking interaction within a short DNA duplex having a C•G base pair at the end (Figure 2-16). These data illustrate the predominance of van der Waals interactions and the hydrophobic interaction, in that large nonpolar aromatic groups stack much more strongly than do natural DNA bases. Nevertheless, there are large differences in the strengths of the stacking interaction between adjacent base pairs (Table 2-5), which are believed to be the result of electrostatic interactions between permanent dipoles of the bases. Consequently, the stacking patterns vary with the helix type and the particular base–base step. Purines stack more avidly than pyrimidines, probably because of their larger surface areas. In B-DNA, the 5′-purine-p-pyrimidine-3′ (Pu–Py) base-pair step has a good stacking overlap and is energetically favorable, whereas the 5′-pyrimidine-p-purine-3′ (Py–Pu) base-pair step has a poor stacking overlap and is thus less stable (Figure 2-17).

Table 2-4. Thermodynamic parameters at 25°C for stacking of two individual nucleic acid bases, as measured for the reaction: unstacked dinucleoside phosphate ↔ stacked dinucleoside phosphate

Dinucleotide phosphate	kcal/mol		
	ΔG	ΔH	T ΔS
ApA	+0.64	−5.31	+5.95
ApU	1.15	−8.39	9.54
GpC	0.55	−7.79	8.34
CpG	0.27	−4.80	5.07
UpU	0.86	−7.79	8.65

Data from R. C. Davis & I. Tinoco, Jr (1968) *Biopolymers* **6**, 230.

Table 2-5. Stacking energies of base pairs

Dinucleotide base pairs	Stacking energies (kcal/mol per stacked pair)
(GC)•(GC)	−14.59
(AC)•(GT)	−10.51
(TC)•(GA)	−9.81
(CG)•(CG)	−9.69
(GG)•(CC)	−8.26
(AG)•(CT)	−6.78
(AT)•(AT)	−6.57
(TG)•(CA)	−6.57
(AA)•(TT)	−5.37
(TA)•(TA)	−3.82

Data from R. L. Ornstein *et al.* (1978) *Biopolymers* **17**, 2341.

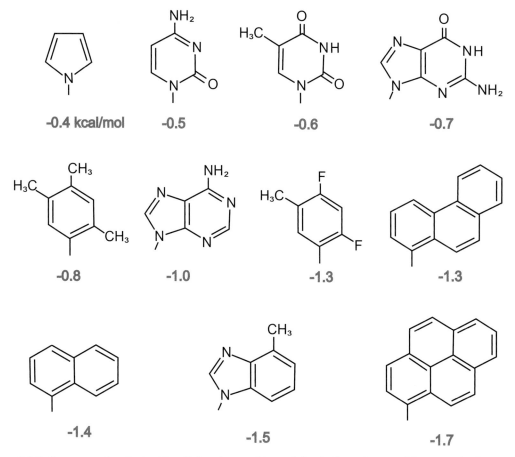

Figure 2-16. Free energies (in kcal/mol) for the stacking of the indicated natural bases and of nonnatural aromatic compounds at the end of a short DNA duplex with a C•G base pair. Data from F. M. Guckian *et al.* (2000) *J. Am. Chem. Soc.* **122**, 2213–2227.

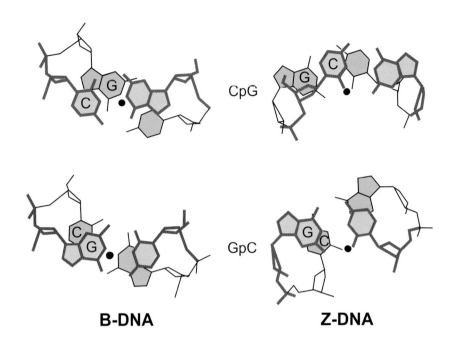

Figure 2-17. Base stacking interactions in B-DNA (*left*) and Z-DNA (*right*) showing pyrimidine–purine step stacking (*top*) and purine–pyrimidine step stacking (*bottom*). The upper base is indicated *blue* with a *gray outline*. The *circle* indicates the helix axis. The pyrimidine–purine steps have very small stacking interactions in B-DNA so these steps are deformed relatively easily.

Nature of base stacking: reference quantum-chemical stacking energies in ten unique B-DNA base-pair steps. J. Sponer *et al.* (2006) *Chemistry* **12**, 2854–2865.

Ab initio determination of the ionization potentials of DNA and RNA nucleobases. D. Roca-Sanjuan *et al.* (2006) *J. Chem. Phys.* **125**, 084302.

Electronic structure of DNA nucleobases and their dinucleotides explored by soft X-ray spectroscopy. Y. Harada *et al.* (2006) *J. Phys. Chem. A* **110**, 13227–13231.

Sensitivity of hydrogen bonds of DNA and RNA to hydration, as gauged by $^1J_{NH}$ measurements in ethanol–water mixtures. M. N. Manalo *et al.* (2007) *J. Biomol. NMR* **37**, 257–263.

Hydration and stability of nucleic acid bases and base pairs. M. Kabelac & P. Hobza (2007) *Phys. Chem. Chem. Phys.* **9**, 903–917.

2.2.B. Double Helices

Double-stranded DNA almost always adopts a helical structure (Figure 2-1) **that is much more rigid and extended than a single strand.** The antiparallel double-stranded structure of DNA was originally inferred by Watson and Crick from diffraction patterns of fibers of natural DNA that were produced by causing it to precipitate. Those studies uncovered two distinct forms of DNA, designated A and B, depending upon whether the fibers were dry (A form) or wet (B form), below or above about 85% relative humidity. Watson and Crick were able to infer from the more simple diffraction patterns of the B form that the DNA has a double-helical structure with planar base pairs perpendicular to the helix axis, and 10 base pairs per turn. Model building produced a structure very close to that accepted today (Figure 2-1) and most DNA *in vivo* appears to be in the B form. On the other hand, further studies of DNA fibers with different nucleotide sequences indicated a variety of double-helical structures (Table 2-6), confirming that even double-stranded DNA does not have a closely defined structure.

Short oligonucleotides can be crystallized, and their detailed structures have been elucidated using X-ray crystallography. **The detailed structures obtained are remarkably variable**, however, indicating that the canonical double helix is not a rigid structure but relatively flexible. The DNA structures in crystals appear to be very subject to crystal-packing forces. Structure determination of DNA molecules by nuclear magnetic resonance (NMR) does not suffer from these crystal constraints, but the traditional NMR approach is inherently more difficult with DNA than with proteins because DNA molecules are long and cylindrical and usually lack the elaborate tertiary structures found in proteins that bring atoms distant in the covalent structure into close proximity. In addition, DNA molecules have a lower density of H atoms. Therefore, the number of short distances of less than 5 Å between H atoms that can be observed by NMR (known as NOEs) is comparatively small in oligonucleotides. NOEs are typically found only between adjacent base pairs, so there are usually few long-range NOEs to define the overall structure. As a result, the structure tends to be defined only locally by NMR, and not very accurately. Fortunately, the orientational information contained in residual dipolar couplings has been very useful in determining accurate DNA structures.

The three most common types of DNA double helices are B-, A- and Z-DNA. Their structures are illustrated in Figure 2-18 and compared in Table 2-7. Details of virtually all the known structures of DNA can be obtained from the **Protein Data Bank (PDB; www.rcsb.org)** and from **http://ndbserver.rutgers.edu**. They can be viewed most readily online using the program **Jmol** at **http://firstglance.jmol.org**.

Table 2-6. Helical forms of DNA and RNA characterized by fiber diffraction

Polynucleotide	Relative humidity (%)	Counterion	Structure[a]	Helix symmetry	Helical rise (Å)	Turn angle (°)	Tilt (°)
Native DNA	75	Na	A	11/1	2.56	32.7	20.3
	92	Na	B	10/1	3.38	36.0	−5.9/ −2.1
	57–66	Li	C	10/1	3.32	38.6	−8.0/ −1.0
Poly (dA–dT)	75	Na	D	8/1	3.04	45.0	−16.4
	<98	Na	A	11/1			
	66	Li	B	10/1			
	56	Li	W	8/2	6.1/2	43.5/ 46.5	15.0/ 16.2
	56	Li	P	24/4	18.55/ 4	36.5– 54.4	18.0– 23.3

Table 2-6. Helical forms of DNA and RNA characterized by fiber diffraction - continued

Polynucleotide	Relative humidity (%)	Counterion	Structure[a]	Helix symmetry	Helical rise (Å)	Turn angle (°)	Tilt (°)
Poly(dG–dC)	<92	Na	A	11/1			
	81	Li	B	10/1			
	56	Li	W	10/2	6.72/2		
	43	Na	Z	6/5	3.63	−30.0	
Poly(dA–dC)•poly(dG–dT)	66	Na	A	11/1			
	66–92	Na	B	10/1			
	66	Na	Z	6/5			
Poly(dA–dG)•poly(dC–dT)	66	Na	C	9/2	3.23	48.0	
	95	Na	B	10/1			
Poly(dA)•poly(dT)	70	Na	B′	10/1	3.29	36.0	7.9/ 1.0
	71		H(A/B)	10/1	3.23		26.5/ 18.0
Poly(dG)•poly(dC)	75	Na	A	11/1	2.56	32.7	
	92	Na	B	10/1			
Poly(dT)•poly(dA)•poly(dT)		Na	A		3.26		7.2/ 9.2
DNA•RNA hybrid	33–92	Na	A	11/1			
Poly(I)•poly(dC)	75	Na	A′	12/1			
Poly(dA)•poly(U)			H(B/A)	11/1	3.06	32.7	13.0
Poly(dI)•poly(C)			H(B/A)	10/1	3.06	32.7	15.0
Poly(A)•poly(U)	<92	Na	A	11/1	2.56	32.7	16.0
Poly(I)•poly(C)	<92	Na	A′	12/1	3.0	30.0	10.0
Poly(U)•poly(A)•poly(U)	72	Na	A	11/1	3.04	32.7	11.7/ 13.6

[a] H, heteromorphic helices, where (A/B) indicates that the first strand is in the A conformation and the second in the B conformation; W, wrinkled helices with dinucleotide repeat; Z, left-handed helix with dinucleotide repeat; P, pleiomorphic helix with hexanucleotide repeat. Data from W. Guschlbauer (1990) in *Concise Encyclopedia of Polymer Science and Engineering* (J. Kroschwitz, ed.), Wiley-Interscience, NY.

Table 2-7. Structural Parameters of A-, B- and Z-DNA in solution

Structural parameter	A-DNA	B-DNA	Z-DNA
Helical sense	Right-handed	Right-handed	Left-handed
Base pairs/turn	11	10.4	12
Rise per base pair (Å)	2.3	3.3	3.7
Helical pitch (Å)	25.4	34	45
Repeat unit (base pairs)	1	1	2
Tilt (°)	19	0	-9
Glycosyl angle	*anti*	*anti*	*anti* at C, *syn* at G
Sugar pucker	C3'-*endo*	C2'-*endo*	C2'-*endo* at C, C3'-*endo* at G
Phosphate conformation $(\alpha/\zeta)(°)$	–88/–44	–40/–98	–146/80 at C, 60/–58 at G
Helical diameter (Å)	25	20	18
Major groove	Narrow and deep	Wide and deep	Flattened
Minor groove	Wide and shallow	Narrow and deep	Narrow and deep

Data from A. H.-J. Wang & H. Robinson (1999) in *Encyclopedia of Molecular Biology* (T. E. Creighton, ed.), Wiley-Interscience, NY.

The specificity of DNA structure clearly arises from the hydrogen bonds between the base pairs, and they also stabilize the double-stranded structure, just as hydrogen bonds stabilize folded protein molecules (Section 11.3.A). Even though they can be replaced by hydrogen bonds to water, they form cooperative networks in double-stranded DNA in which they stabilize each other. The net strengths of individual hydrogen bonds between the base pairs have been measured by comparing the stabilities of analogs of the base pairs lacking the hydrogen-bonding groups: values of between –0.25 and –1.6 kcal/mol per hydrogen bond have been measured for the normal base pairs, so **hydrogen bonds undoubtedly contribute to the stabilities of the double-stranded structures of DNA.** The variations in the degree of stabilization are probably due to variation in the stacking interactions between adjacent base pairs. **Incorrect base pairing has a drastic affect on the stability of the double helix, each incorrect base pair within the interior of the double helix decreasing its stability by 1–4 kcal/mol**; the most important effect is believed to be the need to distort the mismatched pair into a nonstandard geometry, leading to unfavorable conformational interactions in neighboring pairs. Mismatches at the ends of oligonucleotides have little or no effect.

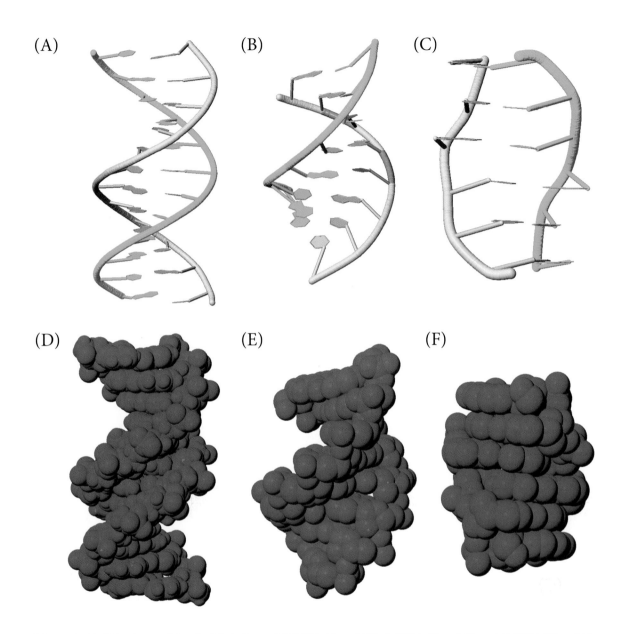

Figure 2-18. Side views of the double helices of B-DNA (A and D), A-DNA (B and E) and Z-DNA (C and F). (A–C) Skeletal models; (D–F) space-filling models. Figures generated from PDB files *1duf*, *115d* and *1dcg*, respectively, using the program Jmol.

A notable feature of all DNA double helices is the presence of a wide **major groove** and a narrow **minor groove** along the surface (Figure 2-1). Both grooves are important because they provide access to the base pairs within the interior of the helix, where proteins and other ligands can recognize the nucleotide sequence. Many ligands are observed to bind in these two grooves, as well as relatively fixed water molecules (Figure 2-19).

The DNA double helices keep the ionized phosphate groups on the outside, in contact with water and counterions, whereas the relatively nonpolar bases are in the interior. Consequently, nucleic acids are very soluble in water but can be precipitated most readily and reversibly by adding nonpolar alcohols, such as isopropanol.

The negative charge associated with each of the phosphate groups makes DNA an extreme polyelectrolyte. These negative charges are generally neutralized by the positive charges of metal ions, polyamines or proteins. Metal ions, such as sodium, potassium or magnesium ions, are used in the screening of the DNA negative charges by interactions with the phosphate O atoms and are crucial for the stability of the DNA structure. This negative charge also repels nucleophilic species, such as hydroxide ions, so phosphodiester bonds are much less susceptible to hydrolytic attack than other esters, such as carboxylic acid esters, which is crucial for maintaining the integrity of the DNA structure and the genetic information stored there. The absence of the 2′-hydroxyl group in DNA further increases its resistance to hydrolysis (Figure 4-2) and probably is the reason why DNA, rather than RNA, tends to be used for storing genetic information.

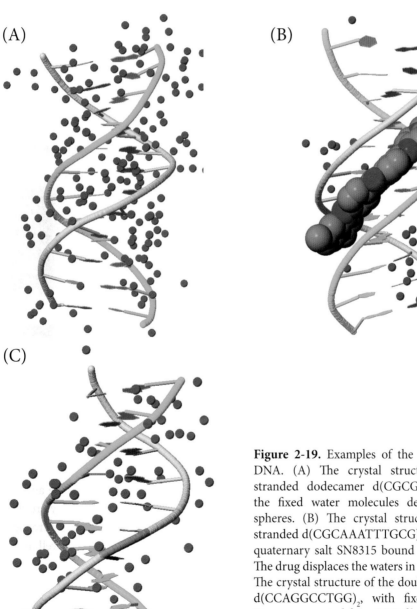

Figure 2-19. Examples of the structures of B-form DNA. (A) The crystal structure of the double-stranded dodecamer d(CGCGAATTCGCG)$_2$, with the fixed water molecules depicted as small red spheres. (B) The crystal structure of the double-stranded d(CGCAAATTTGCG)$_2$, with a quinolinium quaternary salt SN8315 bound to the minor groove. The drug displaces the waters in the minor groove. (C) The crystal structure of the double-stranded decamer d(CCAGGCCTGG)$_2$, with fixed water molecules. Figures generated from PDB files *1fq2*, *1zph* and *2d25* using the program Jmol.

The solvent-accessible surface areas of each nucleotide in seven different double helices are presented in Table 2-8, along with their intrinsic molecular volumes. The intrinsic volumes are similar within 3%, but the accessible surface areas vary by up to 19%. The ionized phosphate groups make up about 47% of the total accessible surface area, the polar groups another 14% and nonpolar atoms about 39%.

Table 2-8. The accessible surface areas of the charged, polar and nonpolar atoms, the overall surface areas, and the intrinsic volumes per base pair for seven nucleic acid duplexes

Duplex	Surface areas (Å²)				Intrinsic volume (Å³)
	Charged	Polar	Nonpolar	Total	
Poly(A)•poly(U)	75.9	36.6	56.2	168.7	272.9
Poly(A)•poly(dT)	72.4	27.6	64.8	164.7	282.3
Poly(dA)•poly(U)	81.0	26.2	66.7	173.8	273.1
Poly(dA)•poly(dT)	78.9	12.3	67.7	158.8	277.2
Poly(dAdT)•poly(dAdT)	74.2	16.4	59.7	150.3	284.1
Poly(dGdC)•poly(dGdC)	78.4	20.5	72.2	171.0	275.4
Poly(dIdC)•poly(dIdC)	71.9	19.1	51.7	142.6	284.0
Average	76 ± 5	23 ± 7	63 ± 8	162 ± 10	281 ± 8

A glossary of DNA structures from A to Z. A. Ghosh & M. Bansal (2003) *Acta Crystallogr. D* **59**, 620–626.

DNA coiled coils. J. L. Campos *et al.* (2005) *Proc. Natl. Acad. Sci. USA* **102**, 3663–3666.

Overview of the structure of all-AT oligonucleotides: organization in helices and packing interactions. L. Campos *et al.* (2006) *Biophys. J.* **91**, 892–903.

Estimation of strength in different extra Watson–Crick hydrogen bonds in DNA double helices through quantum chemical studies. D. Bandyopadhyay & D. Bhattacharyya (2006) *Biopolymers* **83**, 313–325.

Molecular design and synthesis of artificial double helices. Y. Furosho & E. Yashima (2007) *Chem. Rec.* **7**, 1–11.

1. B-DNA

B-DNA is the predominant DNA structure under normal physiological conditions. The B-DNA double helix is right-handed, with its base pairs perpendicular to the helix axis, which passes through the centers of the base pairs (Figure 2-1). The pitch of the helix is determined by the stacking between the adjacent base pairs. The major and minor grooves are roughly equivalent in depth, 8.5 Å and 7.5 Å, respectively, but the width of the major groove is about 12 Å, whereas that of the minor groove only about 6 Å. The bases that define the sequence are more accessible in the major groove, and this is where proteins tend to bind to specific nucleotide sequences (Section 13.2).

The first detailed structure of B-DNA determined by X-ray crystallography (Figure 2-19) was that of a dodecamer with the self-complementary sequence d(CGCGAATTCGCG)$_2$. It exhibited a number of interesting structural features that now appear to be general: the minor groove is narrower at the AATT region of the double helix than at the CGCG ends and is filled by a spine of water molecules that form hydrogen bonds to both the O2 atoms of thymines and the N3 of adenines; the base pairs in the central AATT region of the helix have high propeller twist angles, which enhances stacking of the bases along each strand of the double helix (A•T base pairs are generally more amenable to propeller twisting than G•C base pairs because of being paired by only two hydrogen bonds); finally, the sugar pucker of the deoxyribose ring favors the C2′-*endo* conformation, although conformations ranging from C1′-*exo* to O4′-*endo* are also present. This reflects the flexibilities of B-DNA structures, which depend upon the exact sequence and the environment. In general, runs of As lead to a very narrow minor groove and a high propeller twist, which introduce a bend in the helix axis. Several DNA decamer oligonucleotides with 'mixed' sequences, for example d(CCAGGCCTGG) (Figure 2-19-C), have slightly narrower or wider minor grooves, depending on the sequence. In general, the A•T regions have a narrower minor groove than the G•C regions.

Most DNA molecules in solution have the B-DNA structure, and this structure will be assumed here unless stated otherwise.

Accurate representation of B-DNA double helical structure with implicit solvent and counterions. L. Wang *et al.* (2002) *Biophys. J.* **83**, 382–406.

Sequence-dependent DNA structure: a database of octamer structural parameters. E. J. Gardiner *et al.* (2003) *J. Mol. Biol.* **332**, 1025–1035.

Molecular dynamics simulations of DNA with polarizable force fields: convergence of an ideal B-DNA structure to the crystallographic structure. V. Babin *et al.* (2006) *J. Phys. Chem. B* **110**, 11571–11581.

---

2. A-DNA

B-DNA can be converted to A-DNA under conditions of low hydration and by adding alcohols; the conversion is reversible and occurs on the microsecond time scale. A-DNA was first recognized from fiber X-ray diffraction analysis, where the hydration of the insoluble samples can be varied drastically. The double helix of A-DNA is short and fat, with the base pairs and backbone wrapped further away from the helix axis than in B-DNA (Figure 2-18). The base pairs are tilted significantly relative to the helix axis, by about 19°, and display a minor propeller twist. The equivalent of the major groove of B-DNA is deep and very narrow, with a width of only about 3 Å and a depth of about 13 Å. In contrast, the minor groove has a breadth of 11 Å and a shallow depth of 3 Å.

The A-form of DNA is often observed when short duplexes are characterized by X-ray crystallography. This is now thought to be due to the addition of alcohols to induce crystallization of the DNA. The propensity to form A-DNA is sequence-dependent: **guanine-rich regions readily form A-DNA, whereas stretches of adenine resist it**. This may be due to guanine-rich sequences having a characteristic intrastrand guanine–guanine stacking interaction in the A-DNA double helix. Also, the terminal base pairs from one helix abut the minor groove surface of the neighboring helix in the crystal lattice, minimizing the accessibility of solvent to the wide minor groove. That a low-humidity environment favors formation of the A-type helix indicates that the displacement of surface solvent molecules by hydrophobic base pairs provides a driving force in stabilizing the A-DNA conformation. Complex ions such as the physiological polyamines:

putrescine, $^+H_3N-(CH_2)_4-NH_3^+$

spermidine, $^+H_3N-(CH_2)_4-^+NH_2-(CH_2)_3-NH_3^+$

spermine, $^+H_3N-(CH_2)_3-^+NH_2-(CH_2)_4-^+NH_2-(CH_2)_3-NH_3^+$ (2.4)

cobalt(III)hexamine and neomycin can also stabilize the A-DNA conformation in DNA containing stretches of G•C base pairs. Shorter helices with fewer base pairs convert to A-DNA more readily than do longer helices.

Double-stranded RNA and a number of DNA•RNA hybrids also adopt the A-DNA conformation. The ribose and 2′-deoxyribose sugars are all in the C3′-*endo* conformation, and the 2′-hydroxyl groups of the ribose form hydrogen bonds to adjacent nucleotides in the chain.

The dynamics of the B–A transition of natural DNA double helices. D. Jose & D. Porschke (2005) *J. Am. Chem. Soc.* **127**, 17120–16128.

B-DNA helix stability in a solvent-free environment. E. S. Baker & M. T. Bowers (2007) *J. Am. Mass Spectrom.* **18**, 1188–1195.

---

3. Z-DNA

The Z-DNA structure was unexpected and is unusual in being left-handed. Its name originates from the zig-zag path of the phosphate groups along the DNA backbone. Relative to B- and A-DNA, the Z-DNA double helix is tall and thin, with a diameter of about 18 Å and a helical pitch of about 45 Å (Figure 2-18 and Table 2-7). Z-DNA is favored by alternating Py–Pu sequences, especially (dC–dG)$_n$, and it was discovered in the structure of d(CGCGCG)$_2$. The repeat unit is a CpG dinucleotide, with the glycosyl conformation of cytosine being *anti* and that of guanine being *syn* (Equation 2.1). The sugar pucker for cytosine is C2′-*endo*, while that for guanine is predominantly C3′-*endo* (Figure 2-5). There are 12 base pairs per helical turn, the base pairs are slightly inclined by −9°, and the helical twist angle per dinucleotide repeat is −60°, about −8° for the CpG and −52° for the GpC. The rise per dinucleotide repeat is about 7.4 Å. The Z-DNA conformation has been observed in the crystal structures of several other oligonucleotides, including fragments containing nonalternating Pu–Py sequences and A•T and G•T base pairs.

The single groove of Z-DNA is narrow and deep, with an opening of only 6–7 Å. The equivalent of the major groove is almost nonexistent, being very shallow and exposed to the solvent. Spermine (Equation 2.4) and hydrated metal ions interact with the phosphates and the N7 atoms of guanine bases and neutralize the negative charge. Binding of spermine molecules may help to stabilize the Z-DNA helix, because some are found in the deep groove and bridge it by interacting with phosphate groups on both sides.

DNA is in a dynamic state and forms Z-DNA when circumstances stabilize it, then relaxing to right-handed B-DNA. DNA with alternating dCdG sequences, like poly(dCdG), can interconvert especially easily between the B- and Z-DNA conformations. This is somewhat surprising, because this interconversion requires a substantial change of the purines from *anti* to *syn* conformations (Equation 2.1 and Figure 2-20) plus a change in the hand of the helix. Each time a DNA segment turns into Z-DNA, two B–Z junctions form. There is continuous stacking of bases across this junction, except that one base pair at the junction is broken and the bases on each side are extruded. The tendency to form the Z-conformation is increased by negative supercoiling (Section 3.1), increasing the ionic strength, and by the presence of proteins that bind only to Z-DNA. The left-handedness of Z-DNA is especially adept at relieving the strain of negatively supercoiled DNA, for example during gene transcription.

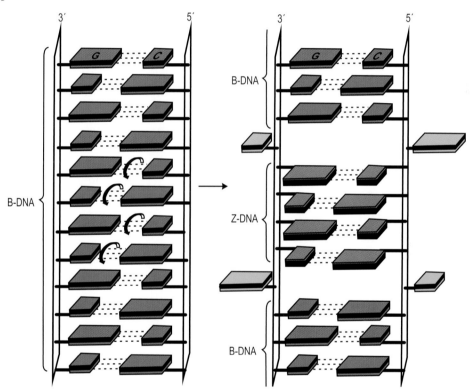

Figure 2-20. Interconversion of B- and Z-DNA requires disrupting the base pairing and flipping the planes of the nucleotide bases. Modified from A. Rich *et al.* (1984) *Ann. Rev. Biochem.* **53**, 799.

Crystal structure of a junction between B-DNA and Z-DNA reveals two extruded bases. S. C. Ha *et al.* (2005) *Nature* **437**, 1183–1186.

The hydration structure of a Z-DNA hexameric duplex determined by a neutron diffraction technique. T. Chatake *et al.* (2005) *Acta Crystallogr. D* **61**, 1088–1098.

Structure of d(TGCGCG)•d(CGCGCA) in two crystal forms: effect of sequence and crystal packing in Z-DNA. S. Thiyagarajan *et al.* (2005) *Acta Crystallogr. D* **61**, 1125–1131.

The rare crystallographic structure of d(CGCGCG)$_2$: the natural spermidine molecule bound to the minor groove of left-handed Z-DNA d(CGCGCG)$_2$ at 10°C. H. Ohishi *et al.* (2007) *Biochem. Biophys. Res. Commun.* **358**, 24–28.

2.2.C. Other DNA Structures

The double-helical structures of DNA described above are not so stable that they predominate under all conditions and with all nucleotide sequences. Certain sequences of DNA form 3-D structures that are not the common A-, B- or Z-DNA double helices.

1. Hoogsteen Base Pairing

Early analysis of nucleic acid base pairs observed that the individual nucleobases cytosine and guanine in water associate to form the Watson–Crick base pair, but that adenine and thymine in isolation preferred to pair differently from that observed in Watson–Crick base pairs. The N7 atom of adenine base paired with N3 of thymine instead of N1 (Figures 2-14 and 2-21). This type of A•T base pair was named the Hoogsteen base pair. The corresponding G•C Hoogsteen base pair is more stable if the cytosine is protonated at the N3 position (which is often depicted as C$^+$; Figure 2-21). An important difference between the two types of base pairs is that the C1′–C1′ distance is only 8.65 Å in the Hoogsteen base pair, significantly shorter than the 10.5 Å of the Watson–Crick base pair. Consequently, Hoogsteen base pairs are not compatible structurally with the B-DNA structure, but they do occur, usually in regions containing significant distortion or near sites of drug intercalation. They are physically possible, but they destabilize the B-DNA structure and are energetically unfavorable.

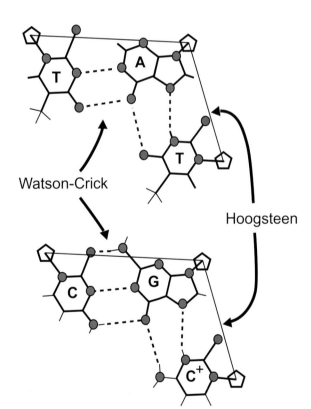

Figure 2-21. Comparison of the Watson–Crick and Hoogsteen base pairs for A•T and G•C. Hydrogen bonds are depicted as *dashed lines*; the two C1′ atoms are depicted as *open pentamers*, and the distance between them is indicated by a *thin line*. N atoms are *blue spheres* and O atoms are *red spheres*.

Certain chemical modifications increase the stability of the Hoogsteen base pair, and modified bases may be used when a Hoogsteen base pair is needed. For example, the modified nucleoside 3-isodeoxyadenosine (iA) forms a stable base pair with thymine using the Hoogsteen conformation: the amino group at the C8 position of 8-amino-adenine can form an additional hydrogen bond with the O2 of thymine. The stable d(CG[iA]TCG)$_2$ duplex illustrates that the iA•T base pair is fully compatible with the normal Watson–Crick base pair.

The Hoogsteen and other alternative base pairs can be generated when certain proteins or drugs bind to DNA (Section 2.5). The antibiotics triostine A and echinomycin bind to –CG– sequences by intercalating (Section 2.5.B) on both sides of a CG dinucleotide base pair. The A•T and G•C$^+$ base pairs adjacent to the antibiotic rings adopt Hoogsteen structures, because this improves the stacking interactions between them and the antibiotic.

The A•T and G•C$^+$ Hoogsteen base pairs have important roles in the structures of nucleic acid triple helices (Section 2.2.C.2). Hoogsteen base pairing is not restricted to DNA but also occurs in transfer RNA (Section 4.2.B), where the nucleic acid bases are often modified and involved in tertiary interactions.

X-ray and NMR studies of the DNA oligomer d(ATATAT): Hoogsteen base pairing in duplex DNA. N. G. Abrescia *et al.* (2004) *Biochemistry* **43**, 4092–4100.

Theoretical study of the Hoogsteen–Watson–Crick junctions in DNA. E. Cubero *et al.* (2006) *Biophys. J.* **90**, 1000–1008.

Hoogsteen base-pairing revisited: resolving a role in normal biological processes and human diseases. G. Ghosal & K. Muniyappa (2006) *Biochem. Biophys. Res. Commun.* **343**, 1–7.

---

2. Triple Helices

Adding increasing amounts of poly(U) to poly(A) first produces the expected 1:1 poly(A)•poly(U) duplex but then a 2:1 complex of three strands is also formed. In the triple-helical structure, the extra strand of poly(U) is bound to the poly(A) strand of the poly(A)•poly(U) duplex using Hoogsteen base pairs (Figure 2-21). The triple helix is important for its possible role in gene transcription and in the design of antisense oligonucleotides (Section 5.3.D).

Triplexes can be either parallel or antiparallel (Figure 2-22). In the parallel motif, a third strand (rich in thymine and cytosine) binds to a purine (adenine and guanine)-rich region within a standard double helix and lies parallel to the purine strand with which it makes contact. In the antiparallel motif, the third strand is rich in guanine and adenine (or thymine) and binds in antiparallel orientation to a purine region of double-helical DNA. In each case, **the third strand lies in the major groove, and the double helix of the other two strands remains largely undistorted**. One example of the 3-D structure of a triple-stranded DNA molecule is shown in Figure 2-23. The third strand forms Hoogsteen base pairs with the purine-rich strand of the duplex. The parallel motif is more stable if the N3 of cytosine is protonated, so it is more stable at low pH; the antiparallel motif is largely insensitive to pH. Triple helices have a greater concentration of ionized phosphate groups than do double helices, so they are more sensitive to the nature and concentration of counterions.

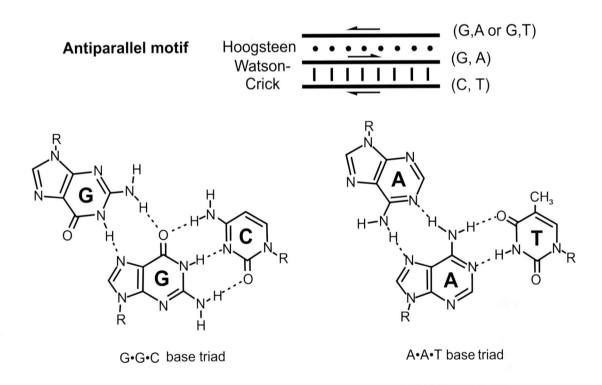

Figure 2-22. Base pairing in triple-helical DNA. The common base triads used in the two types of triplexes in the parallel (*top*) and antiparallel (*bottom*) motifs are described. The orientations of the three strands in the parallel and antiparallel motifs are shown above. In each case, the first and second strands use standard Watson–Crick base pairing, whereas the third strand at the *top* uses Hoogsteen base pairing to the second strand. The T•A•T triad is also found in the antiparallel motif. Adapted from E. T. Kool (1996) *Ann. Rev. Biophys. Biomol. Structure* **25**, 1–28.

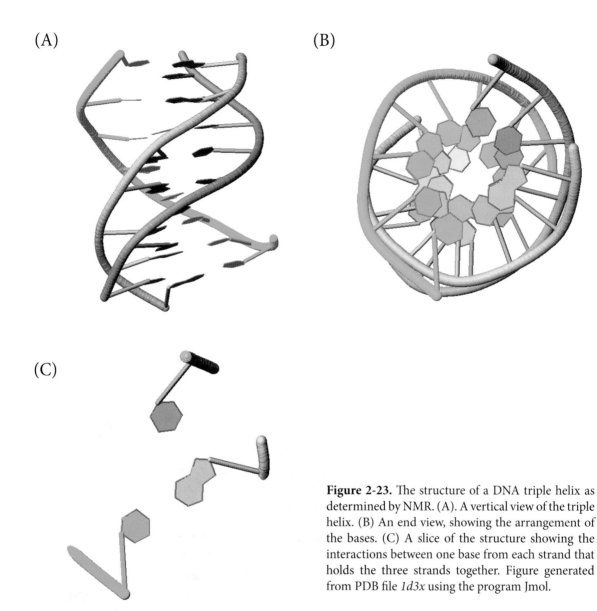

Figure 2-23. The structure of a DNA triple helix as determined by NMR. (A). A vertical view of the triple helix. (B) An end view, showing the arrangement of the bases. (C) A slice of the structure showing the interactions between one base from each strand that holds the three strands together. Figure generated from PDB file *1d3x* using the program Jmol.

Formation of stable triple helices of the (Py)•(Pu)•Py) type is relatively straightforward, but it is more difficult to design oligonucleotides that can bind to a specific DNA duplex using all four bases.

Triplex-forming oligonucleotides: principles and applications. K. M. Vasques & P. M. Glazer (2002) *Quart. Rev. Biophys.* **35**, 89–107.

Effect of divalent cations and cytosine protonation on thermodynamic properties of intermolecular DNA double and triple helices. P. Wu *et al.* (2002) *J. Inorg. Biochem.* **91**, 277–285.

Monitoring denaturation behaviour and comparative stability of DNA triple helices using oligonucleotide–gold nanoparticle conjugates. D. Murphy *et al.* (2004) *Nucleic Acids Res.* **32**, e65.

3. H-DNA: Intramolecular Triple Helices

Stretches of $(GA)_n \cdot (TC)_n$ sequences in genomic DNA are hypersensitive to nucleases during active gene transcription. They undergo a notable change in DNA structure when in a plasmid that is either negatively supercoiled (Section 3.1) or at low pH. The structural change is due to the formation of an intramolecular triple helix for part of the $(GA)_n \cdot (TC)_n$ sequence, with extrusion of a single-stranded loop (Figure 2-24). This structure is known as H-DNA. Other DNA segments with stretches of pyrimidines on one strand and purines on the other have also been shown to adopt the H-DNA structure. The triple helix is formed by the pyrimidine strand and half of the purine strand; the other half of the purine strand is single-stranded. The triple helix is stabilized by Hoogsteen base pairs (Figure 2-21), which accounts for the stabilizing effect of low pH. The single-stranded loop accounts for the susceptibility to nucleases. Negative super-coiling (Section 3.1) introduces the required topology of the chain for the H structure.

Figure 2-24. Schematic model of H-DNA, with an intramolecular triple helix. The homopurine strand is indicated with the *thick red line*, the homopyrimidine strand with a *thin blue line*. The *dashed line* is the half of the homopyrimidine stand incorporated into the triple helix. Adapted from S. M. Mirkin & M. D. Frank-Kamenetskii (1994) *Ann. Rev. Biophys. Biomol. Structure* **23**, 541–576.

When the pyrimidine-rich strand folds back to form a triplex, cytosines from one half of the sequence should interact with G•C but not A•T base pairs in its other half. Conversely, thymines in one half should interact with A•T but not G•C base pairs from the other half. Consequently, a homopurine–homopyrimidine that forms H-DNA should have a sequence that is a mirror repeat.

H-DNA and related structures. S. M. Mirkin & M. D. Frank-Kamenetskii (1994) *Ann. Rev. Biophys. Biomol. Structure* **223**, 541–576.

H-DNA:DNA triplex formation within topologically closed plasmids. S. L. Broitman (1995) *Prog. Biophys. Mol. Biol.* **63**, 119–129.

---

4. Four-stranded Structures: Guanine Quartet

The ends of eukaryotic chromosomes, known as **telomeres**, have DNA sequences that are rich in guanine and contain a short highly repetitive sequence, such as $(TTGGGG)_n$. The guanine-rich strand at the 3′-end exists as a single-strand and is capable of forming a higher ordered DNA structure in which four strands associate, known as the **guanine-** or **G-quartet**. One arrangement of the G-quartet is a four-fold symmetric motif in which the four guanine strands are in parallel orientations (Figure 2-25). The guanine residues are held together by a cyclic head-to-tail network of hydrogen bonds and a nonhydrated metal ion ($K^+ > Na^+ > Cs^+ > Li^+$, depending on its size) is located at the center. The four guanine residues have deoxyribose conformations that alternate between *anti* and *syn*.

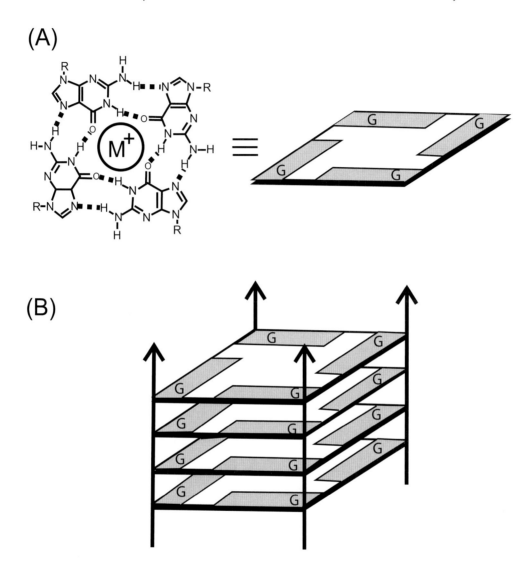

Figure 2-25. Models of the G-quartet. (A) The cyclic array of four guanine bases, each of which is a donor in two hydrogen bonds and an acceptor in another two. At the center is a pocket for interaction with a cation (M). (B) Schematic diagram of how stacks of G-quartets form quadruple helical structures. Adapted from J. R. Williamson (1994) *Ann. Rev. Biophys. Biomol. Structure* **23**, 703–730.

Other four-stranded DNA structures can be formed by tetrads resulting from the association of Watson–Crick base pairs, through either the minor or major groove side of the base pairs. For example, the d(GGGGTTTTGGGG) sequence forms a hairpin, and two such hairpins join to make an antiparallel hairpin dimer, thereby producing a four-stranded helical structure, with the thymines forming loops at either end (Figure 2-26). Structures stabilized by minor-groove tetrads differ from the canonical guanine quadruplex and represent a unique structural motif. For example, the cyclic oligonucleotide dpCCGTCCGT self-associates using intermolecular G•C base pairs and forms a symmetrical dimer stabilized by two G•C•G•C tetrads. The overall 3-D structure is similar to that found in other cyclic and linear oligonucleotides of related sequences, but the relative position of the two base pairs is slipped along the axis defined by the base pairs. This contrasts with other minor-groove G•C•G•C tetrads in which the G•C base pairs are aligned directly. An analogous arrangement of G•C base pairs occurs between the terminal residues of contiguous duplexes in some DNA crystals.

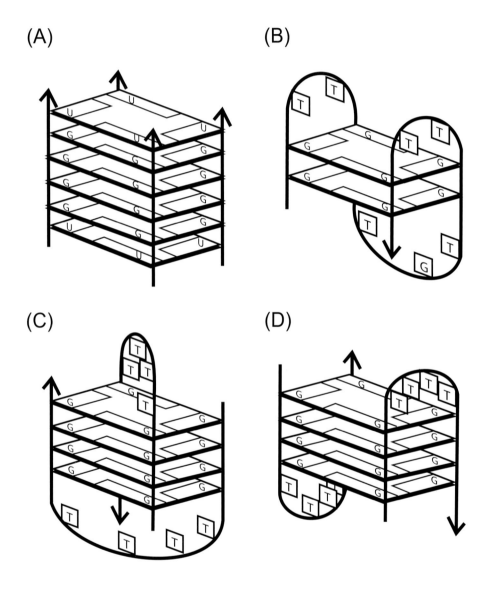

Figure 2-26. Schematic representations of G-quartet structures formed by the following oligonucleotides: (A) ribo(UGGGGU)$_4$; (B) d(GGTTGGTGTGGTTGG); (C, D) d(GGGGTTTTGGGG) with (C) Na$^+$ or (D) K$^+$. Adapted from J. R. Williamson (1994) *Ann. Rev. Biophys. Biomol. Structure* **23**, 703–730.

Telomere end-binding proteins control the formation of G-quadruplex DNA structures *in vivo*. K. Paeschke *et al.* (2005) *Nature Struct. Biol.* **12**, 847–854.

Quadruplex DNA: sequence, topology and structure. S. Burge *et al.* (2006) *Nucleic Acids Res.* **34**, 5402–5415.

Topology variation and loop structural homology in crystal and simulated structures of a bimolecular DNA quadruplex. P. Hazel *et al.* (2006) *J. Am. Chem. Soc.* **128**, 5480–5487.

Four-stranded DNA structures can be stabilized by two different types of minor groove G•C•G•C tetrads. N. Escaja *et al.* (2007) *J. Am. Chem. Soc.* **129**, 2004–2014.

5. *i-Motif*

The complementary DNA strand of the telomere TTGGGG sequence has the sequence CCCCAA. It adopts an unusual structure at low pH: a four-stranded structure, known as the **i-motif**, in which two parallel duplexes intercalate with one another in an antiparallel orientation, greatly extending each chain (Figure 2-27). Each parallel duplex consists of C•C$^+$ base pairs in which one cytosine is protonated.

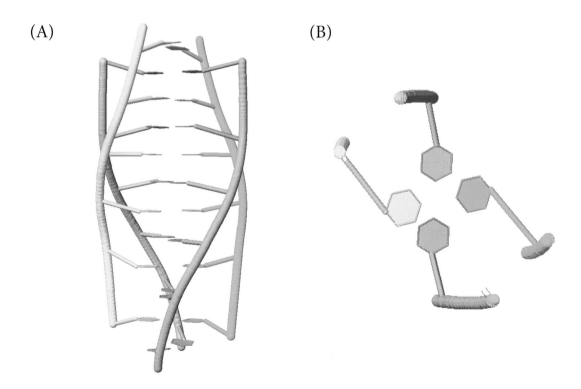

Figure 2-27. The four-stranded i-motif DNA structure determined by NMR. In this case, the sequence of each strand is 5′–d(TCCCCC)–3′, and the cytosine bases are protonated. (A) A vertical view, showing how the base pairs occur between only one pair of strands, but those of the other pair are intercalated between them. Consequently, each oligonucleotide chain is greatly elongated. (B) A slice of the structure showing two base pairs, one from each pair of oligonucleotide chains. The blue and rose bases are hydrogen-bonded, as are the yellow and green, which are above the blue and rose pair. Figure generated from PDB file *225d* using the program Jmol.

Crystal structures of a DNA octaplex with i-motif of G-quartets and its splitting into two quadruplexes suggest a folding mechanism of eight tandem repeats. J. Kondo *et al.* (2004) *Nucleic Acids Res.* **32**, 2541–2549.

DNA octaplex formation with an i-motif of water-mediated A-quartets: reinterpretation of the crystal structure of d(GCGAAAGC). Y. Sato *et al.* (2006) *J. Biochem.* **140**, 759–762.

6. *Inverted Repeat Sequences and Palindromes*

When the sequences of the two strands of DNA are identical, they are said to be 'palindromic'. A palindromic word or text reads the same in both directions. An impressive example is the Latin phrase *Sator arepo tenet opera rotas*, which translates roughly as 'Arepo the sower holds the wheels at work'. In the form:

S A T O R

A R E P O

T E N E T

O P E R A

R O T A S (2.5)

it reads the same in both directions, as well as from top to bottom, and bottom to top. In nucleic acid sequences, the term palindrome is traditionally (although not strictly correctly) applied to double-stranded DNA or RNA (which is two lines) when the nucleotide sequence is the same when read forwards from one strand, then backwards from the other, complementary, strand. In both cases, the sequence is read from 5' to 3', as the two strands are antiparallel. Such sequences also have complementary two-fold rotational symmetry:

$$5'\text{-CCATGG-}3'$$
$$\varepsilon\text{-}\mathrm{\mathsf{DDLV\mathrm{\mathsf{\mathrm{\mathsf{\mathrm{\mathsf{\mathrm{\mathsf{C}}}}}}}}}}\mathrm{\mathsf{C}}\text{-},\mathsf{S}$$
(2.6)

They are more correctly referred to as **inverted repeats**. They occur naturally, especially in **introns**, which interrupt the coding sequence in eukaryotic genes.

The two halves of each strand of a sequence that is said to be palindromic are also complementary to each other, with the potential for the two halves of the same strand to fold back on itself, to form a hairpin structure (Figure 2-28). If it happens in both strands, a cruciform structure is formed. The loops of the cruciform could be any size and any sequence, not necessarily palindromic. Such inverted repeat sequences are very common, being present in sites recognized by restriction enzymes (Section 6.3.A) and genetic repressors and in sequences of transposable elements.

A palindromic repeat sequence adopts a stable fold back structure under supercoiling. A. K. Shukla & K. B. Roy (2006) *J. Biochem.* **139**, 35–39.

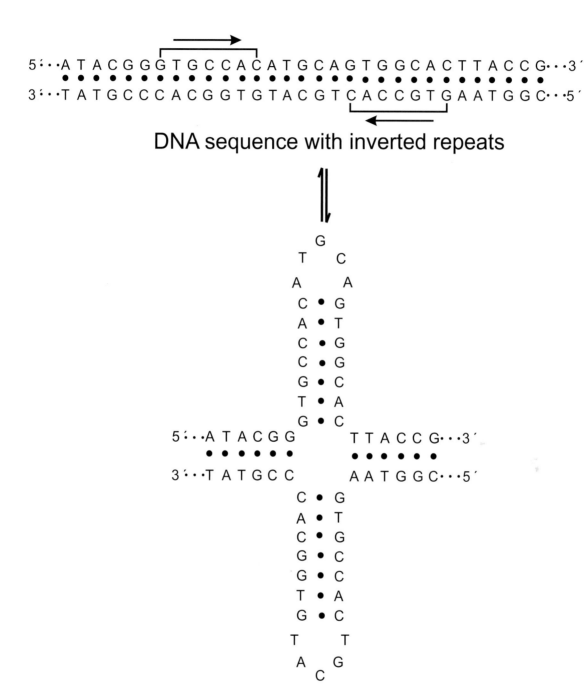

Figure 2-28. Cruciform structure generated from a double helix with inverted repeat sequences. The cruciform is a distorted form that is normally less stable than the linear double helix.

7. Helical Junctions: Cruciforms, Holliday Junctions

The **Holliday junction** is a key intermediate in DNA recombination. Four homologous DNA strands join into a structure that looks like a **cruciform** when observed by electron microscopy. A cruciform structure can also be generated in a double helix if the sequence has inverted repeats (Figure 2-28); it does not happen spontaneously in linear DNA molecules but can be induced by supercoiling (Section 3.1). The detailed structure of the cruciform DNA structure was determined when specific sequences were designed to form several types of so-called **immobile junctions** or **helical junctions**. An example of a four-way junction composed of four oligonucleotides is illustrated in Figure 2-29-A. Each of the four strands is paired with two other strands to form two arms of double helix. The junction is formed because there is no homologous two-fold sequence symmetry flanking the central branch point. Multi-valent metal ions, such as Mg^{2+} and $Co(NH_3)_6^{3+}$, or very high concentrations of monovalent cations, cause the structure to change to a stacked X-shape (Figure 2-29-B). The stacked X-structure is formed by the pairwise coaxial stacking of helices to adopt a right-handed, antiparallel cross with a cross-over angle of about 60° or 120° (Figure 2-29-C). Coaxial stacking can occur in two different ways and produce two different conformers that can interconvert, at rates that depend upon the salt concentration and type. The salt-dependence of these conformational transitions has been traced to the charge on one backbone phosphate group at the junction (Figure 2-29-B). Without the charge on this group, the stacked X-structure occurs in the absence of metal ions.

A number of B-DNA molecules in a crystal lattice have helix–helix packing contacts that are similar to the X-structure.

Happy Hollidays: 40th anniversary of the Holliday junction. Y. Liu & S. C. West (2004) *Nature Rev. Mol. Cell Biol.* **5**, 937–944.

Stereospecific effects determine the structure of a four-way DNA junction. J. Liu *et al.* (2005) *Chem. Biol.* **12**, 217–228.

Solution formation of Holliday junctions in inverted-repeat DNA sequences. F. A. Hays *et al.* (2006) *Biochemistry* **45**, 2467–2471.

The stacked-X DNA Holliday junction and protein recognition. P. A. Khuu *et al.* (2006) *J. Mol. Recognit.* **19**, 234–242.

---

8. Parallel-stranded DNA Duplexes

Double-stranded DNA structures in which the strands are parallel, rather than the normal antiparallel, do not occur naturally, although examples have been encountered in Sections 2.2.C.4 and 2.2.C.5. Other examples have been designed. In one series, the structures involve adenine- and thymine-containing DNA sequences that use nonWatson–Crick base pairs. The stabilities of these duplexes are modest; for example, the melting temperature (Section 5.1) of one comprising 21 nucleotides is 15°C lower than that of the corresponding antiparallel duplex. Another parallel motif using nonWatson–Crick base pairs is known as Π-**DNA**; the d(CGA) sequence has the greatest propensity to form this structure.

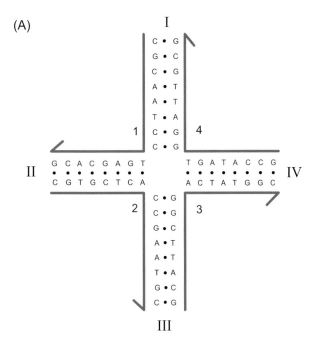

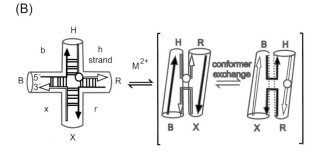

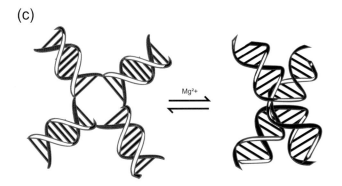

Figure 2-29. A designed stable DNA branched junction. (A) The junction is composed of four strands of DNA, depicted by the *thick lines*, with a *half-arrow* indicating the 3′-end. Each strand is paired with two other strands to form two arms of a double helix, as determined by the indicated base pairs. Adapted from N. C. Seeman & N. R. Kallenbach (1994) *Ann. Rev. Biophys. Biomol. Structure* **23**, 53–86. (B) In the presence of a divalent or trivalent cation, the junctions fold into the stacked X structure, which can have two conformers. The position of the phosphate group that modulates these conformational changes is indicated by the *circle*. (C) Drawing of the 3-D structures of the species described in (B).

An important requirement for a parallel duplex with diverse base pairs is that the two glycosyl bonds within a base pair must come from opposite directions, because of the identical chain polarity. For the normal nucleic acid bases, this can be accomplished using an alternative Watson–Crick base pair conformation, but the A•T and G•C base pairs are then not iso-structural, due to their hydrogen-bonding restrictions. Therefore, it has not been easy to design a stable parallel-strand duplex in which all four normal bases can be incorporated in random order.

Overcoming this difficulty required the use of alternative nucleosides, 2′-deoxyisoguanosine (iG) and 2′-deoxy-5-methylisocytosine (iC), which can form stable atypical Watson–Crick base pairs with the normal 2′-deoxycytosine (C) and 2′-deoxyguanosine (G), respectively. Indeed, oligodeoxynucleotides containing iG and iC can form remarkably stable parallel-stranded duplexes with the complementary (G,C)-containing DNA or RNA strands.

Stretches of parallel-stranded double-helical DNA can occur within antiparallel-stranded Watson–Crick DNA in looped structures or in the presence of sequence mismatches.

The biochemical significance of parallel DNA duplexes. H. M. Buck (2003) *Nucleosides Nucleotides Nucleic Acids* **22**, 1549–1552.

Parallel DNA double helices incorporating isoG or m^{5}isoC bases studied by FTIR, CD and molecular modeling. F. Geinguenaud *et al.* (2005) *Spectrochim. Acta A* **61**, 579–587.

Parallel and antiparallel DNA: fluorescence quenching of ethidium bromide by 7-deazapurines. H. Li & F. Seela (2005) *Nucleosides Nucleotides Nucleic Acids* **24**, 865–868.

2.3. DNA AS A POLYELECTROLYTE: HYDRATION AND COUNTERIONS

Nucleic acids are very soluble in aqueous solution because of their preponderance of ionized phosphate groups, which interact very favorably with water. **Nucleic acids bind twice as much water as do proteins**, some 0.6 g of water per gram of nucleic acid. Of the surface of DNA accessible to solvent, about 45% is made up of phosphate O atoms, 35% of sugar atoms and only 20% of atoms of the nucleobases. DNA is fully hydrated with about 20 water molecules per nucleotide when it adopts the B-structure. The first hydration shell of DNA consists of about 11–12 water molecules per nucleotide that contact the backbone as well as the functional groups of the bases. Water molecules are observed to be most fixed in a 'spine of hydration' in the minor groove that depends upon its width. A narrow minor groove has a single spine of water molecules along its floor, whereas a double ribbon of water molecules is associated with a wide minor groove, bridging the N or O atoms at the edges of the bases to the O4′ atoms of the sugar ring. It is now generally accepted that the hydration shell surrounding the DNA molecule plays an important role in DNA structure and its recognition by proteins and other ligands. Specific water molecules play critical roles in the sequence-specific recognition by proteins, and they modulate the binding of sequence-nonspecific DNA-binding proteins to DNA of random sequences (Section 13.2.A).

The degree of hydration of the double helices is not correlated with the types of atoms exposed to the solvent (Table 2-8), indicating more specific interactions between nucleic acid and water molecules. AT-rich sequences are observed to have a spine of water molecules in the minor groove. In contrast, GC-rich sequences appear to have patches of water molecules in the major groove that would be disrupted by the methyl group of the thymine nucleotide in AT-rich regions. Both types of hydration occur in both the minor and major grooves of poly(dIdC)•poly(dIdC). The minor groove hydration spine of each nucleotide is estimated to contain eight water molecules, the major groove hydration 13. These are a considerable fraction of the 24 water molecules that can fit within the first hydration layer of each nucleotide of a double helix, so some probably involve the second and third layers of water molecules.

A major factor in the solvation of DNA is the single charge on each ionized phosphate group of the backbone, which should be spread amongst the four O atoms. **The polyelectrolyte nature of nucleic acids results in them retaining a significant number of counterions, even when these ions are absent from the bulk solvent.** Being part of the same molecule, the charged phosphate groups are constrained by the covalent bonds to be close to each other, which is unfavorable energetically. The electrostatic repulsions between these ionized phosphate groups are so severe, and the need to screen them with counterions so great, that a number of counterions remain associated with the DNA even in the absence of these ions in the bulk solvent. These bound counterions are said to be 'condensed', in contrast to the other free ions, which form a normal diffuse ionic cloud around the DNA. The sodium ions that are condensed on B-DNA are 76% of the number of phosphate groups, independent of the bulk salt concentration, so **the net charge of the DNA is effectively decreased to –0.24 e per phosphate group. With divalent ions, such as Mg^{2+}, the effective net charge is –0.12 e per phosphate group; it is only –0.08 e for trivalent ions**. Although said to be condensed, these ions are fully hydrated with bound water, associated rather loosely with the DNA, and exchange readily with ions of the bulk solvent. They are located within 7 Å beyond the external surface of the DNA, some 17 Å from the helix axis. Monovalent counterions are generally not observed in oligonucleotide crystal structures and so must be very mobile; divalent cations, such as Mg^{2+}, can be observed crystallographically, but only when they are fixed in position by interacting with two or more groups.

The local concentration of monovalent cations around double-stranded DNA is approximately 1 M. The negative charges on DNA attract H^+ ions as well, so the **pH in close proximity to the DNA molecule is some 2–3 units lower than in the bulk solution**. In terms of the van der Waals dimensions of the atoms, the diameter of B-DNA is only 20 Å **but hydration and electrostatic repulsions cause the effective diameter to be 30 Å at high salt, increasing to 150 Å at 0.01 M salt**. These observations are not unique to DNA but apply generally to any polyelectrolyte of similar charge density.

Cations are usually bound to nucleic acids in a solvated state. They usually interact with DNA via water molecules in their primary solvation shell; direct ion–DNA contacts are probably rare because such interactions would need to compensate for the substantial energetic cost of removing the waters of hydration. On the other hand, Mg^{2+} ions are thought to bind directly to some phosphate groups by losing at least one water from their hydration shells. Solvated ions are usually too large to enter the minor groove, except in the case of AT tracts, where monovalent cations can partition selectively into the minor groove. Solvated ions appear to bind preferentially to guanine bases in the major groove or to the O atoms of phosphate groups. Sites where the ions bind are readily exchangeable: the same site may be occupied by any ion, including more complex molecules, such as spermine (Equation 2.4), or a water molecule. Polyamines like spermine are apparent at specific sites in crystal structures of DNA, but they appear to be disordered by most other techniques.

Divalent cations are much more effective than monovalent, and one Mg^{2+} ion has an influence on the DNA double helix comparable to that of 100–1000 Na^+ ions. Not even K^+ and Na^+ are equivalent, as they produce detectable differences in the structure of the DNA. DNA deformation may be caused by electrostatic collapse around areas of uneven cation density. The release of these counterions can be very significant when they are displaced by binding of a ligand (Section 13.2).

The hydration environment around a DNA double helix, including cation counterions, plays an important role in determining the type of conformation adopted and other properties. For example, right-handed B-DNA with a GC-rich sequence can be converted to left-handed Z-DNA simply by increasing the salt concentration (Figure 2-20).

The partial specific volumes of nucleic acids in water are relatively small, around 0.5–0.6 cm³ g⁻¹. Nucleic acids do not occupy much extra volume in water, indicating that they interact tightly with water molecules. They are influenced remarkably by the solvent composition, especially the salt concentration, because of their high charge density, and they interact differently with the salt and water molecules of an aqueous solution. Consequently, their apparent partial specific volume differs from the true value and varies with the salt concentration (Figure 2-30). This is especially important for nucleic acids because their density (the reciprocal of their partial specific volume) is usually of interest primarily in solutions containing high concentrations of salt, especially of cesium chloride, for density-gradient centrifugation.

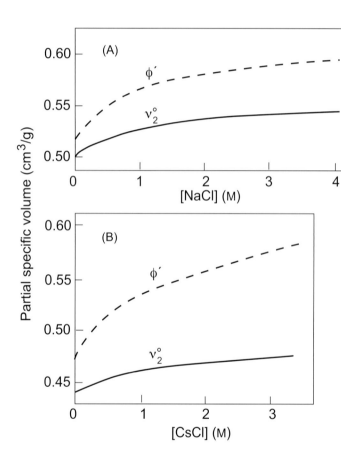

Figure 2-30. The partial specific volume of DNA as a function of the concentration of NaCl (A) or CsCl (B). The apparent partial specific volume is ϕ' (*dashed curve*), which does not account for the interactions with the salt. The true partial specific volume is given by $v°_2$. Data from H. Eisenberg (1976) *Biological Macromolecules and Polyelectrolytes in Solution*, Clarendon Press, Oxford.

The intrinsic compressibility of a nucleic acid is believed to be very small, and the compressibility measured in solution is believed to be primarily that of the hydrating solvent (Table 2-9). The measured values are negative, indicating that the hydration is dominated by interactions with the ionized phosphate groups of the nucleic acids. Differences in compressibilities of the various double helices are attributed to differences in their hydration shells. The more negative the compressibility, the stronger the hydration. The most strongly hydrated is poly(dIdC)•poly(dIdC), poly(dGdC)•poly(dGdC) exhibits intermediate hydration, and the two all-AT duplexes the weakest hydration.

Table 2-9. The apparent molar volumes, and apparent molar adiabatic compressibilities per base pair, for various synthetic DNA duplexes at 25°C

DNA duplex	Apparent molar volume (cm³/mol of base pair)	Apparent molar adiabatic compressibility (10^{-4} cm³/bar/mol of base pair)
Poly(A)•poly(U)	166	−52
Poly(A)•poly(dT)	168	−55
Poly(dA)•poly(U)	173	−41
Poly(dA)•poly(dT)	169	−58
Poly(dAdT)•poly(dAdT)	170	-58
Poly(dGdC)•poly(dGdC)	152	−73
Poly(dIdC)•poly(dIdC)	123	−102

Data from T. V. Chalikian *et al.* (1999) *Biopolymers* **50**, 459–471.

Role of monovalent counterions in the ultrafast dynamics of DNA. S. Sen *et al.* (2006) *J. Phys. Chem. B* **110**, 13248–13255.

The persistence length of DNA is reached from the persistence length of its null isomer through an internal electrostatic stretching force. G. S. Manning (2006) *Biophys. J.* **91**, 3607–3616.

Multiple time scales in solvation dynamics of DNA in aqueous solution: the role of water, counterions, and cross-correlations. S. Pal *et al.* (2006) *J. Phys. Chem.* **110**, 25396–25402.

The DNA structure responds differently to physiological concentrations of K⁺ or Na⁺. B. Heddi *et al.* (2007) *J. Mol. Biol.* **368**, 1403–1411.

Water clustering and percolation in low hydration DNA shells. I. Brovchenko *et al.* (2007) *J. Phys. Chem. B* **111**, 3258–3266.

2.4. DNA FLEXIBILITY AND DYNAMICS: CURVING, TWISTING AND STRETCHING

DNA possesses many of the properties usually associated with a linear polymer, with varying types of both mechanical rigidity and motional flexibility. DNA is unique, however, in the lengths of the chains. A natural DNA molecule will consist of millions to a billion base pairs (bp) and will be millions of times longer than it is wide. A double helix containing 100 million bp will have a covalent length of 3.3×10^8 Å, or 3.3 cm, a diameter of only 20Å (2.0 nm). This extreme elongation presents special topological problems, described in Chapter 3. A further consequence is that such long DNA molecules are readily broken physically by the shear forces generated by standard laboratory manipulations, even just pouring and pipetting; the resulting fragments usually have molecular weights of about 5×10^6, corresponding to about 7500 bp. Passing the solution through a hypodermic syringe needle will produce fragments of about 1500 bp, as will shearing in a homogenizer or blender. Fragments as small as 50–100 bp on average can be generated by sonication.

The motions of DNA in solution can be either large-scale or local: DNA is locally rather stiff, so that a short double helix can be approximated as a rigid rod, but very long DNA molecules are globally flexible. A fully extended DNA double helix might measure nearly 1 m in length, but its flexibility in solution would decrease its average end-to-end distance to only 230 μm, a contraction of more than 4×10^3-fold. DNA helices that are intermediate in length between a rigid rod and a random coil must be treated as 'worm-like chains', which take into account both the local stiffness and the long-range flexibility of the double helix.

The four types of deformation of the linear DNA double helix are curvature, bending, torsion and stretching. The global curvature and bending flexibility are normally expressed as the longitudinal **persistence length**, which is a measure of the resistance of a polymer to lateral bending. It is formally defined as the average projection of the molecular end-to-end distance vector on its initial path vector, in the limit of infinite chain length. Another useful interpretation of persistence length is the distance over which the root mean square (r.m.s.) bend angle in any particular direction is 1 radian (about 57°). A helix of the same length as the persistence length has an r.m.s. end-to-end separation that is 14% shorter than its covalent length.

The persistence length of double-stranded DNA is roughly 150–500 bp, depending upon the ionic strength and the cations present. Consequently, the helix axis of a double-stranded DNA molecule changes direction by 57° over every segment of 150–500 bp. **Normal B-DNA is considerably more flexible than either A-DNA or Z-DNA**: the persistence length of A-DNA is about 1500 bp at moderate ionic strengths. Single-stranded DNA is much more flexible, and its persistence length is only 15Å (1.5 nm) in 2 M NaCl and 30 Å (3 nm), corresponding to 4–9 nucleotides, in 25 mM NaCl.

Double-stranded DNA also displays local torsional stiffness to twisting about its long axis. This is usually expressed as a **torsional persistence length**, that length over which DNA will have an r.m.s. twist deviation of 1 radian (57°). **This torsional persistence length of DNA is about 180 bp**, approximately the same as the longitudinal persistence length. Stretching of the DNA double helix is measured by an **elastic stretch modulus**, S, which when divided by the cross-sectional area of the polymer gives the **Young modulus** of physics. The DNA double helix can be stretched more readily at low ionic strength, probably due to electrostatic repulsions between the phosphate groups.

These deformations of the DNA double helix are believed to be a result of a balance between repulsions between the ionized phosphate groups, attractions between the stacked base pairs and deformation of the covalent structure, although the details are not clear.

The relative concentrations of two sites on a DNA double helix with respect to each other, their **effective concentration**, can be measured by the tendency of the ends of the helix to form a circular molecule (Figure 3-2). The effective concentration reaches a maximum when the two groups are separated by about 500 bp, with a value of about 1×10^{-7} M. The effective concentrations of groups closer together are lower due to the stiffness of the double helix, while those further apart are less likely to encounter each other randomly. When separated by 150 bp, the effective concentration is decreased to about 2×10^{-9} M, but only to about 2×10^{-8} M when separated by 3000 bp. When the two sites are defined in terms of their position on the faces of the double helix, their effective concentrations oscillate in a sinusoidal manner with varying separations, due to the helical periodicity of the double helix and the varying tendencies of different sequences to bend. The phasing over separations of 60–200 bp may alter the effective concentration 5–10-fold.

Certain DNA sequences have abnormal mobilities in gel electrophoresis, suggesting that they have unusual shapes. For example, double-stranded DNA fragments having repeats of $(A)_n$ nucleotides (with $n \geq 4$), separated by another 4–5 nucleotides and phased with the helical repeat, migrate in the gel significantly more slowly than those having mixed sequences. The reason is that the 5′-AAAAA sequence, and its complementary 5′-TTTTT sequence, has an intrinsic bending property that can be demonstrated by the increased efficiency of cyclization of those bent DNA fragments. The molecular basis of the intrinsic bendability of the $(A)_n$ sequence may be due in part to the high propeller twist associated with the A•T base pairs in the $(A)_n$•$(T)_n$ sequence and the presence of only two hydrogen bonds between the bases. The bend in these sequences is relatively smooth, resulting in a curved DNA structure, and the opening (roll) of the bend is towards the minor groove. **Many proteins induce such a smooth bending when they bind to DNA** by having many small single-step bends (Section 13.2.C). **Another type of bent DNA has a sharp kink at a localized site**, usually caused and stabilized by bound proteins.

The large-scale tumbling motion of a DNA duplex of 10–100 bp occurs at a time scale of nanoseconds to microseconds (Figure 2-31). The **correlation times** for tumbling at 0°C of a cylinder the size of an oligonucleotide with 12 nucleotide base pairs are 7.5 ns for rotations about the helix axis and 14 ns for rotations perpendicular to the axis. The latter time increases roughly with the square of the length of the duplex, while the time for the parallel motion increases in proportion to the length. The correlation times at 20°C are about half the values at 0°C. This global flexibility of DNA can account for the behavior of DNA in solution and in gels.

Natural DNA molecules are often closed circles, with no termini. This introduces a further complication that is discussed in Chapter 3.

Electrostatic effects in DNA stretching. A. V. Tkachenko (2006) *Phys. Rev. E* **74**, 041801.

NMR studies of dynamics in RNA and DNA by ^{13}C relaxation. Z. Shajani & G. Varani (2006) *Biopolymers* **86**, 348–359.

Diffusion and segmental dynamics of double-stranded DNA. E. P. Petrov *et al.* (2006) *Phys. Rev. Lett.* **97**, 258101.

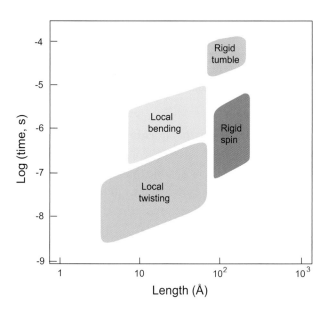

Figure 2-31. Time scales over which DNA duplexes of various lengths move rotationally in various ways. Data from M. Hogan *et al.* (1982) *Proc. Natl. Acad. Sci. USA* **79**, 3518–3522.

2.4.A. Local Flexibility

Local motions are important in many of the biological properties and functions of DNA. That various parts of the DNA structure have considerable flexibility is suggested by the highly anisotropic temperature factors observed in the X-ray crystallography structures of DNA molecules. The average bending angle between adjacent base pairs is approximately 7° under normal conditions, while the average twist angle between adjacent bases is about 3.6°. Enzymes that repair mismatched bases often bind to DNA with the offending base flipped out of the double helix and into the active site of the enzyme; whether this happens, and how frequently or rapidly, in the absence of the enzyme is not known. Many drugs and dyes intercalate into the DNA structure, which requires that adjacent DNA base pairs be separated by up to 6.8 Å (Figure 2-32). This is a large conformational change, involving many torsion angles and sugar pucker rearrangements (Figure 2-5); it is reversible, so DNA must be in constant motion.

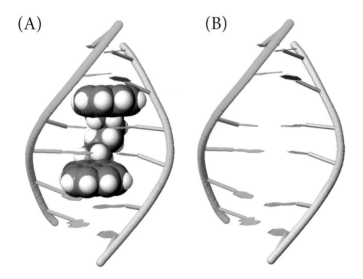

Figure 2-32. The intercalation of double-stranded DNA by the antitumor bisnaphthalimide known as LU-79553. Each strand of the DNA has the sequence 5′–d(ATGCAT). The structure was determined by NMR. (A) The intercalator is shown as a space-filling model but the DNA is depicted with just the nucleobases and a cylinder for the ribose-phosphate backbone. (B) The DNA with the intercalator deleted, illustrating how its structure is distorted. Figure generated from PDB file *1cx3* using the program Jmol.

The biological manipulation and packaging of DNA often depend crucially on local variations in both bending and torsional flexibility, and the probability of DNA untwisting is strongly correlated with a high bending flexibility. The average physicochemical properties of a long DNA molecule approximate those of a thin isotropic homogeneous rod that bends equally in all directions, but more locally DNA behaves as an anisotropic heterogeneous rod that bends primarily only in certain directions. This bending anisotropy is sequence-dependent and to a first approximation reflects both the geometry and stability of individual base steps.

The bendability of DNA is believed to depend primarily upon the stacking energies between neighboring base pairs. DNA sequences with Py–Pu steps in B-DNA are deformed most easily; therefore steps such as CpG, TpA or TpG (= CpA) appear to be more flexible than others and are the preferred binding sites for intercalators and the preferred kinked sites induced by binding of proteins (Section 13.2.C). A change from G•C to A•T base pairs renders that region of the DNA flexible and more susceptible to bending. As a result of its poor stacking, the deformability of a TpA dinucleotide step is inherently much greater than that of an ApT or ApA step. The proximities of the methyl groups of two successive thymines on the opposite strand and of the methyl groups and the phosphate backbone make the helix relatively rigid at an ApA step. In an ApT step, the stacking of the methyl group with the adjacent adenine and the intervening sugar phosphate backbone prevents bending by a roll mechanism. In a TpA step, however, the methyl group projects into the major groove without any significant stacking interactions with either the adjacent adenine or the phosphate backbone, and this step therefore displays greater deformability than the ApA- and ApT-steps. The local dynamics of the CpG site of the dodecamer CGCGAATTCGCG indicate that the deoxyribose of the ninth residue, cytosine, undergoes more motion than the other nucleotides. The greater flexibility of the CpG step has been correlated with the cleavage at this site by the restriction enzyme *Eco*R1 (Section 6.3.C). Therefore, the local dynamics of DNA clearly play important roles in DNA bending induced by proteins, in binding of intercalators, in restriction enzyme action, and undoubtedly in other biological functions.

Single phosphate groups of DNA can be replaced by the double phosphate groups of pyrophosphate or O-ethyl-substituted pyrophosphate without elongating the internucleotide distance: the additional phosphate group is simply flipped out from the sugar-phosphate backbone. This has only a minor impact on the base pairing and stacking, so there are limits to the flexibility of the double helix.

Hydrogen bonding, base stacking, and steric effects in DNA replication. E. T. Kool (2001) *Ann. Rev. Biophys. Biomol. Structure* **30**, 1–22.

Probing single-stranded DNA conformational flexibility using fluorescence spectroscopy. M. C. Murphy *et al.* (2004) *Biophys. J.* **86**, 2530–2537.

The structural basis of DNA flexibility. A. A. Travers (2004) *Philos. Trans. A* **362**, 1423–1438.

2.4.B. Hydrogen Exchange

The technique of hydrogen exchange is very useful in measuring the flexibilities of both nucleic acids and proteins (Section 10.3). The H atoms attached to N and O atoms of the nucleobases can exchange with those of water at rates of roughly $10–10^3$ s^{-1} (i.e. on time scales of $10^{-1}–10^{-3}$ s). Even that attached to C8 of adenine and guanine bases will exchange with measurable rates at elevated temperatures. Such exchange depends upon the environment of the group, especially on whether the H atom is involved in hydrogen bonding.

The imino (–NH–) groups of uracil, thymidine and guanine are sufficiently acidic that their H atom can be removed as a proton by every encounter with a hydroxide ion (HO$^-$); this diffusion-controlled reaction has rate constant k_{OH} of approximately 10^{10} M^{-1} s^{-1}. Weaker bases, such as components of the buffer, can also remove these protons. Acid-catalyzed exchange proceeds by protonation of oxygen followed by deprotonation of the –NH=, to produce the iminol tautomer. Using thymine as an example:

(2.7)

At low pH, acid catalysis in guanine is masked by its ionization and conversion to the conjugate acid. Exchange of an H atom of an amino group of adenine, guanine or cytosine is catalyzed by both acid and base, and there is also a pH-independent rate in the region of neutral pH. The base-catalyzed reaction involves the slow removal of an H ion from an –NH– by a hydroxide ion and its replacement by one from the solvent. The acid-catalyzed reaction involves attachment of H⁺ to a ring nitrogen, followed by removal of one of the amino H atoms (as a proton) by water, to produce as a transient intermediate an abnormal tautomer:

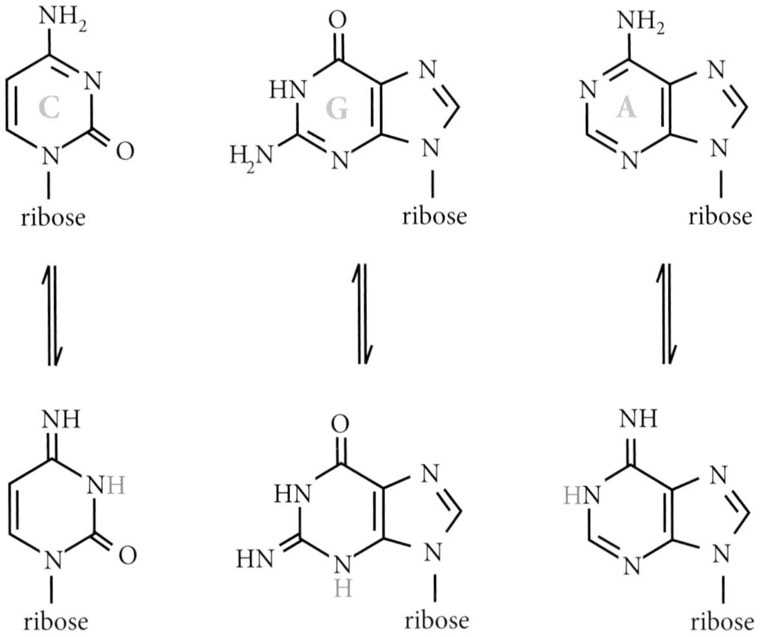

$$\text{(structures of C, G, A and their tautomers)} \tag{2.8}$$

where the H atom from water is shown in red. In the neutral pH region, where the rate is independent of pH, the H atom of the amino group is removed by a hydroxide ion instead of water, to generate the same intermediate. At low pH in each of these cases, acid catalysis is masked by complete ionization and conversion to the conjugate acid (Figure 2-7).

The pH relevant to hydrogen exchange is that within close proximity of the DNA double helix, and this can be quite different from the bulk pH, due to the polyelectrolyte nature of DNA (Section 2.3). Such complications mean that the detailed mechanism by which exchange occurs from a DNA double helix under most circumstances is not certain, as in the case of proteins (Section 10.3). The exception is at high pH, which is possible with DNA because the double helix is stable under these conditions. The exchange rate is determined by the rate at which a group becomes accessible (known as the 'EX1 mechanism'). This has made it possible to measure the rate of base-pair opening in the DNA double helix to be between 2×10^1 and 2×10^3 s^{-1}; slower rates are indicative of unusual structural features. The rates and equilibrium constants of the opening reactions for individual base pairs at various temperatures yield the enthalpies and free energies of the barrier to opening and the open state for each base pair. Neighboring bases one or more positions removed from the opening base pair influence the temperature-dependence of the opening pathway. The rates of the reaction are related directly to the free energies of the open state, which suggests that the transition state in the opening reaction is closer to the native closed state of the base pair than to its open state.

Studies of base pair kinetics by NMR measurement of proton exchange. M. Guéron and J.-L. Leroy (1995) *Methods Enzymol.* **261**, 383–413.

A nuclear magnetic resonance investigation of the energetics of basepair opening pathways in DNA. D. Coman & I. M. Russu (2005) *Biophys. J.* **89**, 3285–3292.

2.5. BINDING OF SMALL MOLECULES

Many small molecules bind to DNA. This usually has significant physiological consequences, and many of these molecules are known as drugs and antibiotics. Most of these molecules bind much more tightly to double-stranded DNA than to single-stranded, so they stabilize the double helix, even if they perturb its structure.

Thermodynamics of nucleic acids and their interactions with ligands. A. N. Lane & T. C. Jenkins (2000) *Quart. Rev. Biophys.* **33**, 255–306.

Synthetic metallomolecules as agents for the control of DNA structure. A. D. Richards & A. Rodger (2007) *Chem. Soc. Rev.* **36**, 471–483.

Volume and hydration changes of DNA–ligand interactions. X. Shi & R. B. Macgregor (2007) *Biophys. Chem.* **125**, 471–482.

2.5.A. Binding to the Minor Groove

Most small molecules that bind to DNA do so in the narrow groove. The narrow groove at A–T sequences in B-DNA provides an excellent binding site for many drugs, due to the greater negative electrostatic potential at the bottom of the minor groove at A–T regions. The N2 amino groups of guanine nucleotides hinder drug binding, both electrostatically and sterically. **Most such drugs replace the spine of hydration water in the narrow minor groove and stabilize the DNA structure, without perturbing its overall conformation significantly** (Figure 2-19-B). The binding energy is due in part to the gain in entropy associated with the displacement of multiple water molecules by one drug molecule.

Binding of distamycin A (a pyrrole-containing antitumor antibiotic) to DNA containing A•T sequences revealed a more complex pattern. Distamycin A can bind DNA not only in a 1:1 drug:duplex complex but also in a 2:1 mode, with two distamycins bound to the minor groove in an antiparallel side-by-side manner. The 2:1 binding mode requires expansion of the minor groove to accommodate the two drug molecules. That it does so reflects the flexibility of B-DNA.

Complexes of the minor groove of DNA. B. H. Geierstanger & D. E. Wemmer (1995) *Ann. Rev. Biophys. Biomol. Structure* **24**, 463–493.

Hydration changes accompanying the binding of minor groove ligands with DNA. N. N. Degtyareva *et al.* (2007) *Biophys. J.* **92**, 959–965.

2.5.B. Intercalation

Virtually any molecule with a sufficiently large aromatic chromophore will bind to DNA double helices by insertion of the flat chromophore between the base pairs (Figure 2-32). Stacking interactions between the intercalator and the adjacent base pairs are believed to be the primary driving force; the intercalators tend to have larger and more electron-deficient chromophores than those of a DNA base pair. Intercalation disrupts the DNA double helix, forcing the base pairs apart by about 3.5 Å and unwinding the helix by 10–25°; yet the stability of the helix is increased because the intercalator binds only to the double helix. The time that different intercalators remain bound (the *residence time*) varies widely, from a few milliseconds to many hours. Binding by intercalation can also be combined with binding in the minor groove by another part of the molecule.

Most of the intercalators are mutagenic and cause frame-shift mutations in which a base pair is inserted or deleted at the site where the intercalator bound to the DNA. These mutations are believed to be caused by the presence of the intercalator during replication of the DNA molecule (Section 6.1), producing mispairing of the nucleotides. Some intercalators, such as daunomycin, are antibiotics and used in cancer chemotherapy.

Variable role of ions in two drug intercalation complexes of DNA. N. Valls *et al.* (2005) *J. Biol. Inorg. Chem.* **10**, 476–482.

Intercalation of organic dye molecules into double-stranded DNA. II. The annelated quinolizinium ion as a structural motif in DNA intercalators. H. Immels *et al.* (2005) *Photochem. Photobiol.* **81**, 1107–1115.

Quantifying force-dependent and zero-force DNA intercalation by single-molecule stretching. I. D. Vladescu *et al.* (2007) *Nature Methods* **4**, 517–522.

---

1. Ethidium Bromide

One of the best-known intercalators is **ethidium bromide** (EB):

(2.9)

because it is widely used in molecular biology laboratories to detect nucleic acids. EB binds with high affinity to double-stranded DNA, with a dissociation constant of only 0.5 μM. The aromatic ring of EB intercalates within the DNA duplex, and the two amino groups hydrogen bond to the phosphate groups across the two DNA strands, which fixes the orientation. Consequently, B-DNA adopts an A-form structure locally and the deoxynucleoside sugar pucker changes from C2′-*endo* to C3′-*endo*. EB does not, however, bind to adjacent sites on the double helix, and there must be at least two base pairs between bound EB molecules. Such 'neighbor exclusion' is common with intercalation and causes binding curves to be more complex than expected with a single population of binding sites.

In aqueous solution, EB is dark yellow in appearance and possesses a broad featureless absorption spectrum, with maximum absorbance at about 480 nm. EB is only weakly fluorescent; its broad featureless fluorescence emission spectrum has a maximum at 617 nm. In aqueous solution, the fluorescence quantum yield is low and the fluorescence lifetime is short because the excited state is quenched by transfer of a proton to an adjacent water molecule. Increasing the EB concentration produces dimerization, a red shift in the absorption maximum, and a decrease in absorbance intensity; the EB dimer is nonfluorescent.

When EB binds to DNA, its absorption spectrum shifts to longer wavelengths, due to stabilization of the excited state, and the fluorescence intensity increases markedly. EB is shielded from solvent molecules and the rate of quenching decreases, increasing both the fluorescence decay time and the quantum yield. The fluorescence decay time of EB is 23 ns when bound to DNA but only about 1.8 ns in water.

The increased fluorescence when EB is bound to DNA provides the most common method for detecting and measuring DNA. The fluorescence is enhanced approximately 30-fold when excitation is at 520 nm and emission at 600 nm. This enhancement is practically independent of the base composition of the DNA. Single-stranded DNA and RNA also stimulate the fluorescence of EB, but to a lesser extent than does duplex DNA.

Appropriately designed dimers of EB show binding affinities and fluorescence enhancements even greater than those of the monomer. One intercalates at a ratio of 1 dimer per 4–5 bp and can detect and quantify picogram amounts of DNA.

EB is a powerful mutagen, extremely toxic by inhalation, ingestion and skin contact, and a suspected carcinogen and reproductive toxin. EB originally found fame as a drug useful in combating trypanosomal, microbial, bacterial and viral infections. All these biological effects are a direct consequence of EB binding to DNA and inhibiting its replication.

Ethidium DNA agarose gel electrophoresis: how it started. P. Borst (2005) *IUBMB Life* **57**, 745–747.

On the electronic structure of ethidium. N. W. Luedtke *et al.* (2005) *Chemistry* **11**, 495–508.

Local conformational changes induced in B-DNA by ethidium intercalation. J. M. Benevides & G. J. Thomas (2005) *Biochemistry* **44**, 2993–2999.

The complex of ethidium bromide with genomic DNA: structure analysis by polarized Raman spectroscopy. M. Tsuboi *et al.* (2007) *Biophys. J.* **92**, 928–934.

---

2. Psoralen Photo Cross-linking

The natural reagent **psoralen**:

(2.10)

intercalates within the DNA double helix and can then undergo chemical reaction with adjacent pyrimidine bases upon being illuminated with UV light. The three-ring structure absorbs UV light with maxima at 220, 250 and 300 nm. Upon irradiation, the 4′–5′ double bond of the psoralen can react with the 5–6 double bond of an adjacent pyrimidine ring, to generate a cyclobutane ring like that depicted in Equation 2.3. This adduct can still absorb UV light in the wavelength range 320–400 nm, and its 3–4 double bond can similarly react with the 5–6 double bond of another adjacent pyrimidine ring. This pyrimidine will be on the opposite strand of the DNA double helix, so the two strands will be cross-linked covalently:

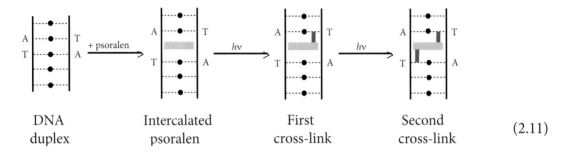

(2.11)

No reaction occurs in the absence of UV light. This reaction can be used in cells, making it useful for studying the *in vivo* structure of DNA.

The most useful forms of psoralen have methyl groups at the 4, 5′ and 8 positions, plus a hydroxymethyl or amino methyl group at the 4′ position. **Anglecins** are angular psoralens that have the five-membered ring fused to the 5–6 bond of psoralen, rather than the 6–7 (or 2′–3′). This angular structure cannot cross-link two bases, but it is a useful reagent for studying the individual cross-linking reactions.

Quantitative analysis of DNA interstrand cross-links and monoadducts formed in human cells induced by psoralens and UVA irradiation. C. Lai *et al.* (2008) *Anal. Chem.* **80**, 8790–8798.

Photoreactivity of furocoumarins and DNA in PUVA therapy: formation of psoralen–thymine adducts. J. J. Serrano-Pérez *et al.* (2008) *J. Phys. Chem.* **112**, 14002–14010.

Modulation of psoralen DNA crosslinking kinetics associated with a triplex-forming oligonucleotide. D. H. Oh *et al.* (2008) *Photochem. Photobiol.* **84**, 727–733.

2.6. CHEMICAL MODIFICATION AS A PROBE OF STRUCTURE

Chemical modification is often used to probe the DNA structure, to determine whether a particular group is accessible to the reagent. Most groups are much less reactive when they are part of a double-helical structure, especially when they are involved in base pairing or bound to proteins (Section 13.1.C).

Diethyl pyrocarbonate (DEPC) carboxyethylates purines at the N7 position when they are single-stranded or in the Z conformation:

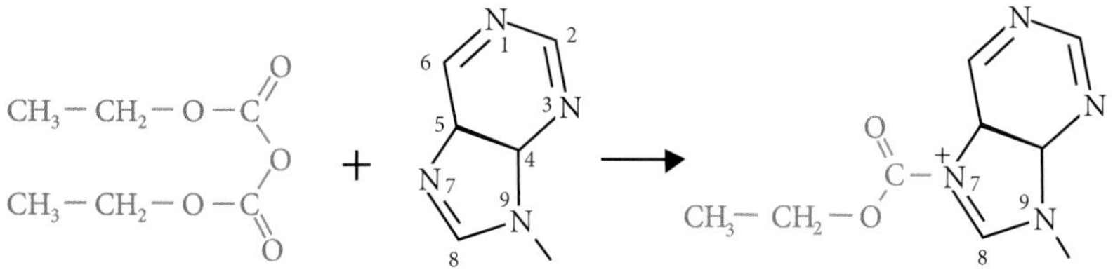

$$\text{Diethylpyrocarbonate} + \text{Purine} \longrightarrow \quad (2.12)$$

The reaction can proceed further at N9 and open the purine ring by expelling C8. **Dimethyl sulfate** methylates the N7 position of guanines (Figure 2-33-C) when in double- or single-stranded structures, but not if it is involved in Hoogsteen hydrogen bonding (Figure 2-21). **Formaldehyde** reacts reversibly with amino groups, just as it does in proteins (Table 7-2), so long as they are not involved in base pairing; but it also produces cross-links (Figure 2-33). **Hydroxylamine** (H_2N–OH) or **ethoxyamine** (H_2N–O–CH_3) converts an amino group to a hydroxyamino (–NHOH) or methoxyamino (–NH–O–CH_3) group (Figure 2-33-B). **Nitrous acid** converts amino groups to hydroxyls (Figure 2-33-A) but also produces cross-links (Figure 2-34).

Bisulfite (HSO_3^-) deaminates cytosine nucleotides to uracil if they are single-stranded (Figure 2-33-D). **Permanganate** or **osmium tetroxide** (OsO_4) will oxidize the C5–C6 double bond of thymine or cytidine residues:

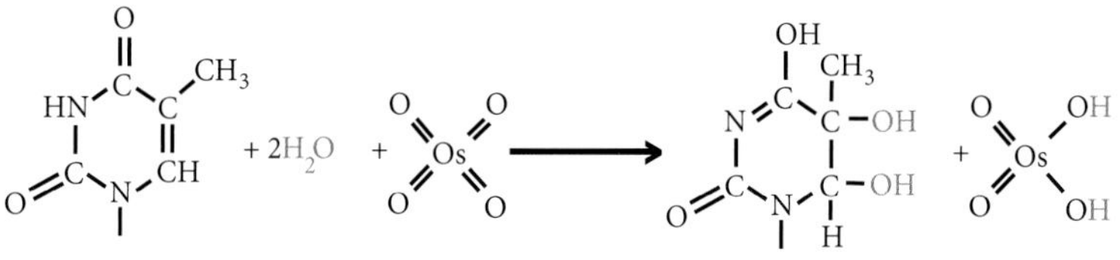

$$\text{Thymine} + 2H_2O + OsO_4 \longrightarrow \quad (2.13)$$

when they are present in single strands. Pyridine increases the rate of reaction with osmium tetroxide and produces an adduct of osmium and pyridine across the double bond. **Bromo-** or **chloro-acetaldehyde** forms fluorescent cyclic etheno derivatives with the base pairing positions of adenines, cytosines and, to a lesser extent, guanines, so long as they are not involved in base pairing:

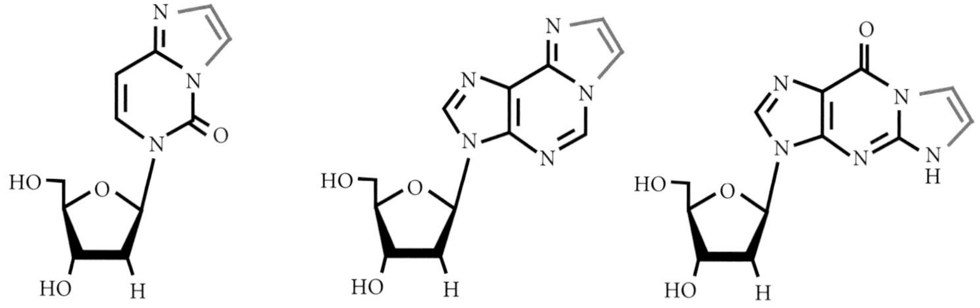

Cytosine   Adenosine   Guanosine   (2.14)

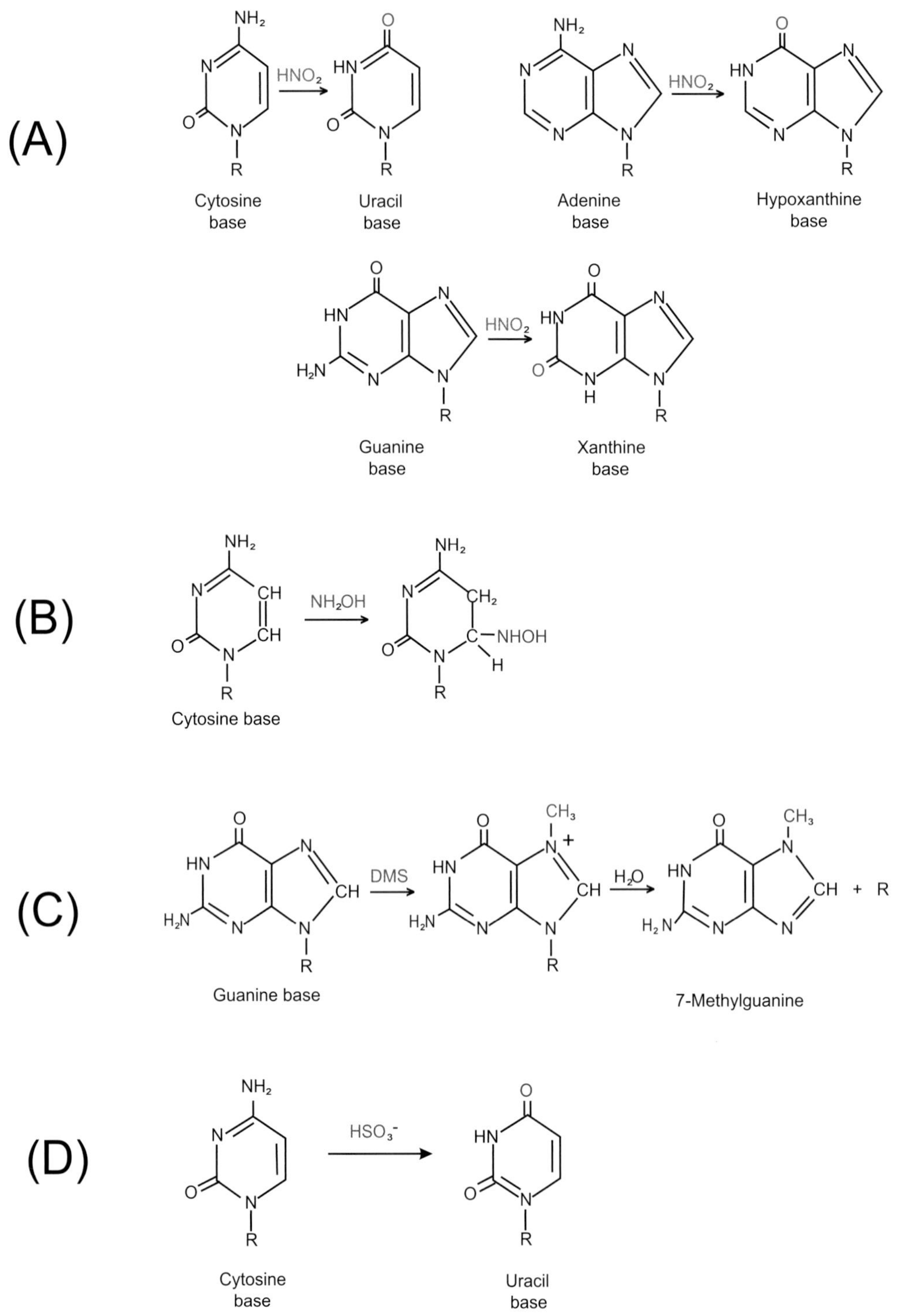

Figure 2-33. Common chemical modifications of DNA bases. The ribose group is depicted by R. (A) Deamination by nitrous acid coverts cytosine to uracil, adenine to hypoxanthine, and guanine to xanthine. (B) Reaction with hydroxylamine, as illustrated here with cytosine. (C) Alkylation of guanine by dimethyl sulfate (DMS); generating the quaternary N atom destabilizes the glycosidic bond and releases the modified base from the deoxyribose sugar. (D) Deamination of cytosine to uracil by bisulfite.

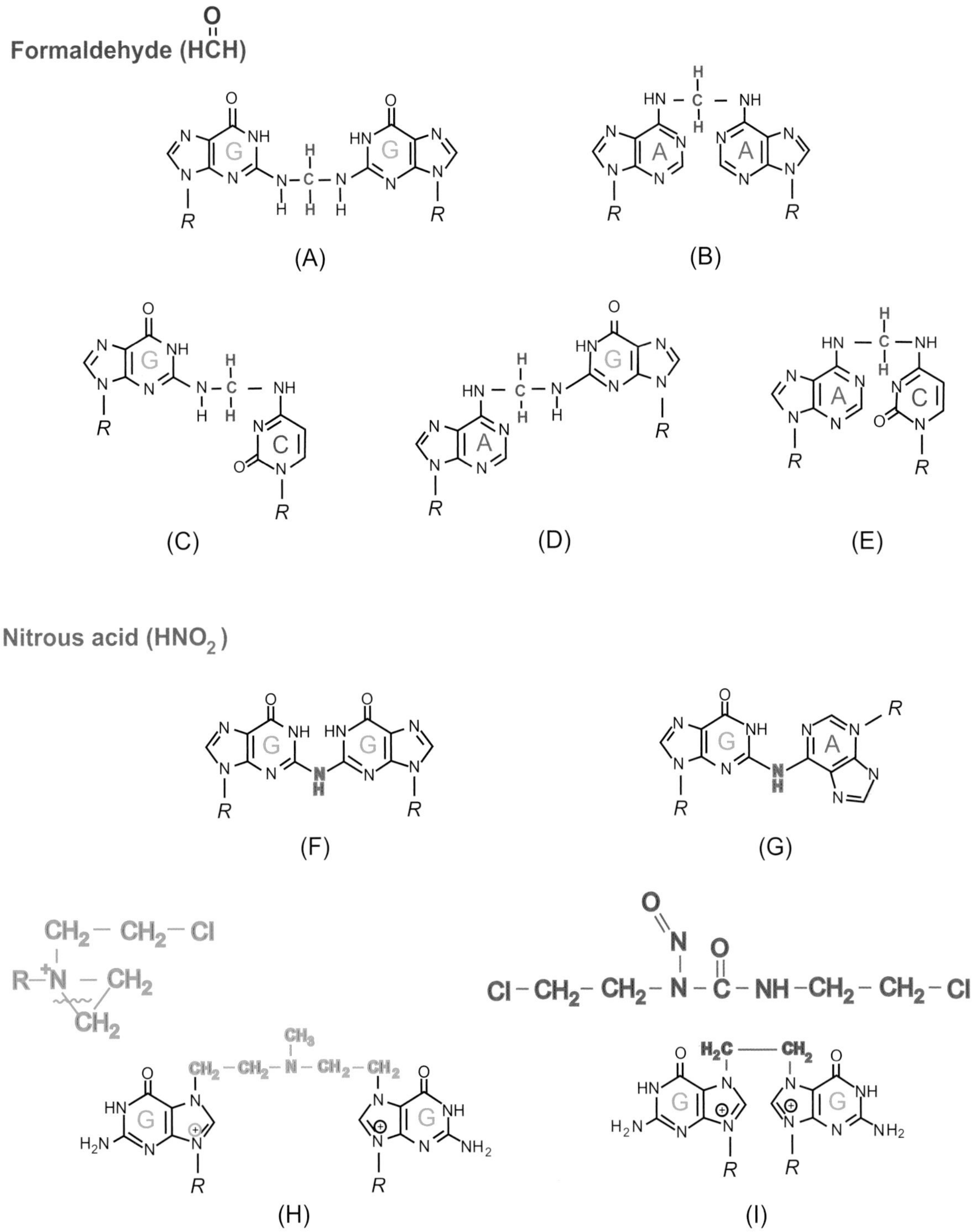

DNA structure   **CHAPTER 2**   83

Figure 2-34. Cross-links produced in double-stranded DNA by formaldehyde ($H_2C=O$), nitrous acid (HNO_2) and two N mustards. The ribose group is depicted by R. Adapted from B. Singer & J. T. Kusmierek (1982) *Ann. Rev. Biochem.* **51**, 655–693.

Nitrosoureas:

$$O=N-\underset{\underset{\displaystyle }{|}}{\overset{\overset{\displaystyle R}{|}}{N}}-\overset{\overset{\displaystyle O}{\|}}{C}-NH_2 \qquad (2.15)$$

will transfer the R group to any accessible O and N atoms of nucleic acids, on the bases, ribose ring and phosphate group.

Chemical modification by **acylation** of amino groups introduced at the 2′-position reports on the local nucleotide flexibility, because unconstrained 2′-amino nucleotides more readily reach a reactive conformation in which the amide-forming transition state is stabilized by interactions between the amine nucleophile and the adjacent 3′-phosphodiester group. The acylation reaction requires that the base lie outside of the helix and in the major groove. Bulged 2′-amine-substituted cytidine nucleotides react approximately 20-fold more rapidly than do nucleotides constrained by base pairing. In contrast, those base-paired but flanked by a 5′- or 3′-bulge react two- or six-fold more rapidly, respectively, than the perfectly paired duplex. The relative lack of 2′-amine reactivity for nucleotides adjacent to a DNA bulge suggests that the structural perturbations of the bulge do not extend significantly into the flanking duplex structure.

Some of these chemical modifications make their adjacent phosphodiester backbones labile to cleavage by the base **piperidine**, so the positions of the modifications in a DNA molecule of defined size can be determined by the lengths of the DNA fragments produced, similar to the approach used in chain-cleavage DNA sequencing (Section 6.4.C). Some chemical probes, as well as some naturally occurring drugs, can cleave DNA by an oxidative mechanism in which an H atom is abstracted from the deoxyribose ring, producing a radical that reacts further with oxidants such as O_2 or H_2O_2, causing chain cleavage. The site of the cleavage can be determined readily from the lengths of the fragments produced. **Hydroxyl radicals** cleave the sugar-phosphate backbone nonspecifically; they are generated spontaneously by ferrous ions in the presence of oxygen and a reducing agent. A complex of ferrous iron with the chelating agent EDTA (Figure 12-7) behaves similarly. Other reagents use bound transition metals to carry out similar reactions. Nonspecific reagents of this type are useful when the DNA molecule is not universally accessible to the reagent, due to being bound to a surface or a protein; the cleavage pattern reflects which nucleotides are accessible, which can indicate the helical periodicity of the double helix. Such nonspecific reagents can be made specific by attaching them covalently to polynucleotides that base-pair with complementary sequences or to agents that intercalate at specific sites; their local concentration can then be much greater than that of the bulk solvent.

Some enzymes are specific for the structure of a polynucleotide. For example, **S1 nuclease** hydrolyzes the backbones of both DNA and RNA when they are single-stranded. It will cleave the single-stranded regions in the loops of cruciforms, in the strand released after triple-strand formation, and in internal loops. It does not cleave where there is a mismatch between a single pair of bases.

Chemical modifications are especially useful in identifying the sites on DNA molecules where specific proteins bind, and thereby protect the DNA from modification (Section 13.1.C). Many of these reactions also produce changes in base-pairing during replication of the DNA and consequently cause genetic mutations, so the reagents responsible are often mutagens and must be handled with care.

Mechanics of DNA flexibility visualized by selective 2′-amine acylation at nucleotide bulges. D. M. John *et al.* (2004) *J. Mol. Biol.* **337**, 611–619.

Photochemical determination of different DNA structures. Y. Xu *et al.* (2007) *Nature Protoc.* **2**, 78–87.

2.6.A. Cross-linking

Chemical groups in DNA that are close in space can be cross-linked using reagents with two reactive groups (Figure 2-34). If two different DNA groups react with the same reagent molecule, they must be in the proximity required by that reagent. This can be a relatively simple probe of the overall structure of the DNA molecule. The reagents used most commonly are *N*-acetyl-*N*′-(*p*-glyoxylbenzoyl)cystamine, bis(2-chloroethyl)-methylamine (also known as 'nitrogen mustard'), *p*-azidophenyl acetimidate and iminothiolane.

~ CHAPTER 3 ~

DNA TOPOLOGY

The linear double-helical structure of DNA described in Chapter 2 is only the first level of DNA structure. DNA molecules can be extremely long, approximately 30 cm for a billion nucleotides, yet they must be accommodated within the nucleus of a cell, the dimensions of which (typically less than 0.1 mm in diameter) are only a tiny fraction of the length of the DNA molecules it must contain; the linear DNA molecules must be compacted extensively. Compaction introduces a second problem: how such a compacted DNA double helix is opened up to serve as a template for its transcription into messenger RNA (Section 6.2). A third problem is even more severe, concerning DNA replication (Section 6.1): how can such a compacted linear molecule be unraveled and separated into its two strands, so that they can be replicated to produce two identical molecules that can then be separated and packaged into two separate nuclei during and after cell division? If the 3 billion base pairs of DNA in the haploid human genome were in one piece (it is actually divided into 23 chromosomes), it would be a linear DNA molecule 1 m long and 20 Å in diameter. To illustrate the physical complexity of such long molecules, consider that this is comparable to a normal piece of string that is 1 mm in diameter being 500 km long! The well-known everyday problem of unraveling a compacted string (even only a few meters in length) illustrates adequately the topological problems that could be encountered within a cell in order to express and duplicate its DNA. Further complications occur with the many DNA molecules that are circular *in vivo* because the termini are joined in normal phosphodiester bonds. Cells have developed very sophisticated machinery to manage these problems of raveling and unraveling DNA. Here we will consider the basic problem of DNA packing: its topology.

Duplex DNA acquires topological properties if organized into a **topological domain**, which is a closed region in which the two strands are linked. The two types of elementary topological domains are (1) **closed circular duplex DNA** (ccDNA), in which the two strands are each covalently continuous, and (2) DNA loops that are closed by the binding of protein molecules (Figure 3-1). Protein-sealed DNA loops can occur locally in parts of DNA molecules that are not closed circular. The principles of circular DNA topology are directly applicable to DNA loops defined by bound protein molecules, and of even greater importance (Chapter 13), but protein molecules can dissociate and then reassociate (Section 13.2). Consequently, most experimental and theoretical work on DNA topological domains has used ccDNA, because it is defined by its covalent bonds. Circular DNA molecules exist in nature as the genomes of bacteria and as plasmids, or they can be prepared easily by using a **DNA ligase** enzyme to link covalently the two ends (Figure 3-2). This occurs much more readily if the two ends of the linear DNA molecule have a few unpaired nucleotides at each end that are complementary to

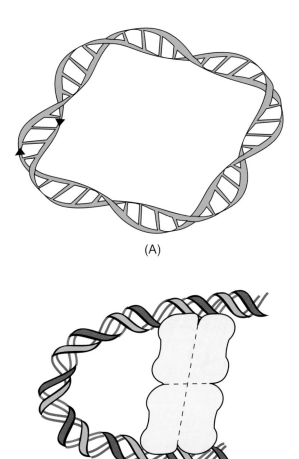

Figure 3-1. Two kinds of elementary topological domains of double-stranded DNA. (A) A closed duplex DNA, in this case a relaxed duplex with no supercoiling. The polynucleotide backbones of the two strands are depicted as the two continuous bands, linked by the ladder-like base pairs. Both strands in this DNA are covalently continuous and the linking number is an integer. There are eight nodes in the projection shown and $Lk = +4$ (Section 3.2). (B) A DNA loop formed by binding of a tetrameric protein. Only the two polynucleotide backbones of the DNA duplex are indicated. In this case, $Lk \approx 6$. The exact value of Lk depends upon the precise location at which the two strands are held together by the protein, and Lk consequently need not be an integer.

each other, known as **sticky ends** (Section 6.3.A). The topology of the resulting circular double helix is altered if one of the ends is rotated relative to the other by one or more full 360° turns before linking the ends.

The reference molecule for DNA topology is the **nicked circular duplex DNA**, in which at least one phosphodiester bond of ccDNA has been hydrolyzed. This destroys the topological domain, because free rotation can take place about the site of the chain scission; such species are designated as **open duplex DNAs**.

Circular DNA molecules like those illustrated in Figures 3-1 and 3-2 are elementary topological domains in which the circular DNA is topologically monomeric and the axis of the DNA is not itself linked to that of any other DNA. DNA can also be organized into higher order, nonelementary topological domains, in which the duplex DNA axes of one or more circular DNAs are themselves linked (Figure 3-3). **Knotted DNA** represents a class in which the duplex axis of a single DNA forms a knot, and it can be produced by site-specific genetic recombination. **Catenated DNA** has the axes of two or more ccDNA submolecules linked and contains one or more higher order topological domains. The submolecules might consist of individual elementary topological domains.

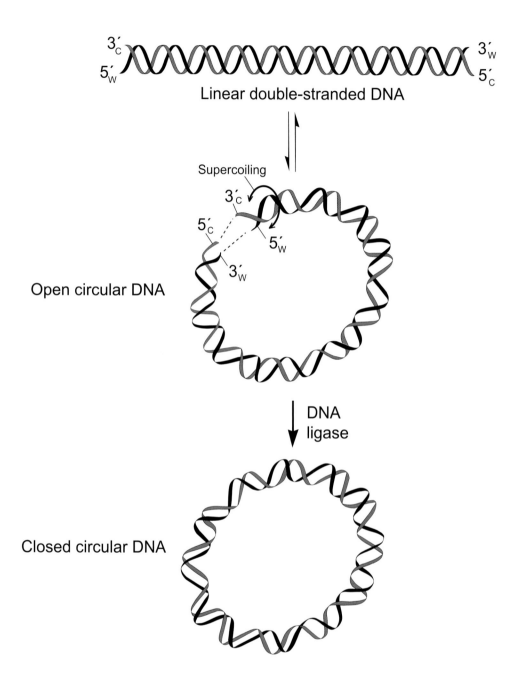

Figure 3-2. Converting a linear double-stranded DNA molecule to a circle. The individual strands in the linear molecule are labeled as W (Watson, *black*) or C (Crick, *red*). Circularization links the opposite ends of each of the W and C strands. It is assisted if the two ends have complementary single strands, known as 'sticky ends'. Supercoiling results if one end is rotated by one or more full turns before the ends are linked. Each twist of 360° changes the linking number by 1. The phosphodiester backbones are completed by the action of a DNA ligase enzyme.

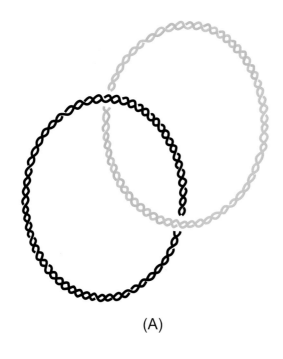

Figure 3-3. Two types of higher order topological domains. (A) Two closed circular DNAs are linked once to form a catenane. The higher order linking number (catenation number) has a value of 1 in this case, because the axes of the two DNA molecules are linked once. (B) The axis of a single DNA molecule is knotted into a trefoil.

Practical methods for supercoiled plasmid DNA production. J. Ballantyne (2006) *Methods Mol. Biol.* **127**, 311–337.

DNA supercoiling and bacterial gene expression. C. J. Dorman (2006) *Sci. Prog.* **89**, 151–166.

DNA topology: dynamic DNA looping. A. Travers (2006) *Curr. Biol.* **16**, R838–840.

Dynamic state of DNA topology is essential for genome condensation in bacteria. R. L. Ohniwa *et al.* (2006) *EMBO J.* **25**, 5591–5602.

Strong effects of molecular topology on diffusion of entangled DNA molecules. R. M. Robertson & D. E. Smith (2007) *Proc. Natl. Acad. Sci. USA* **104**, 4824–4827.

3.1. SUPERCOILING AND SUPERHELICES: TOPOISOMERS

Imagine in Figure 3-2 that one end of a linear DNA duplex is twisted by one or more full turns before the ends are joined. Each twist will introduce and build up elastic strain in the DNA circle that is generated, much as occurs in a rubber band. Consequently, this supercoiling increases the free energy of the DNA. Supercoiling can be either left-handed or right-handed. The number of times the two strands of a ccDNA are coiled about each other cannot be altered without first cleaving at least one of its polynucleotide strands; this can be demonstrated easily with a buckled belt, in which each edge of the belt represents a strand of DNA. The two DNA backbone strands whose interwinding defines a topological domain contain no chemical bonds in common; nevertheless, the two strands cannot be separated except by breaking a covalent bond. This type of connection is sometimes called a **topological bond**.

Circular DNAs often have the duplex coiled about itself, which generates a peculiar twisted appearance. This is a result of **supertwisting**, or **superhelicity**, of the DNA duplex (Figure 3-4); it is often referred to as DNA's 'tertiary structure'. Such superhelical structures arise to lessen the strain introduced in a ccDNA from the way the two individual DNA strands are twisted about each other. Two types of superhelical structure occur and can be reversibly interconverted without strand scission (Figure 3-5). First is the **plectonemic** or **interwound** superhelix, in which the DNA axis winds solenoidally up and down a virtual cylinder, forming a double helix, with smoothly curved connection at the ends of the cylinder. It occurs with purified, monomeric closed circular DNA and is further illustrated in Figure 3-4. Second is the **toroidal** superhelix, in which the DNA axis winds solenoidally around the surface of a torus or a segment of a torus, forming a single helix; it is primarily encountered in DNA associated with other DNA molecules (catenated DNA) or wrapped on a protein surface. These two superhelical forms are different conformations and are interconvertible, but they have opposite handedness. The most simple parameter to describe supercoiling is n, the number of superhelical turns that are apparent in its structure when the molecule lies flat in a plane.

Supercoiled DNA may be converted to a **relaxed circle** simply by cleaving one of its strands at a single site; the DNA backbone of the opposite strand is then free to swivel about its backbone bonds so as to relieve any strain by changing the supercoiling. **The relaxed state of a DNA circle has the lowest free energy and is not considered to be supercoiled**, even though the two strands are wound round each other.

Naturally occurring DNA circles generally have negative supercoiling, i.e. they are **underwound** by about 6%. This is believed to be a result of the way the DNA is packaged (Section 3.7) and it prepares the DNA for physiological processes that require separation of the DNA strands, such as their replication or transcription.

Analysis of chemical and enzymatic cleavage frequencies in supercoiled DNA. H. Tsen & S. D. Levene (2004) *J. Mol. Biol.* **336**, 1087–1102.

DNA axial rotation and the merge of oppositely supercoiled DNA domains in *Escherichia coli*: effects of DNA bends. V. A. Stupina & J. C. Wang (2004) *Proc. Natl. Acad. Sci. USA* **101**, 8608–8613.

Influence of supercoiling on the disruption of dsDNA. S. M. Chitanvis & P. M. Welch (2005) *J. Chem. Phys.* **123**, 124901.

Effect of supercoiling on formation of protein-mediated DNA loops. P. K. Purohit & P. C. Nelson (2006) *Phys. Rev. E* **74**, 061907.

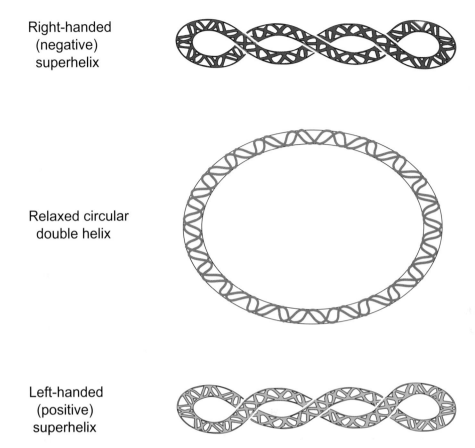

Figure 3-4. Relaxed and supercoiled DNA. The three structures differ in the number of times the two DNA strands are wound about each other (they have different linking numbers) and they cannot be interconverted without breaking one of the strands. The superhelix conformation with the double helices wound around each other is adopted to minimize the strain in the circular double helix. Negatively supercoiled DNA is the form normally present in cells, although the positively supercoiled form is generated transiently during replication and recombination.

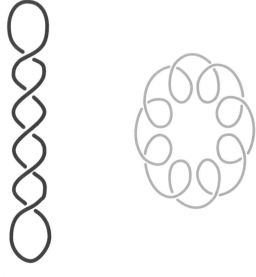

Figure 3-5. The two fundamental types of superhelix winding in superhelical DNA: a plectonemic (or interwound) superhelix (*left*) and a toroidal (or solenoidal) superhelix (*right*). The plectonemic form is the more stable in isolation but the toroidal form is generated by binding to some proteins.

3.2. LINKING NUMBER, *Lk*

Supercoiling is described quantitatively by the **linking number**, *Lk*, which is the number of times that one DNA strand winds about the other and is the most important parameter of DNA elementary topological domains. ***Lk* is a measure of the total number of complete revolutions that either strand makes about the other**; so long as the strands remain intact, *Lk* is a fixed quantity. Two closed duplex DNAs that differ only in *Lk*, but are otherwise identical, are **topoisomers**. They contain exactly the same nucleotide sequence and covalent connections; nevertheless topoisomers with different *Lk*s are distinct and have different properties that depend strongly upon *Lk*.

The linking number has several simple and very useful properties. (1) ***Lk* is an integer for superhelical DNA** because both strands must be closed curves. It need not be an integer for protein-sealed domains, where the geometry can be more flexible. (2) ***Lk* is constant so long as the topological domain remains intact**. *Lk* can be altered only if a topological domain is broken, by cleaving one or both strands of double-stranded DNA or, if appropriate, by the disruption of links to the protein sealing the domain. (3) **The linking number is independent of the DNA geometry** because it is a topological quantity. In particular, *Lk* does not vary with deformation of the trajectory of either strand or with changes in the characteristic duplex geometric quantities (pitch, roll, twist, tilt, propeller twist, etc.). (4) **The linking number is independent of the ordering of the two curves**; consequently, for two DNA strands W and C, $Lk(W,C) = Lk(C,W)$. (5) By convention, **the *Lk* of a closed circular DNA formed by a right-handed double helix is positive**. (6) Tightening the double helix increases *Lk* and is considered to be positive supercoiling.

A topological approach to nucleosome structure and dynamics: the linking number paradox and other issues. A. Prunell (1998) *Biophys. J.* **74**, 2531–2544.

Purification and use of DNA minicircles with different linking numbers. G. Camilloni *et al.* (1999) *Methods Mol. Biol.* **94**, 51–60.

Finite-element analysis of the displacement of closed DNA loops under torsional stress. J. H. White & W. R. Bauer (2004) *Philos. Transact. A* **362**, 1335–1353.

3.2.A. Linking Difference, Δ*Lk*

The **linking difference** is a measure of how the winding of an elementary topological domain of closed duplex DNA deviates from that of its nicked, relaxed circular counterpart. The chemically and biologically interesting properties of a topological domain are actually determined by Δ*Lk*, rather than the linking number *Lk* itself. The linear double helix in Figure 3-2 has the two strands wound around each other in a normal double helix, but supercoiling is introduced by rotating one end relative to the other by one or more turns. The number of turns, which can be negative or positive, determines Δ*Lk*.

The corresponding nicked circular or linear DNAs strictly speaking do not have a linking number but they can be assigned a **pseudolinking number**, Lk_o, which is the number of times the two DNA strands are wound around each other. In contrast to *Lk*, Lk_o need not be an integer, and generally is not, even for homogeneous duplex molecules, as the two strands are not continuous and can have incomplete turns at their ends. Lk_o for relaxed DNA is determined simply by the number of base pairs,

N, and the DNA helical repeat, h_0 base pairs per turn (bp/turn):

$$Lk_o = N/h_0 \qquad (3.1)$$

For linear B-DNA in dilute NaCl at 37°C, h_0 = 10.5 bp/turn on average, although the local value depends upon the base composition. It is also different for noncanonical DNA structures, such as Z-DNA (Section 2.2.B.3) and H-DNA (Section 2.2.C.3), and locally denatured DNAs.

The linking difference is defined as the difference between Lk and Lk_o:

$$\Delta Lk = Lk - Lk_o \qquad (3.2)$$

If $\Delta Lk \neq 0$, the DNA is under a stress that deforms the double helix axis into a superhelix. A relaxed closed duplex tends to be a flat circular molecule, but increasing the supercoiling to $|\Delta Lk| > 1.5$ causes it to adopt a figure 8 type of conformation (Figure 3-4).

1. Relaxed Duplex DNA

Relaxed duplex DNA is that topoisomer closest to nicked circular DNA, whose linking number is the nearest integer to Lk_o. Because Lk_o itself is generally not an integer, a relaxed DNA species usually has a value of ΔLk that is small and fractional, ε, which is simply the difference between the exact value of Lk_o from Equation 3.1 and the nearest integer, so $-0.5 \leq \varepsilon \leq 0.5$. For example, DNA of the plasmid pBR322 has N = 4363 and Lk_o = 415.52 if h_0 = 10.5 base pairs/turn, so the relaxed duplex DNA topoisomer is the one with Lk = 416 and ε = –0.48. In physical terms, ε represents the minimum rotation that is required to bring the 5′ and 3′ ends of a nicked circular DNA together to permit covalent joining by a DNA ligase (Figure 3-2). The value of ε can be determined experimentally from the Boltzmann distribution of topoisomers that forms upon thermal equilibration in the presence of a topoisomerase (Section 3.6.A). This distribution fits well to a Gaussian curve of topoisomer frequency versus ΔLk, and ε is the separation between the center of this distribution and the location of the nearest, most prominent, topoisomer (Figure 3-6).

All naturally occurring closed circular DNAs have $\Delta Lk < \varepsilon$, when the ccDNA is said to be **underwound**. The linking difference is consequently sometimes called the **linking deficiency**, although the term 'linking excess' would be more appropriate for cases in which $\Delta Lk > \varepsilon$ and the DNA is **overwound**. If $\Delta Lk = \varepsilon$, the DNA is said to be **relaxed**, even though it might still contain a small linking difference.

Changing either Lk or Lk_o can change both the sign and the magnitude of ΔLk. The value of Lk can be changed most readily by treatment with a DNA topoisomerase (Section 3.3). In contrast, Lk_o can be changed by altering the temperature, type or concentration of the salt, or by the addition of an intercalating drug (Section 2.5.B). A relaxed duplex DNA can therefore be made to supercoil simply by making a change in its environment, so it is necessary to specify the conditions under which Lk_o is measured. The standard conditions are usually 37°C, 0.2 M NaCl, in the absence of any reagents that affect the DNA twist, when the value of Lk_o is described as Lk_o^0.

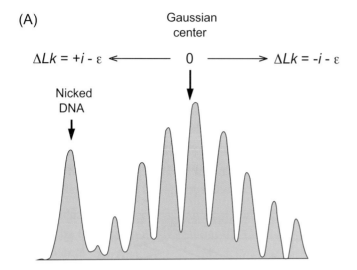
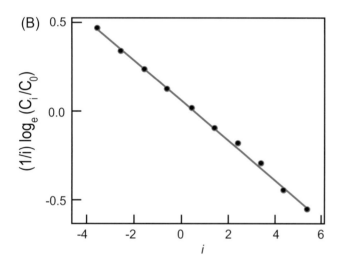

Figure 3-6. Hypothetical analysis of the distribution of topoisomers by gel electrophoresis. (A) Densitometer tracing of the gel, showing the intensity profile as a function of distance. The location of the nicked DNA is indicated, and the other peaks are the topoisomers with different linking numbers. The approximate location of the Gaussian center of the distribution is indicated. (B) Analytical plot of the topoisomer distribution according to Equation 3.15. The ΔLk-values of the topoisomers are $i + \varepsilon$, where $i = 0, \pm 1, \pm 2$, etc.

3.2.B. Superhelix Density, σ

The values of Lk, Lk_o and ΔLk are generally proportional to the DNA length. The supercoiling of two DNA molecules of different lengths can be compared by defining a normalized quantity, the **superhelix density** or **specific linking difference**, σ:

$$\sigma = \frac{Lk - Lk_o}{Lk_o} \tag{3.3}$$

Most naturally occurring circular duplex DNAs are underwound and have values of σ between 0 and –0.1 under standard conditions, but DNAs having σ as low as –0.17 have been prepared *in vitro*. Circular duplex DNAs with relatively large positive σ-values can be prepared using DNA reverse gyrase to increase the linking number (Section 3.3). Both σ and Lk_o are readily measured, so the easiest way to measure Lk is to combine Lk_o with the measured value of either ΔLk or σ. Using the example of pBR322 DNA again, the natural DNA has an average value of σ = –0.06. In this case, the average linking number for this native DNA is 390.58. The linking number of the nearest (most prevalent) topoisomer is 391.

As in the case of ΔLk, the value of σ can be altered by changes in environmental conditions alone, without any change in Lk, such as by changing the temperature. **The temperature coefficient of σ is $\Delta\sigma/\Delta T = 3.1 \times 10^{-4}$ deg^{-1}**. If the temperature is changed from 5° to 37°C, the change in σ is +0.01, which would change ΔLk by +4.16 turns for pBR322 DNA with 4363 base pairs. An increase in temperature thus causes a relaxed DNA to supercoil in the positive sense (a left-handed, interwound superhelix) and reduces the supercoiling of a naturally occurring (underwound) superhelical DNA. Changing the cation species and concentration has similar effects. The parameter pX$^+$ is defined as the negative logarithm of the cation concentration, analogous to the pH. Over the range of ionic strength from 0.05 to 0.3 M, $\Delta\sigma/\Delta pX^+ = 4.47 \times 10^{-3}$ **for the ions Na$^+$, K$^+$, Li$^+$ and NH$_4^+$ and 6.70×10^{-3} for Rb$^+$, Cs$^+$ and Mg^{2+}**. For example, if the potassium ion concentration were to be increased from 0.01 M to 1.0 M, the change in σ should be –0.013, which is comparable to a decrease in temperature of 42°C.

Separation of supercoiled DNA using capillary electrophoresis. D. T. Mao & R. M. Lautamo (2001) *Methods Mol. Biol.* **162**, 333–344.

Separation of large circular DNA by electrophoresis in agarose gels. K. D. Cole & C. M. Tellez (2002) *Biotechnol. Prog.* **18**, 82–87.

Early melting of supercoiled DNA topoisomers observed by TGGE. V. Viglasky *et al.* (2000) *Nucleic Acids Res.* **28**, E51.

3.3. TOPOISOMERASES

DNA topoisomerases are enzymes that catalyze the interconversion of DNA topoisomers. In the process, the DNA strands in a duplex DNA move through each other, to change the linking number. The changes can occur to duplex DNA segments in the same or different molecules, so topoisomerases can also interconvert catenanes and knotted DNA molecules (Figure 3-3). Type I DNA topoisomerases can change the linking number by ±1, by introducing a transient break in one strand, passing the other strand through the break, and resealing the cleaved strand. The topologies of catenanes and knots can be altered by these enzymes only if at least one of the strands contains a nick. Type II topoisomerases break and rejoin both strands at the same time, causing double-stranded DNA to pass through a double-strand break and thereby changing Lk by ± 2. The topology of a catenane or knot can be changed by these enzymes even if the chains are intact.

In the absence of an energy input, topoisomerases catalyze the net interconversion of topoisomers only if new topoisomers have lower free energies than the original; in the presence of such a topoisomerase, the topoisomers become relaxed and the spectrum generated reflects their intrinsic free energies (Figure 3-6). Other topoisomerases can use the energy from hydrolysis of ATP to introduce negative or positive supercoiling, even if this increases the free energy of the ccDNA. **DNA gyrase** is such a type II DNA polymerase and introduces negative supercoiling. In contrast, **reverse gyrase** introduces positive supercoiling, increasing the value of Lk.

The topoisomerases perform these topological tricks by binding the DNA molecule so that the hydroxyl group of a tyrosine residue attacks a DNA phosphate group; this cleaves the backbone of the strand and attaches the cleaved strand to the enzyme. One or both strands are passed through the break, then the cleavage of the polynucleotide chain is reversed, and the intact DNA molecule dissociates from the topoisomerase.

Topoisomerases are crucial components of the cellular machines that catalyze DNA replication and transcription. Without them, the DNA molecules would become hopelessly entangled.

Crystal structure of an intact type II DNA topoisomerase: insights into DNA transfer mechanisms. M. Graille et al. (2008) *Structure* 16, 360–370.

DNA topoisomerases: harnessing and constraining energy to govern chromosome topology. A. J. Schoeffler & J. M. Berger (2008) *Quart. Rev. Biophys.* **41**, 41–101.

Structural studies of type I topoisomerases. N. M. Baker et al. (2009) *Nucleic Acids. Res.* **37**, 693–701.

Topoisomerase II: a fitted mechanism for the chromatin landscape. N. Roca (2009) *Nucleic Acids. Res.* **37**, 721–730.

3.4. TWIST AND WRITHE

The individual conformations of a single topoisomer can be described by two further parameters, the molecule's **twist**, Tw (Section 3.4.A) and its **writhe**, Wr (Section 3.4.B). The twist is the number of complete revolutions that one polynucleotide strand makes about the duplex axis, while the writhe is the number of turns that the duplex axis makes about the superhelix axis. The sum of the two numbers must be the same as the linking number:

$$Lk = Tw + Wr \tag{3.4}$$

This is the most fundamental equation describing DNA topology. It indicates that an increase in twist must be accompanied by a decrease in its writhe (Figure 3-7). A belt wound around a cylinder has a large writhe number but a very small twist; extending the belt decreases the writhe but increases the twist. The linking number of a closed duplex DNA molecule must be an integer, but not the twist and writhe numbers.

The linking number is a topological quantity that remains constant, irrespective of the exact structure and orientation of the molecule, so long as the DNA backbone remains intact. In contrast, the parameters Tw and Wr are geometrical conformational quantities that may change as the DNA bends, twists or kinks.

3.4.A. Twist, Tw

The twist is the number of complete revolutions that one polynucleotide strand makes about the duplex axis in any particular conformation. Twist is a general property of two curves that coil about each other. The twist of one curve (W) about another (C), $Tw(W,C)$, is usually measured by how often W spins about C as one advances along C. In contrast to the linking number, the twist depends upon the ordering of the two curves; except for certain special cases, $Tw(W,C) \neq Tw(C,W)$. By convention, Tw is positive for right-handed duplex turns.

If the DNA axis is a straight line, or if the axis is closed and lies entirely in a plane, the twist is known as Tw_0. For B-DNA in solution, Tw_0 is normally the number of base pairs divided by 10.5, the number of base pairs per turn. In this case, B-form DNA containing 30 base pairs would have a twist of $Tw_0 =$

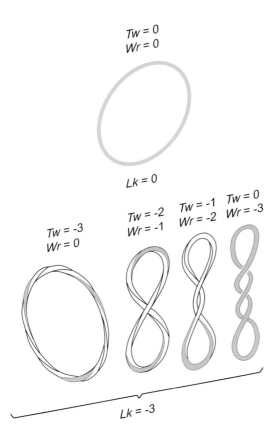

Figure 3-7. Simple illustration of the differences between the linking number (*Lk*), the twist (*Tw*) and the writhe (*Wr*) of a closed duplex DNA molecule. It is represented as a rubber ring of square cross-section, with one surface *colored* and the other three *white*. The twist is apparent by the number of times the visible surface changes color. Adapted from C. R. Calladine & H. R. Drew (1992) *Understanding DNA: the molecule and how it works*, Academic Press, London.

$N/h_0 = 30/10.5 = 2.857$. DNA rarely has a truly linear axis, however, and any out-of-plane motion of the helical axis changes the value of *Tw*. The extent of the deviation depends upon the amount of axial torsion. The twist of any structure can be calculated but it is not straightforward.

Chirality and surface twist of DNA wrapped on protein surfaces. J. H. White & W. R. Bauer (1995) *J. Biomol. Struct. Dynamics* **12**, 815–826.

Is a small number of charge neutralizations sufficient to bend nucleosome core DNA onto its superhelical ramp? G. S. Manning (2003) *J. Am. Chem. Soc.* **125**, 15087–15092.

DNA twisting flexibility and the formation of sharply looped protein–DNA complexes. T. E. Cloutier & J. Widom (2005) *Proc. Natl. Acad. Sci. USA* **102**, 3645–3650.

B-DNA under stress: over- and untwisting of DNA during molecular dynamics simulations. S. Kannan *et al.* (2006) *Biophys. J.* **91**, 2956–2965.

Role of tension and twist in single-molecule DNA condensation. K. Besteman *et al.* (2007) *Phys. Rev. Lett.* **98**, 058103.

Terminal twist-induced writhe of DNA with intrinsic curvature. K. Hu (2007) *Bull. Math. Biol.* **69**, 1019–1030.

3.4.B. Writhe, *Wr*

The writhe (*Wr*), or writhing number, is the number of turns that the duplex axis makes about the superhelix. Writhe is a property of a single closed space curve; in the case of DNA, the appropriate space curve is the duplex axis of the double helix. Writhe **can be calculated only for a closed curve, so *Wr* is not applicable to nicked circular or linear DNAs**. The magnitude of the writhe is directly proportional to *n*, the number of superhelical turns. **The writhe is zero when the DNA's duplex axis is constrained to lie in a plane**. In all other cases some of the winding is always converted into changes in *Tw* and the absolute value of *Wr* is always less than *n*.

The writhe is not a topological quantity so its calculation is *not* independent of the projection chosen; this clearly distinguishes *Wr* from *Lk*. *Wr* is the average of the projected writhing number over all possible projections. Deforming the axis changes the values of the directed writhing numbers, and hence of *Wr*. The easiest way to calculate the writhe is from the difference between the linking number and the twist, using Equation 3.4. This method is indirect but considerably easier than direct methods because both *Lk* and *Tw* are measured much more readily than *Wr*.

A topological invariant to predict the three-dimensional writhe of ideal configurations of knots and links. C. Cerg & A. Stasiak (2000) *Proc. Natl. Acad. Sci. USA* **97**, 3795–3798.

Computation of writhe in modeling of supercoiled DNA. K. Klenin & J. Langowski (2000) *Biopolymers* **54**, 307–317.

Writhe distribution of stretched polymers. S. Sinha (2004) *Phys. Rev. E* **70**, 011801.

Evaluating changes of writhe in computer simulations of supercoiled DNA. R. de Vries (2005) *J. Chem. Phys.* **122**, 064905.

3.5. DNA TOPOLOGY AND GEOMETRY

For any process in which *Lk* is unchanged, any changes in the twist must be matched exactly by changes in the writhe of opposite sign (Equation 3.4 and Figure 3-7):

$$\delta Tw = -\delta Wr \text{ (for constant } Lk\text{)} \tag{3.5}$$

A nicked circular DNA has no writhe, so $Tw_0 = Lk_0$. Introducing this into Equation 3.4 gives:

$$\Delta Lk = \Delta Tw + Wr \tag{3.6}$$

where $\Delta Tw = Tw - Tw_0$ and represents the deviation of the twist from the open circular value for B-DNA. This equation indicates that there are two differences between a relaxed duplex DNA and a nicked circular duplex DNA.

(1) A nicked circular duplex DNA has $Wr = \Delta Tw = 0$, but for the relaxed duplex DNA $Wr + \Delta Tw = \varepsilon$. Consequently, a relaxed duplex DNA can be distorted slightly by some combination of changes in *Tw* and *Wr*. For a perfect elastic rod, all the distortion would go into twist, but it all goes into writhe in the case of DNA.

(2) The twist and writhe remain coupled in a relaxed duplex DNA, as in Equation 3.5. In contrast, for nicked circular DNA ΔLk is not defined and Equation 3.6 does not apply, so in this case ΔTw

and *Wr* are uncoupled and may fluctuate independently. These differences explain why nicked circular DNA often migrates more slowly than relaxed circular DNA in both gel electrophoresis and sedimentation experiments under identical solution conditions.

The important structural parameters for superhelical DNAs have been estimated for closed circular DNAs of 3.5×10^3 and 7.0×10^3 base pairs. The superhelix radius r, measured in Å, depends upon the superhelix density σ (Section 3.2.B) according to the empirical equation:

$$\frac{1}{r} = 0.00153 - 0.268\sigma \tag{3.7}$$

The superhelix pitch is observed to be 54–56°, nearly constant and independent of σ. The length of the superhelix axis is independent of σ but proportional to the contour length of the DNA. The ratio of the superhelix axis length to DNA length is 0.41. The number of supercoils, *n*, is proportional to the absolute value of ΔLk: $n = 0.89 |\Delta Lk|$. The ratio between the twist increment and the writhe is constant, and $\Delta Tw/Wr \approx 3$. Individually, $\Delta Tw = 0.28\ \Delta Lk$ and $Wr = 0.72\ \Delta Lk$, indicating how changes in ΔLk are proportioned between the twist and the writhe.

Some experiments indicate that the superhelix radius is highly salt-dependent, with the opposing duplex segments approaching self-contact at moderate counterion concentrations, but others indicate that the structure remains much more relaxed at high salt, with no evidence of contacts between adjacent duplexes.

The plectonemic form is generally more stable than the toroidal form (Figure 3-5) for ccDNA in isolation. A relaxed circular form generally adopts a figure of 8 structure when ΔLk reaches an absolute value of 1.5–1.9. Thereafter, each increase of ΔLk by 1 usually introduces one more superhelical turn. Supercoiling can also induce large-scale conformational changes, such as local denaturation of the double helix, change from B-DNA to Z- or H-DNA (Section 2.2), or formation of a cruciform structure (Section 2.2.C.7). Increased negative superhelicity can induce adoption of the Z-DNA structure (see Figure 3-11), which made it possible to measure the free energy difference between B- and Z-DNA.

Mathematical modelling of interwound DNA supercoils. D. M. Stump *et al.* (2000) *J. Biomech.* **33**, 407–413.

3.5.A. Experimental Characterization of DNA Topology

Supercoiling affects the overall structure of a DNA molecule and is apparent in this way.

1. Electron Microscopy

The large sizes of DNA molecules make it possible to observe their topologies directly with an electron microscope. However, this requires that the molecules be spread out so that they can be absorbed on the support film in an extended nonaggregated form (Figure 3-8). This is usually accomplished by mixing the negatively charged DNA molecule with a basic protein, such as cytochrome *c*, or low-molecular weight substances, such as benzyl-dimethyl-alkyl-ammonium chloride and ethidium bromide (Section 2.5.B.1). This mixture is spread to form a molecular monolayer at the liquid–air interface, which is then adsorbed onto the support film, stained with uranyl acetate, rinsed in ethanol and air-dried. To improve contrast, the preparation can be rotary-shadowed, usually with platinum and palladium, at an angle of 6–9° from the plane of the metal vapor. Metal shadowing greatly exaggerates the thickness of the DNA molecules but the topology is evident (Figure 3-8).

The linking number of a closed duplex DNA can simply be counted from its structure when the molecule lies flat in a plane. In that case, the writhe is zero, so $Lk = Tw$, and Lk may be measured simply by counting the DNA's duplex turns. Nonplanar structures, however, require other procedures.

Polylysine-coated mica can be used to observe systematic changes in the supercoiled DNA conformation by scanning force microscopy in solution. M. Bussiek *et al.* (2003) *Nucleic Acids Res.* **31**, e137.

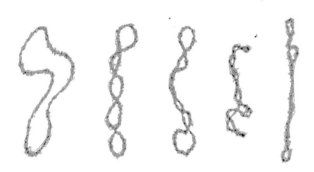

Figure 3-8. Typical results of electron microscopy after rotary shadowing of circular duplex DNA molecules differing in their conformations, from no supercoiling on the *left* to tightly supercoiled on the *right*. The background has been removed.

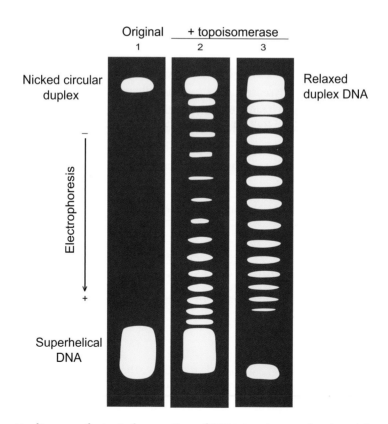

Figure 3-9. Schematic diagram of a typical separation of DNA topoisomers by electrophoresis in an agarose gel. The three lanes show the stepwise relaxation of closed circular DNA by incubation with a topoisomerase. Lane 1 shows the original mixture of superhelical (fastest band) and nicked circular duplex (slowest band) species. Lanes 2 and 3 show the formation of various topoisomers following different times of incubation with the topoisomerase. The final product is relaxed duplex DNA. Adjacent bands contain topoisomers that differ by unity in linking number.

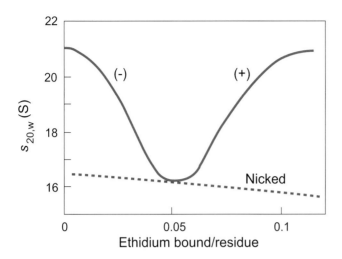

Figure 3-10. Effects on the sedimentation rate of the binding of ethidium bromide to an underwound closed circular DNA from SV40 virus and to the relaxed, nicked circular DNA. The ethidium intercalates between the bases, unwinds the helix, expands the hydrodynamic volume of the molecule and decreases its rate of sedimentation. The s-value of the initially underwound ccDNA approaches that of the relaxed molecule but then increases again as positive supercoils are formed. Data from W. Bauer & J. Vinograd (1968) *J. Mol. Biol.* **33**, 141–171.

2. Gel Electrophoresis to Separate Topoisomers

A supercoiled DNA molecule is more compact than a relaxed DNA molecule of the same length, so supercoiled DNA generally migrates faster when subjected to gel electrophoresis (Figure 3-9) or sedimentation (Figure 3-10). In a population of molecules with a distribution of linking numbers, the various topoisomers will migrate during gel electrophoresis as individual bands (Figure 3-9). Generating all possible values of Lk by the action of a topoisomerase permits the value of Lk to be assigned to each band by simple counting, so long as all the topoisomers can be resolved.

3. Intercalation by Ethidium Bromide

Intercalating agents such as ethidium bromide alter a circular DNA's degree of superhelicity because they cause the DNA double helix to unwind by about 26° at the site of binding (Section 2.5.B). The linking number is unchanged but the decrease in the negative twist of the supercoil is compensated by an increased writhe. Intercalation in a typical underwound ccDNA therefore causes an expansion of its hydrodynamic volume and a decrease in its sedimentation coefficient (Figure 3-10). As more ethidium binds, the sedimentation coefficient decreases to that of relaxed circular DNA, then it increases again as positive supercoils are introduced. If the number of ethidium molecules bound per base at the minimum sedimentation rate is v, the unwinding angle per ligand bound is φ (= 26°) and the average helix repeat in the relaxed form is h_0 (= 10.5 bp/turn), the value of the superhelix density, σ, will be given by:

$$\sigma = -h_0\left(\frac{\varphi}{360°}\right)(2v) = -1.52v \tag{3.8}$$

The molecule described in Figure 3-10 with $v = 0.05$ therefore has $\sigma = -0.076$.

4. Two-dimensional Gel Electrophoresis

Introducing a second dimension and intercalation by ethidium bromide expands the range of topoisomers that can be resolved by electrophoresis and provides information about conformational changes in the DNA molecule. The DNA is first subjected to gel electrophoresis as in Figure 3-9 in a

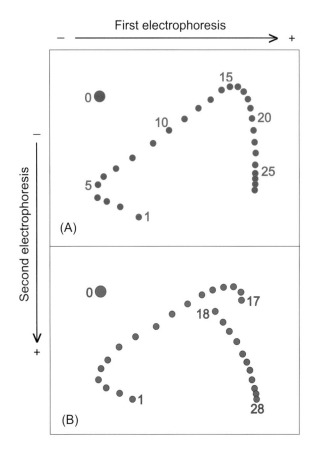

Figure 3-11. 2-D gel electrophoresis of a supercoiled plasmid. The spectrum of topoisomers was generated by incubating a supercoiled plasmid ccDNA with a topoisomerase. The plasmid used in (A) had the polynucleotide d(GC)$_{16}$ inserted in (B). The first electrophoresis was from *left* to *right*. After incubation with ethidium bromide, the second electrophoresis was from *top* to *bottom*. The nicked circular DNA migrates in spot 0. The other spots are numbered from 1 to 28 in order of decreasing *Lk*. Data from J. C. Wang.

slab gel. The separated DNA within the gel is then incubated with ethidium to permit intercalation and reduce the negative writhe (Figure 3-10). The intercalated DNA molecules in the gel are then subjected to electrophoresis at a right angle to that of the first electrophoresis (Figure 3-11). All the topoisomers should be resolved, and their relative mobilities in the first and second dimensions provide information about their topologies.

The two-dimensional (2-D) gel shown in Figure 3-11-A resolved 28 topoisomers that were generated by relaxing a plasmid DNA with a topoisomerase; they are numbered from 1 to 28 in order of decreasing *Lk*. Spot 0 is the nicked circle. The electrophoretic mobilities of isomers 1–4 decreased in both dimensions, indicating that they were positively supercoiled initially. In the original mixture, the relaxed state occurred between topoisomers 4 and 5. The remainder of the topoisomers had increasingly negative supercoiling. The second dimension shows the reduction of negative writhe that results from intercalation. After binding ethidium, topoisomers 1–14 had positive writhe, while 16–28 were still negative. The smooth curve on which these 28 spots lie indicates that no abrupt transitions occurred as a result of the different linking numbers or the change in writhe resulting from intercalation.

The same plasmid was used in Figure 3-11-B, except that it contained a d(GC)$_{16}$ insert, with 32 base pairs alternating between G•C and C•G. The most striking difference is the abrupt change in mobilities of topoisomers 18–28 in the first dimension. A substantial change in overall conformation occurred when *Lk* changed by −1 from that of topoisomer 17. Topoisomers 13–16 migrated in the first dimension at the same approximate rate as those indexed six higher, i.e. 19–22. Their similar migrations suggest that they have the same *Wr*. Therefore, they differ in *Tw* by six, because (Tw_{14}

$- Tw_{20}) = (Lk_{14} - Lk_{20}) - (Wr_{14} - Wr_{20}) = 6 - 0$. This is close to the difference expected if the 32-bp GC insert flipped from right-handed B-DNA, with 10.5 bp/turn, in topoisomers 1–17, to a left-handed Z-helix, with 12 bp/turn, in topoisomers 18–28: the difference in Tw would be expected to be $(32/10.5) + (32/12) = 5.7$.

3.6. ENERGETICS OF SUPERCOILING

A topological domain has associated with it a specific free energy associated with the topological strain, called either the **topological domain free energy** or the **free energy of superhelix formation**. This quantity generally increases with the extent of supercoiling of either chirality; it is generally proportional to the square of ΔLk. The equilibria of all binding reactions that involve changes in either the DNA twist or in the writhe are shifted by this free energy term. Examples are the binding of intercalative drugs, such as ethidium bromide (Section 2.5.B.1), and of proteins (Chapter 13).

The free energy of the topology must be included in any chemical or physical reaction in which the topology changes. The linking number for a DNA topological domain is constant and places an additional constraint upon all chemical reactions and other changes of state; it causes significantly altered thermodynamic quantities, thermal denaturation patterns, transitions to other duplex structures and binding of various reagents. Two opposing tendencies must generally be balanced: (1) minimizing the deviation of the twist, Tw, from that of the corresponding nicked duplex DNA, to reduce the superhelix radius, r, and (2) minimization of the bending distortion that accompanies superhelix formation; electrostatic repulsions between the phosphate groups tend to repel the opposing duplex strands and increase r. The net free energy change due to a change in linking difference, ΔLk, results from some combination of these two factors and is known as the **free energy of superhelix formation**, $\Delta G(\sigma,T)$. Any change in the composition or concentrations of salts affects both factors: the stiffness decreases with increasing ionic strength, but so do the electrostatic repulsions within the duplex. The diameter of the van der Waals surface of B-DNA is only 20 Å, but hydration and electrostatic repulsions cause the effective diameter to be 30 Å at high salt concentrations, increasing to 150 Å at 0.01 M salt. The responses of superhelical structure to changing ionic conditions or temperature are complex.

The free energy due to the topology has been measured, with very similar results, by three independent methods: (1) using the perturbation of the binding constant of intercalating drugs with changes in σ, which can measure ΔG over the entire accessible range of $\pm\sigma$; (2) using the width of the distribution of topoisomers when a closed circular DNA is equilibrated in the presence of a topoisomerase (Figure 3-6), which measures ΔG over a narrow range of $\pm\sigma$ in the vicinity of $\sigma = 0$; and (3) relating the extent of the linking difference to local denaturation of the DNA duplex, which can measure ΔG over a wide range of values of $\sigma < 0$.

The free energy of superhelix formation is described in terms of the superhelix density σ by:

$$\Delta G(\sigma, T) = \frac{NRT}{h_o^2} q(T) \sigma^2 \quad (3.9)$$

where N is the number of base pairs in the DNA, h_0 is the helical repeat of nicked circular or linear duplex DNA (10.5 bp/turn) and $q(T)$ is the free energy coefficient. The last quantity for chains longer than 2400 bp is 1160 bp at 37°C, independent of N and σ. For shorter chains, its value depends upon N:

$$q(37°, N) = 3939 \text{ bp} - 1.1 N \tag{3.10}$$

It decreases at higher temperatures:

$$q(T) = [(0.968 \times 10^6/T) - 1920] \text{ bp} \tag{3.11}$$

The entropy and enthalpy of superhelix formation are independent of temperature but quadratic functions of the superhelix density. For DNA molecules with about 5×10^3 bp, the enthalpy change per mole of base pairs ($\Delta H/N$) has been measured to be 17.3 σ^2 kcal/mol, and the entropy change per mole of base pairs ($\Delta S/N$) is 35 σ^2 cal/mole/deg. The free energy change per mole base pair ($\Delta G/N$) is 6.4 σ^2 kcal/mol at 37°.

The free energy of a supercoiled DNA molecule is usually at a minimum when about 70% of any change in *Lk* is expressed in *Wr* and 30% in *Tw*. Consequently, a decrease in *Lk* causes both right-handed negative supercoiling of the DNA axis and unwinding of the duplex. The energies of the twist deformability are virtually the same for all 10 possible base pair steps but their response is highly nonuniform. In particular, pyrimidine/purine steps are much more flexible than purine/purine steps, followed by purine/pyrimidine steps.

On the origin of the temperature dependence of the supercoiling free energy. J. J. Delrow *et al.* (1997) *Biophys. J.* **73**, 2688–2701.

Effect of polyethylene glycol on the supercoiling free energy of DNA. A. N. Naimushin *et al.* (2001) *Biopolymers* **58**, 204–217.

Link, twist, energy, and the stability of DNA minicircles. K. A. Hoffman *et al.* (2003) *Biopolymers* **70**, 145–157.

3.6.A. Energy Distribution of Topoisomers

The free energy of superhelix formation is best measured with the thermal distribution of topoisomers that results at equilibrium in the presence of a DNA topoisomerase. A circular duplex DNA can be incubated with a topoisomerase, or the appropriate nicked circular DNA can be incubated in the presence of DNA ligase (Figure 3-2), until equilibrium is established. After inactivation of the enzyme, the equilibrium distribution of topoisomers that was produced is separated by gel electrophoresis (Figure 3-6). The energy distribution is determined by the free energy difference between each topoisomer *i* and the hypothetical completely relaxed species at the center of the distribution, with $\Delta Lk = \varepsilon$. The distribution should be described by:

$$c_i = \left(\frac{1}{Q}\right) \exp\left[\frac{\Delta G(i+\varepsilon)}{RT}\right] \tag{3.12}$$

where c_i is the concentration of topoisomer *i*. The factor Q is the partition function, which in this case is simply a constant. Incorporating Equation 3.9 and using the definition of σ, the Boltzmann energy distribution becomes:

$$c_i = \left(\frac{1}{Q}\right)\exp\left[\frac{q}{N}(1+\varepsilon)^2\right] \tag{3.13}$$

It is convenient for analytical purposes to express this equation as a ratio of the concentration of topoisomer *i* relative to that of the most prominent topoisomer, with $i = 0$:

$$\frac{c_i}{c_0} = \exp\left[\left(\frac{q}{N}\right)i(i+2\varepsilon)\right] \tag{3.14}$$

The separation between the maximum of this distribution and the location of the nearest topoisomer should be equal to ε, the difference in winding between a nicked circular DNA and the nearest relaxed, closed DNA. Taking the logarithms of both sides of Equation 3.14 produces:

$$\left(\frac{1}{i}\right)\log_e\left(\frac{c_i}{c_0}\right) = \left(\frac{q}{N}\right)(i+2\varepsilon) \tag{3.15}$$

Plotting the relative concentrations of the various topoisomers on a linear scale should give the expected Gaussian distribution (Figure 3-6-B); the slope gives q/N and the intercept at $i = 0$ yields ε.

Performing such an analysis at different temperatures demonstrates that the distribution of topoisomers is shifted at different temperatures, towards less negative linking numbers at higher temperatures. For example, increasing the temperature from 14°C to 21°C increases the ΔLk of the predominant topoisomers by about 2. The data demonstrate that each base pair of the double helix unwinds by about 0.012° per °C increase in temperature.

Equilibrium distributions of topological states in circular DNA: interplay of supercoiling and knotting. A. A. Podtelezhnikov *et al.* (1999) *Proc. Natl. Acad. Sci. USA* **96**, 12974–12979.

3.6.B. Topology-Dependent Binding of Ligands

The free energy for the binding of ν ligands per base pair to a closed circular DNA, ΔG(ν,σ), is the sum of two contributions: the intrinsic binding free energy, ΔG°(ν), and the associated change in the free energy of superhelix formation, ΔΔG°(σ). The change in free energy upon binding is related to the binding constant K by the usual relationship $\Delta G° = -RT \log_e (K)$. K´ is the intrinsic binding constant, which can be measured by binding to a nicked circular or linear duplex DNA, so $\Delta G°´ = -RT \log_e (K´)$. The change in the superhelix free energy term ΔG°(σ) is:

$$\Delta\Delta G°(\sigma) = [d\Delta G°(\sigma)/d\sigma]\Delta\sigma \tag{3.16}$$

A ligand that alters the duplex rotation by *f* turns upon binding will produce $\Delta\sigma = h_o f/N$. Combining these relationships demonstrates that the binding constant depends upon the superhelix density and the temperature:

$$K(\sigma,T) = K'(T)\exp\left[-q(T)\,\sigma^2\,\frac{2f}{h_0}\right] \tag{3.17}$$

This analysis illustrates how the topology can have very significant effects on K. For example, a protein that binds to a DNA of σ = –0.07 and removes one duplex turn in the process will bind with K = 3K′. In other words, the affinity of such a closed circular DNA for the protein will be approximately three times that of a comparable nicked circular or linear DNA under the same environmental conditions. This effect can be even more pronounced with a supercoiled DNA that has a very high (negative) superhelix density. For example, if σ = –0.17 and one duplex turn is removed in the binding process, K = 600 K′. Consequently, if two otherwise identical DNA molecules of low and high (negative) superhelix density compete for binding to the same protein, the closed DNA of very negative σ will have a 600-fold advantage. Of course, if σ > 0 the effect is reversed and the closed DNA of low σ will bind the more tightly.

The classic example of a reaction that involves a change in the average duplex rotation is the intercalation of ethidium bromide, which unwinds the DNA duplex by 26°, or 0.072 turns of the duplex (Section 2.5.B.1 and Figure 3-10).

Closed circular DNA as a probe for protein-induced structural changes. J. H. White *et al.* (1992) *Trends Biochem. Sci.* **17**, 7–12.

3.7. DNA WRAPPED AROUND THE NUCLEOSOME

The DNA in the chromosomes of eukaryotic organisms is greatly compacted by being wrapped around units of a complex of eight histone proteins: a $H3_2$–$H4_2$ tetramer, flanked by two H2A–H2B dimers. About 166 bp of DNA are wound around the wedge-shaped histone octamer in two left-handed superhelical turns, to form one nucleosome (Figure 3-12). It is a wedge-shaped disk, about 110 Å in diameter and about 55 Å high, with a pseudodyad axis of symmetry close to where the

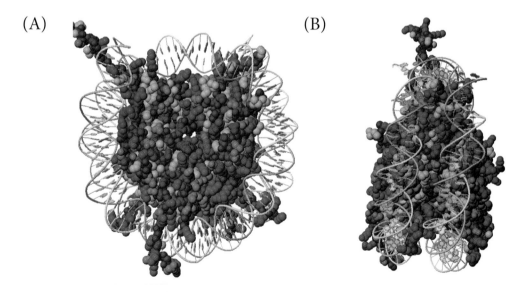

Figure 3-12. The structure of the nucleosome core particle determined by X-ray crystallography. To obtain a homogeneous structure, the histones were synthesized in bacteria to avoid the many post-translational modifications they normally experience, and a synthetic DNA molecule was used so that all the particles had the same DNA sequence. (A) A view down the superhelix axis; (B) a perpendicular view. The DNA is depicted as a skeletal model, whereas the histone proteins are depicted as van der Waals spheres for their atoms. Acidic groups are *red*, basic groups are *blue*, whereas polar but uncharged groups are *mauve*. Figure generated from PDB file *1aoi* using the program Jmol.

bound DNA segment enters and exits. Adjacent nucleosomes are linked by up to 74 bp of DNA, known as the **linker DNA**, plus one molecule of linker histone H1.

The overall helical periodicity of the bound DNA is 10.2 bp/turn but it varies in detail along the DNA. The DNA superhelix is not uniformly bent, due to interactions with the histones; the sharpest distortions are at positions 15 and 40–50 bp on either side of the dyad. Each of the histone dimers has three independent DNA-binding sites (Section 13.3.F). These binding interactions must overcome the free energy cost of bending DNA into the solenoid shape.

The ends of the DNA emerging from the nucleosome can be fixed, which defines a topological domain (Figure 3-13). It is then clear that the DNA wrapped twice around one nucleosome has $\Delta Tw = 0$ and $Wr = -2$. If the histone core is removed and the DNA pulled straight, the writhe is converted to twist, and $\Delta Tw = -2$ and $Wr = 0$.

The helical repeat of 10.2 bp/turn implies that the 146 bp of DNA wrapped on the nucleosome have $Tw = 14.14$. If the DNA duplex had been linearly extended in solution, $Tw_0 = 146/10.54 = 13.85$. Therefore, the surface wrapping and resultant curvature of the DNA axis about the nucleosome increases the twist by only about 2%.

Wrapping DNA in a nucleosome compacts it considerably, but even further compaction occurs when nucleosomes that are adjacent on a DNA molecule interact to form higher ordered structures.

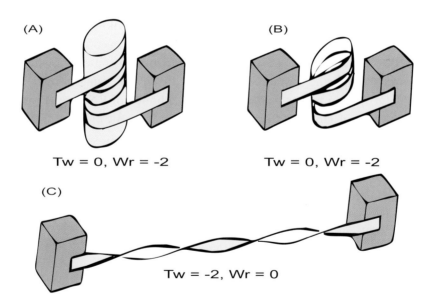

Figure 3-13. The topology of wrapping a molecule like DNA around a histone core to form a nucleosome. In this schematic illustration, the DNA double helix is represented as a simple ribbon, which ignores the double-helical nature of the DNA molecule. (A) The histone core is depicted as the central cylinder, wound with two turns of ribbon, while the two blocks on the *left* and *right* serve to fix the ends of the ribbon. The ribbon is not twisted but has $Wr = -2$. (B) The central cylinder is removed. (C) The ribbon is extended, demonstrating the twist of -2 that results as the writhe goes to zero. Adapted from C. R. Calladine & H. R. Drew (1992) *Understanding DNA: the molecule and how it works*, Academic Press, London.

~ CHAPTER 4 ~

RNA STRUCTURE

The functions of RNA molecules in living organisms range from the storage of genetic information in RNA viruses, defining the structures of macromolecular machines like ribosomes (Section 13.4.F), transferring and regulating the genetic information to the ribosomes, to the enzyme-like activities of ribozymes (Section 14.6). RNA molecules adopt diverse structures in response to these many functional requirements.

RNA has a covalent structure (Figure 4-1) very similar to that of DNA, the only differences being the change from a 2′-deoxyribose sugar in DNA to a ribose sugar in RNA, which involves only a change of a 2′-H atom to a 2′-OH group, and from a methyl group in thymine to an H atom in uracil. The backbone of RNA still consists of 3′ → 5′ phosphodiester bonds, even though an alternative 2′→ 5′ linkage is possible. The alternative linkage is used transiently in some of the functions of RNA, such as the removal and splicing of internal segments during processing of messenger RNA. **The presence of the 2′-OH group makes the phosphodiester bonds of RNA much more susceptible to hydrolysis than those of DNA**, especially at high pH (Figure 4-2). The 2′-hydroxyl group also increases dramatically the variety of three-dimensional (3-D) structures that RNA can adopt. **The bases in RNA molecules are often modified in nature**, much more so than in DNA, and the most common outcomes are illustrated in Figure 4-3. Although the polynucleotide backbone has considerable conformational flexibility, only a relatively small number of conformations are observed in biological RNAs. The 2′-hydroxyl group means that there are vicinal 2′- and 3′-hydroxyl groups at the 3′-ends of RNA molecules that can be oxidized by periodate (IO_4^-) to the 2′,3′-dialdehyde. The dialdehyde is not stable but reacts readily with other groups, especially amines. This reaction can be used to attach the 3′-ends of RNA molecules to a solid support.

The role of DNA in storage and transmittal of genetic information imposes the double-helical structure on most DNA molecules, **whereas RNA molecules are naturally double-stranded only in the RNA viruses**, where they have the same genetic role as DNA. Otherwise, RNA molecules are synthesized as single strands, by transcription of the DNA genome, and they remain single-stranded. Such RNA molecules are not replicated. In fact, **double-stranded DNA is a signal to the cell of a virus infection**, so cells have a special mechanism, known as **RNA interference**, of searching for double-stranded RNA molecules and then responding appropriately by destroying them. **Many RNAs function as unstructured, single-stranded molecules**. For example, the single polynucleotide strand of messenger RNA must be unfolded for the genetic message to be translated; stable structures in the coding region generally inhibit protein biosynthesis, although some structures are important in the regulation of translation (Figure 4-4-C).

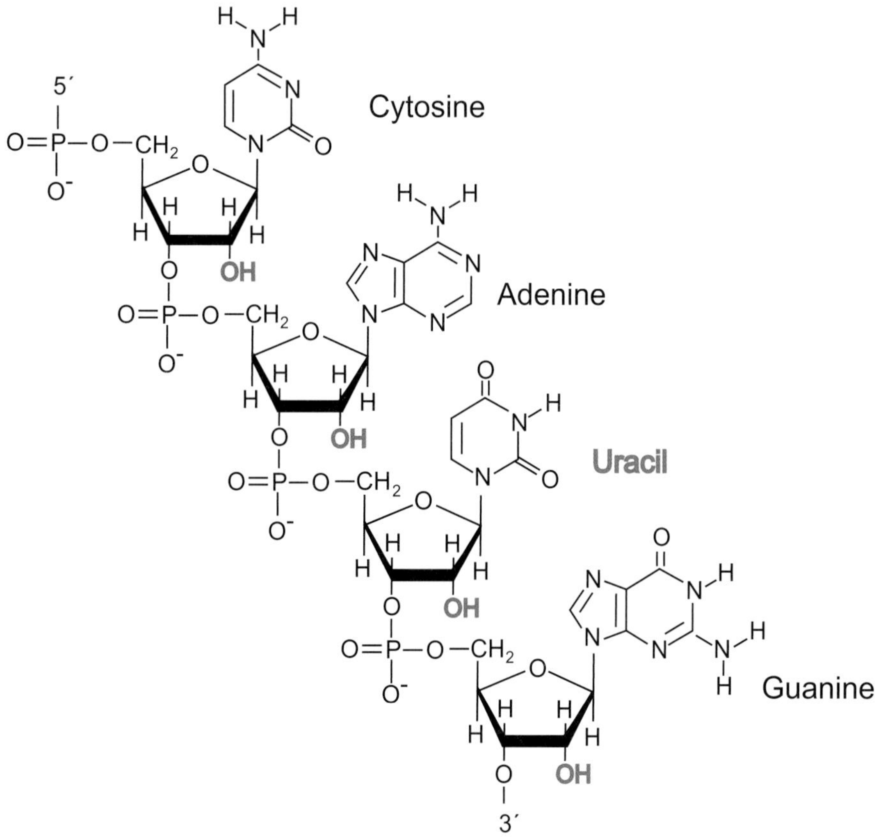

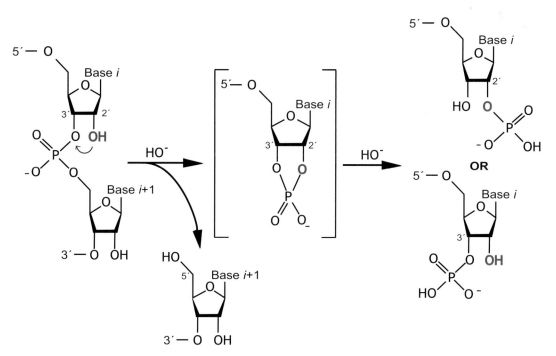

Figure 4-1. The chemical structure of the four primary nucleotides of RNA, showing each of the four bases. The differences from DNA are indicated in *red*.

Figure 4-2. Cleavage of the RNA internucleotide linkage under basic conditions. The 2′-hydroxyl group attacks the phosphate group attached to the adjacent 3′-O atom, to produce the 2′,3′-phospho intermediate (in brackets). After alkaline hydrolysis of the intermediate, the phosphate group can end up on either the 2′- or 3′- O atom.

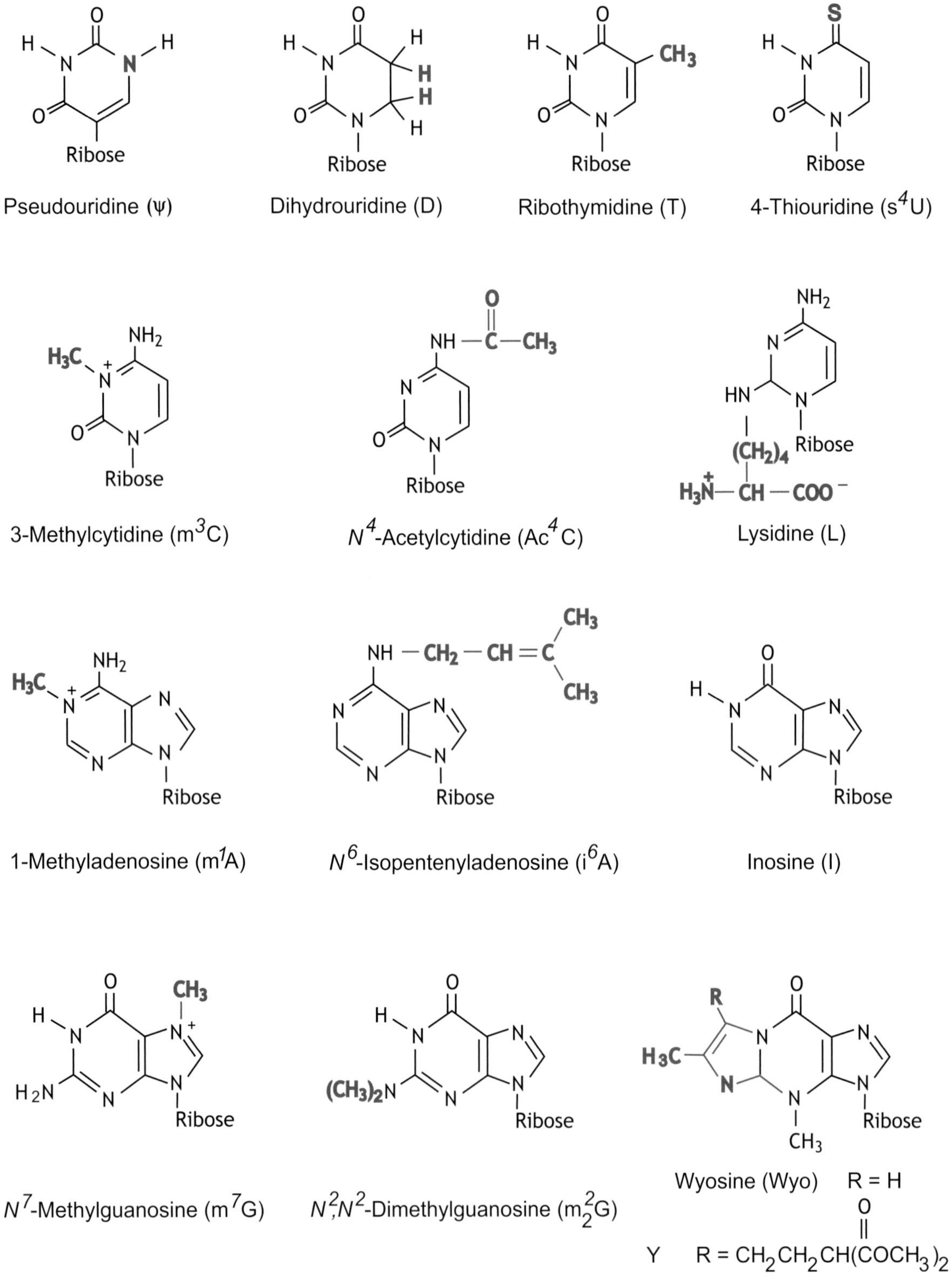

Figure 4-3. The most common modified forms of the nucleotides in tRNA. They may also be methylated at their ribose 2′-positions, and are then indicated by an 'm' after the nucleotide abbreviation, such as Cm, Gm, etc.

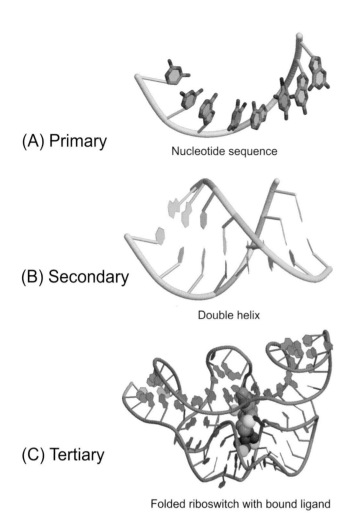

Figure 4-4. Hierarchy of RNA folding. (A) A single strand of RNA and its nucleotide sequence: its primary structure. (B) The A-form double helix of RNA: its secondary structure. (C) The 3-D structure of the thiamine pyrophosphate-specific riboswitch, with bound thiamine pyrophosphate: the tertiary structure. Figure generated from PDB files *1l5d* and *2gdi* using the program Jmol.

Other single-stranded RNA molecules adopt an array of structures to rival proteins in their complexity and functions; some can bind small molecules specifically (Figure 4-4-C) and some even catalyze biochemical reactions (Section 14.6). Many of the coenzymes that are involved in catalysis by proteins are related to the substituents of RNA (Section 14.4.G). **It is thought that RNA was the basis of the first type of living organisms, performing the dual functions of holding the genetic information and catalyzing reactions**. RNA may have performed the present-day functions of both DNA and proteins, which arose later and largely replaced RNA because they are more specialized and better suited to their more specific functions.

It is convenient to describe folded RNA structures in hierarchical terms (Figure 4-4), comparable to those used in describing protein structure (Figure 9-1): primary, secondary, tertiary and quaternary structures. The **primary structure** refers to the sequence of an RNA molecule; **secondary structure** is the regular local structure formed, usually an antiparallel double helix; **tertiary structure** involves interactions between elements of the secondary structure and other groups distant in the sequence; **quaternary structure** is the association of multiple RNA molecules. In many cases, the folding of the RNA molecule involves interactions with one or more protein molecules, which can produce very complex structures, as in ribosomes (Section 13.4.F). In these cases, the common motifs to be described here are incorporated into those structures.

Details of virtually all the known structures of RNA can be obtained from the **Protein Data Bank** (**www.rcsb.org**) and from **http://ndbserver.rutgers.edu**.

The RNA World. R. F. Gesteland & J. F. Atkins, eds (1993) Cold Spring Harbor Press, NY.

NMR spectroscopy of RNA. B. Furtig *et al.* (2003) *Chembiochem.* **4**, 936–962.

RNA structural motifs: building blocks of a modular biomolecule. D. K. Hendrix *et al.* (2005) *Quart. Revs. Biophys.* **38**, 221–243.

NMR methods for studying the structure and dynamics of RNA. M. P. Latham *et al.* (2005) *Chembiochem.* **6**, 1492–1505.

Dynamics-based amplification of RNA function and its characterization by using NMR spectroscopy. H. M. Al-Hashimi (2005) *Chembiochem.* **6**, 1506–1519.

4.1. SECONDARY STRUCTURE OF RNA

The secondary structure of RNA is dominated by the formation of antiparallel double helices stabilized by Watson–Crick base pairs between complementary segments with complementary sequences, comparable to those in DNA (Figure 2-13). The main difference is that uracil (U) bases replace thymine (T), so the canonical base pairs in RNA are C•G, G•C, U•A and A•U. As in DNA, the more stable G•C and C•G base pairs predominate (Table 4-1). Noncanonical base pairs occur much more frequently than in DNA; the most frequent (Table 4-1) is the U•G base pair:

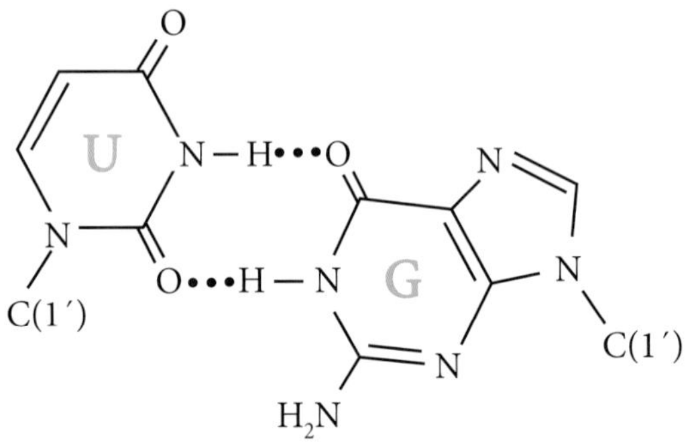

$$\text{U} \cdots \text{G base pair structure} \tag{4.1}$$

It introduces slight distortions into a double-helical structure that are recognized by proteins and other RNAs. RNA structures tend to be more complex than just antiparallel double helices.

RNA double helices, like those of DNA, have an antiparallel right-handed helical conformation. **RNA double helices usually adopt the A-form structure**, however, which differs significantly from canonical B-form DNA double helices in the conformation of the sugar and displacement of the bases from the helical axis (Figure 2-18). In an A-form RNA helix, base-stacking results in nearly perfect parallel orientations of all the bases in the helix. The lengths of hydrogen bonds in RNA are shorter than in DNA and sequence-dependent.

The 2′-hydroxyl group is the major determinant of the conformational and thermodynamic differences between RNA and DNA. Its presence on the ribose sugar in RNA hinders formation of a B-type helix. A-type helices have a deep, narrow major groove that is relatively inaccessible and does

Table 4-1. Frequencies of base pairs in RNA sequences with conserved secondary structures

Base pair	Frequency (%)
G • C	34.3
C • G	28.2
A • U	11.8
U • A	16.9
G • U	3.9
U • G	3.5
Others	1.4

Frequencies in 1081 base pairs in ribosomal and tRNAs and RNase-P, compiled in Table 1 of P. G. Higgs (2000) *Quart. Rev. Biophys.* **33**, 199–253.

not facilitate specific interactions with other molecules. The minor groove exposes the ribose 2′-OH groups, which are good hydrogen bond acceptors and donors. It is also shallow and broad, making it accessible to ligands and other groups, although the bases are not readily distinguished. A network of water molecules spans the minor groove, forming hydrogen bonds to the 2′-OH group. The RNA duplex is less tightly hydrated than the corresponding DNA helix, consistent with the A-type double helix of DNA being favored only at low water concentrations.

RNA double helices are more stiff than those of DNA, with a persistence length (Section 2.4) some 1.5- to two-fold longer than that of DNA. The flexibility patterns of B-DNA and A-RNA are quite different: a few soft essential movements explain most of the natural flexibility of A-RNA, whereas many are necessary for B-DNA. DNA is generally more flexible, but A-RNA is easier to deform in certain ways. The structures of RNA duplexes are not uniform, although they vary less than those of DNA. These variations depend on the sequence and structural context of the helix relative to the global 3-D structure. These local differences lead to very diverse shapes, with profound consequences for recognition by proteins and other ligands.

Unlike DNA, and except for the double-stranded RNA genomes of viruses, **the double helices in natural RNAs are primarily intrastrand**, generated by the single strand doubling back upon itself. The helices are also relatively short, seldom more than 10 base pairs in length, and are interrupted by single-stranded nucleotides forming loop elements: hairpins, bulges and internal loops (Figure 4-5). These, together with the helical junction that is formed when more than two double helices come together, are the secondary structure motifs and the building blocks upon which most complex RNA structures are built.

In many natural folded RNAs, more than half of all nucleotides are incorporated into double-stranded helices (Figure 4-6). **Duplex regions can form through very long-range interactions, between segments distant in the primary structure**, and these interactions are crucial to determine and

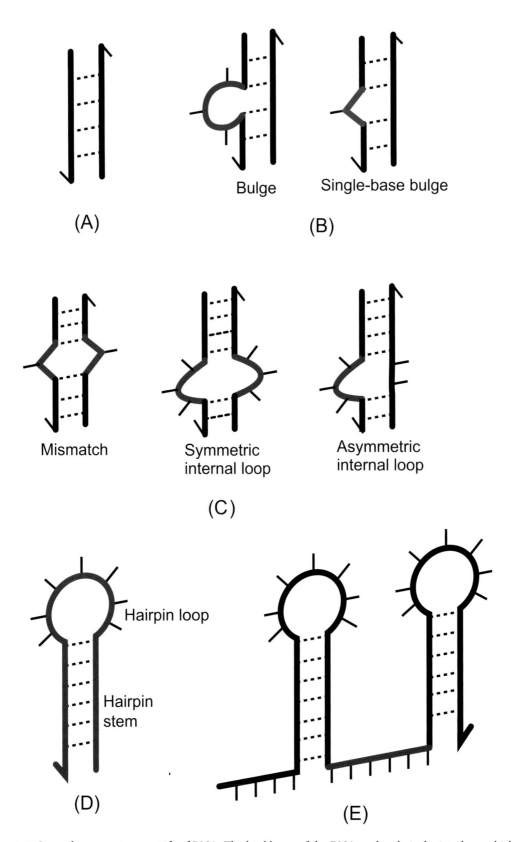

Figure 4-5. Secondary structure motifs of RNA. The backbone of the RNA molecule is depicted as a *thick line*; *dashed lines* indicate complementary antiparallel base pairs, while *thin lines* are unpaired bases. The *half-arrow* indicates the 3′-end of the oligonucleotide. (A) Intramolecular antiparallel double helix; (B) bulges; (C) internal loops within double-stranded segments; (D) hairpin; (E) single-stranded regions between double-stranded hairpins.

stabilize the overall fold of an RNA molecule. For example, opposite strands in double helices within ribosomal RNA are separated by as many as 2000 nucleotides.

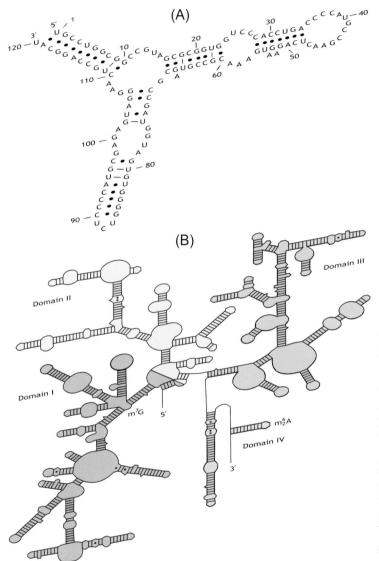

Figure 4-6. Secondary structures of two RNA molecules. (A) The 120-nucleotide 5S RNA of *Escherichia coli*. Watson–Crick base pairs are indicated by *dots*, the common G•U pairs by *dashes*. Adapted from R. R. Gutell *et al.* (1985) *J. Mol. Biol.* **32**, 183. (B) The 1542-nucleotide 16S RNA from *E. coli* ribosomes. The secondary structure was divided up into four domains, indicated by different colors. Two bases that have been modified are indicated. The 3-D structure is shown in Figure 13-37. Adapted from H. F. Noller (1984) *Ann. Rev. Biochem.* **53**, 134.

The relative flexibility of B-DNA and A-RNA duplexes: database analysis. A. Perez *et al.* (2004) *Nucleic Acids Res.* **32**, 6144–6151.

Influence of the 2′-hydroxyl group conformation on the stability of A-form helices in RNA. J. Fohrer *et al.* (2006) *J. Mol. Biol.* **356**, 280–287.

RNA secondary structure design. B. Burghardt & A. K. Hartmann (2007) *Phys. Rev. E* **75**, 021920.

4.1.A. Hairpin Loops

The most common element of RNA secondary structure is the **hairpin** or **stem–loop** (Figure 4-5). For example, bacterial 16S RNA of ribosomes contains approximately 30 hairpins that are phylogenetically conserved (Figure 4-6-B). A hairpin forms when the phosphodiester backbone folds back on itself to form a double-helical tract, called the **stem**, leaving unpaired nucleotides at the turn to form a single-stranded region, called the **loop**. The primary driving force is the base pairing in the stem, which must overcome the opposing requirements to bend the polynucleotide backbone and fix its conformation. Such hairpins can be stable in isolation (Table 4-2).

Table 4-2. The stabilities of isolated RNA hairpins. The nucleotides in the turn are in bold

Sequence	T_m (°C)	$\Delta G°$ (37°C) (kcal/mol)	$\Delta H°$ (kcal/mol)	$\Delta S°$ (cal/mol/°C)
rAC**UUCG**GU[a]	36.8	0.0	−28.9	−93.1
rGGC**ACAA**GCC[b]	62.7	−2.2	−29.2	−86.9
rGGC**GCAA**GCC[b]	71.0	−3.0	−30.3	−88.0
rGGAG**UUCG**CUCC[c]	60.1	−3.4	−48.6	−145.7
rGGAC**GAAA**GUCC[c]	65.9	−4.2	−49.1	−145.0
rGGAC**UUCG**GUCC[c]	71.7	−5.7	−56.5	−163.9
rGGAC**UUCG**GUCC[d]	76.2	−6.3	−55.9	−159.9

[a] In 10 mM sodium phosphate buffer, 0.01 mM EDTA. M. Molinaro & I. Tinoco (1995) *Nucleic Acids Res.* **23**, 3056–3063.

[b] In 100 mM NaCl, 10 mM sodium phosphate buffer, 0.5 mM EDTA. J. SantaLucia *et al.* (1992) *Science* **256**, 217–219.

[c] In 10 mM sodium phosphate buffer, 0.1 mM EDTA. V. P. Antao & I. Tinoco (1992) *Nucleic Acids Res.* **19**, 5901–5905.

[d] In 1 M NaCl, 10 mM sodium phosphate buffer, 0.1 mM EDTA. V. P. Antao & I. Tinoco (1992) *Nucleic Acids Res.* **20**, 819–824.

Most single-stranded loops in natural RNAs vary in length from two to 14 nucleotides. The ribose unit is almost always in the extended 2′-*endo* conformation, rather than the usual 3′-*endo* geometry of A-form RNA helices, presumably to increase the distance that can be spanned by each individual nucleotide. A precise geometry is defined for many loops by hydrogen bonding between nucleotides not involving Watson–Crick base pairing and between bases and backbone functional groups, as well as by stacking interactions between the loop nucleotides. Small hairpin loops contain a high degree of structure, whereas longer loops containing more than seven to eight unpaired nucleotides, such as those present in transfer RNA (Section 4.2.B), preserve extensive base stacking interactions but are generally more poorly structured and thermodynamically less stable. Most loops in ribosomal RNA are four to nine nucleotides in length, perhaps reflecting the need for thermodynamic stability.

Hairpin loops are often recognized by other molecules, especially proteins (Section 13.4). Their binding is often unusual in having a strong entropic contribution to the total binding energy, possibly due to the expulsion of bound water molecules.

Freely diffusing single hairpin ribozymes provide insights into the role of secondary structure and partially folded states in RNA folding. G. Pljevaljcic *et al.* (2004) *Biophys. J.* **87**, 457–467.

Hairpin RNA: a secondary structure of primary importance. P. Svoboda & A. Di Cara (2006) *Cell. Mol. Life Sci.* **63**, 901–908.

Biochemical and thermodynamic characterization of compounds that bind to RNA hairpin loops: toward an understanding of selectivity. J. R. Thomas *et al.* (2006) *Biochemistry* **45**, 10928–10938.

4.1.B. Tetraloops

Tetraloops are hairpin loops consisting of four nucleotides; they are exceptionally common in cellular RNAs. There are two very common families: UNCG and GNRA, where N is any of the four nucleotides and R is a purine. For example, 70% of all tetraloops in 16S ribosomal RNAs from all organisms belong to either of these two families. Other classes include the CUUG, ANYA and (U/A)GGN tetraloops. Tetraloop structures have high thermodynamic stabilities; UNCG is the most stable; capping a double-helical tract with a UNCG tetraloop is equivalent thermodynamically to extending the double-helical stem by two base pairs.

In spite of the differences in their sequences, the tetraloop families have extensive structural similarities (Figure 4-7). In each case, the first and fourth unpaired nucleotides close the loop by forming nonWatson–Crick base pairs, which reduces the number of unpaired nucleotides to only two. Unusual conformations of the backbone are stabilized by hydrogen bonds between its phosphate groups and a base.

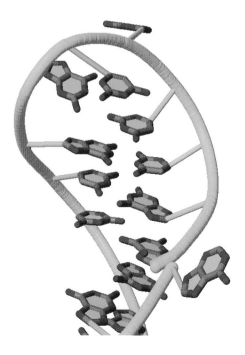

Figure 4-7. The solution structure of a UUCG tetraloop. The nucleobases are depicted in detail, but the ribose–phosphate backbone is depicted only as a cylinder. Figure generated from PDB file *1f7y* using the program Jmol.

Residue specific ribose and nucleobase dynamics of the cUUCGg RNA tetraloop motif by NMR ^{13}C relaxation. E. Duchardt & H. Schwalbe (2005) *J. Biomol. NMR* **32**, 295–308.

Solution structure of a GAAG tetraloop in helix 6 of SRP RNA from *Pyrococcus furiosus*. K. Okada *et al.* (2006) *Nucleosides Nucleotides Nucleic Acids* **25**, 383–395.

4.1.C. Bulges and Internal Loops

Extra unmatched nucleotides (**single-base bulges**) are common structural motifs in the helices of folded RNA molecules, and they often participate in interactions with other molecules and in the formation of tertiary structure. Bulges and internal loops occur when two double-helical tracts are separated by one or more unpaired nucleotides on either one or both strands (Figure 4-5); the former generates a **bulge**, whereas the latter generates an **internal loop**. Equal numbers of bases on each strand produce a **symmetric** internal loop, whereas different numbers of bases generate an **asymmetric loop**. A single base mismatch generates a symmetric internal loop of two nucleotides. Symmetrical internal loops of three nucleotides have the potential of a canonical base pair (middle pair) flanked by noncanonical base pairs on both sides (loop-terminal pair). An **extruded helical single strand** is two or three unpaired bases that are extruded from the main double-helical stack, forming an independent stack of bases. In a **cross-strand stack**, a base on one strand stacks with a base on the opposing strand, rather than stacking with the adjacent bases on its own strand.

The presence of an internal loop or bulge usually reduces the thermodynamic stability of the double helix, but **many internal loops contain G•A base pairs that are noncanonical but contribute to the stability**. Other molecules often recognize such sites, presumably because the unpaired nucleotides are more readily accessible. Such interactions are often associated with conformational transitions in the bulge region, such as flipping out of the bulge base from an intrahelical stack. NonWatson–Crick base pairs readily form within internal loops, while unpaired nucleotides within a bulge may stack within the helix or be bulged outside. **The presence of an internal loop or bulge can induce bending in an RNA helix**; the degree of bending depends on the RNA sequence within the loop and can change upon ligand binding. These motifs consequently are ideal sites for conformational switches, where ligand binding can result in long-range conformational changes.

The NMR structure of an internal loop from 23S ribosomal RNA differs from its structure in crystals of 50S ribosomal subunits. N. Shankar *et al.* (2006) *Biochemistry* **45**, 11776–11789.

The UAA/GAN internal loop motif: a new RNA structural element that forms a cross-strand AAA stack and long-range tertiary interactions. J. C. Lee *et al.* (2006) *J. Mol. Biol.* **360**, 978–988.

Conformational transitions in RNA single uridine and adenosine bulge structures: a molecular dynamics free energy simulation study. A. Bathel & M. Zacharias (2006) *Biophys. J.* **90**, 2450–2462.

4.2. TERTIARY STRUCTURE OF RNA

RNA is much more adept than DNA at forming complex tertiary interactions because of the presence of the 2'-hydroxyl group, which provides each nucleotide with additional hydrogen-bonding potential. In combination with the many polar groups on the bases, a multitude of hydrogen bond interactions are possible. In addition to the unique 2′-hydroxyl group of the RNA backbone, tertiary interactions often involve noncanonical base pairs, unpaired bases and the negatively charged phosphate groups. Nevertheless, secondary structure is the predominant feature of RNA tertiary structures, and RNA secondary structure elements tend to maintain their 3-D structures when excised from very complex tertiary structures. **Two or more secondary structure elements tend to interact, to generate an RNA tertiary structure and define the overall folding of the RNA molecule**. The secondary structure of RNA can orient key residues into positions appropriate for tertiary interactions. For example, the L-shape of all transfer RNAs (Section 4.2.B) is generated by the geometry of the four-way junction that, combined with a conserved length of the double-helical regions, helps to position nucleotides from the T- and D-loops in close proximity to facilitate the tertiary loop–loop base pairs. During formation of the tertiary structure, unpaired bases can twist or flip out of a helical segment and generate unique structural features for recognition by other RNA segments.

Energetically, the secondary structure is the main component of RNA architecture, while the tertiary structure contributes much less to the Gibbs free energy stability of the native folded state. A folded RNA molecule with definite secondary and tertiary structures will normally display two melting peaks, the first, low-temperature, transition corresponding to melting of the tertiary structure, and the second, high-temperature, transition corresponding to melting of the secondary structure (Section 5.2). **RNA structures are less cooperative than are those of proteins** (Section 11.1.B). Therefore, determination of the secondary structure is an essential step in the study of structure–function relationships of an RNA molecule (Section 4.4.A). In other cases, such as the ribosome, the RNA architecture involves a large number of proteins (Section 13.4.F).

Base stacking and hydrogen bonding are often involved in tertiary structure interactions. Base triplets form when an existing base pair becomes involved in another set of hydrogen bonds with a third base (Figure 4-11); this can occur on either the minor groove or major groove side of an RNA double helix. Hydrogen bonds between bases and backbones are observed when double helices are packed together in compact structures (Section 4.2.A.6).

Tertiary structures are often formed by unpaired nucleotides within a secondary structure motif interacting with unpaired nucleotides from another secondary structure module. Any secondary structure motif, hairpin loops, internal loops and bulge loops, can be involved in these interactions. The tertiary contacts can involve intercalation of the bases, base triplets and Watson–Crick base pairing between complementary loop sequences. This type of loop–loop interaction is best illustrated by the tertiary structure of transfer RNA (Section 4.2.B). A transfer RNA initiates folding by forming four hairpin loops using Watson–Crick base pairing, to produce a four-way helical junction. Helical stacking tertiary interactions generate two extended helical domains, which brings into proximity all unpaired residues in the loop regions and facilitates formation of the tertiary base pairing and base triplets that lock the transfer RNA into the L-shaped 3-D structure.

Ions influence the folding of RNA into specific tertiary structures, in that different salts affect the free energy change upon folding. Ions also determine the structure by the way that they are arranged on or near an RNA molecule and occupy various environments. Packing double helices together is opposed by the strong electrostatic repulsions between the negatively charged phosphate groups, unless the negative charges are screened by cations. Divalent Mg^{2+} ions are very effective at stabilizing tertiary structures in RNAs. In most cases, folding of an RNA molecule is so strongly coupled to its interactions with Mg^{2+} that it is difficult to separate the free energies of Mg^{2+}–RNA interactions from the intrinsic free energy of RNA folding. Group I self-splicing introns (Section 4.2.C) have a core of five Mg^{2+} ions that facilitates the close packing of loops into the minor groove of a double helix. The metal-binding sites are generated by base-stacking and hydrogen-bonding interactions between nucleotides.

Non-canonical base pairs and higher order structures in nucleic acids: crystal structure database analysis. J. Das *et al.* (2006) *J. Biomol. Struct. Dyn.* **24**, 149–161.

RNA folding during transcription. T. Pan & T. Sosnick (2006) *Ann. Rev. Biophys. Biomolec. Structure* **35**, 161–175.

Preparation and crystallization of RNA. B. L. Golden (2007) *Methods Mol. Biol.* **363**, 239–257.

NMR studies of RNA dynamics and structural plasticity using NMR residual dipolar couplings. M. Getz *et al.* (2007) *Biopolymers* **86**, 384–402.

4.2.A. Common Structural Motifs

The structures of RNAs can be very variable but certain structural motifs are observed in many different RNA structures.

Three-dimensional motifs from the SCOR, structural classification of RNA database: extruded strands, base triples, tetraloops and U-turns. P. S. Klosterman *et al.* (2004) *Nucleic Acids Res.* **32**, 2342–2352.

Annotation of tertiary interactions in RNA structures reveals variations and correlations. Y. Xin *et al.* (2008) *RNA* **14**, 2465–2477.

---

1. Pseudoknots

Pseudoknots form when a single-stranded region interacts with complementary primary sequences of a hairpin or internal loop by Watson–Crick base pairing (Figure 4-8). A pseudoknot creates an extended helical region through helical stacking of the hairpin double-helical stem with the newly formed loop–loop interaction helix. The pseudoknot is only marginally more stable than the two separate hairpins but its stability can be increased by tertiary interactions such as base triples between unpaired nucleotides in the bridging loops and base pairs within the extended helix.

The presence of a stable pseudoknot structure within the coding region of a messenger RNA, when adjacent to a special heptanucleotide sequence described as 'slippery', causes a shift in the frame with which the sequence of the nucleotide triplets is being read and translated into amino acid residues (Figure 7-21). This type of frame shifting is indispensable in the replication of numerous viral pathogens and is also exploited in the expression of several cellular genes.

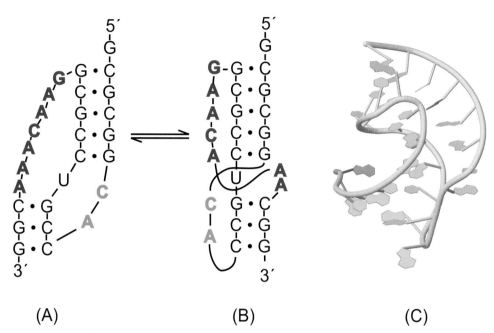

Figure 4-8. An example of a pseudoknot (A), with a schematic illustration of the type of 3-D structure that can be generated (B). An actual example of such a structure is displayed in (C), generated from PDB file *1l2x* using the program Jmol.

Predicting RNA pseudoknot folding thermodynamics. S. Cao & S. J. Chen (2006) *Nucleic Acids Res.* **34**, 2634–2652.

Pseudoknots in RNA secondary structures: representation, enumeration, and prevalence. E. A. Rodland (2006) *J. Comput. Biol.* **13**, 1197–1213.

Tertiary structure of an RNA pseudoknot is stabilized by 'diffuse' Mg^{2+} ions. A. M. Soto *et al.* (2007) *Biochemistry* **46**, 2973–2983.

2. Coaxial Helices: Interhelical Stacking

Nucleotide bases from two separate helices often stack and align their axes to form a pseudocontinuous coaxial helix (Figure 4-9). More than half of RNA structures exhibit statistically significant coaxial packing of helices. Coaxial helices dominate several large RNA structures and are believed to be highly stabilizing. The stacking between helices can occur via a single base or a base-pair bridge between the two helices. Multiple helices can stack in this manner to produce a nearly continuous single helical structure.

Coplanar and coaxial orientations of RNA bases and helices. A. Laederach *et al.* (2007) *RNA* **13**, 643–650.

Figure 4-9. An example of two coaxial helices, formed in this case between two hairpin loops that are depicted in different colors. Figure generated from PDB file *1zci* using the program Jmol.

3. A-minor Motif

One of the most abundant and ubiquitous structural motifs stabilizing RNA tertiary structures is known as the **A-minor motif**. It involves the insertion of the smooth minor-groove edge of an adenine base into the minor groove of a neighboring double helix (Figure 4-10). This occurs preferentially at C•G base pairs, where the fit is especially snug. The base can form hydrogen bonds with one or both of the 2′-hydroxyl groups of those pairs. A-minor motifs stabilize contacts between RNA helices, interactions between loops and helices, and the conformations of junctions and tight turns.

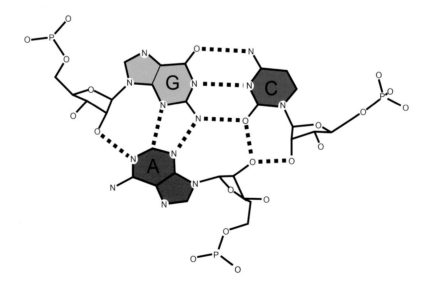

Figure 4-10. Example of an A-minor motif. The adenine nucleotide at the bottom is interacting with a G•C base pair of a double helix, in the minor groove. Besides the hydrogen bonds, there is a very close stereochemical fit between the interacting groups. Only O, N and P atoms are indicated.

RNA tertiary interaction in the large ribosomal subunit: the A-minor motif. P. Nilssen *et al.* (2001) *Proc. Natl. Acad. Sci. USA* **98**, 4899–4903.

Biochemical identification of A-minor motifs within RNA tertiary structure by interference analysis. S. A. Strobel (2002) *Biochem. Soc. Trans.* **30**, 1126–1131.

4. Dinucleotide Platform

A common example of a structural module built upon base-stacking interactions is provided by the **adenosine platform motif** (Figure 4-11). It can also occur with other bases and is known more generally as the **dinucleotide platform**. Two consecutive bases in the same strand at the end of a double helix can form a nonWatson–Crick base pair when within an internal loop, with the bases roughly coplanar. The backbone connecting the two bases is almost perpendicular to the axis of the helix they cap. Consequently, the dinucleotide platform extends the double helix by approximately one unit. The minor groove is enlarged, which facilitates interactions with other parts of the molecule. This generates a platform capable of mediating long-range tertiary interactions, especially the packing of helices in close proximity.

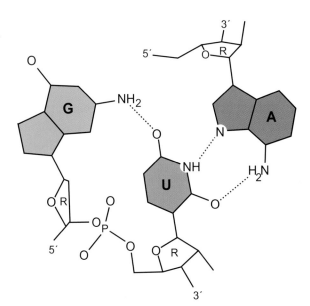

Figure 4-11. The structure of a dinucleotide platform that is also involved in a base triple as observed crystallographically. Hydrogen bonds are indicated by the *dashed lines*; only some of the hydrogen-bonding groups of the bases are shown. R is ribose. The G and U nucleotides are adjacent in the sequence of the polynucleotide chain in the lower *left* and form the dinucleotide platform. The U base is also hydrogen-bonded to an A base from another part of the sequence, to generate a base triple.

5. 'Kissing' Hairpin Loops

Nucleotides in two single-stranded loops can interact by base pairing if their sequences are complementary (Figure 4-12). This structure in 3-D can generate coaxial helices.

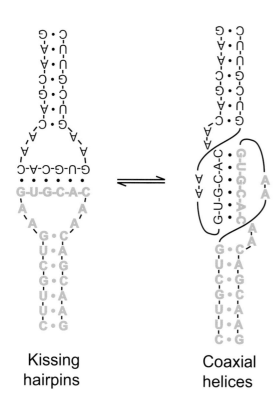

Kissing hairpins Coaxial helices

Figure 4-12. 'Kissing' of two hairpin loops when their single-stranded loops have complementary sequences. Both hairpins have the same sequence here, but that is not necessary. One strand is indicated in *color*. A coaxial stack of helices can be formed by base pairing between the hairpins and their Watson-Crick stems.

6. Ribose Zipper

If two antiparallel strands come into close proximity without being base paired, the ribose rings can interact through hydrogen bonds between consecutive backbone ribose 2'-hydroxyl groups (Figure 4-13). The hydrogen of the 2'-hydroxyl group of the 5'-nucleotide in both strands forms a hydrogen bond to the oxygen of the 2'-hydroxyl and to the base of the 3'-nucleotide on the opposite strand. The hydrogen bond is to the N3 atom when the base is a purine and to the O2 atom when a pyrimidine, so there is no sequence specificity. The zipper is usually limited to two or three nucleotides. Eleven types of ribose zippers can be distinguished on the basis of differences in their ribose–base interactions.

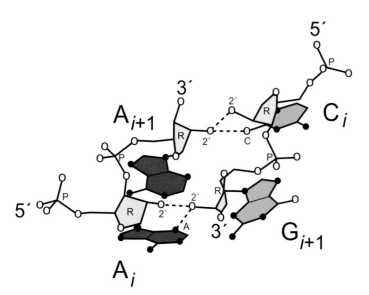

Figure 4-13. Canonical structure of a 'ribose zipper' between nucleotides A–A on one strand and C–G on the other. O atoms are indicated by *open circles*, N atoms by *black circles*. R indicates the ribose ring, while P is the phosphorous atom. The hydrogen bonds are indicated between O2' of the 3'-nucleotide of one strand and O2' and N3 of the base of the 5'-nucleotide of the opposite strand.

Almost all ribose zippers link Watson–Crick double-stranded stems, or stem-like base-paired structures, with loop segments that can be external, internal or junction. Ribose zippers tend to have similar sequences because stacked base triples are formed between consecutive base pairs on the stem or stem-like segment with bases (often adenines) from the loop-side segment.

About two-thirds of the ribose zippers in ribosomes interact with proteins of the complex (Section 13.4.F). Most of these ribosomal proteins bind to the ribose zipper chain segments using basic amino acid residues that form hydrogen bonds to the RNA backbone.

Sequence and structural conservation in RNA ribose zippers. M. Tamura & S. R. Holbrook (2002) *J. Mol. Biol.* **320**, 455–474.

---

7. Uridine Turn

A sharp bend occurs in the phosphate–sugar backbone between the first and second nucleotides of a **uridine turn**, which is also known as a **U turn** or **pi turn** (Figure 4-14). The second base then stacks on a third base. In some cases there is a hydrogen bond between the first and third nucleotides. These motifs serve a variety of functions in different RNA molecules.

Figure 4-14. Structure of a uridine turn at the top of a hairpin. Figure generated from PDB file *1mt4* using the program Jmol.

A U-turn motif-containing stem–loop in the coronavirus 5′-untranslated region plays a functional role in replication. P. Liu *et al.* (2007) *RNA* **13**, 763–780.

The genomic HDV ribozyme utilizes a previously unnoticed U-turn motif to accomplish fast site-specific catalysis. J. Sefcikova *et al.* (2007) *Nucleic Acids Res.* **35**, 1933–1946.

8. Tetraloop/Receptor Interactions

Tetraloops (Section 4.1.B) often interact, via Mg^{2+} ions, with a 'receptor' structure that consists of roughly 10 ribonucleotides in a distorted double helix (Figure 4-15). Interaction of the GAAA tetraloop with its 11-nucleotide receptor is one of the most frequently occurring long-range tertiary interactions in RNAs, and it is important for the stability of the tertiary structure.

Figure 4-15. A tetraloop interacting with its 'receptor'. The tetraloop is *green*, whereas its receptor is *blue*. Figure generated from PDB file *2adt* using the program Jmol.

Site-directed spin labeling studies reveal solution conformational changes in a GAAA tetraloop receptor upon Mg^{2+}-dependent docking of a GAAA tetraloop. P. Z. Qin *et al.* (2005) *J. Mol. Biol.* **351**, 1–8.

Role of metal ions in the tetraloop–receptor complex as analyzed by NMR. J. H. Davis *et al.* (2007) *RNA* **13**, 76–86.

---

9. Roles of Ions

RNA is a polyelectrolyte just like DNA (Section 2.3). In a complex and compact 3-D RNA structure, the sugar–phosphate chain folds several times on itself, and negatively charged phosphate groups inevitably come into close contact. Positively charged ions are necessary to relieve the resulting electrostatic repulsions. Biologically, the most prevalent and efficient cations are magnesium ions and polyamines, such as putrescine, spermidine and spermine (Equation 2.4). Magnesium cations are 'hard' ions, with well-defined electronic structures, that interact favorably with the 'hard' negatively charged O atoms of the phosphate groups. Polyamines like spermine always carry positively charged amino groups, and their flexibilities and small sizes allow them to snuggle into helical grooves and in between helical sugar–phosphate backbones.

Two types of metal-ion interactions occur with RNA: (1) diffuse ions that accumulate near RNA due to the electrostatic field while retaining their hydration sphere, and (2) chelated ions that are in direct contact with RNA at a specific location and may have some waters of hydration displaced by coordination with polar RNA atoms. Such interactions are usually with the ionized phosphate groups but they may also occur with polar groups on the bases. The A-form double helix of RNA has a deep and narrow major groove, and cations are usually observed to be present in both the major and minor grooves. In group I self-splicing introns (Section 4.2.C), an organized core containing five magnesium ions constructs an exterior surface that facilitates the close packing of noncanonical loops into the minor groove of a double helix. Within the interior of this ion core, base-stacking and hydrogen-bonding interactions between nucleotides form specific metal-ion binding sites. Monovalent cations bind to adenosine platforms and to the RNA equivalents of the guanine quartet (Section 2.2.C.4). In many cases the ion-binding site is formed even in the absence of the ion, although it is much less stable, whereas in a few cases large structural rearrangements occur on metal binding.

Varying the magnesium ion concentration during thermal unfolding (Section 5.2) distinguishes between melting of the secondary and tertiary structures. Decreasing concentrations of magnesium ions cause the tertiary structure to melt at decreasing temperatures, whereas the melting temperature of the secondary structure does not change.

Ions and RNA folding. D. E. Draper *et al.* (2005) *Ann. Rev. Biophys. Biomolec. Structure* **34**, 221–243.

Mg^{2+}–RNA interaction free energies and their relationship to the folding of RNA tertiary structures. D. Grilley *et al.* (2006) *Proc. Natl. Acad. Sci. USA* **103**, 14003–14008.

Simulation of Ca^{2+} and Mg^{2+} solvation using polarizable atomic multipole potential. D. Jiao *et al.* (2006) *J. Phys. Chem. B* **110**, 18553–18559.

4.2.B. Transfer RNA Structures

Transfer RNAs (tRNAs) are single RNA chains of 75–93 nucleotides that are present in the cytosol and organelles of all living cells, serving as adapters between the nucleic acid and protein primary structures. Each tRNA is covalently attached to one particular amino acid, and it recognizes the codons of messenger RNA for that amino acid (Figure 7-21); consequently, the attached amino acid is inserted into the growing polypeptide chain when the charged tRNA is bound to the appropriate codon of the messenger RNA that is being translated. A superscript is used to indicate the amino acid accepted and used in protein biosynthesis by each tRNA. Transfer RNAs need to recognize and interact with a number of components of the protein biosynthesis machinery, so they consist of complex folded structures in which some parts have common functions and others are specific for their particular amino acid. Probably because of their complex structures and functions, tRNAs contain many modified versions of the classic nucleotides; more than 80 have been characterized. All such modifications occur after synthesis of the RNA molecule and may be modifications of the base (methylation, thiolation, etc.) or ribose (methylation). Further modifications of the tRNA molecules occur after their biosynthesis (Figure 4-16). Segments can be removed from both ends of the primary transcript, and internal sequences (*introns*) may also be removed. The CCA sequence at the 3′-end, which is linked to the appropriate amino acid, is usually added by a specific enzyme.

Guided by their primary structures, **tRNA molecules fold into cloverleaf-like secondary structures** with well-defined stems and loops that make up the acceptor arm, D-arm and -loop, anticodon arm and loop, and T-arm and -loop (Figure 4-16). The acceptor stem always has seven base pairs and four single-stranded nucleotides, including the absolutely conserved CCA sequence at the 3′-end. The D-stem and loop are of variable length in different tRNAs, whereas the anticodon stem always has five base pairs and the anticodon loop seven nucleotides. The variable region usually has four to five nucleotides but can contain up to 24. Finally, the T-stem always has five base pairs and the T-loop seven nucleotides.

128 CHAPTER 4 RNA Structure

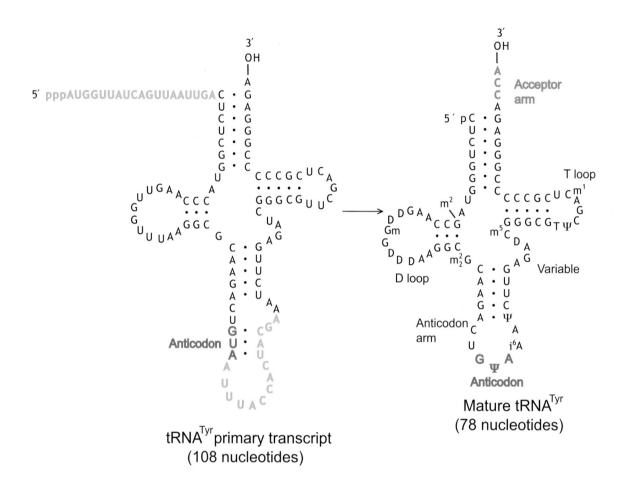

Figure 4-16. The post-transcriptional processing of yeast tRNA^{Tyr}. A 14-nucleotide intron next to the anticodon (*red*) is removed, as is the 19-nucleotide 5′-terminal sequence (green). The sequence –CCA (*blue*) is added at the 3′-end, and several of the bases are modified (Figure 4-3).

Each tRNA has an L-shaped 3-D structure comprising two domains, the acceptor branch (produced by stacking of the acceptor arm and T-arm) and the anticodon branch (the anticodon arm and the D-arm) (Figure 4-17). The angle between both branches is about 90° and the distance between the extremities of the two branches, i.e. between the acceptor end and the anticodon triplet, is 75–80 Å. This 3-D structure is generated by a number of interactions between conserved or semiconserved nucleotides that are distant in the covalent structure but close in the folded conformation. Especially important are those in the D- and T-loops and within the core of the molecule. Some of the nine tertiary interactions present in each tRNA involve triple interactions between three nucleotides simultaneously: 8–14–21, 9–12–23, 10–25–45, 13–22–46.

Transfer RNAs are classified according to the lengths of their variable regions, which can be either four or five nucleotides (class I tRNA) or 10–24 (class II tRNA). The only class II tRNAs are tRNA^{Leu}, tRNA^{Ser} and tRNA^{Tyr} from eubacteria and some organelles, plus that for selenocysteine, the '21st amino acid' (see Section 7.2.O and Figure 4-22). Some tRNAs from mitochondria possess peculiar structural features, such as the absence of one or more stems and loops within the secondary structure and/or the absence of conserved nucleotides. Nonetheless, these tRNAs are thought to fold into the

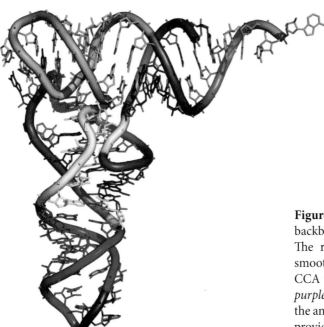

Figure 4-17. The course of the polynucleotide backbone in the 3-D structure of a typical tRNA. The ribose–phosphate backbone is depicted as a smooth rod, with the bases protruding from it. The CCA tail is *orange*, the amino acid acceptor site is *purple*, the D-arm is *red*, the anticodon arm is *blue*, the anticodon is *black*, and the T loop is *green*. Kindly provided by Vossman.

usual L-shaped 3-D structure (Figure 4-17), which is believed to be necessary to fulfill their functions. Over 150 mutations with documented pathogenicity have been identified in the genes for tRNAs within the human mitochondrial genome. The physiological effects of mutationally altering the tRNA sequences are multi-faceted and complex but are ultimately due to the destabilization of certain structural features, perturbing the native tRNA fold that is required for all aspects of its function.

The crystal structure of leucyl-tRNA synthetase complexed with tRNA^{Leu} in the post-transfer-editing conformation. M. Tukalo *et al.* (2005) *Nature Struct. Mol. Biol.* **12**, 923–930.

The crystal structure of tRNA. B. F. Clark (2006) *J. Biosci.* **31**, 453–457.

Crystal structure of a 70S ribosome–tRNA complex reveals functional interactions and rearrangements. A. Korostelev *et al.* (2006) *Cell* **126**, 1065–1077.

Structure of the 70S ribosome complexed with mRNA and tRNA. M. Selmer *et al.* (2006) *Science* **313**, 1935–1942.

4.2.C. Ribozyme Structures

A ribozyme is an RNA molecule capable of catalyzing a chemical reaction by itself in the absence of proteins (Section 14.6). Such catalytic RNAs fold into compact 3-D structures that efficiently activate a single phosphate linkage for a chemical reaction. There are several major types of ribozymes. Some have been identified in **satellite RNAs** from plants and animals, where they participate in the replication of these double-stranded RNAs. A more common category includes the group I and group II **self-splicing introns**. Another category comprises **ribonuclease P**, which is associated with a protein and participates in the processing of tRNAs.

Most of the naturally occurring catalytic RNAs require divalent metal cations. While the metal specificity varies substantially between ribozymes, Mg^{2+} or Mn^{2+} will usually suffice. These metals often have both a structural and catalytic role. The P5abc subdomain of the *Tetrahymena* group I intron dramatically illustrates the roles of divalent metal ions in RNA folding. Three Mg^{2+} ions interact with disparate phosphate groups within the subdomain, which allows the RNA to fold inside-out, in that the phosphates point into the structure and the nucleotide bases point out to the solvent. Some structural metals in RNA can often be substituted with polycations such as spermidine (Equation 2.4) or cobalt hexamine, which emphasizes the importance of charge neutralization for RNA folding.

One of the simplest structures is that of the **hammerhead ribozyme**, which requires only 42 nucleotides for a folded and catalytically active minimal structure. Yet quite dramatically different structures for this minimal ribozyme have been determined crystallographically and by NMR. Each consists of three double-helical stems, but in remarkably different orientations: the first and second helical stems differ in orientation by about 150°. Apparently the crystal lattice induces or selects a more compact structure than predominates in solution, which vividly illustrates the intrinsic flexibility of such RNA molecules. The crystal structures did not explain how catalysis takes place, and it seemed that very substantial changes in conformation would have to occur during the reaction (Section 14.6.B). The structure of a full-length hammerhead ribozyme, with 63 nucleotides (Figure 4-18), however, resolved the inconsistencies. The 50% additional nucleotides produce significant changes throughout the structure, further illustrating the plasticity of RNA structures. This full-length molecule is 1000 times more active than the minimal molecule, and the structure can explain how catalysis takes place (Section 14.6.B).

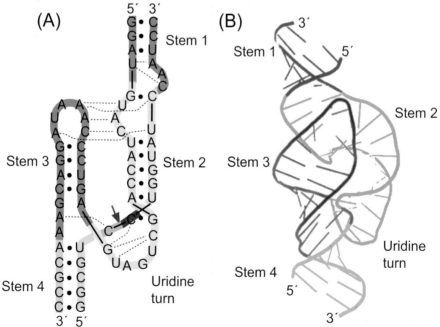

Figure 4-18. The structure of a full-length hammerhead ribozyme consisting of 63 nucleotides in two polynucleotide chains. (A) Schematic diagram of the sequences and secondary structures of the two strands. The colored ribbons and the solid black lines indicate the linear sequences. The dots indicate the base pairs that define the secondary structure. The long-range interactions that determine the tertiary structure are indicated by the dashed lines. The phosphodiester bond that is cleaved by the ribozyme is indicated by the arrow. (B) The 3-D structure of the ribozyme determined crystallographically. The polynucleotide backbone is depicted as the thin cylinder, with the bases depicted as lines off it. Data from M. Martick and W.G. Scott (2006) *Cell* **126**, 309-320.

The **hairpin ribozyme** is a small catalytic RNA comprising two internal loops carried on two adjacent arms of a four-way helical junction. To achieve catalytic activity, the ribozyme folds into a compact conformation that facilitates the formation of tertiary interactions between the two loops (Figure 4-19-A).

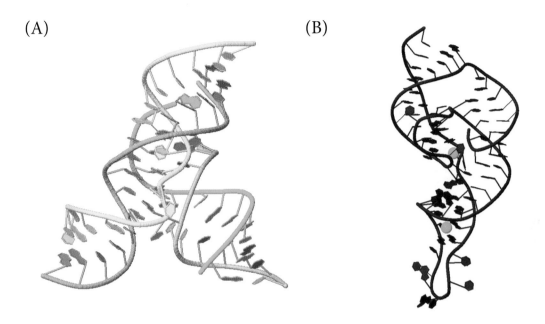

Figure 4-19. Tertiary structures of two types of ribozymes. (A) The crystal structure of a hairpin ribozyme model, comprising four RNA chains. (B) The crystal structure of the precursor of the hepatitis delta virus ribozyme. The *green spheres* are magnesium ions. The bottom of the RNA molecule is distorted because it was bound to a protein in the crystal. Figure generated from PDB files *1x9k* and *1sj3* using the program Jmol.

The **hepatitis delta virus ribozyme** is only somewhat more complex (Figure 4-19-B). It is folded into a double pseudoknot containing five helical stems (P1, P1.1, P2, P3 and P4). The P1 helix is coaxially stacked upon P1.1 and P4, while P2 is stacked on P3. The two stacks are positioned side-by-side, linked by five-strand cross-overs and constrained by the P1.1 pairing.

The **group I and group II introns** are larger and fold into distinct tertiary structures comprising many conserved secondary structural elements. The hallmarks of a group I intron include a common secondary structure of 10 paired segments and several single-stranded joiner (J) segments between the double helices (Figure 4-20). Group II introns also have distinct structures (Figure 4-21). The intron is organized into six double-helical domains (1–6) that originate from a central wheel-shaped loop. Each of the domains has a particular function in the activity of the intron, although D1, D5 and D6 form the intron core. Domain 1 is an extended multi-helical element that includes two looped regions (termed EBS1 and EBS2) that are complementary to two sites in the 5′-exon (termed IBS1 and IBS2). There are also interdomain interactions that allow domain I to serve as the scaffold upon which the intron active site is built. Many of the phylogenetically conserved nucleotides are concentrated in domain 5, which constitutes the catalytic center of the group II intron. Domain 6 includes the branch-point adenosine whose 2′-OH nucleophilically attacks the 5′-splice site during the catalytic action (Figure 14-29). In isolation, domains 5 and 6 together comprise a single double-stranded hairpin (Figure 4-21-B).

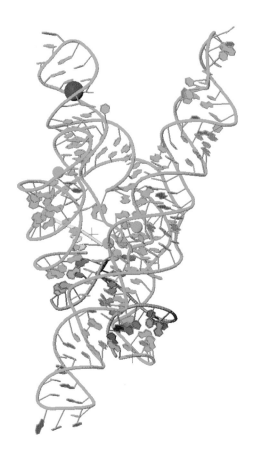

Figure 4-20. The 3-D structure of the *Tetrahymena* group I self-splicing intron. The *blue sphere* is a K$^+$ ion, while the *green spheres* are Mg^{2+} ions. Figure generated from PDB file *1zzn* using the program Jmol.

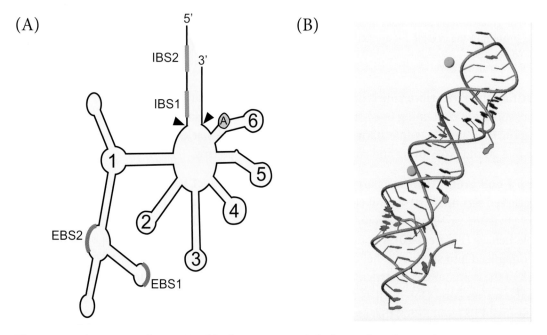

Figure 4-21. Structures of group II self-splicing introns. (A) The predicted secondary structure, with the stem structures being antiparallel base-paired helices. The EBS sequences are complementary in sequence to the IBS sequences; both are indicated as *thick lines*. The circled A in domain 6 is a bulge in the double helix, causing it to bend, and acts as the nucleophile in the first step in splicing. The two *arrowheads* indicate the splice sites. (B) The 3-D structure of domains 5 and 6 in isolation, which comprise a single double-stranded hairpin. Figure generated from PDB data file *1kxk* using the program Jmol.

Ribozymes with new folds and biochemical activities have been found after only limited alterations of the nucleotide sequence (Section 14.6.C). The probability of finding such altered structures increases considerably as the mutational distance from the parental ribozyme increases, suggesting that it must first escape from the 3-D fold of the parent.

The structure–function dilemma of the hammerhead ribozyme. K. F. Blount & O. C. Ulenbeck (2005) *Ann. Rev. Biophys. Biomol. Struct.* **34**, 415–440.

The tertiary structure of the hairpin ribozyme is formed through a slow conformational search. G. Pljevaljcic *et al.* (2005) *Biochemistry* **44**, 4870–4876.

New catalytic structures from an existing ribozyme. E. A. Curtis & D. P. Bartel (2005) *Nature Struct. Mol. Biol.* **12**, 994–1000.

Tertiary contacts distant from the active site prime a ribozyme for catalysis. M. Martick & W. R. Scott (2006) *Cell* **126**, 309–320.

4.3. QUATERNARY STRUCTURE OF RNA

Relatively few examples of the association of RNA molecules to form supramolecular quaternary structures have been well-characterized, but they are relatively important. During splicing of messenger RNA (mRNA) precursors, to remove the introns, for example, the precursors associate with five major ribonucleoprotein (RNP) particles called **small nuclear RNPs** that interact with each other and the mRNA. The dynamic disruption and formation of these RNA–RNA quaternary interactions are essential for RNA splicing. Oligomerization usually depends on the concentrations of cations or polyamines.

The quaternary association of RNA molecules generally occurs by conventional Watson–Crick base pairing. For example, small regulatory RNAs with sequences complementary to other RNA molecules (known as **antisense RNAs**) form intermolecular duplexes during the control of gene expression in both prokaryotes and eukaryotes. Similarly, **guide RNA**s recognize complementary sequences to identify sites where mRNAs are edited after transcription. More complex RNA quaternary structures do not rely exclusively on Watson–Crick pairing but base pairs are still very important. So-called 'kissing-hairpins' (Figure 4-12) form between loop nucleotides in two stem–loop structures with complementary sequences. These structures provide protein recognition sites during the regulation of the copy number of prokaryotic plasmids.

Supramolecular association of RNA molecules by means of nonWatson–Crick interactions is best-characterized in the so-called 'G-quartet structures' (Section 2.2.C.4). These structures form readily *in vitro* for both RNA and DNA sequences that contain stretches of guanidines or uracils, but it is not clear whether they occur *in vivo*.

Regulation of the structure of RNA assemblies through RNA tertiary interactions. R. Ohmori *et al.* (2005) *Nucleic Acids. Symp. Ser.* **49**, 193–194.

4.4. RNA STRUCTURE PREDICTION

The folded structures of RNA molecules are dictated by their nucleotide sequences and are adopted spontaneously, so it should be possible to predict the folded structure from just the nucleotide sequence; the problem is comparable to that of predicting the structures of proteins (Section 9.4). A fundamental difference between RNA and protein structures is that the RNA secondary structure is often stable on its own, whereas that of proteins is at best only marginally stable in the absence of the tertiary structure. Formation of the secondary structure dominates RNA folding, and the tertiary structure forms through relatively weak interactions between pre-formed secondary structure motifs. Consequently, secondary structure is especially easy to be predicted successfully from just the RNA nucleotide sequence, because complementary base pairing in double helices is the most important consideration. Even in the simplest case, however, it is not a trivial exercise because the base pairing in RNA double helices is not restricted to Watson–Crick base pairs and because a sequence of N nucleotides can adopt approximately 1.8^N different secondary structures.

4.4.A. Prediction of Secondary Structure

The most probable secondary structure for a given RNA sequence can usually be identified using computer programs that find the maximum amount of complementary base pairing. The **thermodynamic approach** attempts to find the secondary structure with the lowest free energy, whereas the **phylogenetic approach** uses information available in homologous sequences that have arisen from a common evolutionary ancestor (Section 7.6). Comparison of secondary structures of 16S and 16S-like ribosomal RNA predicted solely on thermodynamic calculations with those derived by the phylogenetic approach demonstrated that the quality of the thermodynamic predictions is variable: the percentages of correctly predicted base pairs ranged between 10% and 90%. Often, the programs produce several solutions that seem equally likely, so identifying the correct one requires additional chemical or biological information.

Revolutions in RNA secondary structure prediction. D. H. Mathews (2006) *J. Mol. Biol.* **359**, 526–532.

Statistical and Bayesian approaches to RNA secondary structure prediction. Y. Ding (2006) *RNA* **12**, 323–331.

Computational RNA secondary structure design: empirical complexity and improved methods. R. Aguirre-Hernandez *et al.* (2007) *BMC Bioinformatics* **8**, 34.

---

1. Thermodynamic Approach

The thermodynamic approach is based on a set of energy parameters for the stabilities of structures formed by short oligonucleotides, plus the observation that the secondary structure of an RNA molecule can be decomposed into elementary motifs that recur in other structures. The sum of the free energies of the secondary elements is assumed to be a reasonable approximation of the total free energy of the RNA, because tertiary interactions are weaker than secondary interactions. The free energy of a given base pair is assumed to be due primarily to hydrogen bonding and base stacking, so it depends only on the immediate neighbors (**nearest-neighbor approximation**). The base pairs considered are the standard Watson–Crick pairs, G•C and A•U, as well as the frequent G•U (Equation 4.1). Stereochemical considerations are also taken into account, such as the requirement that at least three nucleotides occur in a loop between two paired strands (Figure 4-5). Unfortunately, the thermodynamic parameters used are imprecise and incomplete. On the other hand, such calculations have confirmed the expectation that the RNA sequences from organisms that live at higher temperatures have evolved RNA sequences with higher melting temperatures.

Pseudoknots are difficult to predict, because most algorithms are based on decomposition into substructures, which is not possible mathematically with pseudoknots (Figure 4-8). With luck, however, one of the two helices of a pseudoknot might be predicted on the basis of its stability, and inspection of the remaining single-stranded regions might uncover the second helix.

Many computer programs use recursive algorithms based on **dynamic programming** principles, which solve and combine solutions corresponding to subproblems. Each subproblem is solved just once and then saved in a table, avoiding the need to compute the same answer every time the subproblem is encountered. Such algorithms are guaranteed to find the structure that is optimal for a given set of rules. The number of nucleotides, N, can be as large as 10,000. The required computer time, however, is proportional to N^3, and the computer memory required is proportional to N^2. Decomposition into elementary structures makes it possible to consider more complete thermodynamic data, at the price of increased computational time and storage requirements, depending on the way in which energy is assigned to the loops. The free energies of the base pairs and any additional experimental data are included in the energy function, permitting fast computation of the optimal 2-D structure.

Prediction of RNA secondary structure by free energy minimization. D. H. Mathews & D. H. Turner (2006) *Curr. Opinion Struct. Biol.* **16**, 270–278.

A method for finding optimal RNA secondary structures using a new entropy model (vsfold). W. Dawson *et al.* (2006) *Nucleosides Nucleotides Nucleic Acids* **25**, 171–189.

A set of nearest neighbor parameters for predicting the enthalpy change of RNA secondary structure formation. Z. J. Lu *et al.* (2006) *Nucleic Acids Res.* **34**, 4912–4924.

---

2. Phylogenetic Approach

The phylogenetic approach assumes that the function and folding architecture have been conserved during evolution while the nucleotide sequence has changed, producing a family of homologous sequences (Section 7.6); consequently, different versions of an RNA sequence should be compatible with the same secondary structure. The power of this comparative approach stems from the general observation that molecular 3-D architectures change during evolution much more slowly than do sequences (Figure 9-35). A set of homologous sequences with common ancestry and function can pinpoint secondary structure elements that are common to all the sequences.

The first step is to align the sequences, arranging them horizontally so that the lengths of the paired regions juxtapose vertically (Figure 4-22), with inclusion of blanks or gaps in a fashion similar to the alignment of protein sequences (Section 7.6.A.2). The next step consists of searching for nucleotide interactions by measuring correlations between pairs of RNA positions in the alignment, such as compensatory base changes (e.g. an A•U pair changed into a C•G pair). The more compensatory base changes in the sequences, the more probable the secondary structure prediction is correct. The level of ambiguity can always be reduced by using additional sequences. Aligning the sequences

CHAPTER 4 RNA Structure

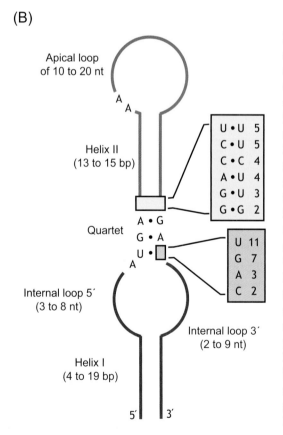

Figure 4-22. Identifying the signal for inserting selenocysteine residues (Section 7.2.O) from the sequences of mRNAs. (A) Sequence alignments of mRNAs coding for various selenocysteine-containing proteins (Gpx, glutathione peroxidases; DI, iodothyronine deiodinase). The base-paired segments, denoted Helix I/I' and Helix II/II' corresponding to the 5'- and 3'-strands, are *underlined*; note the significant number of noncanonical base pairs. The invariant residues in the internal and apical loops are highlighted (*black on red*). At the base of helix II, the quartet made of four nonWatson–Crick pairs that constitute a signature of the SelenoCysteine Insertion Sequence (SECIS element) is boxed and highlighted (*white on blue*). Gaps have been introduced to maximize superpositions between secondary structure elements and sequence similarities. (B) Secondary structure of the SECIS element predicted from the sequence alignment. The central 'quartet' of nonstandard base pairs consists primarily of tandem sheared G•A base pairs involving hydrogen bonds between N2(G) and N7(A), as well as between N3(G) and N6(A). The number of occurrences of the various bases or base pairs at the boxed sites are given in the two boxes (bp, base pairs; nt, nucleotides). Data from R. Walczak *et al.* (1996) *RNA* **2**, 367–379.

and assessing pairs of them should identify the conserved core of the secondary structure. All such methods, however, are plagued with problems of statistical relevance: with only four bases, sequences may be similar purely by coincidence.

The phylogenetic approach is the best method when a set of homologous sequences is available for RNAs with the same biological function. The sequences should be arranged in groups according to the phylogenetic classification or on the basis of their functional properties. The overall power of the approach increases with the diversity of the sequences and the evolutionary distances separating them, while the accuracy of each prediction depends on the number of alterations in each group and subgroup that are related, such as changing both bases of a base pair. A conserved core can often be found by using an iterative stepwise convergent cycle that combines the alignment search and secondary structure assessment, usually by aligning each sequence to a tentative model of the secondary structure.

Use of the information that is present in a large number of homologous sequences is illustrated in Figure 4-22.

Identification and classification of conserved RNA secondary structures in the human genome. J. S. Pedersen *et al.* (2006) *PLoS Comput. Biol.* **2**, e33.

Efficient pairwise RNA structure prediction and alignment using sequence alignment constraints. R. D. Dowell & S. R. Eddy (2006) *BMC Bioinformatics* **7**, 400.

Efficient pairwise RNA structure prediction using probabilistic alignment constraints in Dynalign. A. O. Harmanci *et al.* (2007) *BMC Bioinformatics* **8**, 130.

4.4.B. Prediction of Tertiary Structure

It should be possible, in principle, to predict the 3-D structure of an RNA from just its nucleotide sequence, but it is a formidable problem, comparable to that of predicting the tertiary structures of proteins (Section 9.4). The prediction methods range from very mathematical, relying solely on computer algorithms, to the most pragmatic interpretations of experimental data. A predicted secondary structure is always necessary for construction of a tertiary structure of an RNA molecule.

Tertiary contacts between groups in the tertiary structure can often be inferred through sequence comparisons and chemical modification experiments (Section 2.6), which might indicate the importance of specific atomic positions or uncover some groups that are inaccessible or protected and cannot be explained solely by the secondary structure. Cross-linking experiments can indicate directly which groups are in close proximity in the structure. Such information can be used with the distance-geometry method of generating structures from a set of constraints from NMR data, although it is difficult to compile a sufficient number of constraints and there are problems in choosing the correct chiralities and avoiding knots in the structures. Other computer methods try to find the minimum free energy by simplifying the structure using a **pseudoatom approach**, with either one pseudoatom per helix or one pseudoatom per nucleotide; however, this ignores the asymmetry of the RNA fragments and, more importantly, all the detailed interactions that control RNA folding. A manual approach uses known RNA structures to extract the structure of a fragment of interest. The fragments obtained are then assembled manually on a computer graphics screen, using interactive modeling procedures. Any

3-D structure that results can be refined by restrained energy minimization, molecular mechanics or molecular dynamics procedures. The surface of the final model that is accessible to the solvent can be compared to experimental data on the reactivities of specific groups to chemical reagents.

Modeling should not be considered as a method for generating the correct structure, but primarily as a heuristic tool to help in planning and interpreting experiments. It can suggest new relationships between the various components of a modeled molecule; for example, varying the sequence of a macromolecule in the absence of a 3-D model will be somewhat random and rarely informative. At best, such experiments will confirm the secondary structure of an RNA molecule. The data can still be useful, however, for gradually generating a 3-D structure and suggesting new experiments.

The case of tRNA (Section 4.2.B) is the best example of how such approaches have clarified understanding of the secondary and tertiary structures of RNA molecules: new genome sequences can be searched using precise rules for identifying the sequences of tRNA molecules.

De novo synthesis and development of an RNA enzyme. Y. Ikawa *et al.* (2004) *Proc. Natl. Acad. Sci. USA* **101**, 13750–13755.

Bridging the gap in RNA structure prediction. B. A. Shapiro *et al.* (2007) *Curr. Opinion Struct. Biol.* **17**, 157–165.

The RNAz web server: prediction of thermodynamically stable and evolutionarily conserved RNA structures. A. R. Gruber *et al.* (2007) *Nucleic Acids Res.* **35**, W335–338.

~ CHAPTER 5 ~

DENATURATION, RENATURATION AND HYBRIDIZATION OF NUCLEIC ACIDS

The folded structures of nucleic acids, especially the helices of double-stranded DNA and the folded structures of single-stranded RNA molecules, are disrupted upon exposure to extreme conditions, such as high temperatures, chemical denaturants and extremes of pH. The conditions required to denature a nucleic acid are a quantitative measure of the stability of its structure. In most cases, denaturation of a nucleic acid involves the separation of base-paired double strands, whether two separate strands in double-stranded DNA or intramolecular stems and loops in a single RNA molecule. One of the most important aspects of such structures is that base pairs between guanine and cytosine (G•C and C•G) are generally more stable than those between adenine and thymine (A•T and T•A) or uridine (A•U and U•A) and much more than any other nonstandard (mismatched) base pairs.

In many cases, the original structure can be regained by restoring the conditions. Alternatively, new structures may be generated by adding new polynucleotides with complementary sequences, which can hybridize to the single strands. The feasibility of renaturing nucleic acids depends primarily upon how many independent molecules are generated; the greater the number of different polynucleotide chains generated, the less likely that the original strands can find each other to regenerate complementary structures by base pairing. Because there are only four types of bases in DNA or RNA, regions of some nucleotide sequence complementarity occur even in random sequences, so a variety of base-pairing interactions is possible. The denaturation and renaturation of nucleic acids are central to most techniques in molecular biology.

In many cases, one of the hybridizing strands can be a nucleic acid mimic, such as a protein nucleic acid (PNA; Section 5.4). They can be designed to have a variety of properties to make them useful.

Principles of Nucleic Acid Structure. W. Saenger (1984) Springer, NY.

5.1. DENATURATION OF DOUBLE-STRANDED NUCLEIC ACIDS

The three-dimensional (3-D) structures adopted by nucleic acids, the antiparallel double helices and more complex structures (Chapters 2–4), are only marginally stable and may be disrupted in a variety of ways comparable to those used with proteins (Section 11.1). An exception is high pressures, which can unfold proteins (Section 11.1.G) but do not induce large-scale structural changes in the helical forms of DNA or RNA. Under most conditions, the latter are stabilized by high pressures, but only slightly, indicating that the volume occupied by the double helix in solution is very slightly less than that of the individual strands (Table 2-9).

Effect of hydrostatic pressure on nucleic acids. R. B. Macgregor (1998) *Biopolymers* **48**, 253–263.

Thermodynamics of nucleic acids and their interactions with ligands. A. N. Lane & T. C. Jenkins (2000) *Quart. Revs. Biophys.* **33**, 255–306.

Molecular dynamics applied to nucleic acids. J. Norberg & L. Nilsson (2002) *Acc. Chem. Res.* **35**, 465–472.

Factors regulating thermodynamic stability of DNA structures under molecular crowding conditions. D. Miyoshi *et al.* (2006) *Nucleic Acids. Symp. Ser.* **50**, 203–204.

5.1.A. Methods for Monitoring Denaturation

Different techniques can monitor different aspects of denaturation, whether it be breaking of base pairs or separation of strands. The disruption of base pairing in double-stranded structures is most readily monitored by ultraviolet (UV) absorbance, using the 38% increase in absorbance (**hyperchromism**) at 260 nm (Figure 5-1). Double-stranded nucleic acids exhibit **hypochromicity**, in that the absorbance of the duplex is less than that of the constituent single strands and nucleotides. The shape of the absorbance spectrum does not change, only its intensity, as a result of the absorbing bases being stacked very closely in the double helix so that the electronic systems of neighboring bases interact with each other. Other spectral methods are also useful, especially circular dichroism with relatively small nucleic acids, although the interpretation of the results is not always straightforward. Large changes occur in the hydrodynamic volumes and viscosities of nucleic acids upon their denaturation, especially when the strands separate, but such measurements are not very convenient. Alternatively, unfolding dramatically decreases the electrophoretic mobilities of nucleic acids through gels, and the unfolding and dissociation steps may be visualized directly and easily.

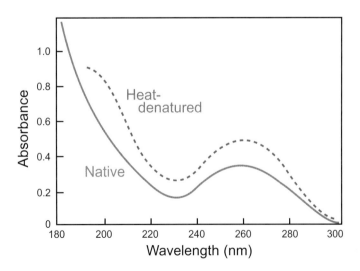

Figure 5-1. Typical UV absorbance of DNA when native (25°C) and when heat-denatured (82°C).

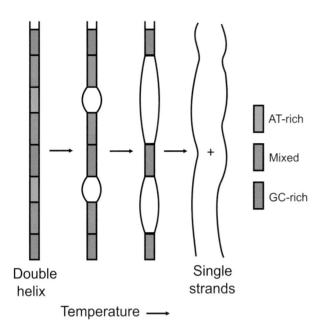

Figure 5-2. The progressive denaturation of a nucleic acid with a nonuniform sequence. The sequence of the double helix (*left*) is divided into regions of about 50 base pairs with three types of sequence, depending upon whether it is AT- or GC-rich; in reality, there is a continuum of sequence variation, from the most AT-rich to the most GC-rich. The final result at high temperatures on the *right* is the two separated and independent strands.

With long DNA molecules, the process of denaturation can be complex (Figure 5-2), but it can be observed almost directly using the technique of **denaturation mapping**. The partly opened intermediate forms are trapped by reacting them with a chemical reagent, such as **formaldehyde** ($H_2C=O$), or its bifunctional version, **glyoxal** ($O=CH-HC=O$); they react chemically with the amino groups of bases and block their participation in base pairing; the open segments cannot then regenerate a double helix. The single- and double-stranded segments can be visualized, after appropriate staining, with an electron microscope (Figure 5-3). The DNA molecules are spread on the surface of the electron microscope grid and made visible by staining with basic proteins, such as cytochrome *c*, and heavy metals. In this way, single- and double-stranded regions can usually be distinguished: the cytochrome *c* molecules bound to double-stranded DNA make it appear somewhat thicker than single-stranded regions, and the double-stranded segment is not quite as kinky. The various single- and double-stranded segments of the DNA molecule are identified simply by their positions relative to the ends of the DNA molecule. This technique of denaturation mapping is useful in physical and genetic studies of very large DNA molecules, such as the genomes of viruses, plasmids, etc.

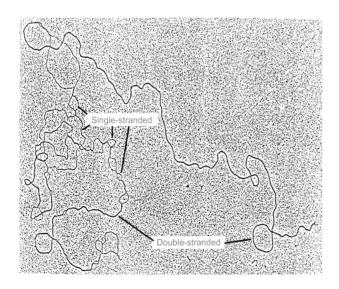

Figure 5-3. Electron micrograph of a DNA molecule that was partially denatured by alkaline pH. Separation of the two strands locally is apparent by a decrease in the thickness of the strand. The sites that are locally denatured coincide with the segments that are relatively deficient in C•G base pairs.

Single- and double-stranded DNAs or RNAs can be distinguished using the enzyme **S1 nuclease**, which hydrolyzes the backbones of only single-stranded nucleic acids. In the case of RNA, the enzyme **RNase V$_1$** hydrolyzes only double-stranded forms. With all such reagents, the possibility that the structure of the nucleic acid molecule will change after being hydrolyzed must be considered, and it is advisable to use these modifications only sparingly, so that no molecules are modified more than once.

Single- and double-stranded molecules can be separated using chromatography on **hydroxyapatite**. Under conditions of low sodium phosphate concentration, both single- and double-stranded DNA bind to hydroxyapatite. At 0.1–0.2 M phosphate buffer, however, the column retains only double-stranded DNA in which at least some 50 adjacent residues are base paired. The double-stranded DNA may be eluted with 0.5 M sodium phosphate. The time–course for dissociation or reassociation may be measured in this way, although it must be kept in mind that this assay does not measure directly base-pair formation, only association or dissociation of complete single strands.

Thermal difference spectra: a specific signature for nucleic acid structures. J. L. Mergny *et al.* (2005) *Nucleic Acids Res.* **33**, e138.

5.1.B. DENATURATION OF DOUBLE-STRANDED DNA

The stability of a double-stranded nucleic acid with a nonuniform base sequence varies by region, and the base-pair composition, especially the fraction of G•C base pairs, determines the stability of each region (Figure 5-2). Under most conditions, regions of the duplex that are composed predominantly of A•T (for DNA) or A•U (for RNA) base pairs are less stable than GC-rich regions (Figure 5-4). As the conditions become increasingly denaturing, the path from fully base-paired duplex to fully-denatured single strands passes through a continuum of intermediate states, where first the AT/AU-rich regions 'melt', forming internal loops, then the regions with increasing GC content, and then finally the most GC-rich regions. The melting transitions of such random-sequence duplexes can be quite broad and the various stages of melting and strand dissociation occur at distinct temperatures.

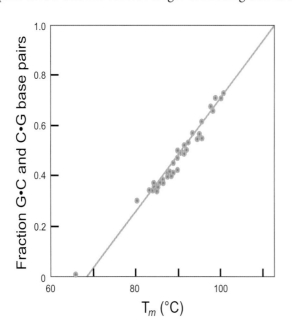

Figure 5-4. Typical variation of the melting temperature, T_m, of various DNAs with their content of G+C nucleotides. The solvent was 0.15 M NaCl and 15 mM sodium citrate (to chelate metal ions).

As the last region melts, the two strands separate (Figure 5-2). Up until that point, the two strands will very quickly regenerate their original structure if the conditions are restored to normal, because this is an intramolecular process. Once the two strands have separated, however, their reassociation is a bimolecular reaction and subject to the concentrations of the two strands and the presence of any other molecules that might compete (Section 5.3).

Thermodynamic and kinetic characterization of the dissociation and assembly of quadruplex nucleic acids. C. C. Hardin *et al.* (2000) *Biopolymers* **56**, 147–194.

Bubbles and denaturation in DNA. T. S. van Erp *et al.* (2006) *Eur. Phys. J. E* **20**, 421–434.

---

1. *Thermal Melting*

Heating is the most common method of denaturing nucleic acids, and their stabilities are often given as their melting temperature, T_m, the temperature at the mid-point of the thermal unfolding transition measured by the breakage of base pairs. **The T_m is largely independent of the length of the polynucleotide strands and their concentration,** even though melting involves dissociation to two single strands, because its value is determined primarily by the pseudo-intramolecular local melting processes that precede melting of the final segment and dissociation of the two strands (Figure 5-2). Only with short oligonucleotides of fewer than 50 base pairs is the T_m dependent upon the length and the concentration. Because strand dissociation occurs only at the last step, different temperatures correspond to the loss of base pairing and to the separation of the strands; the T_m usually reflects the loss of half of the base pairs, and strand separation will take place only at a somewhat higher temperature.

Under typical conditions, the value of T_m for DNA depends primarily upon the salt concentration (see below) and the base composition, because G•C base pairs melt less readily than do A•T. The T_m value is observed to be linearly dependent upon the mole fraction of G•C base pairs, f_{GC} (Figure 5-4), and the T_m in 0.15 M salt may be predicted on just this basis:

$$T_m = (69.3 + 41 f_{GC}) \, °C \tag{5.1}$$

The width of the melting transition of a synthetic DNA with a unique regular nucleotide sequence is quite sharp: the fraction of double helix measured by the hyperchromicity can decrease from 0.75 to 0.25 over a temperature range of 1°C or less. In contrast, the widths of the melting transitions of natural DNAs, with nonregular sequences, measured in the same way are several times broader, because segments approximately 50 base pairs in length with different sequences melt at somewhat different temperatures. Even in the simplest case, the melting process is not readily characterized in terms of thermodynamic parameters, because it is a multi-state reaction, even at equilibrium (Figure 5-2).

Thermal denaturation of double-stranded DNA: effect of base stacking. M. Kohandel & B. Y. Ha (2006) *Phys. Rev. E* **73**, 011905.

Loop dynamics in DNA denaturation. A. Bar *et al.* (2007) *Phys. Rev. Lett.* **98**, 038103.

2. Denaturants

The bases of nucleic acids are relatively hydrophobic (Table 2-3) whereas the sugar–phosphate backbone is very polar. The bases are removed from contact with water within the double helix, and the hydrophobic effect is one of the forces stabilizing the double helix. Consequently, denaturants such as urea and guanidinium chloride that diminish the hydrophobic effect, as well as form hydrogen bonds with the bases (Section 11.1.C), tend to destabilize the double helix. The destabilizing effects of such reagents are independent of their ability to compete in hydrogen bonding but reasonably correlated with their actions in solubilizing the bases (Figure 5-5). **Urea** and **ethylene glycol** both bind preferentially to single-stranded DNA relative to double-stranded DNA and the bulk solution. Urea exhibits a stronger affinity for adenine and thymine bases, while that of ethylene glycol is practically independent of base composition. **Glycine**, **sarcosine** (*N*-methyl glycine) and **glycine betaine**:

$$CH_3-NH-CH_2-CO_2^- \qquad\qquad (CH_3)_3^+N-CH_2-CO_2^-$$

Sarcosine $\qquad\qquad\qquad\qquad$ Glycine betaine $\qquad\qquad$ (5.2)

do not bind preferentially to single strands relative to bulk solution but are excluded from the negatively charged surface of double-stranded DNA. Their preferential binding to single strands relative to double-stranded DNA increases with GC content, demonstrating a stronger affinity for guanine and cytosine bases. Increasing ionic strength diminishes the preferential binding of urea, glycine, sarcosine and glycine betaine, as a result of counterion release upon melting, which is dependent on the water activity and, hence, on co-solute concentration.

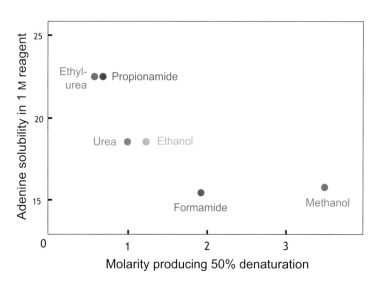

Figure 5-5. Correlation between the potency of six reagents in promoting DNA denaturation and their effects on the solubility of adenine. Data from L. Levine *et al.* (1963) *Biochemistry* **2**, 168.

Formamide ($H-CO-NH_2$) is used most frequently as a denaturant of nucleic acids, because it reduces the T_m of duplexes significantly at moderate concentrations, is readily miscible with water, has low optical absorbance in the UV region and does not react chemically with the DNA.

Some denaturants of nucleic acids, such as **formaldehyde** ($H-CO-H$), react chemically with the amino groups of the bases of the single-stranded form and prevent it regenerating the double helix. Formaldehyde can also produce single-strand breaks and cross-link nucleic acids to proteins (Figure 2-34).

Effect of ethylene glycol, urea, and N-methylated glycines on DNA thermal stability: the role of DNA base pair composition and hydration. L. J. Nordstrom *et al.* (2006) *Biochemistry* **45**, 9604–9614.

The denaturation transition of DNA in mixed solvents. B. Hammouda & D. Worcester (2006) *Biophys. J.* **91**, 2237–2242.

Effect of diethylsulfoxide on the thermal denaturation of DNA. S. A. Markarian *et al.* (2006) *Biopolymers* **82**, 1–5.

Enthalpies of DNA melting in the presence of osmolytes. C. H. Spink *et al.* (2007) *Biophys. Chem.* **126**, 176–185.

3. pH

The stability of standard nucleic acid duplexes is relatively insensitive to pH in the neutral pH region of 5–9 because none of the functional groups present in typical nucleic acids titrate in this region. Modified bases might, however, and the stability would then change. Below pH 5, or above pH 9, the standard duplex is destabilized because of titration of the polar groups on the bases (Figure 2-7). These groups are involved in the hydrogen bonding between pairs of bases, which is altered by their ionization, and the bases acquire a net charge in spite of being buried in the double helix. Low pH decreases the solubility, can cause depurination and strand breakage, and stabilizes triple helices consisting of pyrimidine–purine–pyrimidine and containing cytosine residues (Section 2.2.C.2). High pH, up to about 13, is more benign and is often used to denature nucleic acids, although depurination and strand breakage are still a hazard.

4. Salt Effects

The double helices of DNA and RNA are highly charged due to the ionized phosphate groups of the backbone. Consequently, these polyelectrolytes tend to bind cations to neutralize the charge density, and this condensation of counterions contributes to the conformation and stability (Section 2.3). Nonspecific 'territorial binding' or 'counterion condensation' occurs when the linear charge density of a polymer exceeds a critical value, and it produces an elevated cation concentration in the vicinity of the nucleic acid over that in the bulk solution. The local concentration of Na$^+$ ions on double-stranded DNA can be greater than 1 M, even in the absence of salt in the bulk solvent, and it reduces the effective charge on each phosphate group at pH 7 from –1 to –0.24. Both single- and double-stranded nucleic acids have such condensed counterions, although the linear charge density of the isolated single strands is less than that of the duplex, because the single strands are more free to elongate and diminish the charge density. Consequently, a duplex binds more cations that do the isolated strands. When the strands separate, the excess cations are released into the solution. Greater bulk concentrations of cations make this cation release increasingly unfavorable, so more energy is required to produce dissociation of the duplex, and it is stabilized by high salt concentrations. This effect depends upon the valence of the counterion, with higher valence counterions promoting greater stabilization than monovalent ions at the same concentration. Consequently **the stability of the double-stranded conformation depends markedly upon the ionic strength, and its melting temperature is increased as the ionic strength is increased**.

Anions tend not to interact directly with nucleic acids, due to charge repulsions, and they affect duplex stability only indirectly, according to the Hofmeister series.

The salt effects on the stabilities of natural nucleic acids are independent of their nucleotide sequence and content. The T_m is observed to vary linearly with the logarithm of the total salt concentration, C_s, and the slope is independent of base composition. Consequently, a wide variety of melting data on natural DNAs can be described by the empirical equation:

$$T_m = (16.6 \log[C_s/(1 + 0.7\, C_s)] + 41 f_{GC} + 81.5)\ °C \quad (5.3)$$

where f_{GC} is the fraction of bases that are guanine or cytosine.

Thermodynamic analysis of ion effects on the binding and conformational equilibria of proteins and nucleic acids: the roles of ion association or release, screening and ion effects on water activity. M. T. Record Jr *et al.* (1978) *Quart. Rev. Biophys.* **11**, 103–178.

The molecular theory of polyelectrolyte solutions with applications to the electrolyte properties of polynucleotides. G. S. Manning (1978) *Quart. Rev. Biophys.* **11**, 179–246.

---

5. Prediction of the T_m

The melting temperature of double-stranded DNA is directly proportional to its G+C content, i.e. the fraction of G•C and C•G base pairs (Figure 5-4). The melting temperature may be estimated at neutral pH in the presence of varying salt concentrations (indicated here as the concentration of sodium ion, [Na⁺]) and with varying concentrations of formamide (indicated as the volume %, $\%_f$) for any normal DNA molecule from its length in nucleotides, L, and the mole fraction of G•C and C•G base pairs (f_{GC}):

$$T_m = (81.5 + 16.6 \log([Na^+]/(1 + 0.7[Na^+])) + 41 f_{GC} - 500/L - 0.63\, \%_f)\ °C \quad (5.4)$$

Such equations are valid only for L > 50 base pairs; with shorter strands, the strands separate more readily, so T_m will depend upon the concentrations of the hybridizing strands. Furthermore, the salt effects depend upon the exact length of a short oligonucleotide and its base composition. Equations such as 5.4 will also not be valid if any other components are present that interact differently with the single strands and the duplex.

The stabilities of DNA double helices are independent of G+C content in 3.2 M $(CH_3)_4 N^+\ Cl^-$ or 2.4 M $(CH_3\text{–}CH_2)_4 N^+\ Cl^-$, where the melting temperatures are 94°C and 63°C, respectively. In 5.6 M glycine betaine (Equation 5.2), which has no net charge, the melting temperature is independent of G+C and can be varied from 75°C down to 45°C by varying the KCl concentration from 1 M down to 0.01 M.

Reliable data are available to make reasonably accurate predictions of the T_ms of duplexes involving only short oligonucleotides. They involve primarily the experimentally determined parameters for the various pairs of neighboring base pairs in stabilizing the double helix (Table 5-1), including the enthalpy and entropy changes that permit calculation of the free energy at any temperature, and the effects of base mismatches (Table 5-2). The stabilizing effects of the base pairs must overcome the opposing contribution of the entropy loss on annealing, estimated to be +1.96 kcal/mol at 37°C. A self-complementary oligonucleotide will form intramolecular double helices, and consequently will

Table 5-1. Thermodynamic parameters for DNA nearest-neighbor Watson–Crick base pairs in 1 M NaCl at 37°C

Sequence[a]	$\Delta H°$ (kcal/mol)	$\Delta S°$ (cal/mol/°C)	$\Delta G°$ (kcal/mol)
AA/TT	−7.6	−21.3	−1.00
AT/TA	−7.2	−20.4	−0.88
TA/AT	−7.2	−21.3	−0.58
CA/GT	−8.5	−22.7	−1.45
GT/CA	−8.4	−22.4	−1.44
CT/GA	−7.8	−21.0	−1.28
GA/CT	−8.2	−22.2	−1.30
CG/GC	−10.6	−27.2	−2.17
GC/CG	−9.8	−24.4	−2.24
GG/CC	−8.0	−19.9	−1.84

[a] The slash indicates that the sequences are given in antiparallel orientations. For example, AC/TG means 5′–AC–3′ is base-paired with 3′–TG–5′.

Data from J. SantaLucia & D. Hicks (2004) *Ann. Rev. Biophys. Biomol. Structure* **33**, 415–440.

bind less tightly to another strand, by +0.43 kcal/mol at 37°C. The data in Table 5-2 demonstrate that the order of decreasing stability of the base pairs is: G•C > A•T > G•G > G•T > G•A > T•T > A•A > T•C > A•C > C•C. It is frequently assumed that all mismatches have the same effect on the stability of a heteroduplex, and a common 'rule of thumb' is that 1% of base-pair mismatches decrease the melting temperature by 1°C. The experimental data of Tables 5-1 and 5-2 demonstrate, however, that the different mismatches have effects that vary as much as 2700 in the equilibrium constant for duplex formation and 4.9 kcal/mol in free energy. The two extremes are GGC/CGG and ACT/TCA. Guanine is the most promiscuous base, as it forms the strongest normal pair (G•C) and the strongest mismatches. In contrast, cytosine is the most discriminating base, as it forms the strongest normal pair and the three weakest mismatches.

The thermodynamics of DNA structural motifs. J. SantaLucia & D. Hicks (2004) *Ann. Rev. Biophys. Biomol. Structure* **33**, 415–440.

Table 5-2. Increments in ΔG° (kcal/mol) for single internal matches and mismatches adjacent to Watson–Crick base pairs in 1 M NaCl at 37°C

Sequence[a]	X	Y			
		A	C	G	T
GX/CY	A	0.17	0.81	−0.25	**−1.30**
	C	0.47	0.79	**−2.24**	0.62
	G	−0.52	**−1.84**	−1.11	0.08
	T	**−1.44**	0.98	−0.59	0.45
CX/GY	A	0.43	0.75	0.03	**−1.45**
	C	0.79	0.70	**−1.84**	0.62
	G	0.11	**−2.19**	−0.11	−0.47
	T	**−1.30**	0.40	−0.32	−0.12
AX/TY	A	0.61	0.88	0.14	**−0.88**
	C	0.77	1.33	**−1.30**	0.64
	G	0.02	**−1.30**	−0.13	0.71
	T	**−0.88**	0.73	0.07	0.69
TX/AY	A	0.69	0.92	0.42	**−0.58**
	C	1.33	1.05	**−1.30**	0.97
	G	0.74	**−1.44**	0.44	0.43
	T	**−1.00**	0.75	0.34	0.68

[a] The slash indicates that the sequences are given in antiparallel orientations. For example, AC/TG means 5′–AC–3′ is base-paired with 3′–TG–5′. Watson–Crick pairs are in *bold*.

Data from J. SantaLucia & D. Hicks (2004) *Ann. Rev. Biophys. Biomol. Structure* **33**, 415–440

5.1.C. Double-stranded RNA

Double-stranded RNA is rare in nature but its stability can actually be somewhat greater than that of DNA of the same base composition (but thymine replacing uracil). It adopts an A-form helix rather than the B-form of DNA (Section 2.2.B). The dependence on base composition and formamide concentration is somewhat different, and the equation corresponding to Equation 5.4 is:

$$T_m = (78 + 16.6 \log([Na^+]/(1 + 0.7[Na^+])) + 70 f_{GC} - 500/L - 0.35 \%_f) \,°C \qquad (5.5)$$

5.1.D. DNA•RNA Heteroduplexes

DNA•RNA heteroduplexes also adopt the A-type double helix, like RNA•RNA double helices. They are usually less stable than either homoduplex, except if they have high G+C content; their T_m may be estimated from the equation:

$$T_m = (67 + 16.6 \log[[Na^+]/(1 + 0.7[Na^+])] + 80 f_{GC} - 500/L - 1.0 \%_{f1} - 0.3 \%_{f2}) \,°C \qquad (5.6)$$

where $\%_{f1}$ is the volume percentage of formamide up to 20%, and $\%_{f2}$ is any additional volume percentage. DNA•RNA heteroduplexes are less sensitive to formamide than are the others.

Discovery of the hybrid helix and the first DNA–RNA hybridization. A. Rich (2006) *J. Biol. Chem.* **281**, 7693–7696.

5.1.E. Single-stranded Nucleic Acids

The structures of single-stranded nucleic acids depend crucially upon their nucleotide sequences, in particular their potential for complementary base pairing. Sequences with the potential for substantial base pairing will adopt compact folded structures, like those of transfer RNA and ribozymes (Section 4.2). Other single-stranded RNA or DNA molecules with no significant ability for base pairing between complementary segments will maintain a more unfolded, random coil-like structure. Long nucleic acid polymers with nonuniform sequences, however, invariably have segments with complementary sequences, simply by chance. These single-stranded molecules will consequently form double-stranded hairpins and loops, like those depicted in Figure 5-6. These structures will also unfold at high temperatures, in the presence of denaturants, or at extremes of pH.

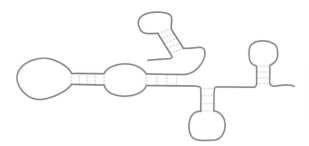

Figure 5-6. Intramolecular base-paired structures formed by nucleic acid single strands. The *long continuous line* is the backbone of the single strand, whereas the *dashed lines* represent complementary base pairs.

Even in the absence of sequence complementarity and base pairing, a wide variety of data indicate that single polynucleotide chains are not random coils (Section 1.3.A) at room temperature. Only at very high temperatures, approaching 100°C, do model polynucleotides approach the properties expected for a polynucleotide random coil. At lower temperatures, the adjacent bases tend to stack and the polynucleotide backbone tends to adopt a helical conformation like it would have in a double helix, but without the other strand. The most vivid illustration of such a structure is that determined by fiber diffraction of polycytosine (poly C). Although superficially similar to one strand of a double helix, the conformation of poly C has only six bases per turn, instead of the 10 or 11 for double helices. Similar structures in other single-stranded nucleic acids are apparent from their circular dichroism spectra.

In solution, each nucleotide of a single polypeptide chain appears to be in equilibrium between a random coil conformation and one in which its base is stacked with adjacent bases (Table 2-5). Such stacking is only slightly cooperative, in contrast to the helix–coil transition in polypeptides (Section 8.4); the equivalent of the cooperativity parameter σ in the Zimm–Bragg model for α-helices (Section 8.4.B.1) is only slightly less than 1 with polynucleotides, whereas it is 10^{-2}–10^{-4} with polypeptide chains. The degree of stacking increases with length of the polynucleotide chain, but only up to about 10 nucleotides; beyond that, the stacking is essentially independent of length. The equilibrium between stacked and flexible conformations shifts with temperature, with flexible conformations predominating near 100°C and stacked conformations predominating only at temperatures lower than 0°C.

RNA denaturation: excluded volume, pseudoknots, and transition scenarios. M. Baliesi *et al.* (2003) *Phys. Rev. Lett.* **91**, 198102.

Random coil phosphorus chemical shift of deoxyribonucleic acids. C. N. Cho & S. L. Lam (2004) *J. Magn. Res.* **171**, 193–200.

Heat capacity changes in RNA folding: application of perturbation theory to hammerhead ribozyme cold denaturation. P. J. Mikulecky & A. L. Feig (2004) *Nucleic Acids Res.* **32**, 3967–3976.

---

1. Physical Stretching

If DNA is stretched physically, it undergoes a large, cooperative structural change at forces greater that 70 picoNewtons (pN), when it stretches to about 1.7 times its crystallographic length. If only the 3′-ends are pulled, the double helix is believed to unwind to a structure resembling a ladder. If only the 5′-ends are pulled, the helix persists but with a large inclination of the base pairs, a narrowing of the minor groove and a thinning of the double helix by roughly 30%. When the molecule is stretched to double its original length, the base pairs rupture and the two strands separate.

5.2. Unfolding and Refolding of Single-stranded RNA Molecules

The unfolding of folded single-stranded RNA molecules, such as transfer RNA and ribozymes (Section 4.2), differs from the denaturation of double helices described above in that it is a solely unimolecular process and does not result in separation of individual polynucleotide chains. It is similar, on the other hand, in that antiparallel double strands, formed by hairpins in this case, are major parts of the structure and can be unfolded in the same way. RNA molecules typically fold across a highly rugged energy landscape, with many energy peaks and minima, and correspondingly there are long-lived folding intermediates, multiple folding pathways and heterogeneous conformational dynamics. Stable secondary structures are believed to be at least partly responsible for the rugged energy landscape of RNA molecules.

Folded RNA molecules can be unfolded and will then refold spontaneously, so the information for their structures resides in their nucleotide sequences. Unfolding can be induced by a gradual change of temperature or a gradual change in the concentrations of the stabilizing Mg^{2+} ions or a denaturant like urea. Unfolding is not very cooperative in the case of RNAs, and intermediates with various structures predominate throughout the transition region between the fully folded and unfolded states. These are equilibrium intermediates, which should have the structures with the lowest free energies under the intermediate conditions. They are readily interconverted and so cannot be assumed to be the intermediates that occur during the kinetic process of unfolding or refolding. In general, the least denaturing conditions cause the tertiary structure to unfold, followed by the various elements of the secondary structure in order of their intrinsic stabilities. Unfolding of tertiary and secondary structures can usually be distinguished because the stability of only the tertiary structure is dependent upon the presence of Mg^{2+} ions. The intermediates usually correspond to isolated double helical hairpins, as they have significant stabilities in isolation (Table 4-2) without being incorporated into higher order structures. A single base pair can stabilize a hairpin by 1–3 kcal/mol, and the secondary structure of a group I self-splicing intron, with 129 base pairs, has been estimated to be 84.7 kcal/mol more stable than the fully unfolded molecule. This is the reason for the much lower cooperativity of unfolding transitions of RNA relative to proteins, where the secondary structure is not stable by itself (Section 11.1).

The equilibrium thermal unfolding of a pseudoknot is illustrated in Figure 5-7. The folded pseudoknot contains an unpaired base, A15, between the two helices and a bulge, A35, in the lower helix. Because of the destabilizing effect of this bulge, the lower helix melts first, and the upper helix is then extended in the intermediate helix by pairing A15 with a previously inaccessible uracil base. At higher temperatures, the more stable helix unfolds.

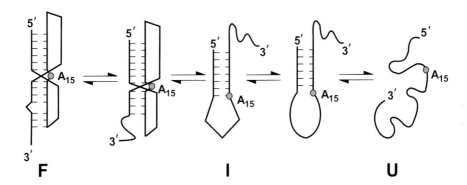

Figure 5-7. Thermal unfolding pathway at equilibrium of a pseudoknot. The *thick line* is the polynucleotide backbone, while the *short thin lines* are bases that are base-paired in the helices. F is the fully folded molecule at low temperatures, I the predominant intermediate, and U the unfolded form at high temperatures. The three intermediates were populated to detectable extents at intermediate temperatures. Nucleotide A15 is unpaired in the folded structure but becomes base-paired in the intermediate after unfolding of the lower helix. Data from C. A. Theimer & D. P. Giedroc (1999) *J. Mol. Biol.* **289**, 1283–1299.

In kinetic studies, the change in structure is caused by rapidly changing the conditions, and the course of the change is followed as a function of time in the new conditions. In this case, the kinetic pathway need not be reversible, and intermediates do not necessarily correspond to low free energy states. RNA molecules are observed to get trapped in metastable states from which it is difficult to escape, reflecting the ruggedness of the energy landscape for a large RNA molecule, with many local energy minima. The folding process occurs over a wide range of time scales: adjacent bases can stack in nanoseconds, individual helices can form in milliseconds, larger secondary structural domains that involve reorganization of certain helices can require seconds, while formation of the fully folded structure can take minutes. The local secondary structure, such as a hairpin loop, forms first, as the two halves of a helix are in close proximity in the unfolded polynucleotide chain. Longer range helices will form more slowly and may only form after previous folding of intermediary short-range helices brings together the two distant halves of the long-range helix. Progressively larger secondary structure domains are generated, and larger domains can form by rearranging and combining some of the smaller elements, but these involve substantial energy barriers between structures, as some helices have to be broken up before other more stable ones can be added (Figure 5-8). As the sizes of domains increase, the height of the energy barrier increases, as does the time required. In general, the transition state occurs early in forming the secondary structure, i.e. the transition state is close to the unfolded polynucleotide, whereas it occurs late in forming the tertiary structure: this latter transition state is close to the fully folded state. Finally, the structure can become frozen in one that does not have the lowest possible free energy but has resulted from the kinetic pathway. Many folded RNA molecules exist in alternative folded conformations. The larger the structure, the greater the discrepancy with the predicted structure of minimum free energy.

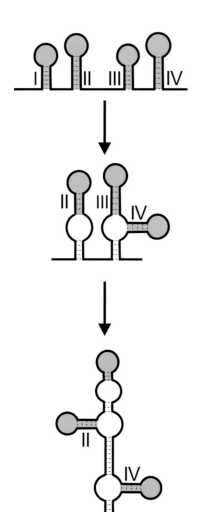

Figure 5-8. Schematic representation of the reorganization of the secondary structure during RNA folding that leads to the formation of progressively larger domains, with base pairing between segments more distant in the sequence. The *thick line* is the polynucleotide backbone, while the *dashed lines* between them represent base pairs in the stem. The *colored* stems have the helices made of segments adjacent in the sequence.

Strategies for RNA folding and assembly. R. Schroeder *et al.* (2004) *Nature Rev. Mol. Cell. Biol.* **5**, 908–919.

RNA folding and unfolding. B. Onoa & I. Tinoco (2004) *Curr. Opinion Struct. Biol.* **14**, 374–379.

Mg^{2+}–RNA interaction free energies and their relationship to the folding of RNA tertiary structures. D. Grilley et al. (2006) *Proc. Natl. Acad. Sci. USA* **103**, 14003–14008.

Concordant exploration of the kinetics of RNA folding from global and local perspectives. L. W. Kwok et al. (2006) *J. Mol. Biol.* **355**, 282–293.

RNA folding during transcription. T. Pan & T. Sosnick (2006) *Ann. Rev. Biophys. Biomol. Struct.* **35**, 161–175.

5.2.A. Transfer RNA Unfolding/Refolding

The unfolding and refolding of transfer RNA molecules has been characterized extensively. The process depends critically upon the salt concentration, particularly the presence of Mg^{2+} ions, which are bound at a few specific locations in the folded tertiary structure; free Mg^{2+} ions stabilize the structure. At high Mg^{2+} concentrations, the tertiary structure is disrupted only at high temperatures, and the remainder of the structure, the secondary structure of double-helical stems, unfolds rapidly thereafter; consequently, **unfolding at high Mg^{2+} concentrations appears to be a single-step, all-or-nothing, cooperative process**. At lower Mg^{2+} concentrations, however, the tertiary structure unfolds much more readily; under such less severe conditions, the various elements of the secondary structure, which include the double-helical stems, but with mismatches, loops and bulges, are sufficiently stable to remain folded, at least transiently. These elements of secondary structure then unfold relatively independently of each other.

Refolding occurs by the reverse of the unfolding process. The individual elements of secondary structure form, and then they interact via unpaired bases to generate the folded tertiary structure. The D-helix is formed in the overall rate-limiting step, so it serves as a nucleus for further folding, which is much faster. It also provides a scaffold for strong Mg^{2+} binding sites. The D-helix is involved in several tertiary interactions with the remainder of the molecule (Figure 4-17). The loop conformations and their interactions can change upon forming the tertiary structure, and a single base pair may change, but the helices do not. In these cases, the secondary structure in the tertiary structure is that which is most intrinsically stable in isolation, and the interactions stabilizing the tertiary structure are significantly weaker. In other cases, however, particularly ribozymes (Section 4.2.C), the secondary structure formed initially is not the same as that in the final tertiary structure, so **rearrangements of the secondary structure can take place during the tertiary folding process**. These can be substantial: in one case, six base pairs are broken and six new ones are formed, a tetraloop is disrupted, and a single-nucleotide bulge is introduced. The rearranged secondary structure is less intrinsically stable than that formed initially, but this is compensated for by the interactions stabilizing the tertiary structure.

Internally mismatched RNA: pH and solvent dependence of the thermal unfolding of tRNA[Ala] acceptor stem microhairpins. E. Biala & P. Strazewski (2002) *J. Am. Chem. Soc.* **124**, 3540–3545.

Study of the influence of metal ions on tRNA[Phe] thermal unfolding equilibria by UV spectroscopy and multivariate curve resolution. M. Vives et al. (2002) *J. Inorg. Biochem.* **89**, 115–122.

Highly conserved modified nucleosides influence Mg^{2+}-dependent tRNA folding. K. N. Nobles et al. (2002) *Nucleic Acids Res.* **30**, 4751–4760.

5.2.B. Ribozyme Unfolding/Refolding

The kinetic folding pathway of the *Tetrahymena* ribozyme (Figure 5-9) includes both Mg^{2+}-dependent and Mg^{2+}-independent steps. The structures of the intermediates on the folding pathway were determined by rapidly hybridizing their single-stranded segments with complementary oligodeoxyribonucleotides, then digesting them with **ribonuclease H**, which is specific for DNA•RNA heteroduplexes. Only segments that were not base paired survive this treatment intact. Both the hybridization and cleavage steps were more rapid than the rates of the folding steps being analyzed. One of the two main structural domains, P4–P6, forms first, and the overall rate-limiting step is a Mg^{2+}-dependent rearrangement preceding stable formation of the second main structural domain, P3–P7. The P3 and P7 helices form in an interdependent manner, and the two domains can be considered kinetic folding units. The rate-limiting step is formation of the triple-helical scaffold, involving strands entering and leaving the shallow groove of P6 and the deep groove of P4. **The intermediates that accumulate kinetically during folding possess some but not all of the fully folded structure**. In particular they have most of the secondary structure and at least some local cooperativity.

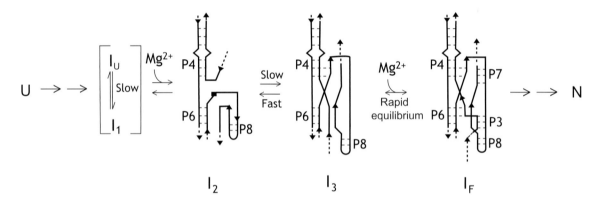

Figure 5-9. Kinetic folding pathway elucidated for the *Tetrahymena* ribozyme. The rate-limiting step in folding is between intermediates I_2 and I_3 and corresponds to the formation of the triple-helical scaffold at the junction between the stacked helices P4 and P6. The two main structural domains of these ribozymes are P4–P6 and P8–P3–P7. Only some of the steps involve binding of magnesium ions. Data from P. P. Zarrinkar & J. R. Williamson (1996) *Nature Struct. Biol.* **3**, 432–438.

Distinct parallel pathways lead the RNA to its native form. The structures of the intermediates populating the pathways are not affected by variation of the concentration and type of background monovalent ions, but they are altered by a mutation that destabilizes one domain of the ribozyme. Starting from different conformational ensembles, but folding under identical conditions, indicated that **the electrostatic environment modulates the molecular flux through different pathways, but the initial conformational ensemble determines the partitioning of the flux**. Misfolding occurs frequently, and kinetically trapped nonproductive intermediates occur. One quasi-native intermediate is only slowly converted to the correctly folded structure.

The folding process of a 115-nucleotide sequence known as SV-11 demonstrates the occurrence of a metastable intermediate (Figure 5-10). The sequence of this RNA is almost a palindrome, and the most stable structure is a single, almost perfect, hairpin. Folding proceeds, however, through a metastable structure involving only local hairpins. The local hairpins form much more readily and rapidly than the single double helix of the hairpin, which would require bringing together the opposite ends of the molecule. The nucleotides that make up the loop in the final hairpin are not involved in such a loop in the metastable intermediate.

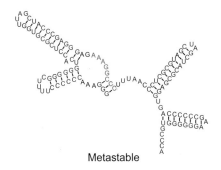

Figure 5-10. The stable and kinetic metastable structures of one strand of the double-stranded RNA molecule SV-11. Its sequence is a nearly perfect palindrome, and the most stable structure is a single, almost perfect, hairpin. The nucleotides in the loop of this structure, –AAAGGCCC– (*red*), have a different structure in the metastable structure. Data from C. K. Biebricher & R. Luce (1992) *EMBO J.* **11**, 5129–5139.

Perturbation of the hierarchical folding of a large RNA by the destabilization of its scaffold's tertiary structure. I. Shcherbadova & M. Brenowitz (2005) *J. Mol. Biol.* **354**, 483–496.

The paradoxical behavior of a highly structured misfolded intermediate in RNA folding. R. Russell *et al.* (2006) *J. Mol. Biol.* **363**, 530–544.

Distinct contribution of electrostatics, initial conformational ensemble, and macromolecular stability in RNA folding. A. Laederach *et al.* (2007) *Proc. Natl. Acad. Sci. USA* **104**, 7045–7050.

5.2.C. Unfolding Using Mechanical Force

Unfolding by pulling the two ends of an RNA molecule apart is simpler than thermal unfolding, because the helices tend to unfold by breaking the base pairs at the end furthest from the loop and closest to the ends being pulled. The folding-energy landscape can be manipulated to control the fate of an RNA molecule: individual RNA molecules can be induced into either native or misfolding pathways by modulating the relaxation rate of the applied force, and they can even be redirected during the folding process to switch from misfolding to native folding pathways.

Mechanical unfolding and refolding of single RNA molecules were originally studied using optical traps attached to the ends of the polynucleotide chain to change suddenly the extension of the molecule. Two methods were traditionally used: 'force-ramp', with the applied force changing continuously, and 'hopping', when the force is held constant and the molecule jumps spontaneously between two different states. The rates of unfolding and refolding are measured directly, but only over very narrow ranges of forces. With a third, 'force-jump', method, the applied force is rapidly stepped to a new value and either the unfolding or refolding event is monitored through changes in the molecular extension. This makes it possible to measure the unfolding and refolding rates independently over a wider range of forces. A 52-nucleotide RNA hairpin with a three-nucleotide bulge exhibits very similar unfolding kinetics and estimated Gibbs free energies when studied using all three methods. The RNA unfolds and refolds without detectable intermediates in constant-force conditions (hopping and force-jump) but it exhibits partially folded intermediates in force-ramp experiments at higher unloading rates. Consequently, folding of even simple RNA hairpins can be more complex than a simple single-step reaction.

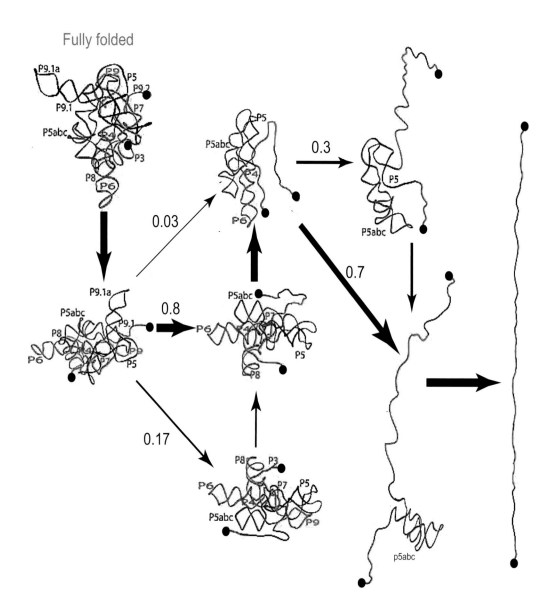

Figure 5-11. Pathway of unfolding of the *Tetrahymena* ribozyme when the two ends are pulled apart mechanically. The flux through the pathway is indicated by the thicknesses of the arrows; the numbers associated with some arrows indicate the probability of that step being followed. Data from C. A. Bustamante (2005) *Quart. Rev. Biophys.* **38**, 291–301.

The *Tetrahymena* ribozyme, with about 400 nucleotides, unfolds through eight intermediates (Figure 5-11). The pathway of unfolding differs for each measurement, presumably because the thermal fluctuations responsible for crossing the kinetic barriers are stochastic. The intermediates were identified by studying mutant forms and subdomains of the RNA, and by adding complementary oligonucleotides that compete with secondary and tertiary interactions by hybridizing to single-stranded segments. The kinetic barriers are caused primarily by tertiary interactions stabilized by Mg^{2+} ions. In the absence of Mg^{2+}, the kinetic barriers disappear and no intermediates are detected.

Single-molecule RNA folding. G. Bokinsky & X. Zhuang (2005) *Acc. Chem. Res.* **38**, 566–573.

Determination of thermodynamics and kinetics of RNA reactions by force. I. Tinoco *et al.* (2006) *Quart. Rev. Biophys.* **39**, 325–360.

Probing the mechanical folding kinetics of TAR RNA by hopping, force-jump, and force-ramp methods. P. T. Li *et al.* (2006) *Biophys. J.* **90**, 250–260.

Modelling RNA folding under mechanical tension. J. R. Vieregg & I. Tinoco (2006) *Mol. Phys.* **104**, 1343–1352.

Force unfolding kinetics of RNA using optical tweezers. I. Effects of experimental variables on measured results. J. D. Wen *et al.* (2007) *Biophys. J.* **92**, 2996–3009.

Real-time control of the energy landscape by force directs the folding of RNA molecules. P. T. Li *et al.* (2007) *Proc. Natl. Acad. Sci. USA* **104**, 7039–7044.

5.3. RENATURATION, ANNEALING AND HYBRIDIZATION

Two nucleic acid strands with complementary base sequences can associate to generate a double helix under the appropriate conditions. This can occur after denaturation of a sample of double-stranded nucleic acid, known as renaturation, or upon mixing two single strands. Another variation is adding a single-stranded molecule, often a short oligonucleotide, with a sequence complementary to one part of the target molecule to a mixture of denatured nucleic acid, to compete with reassociation of the original strands.

Base pairing between complementary strands can be either a relatively straightforward or a very complex process, depending upon the length and sequences of the polynucleotide chains involved. For short oligonucleotides with totally complementary sequences, the reassociation process is fast and complete. When a short double-stranded oligonucleotide is denatured and then re-annealed under nondenaturing conditions, the optical and hydrodynamic properties return quickly to their original values. The rate is determined by the concentrations and frequency of encounter of the complementary strands; once the first base pairs are formed, the pairing of the remaining bases is rapid. For longer chains, however, the process is much more complex, slower and does not necessarily go to completion, because a variety of structures and complexes can be generated initially (Figure 5-6) and hinder the search for the correct complementary partner.

The denaturation of synthetic polymers with repeating sequences is fully reversible when measured by the change in UV absorbance, which indicates that most bases are in intact base pairs in the regenerated double helix. There is hysteresis, however, in the loss and regain of the hydrodynamic properties, in that renaturation does not follow the reverse of the pathway of denaturation; this is due to the formation of branched structures comprising hairpin loops and concatenated aggregates during renaturation. Very long natural DNAs display hysteresis in both their optical and hydrodynamic properties, indicating a failure of a significant fraction of the bases to find their correct partner base and the presence of nonlinear structures.

Before they can base pair, two complementary strands must encounter each other by diffusion, but the presence of other segments with partially complementary sequences will compete. When two long nucleic acid polymers encounter each other, there is only a very small probability of an exact match of many of the base pairs in the initial complex. Instead, small stretches of complementary duplex are formed initially:

(5.7)

The two single strands encounter each other in a bimolecular reaction, so the rate is proportional to their concentrations, and just a few adjacent bases form base pairs. These complexes may dissociate, as they are relatively unstable, or they may grow if the neighboring residues are also complementary:

(5.8)

Zippering of the remainder of the double helix can be very rapid, as it is essentially a unimolecular process. If the remainder of the sequences are not complementary, the initial duplexes formed must dissociate and wait for more productive encounters; this depends upon the temperature being sufficiently high to melt short sequences, but low enough to permit duplex formation. Otherwise, at low temperatures, the unproductive complexes will persist and the most stable base pairing will not be found.

Intramolecular hairpins and internal loops present in the single-stranded nucleic acids (Figure 5-6) present a significant impediment to annealing to another strand. These structures are formed intramolecularly, and thus can be generated much more rapidly than can duplex structures. The first-order rate constant for hairpin formation is 10^5–10^7 s^{-1}, so such structures are formed on the microsecond time scale. The second-order rate constant for oligonucleotide duplex formation is 10^5–10^7 M^{-1} s^{-1}; at typical concentrations used in such experiments, duplexes form generally no faster than in seconds, and often much slower. The competing structures must be unfolded for the annealing process to proceed towards formation of the complementary duplex. The activation energy for such unfolding is large, so they do not dissociate rapidly at temperatures much below the T_m.

Within the double-stranded segments formed in the initial duplexes, defects can occur in the alignment of the base sequences (Figure 5-12), including hairpins and internal loops, as well as mismatches and bulges. Here the activation energy to change them is even greater than for single-stranded structures. These regions may form nucleation complexes that cannot be extended, resulting in large numbers of nonproductive intermediate states. In highly repetitive sequences, large numbers of such structures may be formed initially, but in this case they are readily interconvertible and the proper sequence alignment can be achieved.

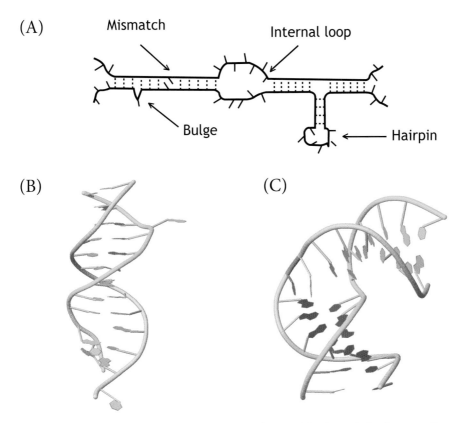

Figure 5-12. Structures of misaligned sequences in nucleic acid duplexes. (A) Schematic illustration of the types of structures formed. The *long continuous line* is the polynucleotide backbone, whereas the *short lines* indicate the bases, most of which are paired in complementary base pairs. (B, C) Crystallographic structures of (B) a single-nucleotide bulge at the *top* and a mismatch at the *bottom*, and (C) a four-nucleotide bulge. (B) and (C) generated from PDB files *1d31* and *1jst* using the program Jmol.

A common strategy to renature nucleic acids efficiently is to incubate them at the T_m of the desired duplex, followed by slow cooling to permit the facile reorganization of the duplex regions, to optimize their sequence alignment. The T_m of the desired duplex may be estimated from the empirical Equations 5.3–5.6. The T_m can be adjusted by manipulating the salt concentration and the amount of formamide; a given T_m can be obtained by many different combinations of the two.

With a short oligonucleotide that is complementary to one segment of a much longer single-stranded nucleic acid, a duplex can form without appreciable competing structures. The process is still hindered, however, by any hairpin structures in the long polynucleotide (Figure 5-6). The two processes differ in that duplex formation is bimolecular, while hairpin formation is unimolecular, so they can be distinguished experimentally by the concentration dependence. The hairpin structure will be favored at low concentrations of the strands, the duplex at high concentrations. The salt concentration also affects these equilibria; the hairpin is favored at low salt concentrations due to its lower charge density. Therefore a combination of salt and oligonucleotide concentrations can be used to generate the desired structure.

Formation of a duplex involves two nucleic acid strands, so their concentrations and lengths also influence the stability of the duplex. There is a complex interplay between length and concentration in determining the stability. The initial nucleation event (Equation 5.7), in which the strands make initial productive contacts, is a bimolecular process that depends upon the strand concentrations. Subsequent base pairing is propagated (Equation 5.8) in a zippering process, in which the nucleated base pairing is extended, which is a pseudo-unimolecular process and occurs with a rate constant of 10^7 s^{-1}. Formally, more than one strand is involved, but the formation of a closed base pair adjacent to a preformed base pair is effectively a monomolecular process analogous to extension of a hairpin.

For short oligonucleotides, the nucleation event dominates complex formation. The concentration dependence of the stability is simple and described adequately by conventional mass action for a biomolecular reaction. Initial encounters probably involve only one or two adjacent base pairs, but this *encounter complex* will be very unstable and dissociate rapidly, unless the adjacent nucleotides can form base pairs as well. If the two sequences are sufficiently complementary, the duplex will extend by a zippering reaction. The rate-limiting step is believed to be formation of the second or third base pair. The apparent rate constant is approximated by:

$$k_2 = k'_N (L_S)^{1/2} / N \tag{5.9}$$

where L_s is the length of the shortest strand participating in duplex formation, N is the total number of base pairs present in nonrepeating sequences, and k'_N is the rate constant for forming the initial nucleus. The value of k'_N is approximately 4×10^5 M^{-1} s^{-1} under optimal conditions. The dependence of k_2 on the square root of L_s rather than L_s is due to the nucleation sites becoming less available as the length of the single strand increases. The value of k_2 is zero at temperatures at or above the melting temperature, because the nucleation complex is not stable at high temperatures but increases at lower temperatures, until it reaches a broad maximum at a temperature about 25°C lower than the melting temperature (Figure 5-13). The rate decreases at lower temperatures, gradually for homopolymers with simple sequences but drastically for normal nucleic acids, due to intramolecular base pairing and the consequent decrease in availability of nucleation sites. The rate of hybridization is maximal at high salt concentrations and gradually decreases at monovalent cation concentrations less than 1 M. Mismatched base pairs affect the stability of the final duplex, but not the rate of forming it, so long as fewer than 10% of the base pairs are mismatched, consistent with the rate-limiting nucleation event involving only about two standard base pairs. It must be remembered that a unique sequence in a large genome (such as the 3 billion base pairs in the human genome) will be present at an extremely low concentration, so hybridization reactions with nucleic acids from an entire genome may reach equilibrium only extremely slowly.

At optimal temperatures, k'_N is about 20–50% smaller for DNA•RNA hybrid formation than for DNA duplexes, depending upon the degree of RNA secondary structure. The rates of forming RNA•RNA duplexes are similarly reduced.

For very long complexes, nucleation accounts for a relatively small part of the total stabilizing free energy, and the concentration dependence of complex formation disappears for long chains. For complexes of intermediate length, both nucleation and propagation make significant contributions to complex formation and stability; the concentration dependence of their stability is reduced relative to that for short complexes of short oligonucleotides, but it does not vanish.

The stability of a long duplex depends upon the fraction of G•C and C•G base pairs (Figure 5-4) but appears to be independent of the actual sequence (Equations 5.3–5.6). This is an illusion, however, that applies only to long duplexes where sequence effects are averaged out over the polymer. For short oligonucleotides, the actual sequence must be taken into account.

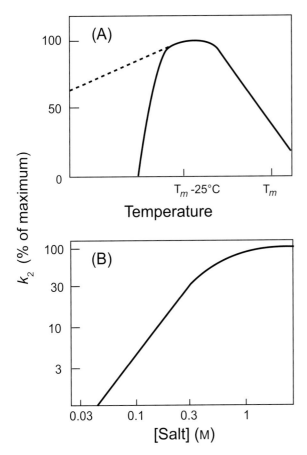

Figure 5-13. Dependence of the rate of hybridization of two complementary nucleic acid strands, k_2, on the temperature (A) and concentration of a monovalent cation (B). The melting temperature of the resulting duplex is T_m. The *solid curves* are those observed with complex natural nucleic acids with very diverse nucleotide sequences, whereas the *dashed curve* in (A) is observed with homopolymers with simple sequences that do not form intramolecular base pairs.

The power of hybridization techniques comes from the dependence of the stability of the hybrid duplex on the degree of complementarity of the constituent strands. Hybridization of nucleic acid strands of different origins produces heteroduplexes *in vitro*, and their stability is a measure of their degree of sequence similarity. The greater the degree of sequence complementarity, the greater the stability of the heteroduplex, so measurement of its stability, by varying the stringency, can give an indication of the degree of sequence identity. When 1% of the bases are mismatched in long strands, the T_m is generally decreased by approximately 1°C, although in short strands this depends upon the nature of the mismatches and their neighboring bases (Table 5-2). There are occasions when the exact sequence of the target DNA or RNA is not known, for example when the sequence of a gene is inferred from the amino acid sequence of its protein product (Figure 7-21). In this case, the base inosine can be used at the uncertain positions in the probe, as it has less stringent base-pairing criteria than adenine, cytosine, guanine and thymine.

Two single-stranded nucleic acid polymers with at least partial complementarity will form duplexes with stretches of double helix, corresponding to the regions of nucleotide sequence complementarity, interleaved with unpaired stretches, corresponding to noncomplementary regions. Two types of singe-stranded loops are introduced by noncomplementarity in the heteroduplex:

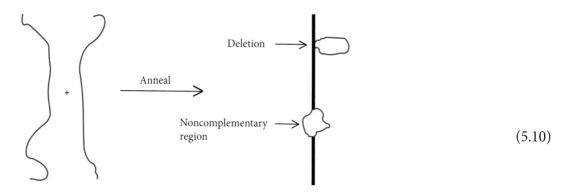

(5.10)

Regions where the bases do not match will produce a loop of nonbase-paired segments in both strands. The smallest loops and those with the same number of nucleotides on each strand are the most stable. On the other hand, the presence of extra nucleotides in only one of the strands, or a deletion in the other, will produce a loop in only that strand.

The thermodynamics of DNA structural motifs. J. SantaLucia (2004) *Ann. Rev. Biophys. Biomolec. Structure* **33**, 415–440.

Statistical thermodynamics and kinetics of DNA multiplex hybridization reactions. M. T. Horne *et al.* (2006) **91**, 4133–4153.

Secondary structure effects on DNA hybridization kinetics: a solution versus surface comparison. Y. Gao *et al.* (2006) *Nucleic Acids Res.* **34**, 3370–3377.

Electrostatic free energy landscapes for nucleic acid helix assembly. Z. J. Tan & S. J. Chen (2006) *Nucleic Acids Res.* **34**, 6629–6639.

Influence of secondary structure on kinetics and reaction mechanism of DNA hybridization. C. Chen *et al.* (2007) *Nucleic Acids Res.* **35**, 2875–2884.

5.3.A. Competing Intramolecular Structures in Individual Single Strands

Formation of a duplex between two single-stranded DNA molecules will be in competition with formation of any intramolecular structures. The most likely intramolecular structures are hairpins, which will be formed whenever the nucleotide sequence has segments with inverted complementarity. They can be formed very rapidly, much more rapidly than the two-stranded duplex under most conditions, so the intramolecular structure will form first and must then be undone in order to generate the duplex, which will be more stable if the two strands are truly complementary. For example, the single-stranded 13-mer 5′-d(GCTACGTAGTCGC) folds to form a hairpin structure with a highly stabilizing central c**GNA**g loop (*bold*) that accommodates a CT mismatched base pair within the double-stranded stem region (*italics*):

$$T\genfrac{}{}{0pt}{}{\diagup A - G - T - C - G - C}{\diagdown G - C - A - T - C - G} \quad (5.11)$$

This loop is especially stable because the guanine and adenine bases form a 'sheared' base-pair conformation stabilized by two hydrogen bonds, and it is stabilized by the base-paired stem as well, so the hairpin of loop and stem is a cooperative structure. The hairpin is in equilibrium with the double-stranded duplex form, which is slow on the nuclear magnetic resonance (NMR) time scale (1–2 s^{-1} at 35°C). The hairpin folds very rapidly (>500 s^{-1}), much more rapidly than the duplex is formed, so it produces a 'kinetic overshoot' that is followed by a much slower equilibration to the equilibrium mixture of hairpin and double-helical duplex, with a rate constant of approximately 0.13 s^{-1} at 25°C at the concentrations used.

Melting of a DNA hairpin is not a cooperative two-state process but involves numerous intermediate structures that are compact, with bases stacked but not base-paired. The melting of such stacking interactions occurs on a time scale of 700 ps to 2 ns, while dissociation of base pairs in the helical segment requires microseconds or longer.

The initial step of DNA hairpin folding: a kinetic analysis using fluorescence correlation spectroscopy. J. Kim *et al.* (2006) *Nucleic Acids Res.* **34**, 2516–2527.

Direct measurement of the full, sequence-dependent folding landscape of a nucleic acid. M. T. Woodside *et al.* (2006) *Science* **314**, 1001–1004.

Structure and folding dynamics of a DNA hairpin with a stabilising d(GNA) trinucleotide loop: influence of base pair mis-matches and point mutations on conformational equilibria. G. D. Balkwill *et al.* (2007) *Org. Biomol. Chem.***5**, 832–839.

DNA folding and melting observed in real time redefine the energy landscape. H. Ma *et al.* (2007) *Proc. Natl. Acad. Sci. USA* **104**, 712–716.

5.3.B. C_0t and R_0t Curves

Monitoring the rate of reassociation of double-stranded DNA is a classic method for evaluating the size and relative sequence complexity of genomic DNA. The duplex DNA of the entire genome is fragmented to moderate-sized pieces, approximately 400 nucleotides long, by physical shearing or cleavage by endonucleases, and the fragments are denatured thermally to produce single-stranded fragments. The reassociation of these fragments is then monitored as a function of time. Being a second-order reaction, the extent to which two complementary DNA strands reassociate should be proportional to C_0t, where C_0 is the initial concentration of nucleotides (not of strands) and t is the time; the degree of reassociation is plotted versus C_0t (Figure 5-14).

If the sequences of all the initial DNA fragments were identical, complementary fragments would encounter each other readily and the mixture would reanneal rapidly, at a low value of C_0t. At the other extreme, reannealing would be very slow and occur only at high C_0t values if the original DNA contained many different sequences; in this case, the concentrations of complementary strands are very low and they encounter each other only infrequently. For a given quantity of DNA, a small genome will have greater concentrations of the same segment than a large genome. The value of C_0t at which the fractional extent of reassociation is 0.5 is used as a measure of the complexity of the genome. One complication with randomly fragmented DNA, with different breakage points, is that complementary segments may hybridize but have sequences at one or both ends that are not present in the other.

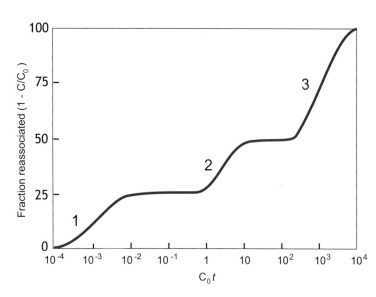

Figure 5-14. Idealized C_0t plot for a DNA that has three distinct components. The first consists of 25% of the DNA that contains about a million copies of repeated DNA and thus reassociates rapidly. The second consists of another 25% that contains 100–1000 copies of repeated DNA. The remaining 50% of DNA consists of single copies of unique sequences; it reassociates most slowly, because the concentrations of the complementary fragments are the lowest.

Such studies on genomic DNA provided the first indications that genomes often contain repetitive DNA, in which the same sequence is present in many copies in the genome. In the fragmented genomic DNA, such DNA segments will be present at higher concentrations than those segments that are present in only one copy per genome, and this fraction of the DNA reanneals at a correspondingly lower C_0t value. The results indicate that different sequences are present in widely varying frequencies throughout the genome, with only a fraction occurring only once per genome.

Hybridization of DNAs from different sources is a ready measure of the average degree of their sequence identity and their genetic and evolutionary relatedness (Section 7.6.A).

A similar approach characterizes the quantity of a specific messenger (m) RNA in a sample by its rate of hybridization to its DNA gene. The DNA gene is radiolabeled and denatured to single strands. An excess of mRNA is then added. The excess mRNA minimizes reannealing of the DNA strands, causes its concentration to remain constant, and causes the hybridization reaction to follow pseudofirst-order kinetics. The amount of DNA•RNA hybrid is measured by separating it from the free single-stranded DNA by hydroxyapatite chromatography or by digesting the single-stranded DNA using S1 nuclease, which does not degrade DNA•RNA hybrids. The time when half of the DNA strands have been converted to DNA•RNA hybrids, $t_{1/2}$, will be inversely proportional to the second-order rate constant for the hybridization, k, and the original concentration of the mRNA of interest, R_0:

$$t_{1/2} = \frac{\log_e 2}{R_0 k} \quad (5.12)$$

The rate constant k is proportional to the size of the mRNA, and its value can be obtained from comparison with the rates of association of comparable pure RNAs and DNAs. Knowing the value of k, the value of R_0 for any sample of mRNA can be determined using Equation 5.12 simply by measuring $t_{1/2}$.

Distortion of quantitative genomic and expression hybridization by Ct-1 DNA: mitigation of this effect. H. L. Newkirk *et al.* (2005) *Nucleic Acids Res.* **33**, e191.

Use of competitive DNA hybridization to identify differences in the genomes of bacteria. O. C. Shanks *et al.* (2006) **66**, 321–330.

DNA–DNA hybridization values and their relationship to whole-genome sequence similarities. J. Goris *et al.* (2007) *Int. J. Syst. Evol. Microbiol.* **57**, 81–91.

5.3.C. Probe Hybridization

A large number of powerful techniques for identification and manipulation of the genetic information present in natural DNA and RNA molecules exploit the ability of complementary nucleic acids to form hybrid duplexes. Most of these hybridization techniques rely on the use of a labeled (usually radioactive) nucleic acid probe to hybridize with a target nucleic acid. If the target is one strand of a double-stranded nucleic acid, it must be denatured first. After removal of any unreacted probe, detection of the remaining probe identifies and quantifies the hybrid duplex; its stability can also provide a measure of the extent of the sequence complementarity between the probe and the target. The probe must have sufficient affinity for its target sequences to resist being removed during the washing procedures.

Locked nucleic acids have been designed to have less conformational flexibility when single stranded and to be restricted to a conformation like that they will adopt when hybridized. Consequently, they have increased affinities for complementary strands. The locking is accomplished by a covalent bridge between the 2′ and 4′ C atoms of the ribose ring. This 'locks' the ribose in the 3′-*endo* conformation, that present in A-form RNA and DNA. Other modifications increase the stabilities of the probes by decreasing their susceptibility to nucleases that would hydrolyze normal nucleic acids. Still others diminish their tendencies to adopt intramolecular secondary structures, which interfere with their hybridization with another molecule; the sequence of a probe should be designed with this in mind.

The length of a specific hybridization probe can be as small as about 15 bases and as long as is necessary, depending upon the complexity of the target sequence and the source of the probe. The minimum length depends upon the size of the target, being the number of bases required to locate the specific sequence uniquely. Long probes have increased sequence specificity, and they can incorporate more label, thereby increasing the signal intensity. They are generally synthesized enzymatically off a template strand of DNA (Section 6.1). Shorter probes of 15–30 nucleotides can be synthesized chemically (Sections 6.6 and 6.7), and the incorporation of reporter groups as label is straightforward. Short probes bind less tightly to the target than do long probes, but weaker binding is often a practical advantage. The stability of the hybrid duplex can then be measured, and it will be sensitive to even single defects of sequence complementarity, including mismatches, deletions, bulges and small internal loops. The binding of a probe is stabilized if it is immediately adjacent to another molecule binding to the same strand, so that there is an additional stacking interaction:

(5.13)

If the probe is shorter than the target, **dangling ends** are produced:

(5.14)

Noncanonical base pairs (mismatches) destabilize the duplex, as do extra or missing nucleotides, which produce a **bulge** in one strand (Figure 5-12):

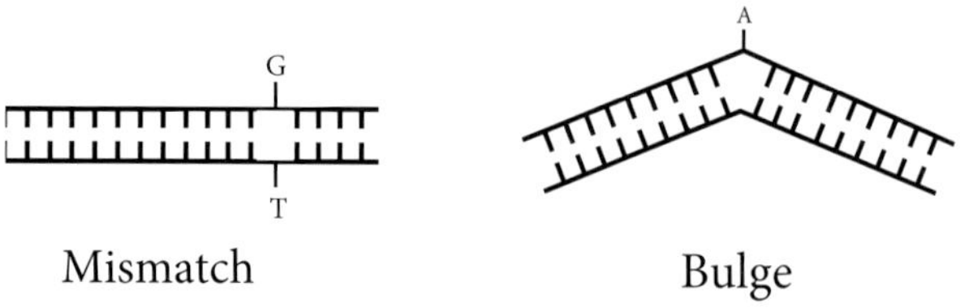

$$\text{Mismatch} \qquad \text{Bulge} \qquad (5.15)$$

If the probe overlaps another oligonucleotide binding to the same strand, this generates a mixture of structures in which base pairing by one or the other in the overlap region is interconverted by **branch migration**:

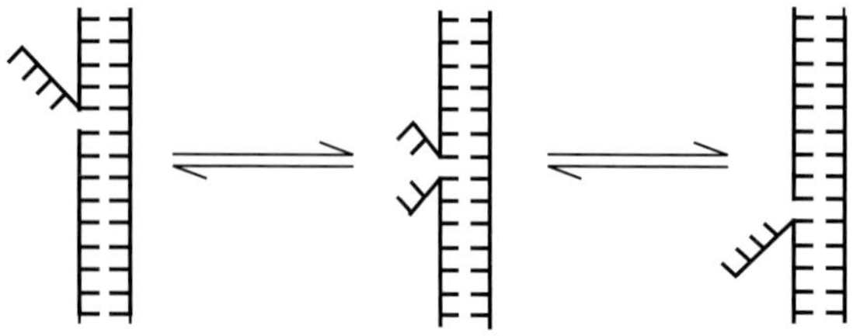

$$(5.16)$$

In many cases, the target nucleic acid is immobilized on a surface, most frequently on nitrocellulose or nylon filters. Immobilization can avoid many competitive binding interactions; for example, if the target was double-stranded, immobilization of the denatured single strands prevents the original duplex from re-forming. In **dot blots** or **slot blots**, the target nucleic acid is deposited on a small area of a blotting matrix; a series of targets can be deposited in different places on the same filter and hybridized simultaneously to a single probe. In **Southern blots**, **Northern blots** and related techniques, target nucleic acids are separated into discrete bands on a gel by electrophoresis. They are then transferred to a filter membrane, which is subjected to a conventional hybridization procedure. Those bands that bind the probe are detected by its label.

Many different probes are attached to solid surfaces in **micro-chip arrays**. After hybridization with complementary nucleic acids in solution, those molecules that did not hybridize are removed by washing and any hybridized molecules detected. The procedures are like those used in ELISA assays.

Padlock probes form especially stable complexes. They are designed so that the sequences at the two ends of the probe are homologous to the two halves of the target sequence and the annealed probe has its termini juxtaposed (Figure 5-15). A phosphodiester bond can then be formed by a DNA ligase between the two ends of the probe, to produce a circular probe molecule. The helical nature of the DNA molecule causes the annealed segment to be wound around the complementary strand. The probe can dissociate only if its backbone is cleaved or if it can slide off an end of the target DNA molecule. It can remain catenated to the target molecule even after denaturing washes that disrupt all base pairing.

RNA labeling and hybridization of DNA microarrays. E. P. Bottinger *et al.* (2003) *Methods Mol. Biol.* **86**, 275–284.

Modified bases in RNA reduce secondary structure and enhance hybridization. H. B. Gamper *et al.* (2004) *Biochemistry* **43**, 10224–10236.

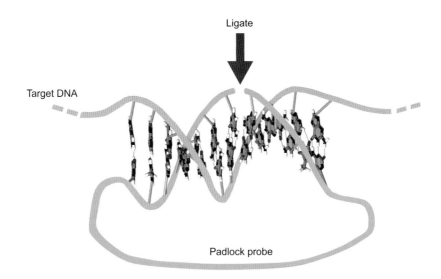

Figure 5-15. A padlock probe designed to be ligated into a circle when it hybridizes to the correct target sequence. The padlock probe is in *blue*, the target DNA sequence in *green*. Only the nucleotides involved in complementary base pairing are shown. The two ends of the linear probe hybridize in juxtaposition on the target sequence. The 5′-phosphate group on one end of the probe is in position to be ligated to the 3′-hydroxyl group at the other end, as indicated by the *arrow*. Once ligated, the padlock probe cannot dissociate, unless it can find its way to an end of the target sequence.

A general method for the synthesis of 2′-O-cyanoethylated oligoribonucleotides having promising hybridization affinity for DNA and RNA and enhanced nuclease resistance. H. Saneyoshi *et al.* (2005) *J. Org. Chem.* **70**, 10453–10460.

The forgotten variables of DNA array hybridization. E. Carletti *et al.* (2006) *Trends Biotechnol.* **24**, 443–448.

Design of LNA probes that improve mismatch discrimination. Y. You *et al.* (2006) *Nucleic Acids Res.* **34**, e60.

1. Stringency

The renaturation or hybridization of nucleic acids depends upon the solution conditions under which they are performed, in particular how they affect the stabilities of nucleic acid duplexes, which is known as the **stringency**. Low-stringency conditions favor duplex formation, while high-stringency conditions destabilize it. Consequently, many duplexes will form under low-stringency conditions, even if not totally complementary in nucleotide sequence, and base-pair mismatches, bulges and hairpins will be tolerated. In contrast, only the most complementary duplexes will form under high-stringency conditions. Different regions of a target nucleic acid can be distinguished on the basis of their sequence complementarity to the probe used.

The stringency conditions are usually adjusted in terms of the temperature, salt concentration and chemical denaturant concentration, as in Equations 5.3–5.6. They are optimized for the particular duplex that is to be formed. Adjustments in stringency can be used to discriminate among nucleic acids with different backbones and to favor a particular type of duplex. For example, use of high formamide concentrations (about 80% by volume) stabilizes RNA•DNA hybrids over DNA•DNA duplexes (Equations 5.4 and 5.6).

Most hybridization solutions contain EDTA to bind divalent cations, a polymer to enhance hybrid formation (presumably by an **excluded volume effect**) and a denaturant, in a buffered salt solution. Chemical denaturants, especially formamide, are frequently used in hybridization reactions to reduce the T_m to an experimentally convenient range. This is particularly important for experiments using RNA, which is readily degraded at high temperature.

Both the rate of formation of the duplex and its stability determine the optimal hybridization conditions. The rate of hybridization usually reaches a maximum at a temperature approximately 25°C below the T_m (Figure 5-13), which can be estimated with Equations 5.3–5.6. The same T_m can be obtained with a variety of different salt and denaturant concentrations.

Optimizing stringency for expression microarrays. J. E. Korkola *et al.* (2003) *Biotechniques* **35**, 828–835.

Specificity assessment from fractionation experiments (SAFE): a novel method to evaluate microarray probe specificity based on hybridisation stringencies. A. L. Drobyshev *et al.* (2003) *Nucleic Acids Res.* **31**, E1-1.

---

2. Analyzing the Extent of Complementarity

How the hybrid duplex is detected and measured depends upon what labels are incorporated into the probe molecule, and a huge number of different methods are used. Radioactive isotopes are the classic means of labeling the probe, but **chemiluminescent** and **bioluminescent** detection schemes are now in common use. These enzymes catalyze reactions that initiate a cascade of reactions that lead to the production of light. They have advantages of speed, cost and safety and are comparable to isotopic techniques in sensitivity. In many cases, a **biotin** label in the probe is detected by streptavidin coupled to alkaline phosphatase or horseradish peroxidase.

The degree of sequence complementarity between two short oligonucleotides can be evaluated by monitoring the change in UV absorbance upon heteroduplex formation (Figure 5-1), or in fluorescence with suitably labeled probes. If a pair of fluorescence donor and acceptor groups are placed in the appropriate positions, so that they come closer together when hybridized to the complementary sequence, the **fluorescence resonance energy transfer (FRET)** between them can be used to give a more specific signal. In **induced fluorescence resonance energy transfer (iFRET)**, the donor is a dye that only fluoresces while interacting with double-stranded DNA and the acceptor is a dye that is covalently linked to the oligonucleotide probe. Hybridization of the probe to its complement induces excitement of the donor dye and subsequent energy transfer to the acceptor dye. The energy transfer reaction (and concomitant hybridization status) can easily be followed by monitoring the fluorescence output of the acceptor dye. **Molecular beacons** are stem–loop hairpin oligonucleotide probes labeled with a fluorescent dye at one end and a fluorescence quencher at the other end; in the free probe with a stem–loop structure, the fluorescent and quenching moieties are in close proximity, so the fluorescence is quenched. When hybridized to a complementary nucleic acid strand, however, they are at opposite ends of the linear molecule and fluorescence is observed. Molecular beacons can differentiate between bound and unbound probes in homogeneous hybridization assays with a high signal-to-background ratio and enhanced specificity compared with linear oligonucleotide probes.

With longer nucleic acids, where some segments may be more complementary than others, the heteroduplex can be subject to digestion by enzymes that degrade only single-stranded regions and leave double-stranded regions intact. The remaining double-stranded fragments can then be isolated by hydroxyapatite chromatography and identified.

Electron microscopy can be used to visualize the regions of sequence homology in long heteroduplexes, after staining with basic proteins and heavy metals. In this way, single- and double-stranded regions can usually be distinguished and the homologous and nonhomologous regions identified (Figure 5-16).

Fluorescent PNA probes as hybridization labels for biological RNA. K. L. Robertson *et al.* (2006) *Biochemistry* **45**, 6066–6074.

Detecting RNA/DNA hybridization using double-labeled donor probes with enhanced fluorescence resonance energy transfer signals. Y. Okamura & Y. Watanabe (2006) *Methods Mol. Biol.* **335**, 43–56.

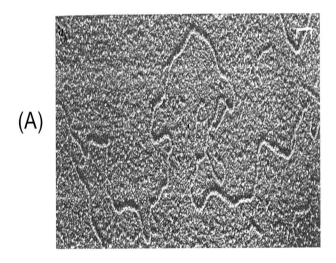

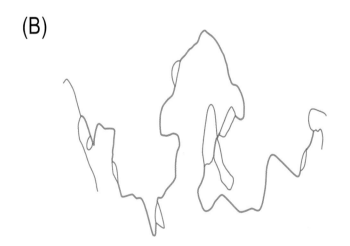

Figure 5-16. Visualization of a partly complementary heteroduplex by electron microscopy. (A) The DNA molecules have been spread on a grid and shadowed. (B) Interpretation of the micrograph. The double-stranded regions are somewhat wider and less kinky than the single-stranded and are indicated in *blue*. The single-stranded regions are in *red*.

Detection of DNA hybridization using induced fluorescence resonance energy transfer. W. M. Howell (2006) *Methods Mol. Biol.* **335**, 33–41.

Fluorescent hybridization probes for sensitive and selective DNA and RNA detection. A. A. Marti *et al.* (2007) *Acc. Chem. Res.* **40**, 402–409.

3. In situ *Hybridization*

Performing hybridization techniques *in situ* makes it possible to visualize the spatial distribution of a target nucleic acid in a tissue, cell or chromosome. There are many variations but all share a set of common steps:

(1) the sample is fixed to preserve its structure;

(2) limited proteinase digestion is carried out, to provide access to the hybridization probe of the target DNA sequence;

(3) the probe is added and hybridization is allowed to proceed under the appropriate conditions;

(4) washing removes unhybridized probe;

(5) the spatial distribution of the immobilized probe is visualized.

Small probes can diffuse into the sample more efficiently than longer probes. This is partially countered, however, by their tendency to be lost in the washing step: the stability of the hybrid duplex involving a short oligonucleotide depends upon its concentration, so the dilution that accompanies the washing step can result in release of some of the probe.

The probe can incorporate radioisotopes or fluorescent groups, but chemiluminescent techniques are now preferred. When detection is by fluorescence (or its quenching), the technique is known as **fluorescence *in situ* hybridization (FISH)**. Visualizing the immobilized probe can be difficult if only a small number of target molecules are present. This has led to the development of techniques that amplify the number of DNA sequences present, usually by the polymerase chain reaction (Section 6.1.C).

Expression analysis of murine genes using *in situ* hybridization with radioactive and nonradioactively labeled RNA probes. A. Chotteau-Lelievre *et al.* (2006) *Methods Mol. Biol.* **326**, 61–87.

Oligonucleotide probes for RNA-targeted fluorescence *in situ* hybridization. A. P. Silverman & E. T. Kool (2007) *Adv. Clin. Chem.* **43**, 79–115.

RNA *in situ* hybridizations on *Drosophila* whole mounts. C. Wulbeck & C. Helfrich-Forster (2007) *Methods Mol. Biol.* **362**, 495–511.

5.4. DNA MIMICS: PEPTIDE NUCLEIC ACIDS

Many practical uses could be imagined for oligonucleotides in medicine, such as in controlling the expression of particular genes by binding to them or their messenger RNAs (Section 6.2.D).

Oligonucleotides are not, however, ideal pharmaceuticals. They are highly charged and thus do not cross cell membranes readily. They are not as stable as would be required, and the sites where they should act are also rich in enzymes (nucleases) that will destroy them. Consequently, much effort has been expended in trying to find analogs that do not have these shortcomings and that, ideally, would bind even more strongly to their targets. The most promising are the peptide nucleic acids (PNA), DNA mimics in which the phosphodeoxyribose backbone of DNA has been substituted by a pseudopeptide backbone to which the normal nucleobases are tethered, to produce an *oligoamide* (Figure 5-17). These DNA mimics can perform many of the functions of RNAs and DNAs. **PNAs of mixed sequences containing all four DNA bases form duplexes with DNA and RNA that obey the Watson–Crick base-pair rules, and two PNAs with complementary base sequences can hybridize into a double-helical structure.**

Only minor modifications of the original amino ethyl glycine backbone can be tolerated; extending any of the three possible linkages of each monomer diminishes the stability of its complexes. The secondary amino group in the nucleobase linker restricts its flexibility and is necessary, because changing it to a tertiary amine inhibits hybridization. Moving a methylene group from the ethyl to glycine moiety has the same effect. On the other hand, substituents may be placed on the α-position of the glycine moiety of the PNA backbone in various ways, so the glycine residue can be replaced by other natural amino acids. So long as the right stereoisomer is chosen, even cyclic substituents are acceptable, such as cyclohexyl at the amino ethyl linker.

PNA is regarded as having great potential for diagnostic and pharmaceutical applications. Other promising DNA mimics are based on a pyrrolidine amide backbone but will not be described here.

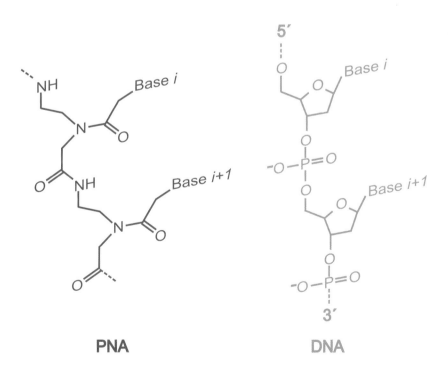

Figure 5-17. Comparison of the chemical structures of DNA and a PNA.

Gene targeting using peptide nucleic acid. P. E. Nielsen (2005) *Methods Mol. Biol.* **288**, 343–358.

Peptide nucleic acids (PNAs), a chemical overview. A. Porcheddu & G. Giacomelli (2005) *Curr. Med. Chem.* **12**, 2561–2599.

Peptide nucleic acid conjugates: synthesis, properties and applications. Z. V. Zhilina *et al.* (2005) *Curr. Top. Med. Chem.* **5**, 1119–1131.

Biological activity and biotechnological aspects of peptide nucleic acid. K. E. Lundin *et al.* (2006) *Adv. Genet.* **5**, 1–51.

Homopolymeric pyrrolidine-amide oligonucleotide mimics: Fmoc-synthesis and DNA/RNA binding properties. T. H. Tan *et al.* (2007) *Org. Biomol. Chem.* **5**, 239–248.

5.4.A. Chemistry and Synthesis

The backbone of PNA is designed to mimic the normal phosphodiester backbone of DNA and RNA and consists of 2-aminoethyl glycine linkages (Figure 5-17). The standard nucleotide bases are connected to this backbone at the amino N atoms by a methylene carbonyl linker. The properties of this polymer are of interest in themselves, but the differences also illustrate some of the important aspects of nucleic acids. In contrast to DNA and RNA, **PNA is neither ionic nor chiral**. The achirality avoids diastereomeric structures and the need to develop synthesis reactions that are stereoselective (Section 1.1.A). Reactive groups on the PNA monomers are protected using either the Boc or Fmoc chemistry that is used in peptide synthesis (Section 7.4). Oligomers can be synthesized on a polystyrene resin using established solid-phase peptide synthesis procedures; they can be released from the support using either anhydrous hydrogen fluoride or trifluoromethanesulfonic acid, then purified chromatographically. Modified forms of PNA can be synthesized readily.

A simple γ-backbone modification preorganizes peptide nucleic acid into a helical structure. A. Dragulescu-Andrasi *et al.* (2006) *J. Am. Chem. Soc.* **128**, 10258–10267.

Peptide nucleic acids with a structurally biased backbone: effects of conformational constraints and stereochemistry. R. Corradini *et al.* (2007) *Curr. Top. Med. Chem.* **7**, 681–694.

Nucleobase modifications in peptide nucleic acids. F. Wojciechowski & R. H. Hudson (2007) *Curr. Top. Med. Chem.* **7**, 667–679.

5.4.B. Hybridization Properties

PNA is a very potent DNA mimic capable of forming both duplex and triplex structures, but with some important differences. **PNA binds equally to DNA and RNA** molecules with the same sequence.

RNA targeting using peptide nucleic acid. E. Nielsen (2006) *Handb. Exp. Pharmacol.* **173**, 395–403.

1. PNA•DNA and PNA•RNA Duplexes

The mixed-sequence PNA pentadecamer H–TGTACGTCACAACTA–NH$_2$ has been studied most extensively. It forms a duplex with complementary DNA that is **antiparallel, with the amino-terminus of PNA at the 3′ end of DNA**. The duplex has a T_m of 70°C, greater than the 53°C of the corresponding DNA•DNA complex. PNA is also able to bind in a parallel orientation, although with a lower affinity (T_m = 56°C). **Mixed-sequence PNA•DNA duplexes generally adopt the antiparallel orientation**. Forming antiparallel PNA•DNA duplexes is faster than forming a corresponding DNA•DNA duplex, whereas forming parallel PNA•DNA complexes is considerably slower. The rates of formation and thermal stabilities of PNA•RNA duplexes are generally greater than for PNA•DNA duplexes. **PNA in complex with DNA or RNA adopts right-handed helical structures** that are probably dictated by the DNA or RNA.

The greater thermal stabilities of PNA•DNA (or PNA•RNA) duplexes than the corresponding DNA•DNA (or RNA•RNA) duplexes are probably due to the absence of electrostatic repulsions between the phosphate groups of the two strands. The stabilities of PNA•DNA hybrids are little affected by changes in ionic strength, except at low ionic strength, where the stability increases; this is believed to be due to a more favorable entropy change upon PNA binding caused by the release of counterions from the DNA polyanion (Section 2.3).

The sequence specificities of PNA oligomers are at least as good as those of DNA: mismatch of a single base pair in the middle of a 15-mer PNA•DNA or PNA•RNA duplex decreases the T_m by 8–20°C.

Cyclohexanyl peptide nucleic acids (chPNAs) for preferential RNA binding: effective tuning of dihedral angle β in PNAs for DNA/RNA discrimination. T. Govindaraju *et al.* (2006) *J. Org. Chem.* **71**, 14–21.

---

2. (PNA)$_2$•DNA Triplexes

PNA oligomers with a high pyrimidine content bind complementary purine DNA by forming predominantly (PNA)$_2$•DNA triple helices, with virtually no duplex. These complexes are very stable, with a T_m > 70°C for decamer PNAs, and have well-defined melting curves. Binding is probably governed by standard Watson–Crick and Hoogsteen base pairing (Section 2.2.C.1) and, as expected, triplex formation involving cytosine in the homopyrimidine PNA is more stable at lower pH (Section 2.2.C.2). The same triplex stoichiometry is observed even at pH 9, however, where cytosine is not expected to be protonated.

3. Strand Invasion: Binding to Double-stranded DNA

PNA is able to dissociate double-stranded DNA, due to the high stability of the (PNA)$_2$•DNA triplexes. For example, homopyrimidine PNA oligomers displace the pyrimidine strands of double-stranded DNA and form (PNA)$_2$•DNA triplexes with the DNA homopurine strand. The DNA pyrimidine strand that was expelled forms a single-stranded loop (Figure 5-18). PNA•DNA complexes formed by strand invasion are both very stable thermodynamically and highly specific. The strand invasion appears to be controlled kinetically, which permits the PNA to search for the target sequence without becoming trapped in incorrect complexes that are thermodynamically stable. Forming a PNA(pyrimidine)$_2$•DNA(purine) strand displacement complex follows pseudofirst-order kinetics, and a single mismatch in a 10-residue PNA reduces the rate of binding by more than a factor of 100.

In the case of PNA•thymine trimers, binding follows somewhat more complex reaction kinetics that suggest the presence of at least one intermediate species. The rate-limiting step appears to be transient opening of a few base pairs in the DNA duplex. The reaction is second-order in PNA concentration, indicating that both PNA strands are involved in the rate-limiting step: the two PNA molecules may be associated loosely with the DNA duplex prior to the base-pair opening step. Intercalators (Section 2.5.B) bound to the starting duplex DNA increase the rate of PNA binding; the intercalation site may facilitate initiation of PNA invasion. The formation of strand displacement complexes is highly dependent on the ionic strength but complexes formed at low ionic strength are generally stable kinetically at salt concentrations up to 0.5 M NaCl.

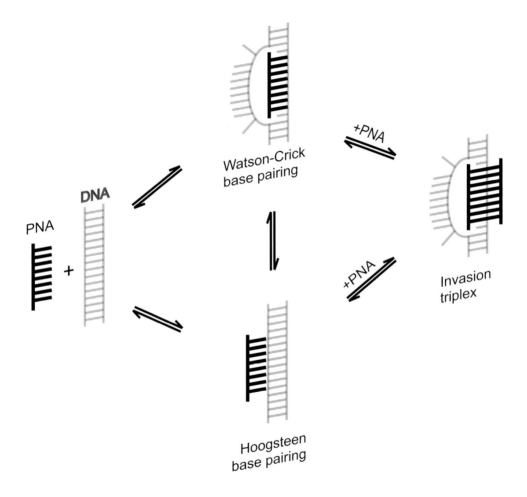

Figure 5-18. Schematic model of the possible mechanisms of PNA_2•DNA triplex formation by strand invasion of a DNA duplex.

Cytosine-rich and alternating thymine–cytosine PNA oligomers bind to their complementary sequence of intact double-stranded DNA by forming PNA•DNA_2 triplexes, which probably involve Hoogsteen base pairing. With purine-rich PNA sequences, duplexes are formed by the invasion into double-stranded DNA at a complementary pyrimidine target (making a PNA•DNA duplex). PNA binds less well to DNA sequences without purine-rich stretches.

Cooperative strand invasion of double-stranded DNA by peptide nucleic acid. T. Sugiyama *et al.* (2005) *Nucleic Acids Symp. Ser.* **49**, 167–168.

End invasion of peptide nucleic acids (PNAs) with mixed-base composition into linear DNA duplexes. I. V. Smolina *et al.* (2005) *Nucleic Acids Res.* **33**, 3146.

Structural diversity of target-specific homopyrimidine peptide nucleic acid–dsDNA complexes. T. Bentin *et al.* (2006) *Nucleic Acids Res.* **34**, 5790–5799.

4. PNA Duplexes and Triplexes

PNA molecules containing complementary sequences form stable antiparallel duplexes with Watson–Crick base paring and very high thermal stability: the T_m of a 10-mer PNA•PNA is 67°C, compared with 51°C for the corresponding PNA•DNA and only 33.5°C for DNA•DNA. The parallel PNA•PNA complex is less stable, with a T_m of 47°C. A PNA_3 triplex can be formed with two pyrimidine decamers and one purine decamer and probably has a helical structure similar to triplex DNA and PNA_2•DNA triplexes (Section 2.2.C.2).

PNA•PNA duplexes have no chiral constituents and probably exist as a racemic mixture of right- and left-handed helices. A chiral amino acid at the carboxyl-terminus of one strand can, however, induce formation of one predominant helical sense over about 10 base pairs, approximately one helical turn, the 'chiral persistence length'. This effect depends on the PNA sequence at the carboxyl terminus, being greatest with a G•C base pair, and on the nature of the terminal chiral amino acid. The D- and L-forms of amino acids on PNA give rise to circular dichroism spectra that are mirror images, indicating that helices of opposite handedness are formed. Whether the side-chain is hydrophobic or hydrophilic also determines the preferred helical sense. Changes in helix chirality occur within minutes at room temperature and follow first-order kinetics. The transition state has a high entropy, as would be expected for a transition from one helical sense to another.

During formation of a PNA•PNA duplex, base pairing occurs rapidly, within less than a second under normal conditions. Some of the PNA•PNA duplexes formed initially undergo a conformational change to the final structure.

5.4.C. Structures of PNA Complexes

PNA•PNA complexes appear to be similar to DNA duplexes in their base-stacking interactions and degree of organization. When bound to RNA or DNA, PNA forms a right-handed helix with a base-pair geometry not very different from that found in B-DNA and A-DNA helices, respectively. In contrast, parallel PNA•DNA and PNA•RNA complexes appear to have different base stacking.

PNA–T_8 hybridized to poly(dA) forms a right-handed PNA_2•DNA triplex in which the base triplets have their planes approximately perpendicular to the helix axis, similar to the standard $(poly(dT))_2$•poly(dA) triplex. The PNA_2•DNA triplex is a remarkably rigid structure, which suggests efficient stacking interactions. Titration of PNA–T_8 added to poly(dA) shows that no duplex stretches are present. Instead, PNA_2•DNA triplexes are distributed evenly on the poly(dA) strand, separated by flexible single-stranded regions. The complex becomes rigid only when the poly(dA) is saturated with 2:1 PNA:DNA bases, completing formation of the triple strand and stacking between each PNA triplex segment.

Antiparallel complementary binding of PNA allows the backbone of the nucleic acid partner to retain a nearly 'normal' conformation: in the RNA•PNA duplex, the RNA strand has an approximately A conformation with C3′-*endo* sugar puckering, whereas the DNA in the PNA•DNA duplex is closer to B-form, with C2′-*endo* sugar puckering. The base-pair positions are, however, more A-like in terms of their helix displacement, although more perpendicular to the helix axis. The PNA•DNA duplex has a pitch of about 13 base pairs, greater than the 10–11 base pairs usually observed with nucleic acid duplexes.

The PNA$_2$•DNA triplex has hydrogen bonds between phosphate groups of the DNA backbone and amide protons in the PNA backbone of the Hoogsteen strand, which could be important for its stability. The two PNA strands are nearly identical and resemble the PNA strand in a PNA•DNA duplex. The PNA$_2$•DNA triplex and the PNA•PNA duplex have relatively wide helices of 26 Å and 28 Å, respectively. They also have a large pitch of 16 and 18 bases per turn, respectively, which is probably the natural conformation of helices with a PNA backbone. The base pairs in the PNA•PNA duplex have stacking overlaps between base pairs that are close to those of A-form nucleic acids. Consequently, PNA helices appear to have a unique structure that only resembles known nucleic acid helices.

The similarity in the base-pair stacking in the PNA•DNA, PNA•RNA and PNA•PNA duplexes indicates that the helical stacking is not influenced markedly by the absence of either backbone charge or chirality in one or both of the strands. The helicity can thus be regarded a result of the stacking interactions and the architecture of the backbone, rather than arising in nucleic acids from repulsions between phosphate groups striving to increase the separation of negative charges along the backbones.

Molecules that intercalate or bind to major grooves in double-stranded DNA (Section 2.5) have no apparent affinity for PNA•DNA complexes, whereas classical binders of the minor-groove bind weakly. No simple ligand able to bind to PNA•PNA duplexes has been found, illustrating perhaps the importance of electrostatic attractions for ligand binding to DNA.

Insights into peptide nucleic acid (PNA) structural features: the crystal structure of a D-lysine-based chiral PNA–DNA duplex. V. Menchise *et al.* (2003) *Proc. Natl. Acad. Sci. USA* **100**, 12021–12026.

Crystal structure of a partly self-complementary peptide nucleic acid (PNA) oligomer showing a duplex–triplex network. B. Petersson *et al.* (2005) *J. Am. Chem. Soc.* **127**, 1424–1430.

~ CHAPTER 6 ~

MANIPULATING NUCLEIC ACIDS

Manipulating nucleic acids is central to molecular biology. Specific segments of DNA or RNA can be identified, isolated, amplified, sequenced and synthesized. Any desired changes can be made to the sequence. The nucleic acid can be inserted into cells and its effect observed. These many techniques make it possible to identify genes and their protein or RNA products, to alter them at will, and to determine their physiological roles.

Nucleic acids have the very great advantage that all those of the same type and size behave almost identically. Consequently, a technique developed with one nucleic acid is very likely to work just as well with any other, in contrast to proteins, where each one is unique. The most important parameter for manipulating a nucleic acid is its length. A second, less important, parameter is its base composition, as nucleic acids with more G•C base pairs denature less readily (Section 5.1.B).

This chapter will not describe all the many techniques that have been devised to manipulate nucleic acids. Instead, it will describe the most fundamental ones that illustrate the physical and chemical principles of them all.

6.1. REPLICATING DNA

The most fundamental characteristic of living organisms is their ability to replicate themselves and multiply. Replication of their genomes is the crucial step. It is easy to envisage **the DNA or, rarely, RNA double helix of the genome simply being dissociated into the individual strands, which then serve as templates for a complementary sequence to be synthesized; the end result is two copies of the original DNA or RNA molecule**. Entire genomes are enormous molecules, however, which means that a multitude of factors are required *in vivo*, especially to prevent these very long molecules from becoming hopelessly entangled (Chapter 3). In particular, **DNA helicases** are required to unwind the duplex DNA ahead of the point of DNA replication. The three distinct stages of genome replication are initiation, elongation and termination; each requires a complex apparatus. Most complex is the initiation step, as replication of the genome must be coordinated with the division of the entire cell. Each cell usually has a single genome, and replicating the genome in the absence of cell division would be deleterious.

This discussion will be limited to much more simple *in vitro* systems, in which a relatively short piece of DNA is replicated using defined enzymes. The first requirement to replicate a double-stranded DNA molecule is to dissociate it into the individual single strands, which can then be used as templates to synthesize a new copy of the missing strand. Although a complex process *in vivo*, in the laboratory it simply requires that the DNA molecules be heated to high temperatures (Section 5.1.B.1).

6.1.A. DNA polymerase

The enzymes that catalyze formation of the complementary strand of a DNA are known as **DNA polymerases**. Each organism has at least one DNA polymerase, but the one used most commonly in molecular biology is DNA polymerase I from *Escherichia coli*. It can synthesize the complementary strand of a DNA single strand in the 5′- to 3′-direction in the presence of all four deoxynucleoside triphosphates (dNTPs) (Figure 6-1). DNA polymerases cannot initiate synthesis, however, so a primer complementary to the site at which replication is to be initiated must be supplied and annealed to the DNA template strand (Figure 6-2). After the DNA polymerase has bound to the primer, the appropriate dNTP binds to the active site of the polymerase, and the polymerase changes conformation. This positions the dNTP so that its α-phosphate group is positioned near the 3′-hydroxyl group of the primer (Figure 6-3) for nucleophilic attack, which produces a phosphodiester bond and releases pyrophosphate (Figure 6-4). The conformational change and nucleophilic attack function in concert to sense that the appropriate Watson–Crick base pairing is occurring and that the correct nucleotide is being introduced. The nascent second strand has now been elongated by one nucleotide. The polymerase then translocates to the next nucleotide of the template strand and attaches the appropriate nucleotide to the growing chain.

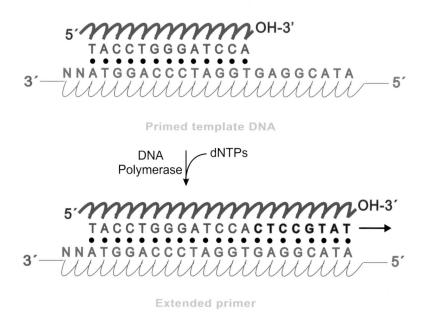

Figure 6-1. DNA chain elongation by DNA polymerase. The *jagged line* represents the DNA backbone, and the bases are indicated by the *letters*; *black circles* represent complementary pairing between the bases. Nucleotides that have been added to the 3′-end of the primer are indicated in *bold*. Addition of a dNTP to the 3′-end of the primer strand produces a new 3′-OH group, to which another dNTP can be added for continued polymerization of the growing chain.

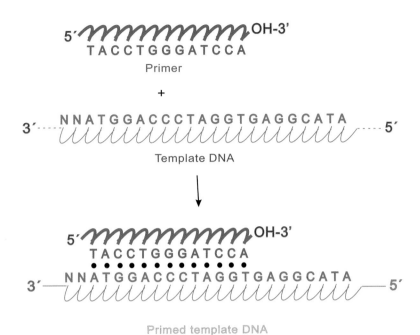

Figure 6-2. Annealing of primer to template DNA. The primer is designed to be complementary in sequence to the site where replication is to be initiated.

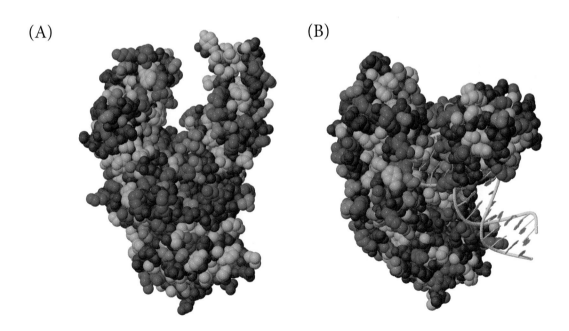

Figure 6-3. Structure of the Klenow fragment of DNA polymerase I. (A) The enzyme from *E. coli* bound to only CTP and pyrophosphate. The atoms of the protein are depicted with their van der Waals radii and colored according to their chemical properties. (B) The enzyme from *Bacillus stearothermophilus* bound to a DNA single strand and a nascent second strand. The DNA is shown in skeletal form, without the details of the backbone. The enzyme in the crystal has added four nucleotides to the original strand, but has been halted by a G•T mismatch. The two upper domains of the enzyme have closed over the DNA molecule. Figure generated from PDB files *1kfd* and *1nkc* using the program Jmol.

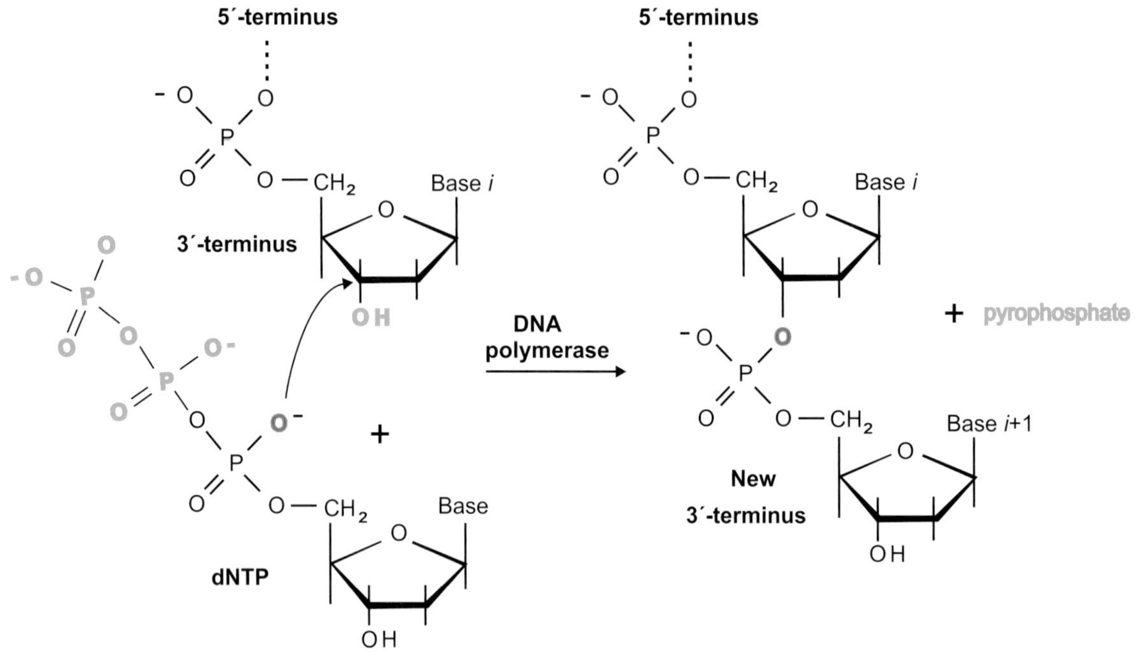

Figure 6-4. Biochemistry of nucleotide chain elongation. DNA polymerase catalyzes the condensation of the appropriate dNTP at the 3′-end of a primed template, releasing pyrophosphate.

Inserting the correct nucleotide is crucial for this process, so the process involves editing and proofreading the sequence of the assembled chain. If an improperly paired base is inserted, the elongation process is interrupted and introduction of the next nucleotide is very slow. The DNA molecule then has time to move to an adjacent site on the DNA polymerase, the exonuclease site, where the incorrect nucleotide is removed. This **3′→5′ exonuclease** activity decreases the frequency of mistakes, to about one mistake in every 50,000 nucleotides added, whereas the error rate is about one per 1000 nucleotides in its absence.

E. coli DNA polymerase I also has a **5′→3′ exonuclease activity** that it uses *in vivo* to degrade certain DNA and RNA fragments. This activity is detrimental to synthesizing DNA copies *in vitro*, and it can be removed by cleaving the DNA polymerase I into two fragments: one of 323 amino acid residues at the N-terminus that comprises the 5′→3′ exonuclease activity, and the other of about 605 residues with the polymerase and 3′→5′ exonuclease activities. The latter is generally known as the **Klenow fragment** and it is used most frequently in molecular biology. The Klenow fragment extends a DNA chain at a relatively modest rate of 50 nucleotides per second. It remains bound to the DNA for about one second on average, before dissociating. Another molecule of DNA polymerase must be bound to continue extending the polynucleotide chain.

Although DNA polymerase I is used extensively in molecular biology to replicate DNA, that is not its main function in nature, which is to fill gaps during the repair and replication of DNA. This function requires the assistance of a DNA ligase to produce a single, intact strand of DNA.

Mechanism and dynamics of translesion DNA synthesis catalyzed by the *Escherichia coli* Klenow fragment. A. Sheriff *et al.* (2008) *Biochemistry* **47**, 8527–8537.

Fingers-closing and other rapid conformational changes in DNA polymerase I (Klenow fragment) and their role in nucleotide selectivity. C. M. Joyce *et al.* (2008) *Biochemistry* **47**, 6103–6116.

Structure and function of 2:1 DNA polymerase•DNA complexes. K. H. Tang & M. D. Tsai (2008) *J. Cell. Physiol.* **216**, 315–320.

6.1.B. DNA ligase

A crucial step in assembling an intact DNA double helix is often the joining of one DNA chain to another, using a **DNA ligase** enzyme. In many cases, this is simply the formation of a phosphodiester bond between two nucleotides that are adjacent in the sequence of a double helix, being brought into proximity by base pairing to a complementary third strand (Figure 6-5). Forming a phosphodiester bond requires energy, which is usually supplied by the hydrolysis of ATP, although in some organisms the coenzyme NAD^+ is used instead.

The ligase reaction requires that the 5'-fragment have a 3'-hydroxyl group and the other strand have a 5'-phosphate group. The two fragments are best abutted without a gap, with no nucleotide missing

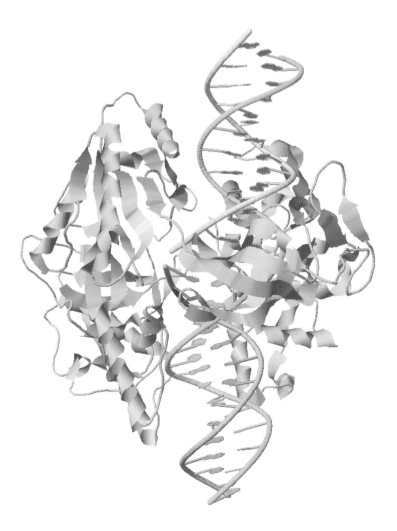

Figure 6-5. Structure of the *E. coli* DNA ligase bound to a nicked DNA double helix. The DNA substrate is indicated in skeletal form, without the details of the polynucleotide backbone. Only the polypeptide backbone of the protein is shown, and as a ribbon, so that the nicked DNA at the active site can be observed. Otherwise, the DNA is in the center of the protein molecule, protected from solvent, and invisible from the outside. Figure generated from PDB file *2owo* using the program Jmol.

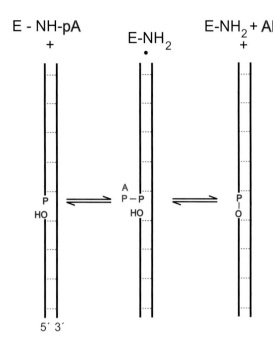

Figure 6-6. The mechanism of action of DNA ligase. On the *left* is a DNA double helix lacking one internal phosphodiester bond; OH is the 3′-hydroxyl, while P is the 5′-phosphate group. The DNA ligase (E) is adenylated on an amino group (Equation 6.1). After the DNA binds to the ligase, the adenylyl group is transferred to the 5′-phosphate. It is then expelled by nucleophilic attack by the 3′-hydroxyl group, generating the phosphodiester bond and releasing AMP and the enzyme. The DNA double helix is now intact.

from the double helix. The reaction proceeds through two intermediates (Figure 6-6). The first is formation of a covalent enzyme–AMP complex, using either ATP or NAD⁺. The adenylyl group (pA) is linked to the ε-amino group of a lysine residue (–NH$_2$) of the enzyme (E), preserving the high energy of the donor, while releasing pyrophosphate (PP$_i$):

$$E\text{-}NH_2 + ATP \rightleftharpoons E\text{-}NH\text{-}pA + PP_i \tag{6.1}$$

This reaction does not require DNA and is reversible, so it can be used to assay the activity of the DNA ligase. In the presence of DNA, the adenylyl group is transferred from the enzyme to the terminal 5′-phosphate of the DNA, but it is then displaced by the 3′-hydroxyl group of the adjacent DNA fragment, forming the phosphodiester bond and releasing AMP. Both strands of the double helix are now intact.

The DNA ligases used most frequently are those from *E. coli*, the thermophile *Thermus aquaticus* and bacteriophage T4.

DNA ligase: a means to an end joining. C. M. Bray *et al*. (2008) *SEB Exp. Biol. Ser.* **59**, 203–217.

Dynamics of phosphodiester synthesis by DNA ligase. A. Crut *et al*. (2008) *Proc. Natl. Acad. Sci. USA* **105**, 6894–6899.

The associative nature of adenylyl transfer catalyzed by T4 DNA ligase. A. V. Cherepanov *et al*. (2008) *Proc. Natl. Acad. Sci. USA* **105**, 8563–8568, E84 and E85.

6.1.C. Polymerase Chain Reaction (PCR)

Once a single strand has been used as a template to synthesize its complementary strand, further

DNA replication requires that the double-stranded molecule be dissociated to its individual strands; they can then be replicated in the same way, to double the number of DNA molecules present. In the laboratory, separating the strands requires only that the DNA be heated to high temperatures. Replication must be carried out promptly, however, to compete successfully with reformation of the original double strands. A primer must be present to hybridize with the single strand and provide a starting point for synthesis of the second strand. When present at high concentrations, the primer competes successfully with renaturation of the two complementary strands.

The polymerase chain reaction (PCR) is somewhat different in that it uses one primer for each of the complementary strands, which delineates the boundaries of the portion of the DNA molecule that is to be replicated. **PCR makes it possible to amplify and produce large quantities of any DNA sequence** and is one of the most useful and powerful techniques in molecular biology. In its most basic form, two oligodeoxyribonucleotides, generally 15–25 nucleotides in length, are used to select a specific target sequence, by being complementary to opposite strands at the two ends of the target sequence (Figure 6-7). Each of the two oligonucleotides serves as a primer for copying one of the two strands of the target sequence using a DNA polymerase enzyme. The two double helices that result are denatured by heating, and the process is repeated. **Each cycle virtually doubles the amount of the desired DNA**. A heat-stable DNA polymerase is used that survives the heat-denaturation of the DNA and is viable for the next cycle.

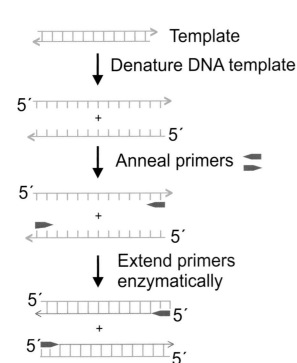

Figure 6-7. One cycle of basic PCR. The DNA of interest is denatured to separate the two strands. Primers complementary to the two ends are annealed to the single strands, and they direct copying of the template by DNA polymerase. Two identical copies of the original molecule result. Another cycle will produce four copies, and the number of molecules nearly doubles for each additional cycle. Other DNA sequences can be present in the original sample but only those that hybridize with the primers will be copied.

Operationally, each cycle consists solely of altering the temperature, from that suitable for annealing of the primers to the DNA and for synthesis by the thermostable polymerase (usually about 72°C) to that required to denature the DNA double helices (usually 94–96°C); simple PCR machines are available to carry out the entire process of several cycles automatically. The temperature for the annealing step is most crucial: a too-high temperature will inhibit annealing, whereas a too-low temperature allows nonspecific priming and the production of spurious amplification products. The optimal temperature for use with primers up to 18 nucleotides long is usually the same as the T_m of the complex and can be predicted by:

$$T = 4°C \times (\text{number of G and C nucleotides}) + 2°C \times (\text{number of A and T}) \qquad (6.2)$$

The two primers used should be designed to have very similar optimal temperatures.

Each cycle doubles the amount of target DNA present: in theory, although not quite in practice. The degree of amplification accomplished depends primarily upon the number of temperature cycles, so **even extremely small amounts of target DNA can be amplified to make substantial quantities**. This exquisite sensitivity is one of the most important attributes of PCR, and makes it possible in theory to detect a single molecule of DNA; on the other hand, it also makes PCR prone to false-positive results due to, for example, exogenous contamination. Contamination ultimately limits the degree of amplification that is practicable. The primer sequences must be chosen with care, and other sequences can be added to their 5′-end to facilitate their subsequent manipulation, such as sites for restriction enzymes to facilitate cloning (Section 6.3).

In **real-time PCR**, the amount of product formed is monitored during the course of the reaction by measuring the fluorescence of dyes or probes introduced into the reaction mixture; the number of amplification cycles required to obtain a particular amount of DNA molecules can be monitored. If the amplification efficiency is known, it is possible to calculate the number of DNA molecules of the amplified sequence that were initially present in the sample. The number of DNA molecules of a particular sequence in a complex sample can be determined with unprecedented accuracy.

Nested primers are complementary to sequences 3′ to the original primers and occur within the amplified sequence. They can be used to carry out a second PCR, to increase the specificity of the reaction and produce increased amounts of product.

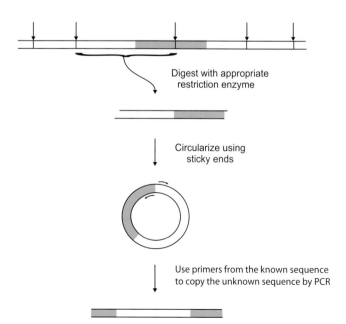

Figure 6-8. Inverse PCR. DNA with only one segment of known sequence (*colored*) is digested with an appropriate restriction enzyme at the sites indicated by *arrows*, to produce pieces that can be circularized by DNA ligase. Basic PCR is then performed using primers complementary to the segment of known sequence, designed so that the DNA polymerase extends the sequence outwards. Consequently, many copies of the restriction fragments containing the segment of unknown sequence result, and their sequences can be determined using normal methods.

Inverse PCR can be used to clone and determine the unknown DNA sequence adjacent to a segment of known sequence (Figure 6-8). The DNA is cleaved into pieces of an appropriate size, which are then converted to circles by DNA ligase (Figure 3-2). The circles containing the known sequence are subjected to PCR using primers complementary to the known sequence so that DNA synthesis occurs outwards from the known sequence. As a result, the segment with unknown sequence is amplified and ready to be sequenced in the normal manner (Section 6.4).

ImmunoPCR is based on chimeric conjugates of specific antibodies linked to nucleic acid molecules. The presence of such an antibody, for example if it is bound to a specific antigen, is detected by amplifying the attached nucleic acid molecule by PCR to generate the signal that is measured. The enormous efficiency of nucleic acid amplification typically leads to a 10^2- to 10^4-fold greater sensitivity than normal immunoassays.

An introduction to PCR techniques. G. Rumsby (2006) *Methods Mol. Biol.* **324**, 75–89.

Introduction to the polymerase chain reaction. Y. M. Lo & K. C. Chan (2006) *Methods Mol. Biol.* **336**, 1–10.

The real-time polymerase chain reaction. M. Kubista *et al.* (2006) *Mol. Aspects Med.* **27**, 95–125.

6.2. PRODUCING RNA

RNA is most commonly produced in nature using DNA as a template, but many viruses have RNA genomes that must be replicated. The enzymes responsible in both instances are known as **RNA polymerases**, being distinguished as either DNA-dependent or RNA-dependent.

6.2.A. RNA Replication: RNA Replicases

Replication of RNA occurs only with the few genomes, primarily those of viruses, that consist of RNA. The RNA-dependent RNA polymerases, or RNA replicases, that replicate viral RNAs are similar in many ways to the polymerases that replicate DNA (Section 6.1.A) but they are not so well characterized, nor are they used routinely in molecular biology. They are often complex molecules containing additional enzymatic activities, such as RNA helicase and others that modify the RNA covalently. Some of their polypeptide chains originate from their host cells. Their most remarkable features are that they (1) are specific for their viral RNA even though they function in the presence of a large excess of host RNAs, (2) amplify their RNA template many-fold during a short infection period, (3) copy the entire template RNA without the use of primers and (4) make about one mistake in every 10^3–10^4 nucleotides incorporated. The error rate is relatively high because they do not have any proofreading activity; this causes RNA viruses to mutate and evolve rapidly.

Alfalfa mosaic virus coat protein bridges RNA and RNA-dependent RNA polymerase *in vitro*. V. L. Reichert *et al.* (2007) *Virology* **364**, 214–226.

In vitro assembly of the Tomato bushy stunt virus replicase requires the host heat shock protein 70. J. Pogany *et al.* (2008) *Proc. Natl. Acad. Sci. USA* **105**, 19956–19961.

6.2.B. Transcription: DNA-dependent RNA Polymerases

The first step in expressing the genetic information in DNA genomes is to produce RNA molecules using the DNA as a template by DNA-dependent RNA polymerases, a process known as **transcription**. The RNA molecules produced may, after modification, function directly, as in the case of transfer RNAs and ribosomal RNAs, or they are messenger RNAs and translated into proteins (Section 7.5.F). The initiation of transcription tends to be where gene expression is regulated, and each gene must be subject to the appropriate control. This requires a very sophisticated apparatus, of which the DNA-dependent RNA polymerase is just a part. As might be expected, these enzymes are physically complex and consist of multiple protein subunits. They usually include a core enzyme that is fully capable of elongation of an RNA chain and its termination but is unable to initiate transcription. The other components are involved in selection of the site at which RNA synthesis will be initiated (known as the **promoter**) and the rate at which it will be transcribed. Very many different proteins are known to be involved and are referred to collectively as **transcription factors**.

Several distinct types of RNA polymerases are often present in an organism. Eukaryotes typically have three that decode the nuclear genome into RNA: DNA-dependent RNA polymerases I, II and III (Pol I, II and III). Pol I is situated exclusively in the nucleolus, where its function is to synthesize ribosomal RNAs. Pol II is responsible for transcribing genes into messenger RNA that will be translated into proteins (Section 7.5.F), plus a few small nuclear RNAs. Pol III is restricted to transcribing a small set of highly expressed, nonprotein-coding (nc) RNA genes, including those for the 5S ribosomal

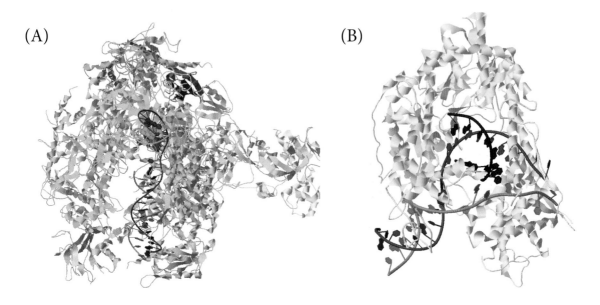

Figure 6-9. DNA-dependent RNA polymerases transcribing DNA. (A) The eukaryotic RNA polymerase complex of 12 polypeptide chains that elongates the messenger RNA. The nucleic acids are depicted in skeletal form, without the details of the polynucleotide backbone. Only the polypeptide backbones of the proteins are depicted, as ribbons, so that the nucleic acids at the active site are visible. Otherwise, they would be buried in the protein and not visible from the outside. (B) The much simpler single-polypeptide chain RNA polymerase from phage T7 transcribing a DNA duplex into RNA. The two DNA strands have been separated within the active site, and the RNA chain being synthesized off one is depicted in *purple*. Figure generated from PDB files *2owo* and *1msw* using the program Jmol.

RNA, transfer RNAs, small nuclear RNAs, microRNAs, short interspersed nuclear element-encoded or transfer RNA-derived RNAs and novel classes of ncRNA that can display significant sequence complementarity to protein-coding genes and might thus regulate their expression. Higher plants have two additional nonessential polymerases, Pol IVa and Pol IVb, that specialize in small RNA-mediated gene silencing pathways.

DNA-dependent RNA polymerases must identify the sequence where transcription should be initiated (the promoter), initiate transcription, translocate and pause along the DNA template, proofread errors, and ultimately terminate transcription at the correct location. Initiation by RNA Pol II requires the assembly of a specific group of general transcription factors at each promoter; the minimal initiation complex contains at least 30 different proteins! The process of elongation is nearly as complex (Figure 6-9-A). *In vivo*, DNA tends to be wrapped in chromatin (Section 3.7), which must be disassembled transiently; this requires the coordinated efforts of Pol II and its associated C-terminal domain kinases, elongation complexes, chromatin-modifying enzymes, chromatin-remodeling factors, histone chaperones (nucleosome assembly factors) and histone variants.

Bacteria have some of the most simple control systems, but even here transcription is complex. The core polymerase consists of four different polypeptide chains with a stoichiometry of $\alpha_2\beta'\beta\omega$; the subunits β' and β are most important catalytically and are similar to the subunits of eukaryotic enzymes. A fully functional enzyme, however, also requires a sigma factor (σ) that is essential for the RNA polymerase to bind to a specific promoter. There are various sigma factors, and bacteria exchange the sigma subunit in the RNA polymerase in response to different environmental stimuli and thereby tune their gene expression according to their current needs. Sigma factors are thought to recognize clearly distinguishable promoter DNA determinants and thereby activate distinct gene sets. The sigma factors themselves are regulated by antisigma factors, which bind and inhibit their cognate sigma factor, and 'appropriators', which deploy a particular sigma-associated RNA polymerase to a specific class of promoters. Adding to the complexity is the regulation of antisigma factors by both anti-antisigma factors, which turn on sigma factor activity, and co-antisigma factors, which act in concert with their partner antisigma factor to inhibit or redirect sigma activity. While the structures and functions of sigma factors are highly conserved, they are sufficiently diverse to enable bacteria to regulate sigma factor activities and respond to the diversity of environmental cues to which the bacterial transcription system has evolved.

The steps in transcription are similar for all DNA-dependent RNA polymerases. Initially the active form of the enzyme binds to the DNA at the appropriate promoter region and separates the two DNA strands there into a 'bubble' of individual strands, consisting usually of about 18 base pairs. Synthesis of a complementary copy of one of the strands is then initiated, but the synthesis is often aborted and the short RNA molecules produced are released. Once a transcript reaches about 8–12 nucleotides, the sigma factor dissociates from the complex and the remaining core enzyme continues extending the transcript without dissociating from the DNA. The 'bubble' of separated strands progresses along the DNA with the polymerase (Figure 6-9). The steps involved in adding each nucleotide to the complementary strands are similar to those involved in DNA synthesis using DNA polymerases (Figure 6-4). At the appropriate point, transcription is terminated and the transcript and polymerase are released.

The complexities of these DNA-dependent RNA polymerases have only been alluded to, but as a result these enzymes are not useful in the molecular biology laboratory to produce RNA from a piece of DNA. This would require all the transcription factors and other ancillary factors to be known for each piece of DNA, which would be impractical and is not feasible at the present time. Fortunately, much more simple enzymes are available from bacteriophages.

RNA polymerase: the vehicle of transcription. S. Borukhov & E. Nudler (2008) *Trends Microbiol.* **16**, 126–134.

RNA polymerase elongation factors. J. W. Roberts *et al.* (2008) *Ann. Rev. Microbiol.* **62**, 211–233.

RNA polymerase II stalling: loading at the start prepares genes for a sprint. J. Q. Wu & M. Snyder (2008) *Genome Biol.* **9**, 220.

Single-molecule studies of RNA polymerase: motoring along. K. M. Herbert *et al.* (2008) *Ann. Rev. Biochem.* **77**, 149–176.

---

1. Single-subunit DNA-dependent RNA Polymerases

The enzymes from bacteriophages T7, T3, SP6 and K11 have been characterized, with that from phage T7 serving as a model. Remarkably, in comparison with the systems described above, the single polypeptide chain of this enzyme is complete and fully functional in initiating, elongating and terminating transcription of any DNA with the appropriate signals in its sequence (Figure 6-9-B). Perhaps not surprisingly, the enzyme is similar structurally to the Klenow fragment of DNA polymerase I (Figure 6-3). The enzyme binds to DNA at a region 6–17 base pairs to the 5′-side of the transcription start site (known as upstream); these base pairs are designated as being at positions −17 to −6. Transcription is initiated when the enzyme binds to sites −5 to +6. Elongation of the transcript requires a substantial change in the structure of the complex of enzyme and DNA.

The phage T7 RNA polymerase does not recognize the transcription signals of the *E. coli* host, nor do the bacterial polymerases recognize the T7 signals. A further useful property of T7 RNA polymerase is that it differs from the *E. coli* enzyme in not being inhibited by the antibiotic **rifampicin**. Consequently, transcription by the host can be blocked while not affecting that by the phage T7 RNA polymerase. The T7 transcription signals are well-known and lie upstream and downstream of the nucleotide sequences to be transcribed; consequently, they can be attached at both ends of a segment of DNA so that it will be transcribed efficiently by the T7 polymerase. A variety of plasmid vectors are available commercially for transcribing any DNA fragment that is inserted into one of them (Figure 6-12).

Mechanism for *de novo* RNA synthesis and initiating nucleotide specificity by T7 RNA polymerase. W. P. Kennedy *et al.* (2007) *J. Mol. Biol.* **370**, 256–268.

Functional architecture of T7 RNA polymerase transcription complexes. D. Nayak *et al.* (2007) *J. Mol. Biol.* **371**, 490–500.

The structure of a transcribing T7 RNA polymerase in transition from initiation to elongation. K. J. Durniak *et al.* (2008) *Science* **322**, 553–557.

T7 RNA polymerase studied by force measurements varying cofactor concentration. T. Thomen *et al.* (2008) *Biophys. J.* **95**, 2423–2433.

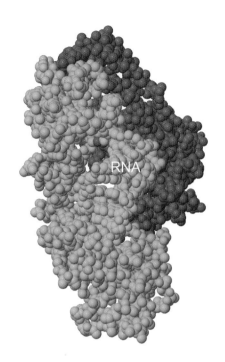

Figure 6-10. The reverse transcriptase of retrovirus HIV-1 binding an RNA pseudoknot. The van der Waals radii of the atoms of all the molecules are shown. The RNA is *green*, whereas the two polypeptide chains of the protein are *blue* and *red*. Figure generated from PDB file *1hvu* using the program Jmol.

6.2.C. Reverse Transcription: RNA into DNA

DNA is not normally produced using RNA as a template, but it does occur with retroviruses that have RNA genomes. These viruses insert their genomes into that of their host cell, which requires that the RNA genome first be converted to one of DNA. This is accomplished by an enzyme produced by the virus, **reverse transcriptase**, which is an 'RNA-dependent DNA polymerase'. It is similar in many respects to other DNA polymerases but is specific for copying RNA (Figure 6-10). The enzyme requires a primer complementary to the RNA strand to be copied. This role is played *in vivo* by a transfer RNA of the host, and the copying process is very complex but results in a double-stranded DNA copy. Reverse transcriptase has a high error rate when transcribing RNA into DNA, in the range of 1 mistake per 17,000–30,000 bases, because it has no proofreading ability. Retroviruses are responsible for human diseases such as AIDS, and currently one of the best ways to fight such a virus is to inhibit its reverse transcriptase activity; such inhibitors tend to be nucleoside analogs (Figure 2-12) and are some of the most powerful anti-AIDS medications available.

Associated with reverse transcriptase is **ribonuclease H** (H for 'hybrid') activity, which is specific for DNA•RNA hybrids and degrades the RNA strand that served as a template. This frees the DNA strand synthesized so that it can be used as a template to synthesize its complementary strand, and reverse transcriptase synthesizes both strands of the DNA copy. Other ribonucleases H can be used to target RNA molecules that have hybridized with a DNA probe.

Reverse transcriptase is used most frequently in molecular biology to make a **complementary DNA** (cDNA) copy of an RNA molecule, so that it can be cloned (Section 6.3.B). The RNA molecule is usually a messenger RNA that would be translated to make a protein (Section 7.5.F). The primer in this case is poly(dT), which is complementary to the poly(A) sequence that is usually at the 3'-ends of messenger RNAs.

Reverse-transcription PCR is PCR (Section 6.1.C) performed on RNA. The first step is the production of a cDNA strand from the RNA, using the enzyme reverse transcriptase rather than DNA polymerase.

Performing quantitative reverse-transcribed polymerase chain reaction experiments. G. Lutfalla & G. Uze (2006) *Methods Enzymol.* **410**, 386–400.

RNA integrity and the effect on the real-time qRT-PCR performance. S. Fleige & M. W. Pfaffl (2006) *Mol. Aspects Med.* **27**, 126–139.

Crystal structure of type 1 ribonuclease H from hyperthermophilic archaeon *Sulfolobus tokodaii*: role of arginine 118 and C-terminal anchoring. D. J. You *et al.* (2007) *Biochemistry* **46**, 11494–11503.

Murine leukemia virus reverse transcriptase: structural comparison with HIV-1 reverse transcriptase. M. L. Coté & M. J. Roth (2008) *Virus Res.* **134**, 186–202.

6.2.D. Antisense Oligonucleotides

A complementary oligonucleotide can hybridize with a target single-stranded DNA or RNA molecule and thereby block its subsequent use. This is especially useful with a messenger RNA (mRNA), which conveys the information from the gene about the amino acid sequence of the protein product of that gene. When expression of this protein is to be blocked in this way, the complementary oligonucleotide is known as an **antisense oligonucleotide**. Formation of the antisense•mRNA heteroduplex in the cell can (1) trigger ribonuclease H activity, leading to mRNA degradation, (2) induce arrest of translation by steric hindrance of ribosomal activity, (3) interfere with mRNA maturation by inhibiting its splicing, or (4) destabilize mRNA precursors in the nucleus; all diminish expression of the target gene. Inhibiting the expression of specific genes helps to elucidate the physiological role of the gene and its product. Such an antisense oligonucleotide might also be useful therapeutically, for example to block an infection by a bacterium or virus. To be useful *in vivo*, however, an antisense oligonucleotide must be much more stable than ordinary nucleic acids, and it must be able to reach the interior of a cell (Section 5.4).

Phosphorothioate oligodeoxynucleotides have been studied extensively because they are resistant to hydrolysis by nucleases. They differ from normal DNA in that one of the nonbridging O atoms in each phosphate group is replaced by a S atom. The melting temperature of a heteroduplex between a phosphorothioate oligonucleotide and RNA is lower than for the DNA•DNA duplex by about 0.5°C for each nucleotide, and some 2.2°C lower than for an RNA•RNA duplex. Consequently, these antisense oligonucleotides must be relatively long, typically at least 17–20 nucleotides, to bind specifically to RNA, and they are not effective if the RNA sequence is involved in a double-stranded structure. Unfortunately, they have been found to bind to and inhibit enzymes involved with nucleic acids, such as DNA polymerase, be immunogenic and have a variety of unexpected side-effects. A key drawback lies in their polydiastereomerism (Section 1.1.A.3) because the various diastereomeric species with the S atoms in different positions exhibit different chemical and biological properties. Another type of molecule that can be useful as an antisense oligonucleotide is a nucleic acid mimic, such as a peptide nucleic acid (Section 5.4).

The phenomenon of antisense also seems to be used naturally, as many genes contain segments on both DNA strands of the genome that are transcribed, and the two oppositely oriented transcripts can overlap. Often, one strand codes for a protein, whereas the transcript from the other strand is nonencoding. Such **natural antisense transcripts** (NATs) can inhibit translation of the sense transcript. NATs are highly prevalent in a wide range of species; for example, around 15% of human protein-encoding genes have an associated NAT. What they do *in vivo* is a mystery.

Catalytic RNAs and DNAs (Section 14.6) can also function as antisense reagents. They cleave target RNAs after base pairing via their antisense flanking arms.

A related antisense phenomenon is **RNA interference**, in which small interfering double-stranded RNA molecules induce RNA degradation through a natural gene-silencing pathway. Being double stranded, they contain both sense and antisense regions of the target RNA. They cause destruction of the target RNA by a poorly understood natural phenomenon in which they are incorporated into the RNA interference (RNAi) targeting complex of RNA and protein that then cleaves the target RNA.

Identifying accessible sites in RNA: the first step in designing antisense reagents. W. H. Pan & G. A. Clawson (2006) *Curr. Med. Chem.* **13**, 3083–3103.

Recent advances in the stereocontrolled synthesis of antisense phosphorothioates. Y. Lu (2006) *Mini Rev. Med. Chem.* **6**, 320–330.

Genome-wide natural antisense transcription: coupling its regulation to its different regulatory mechanisms. M. Lapidot & Y. Pilpel (2006) *EMBO Rep.* **7**, 1216–1222.

Antisense oligonucleotides: target validation and development of systemically delivered therapeutic nanoparticles. C. Zhang *et al.* (2007) *Methods Mol. Biol.* **361**, 163–185.

RNA learns from antisense. D. R. Corey (2007) *Nature Chem. Biol.* **3**, 8–11.

6.3. CLONING

Manipulating DNA is at the very heart of molecular biology. For such an important technique, the physical principle is very simple: DNA is fragmented into pieces of the desired size, then the pieces are inserted into an appropriate **cloning vector** (Figure 6-11) that serves to replicate and store the DNA. The desired fragment is then identified and its cloning vector purified. This **clone** is then the source of pure DNA fragments, and cloning is one method of purifying and isolating a single homogeneous segment of DNA.

The cloning vector need have only sufficient genetic information to permit it to replicate within its host cell. Four basic types are used in the bacterium *E. coli* as a host: plasmids, bacteriophage vectors, cosmids and bacterial artificial chromosomes. All of them have been engineered to accommodate relatively large segments of foreign DNA, without impairing their ability to replicate within their host. **Plasmids** have minimal genomes, just sufficient for them to replicate within the host. They have the advantage of being smaller than the other vectors and of propagating with a high **copy number**, the number of copies of the vector per host cell. **Bacteriophage** vectors can accommodate larger inserts of DNA, up to 120,000 base pairs, and their growth and propagation are straightforward. **Cosmids** combine properties of the preceding two, in that they can replicate as plasmids or they can be packaged as bacteriophages. **Bacterial artificial chromosomes** and **yeast artificial chromosomes** can accept even larger DNA fragments, greater than 300,000 base pairs.

How the clone with the desired DNA fragment is identified depends upon many considerations, but it usually involves hybridization to a single-stranded DNA molecule with a complementary sequence (Section 5.3) or translation to produce a protein with the expected amino acid sequence, reactivity with an antibody or biological activity. Some cloning vectors permit the cloned DNA fragment to be

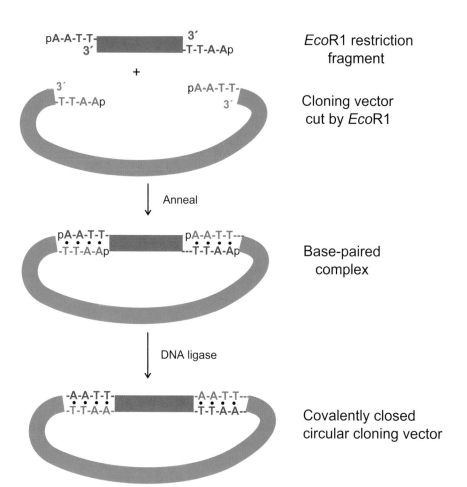

Figure 6-11. Cloning of a DNA fragment into a cloning vector using base pairing between the identical 'sticky ends' of the pair. In this case, both the DNA fragment to be cloned and the cloning vector have been generated by cleavage with the restriction enzyme *Eco*R1 (Equation 6.3). Their complementary sticky ends produce a base-paired complex. The two missing phosphodiester bonds are generated using an enzyme like DNA ligase.

transcribed, translated and expressed as a polypeptide chain (Section 6.3.A); if the product is a folded protein, its biological functions can be assayed.

Strategies for high-throughput gene cloning and expression. L. J. Dieckman *et al.* (2006) *Genet. Eng.* **27**, 179–190.

Enzyme free cloning for high throughput gene cloning and expression. R. N. de Jong *et al.* (2006) *J. Struct. Funct. Genomics* **7**, 109–118.

6.3.A. Expression Vectors

Very often, the cloned DNA is to be transcribed into RNA and translated into a polypeptide chain. Special vectors for this purpose are available that contain all the sequences to signal the initiation and termination of transcription and translation. These signals are at both ends of the cloned sequence, so they are part of the vector and can be used to express any DNA fragment that is cloned in between them. The vector must be inserted into a suitable host cell for expression to occur. The host must provide all the functions necessary for expression; for example, an unprocessed gene will need the introns to be removed from the initial RNA transcript before it can be translated into protein.

A general expression vector for use in prokaryotes is described in Figure 6-12. It is a circular DNA duplex plasmid that contains a sequence serving as an **origin of replication** of the plasmid, so that it can be replicated and maintained within the host. There is usually a **selection marker gene** to ensure that the bacterial host retains the plasmid. This is usually a gene that conveys resistance to an antibiotic, so only cells containing the plasmid can grow in the presence of that antibiotic. A **promoter** sequence provides the site for initiation of transcription; it usually determines the degree to which expression will take place. Eukaryotic promoters usually have a **TATA box** about 30 nucleotides upstream of the transcription start site, with the consensus sequence T–A–T–A–A/T–A–A/T; it is recognized by one of the core transcription factors and positions the RNA polymerase on the DNA. A downstream sequence provides a site for the initiation of translation of the messenger RNA transcript, which must include a site for binding ribosomes and a **start codon**, usually AUG, for initiation of synthesis of the polypeptide chain. In prokaryotes, a **Shine–Dalgarno sequence** is positioned 4–13 nucleotides upstream of the start codon; it is complementary to the 3′-end of an RNA of the ribosome and positions the mRNA and ribosome appropriately. The DNA fragment to be cloned must be inserted just after the start codon and in such a way that the correct triplet reading frame is in phase with the start and stop codons (Figure 7-21). The coding sequence is followed by a **stop codon** to terminate polypeptide chain elongation. Finally there are appropriate signals to stop transcription and add the poly(A) tail to the messenger RNA (which is carried out by a specific enzyme). Certain sequences that have the opposite effect, **antitermination**, can be incorporated to stabilize the RNA polymerase on the DNA and increase elongation of the transcript.

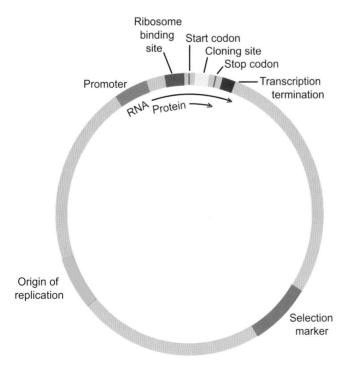

Figure 6-12. A generalized expression vector for prokaryotes. The *circle* represents the DNA duplex of the vector. The origin of replication is for replicating the vector in the host cell. The selection marker is usually a gene conferring on the host resistance to an antibiotic, so that only cells containing the plasmid can grow in the presence of that antibiotic. The promoter specifies initiation of transcription into RNA, which is terminated at the termination signal. The crucial sites on a messenger RNA are the translation site for binding a ribosome and the start and stop codons for producing the polypeptide chain. The DNA of interest is cloned between the latter two sites in such a way that its nucleotide sequence is translated into amino acid residues of the polypeptide chain.

Expression requires that the appropriate RNA polymerase be available. In prokaryotic cells it is usually that from phage T7 and its gene is present on the vector. It is often engineered to be expressed only under certain growth conditions, usually after addition of some inducer molecule. The T7 polymerase is then produced, and it can activate the expression vector. This makes it possible to control the expression of the cloned DNA, which often is deleterious to the host cell.

Baculovirus expression vector system: an emerging host for high-throughput eukaryotic protein expression. B. Shrestha *et al.* (2008) *Methods Mol. Biol.* **439**, 269–289.

pETPhos: a customized expression vector designed for further characterization of Ser/Thr/Tyr protein kinases and their substrates. M. J. Canova *et al.* (2008) *Plasmid* **60**, 149–153.

6.3.B. cDNA Libraries

Messenger RNAs contain the gene sequences of all the proteins that are produced in a cell. They can be readily converted to the amino acid sequences of those proteins because the genetic code is known (Figure 7-21) and because the introns that interrupt the coding sequence in the gene have been removed after transcription. If even only a short segment of the amino acid sequence of the probe is known, it can be used to design a short nucleotide probe that will be complementary to the gene sequence, making it possible to identify the corresponding gene. There is redundancy in the genetic code, however, so the exact sequence of the probe cannot be specified. This can be circumvented by preparing a mixture of probes with all the possible nucleotides represented, so that at least a fraction of the probe molecules will have the correct sequence. Alternatively, less specific bases, such as inosine, can be used where the nucleotide sequence is uncertain.

RNA is much less stable and more difficult to handle than DNA, so the first step in isolating a messenger RNA is to convert it to DNA, using the enzyme reverse transcriptase (Section 6.2.C). Normally, all the messenger RNA of a cell is converted to cDNA and cloned, and the desired molecule is then identified. These steps use the fact that most messenger RNAs have a tract of A nucleotides at their 3′-ends. The population of messenger RNAs can then be isolated by hybridizing them to an affinity column, a solid support containing complementary poly(T) oligonucleotides. All other molecules are removed by washing the solid support, before releasing the messenger RNAs. They are then used as templates for reverse transcriptase to synthesize DNA copies. Oligo-dT is used as the primer, so synthesis is initiated at the 3′-end of the messenger RNA. The resulting double-stranded DNA copy of the messenger RNA can be cloned into an appropriate vector.

Very often a cDNA copy is not complete, due to problems with the reverse-transcription step or because the messenger RNA molecules were not intact but degraded. Even short segments of a few hundred nucleotides can be very useful to identify an expressed gene, however, and they are known as **expressed sequence tags**.

Subtractive hybridization and construction of cDNA libraries. B. Blumberg & J. C. Belmonte (2008) *Methods Mol. Biol.* **461**, 569–587.

Identification of microRNAs and other small regulatory RNAs using cDNA library sequencing. M. Hafner *et al.* (2008) *Methods* **44**, 3–12.

6.3.C. Restriction Enzymes

Manipulating the very long molecules of DNA that occur naturally requires that they be cleaved into smaller fragments of manageable size, but not so small as to be indistinguishable from each other. **Restriction enzymes are endonucleases that recognize and cleave DNA molecules at specific sequences of 4–8 nucleotides**; they are the workhorses of molecular biology. They occur naturally in many bacteria, where they are used to recognize and cleave foreign DNA, such as that introduced by a bacterial virus. The initial fragmentation makes the DNA susceptible to further degradation by general exonucleases. **The host DNA is not cleaved because it has been modified covalently by methylation of specific cytosine or adenosine nucleotides** within the sequence recognized by the restriction enzyme, using a specific methyltransferase enzyme. One class of methyltransferases methylates the C5 carbon of cytosine, to form 5-methyl cytosine, whereas the other class methylates the 4-amino group of cytosine or the 6-amino group of adenine. The endonuclease and the methyltransferase act as a pair; in some cases, the two activities are catalyzed by a single protein molecule. Some bacteriophages have evolved systems to avoid destruction by restriction enzymes of their host by (1) producing proteins that inhibit the restriction enzymes, (2) stimulating the methylation system of the host, (3) modifying their DNA and (4) eliminating from their genomes sites recognized by the restriction enzymes.

Almost all restriction enzymes cleave at DNA sequences that are inverted repeats, in that they have the same sequence on both strands and a dyad symmetry axis (Section 2.2.C.6). Examples of such restriction sequences are the target sites for the restriction enzymes *Eco*R1 (GAATTC) and *Fin*II (CCGG).

Restriction enzymes generally cleave both strands of the DNA molecule, within or near the target sequence. The resulting fragments have a phosphate group at the new 5′-end and a hydroxyl group at the new 3′-end. Often, the bonds cleaved are not at the center of the target sequence, so cleaving between the same sequence on each strand generates fragments with protruding 5′- and 3′-single strands. For example, the *Eco*R1 restriction enzyme cuts between the G and A of the sequence GAATTC:

$$5'-G-A-A-T-T-C-3' \atop 3'-C-T-T-A-A-G-5' \quad \xrightarrow{EcoR1} \quad {5'-G \atop 3'-C-T-T-A-A_p} + {{}_pA-A-T-T-C-3' \atop G-5'}$$

(6.3)

The protruding single strands are known as **staggered**, **sticky** or **cohesive** ends. They are extremely useful, as their base pairing with complementary sequences, perhaps generated by the same restriction enzyme, makes it easy to insert the DNA fragment into another DNA molecule (Figure 6-11). Such manipulations are central to the isolation and cloning of DNA.

Other restriction enzymes cleave at the center of the target sequence and thereby cleave at the phosphodiester bonds on opposite sides of the DNA helix. The resulting fragments do not have single-stranded ends but are known as **blunt-ended**. Lacking any sequence complementarity at the cloning site, they insert into other DNA molecules much less readily but there are no restrictions on what fragments can be cloned.

More than 2600 different restriction enzymes are known, with more than 230 cleavage specificities, and most are available commercially. Different restriction enzymes that recognize and cut at the same position in identical target sequences are referred to as **isoschizomers**. For example, the *Eco*RI and *Rsr*I restriction enzymes both recognize and cut between the G and A of the sequence GAATTC (Equation 6.3). Other restriction enzymes that recognize the same sequence but cut at different positions are known as **neoschizomers**. For example, restriction enzymes *Asp*718I and *Kpn*I both recognize the sequence GGTACC but one cleaves between the two Gs, the other between the two Cs.

Restriction enzymes are very specific for their target sequences under normal physiological conditions and in the presence of Mg^{2+}, but they become less specific under different conditions of pH, salt concentration or Ca^{2+} or Mn^{2+} in place of Mg^{2+}. Any activity towards alternative sequences is known as the **star activity**.

How restriction enzymes became the workhorses of molecular biology. R. J. Roberts (2005) *Proc. Natl. Acad. Sci. USA* **102**, 5905–5908.

The crystal structure of the rare-cutting restriction enzyme *Sda*I reveals unexpected domain architecture. G. Tamulaitiene *et al.* (2006) *Structure* **14**, 1389–1400.

DNA looping and translocation provide an optimal cleavage mechanism for the type III restriction enzymes. N. Crampton *et al.* (2007) *EMBO J.* **26**, 3815–3825.

Use of plasmon coupling to reveal the dynamics of DNA bending and cleavage by single EcoRV restriction enzymes. B. M. Reinhard *et al.* (2007) *Proc. Natl. Acad. Sci. USA* **104**, 2667–2672.

6.3.D. Restriction Maps

To elucidate the structure and sequence of large DNA molecules, it can be useful to **map the positions of restriction sites within the molecule**. This is accomplished by cleaving the large DNA molecule with two or more restriction enzymes that produce a reasonable number of fragments of convenient size. With very large DNA molecules, the restriction enzymes must have target sequences that occur rarely, whereas less specific restriction enzymes are useful with smaller molecules. The fragments resulting from cleavage by either or both restriction enzymes are separated according to their size by gel electrophoresis. The shorter fragments migrate more rapidly, and their mobilities can be compared with those of standard DNA molecules and used to estimate the length of each DNA molecule. This information is then used to construct a **restriction map, which gives the order of the fragments in the original DNA molecule**.

The procedure is illustrated using a short linear molecule of DNA in Figure 6-13. The DNA was digested with the restriction enzymes *Eco*RI and *Bam*H1, which cleave at G/AATTC and G/GATCC sequences, respectively (/indicates the bond at the site of cleavage). Each generated three fragments individually, and five in combination, indicating that the original DNA molecule had two restriction

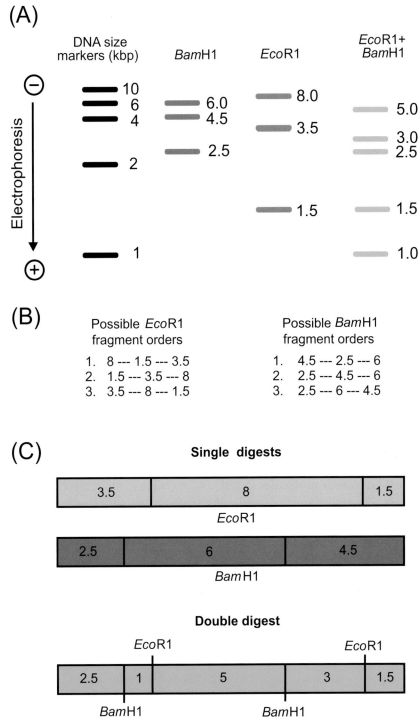

Figure 6-13. Constructing a restriction map. (A) Schematic illustration of the idealized pattern obtained after electrophoresis of the DNA fragments generated by cleavage of a 13,000-base pair DNA molecule with the restriction enzymes EcoR1 and BamH1 individually or together. Electrophoresis was from the *top* to the *bottom*, with the smaller fragments migrating more rapidly. Their sizes are estimated by comparison to the standards in lane 1 on the *left*. The lengths of the fragments are given in thousands of base pairs (kbp). Each restriction enzyme produced three fragments, so there were two restriction sites for each enzyme. (B) All possible orders in the original DNA molecule of the three restriction fragments from each restriction enzyme. The orders (3) from both enzymes combine to reproduce the sizes of the fragments observed with the combined enzymes. (C) The positions of the restriction sites indicated by the data.

sites cleaved by each enzyme. The sizes of the five fragments generated by the two enzymes in combination indicate the positions of these sites within the original DNA molecule. This simple illustration is straightforward, but the procedure is much less so with large molecules containing many restriction sites. In cases of ambiguity, the original DNA molecule can be labeled at one end and the time–course of cleavage by the restriction enzymes monitored. A similar approach using polypeptides is illustrated in Figure 7-14. The labeled molecules all contain one end of the molecule, and fragments corresponding to all the restriction sites should occur at least transiently.

Restriction maps are also important in comparing the genomes of related individuals and species. Genetic mutations can alter the restriction sites or insert or delete genetic material between the sites.

An algorithm for assembly of ordered restriction maps from single DNA molecules. A. Valouev *et al.* (2006) *Proc. Natl. Acad. Sci. USA* **103**, 15770–15775.

6.4. SEQUENCING DNA

A most striking characteristic of natural DNA molecules is their lengths, which can be enormous. The genome of the relatively simple common bacterium *E. coli* is a single double-stranded DNA molecule (closed in a circle) in which each strand contains 4.6 million nucleotides. This corresponds to 9.2 million bits or 1.15 megabytes of information: each residue can be one of four bases and $4 = 2^2$, so each residue contains two bits of information, and 8 bits = 1 byte. DNA molecules from higher organisms can be much larger: each chromosome generally contains a single double-stranded DNA molecule. The human genome consists of approximately 3 billion nucleotides divided amongst 24 molecules (chromosomes). Individual chromosomes can consist of as many as 1 billion nucleotides that, if extended, would stretch 0.33 m in length. Yet **it is possible to determine exactly the covalent structures of these enormous molecules** in terms of their nucleotide sequences, and this is one of the most important accomplishments of molecular biology. The sequence information can then be used to deduce the function of the DNA segment, map its chromosomal location, determine how it might be involved in regulation of gene expression or replication, or elucidate how it interacts with proteins. Furthermore, the nucleotide sequences of the regions coding for protein can be translated, using the genetic code (Figure 7-21), into the amino acid sequences of the proteins, and these protein sequences can be compared with the sequences of all known proteins, giving clues to their functions and evolutionary origins (Section 7.6).

Molecular engineering approaches for DNA sequencing and analysis. X. Bai *et al.* (2005) *Expert Rev. Mol. Diagn.* **5**, 797–808.

Faster, cheaper DNA sequencing. L. Sage (2005) *Anal. Chem.* **77**, 415A–416A.

Towards rapid DNA sequencing: detecting single-stranded DNA with a solid-state nanopore. H. Yan & B. Xu (2006) *Small* **2**, 310–312.

6.4.A. Isolating the DNA Fragments to be Sequenced

DNA sequencing requires that the DNA molecules be relatively short and homogeneous. Initially specific DNA segments were isolated by cloning them into vectors derived from the M13 bacteriophage; these vectors contain a single-stranded DNA chromosome that can accommodate inserts of up to 5000 bases of foreign DNA. The cloning produces homogeneous clones with a single DNA insert. Pure single-stranded DNA can be isolated from this phage easily, and this method is still in common use. Virtually any plasmid can be used in a similar way for DNA sequencing, and essentially any clone of up to about 200,000 bases can be isolated by cloning and then sequenced. DNA is frequently isolated for sequencing by using PCR (Section 6.1.C). Any segments of genomic DNA can be amplified for direct sequencing in a matter of hours.

6.4.B. Chain-termination, Sanger Method

The chain-termination method of DNA sequencing was first described by Sanger in 1977 and is now used almost universally. It involves the enzymatic synthesis of a DNA strand by a DNA polymerase *in vitro*, using the DNA strand to be sequenced as the template (Figure 6-1). Synthesis is initiated by DNA polymerases only at the 3′-end of a primer that is base paired to the DNA template, so synthesis is initiated at only that one site (Figure 6-2). The growing chain is terminated by the occasional random incorporation of a 2′,3′-**dideoxynucleoside triphosphate** (ddNTP), which has no free hydroxyl group and does not permit further DNA synthesis (hence the name *chain-termination*).

Most sequencing applications use as the primer a short synthetic oligonucleotide 18–35 nucleotides in length and complementary in sequence to the template at a unique position. This position is adjacent to, and determines, the region to be sequenced. The sequence of this part of the template must be known in advance; it can, fortunately, be part of the plasmid or virus in which the DNA segment was cloned. The primer is hybridized to the template at an appropriate temperature to ensure that it binds only in the correct position (Figure 6-2). After formation of this duplex, the primer is extended by the DNA polymerase in the presence of the four deoxynucleoside triphosphates: dGTP, dATP, dTTP and dCTP, which are often designated collectively as **dNTPs**.

Extension of the new strand requires a free 3′-hydroxyl group (Figure 6-4). One will be available for continued polymerization of the growing chain so long as normal dNTPs are incorporated (Figure 6-1). The ddNTPs, however, lack a 3′-hydroxyl group, and their incorporation into the DNA terminates chain elongation. Synthesis directed by a sample of template DNA with a unique primer in the presence of all four dNTPs, plus a small concentration of one ddNTP, will produce a population of molecules that have the same 5′-end but different 3′-ends; all will end with the same terminal dideoxynucleotide base, but at different positions. Consequently, the chains synthesized will have a distribution of sizes depending on the sites at which the ddNTP was incorporated (Figure 6-14) and the sizes of the various fragments are determined by the sequence of the template. Four reactions are performed, each with a different ddNTP. The products of the four reactions can be analyzed by electrophoresis on a denaturing polyacrylamide gel, which separates the products by size and can resolve DNA strands differing in length by a single nucleotide. The sequence of the template can be determined from the order of the bands on the gel (Figure 6-15). The shortest oligonucleotides migrate most rapidly, and the sequence can be read off starting there by the presence of bands in the mixtures with the four different ddNTPs. Of course, this sequence is of the strand being synthesized and is the complement of the template strand.

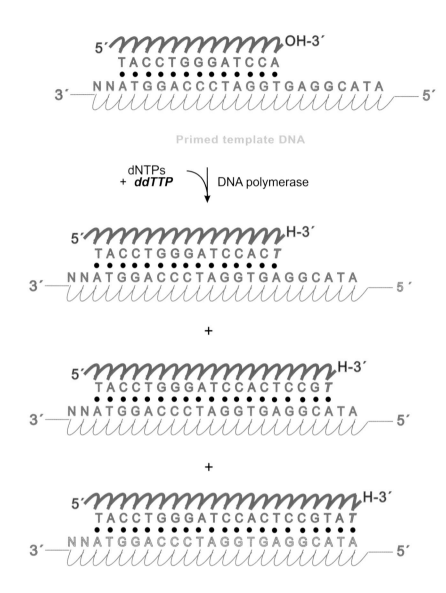

Figure 6-14. Chain-termination by incorporation of a ddNTP. Copying the primed template from Figure 6-2 in the presence of all four deoxynucleoside triphosphate and one dideoxynucleoside triphosphate (ddNTP, in this case ***T***, indicated by *bold italic*) yields a population of extended primer molecules with the same 5′-ends but different 3′-ends lacking a hydroxyl group, depending on the site at which a ddNTP was incorporated. Consequently, the lengths of the extended primer molecules indicate the positions in the sequence of T nucleotides (A on the other strand).

The **DNA polymerase** enzyme is the most critical component. Many DNA polymerases and reverse transcriptases have been used for sequencing. Most have been modified genetically or chemically to eliminate exonuclease activities or improve their utilization of the ddNTPs. The most popular polymerases designed for DNA sequencing are heat-stable because they originate from the thermophile *Thermus aquaticus* and are known as **Taq DNA polymerase**. The 5′–3′-exonuclease activity of natural *Taq* DNA polymerase can degrade the sequencing primers, so it is usually removed. This activity is due to the first 300 amino acid residues at the N-terminus of this enzyme, which can be deleted or a few crucial residues altered. The limited ability of *Taq* DNA polymerase to incorporate ddNTPs is improved more than 10^4-fold simply by changing residue Phe667 to Tyr. The sequencing

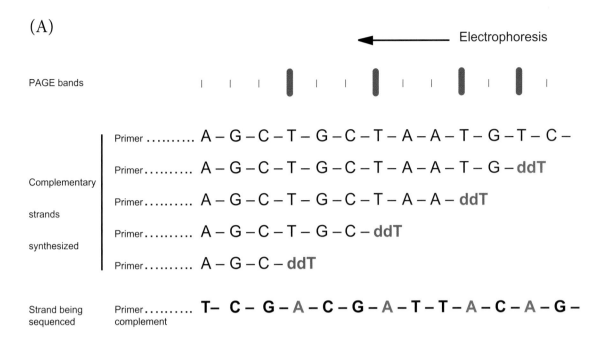

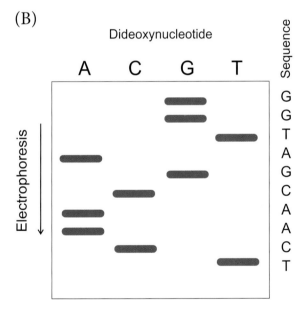

Figure 6-15. Sequencing DNA using the chain-termination method. (A) Complementary strands are synthesized enzymatically from the primer, using the DNA strand as template (Figure 6-1). The complementary strands are elongated until a dideoxynucleotide is incorporated in place of the normal one, which terminates strand elongation (Figure 6-14). A set of complementary strands with their lengths determined by the positions of these nucleotides will result. The same mixtures but using the other three dideoxynucleotides would be analyzed at the same time (B). The complementary strands synthesized are separated by electrophoresis that can resolve strands differing in length by a single nucleotide. The shorter strands migrate more rapidly. The positions of the bands indicate the positions of the four nucleotides in the synthesized DNA strand, and the complete sequence of both strands can be determined.

bands produced should have the same intensity, but those produced by natural *Taq* DNA polymerase vary in intensity more than 15-fold and depend on the nearby sequence. In contrast, those produced by the Tyr667 polymerase vary in intensity less than three-fold, which makes interpretation of the sequencing results much more accurate. The DNA polymerase from phage T7 naturally has a Tyr at the corresponding position and functions similarly, but it is not heat-stable.

Cycle sequencing uses repeated cycles of thermal denaturation and polymerization to increase the amount of product in a DNA sequencing reaction and the sensitivity of the sequencing experiment. During each cycle, the thermostable DNA polymerase extends the annealed primer molecules. The mixture is then heated to 95°C to dissociate the extended primer from the template. The mixture is cooled, allowing another molecule of the excess primer to anneal to the template, and the elongation reactions are repeated. The amount of product DNA increases linearly with the number of cycles, and it can become much greater than the amount of template used originally. The dissociation step enables double-stranded templates to be used directly for sequencing. Cycle sequencing is usually more reliable over a wider range of template concentrations than are other methods. It is used for most large-scale DNA-sequencing projects.

All DNA-sequencing methods require that the strands synthesized be labeled. The label used originally was α-^{32}P dATP. It was simply added to the chain-termination reaction so that the newly synthesized DNA molecules were labeled uniformly with ^{32}P and could be detected by autoradiography. Subsequently, the isotopes ^{33}P and ^{35}S have been used because their radioactive emissions have lower energy and they consequently produce higher resolution autoradiograms. ^{35}S is used in the form of α-thio-dATP, in which an O atom of the phosphate group is replaced by S. Limiting the radioactivity to the terminating ddNTPs offers the advantage that the only DNA chains labeled are those that were elongated and terminated specifically. This diminishes the background radioactivity and produces extremely clean DNA sequencing autoradiograms.

DNA-sequencing methods that use fluorescence rather than radioactivity have been automated and have become essential for large-scale sequencing projects. Either the primers (dye primers) or the dideoxynucleotides (dye terminators) can be made fluorescent and used to label the DNA chains produced by chain-termination. The four bases can be distinguished if they fluoresce with different colors, which makes it possible to use all four in the same sequencing mixture. Fluorescent dye primers can use fluorescence energy transfer to optimize the absorption and emission properties of the label. For example, they might have a fluorescein derivative at the 5′-end that serves as a common fluorescence donor, with the fluorescence acceptor being a rhodamine derivative attached to a modified thymidine within the primer sequence. The spacing between fluorescence donor and acceptor can be varied by the position of the modified thymidine in the primer sequence, altering the efficiency of fluorescence energy transfer. Four distinguishable primers can be devised, one for each of the four bases: all absorb strongly at the same excitation wavelength of 488 nm provided by the fluorescein acceptor, but differ in the wavelength of the fluorescent light they emit; for example, those for the four different nucleotides can emit light at 525, 555, 580 and 605 nm. All four primers can be combined in a single sequencing reaction, improving the sensitivity and accuracy as well as providing the complete sequence of all four nucleotides.

ddNTP terminators labeled with four different fluorescent dyes have also been used extensively for DNA sequencing and are commercially available. Like the radioactive terminators, they have the advantage of labeling only DNA chains that have been elongated and terminated specifically, so background bands are minimized.

Emerging technologies in DNA sequencing. M. L. Metzker (2005) *Genome Res.* **15**, 1767–1776.

Microfabricated bioprocessor for integrated nanoliter-scale Sanger DNA sequencing. R. G. Blazej *et al.* (2006) *Proc. Natl. Acad. Sci. USA* **103**, 7240–7245.

Four-color DNA sequencing by synthesis using cleavable fluorescent nucleotide reversible terminators. J. Ju *et al.* (2006) *Proc. Natl. Acad. Sci. USA* **103**, 19635–19640.

6.4.C. Separating the DNA Fragments by Size

All sequencing methods require the separation of DNA fragments on the basis of their size, and **it must be possible to separate molecules differing in length by a single nucleotide**. A sieving gel is usually necessary to separate the DNA fragments by size; otherwise, DNA molecules of different lengths have identical charge-to-mass ratios and the same electrophoretic mobility in aqueous solution. In the original DNA sequencing experiments using radioactive labeling, separation was by electrophoresis in 4–8% (w/v) cross-linked polyacrylamide gels, containing 7–8 M urea to keep the DNA molecules unfolded. The gels were 0.2–0.4 mm thick and cast between glass plates that were 40–80 cm long; each gel could accommodate up to 96 samples. After electrophoresis for 2–18 hours, the gels were separated from the glass plates and subjected to autoradiography. Usually 200–400 nucleotides of the sequence could be determined from a single sample. These procedures are labor-intensive, however, so there was much incentive to improve the separation methods.

Current separation methods monitor continuously the fluorescence due to the labeled DNA molecules as they migrate electrophoretically past a fixed position on the gel. Noncross-linked gels in capillaries that are 50–100 μm in diameter and 0.4 –0.7 m long provide faster and higher resolution electrophoresis separations, due to the efficient heat transfer of these narrow capillaries. A sequence of 500 or more nucleotides can be determined from a single experiment. A major advantage is that the results are collected and evaluated directly by computer, saving human labor.

End-labeled free-solution electrophoresis overcomes the need for a gel by using a label (or 'drag-tag') to cause the free solution mobility of the DNA to depend upon the size of an oligonucleotide. Attachment of an uncharged drag-tag molecule of a fixed size to various lengths of DNA in the sample selectively slows down smaller DNA chains, because they have less electrophoretic force to pull the drag-tag than do longer DNA molecules. Other variations involve separating the fragments by mass spectrometry.

To obtain the sequence of a very large DNA molecule, such as that of a complete genome, requires short fragments to be sequenced and then reassembled into the complete molecule using overlaps of the sequences of the various segments. A logical approach would be to sequence, say, 400 residues of one segment, then use a short sequence of it to design a primer to sequence the next 400 nucleotides, and so forth, but this is relatively slow. Other approaches to sequencing large genomes fragment the DNA into relatively short pieces of DNA randomly and insert all of them into an appropriate cloning vector; the individual clones are purified and all are sequenced. The original sequence is assembled by finding the same sequence in the various overlapping clones. This approach of random **shotgun sequencing** requires that each nucleotide be sequenced at least twice, and usually many more times, but it can be more efficient than trying to do the sequencing in a rational, logical order.

Recent developments in mass spectrometry for the characterization of nucleosides, nucleotides, oligonucleotides, and nucleic acids. J. H. Banoub *et al.* (2005) *Chem. Rev.* **105**, 1869–1915.

Mass-spectrometry DNA sequencing. J. R. Edwards *et al.* (2005) *Mutation Res.* **573**, 3–12.

DNA sequencing based on intrinsic molecular charges. T. Sakata & Y. Miyahara (2006) *Angew. Chem. Int. Ed. Engl.* **45**, 2225–2228.

Novel design of multicapillary arrays for high-throughput DNA sequencing. A. Tsupryk *et al.* (2006) *Electrophoresis* **27**, 2869–2879.

Scale-up development of high-performance polymer matrix for DNA sequencing analysis. F. Wan *et al.* (2006) *Electrophoresis* **27**, 3712–3723.

6.4.D. Alternative Approaches

Another general sequencing method, known as the **chain-cleavage** or **Maxam–Gilbert** method, was developed early and used extensively. This method also maps the DNA sequence to the sizes of DNA molecules, but it does so by degrading existing DNA chains rather than synthesizing new ones. A sample of purified DNA to be sequenced is first labeled at one end. Labeling the 5'-end is usually done with [γ-^{32}P]ATP and T4 **polynucleotide kinase**. Labeling the 3'-end of either DNA or RNA can be accomplished by adding a radioactive pNp to the 3'-hydroxyl group using the enzyme T4 **RNA ligase**. Then the end-labeled DNA is subjected to a chemical treatment that will break some of the DNA molecules at random, but only at places where one or two bases occurs. Dimethyl sulfate at acid pH methylates both A and G bases, whereas only G is modified at neutral pH (Figure 2-33). Hydrazine ($H_2N–NH_2$) normally reacts with both C and T bases, but only with C in 1.5 M NaCl. These modifications make the adjacent phosphodiester linkages very susceptible to chemical cleavage. The result of each treatment is a population of labeled molecules whose sizes are determined by the sites of cleavage in the sequence. Determination of the sizes in samples cleaved with these four sequence-specific cleavage treatments can yield complete sequence information, as in chain-termination sequencing (Figure 6-15). Unfortunately, it is difficult to find chemical cleavage conditions that consistently give unambiguous, clean cleavage products, and few labeling methods provide labels stable enough for these treatments. For these and other technical reasons, chain-cleavage methods of sequencing are rarely used today.

In the **reverse Sanger** sequencing approach, small amounts of α-thiophosphate forms of the four dNTPs are included while the DNA is copied normally. The α-thiophosphate dNTPs have an O atom of the α-phosphate group replaced by S. They are incorporated readily into the nascent DNA molecules, so a complete polynucleotide chain is synthesized. It is then treated with an exonuclease that removes nucleotides sequentially from the 3-terminus, until an α-S-nucleotide is reached. The lengths of the resistant molecules indicate the positions of the thiophosphate-containing nucleotides, so the use of all four α-thiophosphate dNTPs in separate reaction mixtures generates a sequence ladder.

A very novel method for sequencing DNA relies on the motion of single RNA polymerase molecules when they are transcribing the sequence of a single-stranded DNA molecule into a complementary RNA molecule. When low concentrations of one of the four types of nucleotides limit the rate of transcription, RNA polymerase molecules pause at positions corresponding to that base. This pausing is monitored with an optical trapping apparatus that is capable of resolving base pairs during transcription. Using limiting amounts of each of the four nucleotides, the DNA sequence can be determined from the aligned patterns of pauses recorded from as few as four molecules.

Sequencing end-labeled DNA with base-specific chemical cleavages. A. M. Maxam & W. Gilbert (1980) *Methods Enzymol.* **65**, 499–560.

Single-molecule, motion-based DNA sequencing using RNA polymerase. W. J. Greenleaf & S. M. Block (2006) *Science* **313**, 801.

6.5. SEQUENCING RNA

RNA is much more difficult to handle than DNA, due to its relative instability, single-stranded structure, tendency to adopt folded conformations, and the presence of modified bases. **Consequently, the majority of RNA sequences are obtained by sequencing DNA copies**. The DNA can be either (1) cloned DNA that codes for RNA genes or (2) the cDNA prepared by transcription of the RNA with the enzyme reverse transcriptase (Section 6.2.C). This enzyme synthesizes a cDNA strand using the order of nucleotides in the template RNA strand. The DNA sequences of RNA genes do not, however, provide direct information about all the post-transcriptional modifications of natural RNAs. The initial messenger RNA transcripts are subjected to splicing, cleavage, addition of a 5'-cap moiety, 3'-polyadenylation and RNA editing that alter the sequence. In addition, the primary transcripts of eukaryotic and prokaryotic RNAs are often processed by specific modifications of the nucleotide residues. Roughly 100 modifications of the bases and ribose moieties of the standard four ribonucleotides are known (Figure 4-3) and they tend to be located primarily in the regions of RNAs that are functionally important. Consequently, methods have been developed for sequencing RNA directly and rapidly, comparable to those for DNA sequencing, and special chemical, enzymatic and spectrophotometric techniques have been devised to identify modified nucleotides.

The use of reverse transcriptase for efficient first- and second-strand cDNA synthesis from single- and double-stranded RNA templates. I. E. Tzanetakis *et al.* (2005) *J. Virol. Methods* **124**, 73–77.

Rapid genome sequencing of RNA viruses. T. Mizutani *et al.* (2007) *Emerg. Infect. Dis.* **13**, 322–324.

6.5.A. Direct Sequencing of Oligoribonucleotides

The earliest methods for sequencing RNA were developed for transfer RNAs and relied largely on relatively nonspecific ribonucleases (RNases). The RNA was cleaved into sets of many short oligoribonucleotides using the relatively nonspecific RNases. The fragments produced were separated by column chromatography. Their short sequences were determined from their nucleoside compositions and by stepwise degradation. The overlaps of a large number of these partial sequences could be used to reconstruct the sequence of the original RNA.

An improved method resulted from labeling the RNA molecule at one end and subsequently detecting only molecules containing the label. The 5'-terminus of RNA is usually labeled by removing any existing 5'-phosphate groups of the RNA using the enzyme alkaline phosphatase, followed by complete phosphorylation with [γ-^{32}P]ATP catalyzed by phage T4 polynucleotide kinase. Alternatively, the 3'-end can be phosphorylated by adding the nucleotide cytidine-3',5'[^{32}P] diphosphate using T4 RNA ligase in the presence of ATP. The end-labeled RNA is cleaved partially into random fragments using a nonspecific RNase, and the products separated by two-dimensional **fingerprinting** using electrophoresis and chromatography (Figure 6-16). Only molecules containing the labeled end are observed. A complete series of all the possible fragments can be viewed as a ladder produced by removing nucleotides sequentially from the end of the RNA molecule that is not labeled. The shift in positions of two neighboring fragments upon removing one nucleotide depends on the pK_a values of that nucleotide and identifies it. This approach makes it possible to determine the sequences of 8–15 nucleotides in one experiment.

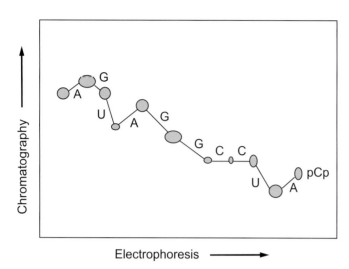

Figure 6-16. Two-dimensional separation of oligoribonucleotides generated by partial cleavage by an endonuclease of an RNA radioactively labeled at its 5′-end. Only those fragments with this radioactivity are detected using autoradiography. The smallest fragments migrate most rapidly during electrophoresis; the smallest in this example is the nucleotide pCp, which would have been at the 5′-end. The shift in mobility upon removal of each 3′-terminal nucleotide gives its identity. The sequence in this hypothetical example would be 5′–CAUCCGGAUGA–3′.

Sequencing methods comparable to the partial cleavage at specific nucleotides of end-labeled DNA (Section 6.4.D) have been adapted to RNA. Enzymatic and chemical procedures are available for cleaving the polynucleotide chain at only certain bases (Table 6-1). The oligoribonucleotides generated are separated by electrophoresis on the basis of their lengths, and only the bands with the end-label are detected by autoradiography. A sample that is nonspecifically cleaved is usually prepared for comparison as a control. The sites of cleavage by the various reagents identify the nucleotide at that position.

Table 6-1. Specificities of cleavage reactions used in RNA sequencing

	Specificity
Enzymatic cleavage by	
Ribonuclease U from *Ustilago sphaerogena*	… ApN …
Ribonuclease T$_1$ from *Aspergillus oryzae*	… GpN …
Ribonuclease A from bovine pancreas	… UpN … and …CpN…
Ribonuclease from *Bacillus cereus*	… UpN … and …CpN …
Ribonuclease Phy M from *Physarum polycephalum*	… ApN … and … UpN …
Chemical cleavage in aniline at pH 4.5 by	
Dimethyl sulfate + NaBH$_4$	… GpN …
Hydrazine	… UpN … > … pCpN …
Hydrazine in 3 M NaCl	… CpN … > … UpN …
Diethyl pyrocarbonate	… ApN … > … GpN …

N depicts any nucleotide. Data from A. Simoncsits *et al.* (1977) *Nature* **269**, 833–836; H. Donis-Keller *et al.* (1977) *Nucleic Acids Res.* **4**, 2527–2538; D. H. Peattie (1979) *Proc. Natl. Acad. Sci. USA* **76**, 1760–1764.

RNA can also be sequenced by methods comparable to the chain-termination method used with DNA (Section 6.4.B). A synthetic oligodeoxyribonucleotide complementary to a known sequence within the RNA molecule is used as primer for DNA synthesis, which is catalyzed by AMV reverse transcriptase in the presence of radioactively labeled dNTPs and one of the four nucleoside 5′-triphosphate dideoxynucleotide analogs (ddNTPs). The sequence is then read from the ladders produced by polyacrylamide gel electrophoresis (PAGE) of the four mixtures using each of the four ddNTPs. The procedure is, however, complicated by the presence of modified nucleotides, because they affect the reverse-transcriptase activity. DNA synthesis is frequently terminated at bulky bases, and the corresponding positions are detected as enhanced bands on the PAGE sequencing ladder. On the other hand, this is a useful indicator of the presence of a modified nucleotide.

Reverse Sanger sequencing is easier with RNA than with DNA (Section 6.4.D). The RNA is copied by reverse transcriptase in the presence of α-S-ribonucleoside triphosphates, in which a phosphate oxygen is replaced by sulfur, that are added to the reaction mixture at a ratio of 1:3 to the normal dNTPs. The phosphorothioate nucleotides are incorporated randomly into the DNA strand synthesized. Four separate mixtures are produced, using the four different α-S-ribonucleoside triphosphates. The full-length DNA transcript is isolated and purified. It is then cleaved with an exonuclease like **snake venom phosphodiesterase**, which removes nucleotides sequentially from the 3′-terminus, until an α-S-nucleotide appears there. The resulting fragments are separated on the basis of their size, and the lengths of the resistant molecules indicate the positions of the thiophosphate-containing nucleotides.

Reverse Sanger sequencing of RNA by MALDI-TOF mass spectrometry after solid phase purification. B. Spottke et al. (2004) *Nucleic Acids Res.* **32**, e97.

Shotgun sequencing of the negative-sense RNA genome of the rhabdovirus Maize mosaic virus. S. E. Reed et al. (2005) *J. Virol. Methods* **129**, 91–96.

6.5.B. Identifying Modified Nucleotides

Modified nucleotides affect the chemical properties of the RNA molecules, especially their modification by enzymes, and complicate the sequencing reactions. The different reactivities of modified nucleotides can cause the fragments terminating there to be irregular, missing or enhanced on the sequencing gels. This provides hints about the presence of modified nucleotides in RNA, but they still need to be identified. Once isolated, modified bases can be identified by thin-layer or column chromatography techniques, comparing them with synthetic standards (Figure 6-17), or by electrospray ionization mass spectrometry.

The nucleotide present at each site of cleavage during sequencing can be labeled specifically and then identified (Figure 6-18). A mixture of oligoribonucleotides is prepared by partial cleavage of the denatured RNA chain. The 5′-hydroxyl groups of the oligoribonucleotides generated are then radiolabeled by T4 RNA kinase and [γ-^{32}P]ATP. After separation by gel electrophoresis according to their size, bands of the different oligoribonucleotides are digested to mononucleotides by ribonuclease T2. Only the terminal nucleotide of each band is ^{32}P-labeled. It can be identified separately by one- or two-dimensional chromatography (Figure 6-17).

Detection and quantification of modified nucleotides in RNA using thin-layer chromatography. H. Grosjean *et al.* (2004) *Methods Mol. Biol.* **265**, 357–391.

A novel method for sequence placement of modified nucleotides in mixtures of transfer RNA. T. M. Wagner *et al.* (2004) *Nucleic Acids Symp. Ser.* **48**, 263–264.

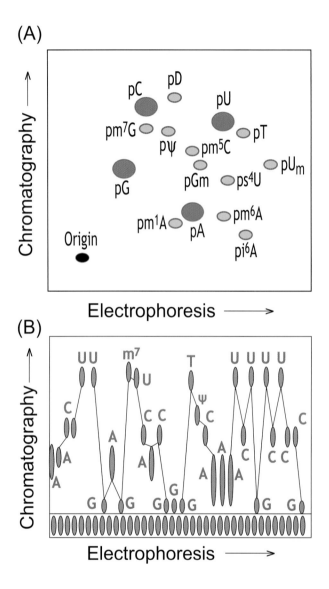

Figure 6-17. Identification of ribonucleotides by chromatography and electrophoresis. (A) Two-dimensional fingerprint of ribonucleoside 5′-phosphates. The major ribonucleotides pA, pU, pG and pC are indicated by the *blue spots*. The others are the minor nucleotides (Figure 4-3): i⁶A, N⁶-isopentenyladenosine; U$_m$, 2′-O-methyluridine; m⁶A, N⁶-methyladenosine; s⁴U, 4-thiouridine; m¹A, 1-methyladenosine; G$_m$, 2′-O-methylguanosine; m⁵C, 5-methylcytosine; Ψ, pseudouridine; D, 5,6-dihydrouridine; m⁷G, 7-methylguanosine; T, 5-methyluridine (thymidine). Data from Y. Kuchino *et al.* (1987) *Methods Enzymol.* **155**, 379–396. (B) Identification of nucleotides at the 5′-ends of random fragments that had been labeled with ³²P at the terminal phosphate group. The fragments had been separated by size by electrophoresis in the horizontal dimension. They were then degraded to individual nucleotides and separated by chromatography in the vertical dimension. The labeled nucleotide was detected by autoradiography. If the electrophoretic bands differ only in their lengths at the 5′-end, the labeled nucleotides will give the sequence of the original RNA molecule, as indicated. Data from J. Stanley & S. Vasilenko (1978) *Nature* **274**, 87–89.

6.6. CHEMICAL SYNTHESIS OF DNA

Many methods in molecular biology require the chemical synthesis of oligonucleotides of defined sequences. One example is the primer that is required for sequencing DNA next to a site with a known sequence (Figure 6-1). Others involve the detection and identification of specific stretches of DNA with a known sequence (Section 5.3.C) and the introduction of changes in nucleotide sequence (Section 6.6.E). Not surprisingly, very powerful methods for synthesizing such oligonucleotides have been developed and it is now a routine procedure, largely performed by automated instruments.

Single-stranded oligonucleotides can be assembled by chemically linking two nucleotides with a phosphodiester bond and then adding individual nucleotides to one of the ends. The process is similar to that used in replicating DNA, except that no enzymes or template are used, so any sequence can

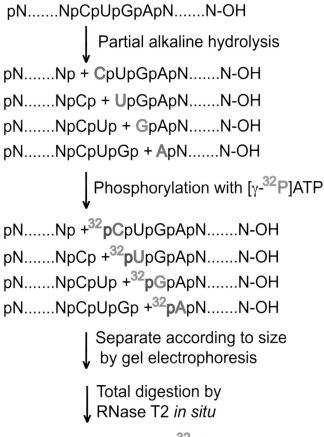

Figure 6-18. The post-labeling method of sequencing of RNA. Limited alkaline hydrolysis of the RNA leads to a mixture of random oligoribonucleotides, which are labeled radioactively at the 5′-end. The fragments are then separated by PAGE according to size. Each fragment is digested completely by ribonuclease T2 to the individual nucleotides. Which nucleotide was labeled, and therefore was at the terminus, is identified by chromatography, as in Figure 6-17-B. A complete set of fragments can provide the sequence of the RNA.

be produced. The process is repetitive, so the efficiency of adding each nucleotide determines what length of oligonucleotide can be produced with a reasonable yield.

DNA nanotechnology: novel DNA constructions. N. C. Seeman (1998) *Ann. Rev. Biophys. Biomol. Structure* **27**, 225–248.

3′-modified oligonucleotides by reverse DNA synthesis. C. D. Claboe *et al.* (2003) *Nucleic Acids Res.* **31**, 5685–5691.

A simple, rapid, high-fidelity and cost-effective PCR-based two-step DNA synthesis method for long gene sequences. A. S. Xiong *et al.* (2004) *Nucleic Acids Res.* **32**, 398.

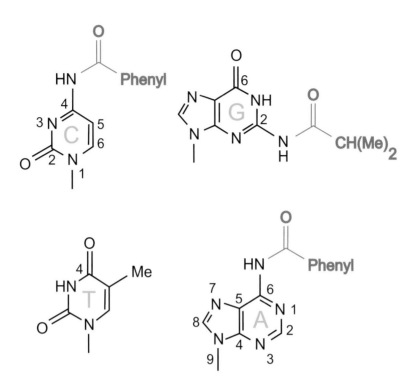

Figure 6-19. The forms of the cytosine (C), guanine (G), thymine (T) and adenine (A) bases used in deoxynucleosides for DNA synthesis. The protecting groups are *colored*. Thymine requires no protecting group.

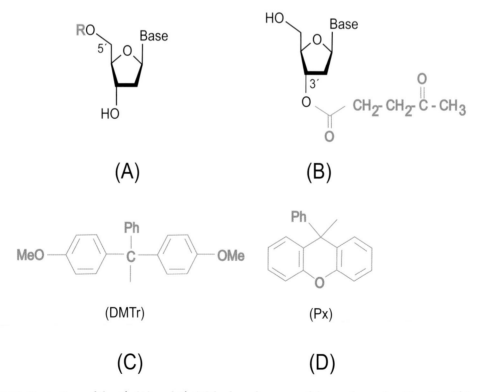

Figure 6-20. Protection of the 5′- (A) and 3′- (B) hydroxyl groups of deoxyribonucleosides. The 5′-hydroxyl is protected with R groups such as (C) (di-*p*-anisyl)phenylmethyl (DMTr) or (D) 9-phenylxanthen-9-yl (Px); Ph is a phenyl group. The protecting groups are *colored*.

6.6.A. Protecting Groups for 2′-Deoxynucleosides

Only one pair of functional groups on the 2′-deoxyribonucleoside building blocks is to be linked covalently, which requires that all other potentially reactive sites be protected reversibly. The protecting groups should be easy to introduce, remain intact throughout the assembly of the DNA sequences and be removed completely under conditions in which the assembled DNA molecule is stable. The nature of the protecting groups is one of the most important factors to be considered in designing a protocol for DNA synthesis. Two other factors to consider are the chemical procedures for generating the phosphodiester linkages between the nucleotides and for purifying the final synthetic DNA molecules.

The glycosidic linkages between the base and the sugar, especially those of purine deoxyribosides, are relatively sensitive to acid. Consequently, the protecting groups used should be removable without the need for very acidic conditions. The functional groups of the adenine, cytosine and guanine bases are generally protected with acyl groups (Figure 6-19). Thymine residues are often left unprotected, although certain procedures may require that their O4 atoms be protected, along with the O6 atoms of *N*-2-acylguanine residues.

The 5′-hydroxyl group is best protected with (di-*p*-anisyl)phenylmethyl (DMTr) or 9-phenylxanthen-9-yl (Px) groups (Figure 6-20). Acyl groups are generally used to protect the 3′-hydroxyl groups (Figure 6-20-B). They have also been used to protect the 5′-hydroxyl group, especially in the solution-phase synthesis of DNA. Both *N*-acyl- and *O*-acyl-protecting groups can be removed at the end of the synthesis by treatment with ammonia, conditions where DNA is very stable.

6.6.B. Coupling Methods

Three different types of coupling methods have been used successfully. Each will be illustrated with a single cycle, in which one nucleotide is attached to another. To synthesize an oligonucleotide of N nucleotides requires that these reaction sequences be repeated $N - 1$ times.

1. Phosphotriester Procedure

All of the internucleotide linkages are protected as phosphotriesters throughout the assembly of the target DNA sequences using this approach, which is the basis for its name. A contrasting method, known as the **phosphodiester approach**, leaves the internucleotide linkages unprotected throughout the assembly of the DNA sequences, but it is now obsolete. The phosphotriester approach couples the free 5′-hydroxyl group of a protected nucleoside or oligonucleotide component to a slight excess of the 3′-(2-chlorophenyl) phosphate moiety of the other nucleoside or oligonucleotide (Figure 6-21). Coupling usually involves a condensing agent such as 1-(mesitylene-2-sulfonyl)-3-nitro-1,2,4-(1*H*)-triazole (MSNT) in anhydrous pyridine solution:

(6.4)

The MSNT also acts as a dehydrating agent and is used in excess.

The fully assembled DNA sequence is unblocked by treatment with an excess of an appropriate oxime (such as E-2-nitrobenzaldoxime) and N^1,N^1,N^3,N^3-tetramethylguanidine:

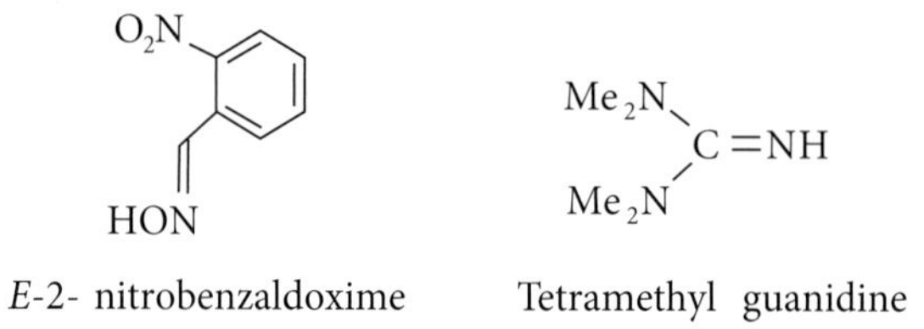

E-2-nitrobenzaldoxime   Tetramethyl guanidine   (6.5)

All of the 2-chlorophenyl-protecting groups are removed without cleaving the internucleotide linkages. The N- and O-acyl-protecting groups (Figure 6-19 and 6-20-B) are subsequently removed by treatment with concentrated aqueous ammonia. The 5′-terminal-protecting group (Figure 6-20-A) is acid-labile and is removed under mild conditions to generate the final DNA sequence. The phosphotriester approach is extremely versatile and can be used in solution or solid-phase and by stepwise addition of mononucleotides or fusing of oligonucleotides.

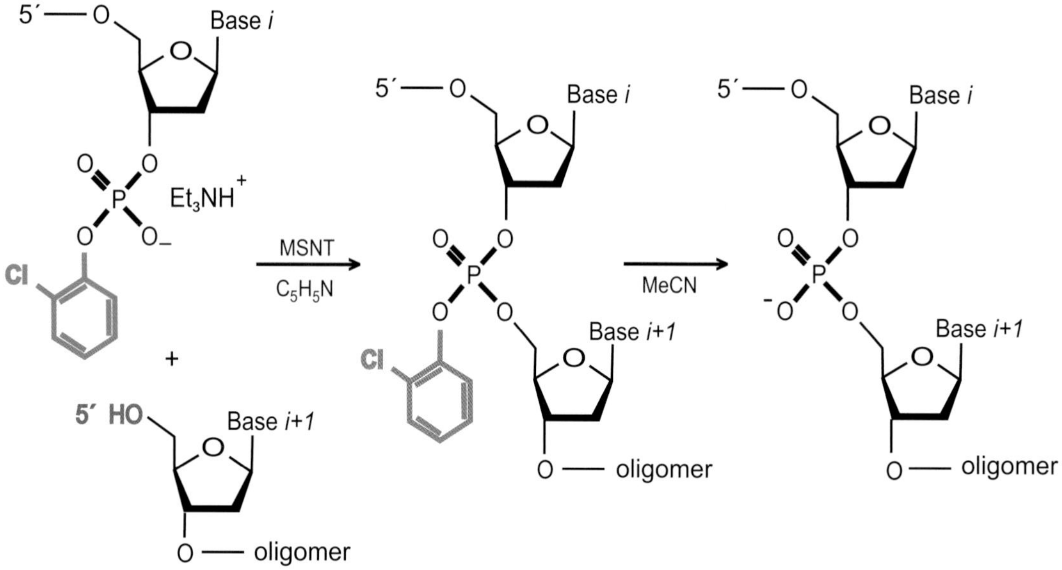

Figure 6-21. The phosphotriester approach to coupling deoxynucleotides. The 5′-reactant is present as the 3′-(2-chlorophenyl) phosphate derivative. Coupling uses a reagent such as MSNT (Equation 6.4) in pyridine, while removal of the protecting groups in the second step uses an oxime and tetramethyl guanidine (Equation 6.5) in acetonitrile (MeCN).

2. Phosphoramidite Procedure

Protected nucleoside 3′-(2-cyanoethyl) N, N-di-isopropylphosphoramidites are used in the phoroamidite approach. In the presence of relatively weak acids, such as 1H-tetrazole, they react readily with the free 5′-hydroxyl group of protected nucleoside or oligonucleotide components (first step in Figure 6-22) to produce protected phosphite triesters. These phosphite triesters are unstable intermediates, so usually they are immediately oxidized by treatment with iodine and a weak base to the corresponding phosphotriesters, as in the second step of Figure 6-22. The 2-cyanoethyl and *N*- and *O*-acyl-protecting groups (Figure 6-19) are removed from the phosphates of the fully assembled DNA sequences by treatment with concentrated aqueous ammonia.

The coupling yields tend to be higher with the phosphoramidite approach than the phosphotriester approach (Figure 6-21) because activated derivatives of phosphorous acid are much more reactive than the corresponding activated derivatives of phosphoric acid. Consequently, **the phosphoramidite approach is widely used for automated solid-phase synthesis of DNA sequences** and in the chemical synthesis of relatively small quantities of the DNA, which are sufficient for most procedures in molecular biology.

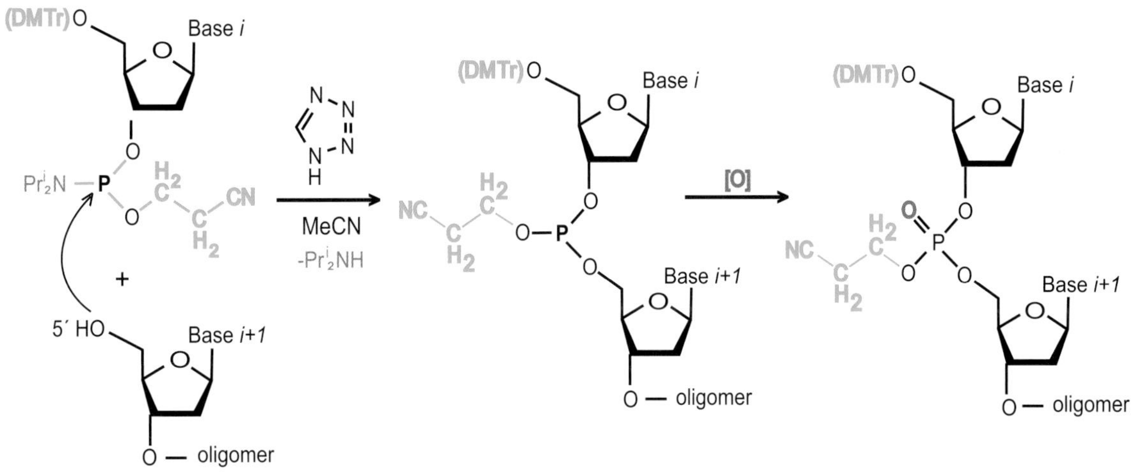

Figure 6-22. The phosphoramidite approach to coupling a deoxynucleotide to the 5′-end of a DNA oligomer (*bottom, left*). The nucleotide to be added is presented as the 3′-(2′-cyanoethyl) *N,N*-di-isopropylphosphoramidite, protected with a DMTr group, and 1*H*-tetrazole in MeCN is a catalyst of the reaction. Pr^i is the isopropyl group and [O] is an oxidant. The protecting groups must be removed from the product of the reaction by treatment with ammonia before another nucleotide can be added to its 5′-end.

Linker phosphoramidite reagents for the attachment of the first nucleoside to underivatized solid-phase supports. R. T. Pon & S. Yu (2004) *Nucleic Acids. Res.* **32**, 623–631.

Internucleotide-linkage formation via the phosphoramidite method using a carboxylic acid as a promoter. M. Tsukamoto *et al.* (2004) *Nucleic Acids Symp. Ser.* **48**, 25–26.

Development of new N-unprotected phosphoramidite building blocks having a silyl-type linker. A. Ohkubo *et al.* (2005) *Nucleic Acids. Symp. Ser.* **49**, 127–128.

---

3. H-Phosphonate Procedure

This approach uses protected nucleoside 3′-(*H*-phosphonate) monoesters (Figure 6-23) as building blocks. It has the advantage that they are easy to prepare and more stable and easier to handle than the building blocks of the phosphoramidite approach. A nucleoside 3′-(*H*-phosphonate) building block reacts readily with the 5′-hydroxyl group of a protected nucleoside or oligonucleotide (the first reaction in Figure 6-23) in the presence of a condensing agent such as pivaloyl chloride in pyridine solution. After completion of the synthesis, the *H*-phosphonate diester groups are oxidized, usually with iodine in the presence of a base in aqueous tetrahydrofuran, to generate unprotected phosphodiester internucleotide linkages, as in the second reaction of Figure 6-23. The *N*- and *O*-acyl-protecting groups can then be removed by treatment with concentrated aqueous ammonia.

The *H*-phosphonate approach is best suited for use in solid-phase synthesis (Section 6.6.D). While it has some merits and the potential to be an important synthetic method, it is more susceptible to side-reactions and is not as efficient in solid-phase synthesis as the phosphoramidite approach.

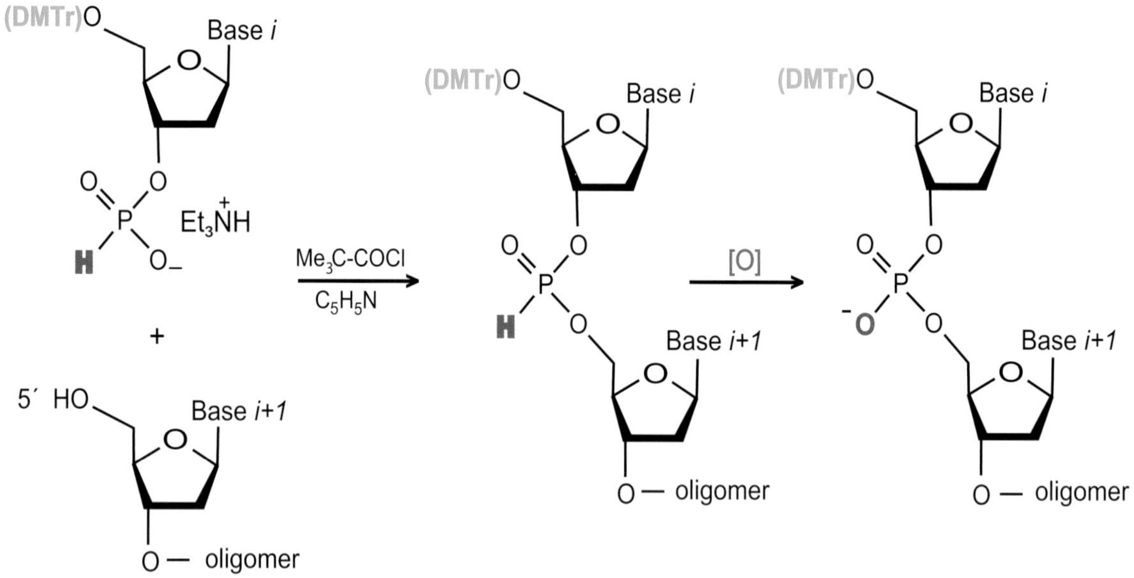

Figure 6-23. The *H*-phosphonate approach to coupling a deoxynucleotide to the 5′-terminus of a DNA oligomer (*bottom, left*). The nucleotide to be added is the 3′-(*H*-phosphonate) monoester (*top, left*) protected with the DMTr group. Coupling occurs in the first step in the presence of a coupling agent, such as pivaloyl chloride (Me$_3$C–CO–Cl) in pyridine (C$_5$H$_5$N). In the second step, the *H*-phosphonate diester is oxidized to give the phosphodiester linkage.

Di- and oligonucleotide synthesis using H-phosphonate chemistry. J. Stawinski & R. Stromberg (2005) *Methods Mol. Biol.* **288**, 81–100.

Developing synthetic methods for bioactive phosphorus compounds using H-phosphonate chemistry: a progress report. T. Johansson *et al.* (2005) *Nucleosides Nucleotides Nucleic Acids* **24**, 353–357.

Stereocontrolled synthesis of H-phosphonate DNA. N. Iwamoto *et al.* (2006) *Nucleic Acids Symp. Ser.* **50**, 159–160.

Efficient preparation of amine-modified oligodeoxynucleotide using modified H-phosphonate chemistry for DNA microarray fabrication. N. K. Kamisetty *et al.* (2007) *Anal. Bioanal. Chem.* **387**, 2027–2035.

6.6.C. Solution-phase DNA Synthesis

Solution-phase synthesis is used primarily when a large quantity of a specific sequence is required. It is laborious, in that it is often necessary to purify the products by chromatography after each coupling step, although this might be no more than filtration through a bed of silica gel. Of the three phosphorylation methods described in Figures 6-21, 6-22 and 6-23, only the phosphotriester approach (Figure 6-21) is very suitable for the synthesis of DNA sequences in solution. This method is very versatile, being useful for both the stepwise addition of individual nucleotides and the coupling of oligonucleotides, in which two or more nucleotide residues are added at a time. It has the advantage that the fully protected intermediates generated are soluble in organic solvents and may therefore be purified by conventional chromatographic techniques. Once all of the protecting groups have been removed, the unprotected DNA sequences generated may, if necessary, be further purified.

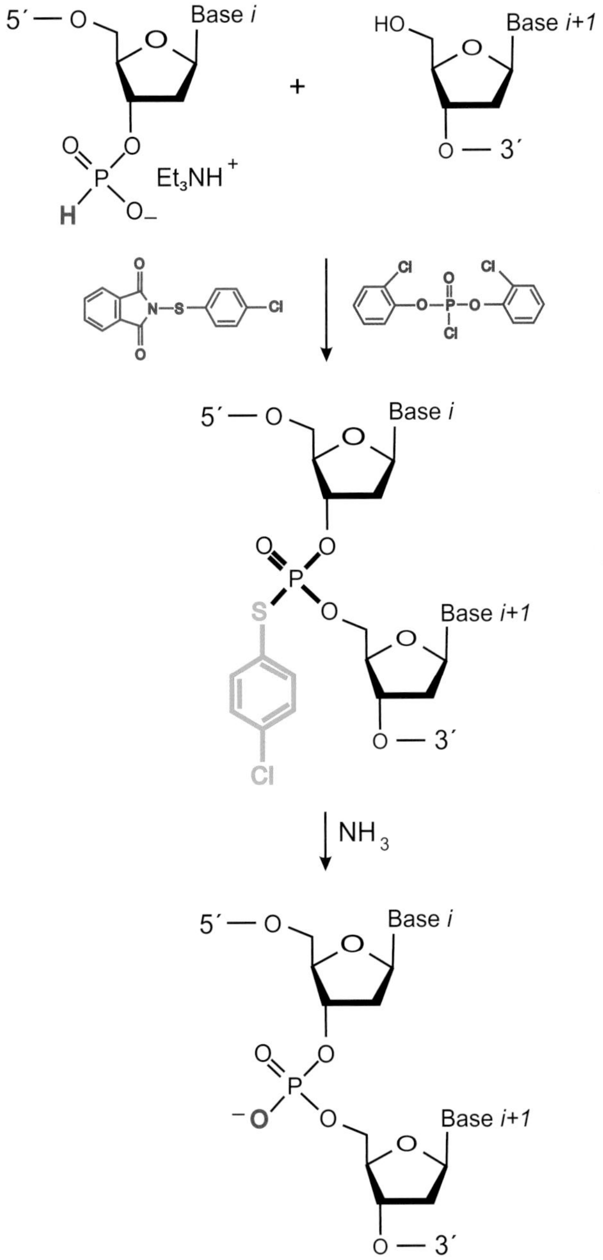

Figure 6-24. One cycle of coupling nucleotides in solution. The 5′-nucleotide is used as the H-phosphonate (Figure 6-23). The second step requires E-2-nitrobenzaldoxime and tetramethylguanidine in acetonitrile, followed by treatment with concentrated aqueous ammonia.

The reaction sequence described in Figure 6-24 is a modified phosphotriester approach involving *H*-phosphonate coupling at low temperature. It produces virtually quantitative coupling, with no side-reactions. The unstable *H*-phosphonate diester intermediates of Figure 6-23 are not isolated but immediately converted at low temperature into the much more stable *S*-(4-chlorophenyl) phosphorothioate intermediates by treatment with the phthalimide derivative described in Figure 6-24. Upon complete assembly of the desired DNA sequence, the *S*-(4-chlorophenyl) phosphorothioate groups are quantitatively converted into phosphodiesters by the oximate treatment described in Figure 6-21.

6.6.D. Solid-phase DNA Synthesis

Solid-phase synthesis is much less laborious than the solution methods and highly preferable, especially if only small (i.e. milligram to gram) quantities of DNA are required. The growing DNA sequence is attached to a solid support, such as controlled-pore glass, and all the reagents that remain after each reaction are washed away. The main advantages of solid-phase synthesis are that (1) it is very rapid, so a DNA sequence of 50 nucleotide residues can easily be produced within 1 day; (2) very high yields of all the chemical reactions are obtained by using large excesses of reagents; (3) the whole process may be fully automated in a DNA synthesizer; and (4) only one purification step is required, of the final product at the end of the synthesis.

All three of the coupling methods described in Figures 6-21, 6-22 and 6-23 have been tried with solid-phase synthesis, and the phosphoramidite approach of Figure 6-22 has emerged as the method of choice because the coupling efficiencies are very high and there are no significant side-reactions. Most commercial automatic synthesizers have been designed specifically for phosphoramidite chemistry.

The main steps in the addition of one nucleotide residue in a synthetic cycle are described in Figure 6-25. The synthesizer uses solutions in acetonitrile of phosphoramidites of each of the four protected deoxynucleosides (Figure 6-19). The solid support is usually porous glass or polystyrene, designated P. To it is attached the desired 3′-terminal nucleoside residue through a succinoyl linker (Figure 6-25); this material is placed in a small column that is inserted into the synthesizer. The first step in the synthetic cycle involves treatment with acid to remove the 5′-*O*-DMTr-protecting group from the 3′-terminal nucleoside residue. The next step is the reaction between the released 5′-hydroxyl group and the appropriate phosphoramidite, catalyzed by 1*H*-tetrazole to give a phosphite triester. Large excesses of both the phosphoramidite and 1*H*-tetrazole are delivered automatically, which can produce a coupling efficiency of up to 98.5%. Any remaining 5′-hydroxyl functions are 'capped' by reaction with a large excess of acetic anhydride to ensure that any molecules that did not react are not involved in subsequent reactions. Consequently, most of the incorrect sequences at the end of the synthesis are of significantly lower molecular weight than the target DNA sequence, which facilitates their removal.

After capping, the phosphite triester group is oxidized to a phosphotriester group, which completes the first synthetic cycle. Removal of the 5′-*O*-DMTr-protecting group produces the final product of Figure 6-25; it is also the first step of the second synthetic cycle that will add a second nucleotide. This cycle is repeated for each successive nucleotide added to the chain. A number of washing steps are involved, but an entire cycle can be completed within 8 minutes.

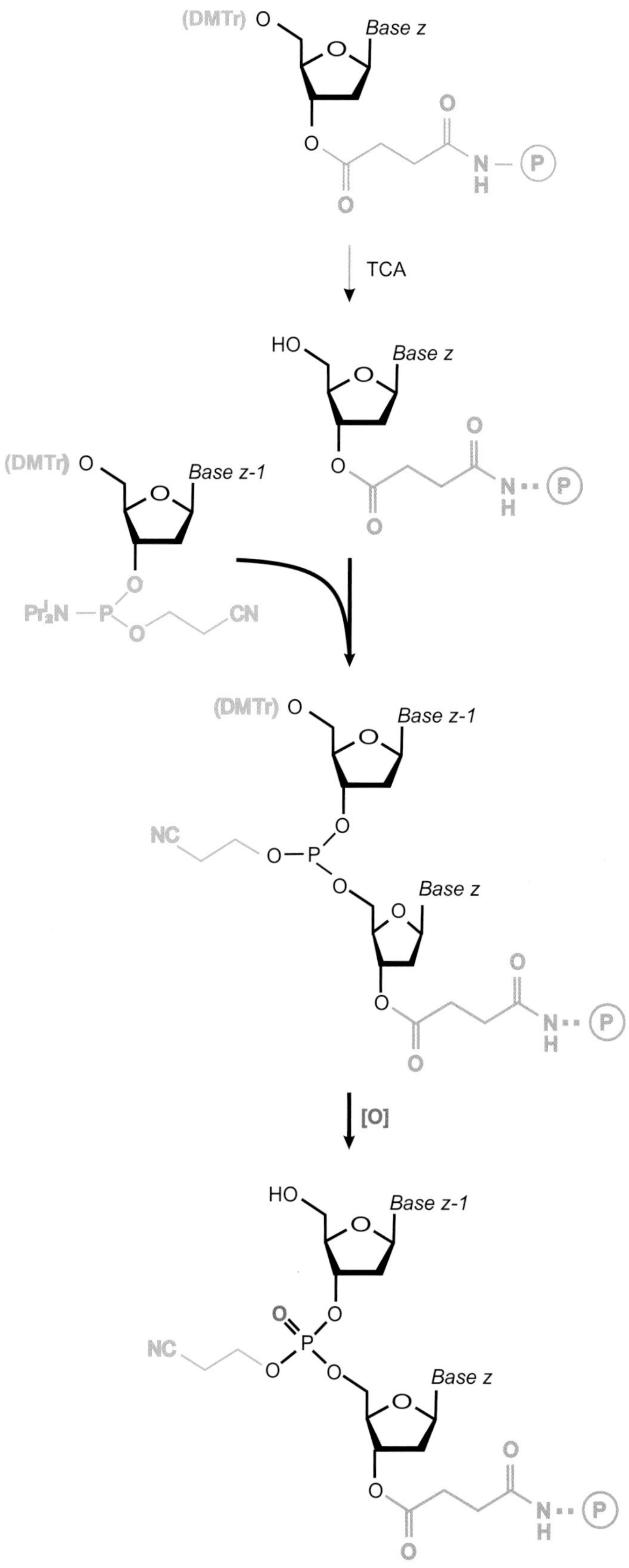

Figure 6-25. One cycle of coupling a nucleotide to the 5′-terminus of a growing DNA chain during synthesis on a solid support (designated by the P in a *circle*). The nucleotide is added as the 3′-(2′-cyanoethyl) *N,N*-diisopropylphosphoramidite protected by the DMTr group (Figure 6-20). TCA is trichloroacetic acid and [O] is an oxidant, such as I_2.

The efficiency of each coupling step can be estimated by measuring spectrophotometrically the amount of colored (di-*p*-anisyl)phenylmethyl (DMTr$^+$) cations that are released at the beginning of each synthetic cycle. If the average coupling efficiency is 98.5%, complete DNA sequences containing 50 and 100 nucleotide residues should be produced with roughly 47% and 22% overall yields, respectively, so DNA sequences containing up to approximately 100 nucleotide residues can be prepared in this way. If the average coupling efficiency is only 97.5%, however, the overall yields would decrease to about 29% and 8%, respectively. Therefore, the coupling procedure needs to be as efficient as possible.

The desired DNA sequence of N nucleotides should be complete after $N - 1$ synthetic cycles but the products must be released from the solid support and all the protecting groups removed. In one unblocking procedure, the 5′-terminal DMTr-protecting group is removed in what would be the first step of synthetic cycle N. The synthetic DNA is then detached from the solid support by treatment with concentrated aqueous ammonia at about 55°C; this also removes all of the 2-cyanoethyl-protecting groups from the internucleotide linkages and all of the *N*-acyl-protecting groups from the bases. The 'capped' truncated sequences are usually removed by column chromatography or PAGE. In another procedure, an ammonolysis step is carried out prior to removal of the 5′-terminal DMTr-protecting group. This procedure is preferable if purification of the crude products is to be carried out by reverse-phase chromatography, because the target DNA sequence is still protected with a 5′-*O*-DMTr group whereas the capped truncated sequences are not; this difference makes it easier to isolate the correct molecules. The DMTr-protecting groups are finally removed by acid hydrolysis under very mild conditions.

6.6.E. Site-directed Mutagenesis

The nucleotide sequence of a fragment of cloned DNA very often needs to be altered. It is possible to change only a single base pair or a few adjacent base pairs, or to create more extensive changes, such as deletions and insertions. There are many procedures for doing this efficiently. In each case, the desired change in sequence must be produced by synthesizing an oligonucleotide with that sequence.

Cassette mutagenesis replaces a restriction fragment from a cloning vector with another fragment with the desired nucleotide sequence (Figure 6-26-A), using restriction enzymes as in Figure 6-11. The new DNA fragment has usually been synthesized chemically. This procedure requires two restriction sites to be available so that the restriction fragment is not too large for DNA synthesis. If not, the alternative method of **enzymatic extension of a mutagenic oligonucleotide** is generally used (Figure 6-26-B). A single-stranded oligonucleotide is synthesized that contains the new sequence flanked on each side by 10–20 nucleotides that are exactly complementary to the DNA strand to be altered. This oligonucleotide base pairs with the original DNA fragment, in spite of the different sequence in the middle. It serves as a primer to initiate copying of the remainder of the original sequence by an appropriate DNA polymerase enzyme. Use of the enzyme thereby avoids the need to synthesize chemically the entire DNA molecule.

Many variations on these methods have been devised, especially using PCR (Section 6.1.C).

Restriction enzyme-mediated integration (REMI) mutagenesis. A. Kuspa (2006) *Methods Mol. Biol.* **346**, 201–209.

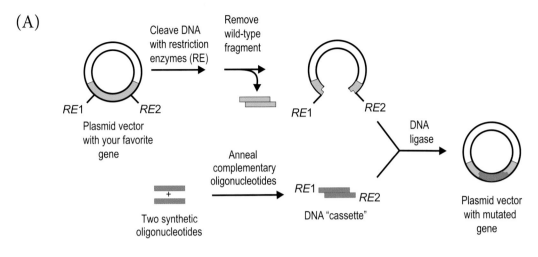

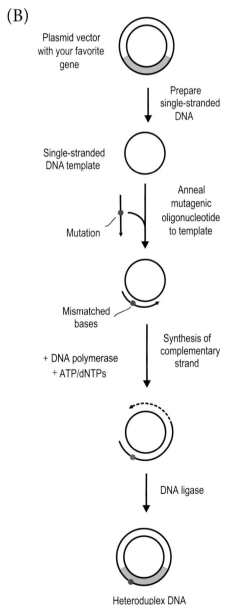

Figure 6-26. Two methods for site-directed mutagenesis. (A) Cassette mutagenesis, in which the DNA sequence to be altered ('your favorite gene') is removed from the plasmid using two restriction enzymes, *RE*1 and *RE*2. It is replaced by a synthetic DNA 'cassette' containing the new sequence. (B) Site-directed mutagenesis by enzymatic extension of a mutagenic oligonucleotide. The single-stranded DNA molecule to be mutated is hybridized by base pairing to an oligonucleotide containing the new sequence, plus flanking sequences complementary to the original DNA molecule. The remainder of the second strand is synthesized enzymatically, using DNA polymerase, ATP and dNTPs. The ends are joined by DNA ligase. The heteroduplex DNA is introduced into the appropriate host, where its replication will produce plasmid molecules with the mutation on both strands.

6.7. CHEMICAL SYNTHESIS OF RNA

The chemical synthesis of RNA is inherently more complicated than that of DNA (Section 6.6) because it requires the use of an additional protecting group, for the 2′-hydroxyl group. With this exception, **the principles of the chemical synthesis of RNA and DNA are virtually the same**. Similar protecting groups are used for the bases and internucleotide linkages, and the same three principal coupling methods (i.e. the phosphotriester, phosphoramidite and *H*-phosphonate approaches; Section 6.6.B) are used in the synthesis of RNA sequences. Likewise, only the phosphotriester approach is suitable for solution-phase synthesis of RNA sequences, and the phosphoramidite approach is again the method of choice for automated solid-phase synthesis.

General deoxyribozyme-catalyzed synthesis of native 3′-5′ RNA linkages. W. E. Purtha *et al.* (2005) *J. Am. Chem. Soc.* **127**, 13124–13125.

Nonenzymatic template-directed RNA synthesis. M. Hey & M. Gobel (2005) *Methods Mol. Biol.* **288**, 305–318.

Synthesis and properties of ENA oligonucleotides targeted to human telomerase RNA subunit. M. Horie *et al.* (2005) *Nucleic Acids Symp. Ser.* **49**, 171–172.

Synthesis, characterization, and biological properties of small branched RNA fragments containing chiral (Rp and Sp) 2′,5′-phosphorothioate linkages. R. Mourani & M. J. Damha (2006) *Nucleosides Nucleotides Nucleic Acids* **25**, 203–229.

Chemical synthesis of RNA including 5-taurinomethyluridine. T. Ogata & T. Wada (2006) *Nucleic Acids Symp. Ser.* **50**, 9–10.

6.7.A. Protecting the 2′-Hydroxyl Group

The protecting group for the 2′-hydroxl group must remain intact until the final unblocking step at the end of the synthesis, and it then must be possible to remove it under very mild conditions without the released 2′-hydroxyl group attacking the adjacent (3′→5′)-internucleotide phosphodiester linkages (Figure 4-2), which will cause their cleavage or migration.

Protecting groups for the 2′-hydroxyl group can be removed by hydrolysis under either basic or acidic conditions. Basic conditions can permit cleavage, but not migration, of the internucleotide linkages (Figure 4-2). The 2′,3′-cyclic phosphate cleavage product formed initially undergoes further hydrolysis to produce a mixture of the corresponding 2′- and 3′-phosphates. Removing the 2′-protecting groups under acidic conditions can produce both cleavage and migration of the internucleotide linkages (Figure 6-27). While cleavage of the internucleotide linkages is clearly undesirable, phosphoryl migration is much more serious in that it produces molecules with one or more 2′→5′-internucleotide linkages. Removing such isomeric contaminants is virtually impossible, even in the case of relatively short oligoribonucleotides. Consequently, acid-labile protecting groups must be removed under conditions sufficiently mild to avoid phosphoryl migration.

Of the 2′-hydroxyl-protecting groups used most widely and successfully, the Thp, Mthp, Ctmp and Fpmp* groups are acetal systems that can be removed under mild acidic conditions:

* *Thp, tetrahydropyran-2-yl; Mthp, 4-methoxytetrahydropyran-4-yl; TBDMS, tert-butyldimethylsilyl; Ctmp, 1-[(2-chloro-4-methyl)phenyl]-4-methoxypiperidin-4-yl; Fpmp, 1-(2-fluorophenyl)-4-methoxypiperidin-4-yl.*

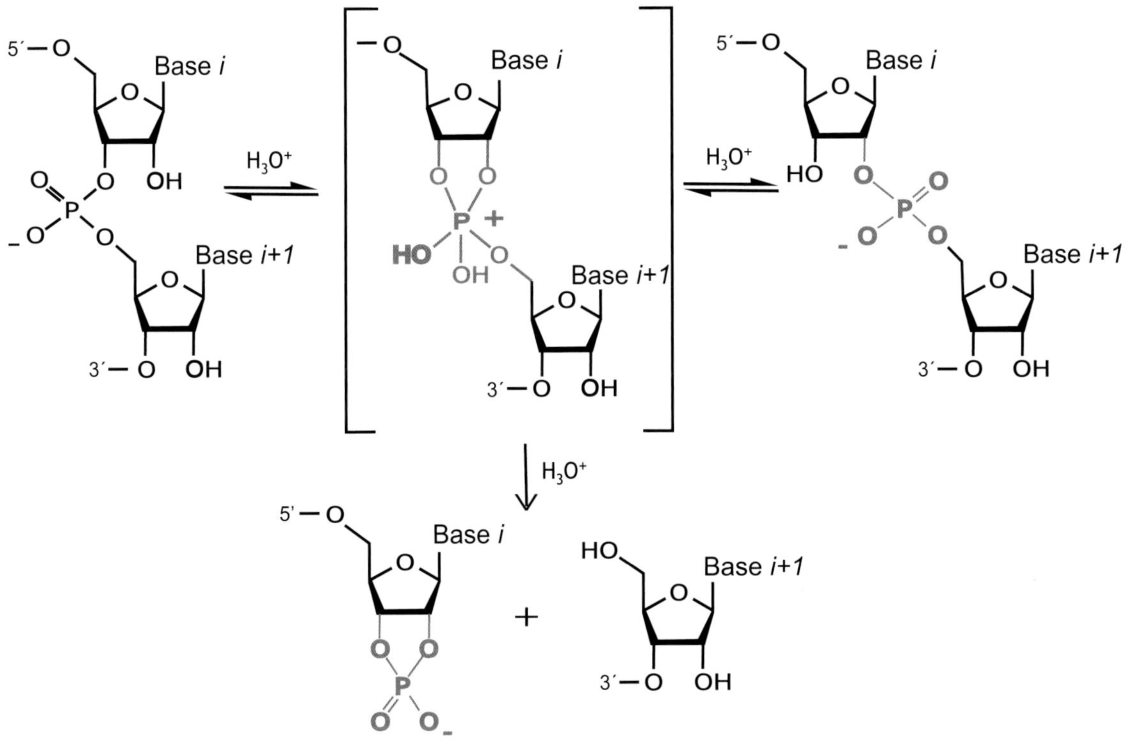

Figure 6-27. Cleavage and migration of the RNA internucleotide linkage under acidic conditions. The migration interconversions are indicated by the horizontal reactions, whereas cleavage occurs in the vertical reaction.

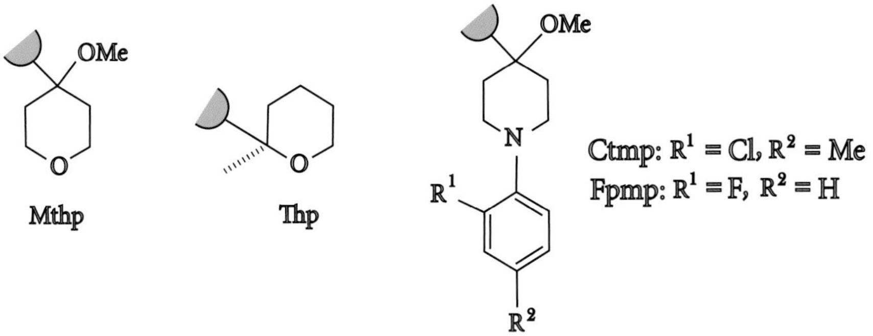

$$\text{Ctmp: } R^1 = Cl, R^2 = Me$$
$$\text{Fpmp: } R^1 = F, R^2 = H \tag{6.6}$$

They are completely stable under the basic conditions of concentrated aqueous ammonia at 55°C that are needed to remove virtually all of the other protecting groups from fully assembled RNA sequences. The 2-nitrobenzyl-protecting group:

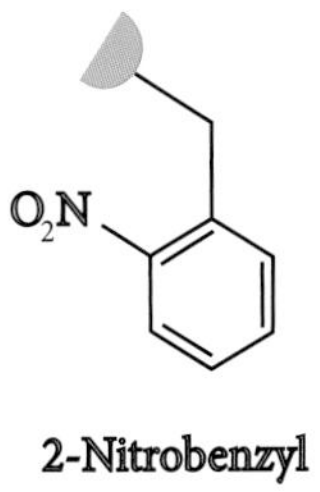

$$\tag{6.7}$$

is removed by photolysis at wavelengths >280 nm. It is also stable to the alkaline treatment. TBDMS:

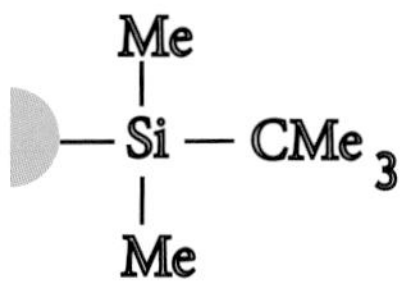

$$\text{—Si(Me)(Me)—CMe}_3$$

TBDMS (6.8)

is used widely in the automated solid-phase synthesis of RNA sequences. It has the advantage that it can be removed in the final unblocking step without using acidic conditions. Under the standard basic unblocking conditions, however, it is partially removed and tends to undergo migration between the 2′- and 3′-hydroxyl groups. Nevertheless, it is used widely in automated solid-phase synthesis.

RNA synthesis using 2′-O-(tert-butyldimethylsilyl) protection. B. S. Sproat (2005) *Methods Mol. Biol.* **288**, 17–32.

Synthesis of RNA using 2′-O-DTM protection. A. Semenyuk *et al.* (2006) *J. Am. Chem. Soc.* **128**, 12356–12357.

A new method for RNA synthesis by use of the cyanoethyl group as the 2′-hydroxyl protecting group. H. Saneyoshi *et al.* (2006) *Nucleic Acids Symp. Ser.* **29**, 125–126.

6.7.B. RNA Synthesis in Solution

As in the case of DNA (Section 6.6.C), solution-phase synthesis is labor-intensive and used only if relatively large quantities of RNA sequences are required.

Relatively long RNA sequences have been synthesized in solution using the phosphotriester approach. In one strategy (Figure 6-28), the 2′-hydroxyl groups are protected with the achiral Mthp group (Equation 6.6) and the methoxymethylene group is used to protect both the 2′- and 3′-hydroxyl groups at the 3′-end that is not to be joined. The 5′-terminal hydroxyl groups are protected with acyl groups, such as the 2-(dibromomethyl)benzoyl (Dbmb) group:

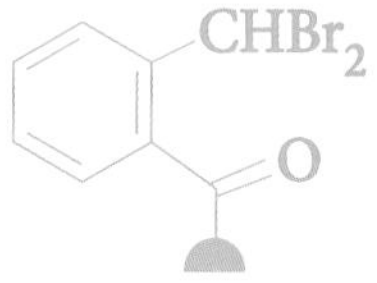

Dbmb (6.9)

The Dbmb groups can be removed under very mild basic conditions. Internucleotide linkages are protected with 2-chlorophenyl groups:

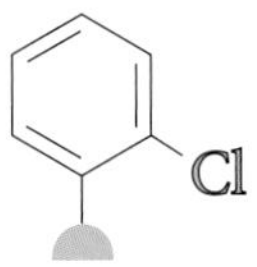

(6.10)

and the bases are protected as:

$$\text{(6.11)}$$

The guanine and uracil residues are protected with aryl groups on O6 and O4, respectively, to prevent the occurrence of side-reactions during the coupling steps. The coupling reactions use MSNT (Equation 6.4) in dry pyridine solution.

Once the fully protected RNA sequences have been assembled, they are unblocked in three steps. First, the 2-chlorophenyl-protecting groups are removed from the internucleotide linkages, and the O6 and the O4 aryl groups are removed from the guanine and uracil residues by treatment with E-2-nitrobenzaldoxime and tetramethyl-guanidine (Equation 6.5). Secondly, all the acyl-protecting groups are removed using concentrated aqueous ammonia. Finally, the Mthp and 2′,3′-terminal methoxymethylene-protecting groups are removed by hydrolysis under mild acidic conditions.

RNA oligonucleotide synthesis via 5′-silyl-2′-orthoester chemistry. S. A. Hatsel *et al.* (2005) *Methods Mol. Biol.* **288**, 33–50.

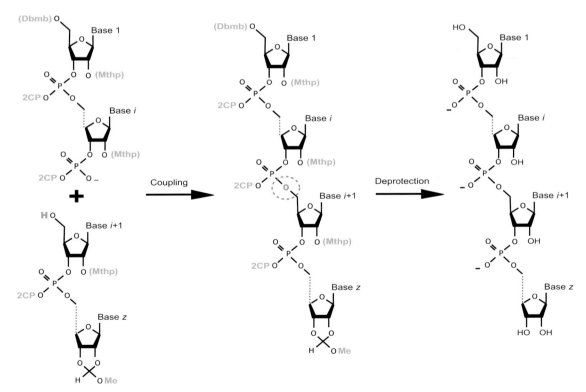

Figure 6-28. Coupling two protected oligoribonucleotides in solution, followed by removal of the protecting groups. The two *dashed* bonds indicate that there are other nucleotides in the chain at these points. The Mthp-protecting group is described in Equation 6.6 and Dbmb in Equation 6.9. 2CP is the 2-chlorophenyl group (Equation 6.10). The methyl group (Me) can be on either the 2′- or 3′-hydroxyl group at the 3′-terminus.

6.7.C. Solid-phase RNA Synthesis

The advantages of the preparation of relatively small (i.e. milligram to gram) quantities of RNA using solid-phase techniques are basically the same as those described with DNA (Section 6.6.D). The protocols involved are also closely similar, apart from the final unblocking procedures. The first step in each cycle is removal of the 5′-terminal protecting group (Figure 6-29, step 1), which must be both rapid and quantitative. The DMTr and Px groups (Figure 6-20) are frequently used as the 5′-protecting

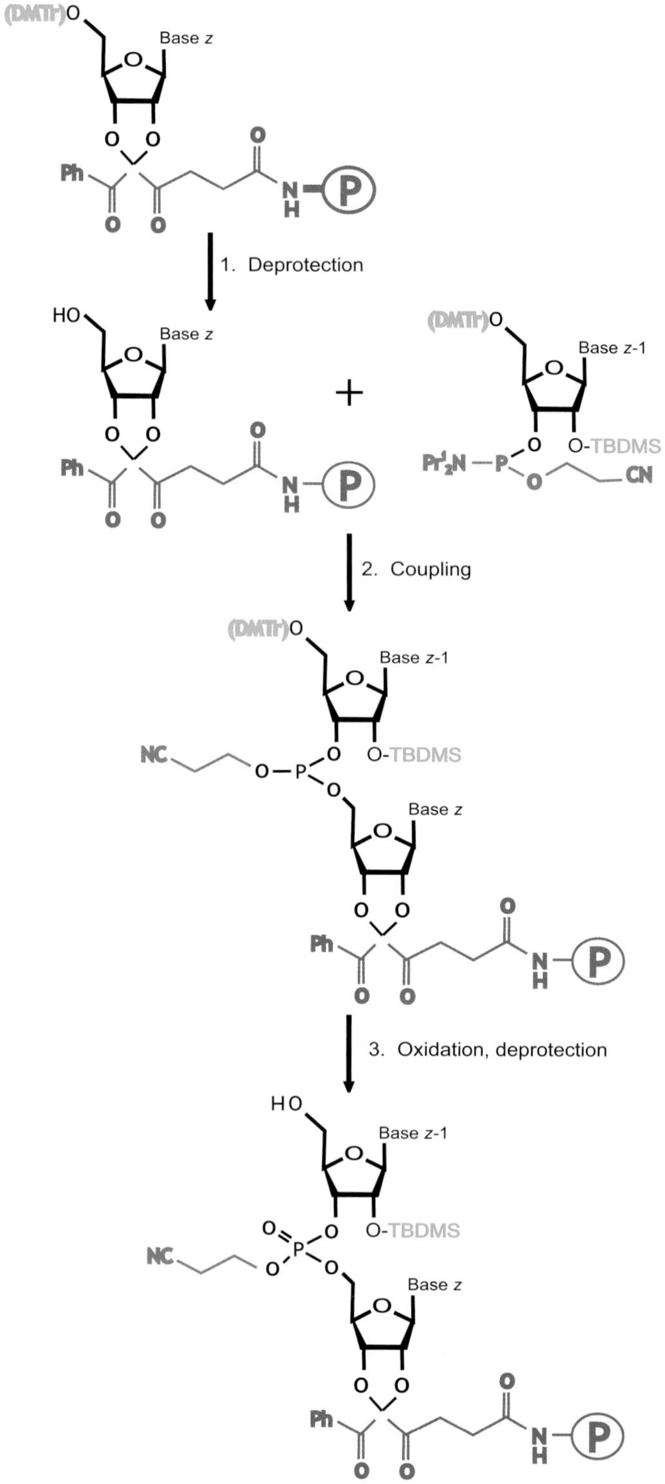

Figure 6-29. One cycle of solid-phase synthesis of RNA. P in a *circle* is the polymer support. The protecting groups DMTr and TBDMS are described in Figure 6-20 and Equation 6.8. The two protecting groups linked to the 2′- and 3′-hydroxyl groups of nucleotide z can be attached to either. The deprotection of the 5′-hydroxyl group of ribonucleotide z (which will be at the 3′-end of the final polymer) in step 1 takes place in 3% trichloroacetic acid in dichloromethane. The coupling in step 2 requires 1H-tetrazole in acetonitrile, followed by acetic anhydride, a weak base and 1-methylimidazole. Step 3 requires 0.1 M iodine in water-pyridine, followed by 3% trichloroacetic acid in dichloromethane. The oligonucleotide is now ready to have another nucleotide added, in another step 2.

groups because they are especially acid-labile. They also have the advantage that it is easy to measure the released groups spectrophotometrically and thereby monitor the efficiency of each coupling step. As in DNA synthesis, the adenine, cytosine and guanine base residues are protected with N-acyl groups (Figure 6-19) and the uracil moieties are left unprotected. The bases of any modified ribonucleosides need to be protected in an analogous manner. 2-Cyanoethyl phosphoramidite building blocks are generally used, as in DNA synthesis (Figure 6-22). The solid support P is usually controlled-pore glass or polystyrene, and the 3′-terminal ribonucleoside residue is usually attached through a 3′- or 2′-succinoyl linker (Figure 6-30). It does not matter which of the two hydroxyl groups is actually linked to the solid support, and the other 2′- or 3′-hydroxyl group is conveniently protected with an acyl group, such as benzoyl.

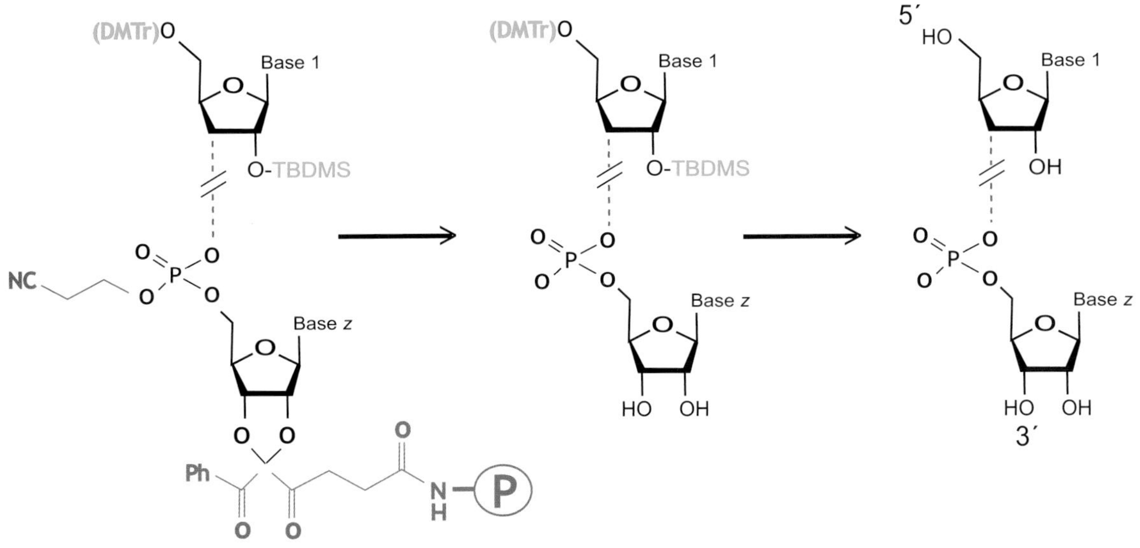

Figure 6-30. Release from the polymer and deprotection of the assembled RNA molecule of z ribonucleotides. The first step takes place in 3% trichloroacetic acid in dichloromethane followed by ammonia in aqueous ethanol; step 2 takes place in triethylamine trihydrofluoride. P is the polymer support.

The 2′-protecting groups must be completely stable to repeated treatment with acid in the steps for removing the DMTr or Px groups at the beginning of each synthetic cycle (Figure 6-29, step 1), as well as to the alkaline conditions used for detachment of the assembled RNA sequences from the solid support and for removal of the N-acyl-protecting groups from the base moieties and of the 2-cyanoethyl-protecting groups from the internucleotide linkages in the penultimate unblocking step (Figure 6-30). TBDMS (Equation 6.8) and Fpmp (Equation 6.6) are the most suitable groups for protection of the 2′-hydroxyl groups, although they require somewhat different methods to release and deprotect the RNA molecule. The RNA sequences protected with Fpmp are partially deprotected by treatment with concentrated aqueous ammonia, without any loss of the Fpmp-protecting groups. This has the considerable advantage that the 2′-protected RNA is completely stable to base-catalyzed hydrolysis and any contaminating ribonucleases and can readily be purified and freed from truncated sequences before it is fully unblocked. Cleavage and migration of the internucleotide linkages during the final unblocking step can be avoided by using mild acidic conditions.

Both methods have produced relatively high-molecular weight RNA sequences. The rates of coupling and coupling efficiencies, however, are somewhat lower than those for the corresponding DNA phosphoramidites.

Recent advances in the high-speed solid phase synthesis of RNA. W. S. Marshall & R. J. Kaiser (2004) *Cur. Opinion Chem. Biol.* **8**, 222–229.

~ CHAPTER 7 ~

POLYPEPTIDE STRUCTURE

Polypeptide chains make up proteins, which are one of the most important classes of biological macromolecules, having both structural and catalytic roles. Polypeptide chains are built up during their biosynthesis from basic building blocks of 21 different amino acids. The amino acids are linked together in a linear polypeptide chain by forming peptide bonds between them, in an order ordained by the nucleotide sequence of the corresponding gene for the protein (Section 7.5.F). These 21 amino acid residues can also be modified post-translationally in very many different ways, to produce even more variety. The amino acid sequence, appropriately known as the **primary structure**, identifies a protein unambiguously, determines all its chemical and biological properties, and specifies (indirectly) the higher levels of protein structure (Chapters 8 and 9).

Chemistry of the Amino Acids. J. P. Greenstein & M. Winitz (1961) John Wiley, NY.

Proteins: structures and molecular properties, 2nd edn. T. E. Creighton (1993) W. H. Freeman and Company, NY.

Structure in Protein Chemistry. J. Kyte (1995) Garland Publishing, NY.

7.1. POLYPEPTIDE CHAINS*

Twenty of the **amino acids** used to synthesize natural proteins have the general structure:

$$\begin{array}{c} R \\ | \\ H_2N-CH-CO_2H \end{array} \tag{7.1}$$

they differ only in the chemical structures of the side-chain, R. The amino group and the carboxyl

---

*A word on nomenclature: a **peptide** is a short polymer of a few amino acid residues with a defined sequence; it usually has properties that are close to those expected from just its constituent amino acids. A **polypeptide** is a longer chain, with more amino acid residues and a defined sequence; it is usually assumed to remain unfolded and to have no special chemical or physical properties. A **polyamino acid** has sequences of varying lengths produced by nonspecific random polymerization of one or a few amino acids. A **protein** is a polypeptide chain with a defined length and amino acid sequence that adopts a specific folded conformation and has special physical and chemical properties.

group give this class of compounds its name. At physiological pH values, both groups are ionized, and this **zwitterion** is the common form of the amino acid in solution. The exceptional amino acid, proline, differs in that its side-chain is bonded to the N atom of the amino group, which is then a secondary amine, and proline is an **imino acid**:

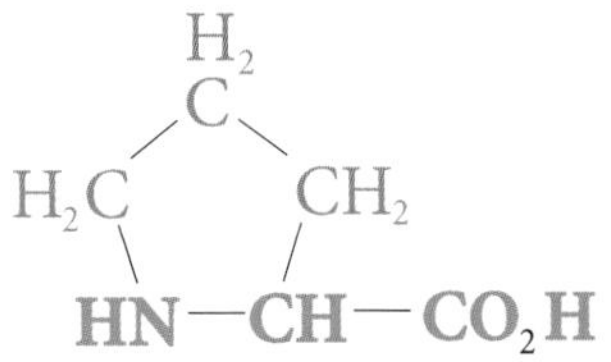

$$\begin{array}{c} H_2C \\ H_2C \quad CH_2 \\ HN-CH-CO_2H \end{array} \qquad (7.2)$$

The central C^α atom is asymmetric in 20 of the amino acids and is always the L-isomer:

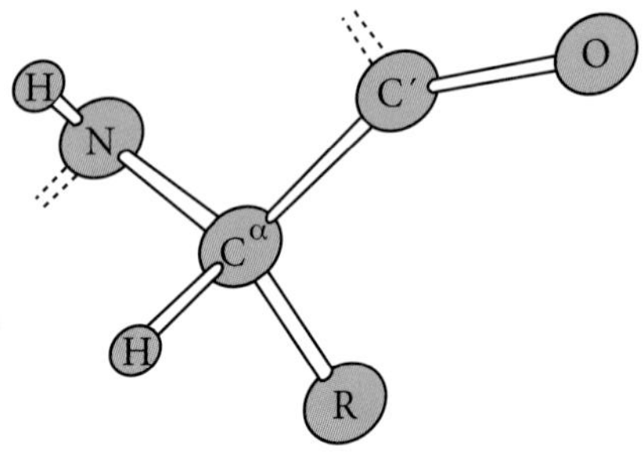

(7.3)

Glycine is the exception, in that its side-chain is simply another H atom, so the C^α atom is not chiral (Section 1.1.A). The L-isomer can be readily identified by looking down from the H atom to the C atom and remembering the "**CORN rule**": the other groups should occur in the clockwise sequence CO, R (side-chain) and N.

Amino acids are linked together by peptide bonds formed by condensation of the α-carboxyl group of one amino acid with the α-amino group of another, producing a dipeptide (Figure 7-1). Repeated condensation of additional amino acids produces tripeptides, tetrapeptides, and so on.

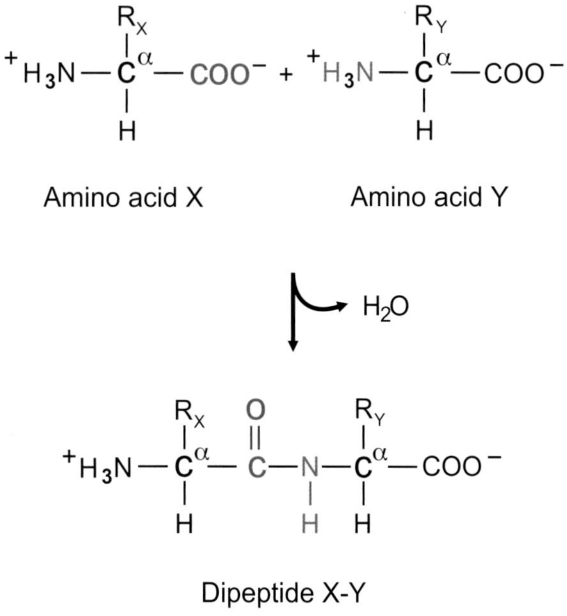

$$^+H_3N-\underset{H}{\overset{R_X}{\underset{|}{C^\alpha}}}-COO^- \; + \; ^+H_3N-\underset{H}{\overset{R_Y}{\underset{|}{C^\alpha}}}-COO^-$$

Amino acid X   Amino acid Y

↓ → H_2O

$$^+H_3N-\underset{H}{\overset{R_X}{\underset{|}{C^\alpha}}}-\overset{O}{\underset{}{\overset{\|}{C}}}-\underset{H}{\overset{}{\underset{|}{N}}}-\underset{H}{\overset{R_Y}{\underset{|}{C^\alpha}}}-COO^-$$

Dipeptide X-Y

Figure 7-1. General scheme for formation of a dipeptide XY from two amino acids, X and Y, by reaction of the carboxyl group of amino acid X with the amino group of amino acid Y. The R group attached to C^α is variable depending on the type of amino acid: R_X and R_Y.

7.2. AMINO ACID RESIDUES

An amino acid incorporated within a polypeptide chain is known as a **residue**. The structures of the side-chains of the 20 normal amino acids used most frequently in protein biosynthesis are described in Figure 7-2. The 21st amino acid, which is used in protein biosynthesis in only a few instances, is selenocysteine, which differs from cysteine only in having a Se atom in place of the S (Section 7.2.O).

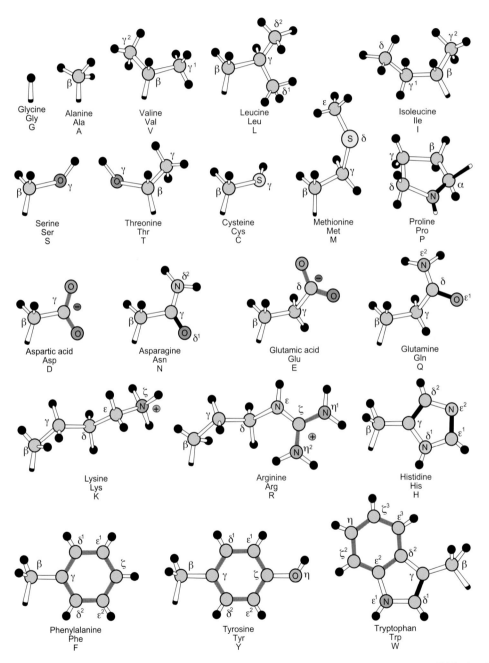

Figure 7-2. The side-chains of the 20 amino acids that occur naturally in proteins. *Small black spheres* are H atoms and *larger green spheres* are C atoms; other atoms are labeled. Double bonds are *black*, and partial double bonds are *gray*. In the case of Pro, the bonds of the polypeptide backbone are included and are *black*. Below the name of the amino acid are the three-letter and one-letter abbreviations commonly used for the residues. Note that Ile and Thr have asymmetric centers in their side-chains and only the isomer illustrated is used biologically.

The central, asymmetric C atom is designated as α, and the side-chain atoms are commonly designated β, γ, δ, ε and ζ from the C$^\alpha$ atom. Chemical groups, however, are often designated by the C atom to which they are bonded: hence, the N$^\zeta$ atom of a Lys residue is said to be part of the ε amino group.

The residues are frequently designated with either three- or one-letter abbreviations (Figure 7-2). The three-letter abbreviations tend to be obvious, but the one-letter abbreviations save space. They are also less likely to be confused; for example, Gln, Glu and Gly are very similar, but not Q, E and G. Amino acid residues in polypeptide chains are properly referred to by changing the ending of their amino acid names, adding or replacing the frequent -ine ending with -yl (for example, alanyl or seryl residues), but residues are often referred to by the name of the amino acid. This complication can be avoided by using the one- or three-letter abbreviations, but they should be used for residues in proteins only, not for the free amino acids. The sequences of amino acids in proteins are usually described using either abbreviation and written from the amino end to the carboxyl end. Starting at the left is the N-terminal residue, which contains the α-amino group and is considered the first residue of the polypeptide chain.

Some relevant properties of individual amino acid resides are given in Table 7-1. Their chemical properties will be described here, usually in terms of their chemical reactivities, for these govern their functions in biological proteins. The reactions of amino acid residues with chemical reagents were extremely important in studying protein structure and function, and much effort was made to find reagents specific for a particular type of residue. Subsequently, however, the roles of individual residues in protein structure and function are usually determined by replacing the residue by another type by manipulating the gene for the protein and producing the altered protein, a process known as **protein engineering** (Section 7.6.C). Nevertheless, understanding the chemical properties of the amino acid residues is essential for understanding their roles in protein structure and function.

Chemical Modification of Proteins. G. E. Means & R. E. Feeney (1971) Holden Day, CA.

Chemical modification. T. Imoto & H. Yamada (1997) in *Protein Function: a practical approach*, 2nd edn (T. E. Creighton, ed.), IRL Press, Oxford, pp. 279–316.

7.2.A. GLYCINE (GLY)

Glycine is the simplest amino acid, with only an H atom for a side-chain. Note that the C$^\alpha$ atom of glycine is not chiral, in contrast to the other amino acids incorporated into proteins, because it is bonded to two H atoms. Consequently, this amino acid residue is not asymmetric.

Table 7-1. Properties of individual amino acid residues

Residue	Mass (Da)	Van der Waals volume[a] (Å^3)	Partial volume in solution[b] (Å^3)	Partial specific volume[b] (cm^3/g)	Frequency in proteins (%)
Ala (A)	71.09	67	86.4	0.732	8.3
Arg (R)	156.19	148	197.4	0.756	5.7
Asn (N)	114.11	96	115.6	0.610	4.4
Asp (D)	115.09	91	108.6	0.573	5.3
Cys (C)	103.15	86	107.9	0.630	1.7
Gln (Q)	128.14	114	142.0	0.667	4.0
Glu (E)	129.12	109	128.7	0.605	6.2
Gly (G)	57.05	48	57.8	0.610	7.2
His (H)	137.14	118	150.1	0.659	2.2
Ile (I)	113.16	124	164.6	0.876	5.2
Leu (L)	113.16	124	164.6	0.876	9.0
Lys (K)	128.17	135	166.2	0.775	5.7
Met (M)	131.19	124	160.9	0.739	2.4
Phe (F)	147.18	135	187.3	0.766	3.9
Pro (P)	97.12	90	120.6	0.748	5.1
Ser (S)	87.08	73	86.2	0.596	6.9
Thr (T)	101.11	93	113.6	0.676	5.5
Trp (W)	186.21	163	225.0	0.728	1.3
Tyr (Y)	163.18	141	190.5	0.703	3.2
Val (V)	99.14	105	136.8	0.831	6.6
Weighted average[c]	119.40	104	137.6	0.703	

[a] Volume enclosed by the van der Waals radii of the atoms.

[b] Increase in volume of water after adding either one molecule or 1 g of residue.

[c] Weighted by frequency of occurrence in proteins to give the value for an average residue in proteins.

7.2.B. Nonpolar Amino Acid Residues (Ala, Leu, Ile, Val)

The side-chains of Ala, Leu, Ile and Val residues are simple hydrocarbons.

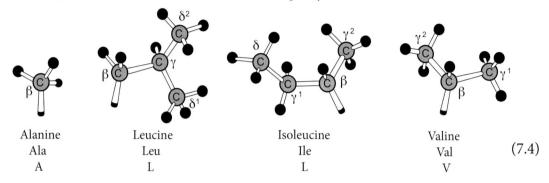

Alanine	Leucine	Isoleucine	Valine	
Ala	Leu	Ile	Val	(7.4)
A	L	I	V	

Consequently, they are nonpolar, inert and chemically unreactive. Note that the side-chain of Ile is chiral and only the one isomer occurs naturally.

7.2.C. Hydroxyl Residues (Ser, Thr)

The side-chains of Ser and Thr residues are dominated by their hydroxyl groups:

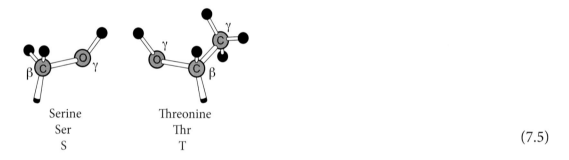

Serine	Threonine	
Ser	Thr	
S	T	(7.5)

These hydroxyl groups are normally no more reactive chemically than that of ethanol, so there are few chemical reactions that can readily and specifically modify Ser and Thr residues in a protein. The only useful general reaction is acetylation with acetyl chloride in aqueous trifluoroacetic acid, which requires relatively harsh conditions. When in the appropriate environment, however, and with a potentially reactive group held in the correct position, the Ser hydroxyl group can function as a potent nucleophile, as in the serine proteinases (Section 15.2.D.1). Specific Ser and Thr residues in certain proteins are subject to various post-translational modifications caused by specific enzymes; in particular, they can be phosphorylated reversibly by specific enzymes, with substantial and important effects on their biological functions (Section 15.2.A).

The hydroxyl group of Ser and Thr is polar and hydrophilic and can function as either a donor or acceptor in hydrogen bonds. The hydroxyl group is sterically able to interact with adjacent polar groups of the polypeptide backbone, which affects its conformation and reactivity. For example, peptide bonds adjacent to Ser and Thr residues are especially susceptible to hydrolysis by acid or metal ions (Section 7.5.A.1).

Note that the Thr side-chain has a center of asymmetry at C^β and that only the one isomer occurs naturally.

Elevated intrinsic reactivity of seryl hydroxyl groups within the linear peptide triads His–Xaa–Ser or Ser–Xaa–His. B. T. Miller & A. Kurosky (1993) *Biochem. Biophys. Res. Commun.* **196**, 461–467.

Susceptibility of the hydroxyl groups in serine and threonine to β-elimination/Michael addition under commonly used moderately high-temperature conditions. W. Li *et al.* (2003) *Anal. Biochem.* **323**, 94–102.

7.2.D. Arginine (Arg)

The Arg side-chain consists of three nonpolar methylene groups capped by the δ guanido group:

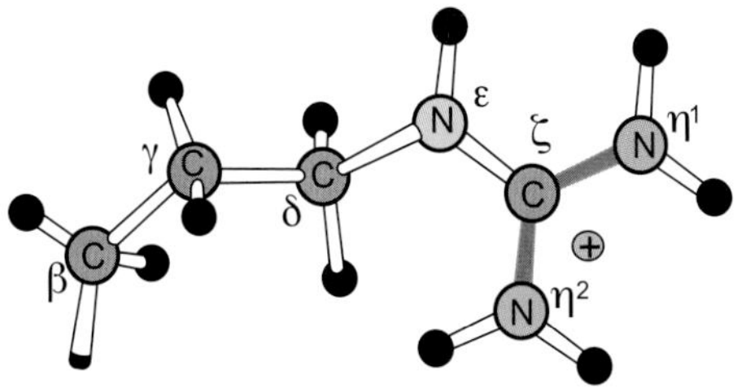

$$\begin{array}{c}\text{Arginine}\\\text{Arg}\\\text{R}\end{array} \tag{7.6}$$

With an intrinsic pK_a value of ~12, the guanido group is very basic and ionized over the entire pH range that can be considered physiological. The positive charge of the guanido group is spread over all four atoms as a result of resonance, which also keeps the group planar:

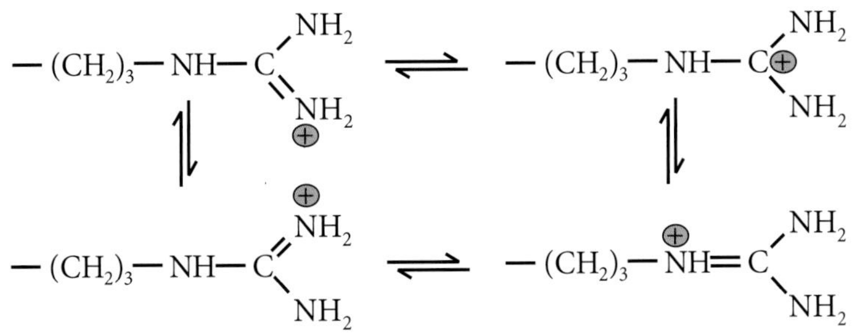

$$\tag{7.7}$$

The guanido group is unreactive in the protonated form, and only very small fractions of the nonionized form are present at physiological pH values. Consequently, Arg residues are not very reactive chemically. Nevertheless, the guanido group of Arg residues can form heterocyclic condensation products with 1,2- and 1,3-dicarbonyl groups, such as those of 2,3-butanedione:

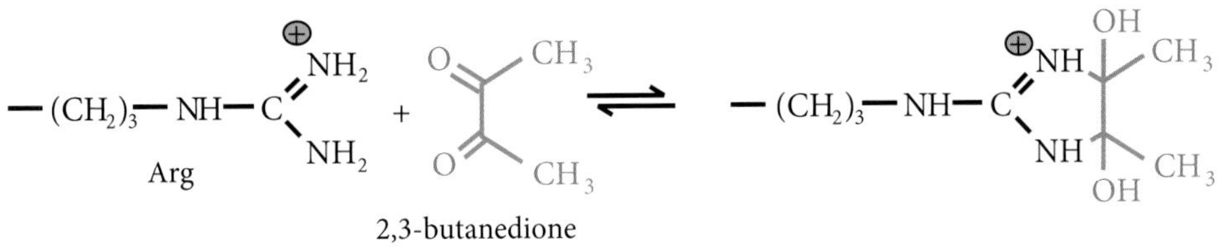

$$\tag{7.8}$$

These interactions occur because the distance between the two carbonyl groups of the reagents closely matches that between the two unsubstituted N atoms of the guanido group. The condensation product can be stabilized by borate, which will complex the adjacent hydroxyl groups.

The guanido group of Arg residues can be cleaved by hydrazine (H_2N–NH_2) to produce a residue (Orn) of the amino acid ornithine:

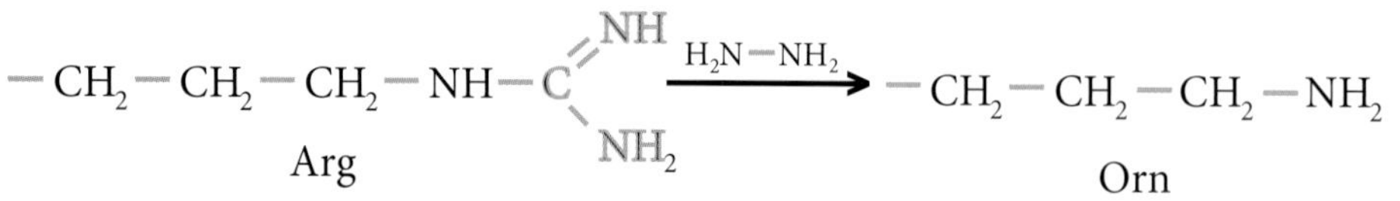

$$-CH_2-CH_2-CH_2-NH-C\begin{matrix}\nearrow NH \\ \searrow NH_2\end{matrix} \quad \xrightarrow{H_2N-NH_2} \quad -CH_2-CH_2-CH_2-NH_2$$

$$\text{Arg} \qquad\qquad\qquad\qquad\qquad\qquad \text{Orn} \qquad\qquad (7.9)$$

A side-reaction, however, can be cleavage of the polypeptide backbone (Section 7.5.E.1).

Functional probing of arginine residues in proteins using mass spectrometry and an arginine-specific covalent tagging concept. A. Leitner & W. Lindner (2005) *Anal. Chem.* **77**, 4481–4488.

Site-selective modifications of arginine residues in human hemoglobin induced by methylglyoxal. Y. Gao & Y. Wang (2006) *Biochemistry* **45**, 15654–15660.

Synthesis of proteins containing modified arginine residues. A. K. Choudhury *et al.* (2007) *Biochemistry* **46**, 4066–4076.

7.2.E. Lysine (Lys)

The Lys side-chain consists of four methylene groups capped by an amino group:

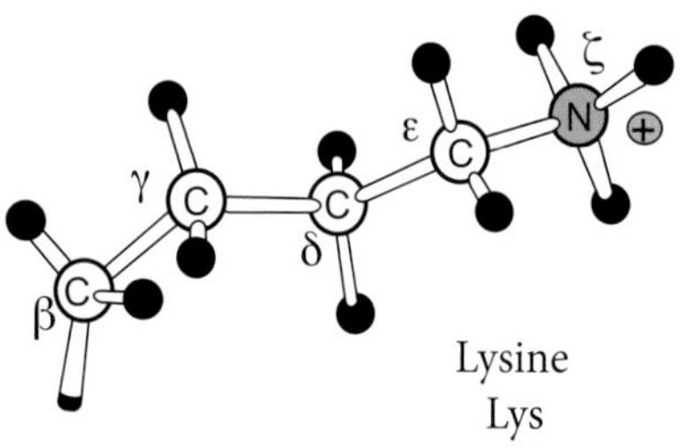

Lysine
Lys
K

$$(7.10)$$

The amino group is ionized under most physiological conditions, as it has an intrinsic pK_a value of ~11.1. The ionized form is unreactive chemically, but a finite fraction of the amino groups are nonionized and they are potent nucleophiles. Consequently, the amino groups of Lys residues readily participate in a wide variety of acylation, alkylation, arylation and amidination reactions (Table 7-2). The α-amino group at the N-terminus of the polypeptide chain is also susceptible to such reactions, but has a pK_a of only 6–8, so it may be modified selectively by controlling the pH of the reaction medium.

Table 7-2. Chemical modifications of amino groups of proteins

Reaction	Reagent	Optimum pH
Acylation	Acetic anhydride	5.5–8
	Citraconic anhydride	8.2
	Maleic anhydride	6–10
	Succinic anhydride	7–10
	Acetylimidazole	>5
	N-Acetylsuccinimide	>4
	N-Hydroxysuccinimide acetate	6.9–8.5
Alkylation and arylation	Iodoacetic acid	7.5–9
	1-Fluoro-2,4-dinitrobenzene	7–11
	2,4,6-Trinitrobenzene sulfonic acid	9.5
Reductive alkylation	Formaldehyde + sodium borohydride	8–10
Amidination	Methyl acetamidate	7–10.5
Carbamylation	Potassium cyanate	>7
Guanidination	1-Guanyl-3,5-dimethylpyrazole nitrate	9.5
	O-Methylisourea	10–11

The number of free amino groups present, and subsequently the extent of their modification with some modifying reagents, can be determined by reaction with **2,4,6-trinitrobenzene sulfonic acid (TNBS)**:

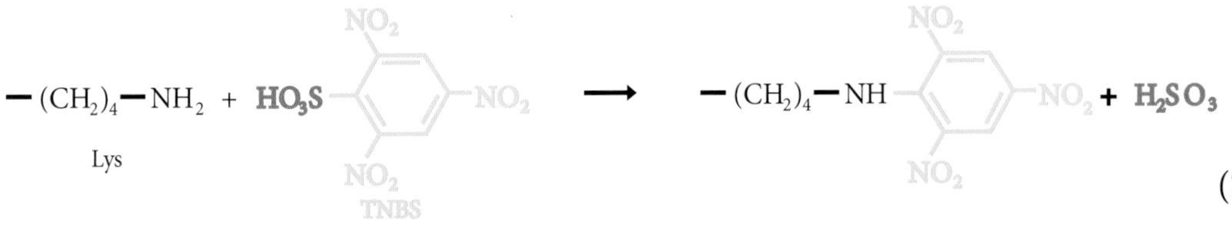

$$-(CH_2)_4-NH_2 + HO_3S-C_6H_2(NO_2)_3 \rightarrow -(CH_2)_4-NH-C_6H_2(NO_2)_3 + H_2SO_3$$

(7.11)

producing trinitrophenyl derivatives with yellow color that can be measured. TNBS also reacts, however, with thiol groups.

The quantification of protein amino groups by the trinitrobenzenesulfonic acid method: a reexamination. P. Cayot & G. Tainturier (1997) *Anal. Biochem.* **249**, 184–200.

1. Acetylation by Anhydrides

Amino groups can be acetylated with a variety of **anhydrides**: acetic, succinic, citraconic (e.g. methylmaleic) and 3,4,5,6-tetrahydrophthaloyl. The reaction occurs rapidly:

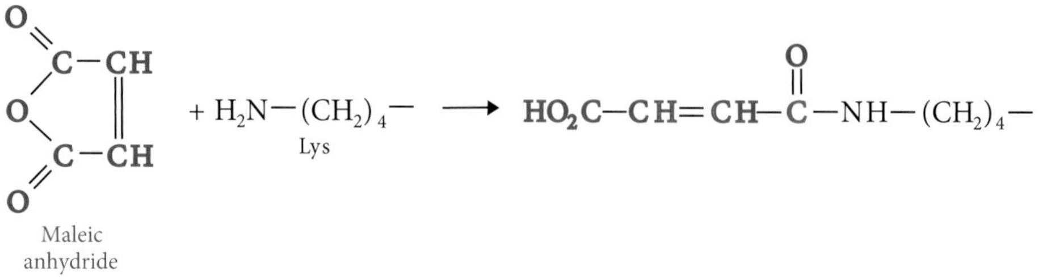

$$\text{Maleic anhydride} + H_2N-(CH_2)_4- \text{(Lys)} \longrightarrow HO_2C-CH=CH-\overset{O}{\underset{\|}{C}}-NH-(CH_2)_4- \tag{7.12}$$

The maleic, citraconic and tetrahydrophthaloyl anhydrides are especially useful for temporary modifications, because the reaction can be reversed readily under acidic conditions.

On-column derivatization and analysis of amino acids, peptides, and alkylamines by anhydrides using capillary electrophoresis. Y. Zhang & F. Z. Gomez (2000) *Electrophoresis* **21**, 3305–3310.

FTIR–ATR spectroscopy for monitoring polyanhydride/anhydride–amine reactions. M. Krishnan & D. R. Flanagan (2000) *J. Control Release* **69**, 273–281.

Reversible fluorescence labeling of amino groups of protein using dansylaminomethylmaleic anhydride. K. Sakata *et al.* (2002) *J. Chromat. B* **769**, 47–54.

---

2. Amidination

The reaction of amino groups with **amidates**:

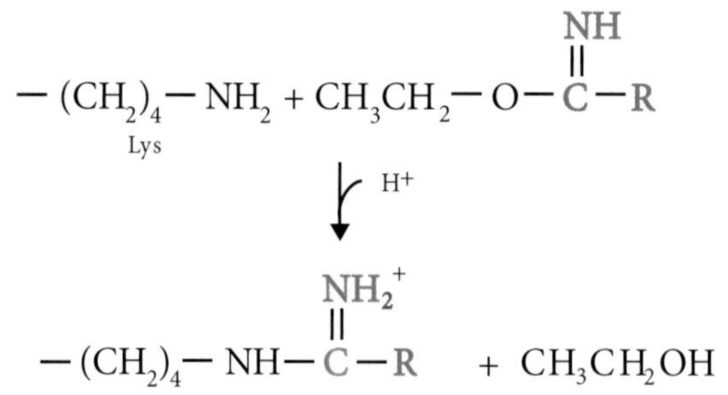

$$-(CH_2)_4-NH_2 \text{ (Lys)} + CH_3CH_2-O-\overset{NH}{\underset{\|}{C}}-R \xrightarrow{H^+} -(CH_2)_4-NH-\overset{NH_2^+}{\underset{\|}{C}}-R + CH_3CH_2OH \tag{7.13}$$

is useful, because the Lys side-chain remains basic and positively charged.

Probing protein tertiary structure with amidination. D. J. Jaenicke *et al.* (2005) *Anal. Chem.* **77**, 7274–7281.

3. Guanidination

Guanidination of amino groups occurs upon their reaction with either **O-methylisourea**

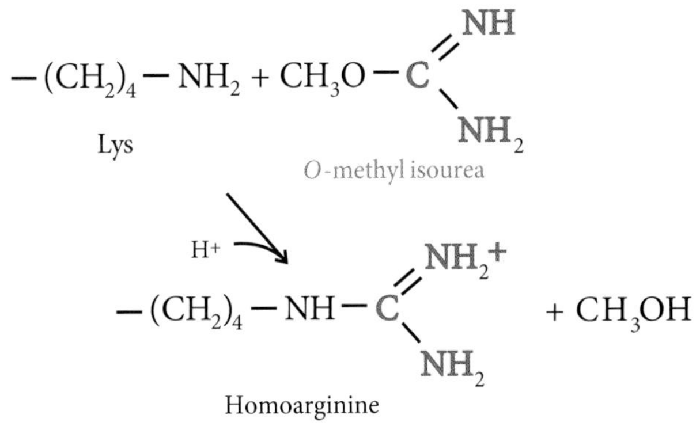

$$-(CH_2)_4-NH_2 + CH_3O-C\underset{NH_2}{\overset{\displaystyle \parallel NH}{}}$$

Lys   *O*-methyl isourea

$$\xrightarrow{H^+} -(CH_2)_4-NH-C\underset{NH_2}{\overset{\displaystyle \parallel NH_2^+}{}} + CH_3OH$$

Homoarginine

(7.14)

or **1-guanyl-3,5-dimethylpyrazol**:

$$R-NH_2 + \text{[1-Guanyl-3,5-dimethylpyrazol]} \xrightarrow{H^+} R-\underset{H}{N}-C\underset{NH_2}{\overset{\displaystyle \parallel NH_2^+}{}} + \text{[3,5-dimethylpyrazole]}$$

Lys

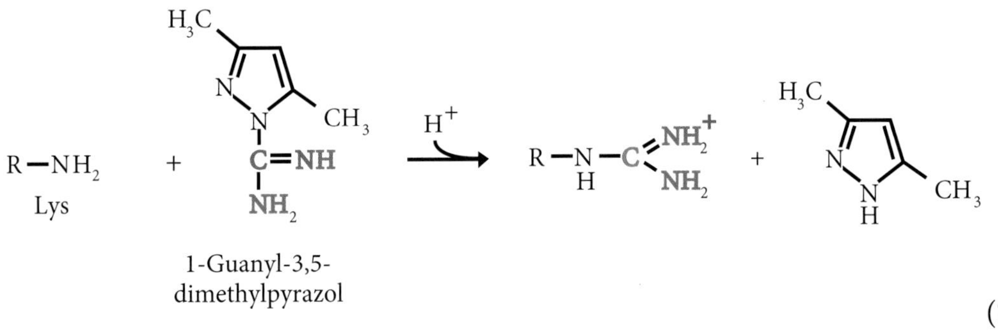

(7.15)

The reaction occurs primarily with the ε amino groups of Lys residues, to convert them to **homoarginine**, and much less with α-amino groups. The modified amino groups retain their positive charge, with increased pK_a; therefore structural effects on the modified polypeptide chains are usually minimal. The function of a protein modified in this way will usually be affected only if the amino group participates intimately in the function.

The reaction is quite specific to amino groups but some side-reactions occur with the thiol and imidazole groups of Cys and His residues, respectively.

Optimization of guanidination procedures for MALDI mass mapping. R. L. Beardsley & J. P. Reilly (2002) *Anal. Chem.* **74**, 1884–1890.

Guanidination chemistry for qualitative and quantitative proteomics. S. Warwood *et al.* (2006) *Rapid Commun. Mass Spectrom.* **20**, 3245–3256.

4. Schiff Base Formation

Lys residues can reversibly form Schiff bases with aldehydes, such as that of the natural cofactor **pyridoxal phosphate** (Section 14.4.G.1) and the chromophore **retinal** in photoreceptor proteins. The Schiff base can be trapped chemically by reduction, for example by **sodium borohydride** (NaBH$_4$):

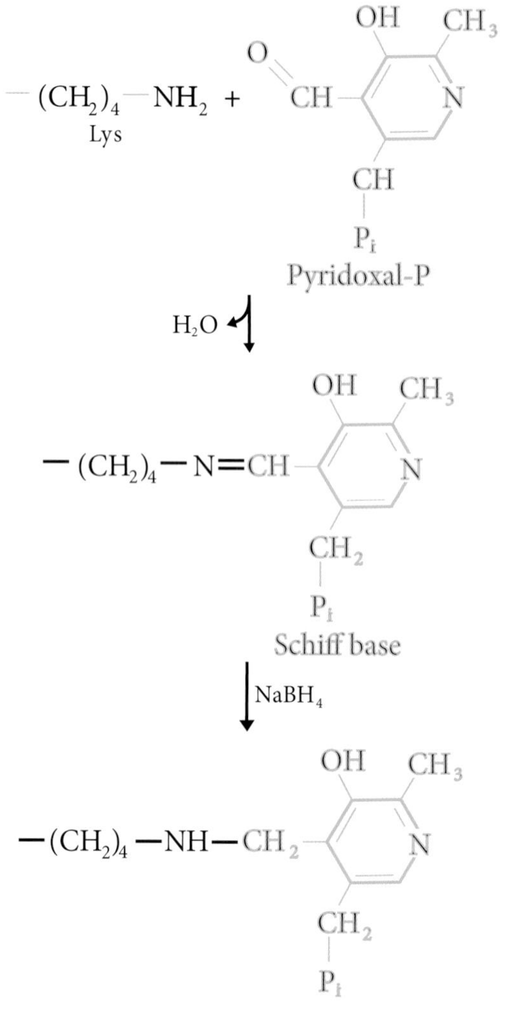

$$\text{—(CH}_2)_4\text{—NH}_2 \text{ (Lys)} + \text{Pyridoxal-P}$$

$$\rightleftharpoons_{H_2O} \text{—(CH}_2)_4\text{—N=CH—[pyridoxal ring]} \quad \text{Schiff base}$$

$$\xrightarrow{NaBH_4} \text{—(CH}_2)_4\text{—NH—CH}_2\text{—[pyridoxal ring]} \tag{7.16}$$

Photochromism and thermochromism of Schiff bases in the solid state: structural aspects. E. Hadjoudis & I. M. Mavridis (2004) *Chem. Soc. Rev.* **33**, 579–588.

Tautomerism of sterically hindered Schiff bases. Deuterium isotope effects on ^{13}C chemical shifts. A. Filarowski *et al.* (2005) *J. Phys. Chem. A* **109**, 4464–4473.

NMR studies of solvent-assisted proton transfer in a biologically relevant Schiff base: toward a distinction of geometric and equilibrium H-bond isotope effects. S. Sharif *et al.* (2006) *J. Am. Chem. Soc.* **128**, 3375–3387.

5. Carbamylation

A widely used modification of amino groups is carbamylation (or carbamoylation) by **cyanate** to form a **homocitrulline** residue:

$$-(CH_2)_4-NH_2 + HN=C=O$$
$$\text{Lys} \quad\quad \text{Cyanate}$$

$$\downarrow$$

$$-(CH_2)_4-NH-\overset{O}{\underset{\|}{C}}-NH_2$$
$$\text{Homocitrulline}$$

(7.17)

This reaction can occur unwittingly in old or poorly maintained solutions of urea (Section 11.1.C), which generate cyanate spontaneously.

Carbamylation of proteins in 2-D electrophoresis: myth or reality? J. McCarthy *et al.* (2003) *J. Proteome Res.* **2**, 239–242.

7.2.F. Histidine (His)

The side-chain of His residues consists primarily of an imidazole group:

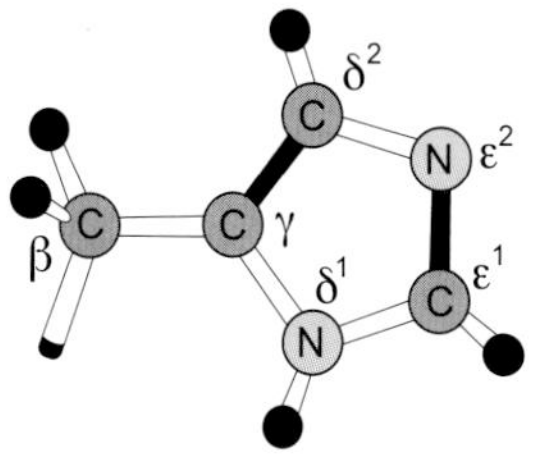

Histidine
His
H

(7.18)

The two N atoms of the His side-chain are designated here as δ^1 and ϵ^2 but they are also known as π and τ, respectively, or as N1 and N3. The latter designation is often ambiguous, because biochemists usually assign the number 1 to the N atom adjacent to C^γ, whereas organic chemists tend to reverse the numbers.

The chemical properties of imidazole make the His residue an extremely effective nucleophilic catalyst. First, imidazole is an amine, which in terms of its basicity is much more reactive than a hydroxide ion. Secondly, it is a tertiary amine, which is intrinsically more nucleophilic than primary or secondary amines. Normally the greater intrinsic reactivity of tertiary amines is nullified by increased steric hindrance, but the atoms bonded to the two N atoms of imidazole are held back by the five-membered ring and produce relatively little steric hindrance. Thirdly, imidazole is one of the strongest bases that can exist at neutral pH because it ionizes with a pK_a value near 7. A weaker base would be less reactive as a nucleophile, whereas a stronger base would be less reactive because it would be protonated to a greater extent at neutral pH.

The N atom without the H atom of the imidazole ring is a nucleophile and a hydrogen bond acceptor, while the other N atom is an electrophile and a donor for hydrogen bonding. This is almost the chemical equivalent of being ambidextrous, so the nonionized His side-chain is extremely versatile and is often involved directly in protein function. The H atom can be on either the δ^1 or ϵ^2 N atom, so imidazole can exist as either of two tautomers (Section 1.1.C). When free in solution, the imidazole group has the H atom predominantly on the ϵ^2 N atom, which has a pK_a value ~0.6 pH unit higher than that of the δ^1 atom. The relative pK_as of the two N atoms can vary with their local environment, however, and both forms are found in folded proteins.

The nucleophilicity of the nonionized His side-chain is destroyed when it is protonated at the second N atom. This occurs readily, with a pK_a value near 7. Resonance ensures that the positive charge is shared by the two N atoms:

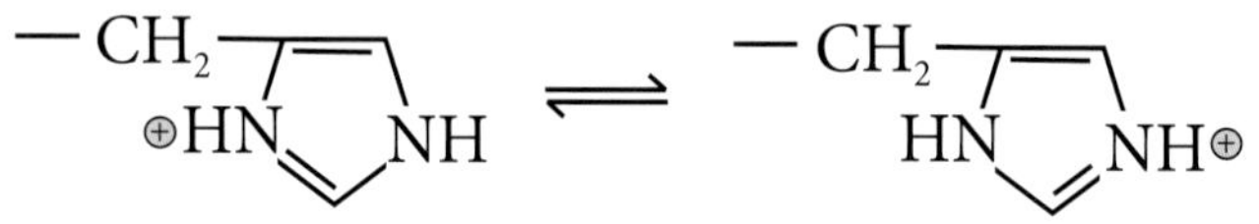

$$(7.19)$$

The $C^{\epsilon 1}$ atom, between the two N atoms, can exchange its H atom slowly with aqueous solvents (Section 10.3). This suggests that there is a small probability of it being deprotonated at least transiently:

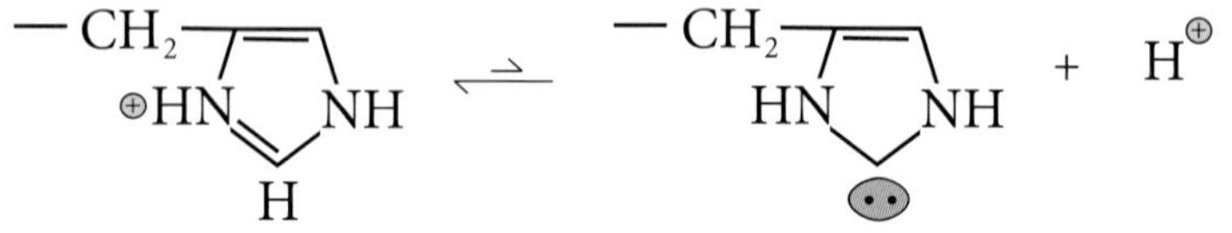

$$(7.20)$$

His residues are especially useful in nuclear magnetic resonance (NMR) studies of proteins because the H atom on the $C^{\epsilon 1}$ atom is usually well-resolved from the multitude of resonances of the other H atoms in proteins. Protonation of the imidazole group usually shifts its resonance by ~1 p.p.m. to a lower field strength, which makes it relatively easy, even in large proteins, to measure the pK_a values of individual His residues.

The imidazole group has an intrinsic affinity for metal ions, especially zinc, iron and copper. This is frequently used in proteins to bind such metal ions (Section 12.2).

In principle, the imidazole group is capable of undergoing numerous chemical reactions, but most of them occur much more readily with amino and thiol groups. Consequently, there are very few reactions to modify His residues specifically in proteins.

Structure characterization of functional histidine residues and carbethoxylated derivatives in peptides and proteins by mass spectrometry. M. Kalkum *et al.* (1998) *Bioconjug. Chem.* **9**, 226–235.

pK values of histidine residues in ribonuclease Sa: effect of salt and net charge. B. M. Huyghues-Despointes *et al.* (2003) *J. Mol. Biol.* **325**, 1093–1105.

7.2.G. Acidic Residues (Asp, Glu)

The side-chains of Asp and Glu residues are dominated by their carboxyl groups:

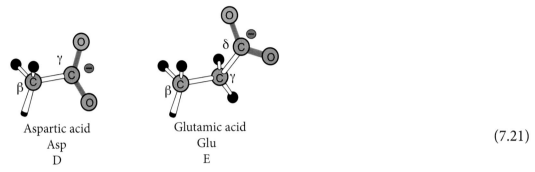

$$\begin{array}{cc} \text{Aspartic acid} & \text{Glutamic acid} \\ \text{Asp} & \text{Glu} \\ \text{D} & \text{E} \end{array} \qquad (7.21)$$

The intrinsic pK_a value is close to 3.9 in Asp and 4.3 in Glu, so these residues are ionized and very polar under physiological conditions.

Carboxyl groups are not very reactive, and catalysts or activation processes are required for them to react chemically. Esterification and amidation are the usual chemical modifications. Trialkyloxonium salts, such as triethyloxonium tetrafluoroborate, are commonly used for mild esterification of carboxyl groups in proteins. Diazoacetyl compounds react with carboxyl groups at acidic pH to form esters:

$$\begin{array}{c} \quad\quad\quad\quad\quad\quad\quad\; \text{O} \\ \quad\quad\quad\quad\quad\quad\quad\; \| \\ -\text{CO}_2\text{H} + \text{N}_2\text{CH}-\text{C}-\text{NH}-\text{R} \\ \quad\quad\quad\quad\quad\quad \text{Diazo amide} \\ \downarrow \\ \text{O} \quad\quad\quad\; \text{O} \\ \| \quad\quad\quad\; \| \\ -\text{C}-\text{O}-\text{CH}_2-\text{C}-\text{NH}-\text{R} + \text{N}_2 \end{array} \qquad (7.22)$$

Other reagents, such as acid halogen compounds and ethylenimine (Equation 7.31), have also been used to esterify the carboxyl groups in proteins.

Amidation of carboxyl groups with simple amines is possible after activation of the carboxyl groups with water-soluble **carbodiimides**, such as 1-ethyl-3-(3-dimethylaminopropyl)carbodiimide. At a pH of about 5, the reaction proceeds as:

$$\begin{array}{c} \text{O} \\ \| \\ \text{R}^1-\text{COH} + \text{R}^2\text{N}=\text{C}=\text{NR}^3 \\ \text{Carboxyl} \quad\quad\quad \text{Carbodiimide} \\ \downarrow\; \text{H}^+ \end{array}$$

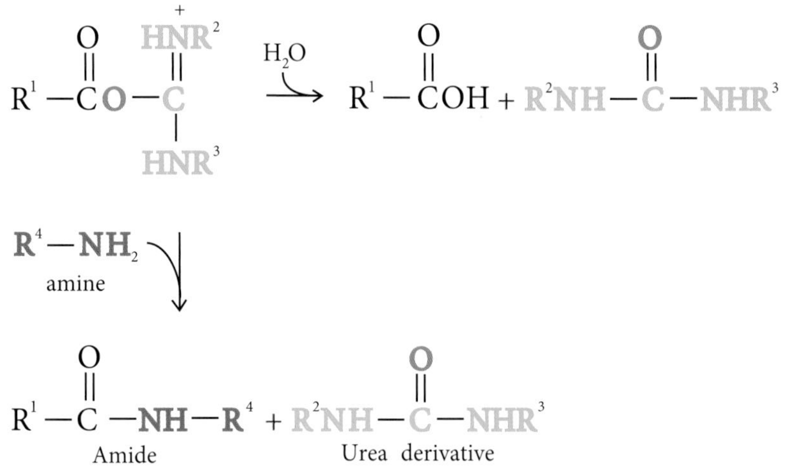

$$R^1-\overset{O}{\underset{\|}{C}}-O-\overset{\overset{+}{H}NR^2}{\underset{\underset{HNR^3}{|}}{C}} \xrightarrow{H_2O} R^1-\overset{O}{\underset{\|}{C}}OH + R^2NH-\overset{O}{\underset{\|}{C}}-NHR^3$$

$$R^4-NH_2 \text{ amine} \searrow$$

$$R^1-\underset{\text{Amide}}{\overset{O}{\underset{\|}{C}}-NH-R^4} + \underset{\text{Urea derivative}}{R^2NH-\overset{O}{\underset{\|}{C}}-NHR^3} \qquad (7.23)$$

Carboxyl groups have a weak intrinsic affinity for Ca^{2+} ions, so Asp and Glu residues are used in this way in many calcium-binding proteins (Section 12.3).

Asp and Glu residues differ only in having one or two methylene groups. They might then be expected to have very similar chemical and functional properties in proteins, but they do not. The different lengths of the side-chains cause their carboxyl groups to interact very differently with the polypeptide backbone, so they have different effects on the conformation and chemical reactivity of the peptide backbone. The Asp carboxyl group interacts more readily with its adjacent polypeptide backbone than does that of Glu. This interaction causes the chain to be cleaved relatively easily at Asp residues (Section 7.5.B.2). Peptide bonds between –Asp–Pro– residues are particularly labile in acid, because the Asp carboxyl group interacts with the adjacent Pro residue's unique tertiary N atom.

Studies on the mechanism of aspartic acid cleavage and glutamine deamidation in the acidic degradation of glucagon. A. B. Joshi *et al.* (2005) *J. Pharm. Sci.* **94**, 1912–1927.

Capillary electrophoresis analysis of hydrolysis, isomerization and enantiomerization of aspartyl model tripeptides in acidic and alkaline solution. S. De Boni & G. K. Scriba (2007) *J. Pharm. Biomed. Anal.* **43**, 49–56.

Conformation-dependent racemization of aspartyl residues in peptides. K. Kuge *et al.* (2007) *Chemistry* **13**, 5617–5621.

7.2.H. Amide residues (Asn, Gln)

The side-chains of Asn and Gln residues are the same as those of Asp and Glu, except that the carboxyl group of each has been converted to the amide:

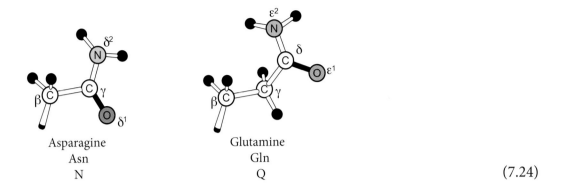

Asparagine
Asn
N

Glutamine
Gln
Q

(7.24)

The amide side-chain does not ionize and is not very reactive chemically. It is polar, however, and can serve as both a donor and acceptor of hydrogen bonds.

When Gln residues are at the N-terminus of a peptide chain, they spontaneously cyclize:

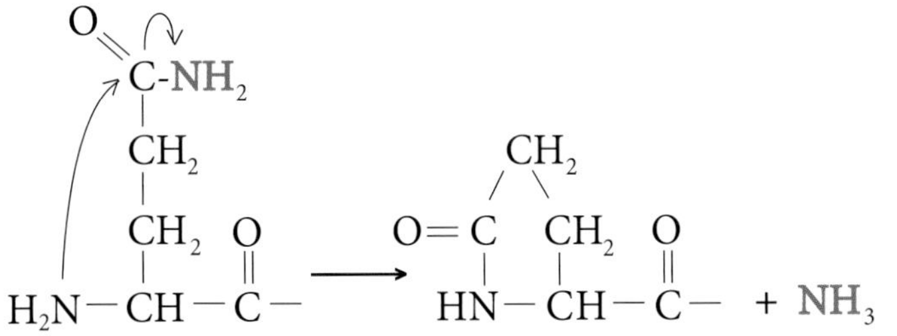

$$\begin{array}{c}\text{O} \\ \diagdown \\ \text{C-NH}_2 \\ | \\ \text{CH}_2 \\ | \\ \text{CH}_2 \quad \text{O} \\ | \quad || \\ \text{H}_2\text{N-CH-C-} \end{array} \longrightarrow \begin{array}{c} \quad\text{CH}_2 \\ \diagup\quad\diagdown \\ \text{O=C}\quad\text{CH}_2\quad\text{O} \\ |\qquad|\quad || \\ \text{HN-CH-C-} \end{array} + \text{NH}_3$$
(7.25)

The resulting residue of **pyrrolidone carboxylic acid** makes the N-terminus unreactive in most procedures for sequencing proteins, such as the Edman degradation (Section 7.5.E.2).

The amide group is labile at extremes of pH and at high temperatures, and Asn and Gln residues can deamidate to Asp and Glu.

Reactions of the amide side-chains of glutamine and asparagine *in vivo*. F. Wold (1985) *Trends Biochem. Sci.* **10**, 4–6.

---

1. Deamidation

Loss of the amide amino group, **deamidation**, occurs spontaneously in proteins at Asn and Gln residues, which can cause heterogeneity, instability and defects in the function of the protein. The amide bond is relatively labile, especially in acidic media, and is easily hydrolyzed chemically. The rate of this reaction is sensitive to the environment and depends upon the conformation of the polypeptide chain. Exposed amides are more readily deamidated than those buried in the interior of a protein; denatured proteins often have high potential for deamidation, because the polypeptide chain is flexible and the amide group is free in solution. The rate of deamidation depends upon the amino acid sequence and can vary nearly 100-fold, with half-lives from 3.3 to 277 days measured in model peptides.

The Asn amide group is especially labile because its side-chain is sterically suited to interact with the –NH– group of the following residue in the polypeptide chain. It can form transiently a cyclic succinimidyl derivative (Figure 7-3), which can undergo racemization and hydrolysis to cleave the polypeptide chain. It can also produce a mixture of D- and L-isomers of Asp and **isoAsp** residues, in which the backbone peptide bond involves the side-chain carboxyl group instead of the usual α-carboxyl. Racemization can also occur without deamidation. Asn residues deamidate up to 50 times more rapidly if the following residue is Gly, because the absence of a side-chain favors formation of the **succinimide**. Consequently, –Asn–Gly– sequences deamidate especially easily. Asn–Gly peptide bonds can be readily cleaved by incubation with **hydroxylamine**, which can react with the succinimide ring instead of water. The same reactions can also occur with the Asp–Gly sequence, although less rapidly (Figure 7-3).

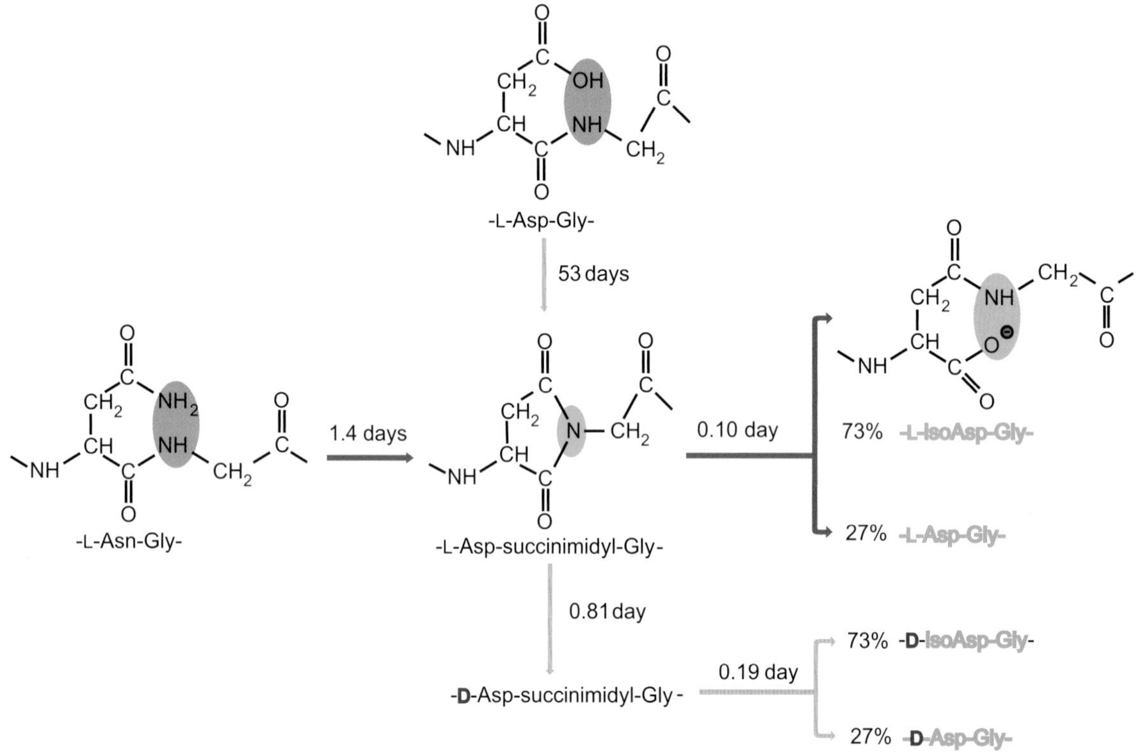

Figure 7-3. Spontaneous formation of peptide succinimides from Asp and Asn residues and the products of hydrolysis. The times shown for the various reactions are the half-times at 37°C and pH 7.4 for the model peptides Val–Tyr–Pro–Asn–Gly–Ala and Val–Tyr–Pro–Asp–Gly–Ala. The most rapid sequence is from *left* to *right*. When the Gly residue is replaced by Leu, the rate of succinimide formation is 50 times slower. Data from S. Clarke (1987) *Int. J. Peptide Protein Res.* **30**, 808–821.

Prediction of primary structure deamidation rates of asparaginyl and glutaminyl peptides through steric and catalytic effects. N. E. Robinson & A. B. Robinson (2004) *J. Pept. Res.* **63**, 437–448.

Reaction mechanism of deamidation of asparaginyl residues in peptides: effect of solvent molecules. S. Catak *et al.* (2006) *J. Phys. Chem. A* **110**, 8354–8365.

7.2.I. Cysteine (Cys)

The most reactive group in the normal amino acids is the thiol group of the Cys side-chain:

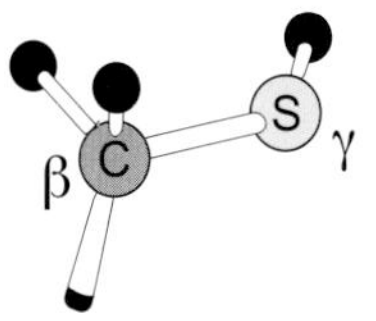

Cysteine
Cys
C
(7.26)

It undergoes numerous chemical reactions with a variety of reagents. A most important property is its tendency to ionize, to the thiolate anion $-S^-$:

$$-SH \leftrightarrow -S^- + H^+ \quad (7.27)$$

The pK_a value of a typical exposed Cys thiol group is 8.7 but values as low as 3.5 and > 11 are known in proteins, depending on their environment. Only the ionized thiolate anion is chemically reactive normally, and only it absorbs light in the near-ultraviolet (UV) region (about 250 nm). The thiolate anion is known as a soft nucleophile, poorly solvated and highly polarizable, with vacant d-orbitals and nucleophilic power much greater than would be predicted from its basicity.

In the nonionized –SH form, the thiol group is not very polar or reactive. Thiol groups have only very weak hydrogen-bonding capabilities, and they may be buried and made inaccessible to solvent water without the need for any hydrogen bonding.

Thiol groups undergo too many chemical reactions to catalog completely. Only the most important are described here.

Detection and quantitation of biological sulfhydryls. A. Russo & E. A. Bump (1988) *Methods Biochem. Anal.* **33**, 165–241.

Reflections on the role of the thiol group in biology. N. Haugaard (2000) *Ann. NY Acad. Sci.* **899**, 148–158.

Counting the number of disulfides and thiol groups in proteins and a novel approach for determining the local pK_a for cysteine groups in proteins *in vivo*. E. Bellacchio *et al.* (2001) *J. Synchrotron Radiat.* **8**, 1056–1058.

---

1. Alkylation of Thiol Groups

Thiolate anions react rapidly with alkyl halides, especially **iodoacetamide** and **iodoacetate**, to generate very stable adducts:

$$-S^- + I-CH_2CONH_2 \rightarrow -S-CH_2CONH_2 + I^- \quad (7.28)$$

$$-S^- + I-CH_2CO_2^- \rightarrow -S-CH_2CO_2^- + I^- \tag{7.29}$$

The reaction occurs by the thiolate anion attacking and displacing the I atom of iodoacetamide or iodoacetate in a nucleophilic reaction. The rate decreases at low pH values, where the thiol group is not ionized, and the apparent pK_a value of the thiol group can be estimated from the pH-dependence of the rate of the reaction. There have been exceptional cases where the results have been misleading, but they have been observed in proteins that catalyze thiol–disulfide exchange; the protein may modify the energy of the transition state for the reaction. A fully ionized cysteine thiol group, with a pK_a of ~8.7, reacts with iodoacetamide with a second-order rate constant at 25°C of approximately 25 s^{-1} M^{-1}. Iodoacetate reacts at about one-third of this rate. The resulting groups are commonly known as **carboxamidomethyl** and **carboxymethyl**, respectively. They are stable to the conditions routinely used to hydrolyze proteins to their substituent amino acids, except that the carboxamidomethyl group gets hydrolyzed to the carboxymethyl (Section 7.5.A.1).

It is important to realize that protons are effectively released from the –SH group by the reaction. Frequently, high concentrations of a thiol reagent are used to reduce all Cys residues, and then a large excess of iodoacetate or iodoacetamide is added to react with both the protein and reagent. Unless very high concentrations of a buffer are present, or a base is added to raise the pH, the pH will drop and the rate of the reaction will slow, or even stop before completion.

Both iodoacetamide and iodoacetate also react with amino groups, but at lower rates and primarily at very alkaline pH values where the amino groups are not ionized. Both Lys and His residues can be modified in this way. The two reagents can also react with the S atom of Met residues, to generate the positively charged sulfonium salt (Equation 7.47), but at a much lower rate. This rate is essentially independent of pH, however, so it can become the predominant modification reaction of a protein at very low pH values.

Many variants of iodoacetate and iodoacetamide have been devised by adding further groups with absorbance and fluorescence properties that serve as reporters of structure in proteins.

Reactivity and ionization of the active site cysteine residues of DsbA, a protein required for disulfide bond formation *in vivo*. J. W. Nelson & T. E. Creighton (1994) *Biochemistry* **33**, 5974–5983.

---

2. Thiol Addition Across Double Bonds

Thiolate anions are sufficiently nucleophilic to add across C=C double bonds, as in maleic anhydride. ***N*-Ethylmaleimide** is the classic reagent that is used most frequently to modify thiol groups:

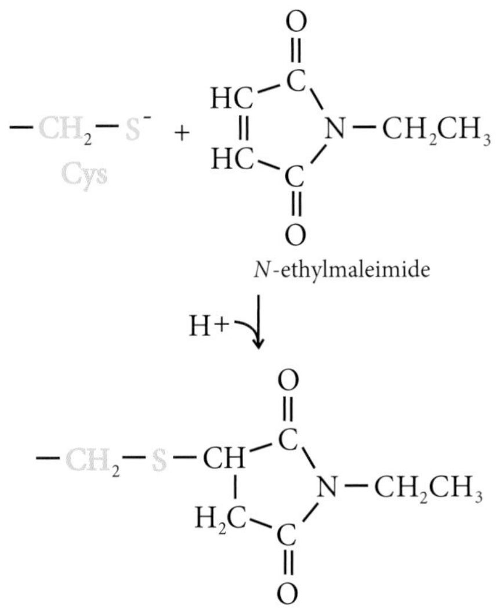

$$\text{—CH}_2\text{—S}^- + \underset{N\text{-ethylmaleimide}}{\begin{array}{c}\text{HC}\overset{\displaystyle O}{\underset{\displaystyle \|}{-C}}\\ \|\quad\quad\text{N—CH}_2\text{CH}_3\\ \text{HC}\underset{\displaystyle \|}{-C}\\ O\end{array}}$$

$$H^+ \downarrow$$

$$\text{—CH}_2\text{—S—CH}\overset{\displaystyle O}{\underset{\displaystyle \|}{-C}}\\ |\quad\quad\text{N—CH}_2\text{CH}_3\\ \text{H}_2\text{C}\underset{\displaystyle \|}{-C}\\ O \tag{7.30}$$

Thiol groups can also open the ring of **ethylenimine**:

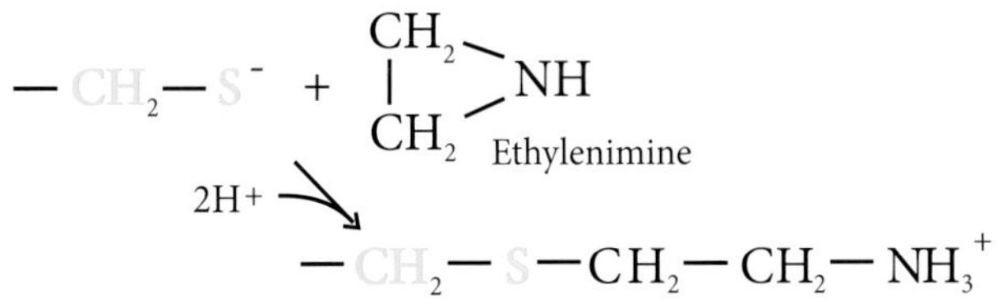

$$\text{—CH}_2\text{—S}^- + \underset{\text{Ethylenimine}}{\begin{array}{c}\text{CH}_2\\ |\quad\searrow\text{NH}\\ \text{CH}_2\nearrow\end{array}}$$

$$2H^+ \searrow$$

$$\text{—CH}_2\text{—S—CH}_2\text{—CH}_2\text{—NH}_3^+ \tag{7.31}$$

With Cys residues in proteins, the resulting side-chain is now positively charged and the proteolytic enzyme trypsin will cleave the following peptide bond (Section 7.5.B.1).

3. Binding of Metal Ions

Thiol groups form complexes of varying stabilities with a variety of metal ions. The most stable are those with divalent mercury, Hg^{2+}, but its divalency means that complexes with a variety of stoichiometries are formed. Consequently, univalent organic mercurials of the type R–Hg$^+$ tend to be used instead, because they more reproducibly form 1:1 complexes with thiol groups. Thiol complexes with silver are less stable, but univalent Ag^+ reacts stoichiometrically and can be used to titrate thiol groups. Copper, iron, zinc, cobalt, molybdenum, manganese and cadmium ions all form various complexes with thiol groups.

Coordination of heavy metals by dithiothreitol, a commonly used thiol group protectant. A. Krzel *et al.* (2001) *J. Inorg. Biochem.* **84**, 77–88.

---

4. Oxidation of Thiol Groups

Thiol groups are readily oxidized by oxygen. The reaction occurs especially rapidly in the presence of trace amounts of metal ions, such as Cu^{2+}, Fe^{2+}, Co^{2+} and Mn^{2+}, and the complex of metal and thiol may be the actual reactant with oxygen. Various oxidation states are possible, but some are intrinsically unstable. Only two oxidized forms of thiol groups are generally encountered: **disulfide** (–S–S–) and **sulfonic acid** (–SO$_3^-$). The end-product of air oxidation is normally the disulfide:

$$2\text{-SH} + 1/2\, O_2 \rightarrow \text{-S-S-} + H_2O \qquad (7.32)$$

As suggested by the participation of only half a molecule of O_2, the reaction mechanism is complex. Many extracellular proteins have their Cys residues paired in disulfide bonds.

More potent oxidizing agents produce sulfonic acids: **cysteic acid**, with the CH_2–SO_3^- side-chain. It is generated by **performic acid** (HCO_3H) from either the thiol or disulfide forms of Cys residues. The reaction undoubtedly proceeds through the intermediate **sulfenate** (–SO$^-$) and **sulfinate** (–SO$_2^-$) oxidation states, but they are generally unstable and not normally detectable. They can, however, be specifically stabilized in a protein structure.

Global methods to monitor the thiol–disulfide state of proteins *in vivo*. L. I. Leichert & U. Jakob (2006) *Antioxid. Redox Signal.* **8**, 763–772.

---

5. Disulfide Bonds

The disulfide bond can be rotated by 360°, but this affects its free energy. It is most stable with a C–S–S–C torsion angle of ±90°:

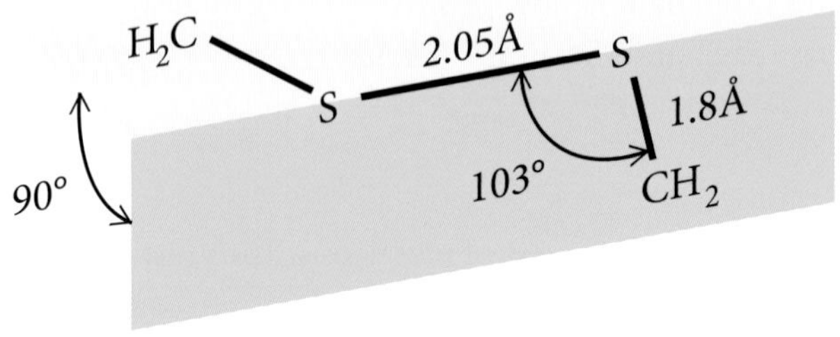

$$\qquad (7.33)$$

An angle of 0° or ±180° is believed to increase the free energy by ~6.7 kcal/mol (28 kJ/mol), which would correspond to the free energy barrier to complete rotation about the disulfide bond. Disulfide bonds in cyclic five-membered rings, such as that of **lipoic acid**, have a torsion angle of only ±30° and are strained by ~3.7 kcal/mol. On the other hand, the disulfide bonds in cyclic six-membered rings, such as that in the disulfide form of dithiothreitol (Section 7.2.1.7), have torsion angles of ~±60°, but there are few indications that they are strained.

The disulfide bond is a stable covalent bond yet it can be broken easily by an ionized thiol group, in the **thiol–disulfide exchange reaction** (Section 7.2.1.6). This reduces the disulfide bond to two thiol groups. Protein disulfide bonds (depicted here as P_S^S) can be reduced by reacting them with an excess of a thiol reagent, RSH, such as β-**mercaptoethanol** (HO–CH$_2$–CH$_2$–SH) or dithiothreitol (Equation 7.43):

$$P_S^S + 2\,RSH \leftrightarrow P_{SH}^{SSR} + RSH \leftrightarrow P_{SH}^{SH} + RSSR \tag{7.34}$$

Disulfide bonds can also be reduced chemically by many other reagents, including **borohydride**, tributylphosphine and tris-(2-carboxyethyl)-phosphine. The **phosphines** are unusual in that they function at acidic pH, where the disulfide bond and thiol groups tend not to react, nor do they react with many of the reagents that modify thiol groups; consequently, reduction of the disulfide bond and alkylation of the resulting thiol groups can be carried out simultaneously.

Nucleophiles such as **cyanide**, **sulfite** and **hydroxide** ion can also cleave disulfide bonds reversibly:

$$RS^- + RSCN \underset{CN^-}{\rightleftarrows} R-S-S-R \underset{}{\overset{SO_3^-}{\rightleftarrows}} RSSO_3 + RS^-$$

$$\downarrow OH^-$$

$$RSOH + RS^- \tag{7.35}$$

It is difficult, however, to drive these reactions to completion simply by adding excess reagent. The reaction can be pulled to completion by adding a second reagent that reacts with the thiol group generated. In the case of **nitrothiosulfobenzoate** (NTSB) the reaction can go to completion, and one mol of the colored 5-thio-2-nitrobenzoic acid (TNB) product is generated for each mole of disulfide bond originally present:

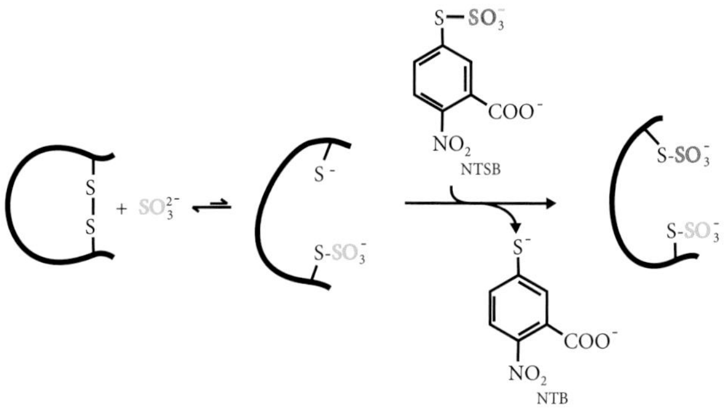

$$\tag{7.36}$$

This is the most convenient assay for protein disulfide bonds, so long as thiol groups are not also present (they can be blocked irreversibly prior to measuring the disulfide bonds).

The number of different disulfide bonds that can be formed by randomly pairing Cys residues increases dramatically with the number of Cys residues (Table 7-3). Which particular Cys residues in a protein are linked by disulfide bonds can be determined chemically by various types of peptide mapping (Section 7.5.C) and diagonal techniques (Section 7.5.D.2).

Protein disulfide bond determination by mass spectrometry. J. J. Gorman *et al.* (2002) *Mass Spectrom. Rev.* **21**, 183–216.

Disulfide bonds, their stereospecific environment and conservation in protein structures. R. Bhattacharyya *et al.* (2004) *Protein Eng. Des. Sel.* **17**, 795–808.

Formation of disulfide bonds in proteins and peptides. G. Bulaj (2005) *Biotechnol. Adv.* **23**, 87–92.

6. Thiol-Disulfide Exchange

Thiol groups and disulfide bonds undergo a spontaneous chemical reaction in which the ionized thiolate anion displaces one S atom of the disulfide bond in an S_N2 type of reaction:

$$-S_{nuc}^{-} + S_c - S_{lg} \leftrightarrow S_{nuc} - S_c + {}^{-}S_{lg}- \tag{7.37}$$

The three S atoms involved in the reactions are generally labeled as the nucleophile (nuc), central (c) and leaving group (lg). The thiolate anion nucleophile attacks one end of the disulfide bond, ideally along its axis, and that S atom becomes the central atom. The transition state for the reaction is believed to have the three atoms linear and equally spaced. The negative charge of the thiolate anion is spread symmetrically, with more on the terminal atoms than on the central one.

Table 7-3. The number of ways in which $2n$ sulfhydryl groups can combine to form n disulfide bonds

n	Number of combinations
1	1
2	3
3	15
4	105
5	945
6	1.0×10^4
7	1.4×10^5
8	2.0×10^6
9	3.4×10^7
10	6.5×10^8
11	1.4×10^{10}
12	3.2×10^{11}
13	7.9×10^{12}
14	2.1×10^{14}
15	6.2×10^{15}
16	1.9×10^{17}
17	6.3×10^{18}
18	2.2×10^{20}
19	8.2×10^{22}
20	3.2×10^{24}
21	1.3×10^{25}
22	5.6×10^{26}
23	2.5×10^{28}
24	1.1×10^{30}
25	5.8×10^{31}

The ionized form of the thiol group is the reactive species, so the rate of the reaction varies with the pH below the pK_a of the thiol group. The reaction can be quenched by acidification, making the thiolate anion insignificant. Alternatively, such mixtures can be trapped irreversibly by rapidly reacting all free thiol groups with reagents such as iodoacetamide (Equation 7.28), iodoacetate (Equation 7.29) and or N-ethylmaleimide (Equation 7.30). The thiol and disulfide species present in both equilibrium and kinetic trapped mixtures can be analyzed chemically, for example by high performance liquid chromatography (HPLC) under acidic conditions.

The rate of reaction depends on the affinity for electrons of each of the three S atoms; this is conveniently measured by their pK_a values when they are thiol groups. These three pK_a values can reliably predict the second-order rate constant for the intermolecular reaction between model compounds. The relationship can be expressed relative to the concentration of thiolate anion (k_{RS^-}) or the total amount of thiol group present (k_{obs}):

$$\log k_{RS^-} \text{ (s}^{-1}\text{ M}^{-1}\text{)} = 4.5 + 0.59\, pK_a^{nuc} - 0.40\, pK_a^{c} - 0.59\, pK_a^{lg} \tag{7.38}$$

$$\log k_{obs} = 4.5 + 0.59\, pK_a^{nuc} - 0.40\, pK_a^{c} - 0.59\, pK_a^{lg} - \log(1 + 10^{(pKanuc - pH)}) \tag{7.39}$$

With any mixture of different thiol groups and disulfide bonds, more than one type of reaction will take place, involving all three S atoms; the combined rate will be observed. The reaction is more simple using a mono- or di-oxide form of the disulfide; only the nonoxidized S atom undergoes thiol–disulfide exchange:

$$\text{R}-\overset{\overset{\displaystyle O}{\|}}{\text{S}}-\text{S}-\text{R} + \text{X}-\text{SH} \longrightarrow \text{R}-\overset{\overset{\displaystyle O}{\|}}{\text{S}}\text{H} + \text{X}-\text{S}-\text{S}-\text{R} \tag{7.40}$$

$$\text{R}-\overset{\overset{\displaystyle O}{\|}}{\underset{\underset{\displaystyle O}{\|}}{\text{S}}}-\text{S}-\text{R} + \text{X}-\text{SH} \longrightarrow \text{R}-\overset{\overset{\displaystyle O}{\|}}{\underset{\underset{\displaystyle O}{\|}}{\text{S}}}\text{H} + \text{X}-\text{S}-\text{S}-\text{R} \tag{7.41}$$

Consequently, the reaction stops at this stage and will go to completion.

The maximum rate is predicted to occur with a thiol group that has a pK_a the same as the pH of the reaction. If the pK_a value is greater, there will be less ionized form of the thiol group; if the pK_a is lower, the thiol will be ionized but is less nucleophilic.

The reverse reaction is also governed by Equations 7.38 and 7.39, making it possible to predict the equilibrium constant for the reaction. It will depend upon the pH if the attacking and leaving S atoms have different pK_a values and vary over the pH interval between the two thiol pK_a values. The equilibrium favors the thiol group with the lower pK_a value, the opposite to what would have been predicted from just the effect of ionization of the two thiol groups.

Strain in the disulfide bond increases its rate of reaction. For example, the strained disulfide bond of lipoic acid reacts considerably more rapidly than otherwise expected, suggesting ~3.8 kcal/mol of conformational strain in its five-membered ring. The thiol–disulfide exchange reaction is inhibited if the thiol or disulfide groups are buried and inaccessible, or if there are bulky substituents adjacent to the S atoms. Positive charges near the disulfide bond will attract the attacking thiolate anion and increase the rate of the reaction, whereas negative charges have the opposite effect. Formation of a mixed disulfide with opposite charges close on the two moieties is favored, while that with like charges is disfavored. Such electrostatic effects are substantial in peptides only when the charged groups are on adjacent residues, and their magnitude can be decreased in magnitude by electrostatic screening with high salt concentrations.

Thiol–disulfide exchange is one of the most specific chemical reactions available in molecular biology, as thiol groups and disulfide bonds tend not to react with other groups on proteins or nucleic acids. It is routinely used in molecular biology to reduce protein disulfide bonds using reagents like β-mercaptoethanol (Equation 7.34). Two sequential thiol–disulfide exchange reactions are necessary, proceeding through a mixed disulfide between the reagent and the protein. These reactions are readily reversible and can be used in reverse to add disulfide bonds to a protein (Section 11.4.C).

Theoretical insights into the mechanism for thiol/disulfide exchange. P. A. Fernandes & M. J. Ramos (2004) *Chemistry* **10**, 257–266.

---

7. Dithiothreitol, Dithioerythritol

Dithiothreitol (DTT or DTT_{SH}^{SH}) and **dithioerythritol** (DTE) each have two thiol groups and closely related structures:

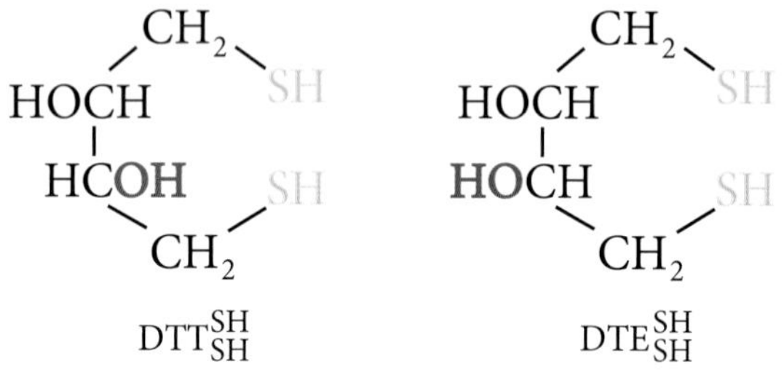

$$\begin{array}{cc}
\text{DTT}_{SH}^{SH} & \text{DTE}_{SH}^{SH}
\end{array}$$
(7.42)

They were introduced by Cleland to improve the potency of reagents like mercaptoethanol with single thiol groups, and they are often known as **Cleland's reagent**. They are potent reductants of disulfide bonds (e.g. RSSR) because they form a stable, cyclic, intramolecular disulfide bond, abbreviated here as DTT_S^S and DTT_S^S:

$$DTT_{SH}^{SH} + RSSR \leftrightarrow DTT_S^S + 2\,RSH \tag{7.43}$$

The equilibrium constant for the reaction is given by:

$$K_{eq} = \frac{DTT_S^S[RSH]^2}{DTT_{SH}^{SH}[RSSR]} \tag{7.44}$$

With a typical model RSSR disulfide bond, like that of glutathione or mercaptoethanol, K_{eq} has the value 200 M at high pH and 2800 M at low pH values. The pH dependence over the range of ~8–11 is due to differences in the pK_a values of the thiol groups involved. The first thiol group of DTT_{SH}^{SH} to ionize has an apparent pK_a of ~9.2, while that of the second is ~10.2. The difference between these two values is exaggerated because it includes a statistical factor: either of the two thiol groups can ionize first, and either of the two thiolate anions can pick up the first proton. Consequently, the first and second equilibrium constants should differ by a factor of four, which would introduce a difference in observed pK_a values of 0.6 pH units even if the two thiol groups have identical intrinsic tendencies to ionize. Correcting for this, the difference in apparent pK_a values of 1.0 pH units indicates that there is little electrostatic interaction between the two thiol groups; ionization of one increases the pK_a of the other by only 0.4 pH units.

The equilibrium constant of Equation 7.44 has the unit of concentration because the disulfide bond of DTT is intramolecular and reducing it produces only a single molecule. In contrast, reduction of the disulfide bond of RSSR generates two RSH molecules, so its disulfide bond can be considered 'intermolecular'. The equilibrium constant can therefore be interpreted in terms of the **effective concentration** of the two thiol groups of DTT_{SH}^{SH} relative to each other. Correcting for ionization effects indicates that it has the relatively large value of ~120 M. This is believed to be due to the two thiol groups being kept in reasonable proximity and forming a disulfide bond in a favorable six-membered ring structure. On the other hand, the disulfide bond of DTT_S^S has a CSSC torsion angle of only ~60°, rather than the most favorable 90° (Equation 7.33), so it appears to be somewhat strained. Nevertheless, there is only a small change in enthalpy of no more than 0.5 kcal/mol when DTT_{SH}^{SH} reduces model linear disulfide bonds (Equation 7.43), so the energetic strain appears to be minimal.

The dithiol and disulfide forms of DTT may be distinguished readily because only the disulfide form absorbs near-UV light, with a maximum at about 280 nm. The dithiol form tends to bind many heavy metal ions.

Direct measurement of the equilibrium between glutathione and dithiothreitol by high performance liquid chromatography. M.-H. Chau & J. W. Nelson (1991) *FEBS Lett.* **291**, 296–298.

---

8. Ellman's Reagent

The reagent 5,5′-dithiobis-(2-nitrobenzoic acid), **DTNB**, is also known as **Ellman's reagent**. It is one of the most useful reagents for measuring thiol groups quantitatively and easily, and many variants of Ellman's reagent have been devised for special purposes. At alkaline pH, thiol groups undergo thiol–disulfide exchange with the disulfide bond of DTNB, to produce 5-thio-2-nitrobenzoic acid (TNB) and a mixed disulfide:

$$-CH_2-S^- + \underset{\text{DTNB}}{O_2N-\underset{HO_2C}{\bigcirc}-S-S-\underset{}{\bigcirc}\overset{CO_2H}{-}NO_2} \;\rightleftharpoons\; -CH_2-S-S-\underset{}{\bigcirc}\overset{CO_2H}{-}NO_2 \;+\; \underset{\text{TNB}}{{}^-S-\underset{CO_2H}{\bigcirc}-NO_2} \quad (7.45)$$

The appearance of TNB is measured by its intense color, with an absorption maximum at 409.5 nm in normal aqueous solution. In 6 M guanidinium chloride (GdmCl), which is often used to unfold proteins and to make their thiol groups accessible (Section 11.1.C), it is shifted to 421 nm. Nevertheless, the reaction for historical reasons is generally monitored at 412 nm, where the molar absorbance coefficient at 25°C is 14,150 M^{-1} cm^{-1} in standard buffers and 13,700 in 6 M GdmCl; these values vary with temperature.

Use of an excess of Ellman's reagent ensures that the reaction goes essentially to completion. This is assisted because the pK_a of the thiol group of TNB is only ~4.5, considerably lower than the values near 9 of normal Cys thiol groups; thiol–disulfide exchange favors the more acidic thiol group. One mol of TNB is produced per mol of original thiol group, irrespective of whether the original thiol groups react only with the Ellman's reagent or also with the mixed disulfide.

Molar absorption coefficients for the reduced Ellman reagent: reassessment. P. Eyer *et al.* (2003) *Anal. Biochem.* **312**, 224–227.

7.2.J. Methionine (Met)

Met residues have a long side-chain that is unbranched, nonpolar and relatively unreactive chemically:

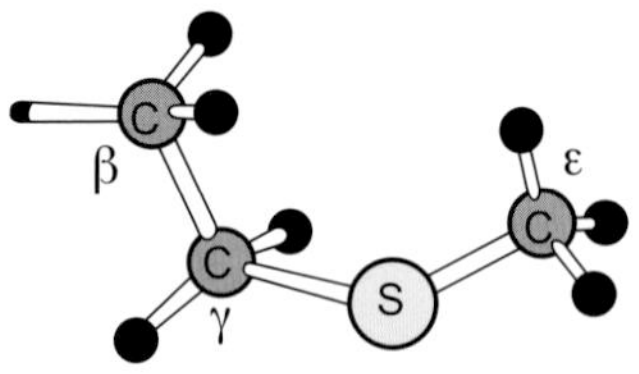

Methionine
Met
M
(7.46)

The S atom is only slightly nucleophilic. At acidic pH, however, it can become the most potent nucleophile in proteins because it cannot be protonated, unlike other nucleophiles in proteins. Under acidic conditions, it can be selectively modified by forming **sulfonium salts** with alkylating agents, such as iodoacetate and methyl iodide:

$$-(CH_2)_2-S-CH_3 \xrightarrow{CH_3I} -(CH_2)_2-\overset{+}{S}(CH_3)-CH_3 + I^- \qquad (7.47)$$

This reaction can be reversed by thiol reagents; the methyl group removed is equally likely to be the original one or that introduced by the methyl iodide, so this reaction using labeled methyl iodide makes it possible to introduce a radioisotope label in 50% of the residues.

The S atom is also susceptible to oxidation by air or by more potent oxidants, such as peroxides. The **sulfoxide** is formed first, followed by the **sulfone**:

$$-(CH_2)_2-S-CH_3 \xrightarrow{[O]} -(CH_2)_2-\underset{}{\overset{O}{\underset{\|}{S}}}-CH_3 \xrightarrow{[O]} -(CH_2)_2-\underset{\underset{\|}{O}}{\overset{\overset{\|}{O}}{S}}-CH_3 \quad (7.48)$$

Any such oxidation abolishes the reactivity with alkylating reagents. The first step, but not the second, can be reversed by thiol groups. Oxidized Met residues in peptide thioesters can be reduced rapidly with NH_4I to the corresponding sulfide by using Me_2S as coreductant.

The Met side-chain reacts with the reagent CNBr, which provides a useful method for cleaving polypeptide chains at Met residues (Section 7.5.B.2).

Cyclic oxidation and reduction of methionine residues of proteins in antioxidant defense and cellular regulation. E. R. Stadtman (2004) *Arch. Biochem. Biophys.* **423**, 2-5.

A comprehensive picture of non-site specific oxidation of methionine residues by peroxides in protein pharmaceuticals. J. W. Chu *et al.* (2004) *J. Pharm. Sci.* **93**, 3096–3102.

The reduction of oxidized methionine residues in peptide thioesters with NH_4I–Me_2S. C. P. Hackenberger (2006) *Org. Biomol. Chem.* **4**, 2291–2295.

7.2.K. Phenylalanine (Phe)

The side-chain of a Phe residue is dominated by the aromatic ring:

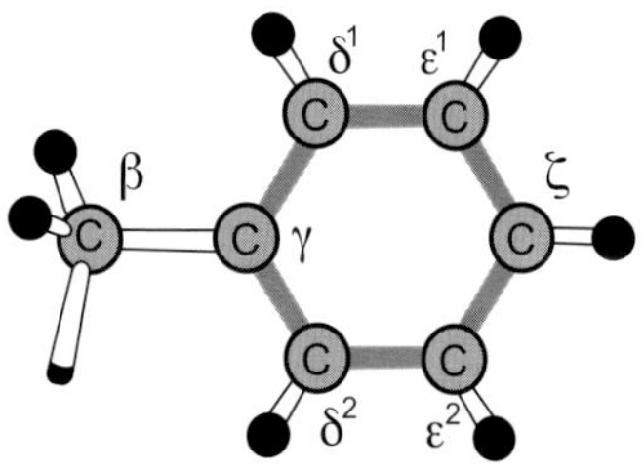

(7.49)

It is similar chemically to benzene and toluene, so it is chemically reactive only under extreme conditions. The Phe side-chain is nonpolar, other than the tendency of the electrons of the peripheral H atoms to be drawn into the aromatic ring. The Phe residue does absorb near-UV light, but much less than the other aromatic residues Tyr and Trp, which normally overwhelm the contribution of Phe residues to the spectral properties of proteins.

7.2.L. Tyrosine (Tyr)

The Tyr side-chain consists primarily of a phenolic group:

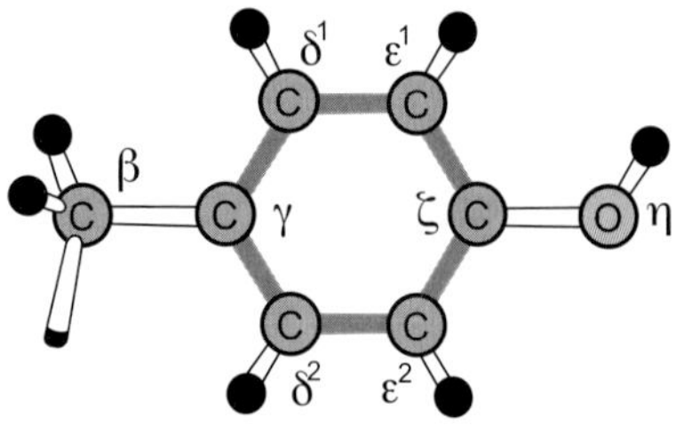

$$\text{Tyrosine} \quad \text{Tyr} \quad \text{Y} \tag{7.50}$$

This phenolic group gives Tyr residues spectral properties that are especially useful (Table 7-4). The hydroxyl group makes this residue more polar than the otherwise similar Phe residue and the aromatic ring much more reactive in electrophilic substitution reactions. Usually these reactions occur at the symmetrical positions, designated here as ε^1 and ε^2 but frequently numbered as 3 and 5, respectively. Tyr side-chains can be readily nitrated and iodinated:

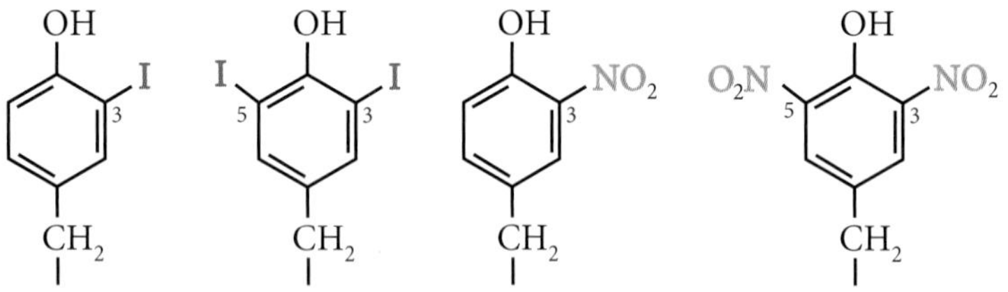

$$(7.51)$$

Table 7-4. UV absorbance properties of tyrosine and several derivatives

Amino acid	pK_{app} of –OH group	Nonionized –OH		Ionized –OH	
		λ_{max} (nm)	Molar absorbance (M^{-1} cm^{-1})	λ_{max} (nm)	Molar absorbance (M^{-1} cm^{-1})
Tyrosine	10.1	274.5	1400	293	2400
ε-Iodotyrosine	8.2	283	2750	305	4100
$\varepsilon^1,\varepsilon^2$-Diiodotyrosine	6.5	287	2750	311	6250
ε-Nitrotyrosine	7.2	360	2790	428	4200
ε-Aminotyrosine	10.0[a]	275	1600	320	4200
O-Acetyltyrosine	–	262	262	–	–

[a] The pK_{app} of the aromatic amino group is approximately 4.8. Data from A. N. Glazer (1976) in *The Proteins*, Vol. 2, 3rd edn (H. Neurath & R. L. Hill, eds), Academic Press, NY, pp. 1–103.

Nitration can be produced by reaction with the reagent **tetranitromethane**, $C(NO_2)_4$, but it also occurs naturally by reaction with the physiologically important reagent **nitric oxide**. 3-Nitrotyrosine is colored, with an absorbance maximum at 428 nm at alkaline pH (Table 7-4), and can be used as a spectrophotometric reporter group. The nitrotyrosine residue can be reduced by sodium hydrosulfite to an **aminotyrosine** residue, which can then be altered by further chemical modifications.

The Tyr hydroxyl group has an intrinsic pK_a of ~11.1, so it ionizes only at very alkaline pH values. It can function in the nonionized form as either a donor or acceptor of hydrogen bonds. The hydroxyl group can be acetylated by reaction with acetic anhydride:

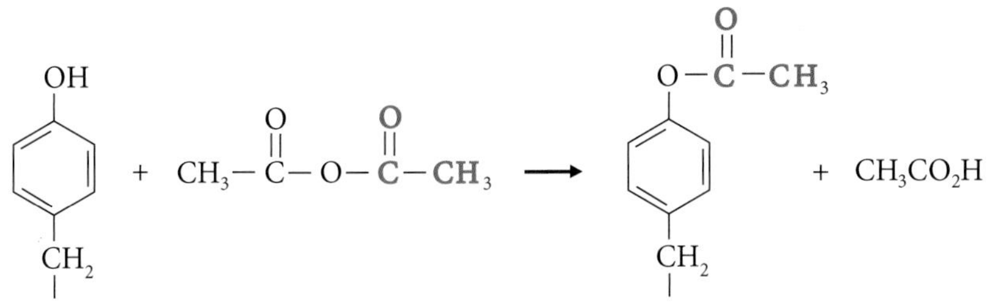

$$\text{Tyr-OH} + CH_3-\overset{O}{\underset{\|}{C}}-O-\overset{O}{\underset{\|}{C}}-CH_3 \longrightarrow \text{Tyr-O-}\overset{O}{\underset{\|}{C}}-CH_3 + CH_3CO_2H \tag{7.52}$$

NO-dependent protein nitration: a cell signaling event or an oxidative inflammatory response? F. J. Schopfer *et al.* (2003) *Trends Biochem. Sci.* **28**, 646–654.

Protein tyrosine nitration in hydrophilic and hydrophobic environments. S. Bartesaghi *et al.* (2006) *Amino Acids* **32**, 501–515.

7.2.M. Tryptophan (Trp)

The Trp side-chain consists largely of an indole group:

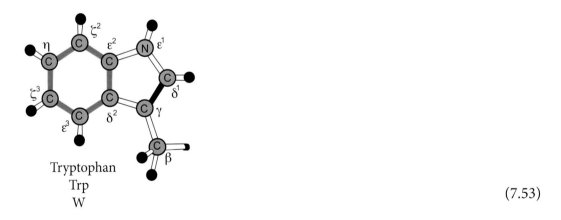

Tryptophan
Trp
W

(7.53)

It is the largest and most complex of all the normal amino acids. The five-membered ring is susceptible to oxidation and various other chemical reactions. **Iodine** and ***N*-bromosuccinimide** (NBS) oxidize the Trp indole ring to that of **oxindolealanine**:

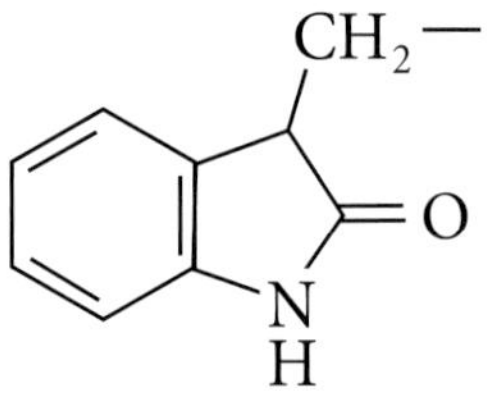

$$\text{(structure: 3-substituted 2-oxindole with CH}_2\text{— side chain)} \tag{7.54}$$

Ozone opens the indole ring to that of **N-formylkynurenine**:

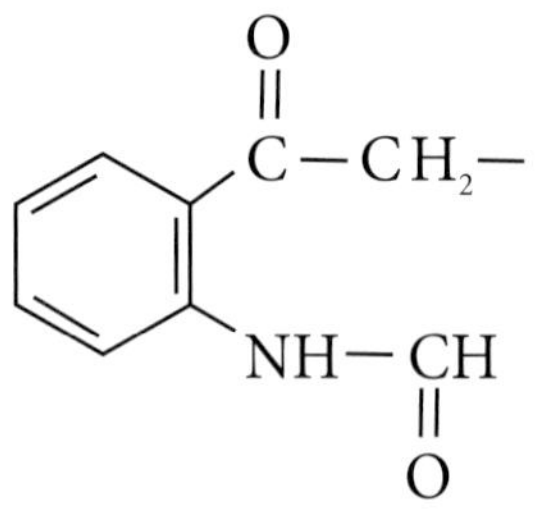

$$\text{(structure: 2-(formylamino)phenyl ketone with —CH}_2\text{— side chain)} \tag{7.55}$$

Peptide bonds following Trp residues can be cleaved with varying efficiencies by several chemical reagents, especially **iodosobenzoic acid** and **BNPA-skatole**. Most of these procedures have the disadvantages of side-reactions and oxidizing Cys and Met residues.

Except for the tendency of the electrons of the peripheral H atoms to be drawn into the aromatic ring, the Trp side-chain is largely nonpolar and it is generally considered the most hydrophobic of all the amino acid residues (Table 7-6). The N atom is usually involved in hydrogen bonding, but only weakly.

The absorbance of near-UV light by the indole ring, plus its fluorescence, makes Trp residues extremely useful in spectral studies of protein structure. In addition, there are relatively few Trp residues in proteins; they often have only one.

Interfacial tryptophan residues: a role for the cation-pi effect? F. N. Petersen *et al.* (2005) *Biophys. J.* **89**, 3985–3996.

Modification of tryptophan and tryptophan residues in proteins by reactive nitrogen species. F. Yamakura & K. Ikeda (2006) *Nitric Oxide* **14**, 152–161.

A matrix-assisted laser desorption/ionization compatible reagent for tagging tryptophan residues. C. Li *et al.* (2006) *Eur. J. Mass Spectrom.* **12**, 213–221.

7.2.N. Imino Acid (Pro)

The amino acid residue Pro is unique in that its side-chain is bonded covalently to the N atom of the peptide backbone, indicated by the black bonds:

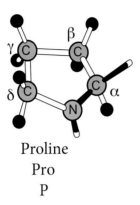

Proline
Pro
P   (7.56)

Consequently, the peptide backbone at Pro residues has no amide H atom for use as a donor in hydrogen bonding or in resonance stabilization of the peptide bond of which it is part (Equation 8.1). The cyclic five-membered ring is not planar but is invariably puckered, with the C^γ atom displaced ~0.5 Å from the plane defined by nearly coplanar C^α, C^β, C^δ and N atoms. This ring imposes rigid constraints on rotation about the N–C^α bond of the backbone (Section 8.1). The side-chain atoms of Pro residues are nonpolar and chemically inert. Pro residues are also unique in that their preceding peptide bond has a significant intrinsic tendency to adopt the *cis* conformation, generally ~10% of the time in model peptides (Section 8.2).

The flexibility in the proline ring couples to the protein backbone. B. K. Ho *et al.* (2005) *Protein Sci.* **14**, 1011–1018.

7.2.O. Selenocysteine (Sec)

The amino acid **selenocysteine** is incorporated only rarely in proteins. It differs from Cys only in having the S atom of Cys replaced by a Se atom:

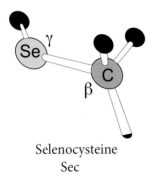

Selenocysteine
Sec   (7.57)

This **selenol** group is an essential component in the active sites of a few important enzymes, such as glutathione peroxidase and formate dehydrogenase. A normal selenol group ionizes with a pK_a of 5.2, so it would be fully ionized at physiological pH. This contrasts with Cys thiol groups, which have considerably higher pK_a values and normally are only slightly ionized at pH 7. The ionized selenol group is a stronger nucleophile than an ionized thiol group. Consequently, enzymes containing Sec residues are much less active if these residues are replaced by Cys.

Selenocysteine. T. C. Stadtman (1996) *Ann. Rev. Biochem.* **65**, 83–100.

Selenocysteine in proteins: properties and biotechnological use. L. Johansson *et al.* (2005) *Biochim. Biophys. Acta* **1726**, 1–13.

7.2.P. Physical Properties and Hydrophobicities of Amino Acid Residues

The van der Waals and partial volumes of the amino acid residues present in proteins are given in Table 7-1; the accessible surface areas of the amino acid residues and their various parts are given in Table 7-5.

Table 7-5. Accessible surface areas (Å^2) of amino acid residues in a Gly–X–Gly tripeptide in an extended conformation

Residue	Total	Main-chain atoms	Side-chain atoms Total	Nonpolar atoms	Polar atoms
Ala	113	46	67	67	
Arg	241	45	196	89	107
Asn	158	45	113	44	69
Asp	151	45	106	48	58
Cys	140	36	104	35	69
Gln	189	45	144	53	91
Glu	183	45	138	61	77
Gly	85	85			
His	194	43	151	102	49
Ile	182	42	140	140	
Leu	180	43	137	137	
Lys	211	44	167	119	48
Met	204	44	160	117	43
Phe	218	43	175	175	
Pro	143	38	105	105	
Ser	122	42	80	44	36
Thr	146	44	102	74	28
Trp	259	42	217	190	27
Tyr	229	42	187	144	43
Val	160	43	117	117	

Data from S. Miller *et al.* (1987) *J. Mol. Biol.* **196**, 641–656.

Partial molar volumes of proteins: amino acid side-chain contributions derived from the partial molar volumes of some tripeptides over the temperature range 10–90 degrees C. M. Hackel *et al.* (1999) *Biophys. Chem.* **82**, 35–50.

1. Hydrophilicities

The hydrophilicities of the amino acid side-chains have been measured by their partition coefficients between the vapor phase and water, using model compounds in which the main chain is replaced by an H atom. For example, CH_4 would be the model for the Ala side-chain, toluene for Phe. A model for the peptide backbone is *N*-methylacetamide:

$$CH_3-\underset{\underset{\displaystyle}{\|}}{\overset{\overset{\displaystyle O}{\|}}{C}}-NH-CH_3 \qquad (7.58)$$

Ionized molecules have negligible tendency to vaporize, so the partition coefficients measured for the corresponding nonionized molecules are corrected to the fraction of nonionized form present at pH 7. The measured hydrophilicity values of the amino acid side-chains, normalized so that of Gly is zero, are given in Table 7-6.

Molecules with polar hydrogen bond donors or acceptors strongly prefer the aqueous environment, because they form hydrogen bonds to water approximately as well as do other water molecules. Maximum hydrophilicity is observed with ionized molecules and with those that can act as both donor and acceptor in hydrogen bonds with water. The peptide bond is as hydrophilic as the side-chains of Asn and Gln residues. In contrast, nonpolar molecules do not interact as favorably with water and have relatively low solubilities. Nevertheless, they all are more hydrophilic than the H atom of the reference Gly residue.

2. Hydrophobicities

The hydrophobicities of the individual amino acid side-chains have been measured experimentally in a variety of ways, using free amino acids, amino acids with the amino and carboxyl groups blocked, and analogs of the side-chain with the backbone replaced by a H atom, and by measuring their partitioning into a variety of nonpolar solvents, including ethanol, octanol, dioxane and cyclohexane. The hydrophobic interaction is defined here as the free energy of transfer from water to a nonpolar liquid, ΔG_{tr}, and the more hydrophobic molecules have the more negative hydrophobicities. The absolute values of ΔG_{tr} are not very relevant, only their relative values, so the side-chain hydrophobicities are obtained by subtracting the value measured for Gly. Unfortunately, the hydrophobicity values measured in the various ways vary substantially, so several representative scales are given in Table 7-6.

The apparent hydrophobicities of the amino acid side-chains vary enormously, depending primarily upon whether or not there are polar groups present. Ionized and polar side-chains interact strongly with water and have much lower solubilities in nonpolar solvents, due to the unfavorable energetics of placing a polar group in a nonpolar environment. The magnitude of this effect varies enormously, depending upon the solvent and the molecule, and is probably the main source of the variation in hydrophobicity scales. Within a nonaqueous solvent, the polar groups of the side-chains will, when

Table 7-6. Relative hydrophilicities and hydrophobicities (kcal/mol) of amino acid side-chains

Amino acid residue	Hydrophilicity[a]	Hydrophobicity			
		Side-chain analogs[a]	Amino acids[b]	N-acetyl amides[c]	Calculated[d]
Arg	−22.31	15.86	3.0	1.84	3.95
Asp	−13.34	9.66	2.5	1.40	3.81
Glu	−12.63	7.75	2.5	1.17	2.91
Asn	−12.07	7.58	0.2	1.09	1.91
Lys	−11.91	6.49	3.0	1.81	2.77
Gln	−11.77	6.48	0.2	0.39	1.30
His	−12.66	5.60	−0.5	−0.23	0.64
Ser	−7.45	4.34	0.3	0.07	1.24
Thr	−7.27	3.51	−0.4	−0.48	1.00
Tyr	−8.50	1.08	−2.3	−1.75	−1.47
Gly	0	0	0	0	0
Pro	−1.4	−0.96	−0.99		
Cys	−3.63	−0.34	−1.0	−2.80	−0.25
Ala	−0.45	−0.87	−0.5	−0.57	−0.39
Trp	−8.27	−1.39	−3.4	−4.11	−2.13
Met	−3.87	−1.41	−1.3	−2.24	−0.96
Phe	−3.15	−2.04	−2.5	−3.26	−2.27
Val	−0.40	−3.10	−1.5	−2.22	−1.30
Ile	−0.24	−3.98	−1.8	−3.28	−1.82
Leu	−0.11	−3.98	−1.8	−3.10	−1.82

[a] Hydrophilicity was measured by the partition coefficient (K_D) of the model for each side-chain (backbone replaced by H atom) from vapor → water, hydrophobicity from water → cyclohexane. For ionizing side-chains, the values were corrected for the fraction of each side-chain that was ionized at pH 7. Both scales were normalized to zero for the value for Gly. Data from A. Radzicka & R. Wolfenden (1988) *Biochemistry* **27**, 1664–1670.

[b] Some values were measured from the relative solubilities of the amino acids in water and ethanol or dioxane. Data from Y. Nozaki & C. Tanford (1971) *J. Biol. Chem.* **246**, 2211–2217. Other values were extrapolated from these data; M. Levitt (1976) *J. Mol. Biol.* **104**, 59–107.

[c] Measured from the partition coefficient between water and octanol of the N-acetyl amino acid amides. Data from J. Fauchère & V. Pliska (1983) *Eur. J. Med. Chem.* **18**, 369–375.

[d] Calculated from the hydrophobicities of the individual groups that make up each side-chain, using data for the partition coefficient between water and octanol of many model compounds.

possible, make electrostatic and hydrogen bond interactions with other polar groups, such as of the peptide backbone and any of the solvent; this includes any water that is also present, which can be substantial in solvents such as octanol. Such intramolecular polar interactions will cause molecules with polar groups to appear to be more hydrophobic than they really are. Perhaps for that reason, the most extreme values of hydrophobicity have been measured using models for just the side-chain, without the polypeptide backbone, and the apolar solvent with the least polar nature, cyclohexane (Table 7-6). Because of the widely varying hydrophobicities of amino acid side-chains, the more neutral term hydropathy is often used to describe their relative preferences for aqueous and nonpolar environments.

The diversity of polar and nonpolar groups in amino acid side-chains and in the polypeptide backbone makes it advisable to consider the individual groups rather than the side-chain as a whole and to consider simply their hydration by water (Table 7-7). These values can then be used to calculate the thermodynamics of hydration of the amino acid side-chains (Table 7-8). The free energy of transfer from water to nonpolar solvents of the side-chains is correlated with the surface area, but only of the nonpolar side-chains, and there are remarkably large differences between the hydrophobicities measured in different ways (Figure 7-4-A). The values measured for the transfer of the other amino acids are dominated by their polar groups. That solvation by water is the predominant factor is illustrated by the excellent correlation of the surface areas of all the amino acid side-chains with their free energy of transfer from vapor to cyclohexane (Figure 7-4-B); probably the polar groups are equally uncomfortable in both phases. The heat capacities of aqueous solutions of the side-chain analogs are directly proportional to their nonpolar accessible surface areas (Figure 7-5), except that those side-chains with ring structures and with an S atom give slightly lower values. Polar groups have opposite, but smaller, effects on the heat capacity. The unusually large heat capacities of aqueous solutions of nonpolar molecules and the anomalous thermodynamics of the hydrophobic effect arise from the interactions of water with nonpolar atoms.

The **hydrophobic moment** of a polypeptide chain is calculated from the vector sum of the contributions of each of the amino acid residues. This is given by a vector pointing from the C^{α} atom to the center of the side-chain, with a length proportional to the hydrophobicity of the side-chain. This parameter can often account for the architecture and interactions of large molecules like proteins.

Evaluation of methods for measuring amino acid hydrophobicities and interactions. K. M. Biswas *et al.* (2003) *J. Chromatogr. A* **1000**, 637–655.

A linear function for the approximation of accessible surface area of proteins. R. G. Kim & C. Y. Choi (2006) *Protein Pept. Lett.* **13**, 549–553.

Solvation free energy of amino acids and side-chain analogues. J. Chang *et al.* (2007) *J. Phys. Chem. B* **111**, 2098–2106.

Table 7-7. Thermodynamic parameters for the hydration by water at 25°C of each 1 Å² of accessible surface area of the nonpolar and polar parts of polypeptide chains

Surface	Enthalpy (J/mol Å²)	Entropy (J/K mol Å²)	Free energy (J/mol Å²)	Heat capacity (J/K mol Å²)
Aliphatic	−122	−0.578	50	2.22
Aromatic	−148	−0.319	−53	1.23
Polar part of				
Arg	−827	−0.478	−685	−0.20
Asn	−894	−0.654	−699	−0.96
Asp	−715	−0.469	−575	−1.35
Cys	−271	−0.402	−151	−3.88
Gln	−703	−0.591	−527	−0.19
Glu	−562	−0.436	−432	−0.52
His	−1128	−0.693	−922	−1.28
Lys	−714	−0.482	−570	−1.57
Met	−473	−0.412	−350	−3.88
Ser	−1045	−0.983	−752	−1.30
Thr	−1287	−1.053	−972	−1.25
Trp	−1161	−0.693	−954	3.78
Tyr	−854	−0.415	−730	0.08
−CO−NH−	−1702	−1.026	−1396	−1.64

To convert Joules (J) to calories, divide by 4.16.

Data from P. L. Privalov & G. I. Makhatadze (1992) *J. Mol. Biol.* **224**, 715–723 and (1993) *J. Mol. Biol.* **232**, 660–679; G. I. Makhatadze & P. L. Privalov (1993) *J. Mol. Biol.* **232**, 639–659.

Table 7-8. Thermodynamic parameters for the hydration by water at 25°C of the side-chains of the amino acid residues

Residue	Enthalpy (kJ/mol)	Entropy (J/K mol)	Free energy (kJ/mol)	Heat capacity (J/K mol)
Ala	−8.28	−40.3	3.72	166.7
Arg	−99.38	−102.6	−68.80	273.4
Asn	−67.05	−70.6	−46.02	88.8
Asp	−47.33	−54.9	−30.96	89
Cys	−22.95	−48.0	−8.65	237.6
Gln	−70.44	−84.5	−45.27	180.2
Glu	−50.72	−68.8	−30.21	179
His	−69.63	−74.0	−47.58	179.6
Ile	−17.12	−79.5	6.57	402.3
Leu	−17.12	−79.5	6.57	381.7
Lys	−48.78	−91.9	−21.39	249.8
Met	−34.62	−85.3	−9.19	175.9
Phe	−25.31	−62.0	−6.83	383
Pro	−10.17	−41.7	2.25	177.7
Ser	−42.99	−60.8	−24.87	81.2
Thr	−45.05	−72.2	−23.52	184.5
Trp	−58.79	−57.2	−41.73	458.5
Tyr	−57.37	−70.3	−36.43	301.7
Val	−13.73	−65.6	5.82	314.4
−CO−NH−	−59.57	−35.9	−48.87	−57.4

To convert Joules (J) to calories, divide by 4.16.

Data from P. L. Privalov & G. I. Makhatadze (1992) *J. Mol. Biol.* **224**, 715–723 and (1993) *J. Mol. Biol.* **232**, 660–679; G. I. Makhatadze & P. L. Privalov (1993) *J. Mol. Biol.* **232**, 639–659 and (1995) *Adv. Protein Chem.* **47**, 307–425.

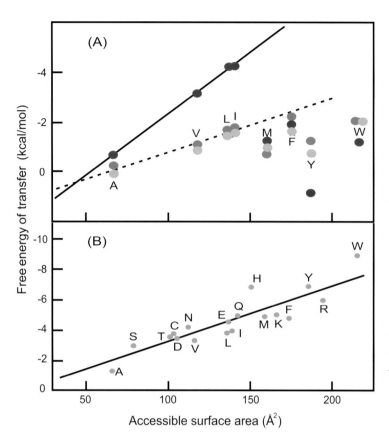

Figure 7-4. Relationship between the accessible surface areas of the nonpolar amino acid side-chains and their free energies of transfer (A) from water to nonaqueous solvent and (B) from vapor to cyclohexane. The free energies of transfer in (A) are from Table 7-6; *red points* represent measurements with the side-chain analogs, using cyclohexane as the nonpolar solvent; the *green points* were obtained with the free amino acids and ethanol and dioxane as the nonpolar solvent; the *blue points* were calculated from the hydrophobicities of the parts of each side-chain. The slope of the *solid line* in (A) is 43 cal/Å², for the *dashed line* 20 cal/Å². The slope of the line in (B) is 41 cal/Å². Adapted from T. E. Creighton (1993) *Proteins: structures and molecular properties*, 2nd edn, W. H. Freeman, NY, p. 161.

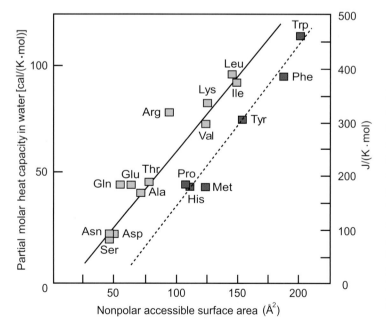

Figure 7-5. Correlation of the heat capacities in aqueous solution at 25°C of analogs of the amino acid side-chains with the accessible surface areas of their nonpolar atoms. The *upper straight line* fits all the side-chains (*green points*) except those with ring structures and the sulfur-containing Met (*blue points, dashed line*). The slope of the *upper line* is 0.72 cal/°C mol Å² (300 J/K mol nm²). Data from G. I. Makhatadze & P. L. Privalov (1990) *J. Mol. Biol.* **213**, 375–384.

7.3. PROTEIN DETECTION

Many methods have been developed for detecting and measuring polypeptide chains and proteins in solution, in an electrophoresis gel or on a blotting membrane. The classic **Kjeldahl method** detects all the N atoms of proteins as ammonia after heating them with sulfuric acid; it is used only with very complex samples containing many different types of molecules. More widely used colorimetric methods are the **Lowry**, **biuret** and **ninhydrin** assays. The **UV absorbance** method uses the absorbance of Trp and Tyr residues at 280 nm and is a very convenient method for determining the protein content of pure samples when the protein is the only material present that absorbs UV light and its molar absorbance (extinction coefficient) is known. Convenient and sensitive methods for the detection and measurement of proteins use the binding of dyes, such as **Coomassie Brilliant Blue** and **Ponceau S**, or reaction with a dye that becomes fluorescent, such as **Fluorescamine**.

Many of these methods can be employed for the detection of proteins on a membrane or in a polyacrylamide gel. The most sensitive detection method is **silver staining**, which can detect 2–5 ng of protein in an electrophoresis band and also detects other molecules. For most of these assays, the responses of different proteins vary quantitatively, depending on their amino acid composition, which makes quantitative measurement difficult.

7.3.A. Biuret Reaction

Proteins form a blue color in alkaline solution containing Cu^{2+} ions. The color is due to a **biuret**-like compound that is generated by losing protons from the N atoms of the peptide bonds:

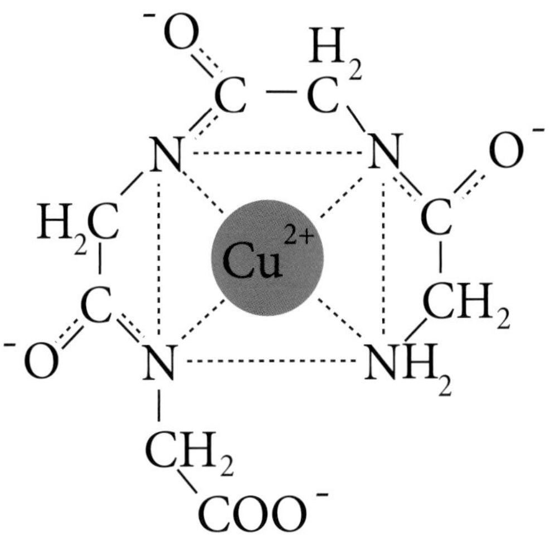

$$\text{(structure of Cu}^{2+}\text{ biuret-like complex)} \tag{7.59}$$

This biuret reaction is not very sensitive but it has the advantage that different proteins produce similar amounts of color. Consequently it is useful for measuring the protein contents of widely varying heterogeneous samples. A variety of modifications of the original biuret reaction have made it possible to measure 50 μg to 1 mg of protein.

Optimization of the conditions for biuret complex formation for the determination of peptides by capillary electrophoresis with ultraviolet detection. A. J. Gavron & S. M. Lunte (2000) *Electrophoresis* **21**, 2067–2073.

Cross-reactivity of amino acids and other compounds in the biuret reaction. G. L. Hortin & B. Meilinger (2005) *Clin. Chem.* **51**, 1411–1419.

7.3.B. Lowry Assay

The Lowry method is a sensitive colorimetric protein determination method that can measure 1–20 μg of protein. It is a combination of the phenol and biuret methods. The **Folin and Ciocalteu phenol reagent** reacts with phenols; it contains sodium molybdate (VI), sodium tungstate (VI) and phosphate; the active constituents are believed to be:

$$3\ H_2O \cdot P_2O_5 \cdot 13\ WO_3 \cdot 5\ MoO_3 \cdot 10\ H_2O$$

$$3\ H_2O \cdot P_2O_5 \cdot 13\ WO_3 \cdot 4\ MoO_3 \cdot 10\ H_2O \tag{7.60}$$

This reagent is added to the protein sample after the biuret reaction (Section 7.3.A) and the absorbance generated at 750 nm is proportional to the amount of protein present.

The color response of different proteins is variable, however, depending upon the amino acid composition. The side-chains of Tyr, Trp, His and Asn residues also react in isolation, producing blue color, but Pro residues prevent the adjacent peptide backbone from reacting. The greatest problem with the assay, however, is the wide variety of nonprotein substances that interfere, either producing a blue color themselves or inhibiting color development by the protein. This can be circumvented most readily by removing the interfering substances before analysis.

Modification of the Lowry assay to measure proteins and phenols in covalently bound complexes. A. L. Winters & F. R. Minchin (2005) *Anal. Biochem.* **346**, 43–48.

7.3.C. Ninhydrin

Ninhydrin reacts with free amino groups, including those in amino acids, peptides and proteins. Heating under mildly acidic conditions generates a blue-purple color, with an absorption maximum at 570 nm, known as 'Ruhemann's purple':

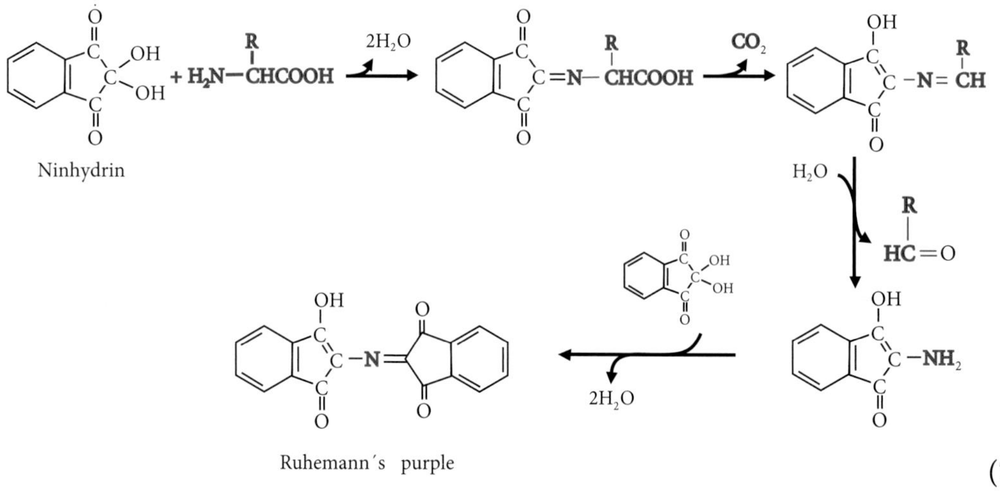

$$\tag{7.61}$$

Imino acids such as proline and hydroxyproline react differently and generate color with an absorption maximum at 440 nm.

This reaction is unusual in that at pH 5.5 it results in the formation of the same soluble chromophore by all primary amines that react, be they amines, amino acids, peptides, proteins or even ammonia. The chromophore is not chemically bound to the protein or other insoluble material, so it is not lost when the insoluble substrate is removed by centrifugation or filtration after the reaction is completed.

Applications of the ninhydrin reaction for analysis of amino acids, peptides, and proteins to agricultural and biomedical sciences. M. Friedman (2004) *J. Agric. Food Chem.* **52**, 385–406.

Ninhydrin as a reversible protecting group of amino-terminal cysteine. C. T. Pool *et al.* (2004) *J. Peptide Res.* **63**, 223–234.

The development of novel ninhydrin analogues. D. B. Hansen & M. M. Jouillie (2005) *Chem. Soc. Rev.* **34**, 408–417.

7.3.D. Fluorescamine

Fluorescamine itself is nonfluorescent but it reacts readily with primary amino groups to form highly fluorescent compounds:

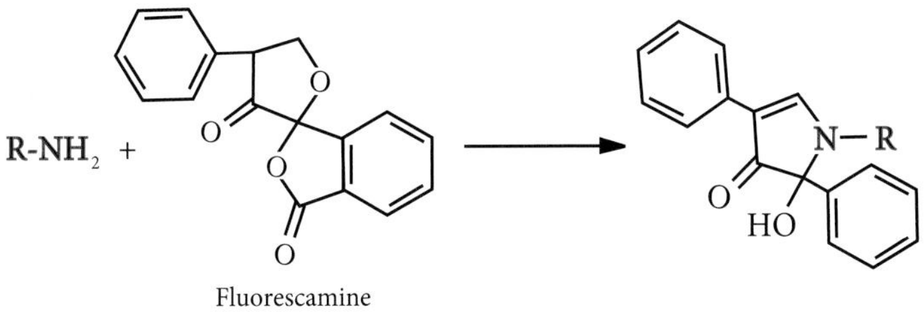

$$\text{R-NH}_2 + \text{Fluorescamine} \longrightarrow \text{product} \tag{7.62}$$

The fluorescence of the products has an excitation maximum at 390 nm and an emission maximum at 475 nm. Excess fluorescamine reacts with water and is destroyed; the product formed upon reacting with ammonia is only very slightly fluorescent. It is 10–100 times more sensitive than the ninhydrin reaction. Consequently, fluorescamine is ideal for detecting amino groups, especially in proteins, peptides and amino acids. The fluorescence yield varies with the amino group, however, so it is not useful for quantitative measurements.

Fluorescamine has low solubility and stability in water, but it can be substituted by ***o*-phthalaldehyde**, which reacts similarly with primary amino groups and gives highly fluorescent products in the presence of mercaptoethanol ($HS-CH_2CH_2-OH$):

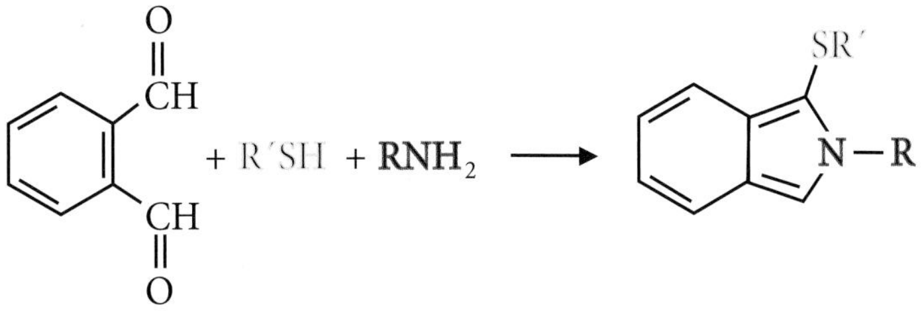

$$\text{o-phthalaldehyde} + R'SH + RNH_2 \longrightarrow \text{product} \tag{7.63}$$

The determination of polyamines and amino acids by a fluorescamine-HPLC method. K. J. Hunter & A. H. Fairlamb (1998) *Methods Mol. Biol.* **79**, 125–130.

Postcolumn derivatization of peptides with fluorescamine in capillary electrophoresis. R. Zhu & W. T. Kok (1998) *J. Chromatogr. A* **814**, 213–221.

Enhancement of the fluorescence and stability of o-phthalaldehyde-derived isoindoles of amino acids using hydroxypropyl-beta-cyclodextrin. B. D. Wagner & G. J. McManus (2003) *Anal. Biochem.* **317**, 233–239.

7.3.E. Coomassie Brilliant Blue

The staining procedure used most frequently for proteins in polyacrylamide gels is Coomassie Brilliant Blue. It can detect about 0.1 µg of protein in an electrophoresis band. Coomassie Brilliant Blue is commonly available in two different forms, R250 and G250:

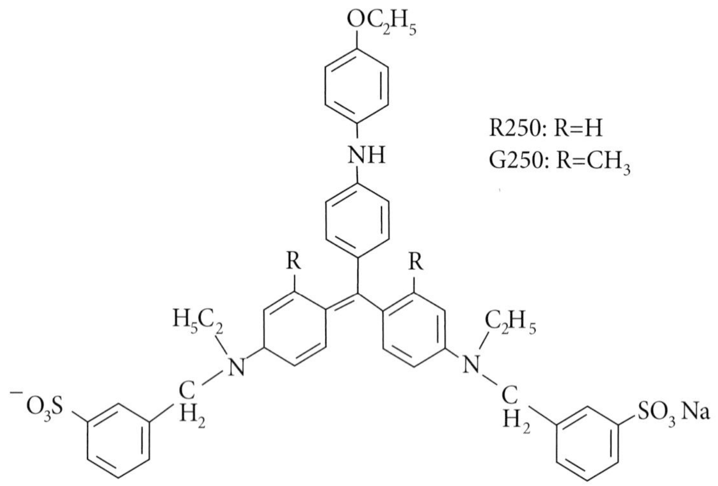

R250: R=H
G250: R=CH$_3$

(7.64)

Although the two dyes are structurally very similar, they require different staining procedures and should not be interchanged in any particular protocol.

These two dyes do not react chemically with proteins but merely form noncovalent complexes. The interaction with proteins is believed to be primarily ionic, involving the acidic sulfonate groups on the dye and basic groups on the protein, but nonpolar van der Waals forces are probably also involved. Consequently, the dyes do not bind equally to all proteins, so two samples or electrophoretic bands with the same amount of blue color need not contain the same amount of protein.

Coomassie Brilliant Blue G250 is often used to measure the amount of protein in solution, known as the **Bradford assay**. It uses the shift in the absorbance spectrum of the dye when it binds to protein; the absorbance maximum changes from 465 to 595 nm under certain acidic conditions. This assay is widely used, due to its simplicity, although different proteins produce somewhat different responses.

Blue silver: a very sensitive colloidal Coomassie G-250 staining for proteome analysis. G. Candiano *et al.* (2004) *Electrophoresis* **25**, 1327–1333.

The Coomassie chronicles: past, present and future perspectives in polyacrylamide gel staining. G. B. Smejkal (2004) *Expert Rev. Proteomics* **1**, 381–387.

Sensitive, quantitative, and fast modifications for Coomassie Blue staining of polyacrylamide gels. R. Westermeier (2006) *Proteomics* **6** (Suppl. 2), 61–64.

7.3.F. Ponceau S

Ponceau S is a red dye:

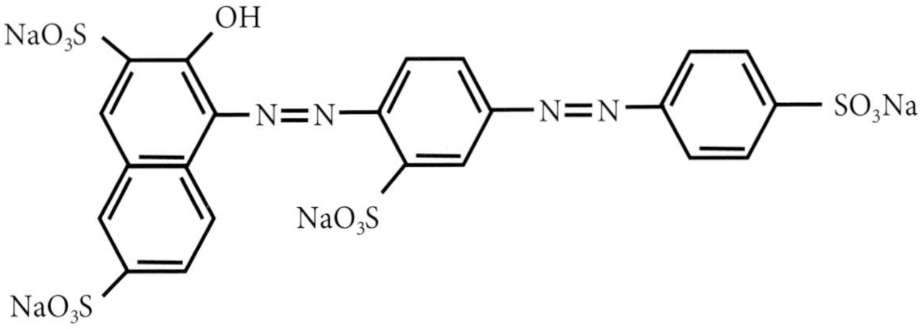

$$\text{[structure of Ponceau S: naphthalene with NaO}_3\text{S, NaO}_3\text{S, and OH substituents, linked via N=N to a benzene ring with NaO}_3\text{S, linked via N=N to a benzene ring with SO}_3\text{Na]} \tag{7.65}$$

It is commonly used to stain proteins on blotting membranes. Stained proteins absorb the dye, so they appear red on a pink background. The dye binds reversibly to proteins, so it can be washed out; it also is not very sensitive, requiring 0.2 μg of protein in a normal spot to be visible. Ponceau S is used for staining proteins primarily because the procedure is simple and rapid.

Ponceau S as a dye for quantitative protein assay. E. Gracia & F. Fernandez-Belda (1992) *Biochem. Int.* **27**, 725–733.

Protein determination by ponceau S using digital color image analysis of protein spots on nitrocellulose membranes. S. V. Bannur *et al.* (1999) *Anal. Biochem.* **267**, 382–389.

7.4. PEPTIDE SYNTHESIS

The chemical synthesis of peptides involves the sequential addition of amino acids to a growing peptide chain, using chemical coupling procedures. Reaction must occur only between the α-amino group of one reactant and the α-carboxyl group of the other; consequently, the α-carboxyl group of the first reactant and the α-amino group of the second must be protected, as must the other potentially reactive groups on the side-chains of the amino acids. All these protections must be reversible, so they can be removed when desired. The blocking groups on the side-chains will be removed only after the polypeptide chain is assembled, so they must be stable under conditions where the terminal α-blocking groups are removed. One of the latter must be removed after addition of each amino acid residue, to liberate the terminal residue of the elongated peptide for addition of the next residue. All of these reactions must go very nearly to completion; **with a repetitive procedure like linking together 100 amino acid residues, any errors accumulate very rapidly**. For example, even if each residue is added correctly 99% of the time in each step, the final yield of correct product of 100 amino acid residues will be only 37% (= 0.99^{99}).

Nevertheless, the chemical synthesis of small proteins is feasible. The development of solid-phase procedures and automated synthesizer equipment has reduced the labor and errors associated with repetitive manual procedures. Purification technologies have minimized the problems caused by the accumulation of byproducts resulting from synthetic difficulties in chain assembly and deprotection of side-chains. The repetitive nature of the process can be circumvented by preparing smaller peptides and then linking them together (Section 7.4.A.1).

Chemical synthesis is not limited to the 20 amino acids of the L-isomer that are used to synthesize natural proteins. Any type of amino acid can be used, including the D-isomer.

Chemical synthesis of proteins. B. L. Nilsson *et al.* (2005) *Ann. Rev. Biophys. Biomolec. Structure* **34**, 91–118.

Fundamentals of modern peptide synthesis. M. Amblard *et al.* (2005) *Methods Mol. Biol.* **298**, 3–24.

7.4.A. Chemistry of Polypeptide Chain Assembly

The peptide chain is routinely assembled stepwise from the C- to the N-terminus because this best preserves the chiral integrity of the component amino acids. The amino group of an amino acid is protected with suitable groups, such as the widely used urethanes:

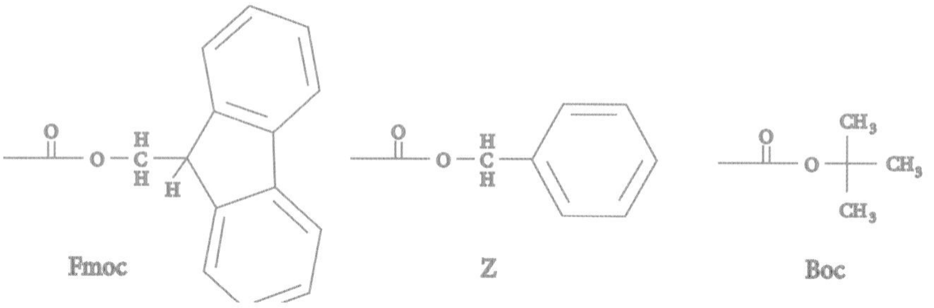

$$\text{Fmoc} \qquad \text{Z} \qquad \text{Boc} \tag{7.66}$$

where Fmoc = fluorenylmethyloxycarbonyl, Z = carbobenzoxy or benzyloxycarbonyl and Boc = *tert*-butyloxycarbonyl.

Peptide bonds do not form spontaneously between carboxyl and amino groups, so one has to be activated; almost invariably it is the carboxyl group. The driving force in carboxyl activation is usually augmentation of the electrophilic character of the C atom. It normally has a low electron density, and this is decreased even further by a negative inductive effect of the activating substituent, X:

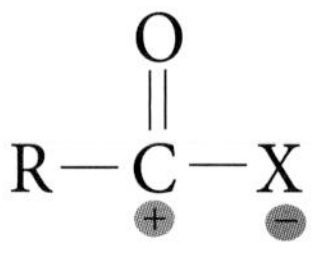

$$R-\overset{O}{\underset{\oplus}{C}}-\overset{}{\underset{\ominus}{X}} \tag{7.67}$$

The electrophilic center permits attack by a nucleophilic nonionized amino group:

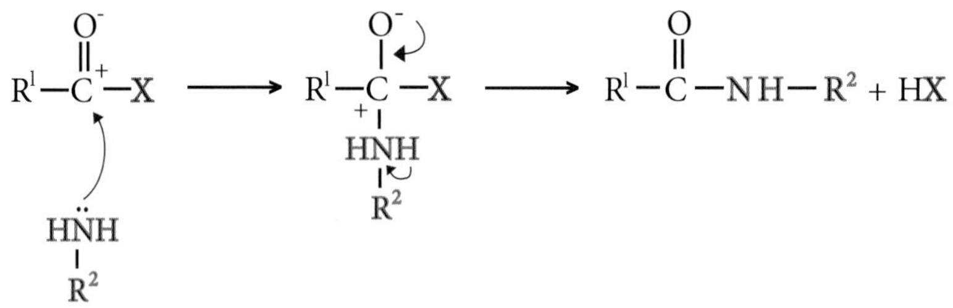

$$R^1-\overset{+}{\underset{\underset{\underset{R^2}{|}}{\underset{H\ddot{N}H}{}}}{\overset{\overset{O^-}{\|}}{C}}}-X \longrightarrow R^1-\underset{\underset{\underset{R^2}{|}}{\underset{HNH}{+}}}{\overset{\overset{O^-}{|}}{C}}-X \longrightarrow R^1-\overset{\overset{O}{\|}}{C}-NH-R^2 + HX \quad (7.68)$$

Examples of activated carboxyl groups are (1) azides, $-CO-N_3$; (2) acid chlorides, $-CO-Cl$; (3) mixed anhydrides, $-CO-O-CO-R$; and (4) activated esters, $-CO-OR'$, where R' is often *p*-nitrophenyl.

Coupling reagents can also be used to produce a peptide bond between free amino and carboxyl groups. These reagents usually generate transient intermediates like those just described, i.e. acid chlorides, active esters, etc. The most successful coupling reagents are the **carbodiimides**, especially dicyclohexylcarbodiimide. They react with free carboxyl groups to make them reactive towards amines (Equation 7.23). Incorporation of several sequential amino acids that are sterically hindered (valine, isoleucine, aminoisobutyric acid, etc.) often requires the use of special coupling reagents, such as acid fluorides, to provide reasonable reaction rates and efficient coupling yields.

Activation of the carboxyl group can cause racemization of the activated residue, because it enhances transient loss of the H atom on the C^α atom. Special procedures for coupling peptides that minimize this have been developed; Gly and Pro residues are used whenever possible as the C-terminal residue in fragment coupling, because of their resistance to racemization.

Chemical or enzymatic ligation of peptide fragments offers an alternative method for generating a peptide bond. Proteinase enzymes that normally cleave polypeptide chains (Section 7.5.B.1) can be induced, by choosing the appropriate conditions, to synthesize peptide bonds by the reverse reaction. Most important is to use nonaqueous solutions, for water is required for peptide bond cleavage but is produced in the reverse reaction.

After the addition of a protected amino acid, its protected amino group must be liberated for the next reaction while retaining the protecting groups on the side-chains of the growing peptide chain. This is known as **orthogonal protection**, in which the α-amino group is protected with a group that can be removed selectively by a mechanism that does not remove the groups protecting the side-chains. Most commonly used is the Fmoc group (Figure 7-6), which can be removed repetitively from the α-amino groups under basic conditions, combined with acid-labile protecting groups, such as Boc or OtBu, on the side-chains that must remain throughout chain elongation.

Maximal protection, in which all the potentially reactive groups, except for the amino and carboxyl groups to be coupled, are blocked, has the disadvantage that it is often difficult to remove all the protecting groups after completion of the synthesis. By careful control of reaction conditions (solvent, pH, etc.), some of the side-chain protection can be eliminated, thereby minimizing problems associated with removal of the protecting groups.

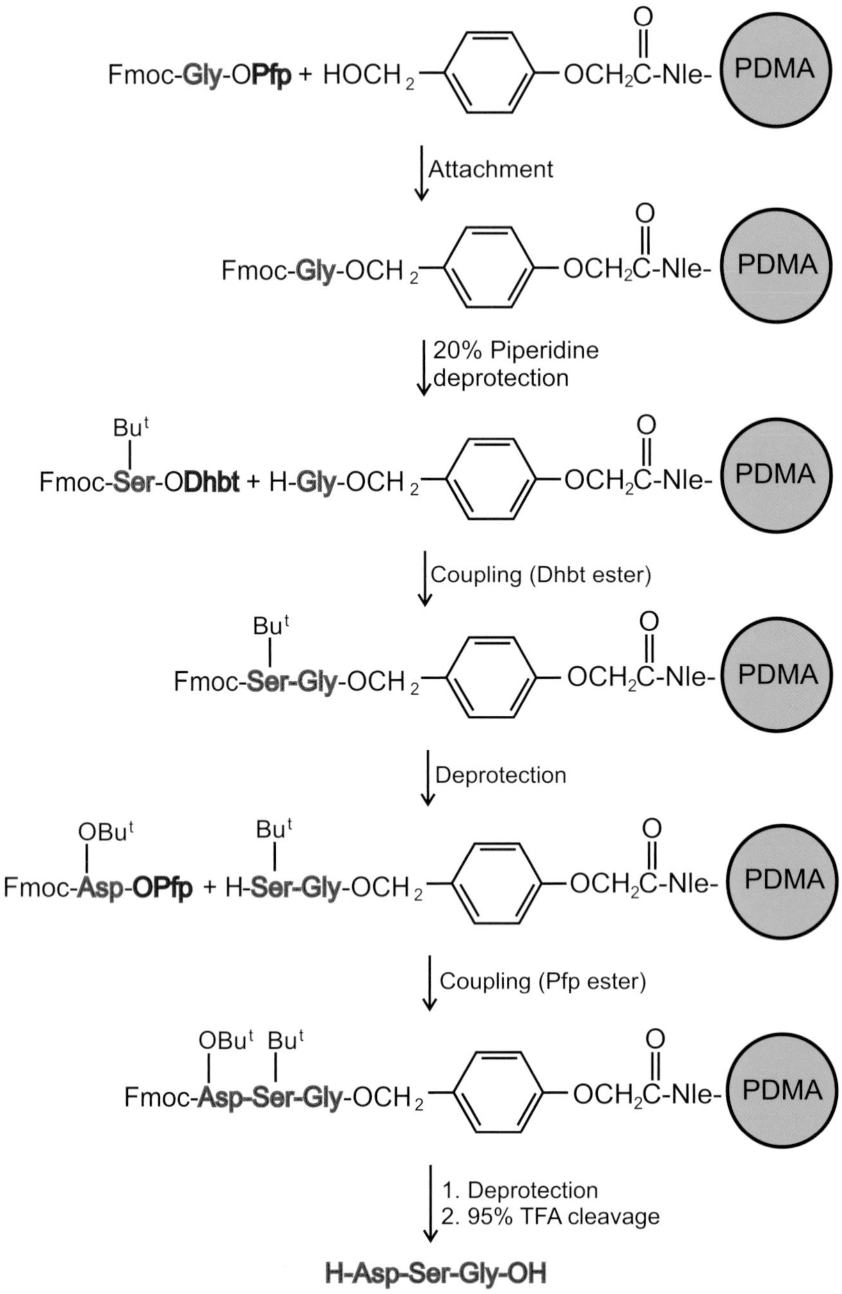

Figure 7-6. Synthesis of the tripeptide Asp–Ser–Gly using the Fmoc strategy. The solid-phase resin is cross-linked polydimethylacrylamide (PDMA). To this is coupled one residue of an unnatural amino acid, norleucine (Nle), which serves as an internal standard when monitoring the synthesis by hydrolyzing a portion of the resin, followed by amino acid analysis. The aromatic moiety attached to the amino group of the Nle residue is an acid-labile linkage agent, from which the assembled chain will be cleaved. The first amino acid is attached to the linkage agent through an ester bond. The amino acids are added sequentially to the resin after activating their carboxyl groups, in this case either as the pentafluorophenyl (Pfp) ester or that derived from 3,4-dihydro-3-hydroxy-4-oxo-benzotriazine (Dhbt). The amino acid side-chains are protected with acid-labile groups such as *tertiary*-butyl (Bu^t), its ester (OBu^t) or urethane (Boc). The amino groups of the amino acids are protected by the base-labile Fmoc group, which is removed by piperidine. When the last amino acid has been added, the Fmoc group is removed and the peptide is released from the resin and the side-chain blocking groups by treatment with trifluoroacetic acid (TFA).

Recent trends in protease-catalyzed peptide synthesis. C. Lombard *et al.* (2005) *Protein Peptide Lett.* **12**, 621–629.

Emerging methods in amide- and peptide-bond formation. J. W. Bode (2006) *Curr. Opinion Drug Discov. Devel.* **9**, 765–775.

Aryldithioethyloxycarbonyl (Ardec): a new family of amine protecting groups removable under mild reducing conditions and their applications to peptide synthesis. M. Lapeyre *et al.* (2006) *Chemistry* **12**, 3655–3671.

Fluorous (trimethylsilyl)ethanol: a new reagent for carboxylic acid tagging and protection in peptide synthesis. S. Fustero *et al.* (2006) *J. Org. Chem.* **71**, 3299–3302.

1. Chemical Ligation of Peptide Fragments

The synthesis of large peptides (or small proteins) containing 60–100 residues has been dominated by fragment condensation in solution with maximal protection of side-chains. The desired protein sequence is divided into peptide segments of ~15 residues, which is about the maximum length of peptide that can be prepared easily by stepwise addition of amino acids. Once all the fragments have been prepared, they are combined pairwise to generate segments of about 30 residues. These are then combined pairwise to generate fragments of ~60 residues, and this fragment condensation continues until the desired sequence is obtained. The polypeptide chain is finally deprotected, purified and induced to fold (Figure 7-7).

Frequently, unprotected fragments are coupled selectively using **chemical ligation**. The best-developed strategy for this approach uses an N-terminal Cys residue on one fragment, with a special reactive C-terminal group on the other peptide (Figure 7-8). Two unprotected peptide segments are prepared and then combined in aqueous solution. The C-terminal fragment has a Cys residue at its N-terminus, and the other peptide fragment has a thioester at its C-terminus. The thioester is displaced by the thiolate anion of the N-terminal Cys residue, followed by an acyl migration, through formation of a five-membered ring, to generate the desired stable peptide bond. If the S atom of a Cys residue within the peptide chain is involved, the thioester formed can be displaced by other thiols; eventually, the fragment migrates to the N-terminal Cys, when the irreversible rearrangement can occur. An alternative strategy uses an N-terminal β-bromoalanine of one fragment and the C-terminal thioester of the other to give the same covalent thioester intermediate by thioesterification.

This procedure is useful only when the protein has Cys residues in useful positions in the polypeptide chain. Attempts are being made to permit coupling at other residues. Some success has been achieved at X-Gly, Gly-X and X-His sites.

Synthesis of proteins by native chemical ligation using Fmoc-based chemistry. J. A. Camarero & A. R. Mitchell (2005) *Protein Peptide Lett.* **12**, 723–728.

Protein chemical ligation as an invaluable tool for structural NMR. A. Shechtman (2005) *Protein Peptide Lett.* **12**, 765–768.

Insights into the mechanism and catalysis of the native chemical ligation reaction. E. C. Johnson & S. B. Kent (2006) *J. Am. Chem. Soc.* **128**, 6640–6646.

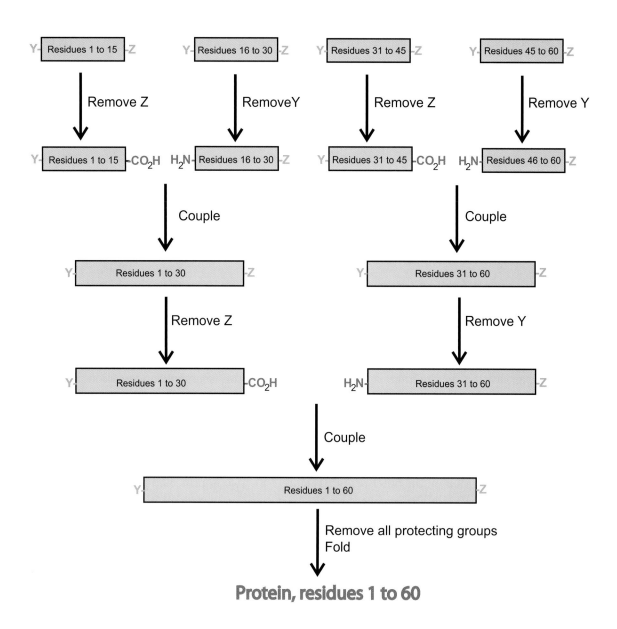

Figure 7-7. Solution fragment condensation synthesis of a hypothetical 60-residue protein using a maximal protection strategy. The 15-residue initial peptides shown at the *top* are prepared by stepwise addition of urethane-protected amino acids to the protected C-terminal residue. To minimize racemization upon fragment coupling, it is preferable that the C-terminal residue be Gly or Pro.

7.4.B. Solution or Solid-Phase?

The most fundamental question about peptide synthesis is whether the polypeptide chain being synthesized is to be in solution or attached to a solid matrix. The problem is similar to that discussed previously concerning the synthesis of DNA (Section 6.6) and RNA (Section 6.7).

The first approaches to peptide synthesis were based on solution methods, with isolation and characterization of each intermediate. This generated well-defined products after each step, with confidence in their final structures, at the price of substantial labor and losses in purifying each intermediate.

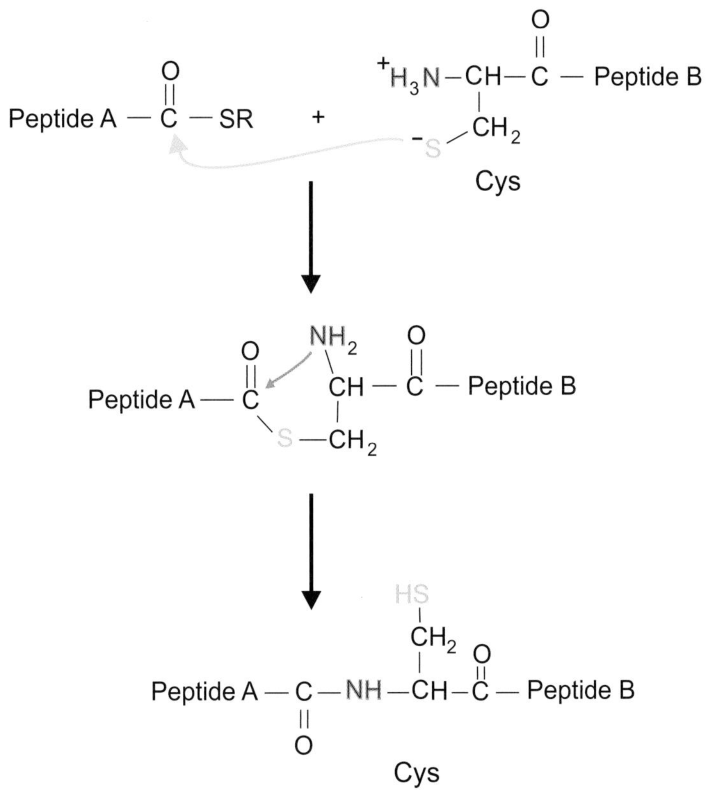

Figure 7-8. Native chemical ligation of two unprotected peptide fragments using thiol–thioester exchange with ligation at the N-terminal Cys residue. The first step is the specific reaction between the C-terminal thiol ester of peptide A and the N-terminal thiol group of peptide B. The second step is the spontaneous rearrangement of the acyl group to form the desired peptide bond.

The solid-phase approach uses a polymeric protecting group, which attaches the polypeptide chain to a solid support (Figure 7-9). The use of excess reagents can force reactions to virtual completion, and separation of the polymeric product from excess reagents by filtration and washing is almost trivial. Intermediates are not isolated. Consequently, the process can be automated, and instruments are now available commercially that will assemble polypeptides of any desired sequence without human intervention at a rate approaching 100 residues per day. Once assembled, the polypeptide chain is released from the polymer and deprotected.

The two chemical approaches used most frequently with solid-phase procedures are known as **Boc** and **Fmoc**. The Boc strategy (Figure 7-10) is often used with a cross-linked polystyrene support and a benzyl ester linkage to the polymer requiring a strong acid, such as hydrogen fluoride (HF), for deprotection. The Fmoc procedure (Figure 7-6) uses an acid-labile linkage, like that of *p*-methoxybenzyl ester, to the resin and a base-labile Fmoc group to protect the amino group of the added amino acids. Protection of the side-chains can be acid-labile, like the resin linkage, and cleavage using acid will produce the free, unprotected peptide. Alternatively, more stable protecting groups on the side-chains will produce the protected peptide, perhaps for use in fragment condensation after purification and characterization. In this case, the protecting groups are finally removed with a strong acid, such as HF. The purity of the final product depends on complete reaction at each synthetic step and minimization of side-reactions during the assembly of the oligomeric peptide. Unraveling any problems is hampered by the growing

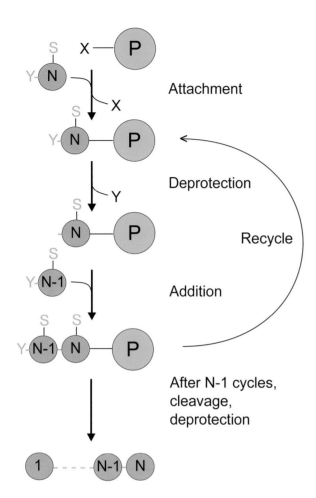

Figure 7-9. General scheme for solid-phase peptide synthesis. The *blue circles* are amino acids, the *green circle* is the polymer to which the nascent chain is attached; attachment is usually through the α-carboxyl group of the C-terminal residue of the desired peptide sequence. The chain is usually extended stepwise from the C-terminus (residue N) towards the N-terminus (residue 1). In the addition step, the α-carboxyl group of the amino acid being added must react only with the α-amino group of the nascent chain. Y is the group protecting the α-amino group of the amino acid being added, which must be removed for addition of the next amino acid. S is the group protecting the side-chain of each residue when necessary. Once the polypeptide chain has been completed after N – 1 cycles, it is cleaved from the resin and all the remaining protecting groups removed.

polypeptide chain being attached to the polymeric support, which limits the use of normal methods of characterizing intermediates. The majority of undesirable side-products in the final product are usually due to incomplete deprotection of the side-chains. Nevertheless, average reaction yields are estimated to be >99.5% for each residue added, and synthesis of peptides in the 50- to 100-residue range is routine.

On the other hand, unambiguous synthesis of larger peptides and small proteins is best accomplished by assembly of fragments that have been purified and fully characterized (Section 7.4.A.1). This prevents the accumulation of side-products with only minor structural differences that can be difficult to remove in the final mixture.

Methods and protocols of modern solid phase peptide synthesis. M. Amblard *et al.* (2006) *Mol. Biotechnol.* **33**, 239–254.

7.4.C. Peptide Libraries

The goal of classical peptide synthesis is to synthesize a single peptide, with a defined sequence, in pure form. A totally different approach, however, is to synthesize complex mixtures of peptides containing molecules with all possible sequences, known as a **peptide library**. These mixtures are then screened to detect that subset of peptides that has the desired biological properties.

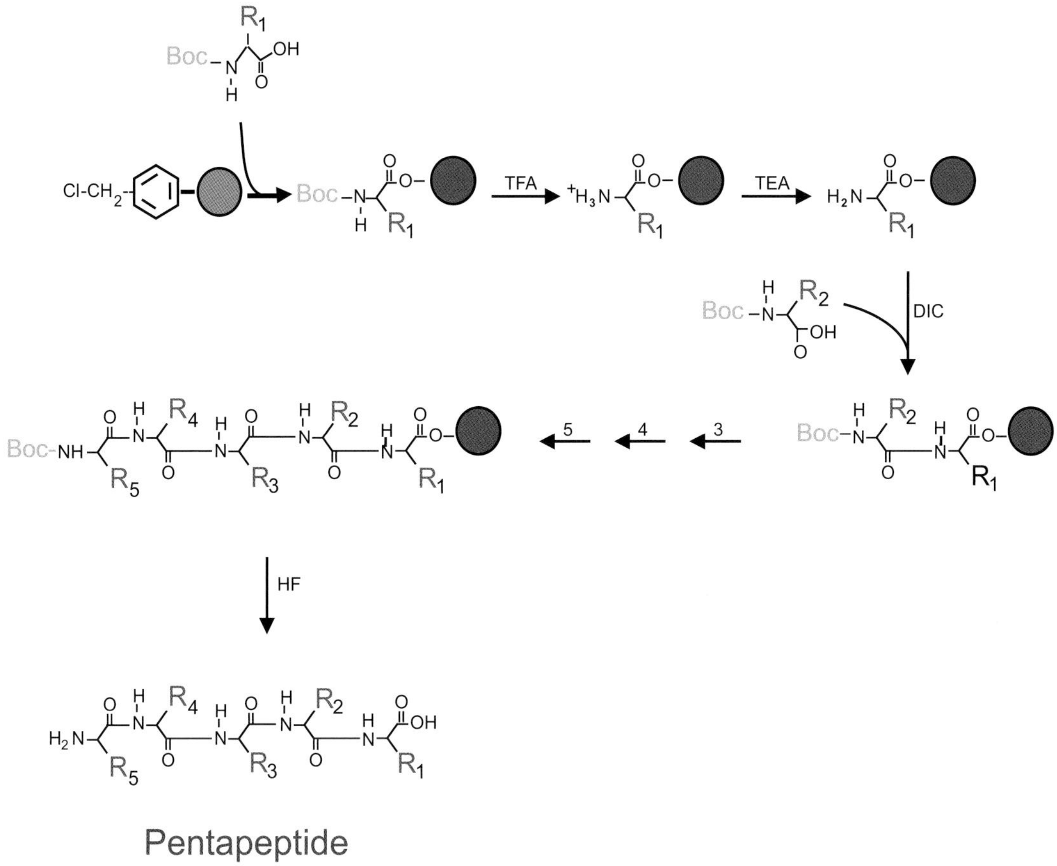

Figure 7-10. The Boc strategy for peptide synthesis. The side-chains of the amino acids are designated as R_i. The first C-terminal residue has its amino group protected with the *t*-butyloxycarbonyl (Boc) group and is coupled to the solid support (*orange sphere*) through its carboxyl group. Subsequently, the benzyl group is included in the solid supports (*red*). The Boc group is removed with trifluoroacetic acid (TFA), which is then neutralized with a base such as triethylamine (TEA). The next amino acid, with the amino group blocked by Boc, is added, using a suitable coupling reagent such as diisopropylcarbodiimide (DIC). Chain elongation is repeated until the desired sequence is obtained. The peptide chain is released from the solid support by cleavage with hydrogen fluoride (HF).

The synthesis of peptide mixtures is relatively easy; a mixture of amino acids, rather than just a single one, is added in each step of the synthesis. A huge number of peptides is possible: even if only the 20 natural amino acids are used, a peptide of n residues can occur with 20^n different sequences; with only 10 residues, this is 1×10^{13} different sequences. A sample of manageable size is not going to contain more than 10^{13} peptides, so it is difficult to prepare a complete library of sequences with more than 10 residues. The final heterogeneous peptide product is not purified but is used as the mixture. The difficult part of such a combinatorial approach is devising the selection procedure to identify those sequences with the desired properties, which depends entirely on those properties.

Identifying the active peptides is facilitated if they are attached to beads, like those used in solid-phase synthesis, and each bead contains a single type of peptide. Such one-bead one-compound peptide libraries are prepared by dividing the beads into 20 portions and attaching a single, different, amino acid to each portion. These portions are then combined, mixed, divided again into portions and a further round of amino acid addition carried out. Consequently, each bead carries a single type of peptide. Each bead may contain several nanomoles of peptide, so assays can be performed directly on the resin beads. Individual beads identified as having the desired properties are then subjected to peptide sequencing (Section 7.5) to determine the sequence of the active peptide. Alternatively, the peptides may be released partially from the bead but constrained to remain close to the original bead, such as by embedding in a soft agar medium. Zones of biological activity identify which beads contain the active peptide.

Peptide libraries can also be constructed by using cells to biosynthesize proteins from genes with random nucleotide sequences added at one end (Section 7.6.C). The proteins are usually the coat proteins of bacterial viruses, and the random segments are designed to be readily accessible ('displayed') in the assembled virus. Any virus with a sequence with the desired properties can be selected and amplified, and the amino acid sequence identified by sequencing the genome of the virus (Section 7.5.F).

Peptide libraries: at the crossroads of proteomics and bioinformatics. B. E. Turk & L. C. Cantley (2003) *Curr. Opin. Chem. Biol.* **7**, 84–90.

Combinatorial solid phase peptide synthesis and bioassays. D. S. Shin *et al.* (2005) *J. Biochem. Mol. Biol.* **38**, 517–525.

T7 lytic phage-displayed peptide libraries exhibit less sequence bias than M13 filamentous phage-displayed peptide libraries. L. R. Krumpe *et al.* (2006) *Proteomics* **6**, 4210–4222.

7.5. PEPTIDE AND PROTEIN SEQUENCING

The sequence of amino acid residues within a polypeptide chain can be determined by chemical methods. To do this with a large protein requires a considerable amount of effort, however, and the preferred method with natural proteins now is to determine the sequence of the corresponding **gene**, which can be converted to the protein sequence using the genetic code (Section 7.5.F). The gene sequence of the gene is considerably easier to determine, so long as the gene has been identified and isolated, but it gives only the amino acid sequence of the polypeptide as originally synthesized; many polypeptide chains are modified covalently after their biosynthesis, and such post-translational modifications can be determined only by chemical analysis of the polypeptide chain. Moreover, information about at least part of the amino acid sequence of the protein is often required in order to identify its gene.

7.5.A. Amino Acid Analysis

The first step in any sequencing procedure, and at many subsequent stages, is to determine what amino acid residues are present in a peptide or polypeptide chain. The amino acid compositions of proteins are routinely determined by completely hydrolyzing the peptide bonds of the polypeptide chain and then measuring the quantities of the constituent amino acids that are released (Figure 7-11).

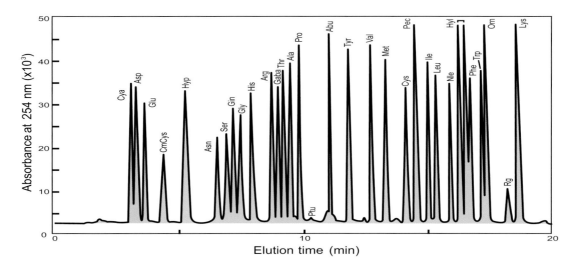

Figure 7-11. Chromatographic separation of amino acids after derivatization with phenylisothiocyanate. A sample containing 250 pmol of each amino acid was separated on a 3.9 × 300-mm Picotag reverse-phase column. Elution was with a convex gradient from eluent A (0.14 M sodium acetate, 0.05% triethylamine, 6% acetonitrile, pH 6.4) to 54% A, 46% eluent B (60% acetonitrile in water). To conserve space, the three-letter abbreviations for amino acid residues are used here for the standard amino acids, except that in this case Cys was the disulfide-bonded cystine. The nonstandard amino acids are: CmCys, carboxymethyl cysteine; Hyp, hydroxyproline; Gaba, γ-amino-butyric acid; Ptu, phenylthiourea (generated from ammonia by derivatization); Abu, α-amino-butyric acid; Nle, norleucine; Hyl, hydroxylysine; Orn, ornithine; Rg, contaminant from reagents. Data from S. A. Cohen & D. J. Strydom (1988) *Anal. Biochem.* **174**, 1–16.

Amino acid analysis. An overview. M. I. Tyler (2000) *Methods Mol. Biol.* **159**, 1–7.

Validation of amino acid analysis methods. A. J. Reason (2003) *Methods Mol. Biol.* **211**, 181–194.

---

1. Peptide Bond Hydrolysis

The traditional method of hydrolyzing polypeptide chains has been to incubate them in 6 M HCl at ~110°C for 24–72 hours in the absence of oxygen. More modern methods use other acids, higher temperatures and shorter periods of time. Most peptide bonds hydrolyze at similar rates, but those between the large nonpolar amino acid residues, particularly Val, Leu and Ile, are hydrolyzed more slowly and require longer hydrolysis times or the addition of organic acids such as **trifluoroacetic acid**. Hydrolysis is presumably hindered sterically by the bulky side-chains.

Any chemical procedure that hydrolyzes the peptide bonds of the backbone will also hydrolyze the chemically similar amide side-chains of Asn and Gln residues, to produce the amino acids aspartic acid and glutamic acid, respectively. It is feasible, although not convenient, to measure the total number of Asn and Gln residues by measuring the amount of ammonia released during the hydrolysis, but otherwise it is not possible to distinguish between Asp and Asn and Glu and Gln after acid hydrolysis of the polypeptide chain. In this case, it is common practice to designate such uncertain residues by the three-letter abbreviations Asx and Glx, or by the one-letter abbreviations B and Z, respectively.

Trp residues are usually destroyed completely by acid hydrolysis, probably as a result of the reaction with chlorine, which is produced by oxidation of HCl. These residues can be protected by the addition of thiol or sulfonic acid compounds or of phenol to scavenge the chlorine. Tyr residues are also susceptible to chlorination, but they are usually lost only partially. The thiol groups of Cys residues are oxidized and the amino acid partially destroyed by acid hydrolysis; this residue is best analyzed after performic acid oxidation of the protein to convert all the Cys residues to cysteic acid (Equation 7.77). If a protein is hydrolyzed to amino acids under conditions where a disulfide bond persists, the result is two cysteine amino acids linked by a disulfide bond, which historically is known as the amino acid **cystine**. Consequently, Cys residues linked by disulfide bonds are frequently referred to as '**1/2-cystine**'.

Some of the problems with acid hydrolysis can be overcome by using other procedures, such as hydrolysis by alkali or proteinases. Other amino acids, notably serine and threonine, are destroyed by alkaline hydrolysis, however, and total proteinase digestion to amino acids is not straightforward. Consequently, acid hydrolysis remains in common use.

Hydrolysis of samples for amino acid analysis. I. Davidson (2003) *Methods Mol. Biol.* **211**, 111–122.

The effect of hydrolysis time on amino acid analysis. A. J. Darragh & P. J. Moughan (2005) *J. AOAC Int.* **88**, 888–893.

---

2. Quantifying Amino Acids

The identities and quantities of the various amino acids present in protein hydrolyzates are normally determined by automated amino acid analyzers (Figure 7-11). The amino acids are separated by column chromatography and measured quantitatively as they emerge from the column. Traditional methods used ion-exchange chromatography of the free amino acids, followed by detection with ninhydrin (Section 7.3.C) or fluorescent reagents like fluorescamine (Section 7.3.D). Proline does not react in the usual manner with such reagents, due to the absence of an amino group, so special procedures are required to detect it. More rapid and sensitive methods now predominate, in which the amino acids are reacted with suitable reagents prior to the chromatographic separation, rather than after. A popular method is to react the amino acids with **phenylisothiocyanate** (Equation 7.80) and then to separate the colored derivatives by reverse-phase chromatography. With this procedure, a complete quantitative amino acid analysis can be carried out in just a few minutes with only picomole quantities of amino acids (Figure 7-11). The method is calibrated by running standards with known amounts of each of the amino acids.

The relative numbers of aromatic residues (Phe, Tyr and Trp) in intact proteins and peptides can usually be determined from the UV absorbance spectrum of the purified protein. The conditions must be such that the polypeptide chain is fully unfolded so that its spectrum is the sum of its constituent residues.

Amino acid analysis does not give the number of residues of each amino acid per polypeptide chain directly, only the molar ratios of the various amino acids. The true molecular weight of the polypeptide chain, in the absence of any nonamino acid moieties, must be known for the amino acid analysis results to be converted to the number of residues of each amino acid per chain.

Recent advances in amino acid analysis by capillary electrophoresis. V. Poinsot *et al.* (2006) *Electrophoresis* **27**, 176–194.

Validation of a reversed-phase HPLC method for quantitative amino acid analysis. M. P. Bartolomeo & F. Maisano (2006) *J. Biomol. Tech.* **17**, 131–137.

3. Counting Residues

The polypeptide chains of natural proteins have integral numbers of each of the 20 amino acids. Most currently accepted methods of amino acid analysis produce only the ratio of moles of amino acid per mole of protein. Experimental error and uncertainty about the molecular weight of the protein mean that this value is rarely found to be close to an integer. The procedure for counting residues determines the integral number of residues of certain amino acids independently of any other property of the polypeptide chain, including its molecular weight. This information may then be combined with the results of amino acid analysis to generate a more accurate value for the molecular weight of the protein and the number of other amino acid residues. This approach can also provide information about the net charge on a protein.

The N residues of one type of amino acid are modified gradually and specifically to generate a complete spectrum of molecules with 0, 1, 2, 3,..., N of the groups modified. A single modification reaction can be used, varying the average extent of the reaction; alternatively, the competition with a related, but distinguishable, reagent can be used and the modification by both reagents taken to completion. The spectrum of species produced is determined by separating them using a technique that is sensitive only to the number of groups modified by the particular reagent. The number of species is simply counted.

Modifications that alter the charge of a residue are most useful. The number of modifications introduced into the protein is then determined by separating the various species by ion-exchange chromatography, electrophoresis or isoelectric focusing. The separation must only be sensitive to the number of groups modified in charge, not which particular ones in the polypeptide chain; this can usually be accomplished by carrying out the modification and separation on unfolded molecules, where differences between the N residues are minimized. There should be $N + 1$ species present after the modification.

Reaction with a reagent like iodoacetic acid introduces an acidic group onto the thiol groups of Cys residues (Equation 7.29). The closely related iodoacetamide is neutral and does not introduce a charged group (Equation 7.28). Competition of the two reagents for thiol groups can be used by varying the ratio of iodoacetate to iodoacetamide. Reaction of the thiol groups is taken to completion. The number of acidic groups introduced is determined by electrophoresis under denaturing conditions (Figure 7-12-A). The result will give the number of total Cys residues or of thiol groups, depending upon whether any disulfide bonds present were reduced before the counting procedure. The difference between the two results can provide the number of disulfide bonds that were present originally.

Amino groups can be modified specifically by reacting them with a reagent like maleic or **succinic anhydride** (Equation 7.12), which replaces a basic amino group with an acidic group. Consequently, the net charge of the protein can change by up to two units for each amino group reacted. The extent of the reaction is varied by adding different amounts of reagent to the unfolded protein (Figure 7-12-B). This procedure counts the number of Lys residues and the amino group at the N-terminus.

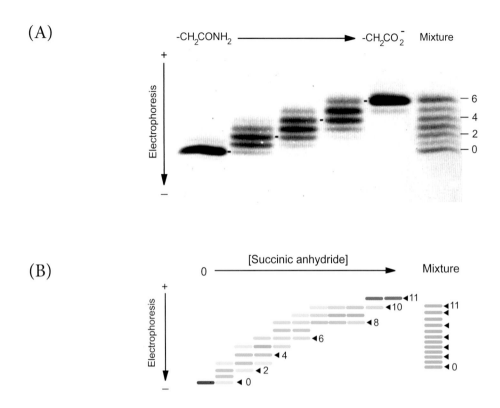

Figure 7-12. Counting thiol and amino groups in proteins. (A) Counting the six Cys residues of reduced bovine pancreatic trypsin inhibitor (BPTI). The thiol groups of the reduced protein were reacted with the neutral iodoacetamide (1st lane), acidic iodoacetic acid (5th lane) and mixtures of the two at ratios of 1:1, 1:3 and 1:9 in lanes 2, 3, and 4, respectively. Lane 6 contained a mixture of equal portions of the samples applied to lanes 1–5. Electrophoresis of the basic protein was in 8 M urea from *top* to *bottom*. The electrophoretic mobility was decreased in proportion to the number of acidic carboxymethyl groups on each molecule; the number for each of the electrophoretic bands is indicated on the *right*. The competition between the two reagents indicates that iodoacetamide reacts three times more rapidly than does iodoacetic acid under the conditions used here. Adapted from T. E. Creighton (1980) *Nature* **284**, 487–489. (B) Counting the amino groups of bovine ribonuclease A by electrophoretic separation of the mixtures produced by progressive succinylation. The original unmodified protein is on the *left* and the degree of succinylation increases to the *right*. Electrophoresis was in 8 M urea at pH 3.6 from *top* to *bottom*. The separate lane on the *far right* is of a mixture obtained by combining all the individual samples; in this case, electrophoresis was at the slightly lower pH of 3.45. Alternating bands are marked by *arrows*, with the number of succinyl groups indicated for a few of the bands. The results confirm the presence of 11 amino groups in ribonuclease A. Original data from M. Hollecker & T. E. Creighton (1980) *FEBS Lett.* **119**, 187.

The number of groups modified is counted by the number of new bands generated that are apparent by electrophoresis. The electrophoretic mobility is proportional to the net charge of the protein molecule, which is changed gradually by the modification of specific residues. The extent of modification that is necessary to give the protein zero electrophoretic mobility therefore indicates the net charge of the original protein molecule. Alternatively, the stepwise modification can be used to establish a scale for electrophoretic mobility as a function of changes in net charge.

Similar approaches should be feasible for other amino acids for which specific modifications are possible, but specific protocols have not been developed.

Counting integral numbers of amino acid residues per polypeptide chain. T. E. Creighton (1980) *Nature* **284**, 487–489.

Counting integral numbers of amino groups per polypeptide chain. M. Hollecker & T. E. Creighton (1980) *FEBS Lett.* **119**, 187–189.

7.5.B. Fragmentation of a Protein into Peptides

The crucial step in protein sequencing is to cleave the protein into a defined set of peptides sufficiently short that their sequences can be determined. The procedure is repeated, cleaving the polypeptide chain at different positions. The overlaps in the sequences of these peptides permit the entire sequence of the polypeptide chain to be inferred.

A variety of techniques have been developed for cleaving polypeptide chains specifically (Table 7-9). Proteolytic enzymes are used most commonly, but there are also chemical methods that are specific for peptide bonds adjacent to certain types of residues.

Before cleavage, Cys residues are often alkylated with [^{14}C]iodoacetic acid (Equation 7.29) or **4-vinylpyridine** to prevent disulfide formation and peptide aggregation, and to promote convenient detection of Cys-containing peptides. After cleavage, the peptides produced can be derivatized with fluorescent tags for detection. In general, however, separation problems may be introduced by pre-column derivatization, and the digest is often analyzed directly. A practical example of capillary HPLC peptide mapping (Figure 7-13) illustrates how *in vivo* oxidation of a recombinant protein hormone after its intravenous administration can be detected.

Selective cleavage and modification of peptides and proteins. T. F. Spande *et al.* (1970) *Adv. Protein Chem.* **24**, 97–260.

1. Proteolytic Enzymes

Proteinases are enzymes (Chapter 14) that hydrolyze one or more of the peptide bonds of the polypeptide backbone of a protein molecule, often only at specific types of amino acid residues (depicted here as X):

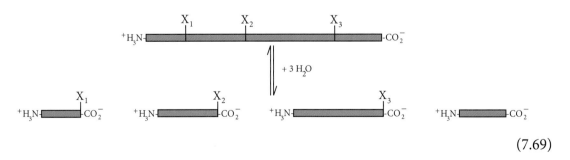

(7.69)

Table 7-9. Methods for cleaving polypeptide chains at specific residues

Peptide bond cleaved[a]	Enzyme or procedure
Ala–Yaa	Elastase, bromelain
Arg–Yaa	Trypsin, endoproteinase Arg-C, clostripain
Asn–Gly	Hydroxylamine
Asp–Yaa	V-8 proteinase
Asp–Pro	Mild acid
Xaa–**Asp**	Asp-N proteinase
Xaa–**Cys**	Cyanylation
Glu–Yaa	V-8 proteinase
Gly–Yaa	Elastase
Leu–Yaa	Pepsin
Xaa–**Leu**	Thermolysin
Lys–Yaa	Trypsin, endoproteinase Lys-C, bromelain
Met–Yaa	CNBr
Phe–Yaa	Chymotrypsin, pepsin
Xaa–**Phe**	Thermolysin
Pro–Yaa	Prolylendopeptidase
Trp–Yaa	Iodosobenzoic acid, chymotrypsin
Tyr–Yaa	Chymotrypsin, bromelain

[a] Xaa and Yaa can be almost any amino acid, except for Pro in many instances. Cleavage is C-terminal to residue Xaa, N-terminal to Yaa.

From T. E. Creighton (1993) *Proteins: structures and molecular properties*, 2nd edn, W. H. Freeman, NY, p. 38.

Proteinases are generally classified according to the hydrolytic mechanism that they use for peptide bond cleavage:

(1) The **carboxyl proteinases** have the carboxyl groups of two Asp residues at their active site, which activate the water molecule that will cleave the peptide bond.

(2) The **serine proteinases** use a unique **catalytic triad** of an Asp carboxyl group, a His imidazole group and a Ser hydroxyl group.

(3) The **thiol proteinases** use the thiol group of a Cys residue.

(4) The **zinc proteinases** have a zinc ion as part of the active site.

Proteinases are also classified on the basis of where they act on a protein chain. **Exoproteinases** remove amino acids sequentially from the end of a chain, **aminopeptidases** from the amino terminus and **carboxypeptidases** from the carboxyl terminus. **Endoproteinases** cleave bonds in the middle of a protein chain.

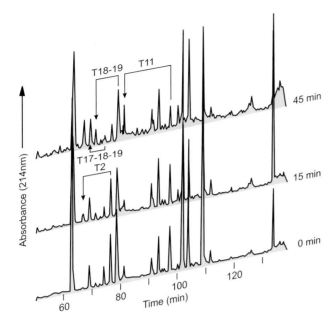

Figure 7-13. Peptide mapping to demonstrate the oxidation of Met residues *in vivo*. HPLC analysis is shown of the peptides produced by trypsin from recombinant human growth hormone (*lower profile*) and of the same protein recovered from rat serum taken 15 min (*middle profile*) and 45 min (*upper profile*) after an intravenous injection. *Arrows* link Met-containing peptides (*right* in each pair) with their corresponding oxidized forms (*left* in each pair). The oxidized form increases with time *in vivo*. Data from J. E. Battersby *et al.* (1995) *Anal. Chem.* **67**, 447–455.

Proteinases vary in their specificity for peptide bonds adjacent to certain residues; some cleave the peptide bond preceding the specificity residue but most cleave the following peptide bond. **Trypsin** is one of the most specific proteolytic enzymes, cleaving only peptide bonds following Lys or Arg residues. Cleavage by trypsin can be restricted to Arg residues by reversibly blocking the ε amino groups of Lys residues by maleylation or citraconylation (Equation 7.12). Trypsin will also cleave after Cys residues if the thiol group has been reacted with ethylenimine to produce a basic side-chain (Equation 7.31). Alternatively, the enzymes **clostripain** and **submaxillary proteinase** have natural specificity only for Arg residues, while **endoproteinase Lys-C** and **lysylendopeptidase** are specific for Lys residues. The **V8 proteinase** from *Staphylococcus aureus* is complementary to these enzymes in that it cleaves only after the acidic residues Asp and Glu. **Chymotrypsin** is fairly specific for the peptide bonds following the aromatic residues Tyr, Phe and Trp but it also tends to cleave after other nonpolar residues, especially Leu. **Thermolysin** preferentially cleaves the peptide bond preceding Leu, Val, Ile and Met residues. **Prolylendopeptidase** cleaves after Pro residues, and **Asp-N proteinase** cleaves before Asp residues.

Common features of the four types of protease mechanism. L. Polgár (1990) *Biol. Chem. Hoppe-Seyler* **371** (Suppl.), 327–331.

The structure and function of the aspartic proteinases. D. R. Davies (1990) *Ann. Rev. Biophys. Biophys. Chem.* **19**, 189–215.

Trypsin cleaves exclusively C-terminal to arginine and lysine residues. J. V. Olsen *et al.* (2004) *Mol. Cell. Proteomics* **3**, 608–614.

2. Chemical Methods of Cleavage

The most widely used chemical method of cleaving specific peptide bonds uses **cyanogen bromide (CNBr)** to cleave the peptide bond after Met residues. The Met S atom is sufficiently nucleophilic (but not if it is oxidized; Equation 7.48) for it to react with CNBr:

$$-CH_2-CH_2-S-CH_3 + CNBr \longrightarrow -CH_2-CH_2-\overset{+}{\underset{|}{S}}-C\equiv N + Br^- \quad (7.70)$$
$$\hspace{6cm} CH_3$$

The stereochemistry of the Met side-chain favors the intramolecular rearrangement of the sulfonium salt:

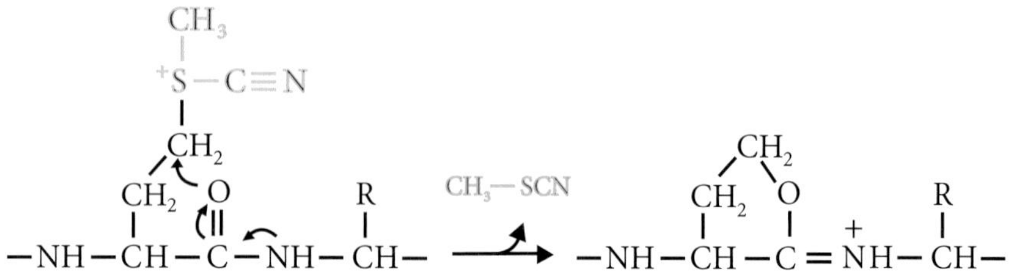

$$(7.71)$$

The iminolactone generated is hydrolyzed by water, which cleaves the polypeptide chain:

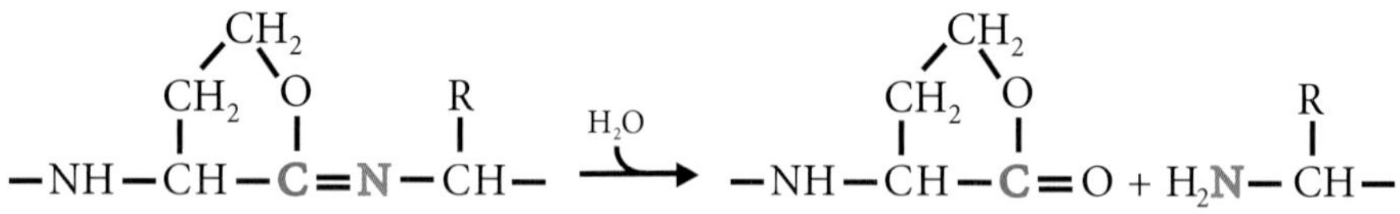

$$(7.72)$$

The original Met residue has been converted to a **homoserine (Hse) lactone** residue at the C-terminus of the N-terminal fragment produced. The lactone is hydrolyzed reversibly to the free acid:

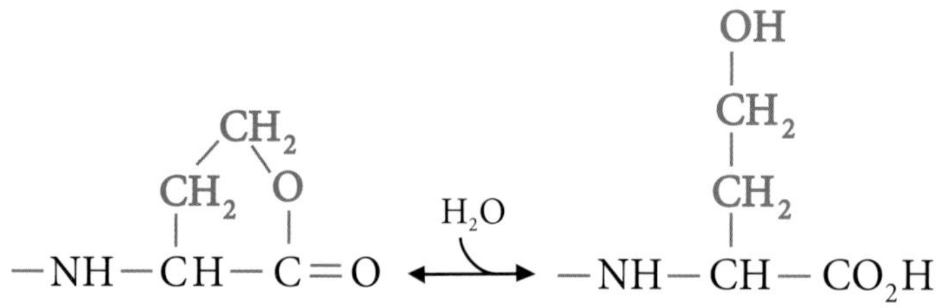

$$(7.73)$$

The two forms differ in net charge, so this reversible equilibrium can complicate separation of the peptides generated by CNBr. A further complication is that Met residues preceding Thr, Ser and Cys residues are often cleaved incompletely.

Cyanylation with 2-nitro-5-thiocyanobenzoic acid can be used to cleave peptide bonds preceding Cys residues:

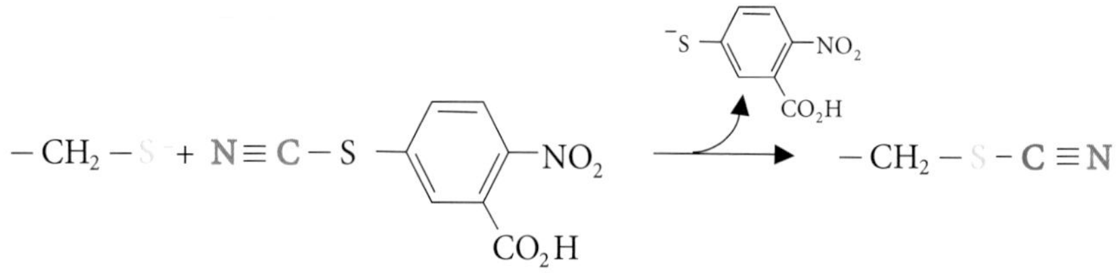

$$(7.74)$$

Incubation at alkaline pH causes peptide bond cleavage:

$$\underset{\begin{array}{c}R\\|\\-CH-\end{array}}{}\underset{\begin{array}{c}O\\||\\C\end{array}}{}-NH-\underset{\begin{array}{c}\overset{\displaystyle N}{\overset{|||}{C}}\\|\\S\\|\\CH_2\\|\\CH\end{array}}{}\underset{\begin{array}{c}O\\||\\C\end{array}}{}-NH-\xrightarrow{OH^-}\underset{\begin{array}{c}R\\|\\-CH-\end{array}}{}CO_2^- + HN-\underset{\begin{array}{c}HN\diagdown\,_S\\\diagup\,C\diagdown\\CH_2\quad\diagdown\\\diagup\\CH\end{array}}{}\underset{\begin{array}{c}O\\||\\C\end{array}}{}-NH- \quad (7.75)$$

This reaction has the disadvantage that the new N-terminus generated is blocked and not amenable to sequence analysis by the Edman procedure (Section 7.5.C).

Peptide bonds following Trp residues can be cleaved with varying efficiencies by **iodosobenzoic acid** and **BNPA-skatole**. These procedures have the disadvantage, however, of side-reactions and oxidizing Cys and Met residues.

Peptide bonds adjacent to Asp, Ser and Thr residues are especially susceptible to **acid hydrolysis** because the polar groups of the side-chains of these residues can interact with the adjacent peptide bond of the backbone. In particular, **Asp–Pro** bonds are cleaved readily by dilute acid, probably because of the stereochemical ability of the Asp side-chain carboxyl group to interact chemically with the unique tertiary N atom of the Pro residue (Figure 7-3).

Asn–Gly sequences can be cleaved by hydroxylamine, due to the tendency of such Asn residues to form succinimide derivatives transiently and to deamidate by reacting with water (Figure 7-3). If the succinimide ring is opened by reaction with hydroxylamine instead of water, peptide bond cleavage results. The original Asn residue is now an Asp at the new C-terminus of the site of cleavage, but it has a **hydroxamate** group (–CO–NHOH) in place of either the α- or β-carboxyl group.

Current developments in chemical cleavage of proteins. K. K. Han *et al.* (1983) *Int. J. Biochem.* **15**, 875–884.

A method for C-terminal sequence analysis in the proteomic era (proteins cleaved with cyanogen bromide). B. Samyn *et al.* (2006) *Nature Protocols* **1**, 318–323.

7.5.C. Peptide Mapping

Peptide mapping is a relatively simple method for rapidly comparing the amino acid sequences of two related proteins and a very sensitive way of detecting differences between closely related proteins. The peptides produced by a specific cleavage method are separated into a one- (1-D) or two-dimensional (2-D) **peptide map** and the maps of related proteins are compared. A single amino acid difference between two peptides, due to either gene mutation or post-translational modification, will usually alter the physical properties of the peptide in which that difference occurs, which will change the position of that peptide on the peptide map. Differences of a single amino acid residue can be found rapidly. Classical peptide mapping uses 2-D fingerprinting by chromatography and electrophoresis on paper sheets (see Figure 7-15); such methods are still useful when the amounts of protein available are sufficient, but molecular biology now relies on more sensitive approaches.

Comparing proteins available only in very small amounts is possible with modern separation techniques, such as capillary **HPLC** (Figure 7-13), **capillary electrophoresis**, high-resolution 2-D **gel electrophoresis** and **mass spectrometry**. With mass spectrometry, the mapping, detection and analysis are carried out in one step. Using entire peptide mixtures or pre-selected fragments, direct determination of the total masses of peptide fragments from a protein gives a unique **mass fingerprint** and is now the mapping method of general choice (see Figure 7-20).

The approximate positions in a polypeptide chain of amino acids for which specific polypeptide chain cleavage reactions are available (Table 7-9) can be determined relatively easily. The polypeptide chain is labeled in some way at one of its two ends (Figure 7-14). The cleavage reaction is then carried out partially, so that only a fraction of the molecules are cleaved at each target position. The resulting mixture of cleaved and uncleaved molecules is sorted according to length by SDS polyacrylamide gel electrophoresis, but only the peptides containing the label at one end are detected. The lengths of these fragments indicate the positions of the cleavage sites throughout the polypeptide chain.

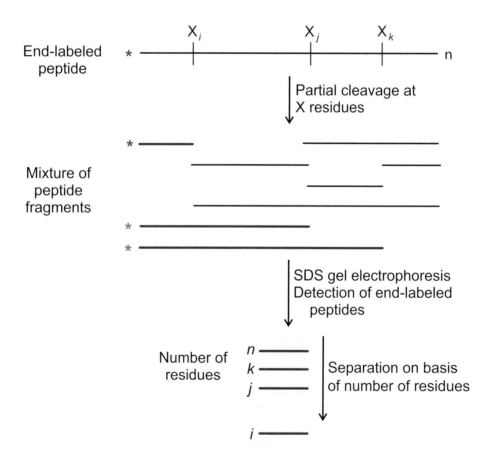

Figure 7-14. Mapping the positions of specific residues along the polypeptide chain by end-labeling and partial cleavage at the residues. The polypeptide chain is labeled at one end, indicated by the *asterisk*; the label can be radioactive or fluorescent, or be bound specifically by some antibody. Partial cleavage at the residues of interest generates a wide variety of peptides. The peptides are separated according to their size, usually by SDS gel electrophoresis, giving the lengths of the peptides. The labeled fragments give the positions of the cleavage points relative to the labeled end. From T. E. Creighton (1993) *Proteins: structures and molecular properties*, 2nd edn, W. H. Freeman, NY, p. 40.

A general procedure for the end labeling of proteins and positioning of amino acids in the sequence. D. G. Jay (1984) *J. Biol. Chem.* **259**, 15572–15578.

Determination of the relative positions of amino acids by partial specific cleavages of end-labeled proteins. R. A. Jue & R. F. Doolittle (1985) *Biochemistry* **24**, 162–170.

Peptide mapping. T. Bergman & H. Jörnvall (1999) in *Encyclopedia of Molecular Biology* (T. E. Creighton, ed.), Wiley-Interscience, NY, pp. 1797–1801.

7.5.D. Diagonal Maps

Those peptides from a protein that contain a particular type of amino acid, such as all the peptides containing Cys residues, can be identified, isolated and purified selectively using so-called **diagonal techniques**. They use a change in the properties of peptides produced by the selective modification of the amino acid residues of interest. When a mixture of peptides generated for peptide mapping is separated in one dimension and then subjected to the same separation procedure a second time, but at right angles to the first, all the peptides will have the same mobilities in both dimensions and consequently will lie on a diagonal line (Figure 7-15). Electrophoresis or chromatography can be used for the separation, but it should be carried out on a 2-D medium, such as paper, thin-layer plates or gels. Certain peptides will have a different mobility in the second dimension if they are modified chemically after the first electrophoresis. The entire mixture of peptides is treated to modify all the residues of a particular type in a manner that changes their separation properties. They will then have different properties in the second dimension, so the peptides containing these residues will lie off the diagonal of all the other peptides. These peptides are readily identified and have been separated from the other peptides.

Unfortunately, the 2-D separations that are best suited for diagonal techniques are no longer in common use, having been supplanted by more modern techniques that require less material but operate in a single dimension. The techniques developed for diagonal methods can still be used with 1-D separations but the fractions generated by the first separation must be separated again individually.

1. Isolating Peptides Containing Certain Amino Acids

Peptides containing **Cys** residues can be purified selectively by first blocking their thiol groups by reaction with iodoacetic acid (Equation 7.29). A 2-D diagonal peptide map is then prepared by electrophoresis at about pH 3.5. The peptides are exposed to performic acid vapor between the two separations, which oxidizes the S atom of the modified Cys residues to the sulfone:

$$-CH_2-S-CH_2-CO_2^- \quad \xrightarrow{[O]} \quad -CH_2-\overset{\overset{O}{\|}}{\underset{\underset{O}{\|}}{S}}-CH_2-CO_2^- \qquad (7.76)$$

Carboxymethyl-Cys

This modification decreases the pK_a value of the carboxyl group, so peptides containing carboxymethyl-Cys residues are somewhat more acidic in the second dimension; they lie to one side of the diagonal of unmodified peptides (Figure 7-15-A). Alternatively, the thiol groups can be reacted initially with *N*-ethylmaleimide (Equation 7.30); the introduced *N*-ethyl succinimide group can then be hydrolyzed to the succinamic acid by treatment with ammonia vapor after the first separation.

Analogous techniques have been developed for a few other residues, using appropriate chemical modifications specific for those residues.

(1) The amino groups of **Lys**-containing peptides are initially blocked with trifluoracetyl or maleyl groups by treatment with the corresponding anhydride, which introduces acidic groups (Equation 7.12). These groups are easily removed after the first electrophoresis by acidification, changing the electrophoretic mobilities of these peptides in the second dimension.

(2) The S atoms of **Met** residues are first alkylated with iodoacetamide to produce the charged sulfonium derivative (Equation 7.70). After the first electrophoresis, heating causes the polypeptide chain to be cleaved at this derivative; the reaction is analogous to that involved in cleavage by cyanogen bromide (Equation 7.72).

(3) The side-chains of **His** residues are first blocked by reaction with fluorodinitrobenzene; after the first electrophoresis, they are regenerated by exposure to a volatile thiol-containing reagent, such as β-mercaptoethanol, which removes the dinitrophenyl groups.

(4) The guanido groups of **Arg** residues are first blocked by reaction with cyclohexanedione (Equation 7.8); after the first electrophoresis, they are regenerated at alkaline pH.

(5) **Trp** residues can be modified after the first separation by reaction with *o*-nitrophenylsulfenyl chloride.

A diagonal procedure for isolating sulfhydryl peptides alkylated with *N*-ethylmaleimide. H. Gehring & P. Christen (1980) *Anal. Biochem.* **107**, 358–361.

---

2. Identifying Disulfide Bonds

Diagonal techniques were introduced initially to determine which pairs of Cys residues in a protein are linked by disulfide bonds. Using conditions where the disulfide bonds are stable and cannot rearrange or be reduced, the protein is cleaved into peptides that are separated in the first dimension. Any peptides linked by one or more disulfide bonds migrate together (Figure 7-15-B). Before the second separation, the disulfide bonds are cleaved by exposure to performic acid, which converts each of the Cys residues to cysteic acid:

$$R^1-S-S-R^2 \xrightarrow{[HCO_3H]} R^1-SO_3^- + R^2-SO_3^- \tag{7.77}$$

The peptides that were linked by disulfide bonds originally are now independent and more acidic. They migrate differently in the second dimension and consequently lie off the diagonal of unmodified peptides. Their common mobility in the first dimension indicates which were linked by disulfide bonds.

Detection and mapping of widespread intermolecular protein disulfide formation during cardiac oxidative stress using proteomics with diagonal electrophoresis. J. P. Brennan *et al.* (2004) *J. Biol. Chem.* **279**, 41352–41360.

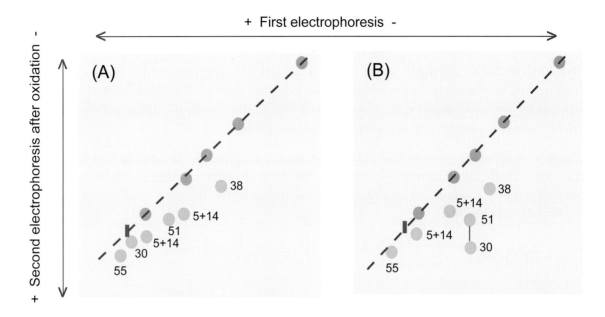

Figure 7-15. Diagonal electrophoresis and the isolation and identification of Cys residues modified by reaction with iodoacetate and linked by disulfide bonds. The proteins were digested with trypsin, followed by chymotrypsin, and the resulting peptides were separated by paper electrophoresis at pH 3.5 in the horizontal direction (anode at *left*); the *dark blue rectangle* marks the origin. After exposure to performic acid (HCO_3H) vapor, the electrophoresis was repeated in the vertical direction (anode at the *bottom*). The peptides were visualized by staining with ninhydrin. The diagonal indicated by the *dashed line* is defined by peptides that were not altered by the performic acid and consequently had the same mobility in both dimensions; many of these peptides have migrated off the paper at the *upper right*. (A) Reduced bovine pancreatic trypsin inhibitor (BPTI) in which the six Cys residues (5, 14, 30, 38, 51, 55) had been blocked by reaction with iodoacetate. The six peptides containing these residues define a second diagonal, below the first, because they were slightly more acidic in the second dimension due to their oxidation to the sulfone. The peptide containing both Cys5 and Cys14 is present in two forms due to partial cleavage. (B) The major one-disulfide intermediate in the BPTI disulfide folding pathway (Figure 11-26) that had been trapped by reacting all four free Cys thiol groups with iodoacetate. The two peptides containing Cys30 and Cys51 are absent from the second diagonal of modified Cys residues and had the same, more rapid mobility in the first dimension, indicating that they were originally linked by an intramolecular disulfide bond between these two Cys residues. The absence of these two peptides from the second diagonal indicates that essentially all of the molecules had this disulfide bond. Data from T. E. Creighton (1974) *J. Mol. Biol.* **87**, 603–624.

7.5.E. Sequencing

The procedures described here are capable of determining the entire amino acid sequence of most proteins. Alternatively, the amino acid sequence of just part of the protein is determined and then used to design an oligonucleotide that should be complementary in nucleotide sequence to one part of the sequence of the gene that encodes the protein. The gene is usually identified and cloned by its hybridization to such an oligonucleotide probe (Section 5.3.C). Protein sequencing techniques are also necessary to determine what post-translational modifications occur to the initial polypeptide chain generated from its gene sequence.

The ABC's (and XYZ's) of peptide sequencing. H. Steen & M. Mann (2004) *Nature Rev. Mol. Cell Biol.* **5**, 699–711.

1. Amino-terminal and Carboxyl-terminal Residues

Most methods for identifying N- and C-terminal residues rely upon the terminal amino or carboxyl group having normal chemical properties and not being blocked covalently. N-terminal pyrrolidone carboxylate groups generated spontaneously from Gln residues at the N-terminus (Equation 7.25) can be removed by the enzyme **pyroglutamyl aminopeptidase**.

Chemical methods to identify the N-terminal residue of a protein or peptide involve labeling chemically all the amino groups, hydrolyzing the peptide to the constituent amino acids, then determining which amino acid is labeled on the α-amino group. The most widely used procedure employs **dansyl chloride**:

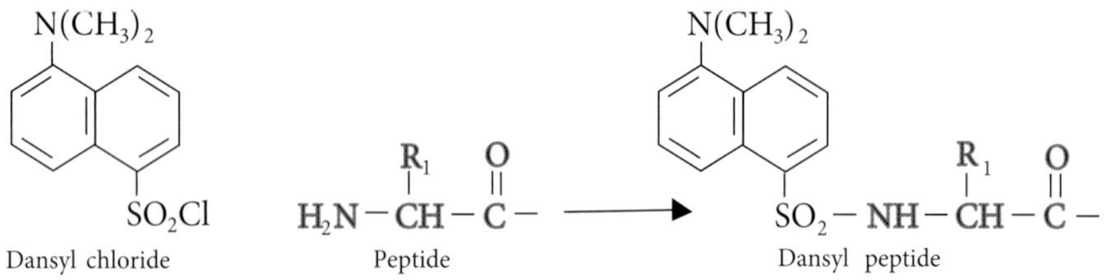

$$\text{Dansyl chloride} + \text{Peptide} \longrightarrow \text{Dansyl peptide} \tag{7.78}$$

The product of reaction with the α-amino group withstands hydrolysis of the peptide to the amino acids. The labeled amino acids are detected by the fluorescence of the dansyl derivative and can be identified by chromatographic techniques in which each amino acid derivative has a characteristic mobility.

The α-amino groups are reacted with **cyanate** in another procedure. Upon acid hydrolysis of the peptide to amino acids, the N-terminal residue is converted to the hydantoin:

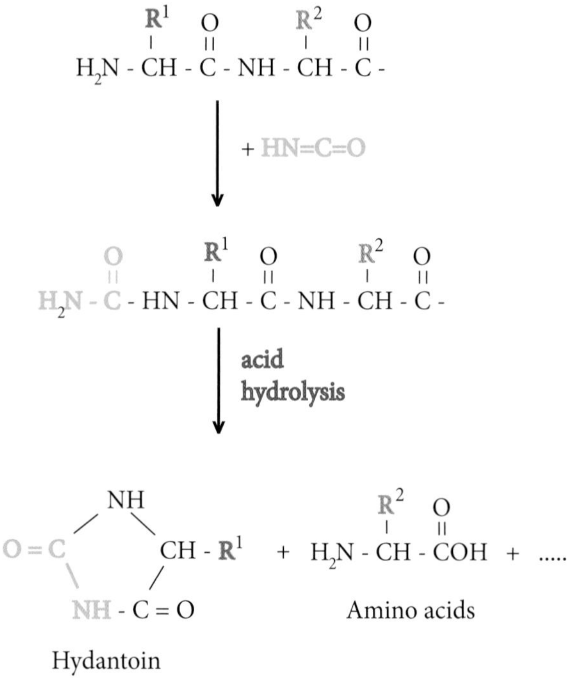

$$\tag{7.79}$$

The only amino acid present as the hydantoin is the one that was at the N-terminus.

C-terminal residues can be identified as described in Section 7.5.E.3.

Short sequences of residues at either the amino or carboxyl ends of peptides can often be determined by using **exopeptidases** that liberate single amino acids sequentially from the ends of the chain; examples are **leucine aminopeptidase** and **carboxypeptidase**. The second residue can be released only after the first has been removed, and so forth, so the order of their release gives the order of the amino acids in the sequence. The amino acids liberated are determined by conventional amino acid analysis as a function of time, and the order of the residues is determined only if they are removed sequentially with different time–courses. This does not always occur, however, because the various amino acids are removed at varying rates. For example, if the second residue can be cleaved by the exopeptidase much more rapidly than the first, the two will appear to be released simultaneously, at the rate of the first; it will be impossible to determine the relative order of these two amino acids.

C-terminal sequencing of peptide hormones using carboxypeptidase Y and SELDI-TOF mass spectrometry. D. R. Cool & A. Hardiman (2004) *Biotechniques* **36**, 32–34.

C-terminal ladder sequencing of peptides using an alternative nucleophile in carboxypeptidase Y digests. A. Hamberg et al. (2006) *Anal. Biochem.* **357**, 167–172.

---

2. Sequencing from the N-terminus: the Edman Degradation

The **Edman degradation** procedure for determining the sequence of amino acid residues from the N-terminus (Figure 7-16) is one of the most successful and widely used methods in protein chemistry. It removes and identifies one amino acid at a time from the amino end of a peptide and can be repeated many times to determine the sequence of amino acids.

The terminal α-amino group must be free so that it can react with phenylisothiocyanate in alkaline medium to give the phenylthiocarbamyl (PTC) derivative:

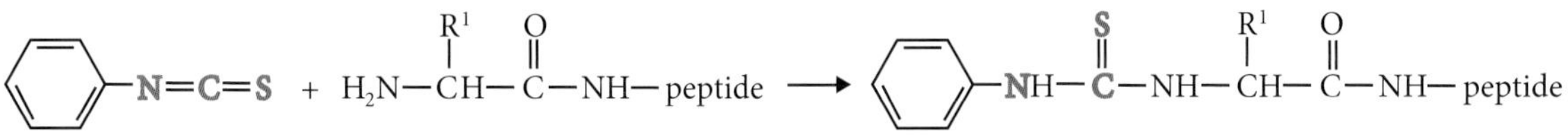

$$\text{Ph-N=C=S} + \text{H}_2\text{N-CH}(R^1)\text{-C(=O)-NH-peptide} \longrightarrow \text{Ph-NH-C(=S)-NH-CH}(R^1)\text{-C(=O)-NH-peptide} \tag{7.80}$$

The PTC group causes the first peptide bond to be relatively unstable, and it is hydrolyzed under acidic conditions where the other peptide bonds are stable:

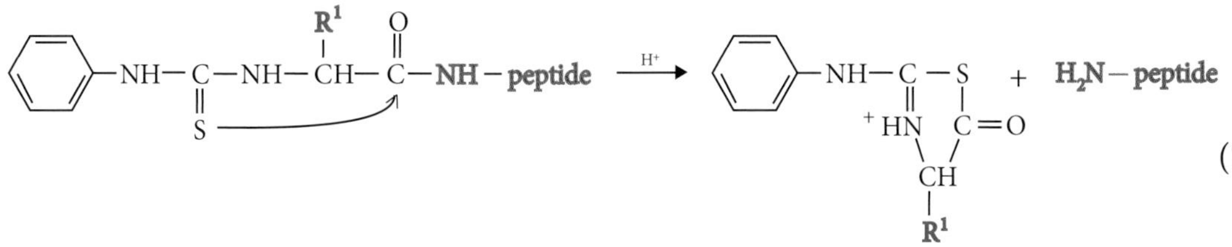

$$\text{Ph-NH-C(=S)-NH-CH}(R^1)\text{-C(=O)-NH-peptide} \xrightarrow{H^+} \text{Ph-NH-C(-S-)(=}^+\text{HN-)} \text{CH}(R^1)\text{-C=O cyclic} + \text{H}_2\text{N-peptide} \tag{7.81}$$

The cyclic derivative of the N-terminal residue that results rearranges in aqueous solution to the phenylthiohydantoin (PTH) derivative:

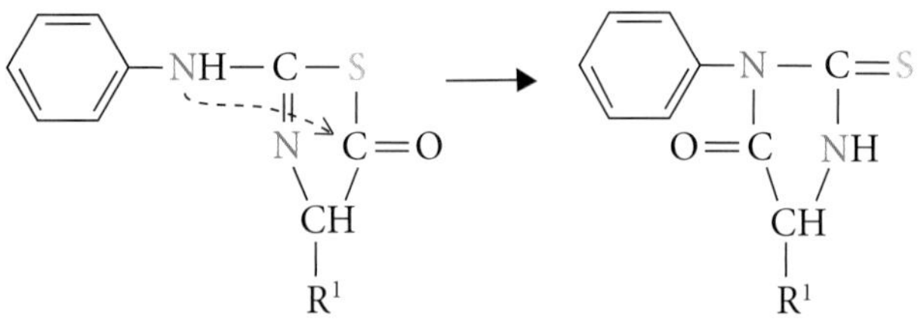

$$\begin{array}{c}\text{Ph-NH-C-S}\\ \text{N-C=O}\\ \text{CH}\\ \text{R}^1\end{array} \longrightarrow \begin{array}{c}\text{Ph-N-C=S}\\ \text{O=C-NH}\\ \text{CH}\\ \text{R}^1\end{array} \qquad (7.82)$$

The amino acid that was at the N-terminus and released as the PTH derivative is identified by chromatographic techniques. Alternatively, the N-terminal residue on the peptide after each cycle of Edman degradation can be determined using dansyl chloride (Equation 7.78).

The great virtue of the Edman method is that it leaves the polypeptide chain altered only by the loss of the N-terminal residue. This shortened peptide is ready for a second cycle of the same procedure, to identify the second residue in the original sequence. This procedure is repeated sequentially to determine the amino acid sequence; generally about 30 residues may be determined in this way, but as many as 70 are possible in some cases.

As with all repetitive procedures, the primary difficulty with the Edman procedure is that errors are cumulative. If at any stage a significant amount of unreacted N-terminal residue remains on the peptide, it will appear in the next cycle, and this error will be propagated in all subsequent steps. Gradually, the amount and diversity of the background material rises until the correct residue released can no longer be identified unambiguously. Another contribution to the background is random breakage of the polypeptide chain under the relatively harsh conditions used. Side-reactions gradually prevent polypeptide molecules from participating in the reaction, and some of the polypeptide is lost by being soluble in some of the organic solvents used for extraction. The latter problem is alleviated in automated instruments by attaching the polypeptide chain covalently to a solid support. There is no ideal method of making the attachment but proteins are often isolated by blotting them electrophoretically onto membranes; they are firmly attached to such membranes and may be subjected directly to sequence analysis.

In gas-phase sequenators, the sample is embedded in a film on the surface of a thin disk. Some of the reagents are delivered as vapor, which minimizes loss of the polypeptide and contamination of the sample. As little as 5–10 picomoles of protein are required for analysis in this way. The efficiency of an average cycle can be as great as 98%, which permits up to 70 residues to be sequenced.

N-terminal protein sequencing for special applications. P. J. Jackson (2003) *Methods Mol. Biol.* **211**, 287–300.

Identification of seven modified amino acids by Edman sequencing. D. Brune et al. (2006) *J. Biomol. Tech.* **17**, 308–326.

Edman sequencing of proteins from 2D gels. S. Komatsu (2007) *Methods Mol. Biol.* **355**, 211–217.

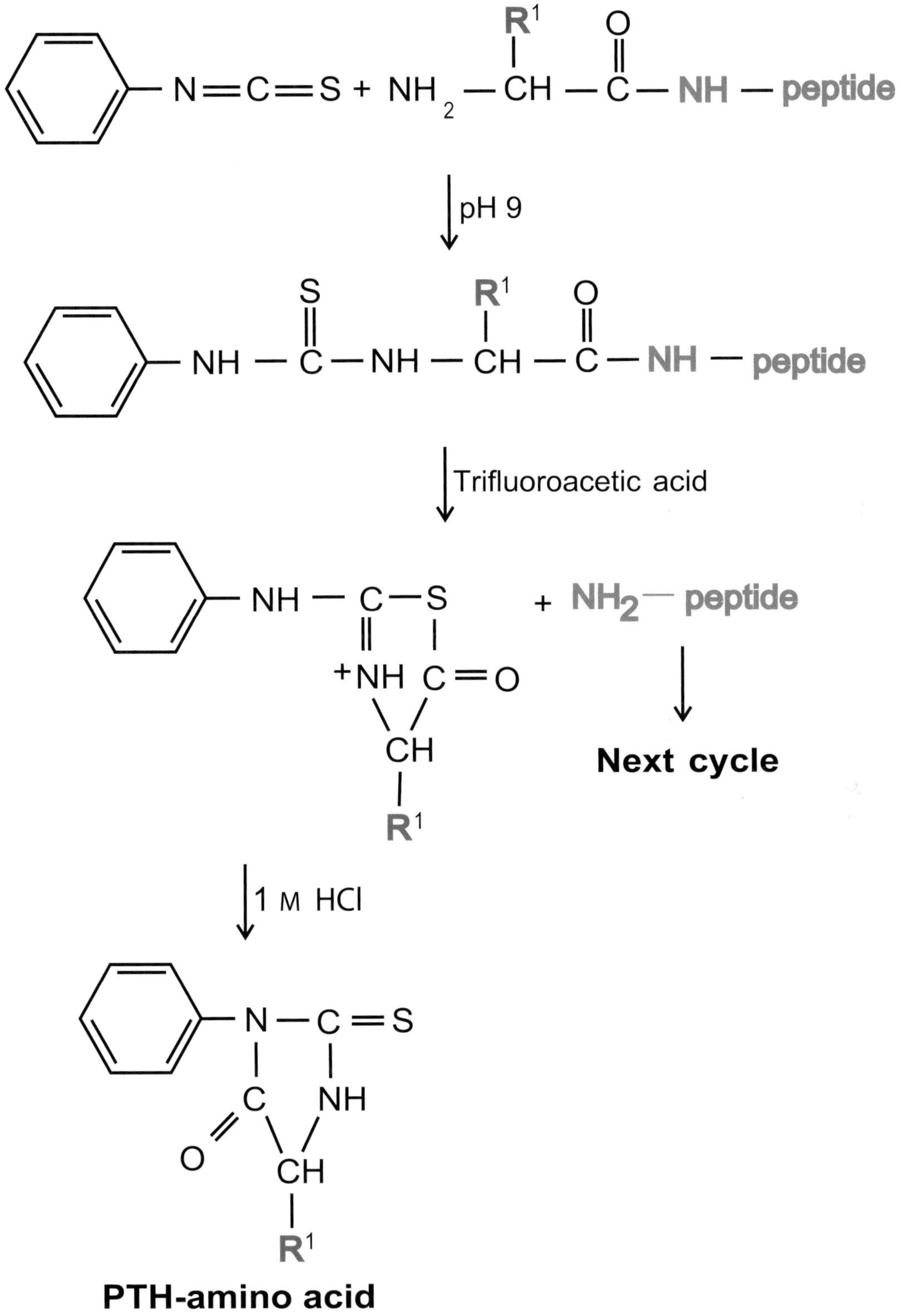

Figure 7-16. One cycle of the Edman degradation procedure for sequencing peptides from the N-terminus. Note that the N-terminal residue is removed as a PTH amino acid derivative, producing the remaining, shortened peptide for determination of the new N-terminal residue in the next cycle. Adapted from T. E. Creighton (1993) *Proteins: structures and molecular properties*, 2nd edn, W. H. Freeman, NY, p. 33.

3. Sequencing from the C-terminus

Developing a method comparable to the Edman degradation for sequencing from the C-terminus was much more difficult, but it is now possible to obtain sequences of a few residues from the C-termini of peptides and proteins (Figure 7-17). The protonated C-terminal carboxyl group reacts with diphenylphosphoroisothiocyanatidate, which then cyclizes to the peptidylthiohydantoin. This

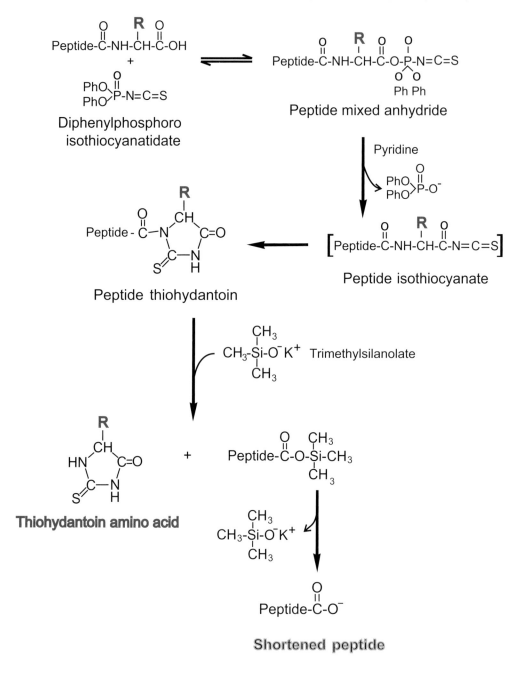

Figure 7-17. Chemical sequencing from the C-terminus of a peptide or protein. The protonated C-terminal carboxyl group reacts with diphenylphosphoroisothiocyanatidate, which cyclizes to the peptide thiohydantoin. Treatment with potassium trimethylsilanolate liberates the thiohydantoin amino acid and leaves a shortened peptide ready for the next cycle.

moiety is cleaved by trimethylsilanolate, to generate a peptide shortened by one residue and ready for the next cycle. The thiohydantoin derivative of the original C-terminal residue is identified chromatographically.

Automated methods for C-terminal protein sequencing. J. M. Bailey & C. G. Miller (2004) *Methods Mol. Biol.* **64**, 259–269.

C-terminal sequencing method for proteins in polyacrylamide gel by the reaction of acetic anhydride. K. Miyasaki & A. Tsugita (2006) *Proteomics* **6**, 2026–2033.

---

4. Sequencing by Mass Spectrometry

Mass spectrometry (MS) is now the most powerful technique for determining the covalent structures of proteins. The procedure is not hindered by the presence of blocking groups on the terminal amino and carboxyl groups like the sequencing techniques described above, and it can even identify the blocking groups. It is also able to identify atypical amino acid residues and the various post-translational modifications that often occur in proteins. Consequently, it has been developed into a most useful and powerful technique for sequencing directly peptides of up to 15 residues, for identifying the residues released by the sequencing techniques described above, and for checking the covalent structure deduced by other methods.

In some approaches, fragmentation of the polypeptide chain occurs within the mass spectrometer, and the molecular weights of the fragments produced are used to infer the covalent structure of the original polypeptide chain (Figure 7-18). The fragmentation patterns of a peptide induced by bombardment with electrons or neutral atoms, such as Ar atoms, are potentially very complex, but the peptide backbone tends to fragment primarily at the peptide bond. Other fragmentations also occur, and the side-chains may also be cleaved. Fortunately, the usual fragmentation patterns for

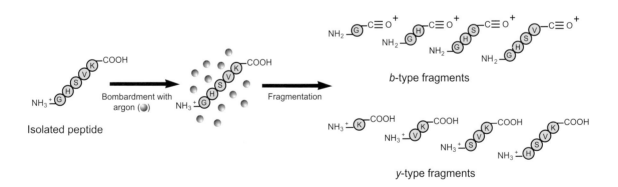

Figure 7-18. Sequence analysis of a peptide by its fragmentation in a tandem mass spectrometer. The ionized peptide is isolated in the first region of the mass spectrometer, followed by collisional dissociation with an inert gas in the second region. The fragment ions produced are separated according to their mass in the third region. Peptide fragmentation produces predominantly the *b*- and *y*-type ions depicted, although others are possible.

each type of residue are known, so all the ion fragments produced with a peptide may be used to determine its sequence. In **tandem mass spectrometry (MS/MS),** a second spectrometer is used on a selected ion from the first spectrometer to analyze the fragmentation products produced within the spectrometer by electron or atom bombardment.

Another biopolymer sequencing method, **ladder sequencing**, has been developed for both proteins and oligonucleotides. This method uses MALDI-MS in combination with enzymatic or chemical digestion to generate sequence-specific ladders of proteins and oligonucleotides. For example, ladder sequencing of a protein involves carrying out the Edman degradation sequentially (Section 7.5.E.2) but with only partial blocking at each step, so that a fraction of the molecules becomes impervious to further degradation. The final result of a number of cycles is a **protein ladder** of all the forms of the polypeptide chain lacking 0, 1, 2, 3, … amino acid residues from the N-terminus. The mixture of peptides is analyzed by MALDI-MS (Figure 7-19). The decrease in molecular weight after each cycle of the Edman degradation identifies the amino acid that was removed. In this way, the N-terminal sequence of the polypeptide chain can be 'read off' from the spectrum of masses.

MS techniques are often used in conjunction with the usual techniques for fragmentation of the protein into various smaller peptides. The molecular masses of the spectrum of peptides produced by various methods to fragment the original polypeptide chain (Section 7.5.B) can provide very useful information on how smaller peptides of known sequence are linked together in the primary structure of the intact polypeptide chain. In fact, the spectrum of peptides produced from a protein by a specific fragmentation process, such as digestion by the proteinase trypsin, is so diagnostic of a protein that it can be used to identify an unknown protein if its sequence is already known (Figure 7-20). This is very often the case now, as the sequences of all the genes of many organisms are known, so the primary structures of their gene products are also known. The molecular masses of the peptide fragments from an unknown protein are compared with all those expected from known proteins in the databases of protein sequences; a match usually indicates that the unknown protein is identical with the matched protein in the database.

Mass spectral analysis in proteomics. J. R. Yates (2004) *Ann. Rev. Biophys. Biomolec. Structure* **33**, 297–316.

Challenges in mass spectrometry-based proteomics. J. Reinders *et al.* (2004) *Proteomics* **4**, 3686–3703.

Shotgun protein sequencing by tandem mass spectra assembly. N. Bandeira *et al.* (2004) *Anal. Chem.* **76**, 7221–7233.

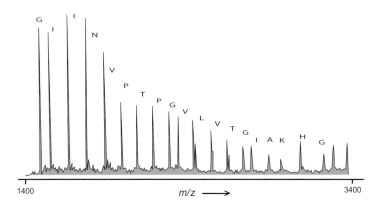

Figure 7-19. Ladder sequencing of proteins by mass spectrometry. Shown is the mass spectrum of a mixture of peptides generated by sequential partial removal of amino acids from the N-terminus of the peptide. The amino acid removed is identified from the consequent shift in mass. Data from B. T. Chait *et al.* (1993) *Science* **262**, 89.

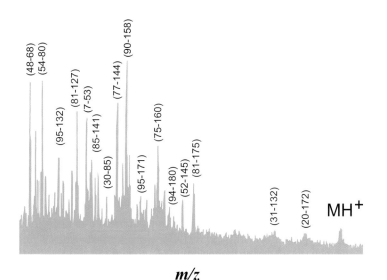

Figure 7-20. Typical example of the identification of a protein from the mass spectrum of peptides produced by digestion of the protein by a proteinase of known specificity. The molecular weights of the peptides produced are diagnostic of their sequence, and searching sequence databases for a match can identify the protein unambiguously.

Protein sequencing by mass analysis of polypeptide ladders after controlled protein hydrolysis. H. Zhong *et al.* (2004) *Nature Biotech.* **22**, 1291–1296.

Application of mass spectrometry in proteomics. I. C. Guerrera & O. Kleiner (2005) *Biosci. Rep.* **25**, 71–93.

Electron capture dissociation mass spectrometry in characterization of peptides and proteins. R. Bakhtiar & Z. Guan (2006) *Biotechnol. Lett.* **28**, 1047–1059.

7.5.F. Protein Sequences from Gene Sequences

The amino acid sequence of a protein is determined in nature by the nucleotide sequence of its **gene**. The DNA corresponding to the gene (Chapter 2) is first transcribed into the complementary **messenger RNA** (Section 6.2.C), and its sequence of nucleotides is read off, three nucleotides (one **codon**) at a time, using the **genetic code** (Figure 7-21) to specify the amino acid sequence of the polypeptide chain.

The region of the gene coding for the polypeptide chain must be identified first (Figure 7-22). Each amino acid is specified by a triplet of nucleotides, so any gene sequence can be translated in three different **reading frames**, which will produce three very different amino acid sequences. The coding region may also be interrupted by **introns**, segments that are removed from the messenger RNA copy of the gene before it is translated. Such sequences have been removed from mature messenger RNA, so it is easiest to obtain a protein sequence from it, usually as a complementary DNA copy (Section 6.2.C).

The coding sequence and correct reading frame can usually be identified as the longest segment lacking the termination codons for cessation of the polypeptide chain (UAA, UAG and UGA), known as an **open reading frame**. Such termination sequences should appear on average every 21 nucleotides in a random sequence, but they appear only at the ends of coding sequences that are much longer. The genetic code is also redundant, in that more than one codon specifies most amino acids, but each organism usually favors certain codons over others within coding regions; this can also be a clue as to the coding region.

First position	Second position				Third position
	U	C	A	G	
U	Phe Phe Leu Leu	Ser Ser Ser Ser	Tyr Tyr Terminate Terminate	Cys Cys Terminate Trp	U C A G
C	Leu Leu Leu Leu	Pro Pro Pro Pro	His His Gln Gln	Arg Arg Arg Arg	U C A G
A	Ile Ile Ile Met	Thr Thr Thr Thr	Asn Asn Lys Lys	Ser Ser Arg Arg	U C A G
G	Val Val Val Val	Ala Ala Ala Ala	Asp Asp Glu Glu	Gly Gly Gly Gly	U C A G

Figure 7-21. The genetic code of protein biosynthesis. The nucleotides of the 64 possible triplet codons are those on the messenger RNA.

The N-terminus of the polypeptide chain is more readily recognized, as it is always a Met residue, which is always encoded by the codon AUG (Figure 7-21). The nucleotide sequence prior to the start codon also contains certain sequences that tell the cell translation machinery to start translating at that point. These sequences depend upon the cell and the gene.

An amino acid sequence obtained from a gene sequence must be considered tentative, as a number of complications can occur during translation of a messenger RNA. A variant genetic code is used in some organelles and species. A specific codon might be translated atypically, as in the case of UGA codons that code for selenocysteine (Section 7.2.O) instead of chain termination (Figure 7-21). The adjacent nucleotides determine which way the codon is translated. A specific protein is also involved in the recognition of a UGA codon coding for selenocysteine. In other cases, some messenger RNAs are 'edited' by insertion of nucleotides that were not present in the gene sequence. In a few other cases, a nucleotide can be skipped during translation, shifting the reading frame. Fortunately, such complications are rare.

1. Post-translational Modifications

The polypeptide chain that results from translation of the gene sequence is often modified by various post-translational alterations, and a great variety of modified amino acid residues are known. Many of these arise by spontaneous chemical reactions and occur in only a few copies of each protein. Much more important are those that occur specifically and in virtually all molecules of the protein; they are invariably produced by specific enzymes. Whether or not such a modification occurs depends in part upon whether or not the protein comes into contact with the modifying enzymes. Most of

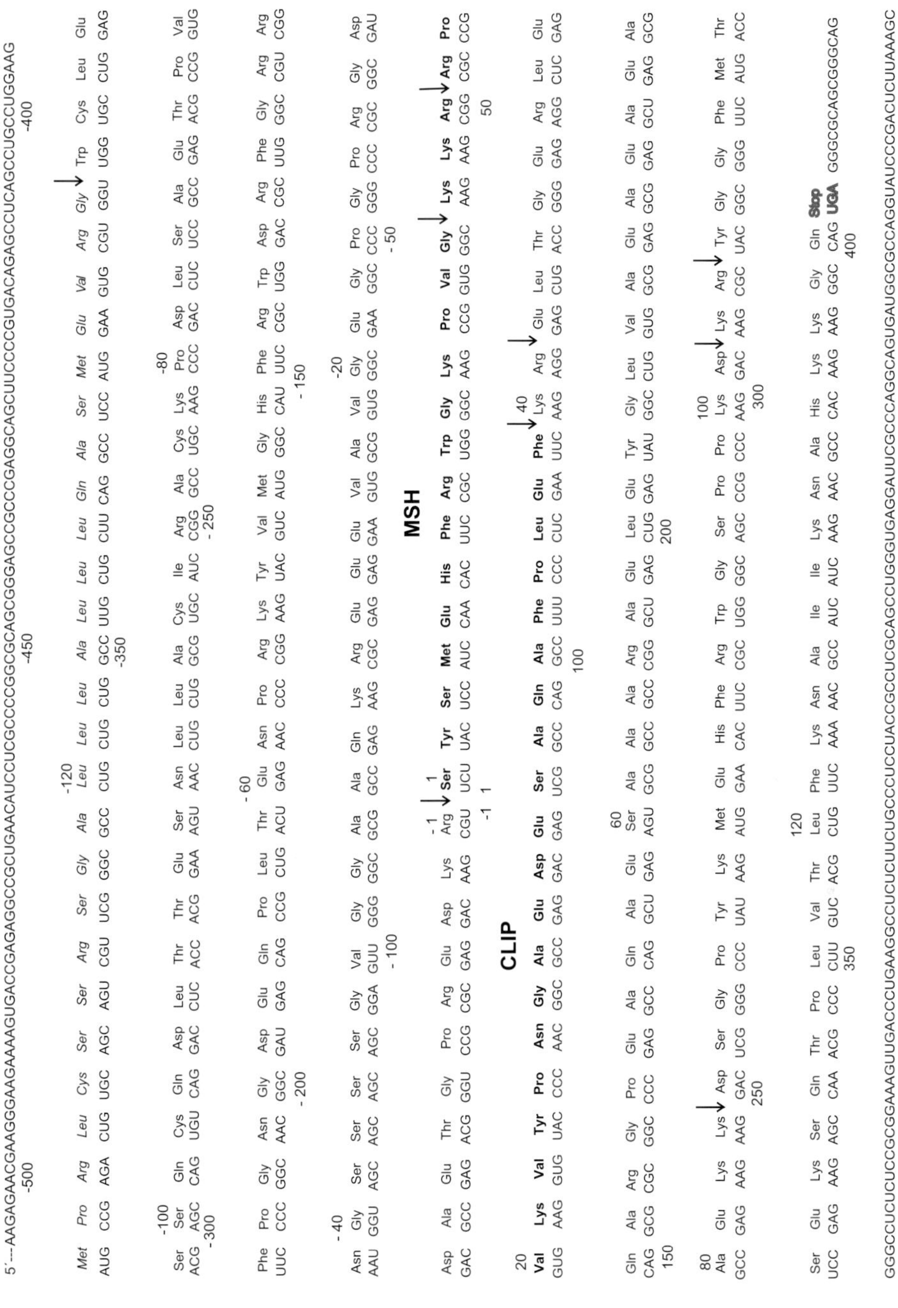

Figure 7-22. Example of determining the amino acid sequence of a protein from the nucleotide sequence of its messenger RNA (mRNA), or complementary DNA (cDNA) copy. The mRNA sequence was determined from the sequence of cloned cDNA for the hormone precursor pre-pro-opiomelanocortin. The inferred sequence of the protein synthesized from the mRNA is indicated above the nucleotide sequence, starting at the initiating Met residue. The numbering systems for both the nucleotides (*bottom*) and amino acid residues (*top*) begin at the N-terminus of the mature hormone, adrenocorticotropic hormone (*bold*); preceding nucleotides and residues have negative numbers. The first 26 residues of the precursor protein comprise the signal peptide (*italics*) that directs the polypeptide chain into the endoplasmic reticulum. The mature hormone is released from the precursor polypeptide chain by proteolytic cleavages at the *arrows* and it is further cleaved into melanocyte-stimulating hormone (MSH) and corticotropin-like intermediate lobe peptide (CLIP). As in this case, processing of peptide hormone precursors usually occurs at sequential Lys and Arg residues. At the 3′-end of the mRNA is a portion of the usual poly-A tail.

Table 7-10. Cellular sites of major post-translational modifications of proteins

Cytoplasm
- Removal of initiating Met residue
- Acetylation of α-NH_2 groups
- Myristoylation of α-NH_2
- O–glycosylation with GlcNAc
- Addition of palmitoyl groups
- Proteolytic processing of virus polyprotein

Endoplasmic reticulum
- Removal of signal peptide
- Core glycosylation of Asn residues
- Addition of palmitoyl and glycosyl phosphatidylinositol groups
- Carboxylation of Glu residues
- Hydroxylation of Pro and Lys residues of procollagen
- Disulfide bond formation

Golgi apparatus
- Modification of N-glycosyl groups
- O-Glycosylation with GalNAc
- Sulfation of Tyr residues

Secretory vesicles and granules
- Amidation of α-CO_2H group
- Proteolytic processing of some precursors

Mitochondria, chloroplasts
- Removal of signal peptide

them are localized in specific parts of the cell (Table 7-10). These covalent modifications must usually be identified by chemical analysis of the mature protein, although the occurrence of some can be predicted from just the local amino acid sequence.

The Met residue that initiates translation and occurs at the N-terminus of the nascent polypeptide chain is often removed by a specific enzyme, depending upon the nature of the second residue; large, hydrophobic or charged residues inhibit its removal. Polypeptide chains are often modified by removing stretches of amino acid residues by the action of proteolytic enzymes (Sections 7.5.B.1 and 15.2.D). An extreme example is illustrated in Figure 7-22. Most frequent is the removal of the **signal peptide** at the N-terminus that causes the nascent **pre**-polypeptide chain to be extruded into the endoplasmic reticulum or other membrane of a cell, and thus to be exported from the cytoplasm. Many proteins, known as **pro**-proteins, are activated by proteolytic cleavage, and this is an important mechanism of ensuring that they are active only under the right circumstances (Section 15.2.D).

The N-terminal amino group is often modified by **acetylation**, or addition of the **myristoyl** fatty acid in the case of N-terminal Gly residues:

$$H_3C-\overset{O}{\underset{\|}{C}}-NH- \qquad H_3C-(CH_2)_{12}-\overset{O}{\underset{\|}{C}}-NH-$$

Acetylated   Myristoylated   (7.83)

The terminal carboxyl group of a polypeptide chain can be linked to very complex **glycosylphosphatidylinositols**, which anchor the protein in a membrane (Section 9.2). C-terminal Cys residues can have comparable farnesyl and geranylgeranyl groups attached to the side-chain thiol group. In a similar way, palmitoyl groups can be attached to the side-chains of internal Cys residues. A **C-terminal amide** group is found at the C-termini of many peptide hormones; this group is the result of modifying a Gly residue that was originally at the C-terminus.

Glycosylation is one of the most frequent post-translational modifications. The attachment of carbohydrates is a very complex process that uses at least eight different sugar monomers, and they are linked together in a variety of covalent bonds. Glycosylation of Asn residues is known as ***N*-glycosylation**, as the sugar is attached to the N atom of the Asn side-chain. Asn residues that are glycosylated always occur in the sequence Asn–Xaa–Ser, Asn–Xaa–Thr or Asn–Xaa–Cys, where Xaa is any amino acid residue other than Pro. Consequently, N-glycosylation sites can usually be identified simply from the amino acid sequence, but only if the protein traverses the endoplasmic reticulum (Table 7-10). The side-chain O atoms of Ser and Thr residues can be subject to ***O*-glycosylation** by the covalent attachment of *N*-acetylgalactosamine groups.

The side-chains of many Pro and Lys residues of the structural protein collagen (Figure 8-16) have **hydroxyl groups** added (Equations 8.10 and 8.11). These modifications are specific to collagens and are important for stabilizing their triple-helix structures (Section 8.5.C). Proteins involved in blood clotting have **carboxyl groups** added to the side-chains of some of their Glu residues, to improve their affinities for Ca^{2+} ions (Section 12.3.B)

A physiologically most important post-translational modification is the **phosphorylation** of the hydroxyl groups of Ser, Thr and Tyr side-chains. In contrast to most other modifications, phosphorylation is reversible, and it is an important regulator of the activities of many proteins (Section 15.2.A).

Proteins that are exported from the cytoplasm very often form **disulfide bonds** between pairs of their Cys residues (Section 7.2.I.5). Which pairs of Cys residues are linked in this way depends upon which are brought into the appropriate proximity by the folding of the polypeptide chain. Disulfide bonds are more stable outside the relatively reducing environment of the cytosol, and they then provide stabilization to the folded conformation. Although disulfide bonds can be generated chemically (Section 11.4.C), they are introduced by protein catalysts *in vivo*.

The characterization of protein post-translational modifications by mass spectrometry. R. E. Schweppe *et al.* (2003) *Acc. Chem. Res.* **36**, 453–461.

A newly discovered post-translational modification: the acetylation of serine and threonine residues. S. Mikherjee *et al.* (2007) *Trends Biochem. Sci.* **32**, 210–216.

Modificomics: posttranslational modifications beyond protein phosphorylation and glycosylation. J. Reinders & A. Sickmann (2007) *Biomol. Eng.* **24**, 169–177.

7.6. PRIMARY STRUCTURES OF NATURAL PROTEINS: EVOLUTION AT THE MOLECULAR LEVEL

Natural proteins vary enormously in both the lengths and amino acid sequences of their polypeptide chains. They may be as long as 25,000 amino acid residues, or as short as 50, but **most protein polypeptide chains contain between 200 and 500 residues**. The average molecular weights of proteins from prokaryotic and eukaryotic cells have been estimated to be 24,000 and 31,700, respectively, which correspond to about 212 and 280 amino acid residues. These averages are weighted according to the abundance of the various proteins. Estimates using individual proteins have given an average length of about 300 residues in prokaryotes and about 450 in eukaryotes.

The number of known protein sequences is very large and growing constantly; compilations of protein sequences are in the form of computer **databases** that are accessible over the World Wide Web. The most widely available are listed in Table 7-11. The **GenBank**, **EMBL** and **DDBJ** databanks contain primarily nucleotide sequences, from which protein sequences are derived. They have virtually the same content because they exchange data daily. The main protein sequence databases are **SWISS-PROT** and the Protein Identification Resource (**PIR**). They are not identical; the entries in SWISS-PROT are especially well-documented, whereas PIR is classified into superfamilies of related proteins. The Protein Research Foundation database (**PRF**) also contains peptide and protein sequences determined directly. Each sequence database is a collection of entries, each with a unique 'accession number' and consisting of a definition of the sequence, its source, literature references, and other text information, followed by a table describing the biological features of the sequence data and the sequence itself.

Although each protein sequence is unique, **the 20 amino acids are used with very similar frequencies in virtually all proteins** (Table 7-1); only small proteins or those with unusual structures differ significantly from the usual composition. Membrane proteins (Section 9.2) have somewhat higher levels of nonpolar amino acids (Leu, Ile, Val, Met, Phe and Tyr) and slightly lower levels of those

Table 7-11. World Wide Web addresses for the major sequence databases

Database	Address
GenBank	www.ncbi.nlm.nih.gov/Genbank
EMBL	www.ebi.ac.uk
DNA Data Bank of Japan	www.ddbj.nig.ac.jp
SWISS-PROT	www.expasy.ch
Protein Information Resource (PIR)	pir.georgetown.edu
Protein Research Foundation (PRF)	www.prf.or.jp

normally ionized (Lys, Arg, Glu, Asp and His). The most atypical compositions are usually those of fibrous proteins, such as collagen, with its preponderance of Gly and Pro residues, which have regular conformations and repetitive amino acid sequences (Section 8.5).

With 20 different amino acid residues possible at each of 200–500 positions, **the number of amino acid sequences possible is astronomical** (20^n for a polypeptide chain of n residues). It is thought that approximately 1% of random amino acid sequences could produce a protein that would have structural properties like those found to be necessary for biological function (Chapter 9). Consequently, an immense number of different proteins might be possible, a number too large to comprehend, and even single copies of all such proteins could not coexist in this universe.

The lengths of the polypeptide chains and the distributions of amino acids within known sequences of nonfibrous proteins appear to be nearly random, with no obvious signs of nonrandomness. This is an initial hint that the principles of protein structure and function are neither simple nor universal.

The primary structure of a protein may appear to be random, but it certainly is not. First, it reflects the evolutionary history of the protein and its gene. So many different protein sequences are possible that **whenever two proteins are found to have amino acid sequences that are significantly similar, this is not a coincidence, and they undoubtedly arose from a common ancestor during evolution** and had the same sequence then. Over many years each sequence has been altered by mutations and refined by evolutionary selection. Secondly, the amino acid sequence determines all aspects of the structure and function of each protein.

Simulating protein evolution in sequence and structure space. Y. Xia & M. Levitt (2004) *Curr. Opin. Struct. Biol.* **14**, 202–207.

Converging on a general model of protein evolution. J. T. Herbeck & D. P. Wall (2005) *Trends Biotechnol.* **23**, 485-487.

Darwinian evolution can follow only very few mutational paths to fitter proteins. Weinreich D.M. et al. (2006) *Science.* **312**, 111-114.

Mechanisms of protein evolution and their application to protein engineering. M. E. Glasner *et al.* (2007) *Adv. Enzymol.* **75**, 193–239.

7.6.A. Homologous Genes and Proteins

The present-day sequences of nucleic acids and proteins, and their three-dimensional (3-D) structures, reflect their evolutionary origin and that of the species in which they occur. **Similar amino acid or nucleic acid sequences imply evolutionary descent from a common ancestor** and it is implausible to imagine that similar long sequences could have arisen by any mechanism other than evolutionary divergence. Consequently, nucleic acid and protein sequences are extremely useful in reconstructing evolution (Figure 7-23) and evolutionary considerations are equally as useful in understanding protein structure and function. The probable amino acid sequence of a protein at any point along the phylogenetic tree, at any time during evolution, can be inferred from the present-day sequences, including the amino acid changes that must have taken place.

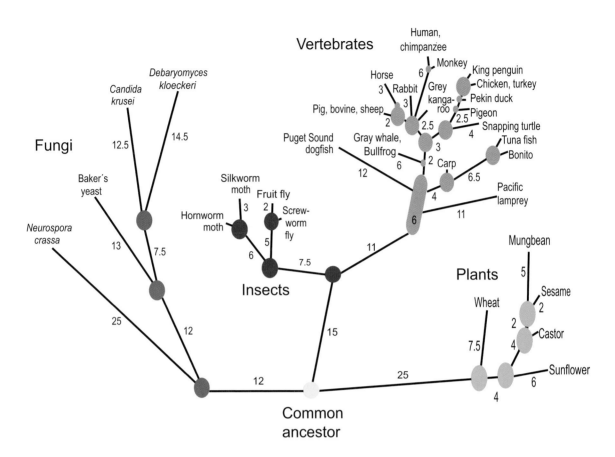

Figure 7-23. Phylogenetic tree constructed from the sequences of cytochromes *c* by minimizing the number of amino acid residue changes that occurred throughout the tree. The length of each branch is proportional to the indicated number of changes of amino acid residues that would have been necessary. The branch points within the *elongated oval* were not adequately defined by the sequences. Data from M. O. Dayhoff (1972) *Atlas of Protein Sequence and Structure*, National Biomedical Research Foundation, DC.

Nucleic acid or protein sequences that have evolved by divergence from a common ancestor are said to be **homologous**. Homologous proteins and nucleic acids start out identical within a population of interbreeding individuals, until the population splits into two and becomes two distinct species that no longer interbreed. After that, the populations, with their genes and proteins, become independent of each other and gradually diverge by the evolutionary processes of mutation and selection. Depending primarily upon the time since they diverged, homologous sequences can be identical, similar to varying extents, or even not recognizably similar if the similarities have disappeared due to extensive divergence. Homology almost invariably extends to the level of the 3-D structures of the proteins, and such structures are almost always better conserved than the sequence (Section 9.3).

Proteins and genes are either homologous or not. It is not strictly correct, although the practice is common, to describe two proteins that were initially identical but now, for example, have sequences only 50% identical, as being '50% homologous'. Instead, this type of 50% identity spread throughout the sequence should be recognized as implying total homology. Only when different parts of a protein or gene sequence have different origins (which happens frequently with large, mosaic proteins comprised of multiple domains; Figure 7-24) is it correct to speak of partial homology. For example, the gene for the low-density lipoprotein (LDL) receptor appears to have obtained 34% of its nucleotides from

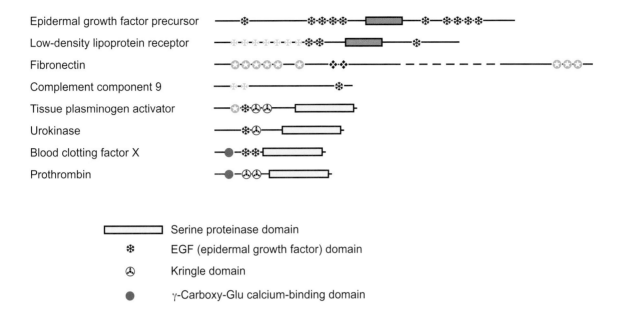

Figure 7-24. Mosaic structures of some vertebrate proteins. Each type of symbol represents members of a set of homologous segments of polypeptide chain, which in these examples are also independent structural units, or domains, of the protein. Those with known structural or function properties are identified at the *bottom*. Adapted from T. E. Creighton (1993) *Proteins: structures and molecular properties*, 2nd edn, W. H. Freeman, NY, p. 133.

an ancestral gene for complement component C9, 48% of its nucleotides from an ancestral gene for epidermal growth factor (EGF) and 18% from elsewhere. Consequently, the LDL receptor is truly 34% homologous to C9 and 48% homologous to EGF.

The only alternative explanation to divergence to explain similarities in sequence or structure is evolutionary **convergence**, where two or more initially unrelated sequences have become similar for functional reasons, under the pressure of evolutionary selection. Convergence is a fairly common evolutionary phenomenon at the macroscopic level, and may be encountered at the level of protein 3-D structure (Section 9.3), but **there are no instances where the sequences of nucleic acids or proteins have been shown conclusively to have become substantially similar by convergence**.

The evolutionary aspects of protein structure are extremely powerful in understanding any protein. The very first step in characterizing a new gene or protein is to find out with what others it is homologous. This provides invaluable clues regarding the origin and function of the new gene or protein.

On the conservation of protein sequences in evolution. B. Kisters-Wolke *et al.* (2001) *Trends Biochem. Sci.* **25**, 419–421.

Genome and protein evolution in eukaryotes. R. R. Copley *et al.* (2002) *Curr. Opin. Chem. Biol.* **6**, 39–45.

Gene family phylogenetics: tracing protein evolution on trees. J. Thornton (2002) *EXS* **92**, 191–207.

An integrated view of protein evolution. C. Pal *et al.* (2006) *Nature Rev. Genet.* **7**, 337–348.

1. Detecting Sequence Homology

Aligning two or more sequences is an essential step in determining whether they are truly homologous, and numerous procedures and computer programs are available for this purpose. Most such programs are designed to compare a new sequence with all the other known sequences in the sequence databases (Table 7-11). To do this on a reasonable time scale is not a trivial accomplishment, and the most popular programs are exceedingly sophisticated.

Sequences that have most of their residues identical are almost certainly homologous, but only slight similarities present a problem in deciding whether they are significant. With distantly related sequences, where many changes have occurred during their divergence, the number of amino acid identities may be almost insignificant. It is difficult to give simple rules for the degree of similarity necessary to demonstrate unambiguously that two protein sequences are homologous, as this depends upon the lengths of the sequences and their amino acid compositions (Figure 7-25). If all 20 amino acids occurred with the same frequencies, random unrelated sequences would be expected to be identical at only 5% (1/20) of their residues, only somewhat higher (about 6%) when the amino acids occur with their observed variable frequencies (Table 7-1). In addition, however, it is necessary to allow some insertions and deletions (**indels**) of residues when comparing homologous proteins, and overzealous use of indels can make even unrelated sequences become very similar. Nevertheless, for long sequences with typical amino acid compositions and with only limited introduction of insertions and deletions, **an identity of greater than 20% is usually sufficient to indicate homology**.

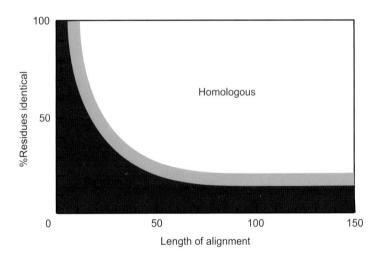

Figure 7-25. Approximate relationship between the fraction of amino acid residues that are identical and the length of the sequence comparison, for deciding whether any sequence similarities indicate homology. For long sequences, more than 20% of the residues must be identical, whereas even identical pentapeptide sequences are not significant. The dividing line between the two areas is not defined precisely.

It is helpful with distantly related proteins to consider also the nature of the amino acid residues and the ease with which they can replace each other by simple genetic mutations that alter one nucleotide of a gene (Figure 7-21). Analysis of closely related proteins demonstrates that similar amino acids replace each other most frequently (Section 7.6.A.4). A '**similarity index**' matrix can be used to weight each alignment of two different amino acid residues. A variety of matrices have been devised for this purpose, but that used most frequently is semi-empirical and based upon what replacements have actually occurred during the evolution of closely related proteins (Figure 7-26). The observed replacements are quite different from what would be expected from just the probability of random mutations of the genetic code (Figure 7-21); instead, they reflect the chemical properties of the amino acid residues and their frequencies of occurrence in proteins (Table 7-1). From all these considerations,

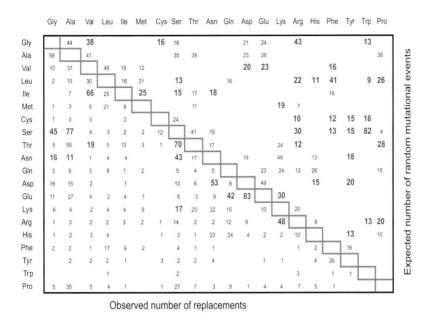

Figure 7-26. Relative frequencies of amino acid replacements in 1572 examples of closely related proteins that are observed (*bottom, left*) and those expected for random single-nucleotide mutations (*top, right*). The greatest discrepancies between the observed and random replacements are shown in *bold*. Replacements involving chemically similar amino acids are generally observed to be much more frequent than expected for random mutations.

a **mutation data** (MD) or **point accepted mutation** (PAM) matrix can be prepared for evaluating the significance of each alignment. One PAM is equivalent to a unit of evolutionary divergence in which 1% of the amino acid residues of a protein have been changed. Pairs of closely related sequences have been used to collect mutation frequencies corresponding to 1 PAM, then extrapolated to a distance of 250 PAMs. One example of a PAM250 matrix is given in Figure 7-27.

To estimate the significance of any similarity between two protein sequences, it should be compared with the values obtained in the same way with randomly permuted versions of one of the sequences in which the exact length and amino acid composition of the protein is preserved. The statistical variation of the random comparisons provides a measure of the significance of any observed similarity.

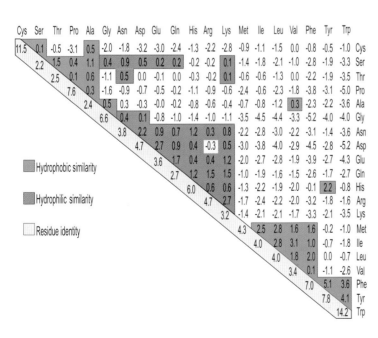

Figure 7-27. The PAM250 score matrix for aligning sequences and conducting homology searches. The value given for each pair of amino acid residues is the natural logarithm of the frequency with which that pair of residues has been changed during the evolution of proteins of known sequence. Negative numbers indicate changes that are disfavored; positive numbers indicate favored exchanges and are *shaded*. The highest residue exchange scores are between bulky aromatic residues (Phe, Tyr, Trp) and are stronger than exact matches of highly mutable residues such as Ser. The greatest mismatches are between bulky hydrophobic residues and small or negatively charged residues. Data from S. A. Benner *et al.* (1994) *Protein Eng.* **7**, 1323–1332.

An effective way of searching for significant similarities between two sequences is to generate a **diagonal dot-plot** comparison of the two (Figure 7-28). A short segment of one sequence (perhaps 1–30 residues) is compared with all possible segments of the same length of the second sequence. All possible segments of the first sequence are compared in this way, to generate a matrix of all segments of one sequence versus all those of the second. The similarities between each pair of segments are scored by the number of identities between the aligned residues, the number of mutations necessary to convert one to the other, or weighting the differences between them. If the similarity between any two segments is greater than some threshold value, a positive score is registered in the comparison matrix. This is exhibited as a dot when the matrix is displayed with the sequence of one protein aligned vertically, the other horizontally. Alternatively, the magnitude of the score can be entered and only the highest scores displayed (Figure 7-28). Homology is especially evident to the eye as a diagonal run of dots or high scores. One of the advantages of this method is that insertions and deletions of amino acid residues do not interfere with the detection of homologies but merely cause the diagonal to break and shift. Also, all possible comparisons of the sequences are made, so that no preconceptions about the alignment of the sequences are necessary. The significance of any similarities can be judged by comparison with other regions of the dot-plot, and the scoring threshold can be raised or lowered to search for very weak homologies.

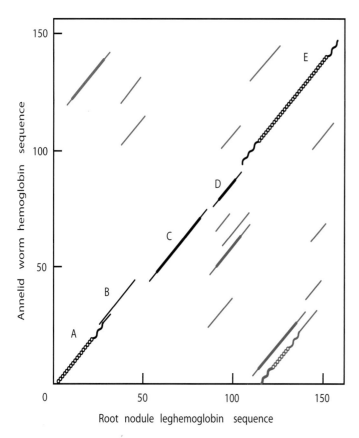

Figure 7-28. Diagonal dot-plot comparison of the amino acid sequences of the distantly related annelid worm hemoglobin and root nodule hemoglobin, which have no more than 20% of their residues identical. Special comparison methods were used to detect this very marginal similarity. The comparison of each alignment calculated the probability with which the various pairs of aligned residues replace each other in other homologous proteins (Figure 7-26) and the similarity in five physical parameters of the amino acid residues. The results from window lengths of between 7 and 25 residues of each protein were overlaid. The most significant similarities found are indicated by the *diagonal lines*, in the following order of increasing significance: *thin lines < thick lines < wavy lines < circles*. The diagonal segments labeled A through E correspond to the five structurally homologous segments that are known from the crystal structures of the two proteins (Figure 9-34). Data from P. Argos (1987) *J. Mol. Biol.* **193**, 385–396.

Multi-species sequence comparison: the next frontier in genome annotation. I. Dubchak & K. Frazer (2003) *Genome Biol.* **4**, 122.

The limits of protein sequence comparison? W. R. Pearson & M. L. Sierk (2005) *Curr. Opinion Struct. Biol.* **15**, 254–260.

Homology assessment and molecular sequence alignment. A. J. Phillips (2006) *J. Biomed. Inform.* **39**, 18–33.

2. Aligning Homologous Sequences

Aligning two or more homologous sequences is straightforward only if they are very similar. The criterion generally used is to minimize the number of genetic events needed to account for the differences between the sequences, the principle of maximum parsimony or minimal mutations. The assumption that this will be the correct alignment is most valid when there have been the fewest mutational events. With increasing numbers of mutations, the possibility of multiple mutations at some sites becomes significant, and reverse mutations might have occurred, which will not be apparent from the contemporary sequences. Consequently, **the greater the number of differences between two sequences, the greater the probability that an alignment of their sequences is not correct**.

The greatest complication in aligning sequences arises from the need to allow for insertion and deletion of amino acid residues. The introduction of gaps diminishes the significance of any alignment, so a penalty needs to be assigned to each gap introduced when assessing the significance of the alignment. The possibility of introducing gaps also increases greatly the computation complexity. The widely used database search tools BLAST and FASTA search initially for the highest scoring matched regions without allowing for gaps. They can then examine the most likely alignments by permitting the introduction of gaps.

A pair of sequences is usually aligned by using a 2-D matrix in which every residue in one sequence is scored against every residue in the other. The algorithm starts in the top left corner of the matrix and checks the various choices at each location; it finishes in the lower right corner. At each location, it decides whether to match the two residues and continue aligning from the previously aligned residue pair, or to pay the penalty and insert a one-residue gap into either sequence. Scores for matching residues are taken from matrices such as PAM250 (Figure 7-27). The best local alignment is taken as the highest-scoring continuous positive path through the matrix. The standard Needleman–Wunsch algorithm will find the optimal full-length alignment for a pair of sequences.

Sequence alignments benefit greatly from having multiple sequences that are known to be homologous. A new sequence might have substantial similarities to just a few of the multiple sequences, which would be missed if the new sequence were compared with only a single sequence. Consequently, alignments of multiple sequences tend to be much more accurate. Multiple sequence alignments also indicate which residues are most conserved, and therefore most likely to be important functionally. Aligning more than three or four sequences simultaneously is, however, a daunting computational process, and computer algorithms that are rigorously correct have been impractical to implement.

The sequences of distantly related proteins can be aligned most accurately if the 3-D structures of one or more of the proteins are known (Section 9.3). The 3-D structure of a protein is more conserved during evolution than its primary structure, and it places restrictions on the amino acid changes that can occur. Amino acid residues occupying equivalent positions in the structure are aligned in the sequence.

A study on protein sequence alignment quality. A. Elofsson (2002) *Proteins* **46**, 330–339.

Multiple sequence alignment. R. C. Edgar & S. Batzoglou (2006) *Curr. Opinion Struct. Biol.* **16**, 368–373.

State of the art: refinement of multiple sequence alignments. S. Chakrabarti *et al.* (2006) *BMC Bioinformatics* **7**, 499.

3. Orthologous/Paralogous Genes and Proteins

The most obviously related genes, and their corresponding proteins, are those that occupy the same locations in the genetic material, and their proteins serve the same function, in different species of organisms. For example, it is possible to find comparable hemoglobins, cytochromes *c*, glyceraldehyde phosphate dehydrogenases, DNA polymerases, histones, etc., in a wide variety of organisms and to compare their gene and protein sequences. Excellent examples are the cytochromes *c* from various species, which have very similar sequences and essentially the same function in each species (Figure 7-29). Such genes and proteins are said to be orthologous. They diverged at the same time as the species in which they occur diverged from their common evolutionary ancestor. Therefore, the phylogenetic tree implied by these protein or gene sequences should be the same as that for the species (Figure 7-23).

```
                                 1              10              20              30              40              50
Human and chimpanzee                     GDVEKGKKIFIMKCSQCHTVEKGGKHKTGPNLHGLFGRKTGQAPGYSYTA
Pig, bovine, and sheep                   GDVEKGKKIFVQKCAQCHTVEKGGKHKTGPNLHGLFGRKTGQAPGFSYTD
Grey kangaroo                            GDVEKGKKIFVQKCAQCHTVEKGGKHKTGPNLNGIFGRKTGQAPGFTYTD
Chicken and turkey                       GDIEKGKKIFVQKCSQCHTVEKGGKHKTGPNLHGLFGRKTGQAEGFSYTD
Snapping turtle                          GDVEKGKKIFVQKCAQCHTVEKGGKHKTGPNLNGLIGRKTGQAEGFSYTE
Puget Sound dogfish                      GDVEKGKKVFVQKCAQCHTVENGGKHKTGPNLSGLFGRKTGQAQGFSYTD
Pacific lamprey                          GDVEKGKKVFVQKCSQCHTVEKAGKHKTGPNLSGLFGRKTGQAPGFSYTD
Garden snail                             GZAZKGKKIFTQKCLQCHTVEAGGKHKTGPNLSGLFGRKQGQAPGFAYTD
Screw-worm fly               GVPAGDVEKGKKIFVQRCAQCHTVEAGGKHKVGPNLHGLFGRKTGQAAGFAYTN
Tobacco hornworm moth        GVPAGNADNGKKIFVQRCAQCHTVEAGGKHKVGPNLHGFFGRKTGQAPGFSYSN
Candida krusei             PAPFEQGSAKKGATLFKTRCAQCHTIEAGGPHKVGPNLHGIFSRHSGQAEGYSYTD
Rust fungus                  GFEDGDAKKGARIFKTRCAQCHTLGAGEPNKVGPNLHGLFGRRSGTVEGFSYTD
Rape and cauliflower     ASFDEAPPGNSKAGEKIFKTKCAQCHTVDKGAGHKQGPNLNGLFGRQSGTTAGYSYSA

                                        60              70              80              90             100
Human and chimpanzee                ANKNKGIIWGEDTLMEYLENPKKYIPGTKMIFVGIKKKEERADLIAYLKKATNE
Pig, bovine, and sheep              ANKNKGITWGEETLMEYLENPKKYIPGTKMIFAGIKKKGEREDLIAYLKKATNE
GrEy kangaroo                       ANKNKGIIWGEDTLMEYLENPKKYIPGTKMIFAGIKKKGERADLIAYLKKATNE
Chicken and turkey                  ANKNKGITWGEDTLMEYLENPKKYIPGTKMIFAGIKKKSERVDLIAYLKDATSK
Snapping turtle                     ANKNKGITWGEETLMEYLENPKKYIPGTKMIFAGIKKKAERVDLIAYLKDATSK
Puget Sound dogfish                 ANKSKGITWQQETLRIYLENPKKYIPGTKMIFAGLKKKSERQDLIAYLKKTAAS
Pacific lamprey                     ANKSKGIVWNQETLFVYLENPKKYIPGTKMIFAGIKKEGERKDLIAYLKKSTSE
Garden snail                        ANKGKGITWKNQTLFEYLENPKKYIPGTKMVFAGLKBZTERVDLIAYLZZATKK
Screw-worm fly                      ANKAKGITWQDDTLFEYLENPKKYIPGTKMIFAGLKKPNERGDLIAYLKSATK
Tobacco hornworm moth               ANKAKGITWQDDTLFEYLENPKKYIPGTKMVFAGLKKANERADLIAYLKQATK
Candida krusei                      ANKRAGVEWAEPTMSDYLENPKKYIPGTKMAFGGLKKAKDRNDLVTYMLEASK
Rust fungus                         ANKKAGQVWEEETFLEYLENPKKYIPGTKMAFGGLKKEKDRNDLVTYLREETK
Rape and cauliflower                ANKNKAVEWEEKTLYDYLLNPKKYIPGTKMVFPGLKKPQDRADLIAYLKEATA
```

Figure 7-29. Alignment of the amino acid sequences of some cytochromes *c* from various eukaryotes that are believed to be orthologous. The one-letter code (Figure 7-2) is used; Z is either E or Q (Glu or Gln), B is either D or N (Asp or Asn). Residues that are identical to those in the human protein are in *blue*; chemically similar residues (S,T; I,V,L,F,M; F,Y,W; D,E; K,R) are in *green*.

The more closely related the organisms, the shorter the time since they diverged from a common evolutionary ancestor and the more closely related are the sequences of their orthologous genes and proteins (Figure 7-30). Closely related species have closely related proteins. For example, the cytochromes *c*, hemoglobin α-, β- and γ-chains, and fibrinopeptides of chimpanzees are identical to those of humans, while their myoglobins and hemoglobin δ chains each have one amino acid difference. As less closely related species are compared, the differences between their genes and proteins increase (Figure 7-30). In fact, after their separation, genes appear to acquire differences at a constant rate, which depends upon the protein (Table 7-13). **The number of amino acid differences between two orthologous protein sequences is usually proportional to the distance they are separated on the phylogenetic tree** (Figure 7-23). The differences in the sequences of homologous proteins indicate the degree of flexibility that is possible in these proteins while maintaining the same function.

	Chimpanzee	Sheep	Rattlesnake	Carp	Snail	Moth	Yeast	Cauliflower	Parsnip
Human	0	10	14	18	29	31	44	44	43
Chimpanzee		10	14	18	29	31	44	44	43
Sheep			20	11	24	27	44	46	46
Rattlesnake				26	28	33	47	45	43
Carp					26	26	44	47	46
Garden snail						28	48	51	50
Tobacco hornworm moth							44	44	41
Baker´s yeast (iso-l)								47	47
Cauliflower									13

Figure 7-30. Number of amino acid differences between cytochromes *c* of a few representative species. These proteins all consists of at least 104 amino acid residues, so each pair has at least half their residues identical.

In contrast, genes and proteins that are not genetically and functionally equivalent are said to be **paralogous**; they originated by duplication of a gene into two copies within the same genome (Section 7.6.B.1). With two or more copies of a particular gene within the same genome, they are free to evolve independently and acquire different functions. **Homologous genes and proteins that are present in the same genome must be paralogous**. Excellent examples are the various polypeptide chains of hemoglobins and myoglobin from humans (Figure 7-31). At least some of the amino acid changes that have occurred between paralogous proteins must account for the differences in their functions.

The sequences of orthologous genes can be used to infer the evolutionary tree by which the species arose during evolution, because orthologous genes diverged at the same time as the species in which they occur (Figure 7-23). In contrast, paralogous genes can originate at any time, so **the phylogenetic tree implied by comparison of paralogous sequences will not be the same as that of the species involved**. This is a powerful method for distinguishing between orthologous and paralogous sequences in the absence of other information about their functions.

Tree pattern matching in phylogenetic trees: automatic search for orthologs or paralogs in homologous gene sequence databases. J. F. Dufayard *et al.* (2005) *Bioinformatics* **21**, 2596–2603.

Human and orthologous gene nomenclature. M. W. Wright & E. A. Bruford (2006) *Gene* **369**, 1–6.

Modeling amino acid substitution patterns in orthologous and paralogous genes. G. C. Conant *et al.* (2007) *Mol. Phylogenet. Evol.* **42**, 298–307.

4. *Nature of Amino Acid Sequence Differences*

Closely related proteins generally differ only by replacements of one amino acid by another at a few positions within the polypeptide chain (Figure 7-30). Less frequent are differences in the total number of residues in the polypeptide chain, owing to the deletion or insertion of residues within the polypeptide chain or at either end. With more distant relationships, the number and nature of the sequence differences can vary greatly.

Wide variation of amino acid residues occurs at some positions (Figures 7-29 and 7-31). At others, only amino acids with chemically similar side-chains are found, and no variation is found at certain positions. There obviously have been constraints on the evolutionary changes in sequence that have occurred, but the variation that has occurred is perhaps the more amazing; orthologous proteins have

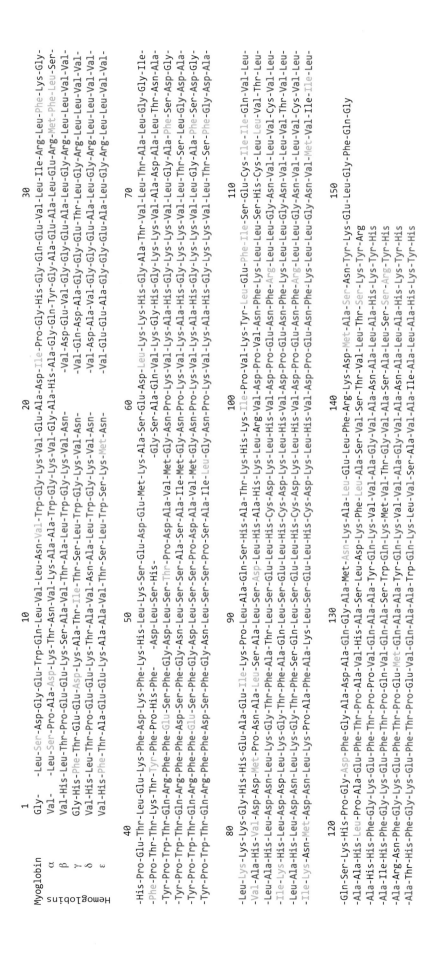

Figure 7-31. Amino acid sequences of members of the human globin family: myoglobin and the hemoglobin polypeptide chains, which are all paralogous. Positions where three or more of the amino acid residues are identical are in *blue*; chemically similar residues are in *green*.

the same function, often with indistinguishable properties, yet their sequences can vary markedly. The greater divergence observed between proteins from distantly related species usually reflects primarily the greater time since the species separated, during which mutational change has been permitted. With paralogous comparisons, the differences tend to be greater, due to the proteins having acquired different properties and functions.

The constraints on change in the amino acid sequences of orthologous proteins are a result of the requirement that these proteins maintain their function. **The more important an amino acid residue, the lower its probability of changing during evolutionary divergence**. Constraints on evolutionary divergence can also be apparent within the gene sequences coding for closely related proteins. Nucleotide changes at the third position of a codon that do not alter the amino acid specified (Figure 7-21) occur more frequently than those that do. Genomic sequences that do not code for protein change more rapidly than those that do. Flanking sequences of a gene that are conserved are usually those involved in regulation of expression of the gene.

In general, the amino acid replacements that occur during protein divergence are very nonrandom, both in the degrees to which different residues change and in the amino acids that replace each other. The most prevalent replacements occur between amino acids with similar side-chains: Gly/Ala, Ala/Ser, Ser/Thr, Ile/Val/Leu, Asp/Glu, Lys/Arg, Tyr/Phe, and so forth (Figure 7-26). The numbers of observed replacements also reflect the frequency with which each amino acid occurs in proteins. Correcting for this gives the normalized values for the relative mutabilities of the different amino acid residues (Table 7-12), which vary more than seven-fold. The acidic and hydrophilic residues are those changed most frequently, whereas the large, hydrophobic residues are replaced least often. The other conserved residues are Gly, Pro and Cys.

This bias by which amino acids are mutated to varying extents and replaced by nonrandom amino acids does not arise from the nature of the genetic code or from the types of mutations that occur. The amino acid replacements expected with random single nucleotide mutations are shown in the upper half of Figure 7-26; this distribution is remarkably different from that observed between homologous proteins, shown in the lower half of the figure.

The bias in amino acid replacements reflects the role of selection in ensuring that only those mutations that do not disrupt the structure and function of the protein are permitted to survive. Consequently, chemically similar replacements are found most frequently. Once the structure and function of a protein are understood, the reasons for the observed variation of the primary structure during evolution usually become clear (Section 9.3).

Detecting compensatory covariation signals in protein evolution using reconstructed ancestral sequences. K. Fukami *et al.* (2002) *J. Mol. Biol.* **319**, 729–743.

Mutual information in protein multiple sequence alignments reveals two classes of coevolving positions. G. B. Gloor *et al.* (2005) *Biochemistry* **44**, 7156–7165.

Protein evolution: causes of trends in amino-acid gain and loss. L. D. Hurst *et al.* (2006) *Nature* **442**, E11–12.

An empirical codon model for protein sequence evolution. C. Kosiol *et al.* (2007) *Mol. Biol. Evol.* **24**, 1464–1479.

Table 7-12. Relative variabilities of amino acid residues during divergence of homologous proteins

Residue	Variability[a]
Asn	100
Ser	90
Asp	79
Glu	76
Ala	75
Thr	72
Ile	72
Met	70
Gln	69
Val	55
His	49
Arg	49
Lys	42
Pro	42
Gly	37
Tyr	31
Phe	31
Leu	30
Cys	15
Trp	13

[a] The number of times that a given amino acid residue changed during the evolution of various proteins was divided by the number of times that the residue occurred. These values were then normalized by setting the largest one to 100. Data from M. O. Dayhoff (1978) *Atlas of Protein Sequence and Structure*, Vol. 5, Suppl. 3, National Biomedical Research Foundation, DC.

5. Rates of Divergence

Different proteins have evolved at different rates (Table 7-13). For example, humans and rhesus monkeys differ in 1% of the residues of their cytochromes *c* but in 3–5% of the residues of their hemoglobin α- and β-polypeptide chains and in 30% of the residues of their fibrinopeptides. A similar ordering of the relative degrees of variation of these proteins is observed when other species are compared.

Table 7-13. Rates of evolutionary change of some proteins

Protein	Accepted point mutations/100 residues/10^8 years
Histones	
H4	0.25
H3	0.30
H2A and H2B	1.7
H1	12
Fibrous proteins	
Collagen (α-1)	2.8
Crystallin (αA)	4.5
Intracellular enzymes	
Glutamate dehydrogenase	1.8
Triosephosphate isomerase	5.3
Lactate dehydrogenase M	7.7
Electron carriers	
Cytochrome c	6.7
Plastocyanin	14
Ferredoxin	17
Hormones	
Glucagon	2.3
Corticotropin	4.2
Insulin A and B chains	7.1
Lutropin α-chain	14
Prolactin	20
Growth hormone	25
Lutropin β-chain	33
Insulin C peptide	53
Oxygen-binding proteins	
Myoglobin	17
Hemoglobin β-chain	30
Secreted enzymes	
Trypsinogen	17

Table 7-13. Rates of evolutionary change of some proteins - continued

Protein	Accepted point mutations/100 residues/10^8 years
Ribonuclease A	43
Immunoglobulins	
κ chains (V region)	100
κ chains (C region)	111
γ-chains (V region)	143
Snake venom toxins	
Long neurotoxins	111
Short neurotoxins	125
Other proteins	
Parvalbumin	20
Albumin	33
α-Lactalbumin	43
Fibrinopeptide A	59
Casein κ chain	71
Fibrinopeptide B	91

Data from A. C. Wilson *et al.* (1977) *Ann. Rev. Biochem.* **46**, 573–639.

The rate of evolutionary change of any protein or gene appears to serve as an **evolutionary clock** (Figure 7-32), with the clock for each gene or protein ticking at a particular rate in all species in which it occurs. This makes amino acid sequences extremely valuable for reconstructing the process of evolution. **The number of differences depends upon the time since divergence, and it is generally independent of the evolutionary path the gene or protein followed**. This is evident from the similarities in the number of differences when comparing proteins from animals with those from other living organisms, such as plants. The number of differences is very similar irrespective of which animals or plants are compared (Figure 7-30).

The greatest exceptions to constant rates of divergence are those instances where changes in the function of the protein have occurred. For example, the protein α-crystallin has evolved at an anomalously high rate in the blind mole rat. This protein is a major structural component of the eye lens, and its function was lost when the blind mole rat lost its sight during its evolution. The greatly increased rate of divergence since then presumably reflects the drastic easing of functional constraints on the changes that are permitted in this protein in the blind mole rat.

Functional genomic analysis of the rates of protein evolution. D. P. Wall *et al.* (2005) *Proc. Natl. Acad. Sci. USA* **102**, 5483–5488.

Family-specific rates of protein evolution. H. Luz & M. Vingron (2006) *Bioinformatics* **22**, 1166–1171.

Structural determinants of the rate of protein evolution in yeast. J. D. Bloom *et al.* (2006) *Mol. Biol. Evol.* **23**, 1751–1761.

The quest for the universals of protein evolution. E. P. Rocha (2006) *Trends Genet.* **22**, 412–416.

Protein evolution is faster outside the cell. K. Julenius & A. G. Pedersen (2006) *Mol. Biol. Evol.* **23**, 2039–2048.

6. Roles of Selection

How can we explain the wide range of evolutionary rates in different proteins (Table 7-13) and the nearly constant rate with time for each protein? The most plausible explanation is that **the observed evolutionary changes that have occurred are largely due to neutral mutations that have not affected significantly the function of the protein or the fitness of its gene**. Having little or no effect on function, **neutral mutations** would not have been selected for or against, but would have arisen by the inevitable process of mutation along the lineage of each gene. If truly neutral mutations are possible, each gene must acquire them during its evolution, so **every gene and protein will acquire neutral mutations at the rate at which such mutations occur in a single copy of its gene**.

This is not to say that natural selection has not been important, as it certainly selected against adverse mutations (Section 7.6.A.6.a) and there are a few instances where it appears to have had a positive effect (Section 7.6.A.6.b).

Functional and nonfunctional mutations distinguished by random recombination of homologous genes. H. Zhao & F. H. Arnold (1997) *Proc. Natl. Acad. Sci. USA* **94**, 7997–8000.

Sequence divergence, functional constraint, and selection in protein evolution. J. C. Fay & C. I. Wu (2003) *Ann. Rev. Genomics Hum. Genet.* **4**, 213–235.

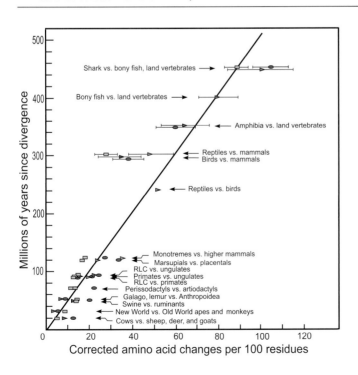

Figure 7-32. The constant rate of evolutionary divergence of the amino acid sequences of the globins. *Triangles*, hemoglobin α-chains; *circles*, hemoglobin β-chains; *squares*, myoglobins. The number of amino acid substitutions that occurred is plotted against the time since divergence of the species that has been deduced from the fossil record. The older divergence times are the least accurate, and horizontal error bars are given that extend over ±2 standard deviations, indicating the 95% confidence limit. Data from N. Nei (1987) *Molecular Evolutionary Genetics*, Columbia University Press, NY.

a. Neutral Mutations and Negative Selection

According to the **neutral mutation hypothesis**, the constant rate of evolutionary divergence of each protein would be the same as its particular neutral mutation rate per gene copy, which is the total mutation rate per gene times the fraction of mutations that are effectively neutral. Assuming for simplicity that the total mutation rate is the same for each codon of a gene, **the neutral mutation rate would differ for different genes because of the different fractions of mutations that are effectively neutral**; each gene or protein would differ in how much variation in amino acid sequence is possible without an effect on its function. If the precise amino acid sequence is not critical for the function of the protein, a large fraction of the total mutations would be neutral, and the protein would evolve rapidly. **Fibrinopeptides** are such examples, as they appear to function primarily to block the aggregation of the precursor protein, fibrinogen. They are cleaved proteolytically from the amino ends of two of the three fibrinogen polypeptides in the first step of blood clotting and play no further known role. As a consequence of their removal, the fibrinogen is converted to fibrin, which aggregates and forms the basis of the blood clot. The only known functional constraints on the amino acid sequences of the fibrinopeptides are a carboxyl-terminal Arg residue, required for proteolytic cleavage by thrombin, and a somewhat acidic net charge, probably to inhibit aggregation of the precursor, fibrinogen. Within these minor limitations, many different amino acid sequences are functional, which would explain why changes during evolution have occurred at relatively rapid rates in these protein segments (Table 7-13).

At the other extreme, proteins for which very few amino acid replacements are acceptable evolve at very slow rates. An example is **cytochrome c**, which must interact with a number of other proteins in its biological function of transferring electrons. Evolutionary variation has occurred at only some sites (Figure 7-29), which do not play a crucial role in this protein's biological function. This protein has changed at a moderately low rate (Table 7-13).

Generally, **the degree of change within a protein's amino acid sequence is found to be inversely proportional to the biological importance of each residue**. Most variable are those residues that occur on the surface of the protein and are not involved in functional interactions with other molecules. Most conserved are those amino acid residues that are involved most directly in the biological function of the protein. The same considerations apply to the gene sequence.

Proinsulin is a good example, with widely different rates of divergence of the A, B and C segments of its polypeptide chain. The C peptide portion has evolved at a rate that is seven times more rapid than that of the remainder of the hormone, the A and B chains, which make up the functional hormone, insulin (Table 7-13). The C peptide is removed proteolytically from the middle of the proinsulin polypeptide chain, after it has folded to its correct conformation. The primary role of the C peptide appears to be to link the A and B segments covalently, to ensure correct intramolecular folding of the protein; it has no other known role, and other cross-links are able to function in the refolding of insulin *in vitro*. **The seven-fold greater rate of divergence of the C peptide than of the A and B chains therefore reflects the fewer constraints on its precise amino acid sequence** relative to the functional parts of the hormone.

The available data indicate that the type of changes at the molecular level that have occurred during evolution are those least likely to have functional consequences and least likely to have been selected. The occurrence of primarily nonfunctional changes is most readily explained as the result of the accumulation of neutral mutations. **The role of natural selection at the molecular level seems to be primarily negative**, weeding out the deleterious mutations that affect function.

b. Positive Selection for Functional Mutations

Of course, functional change has occurred during evolution, as evidenced by the diversity of living organisms. However, this diversity is often not evident at the molecular level, in that orthologous proteins with the same function from different species usually have very similar functional properties. Some exceptions are known: the **hemoglobins** of vertebrates are orthologous but vary widely in the ways that their oxygen-binding properties are regulated (Section 12.5.B). For example, fish hemoglobins are used for respiration in the normal way but they also secrete oxygen into the swim bladder and eye in order to alter the buoyancy of the fish. This novel release of oxygen occurs in response to a decrease in the pH; it is known as the **Root effect** and does not occur with hemoglobins from nonfish species. In another example, crocodiles are able to stay under water for as long as an hour because their hemoglobins have evolved to liberate oxygen only when absolutely required. Some birds are able to fly at very high altitudes because their hemoglobins have very high oxygen affinities. These are just a few examples of the ways that hemoglobins have evolved to be appropriate to their environments, and these evolutionary changes would be expected to have been hastened by natural selection. All of these functional differences can, however, be attributed to mutational alterations of just a few residues, and the majority of the evolutionary divergence that has occurred in the hemoglobins is believed to be largely neutral.

There are remarkably few other instances where natural selection obviously had a positive effect, in selecting *for* favorable mutations. This may be because such changes in functional properties of a protein will involve only a few residues and not be readily apparent at the sequence or structural level, because the majority of changes that have occurred will be of the neutral type. It may also have operated during a relatively short evolutionary time period. Positive selection will produce effects opposite to those of negative selection: functional residues are changed most frequently, as are nucleotides that cause amino acid replacements. Consequently, positive selection will be apparent from the overall sequence changes only when it has predominated over negative selection and neutral changes. This is usually not the case, unless the positive selection has occurred rapidly and recently, before neutral mutations can predominate.

The most dramatic known instance is in certain protein inhibitors of proteolytic enzymes (Section 7.5.B.1). These **proteinase inhibitors** act by binding at the active site of the proteolytic enzyme and block its access to substrate. The evolutionary variation observed with some of the proteinase inhibitors and their genes is just the opposite to that normally observed with neutral mutations, in that the functionally important regions have changed the most. The most variable parts of the genes are those coding for the protein, where most of the nucleotide replacements change the encoded amino acid. Those amino acid residues known to interact directly with the proteolytic enzymes have changed the most. Some of the inhibitors have been shown to be specific for different proteinases. A corresponding **hypervariability** of the active site regions of certain proteolytic enzymes has also been detected, so some proteinases and their inhibitors may be coevolving by positive selection. It may be significant that snake toxins have evolved at a very high rate, higher even than that of fibrinopeptides (Table 7-13), possibly because they must adapt to functional evolution in their targets.

Evolution to produce the myriad of living organisms present today must have occurred by many instances of adaptive evolution due to positive selection, but specific instances other than those just described have been difficult to identify. Most evolutionary divergence of proteins is probably of the neutral variety and of no functional significance, and it probably masks the functional changes.

Adaptive evolution of genes and gene families. W. J. Swanson (2003) *Curr. Opinion Genet. Dev.* **13**, 617–622.

Positive selection in the human genome inferred from human-chimp-mouse orthologous gene alignments. A. G. Clark *et al.* (2003) *Cold Spring Harbor Symp. Quant. Biol.* **68**, 471–477.

Adaptive evolution and functional divergence of pepsin gene family. V. Carginale *et al.* (2004) *Gene* **333**, 81–90.

The genomic rate of adaptive evolution. A. Eyre-Walker (2006) *Trends Ecol. Evol.* **21**, 569–575.

Detecting natural selection at the molecular level: a reexamination of some 'classic' examples of adaptive evolution. L. Nunney & E. L. Schuenzel (2006) *J. Mol. Evol.* **62**, 176–195.

7.6.B. Gene Rearrangements and the Evolution of Protein Complexity

Genetic, biochemical and morphogenetic complexity have all increased during evolution. It is difficult to imagine how useful genes could arise *de novo* at the required rate but, once some functional genetic material is present, it would seem easiest to increase its complexity by duplicating the pre-existing genes. With an additional copy of a gene available in the genome, one of the paralogous genes may continue to provide the original function, while the other could accumulate mutations that alter this function. If the second gene were to evolve eventually to serve yet another useful function, it would tend to be retained within the genome and to be passed on to further generations.

How big is the universe of exons? R. L. Dorit *et al.* (1990) *Science* **250**, 1377–1382.

Do orthologous gene phylogenies really support tree-thinking? E. Bapteste *et al.* (2005) *BMC Evol. Biol.* **5**, 33.

1. Gene Duplications

Many proteins from the same organism are found to be homologous and obviously are the products of gene duplication; their genes are said to comprise a **gene family** and to be paralogous. The most striking and well-studied gene family is that of the globins of higher organisms, which includes single-chain myoglobin and the various polypeptide chains of tetrameric hemoglobins. All members of this family are present in the same organism, and they have functions that are similar, although distinct: myoglobin stores oxygen in muscle, whereas the various hemoglobins transport oxygen within erythrocytes at various stages of the life cycle (Section 12.5.B). The similarities between the amino acid sequences of these polypeptide chains (Figure 7-31) are sufficiently striking to leave little doubt that they must have arisen by gene duplications of a common ancestor, and they also have very similar 3-D structures (Figure 9-34).

The probable sequence of events may be inferred from the relative similarities of the sequences. The myoglobin sequence is least like the others, so it probably diverged first by duplication of the ancestral globin gene, to give the myoglobin and ancestral hemoglobin genes. The latter probably duplicated again at a later time and gave rise to the α-globin gene and the ancestor of the β-type genes (β, γ, δ and ε). The next duplication of the β-type, at least 200 million years ago, would have produced the line to the γ- and ε-genes plus that to the β- and δ-genes; both lines duplicated further at a later date (about 100 million and 40 million years ago, respectively) to produce each pair of genes. More recently, two human genes for γ-chains appeared, ᴳγ and ᴬγ, which produce γ-chains that differ only in having either Gly or Ala at position 136. There are also two human α-genes but they produce identical α-polypeptide chains.

It is often assumed that a newly sequenced gene or protein will have the same type of function as any other protein to which it is found to be homologous. Such preservation of function is not necessary, however, and varying degrees of functional divergence are known. For example, the proteinases of the trypsin family catalyze the same basic reaction but with markedly different substrate specificities. **Haptoglobin** is homologous with the trypsin proteinases and undergoes similar proteolytic processing and activation (Section 15.2.D.1) but it has no proteolytic activity; its physiological function is to bind αβ complexes of hemoglobin chains. **Lysozyme** from hen eggs is homologous with α-**lactalbumin** from bovine milk: of 123 residues, 47 are identical and many others are physically similar, and the two proteins have very similar structures. Yet lysozyme functions to degrade the polysaccharides of bacterial cell walls, whereas α-lactalbumin has no such activity. Instead, α-lactalbumin regulates lactose synthesis by binding to galactosyl transferase, a membrane-bound enzyme. The latter enzyme normally glycosylates proteins in the Golgi apparatus but it synthesizes lactose when complexed with α-lactalbumin. The only apparent similarity in the functions of these two proteins is that they both act in reactions involving saccharides, although there is no evidence that α-lactalbumin actually interacts directly with the saccharide. **Ovalbumin**, a seemingly functionless food storage protein of egg white, is homologous to the **serpin** class of proteinase inhibitors, as is **angiotensinogen**, the precursor to the hormone angiotensin; neither has proteinase inhibitor activities, nor do the serpins act as hormones.

In other cases, proteins have developed auxiliary functions. The most striking examples are several enzymes (e.g. lactate dehydrogenase, argininosuccinate lyase, aldose reductase, enolase and glutathione transferase) that have been duplicated and adapted to serve as structural proteins of the eye lens, or have simply added this role as an additional function.

Gene duplications: the gradual evolution of functional divergence. J. F. Brookfield (2003) *Curr. Biol.* **13**, R229–230.

Functional divergence prediction from evolutionary analysis: a case study of vertebrate hemoglobin. S. Gribaldo et al. (2003) *Mol. Biol. Evol.* **20**, 1754–1759.

Duplication and divergence: the evolution of new genes and old ideas. J. S. Taylor & J. Raes (2004) *Ann. Rev. Genet.* **38**, 615–643.

---

2. Protein Elongation by Intragene Duplication

Duplication of the genetic material within a gene will cause the gene to be elongated and can cause at least part of the amino acid sequence to be duplicated. The duplicated portions can then diverge and accumulate mutational changes. Such gene extension by duplication is often most evident in diagonal dot-plots of the sequence (Figure 7-28), where two or more diagonals of homology are observed for a single stretch of sequence.

A striking example of gene elongation by duplication is found with the **ferredoxins**, where the homology between the two halves of the amino acid sequence is very apparent (Figure 7-33). In **transferrin**, the two 340-residue halves are 50% identical in sequence and similar in structure (Figure

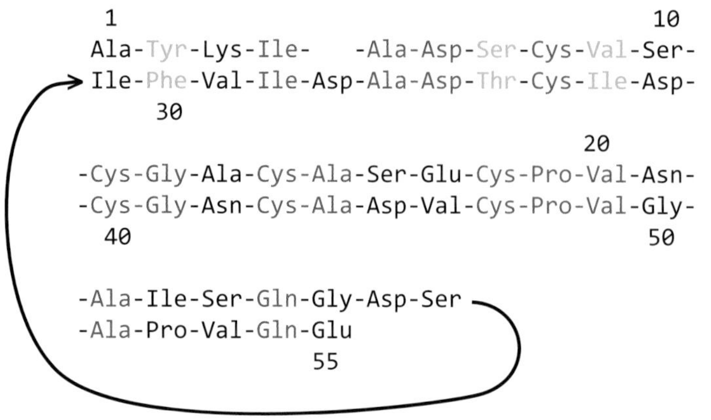

```
       1                            10
   Ala-Tyr-Lys-Ile-   -Ala-Asp-Ser-Cys-Val-Ser-
 ⟶ Ile-Phe-Val-Ile-Asp-Ala-Asp-Thr-Cys-Ile-Asp-
       30
                              20
   -Cys-Gly-Ala-Cys-Ala-Ser-Glu-Cys-Pro-Val-Asn-
   -Cys-Gly-Asn-Cys-Ala-Asp-Val-Cys-Pro-Val-Gly-
       40                      50

   -Ala-Ile-Ser-Gln-Gly-Asp-Ser
   -Ala-Pro-Val-Gln-Glu
                55
```

Figure 7-33. Homology between the two halves of the primary structure of *Clostridium pasteurianum* ferredoxin. Identical residues are in *blue*, similar residues *green*.

12-11) and each half binds a single Fe atom. **Serum albumin** has a triplicated structure derived from a simpler 190-residue form.

Protein repeats: structures, functions, and evolution. M. A. Andrade *et al.* (2001) *J. Struct. Biol.* **134**, 117–131.

---

3. Gene Fusion and Division

Many proteins, especially those unique to vertebrates, have mosaic structures in which various segments appear to have had different origins (Figure 7-24). One segment might be homologous to one protein family, another to a different family. Such proteins give the impression of having been assembled by stringing together modules, presumably by fusing together their individual genes. Each of the modules usually corresponds to an entire structural and functional domain of a protein. For example, a calcium-sensitive proteinase was apparently constructed by fusing the genes for a calcium-binding protein and a thiol proteinase.

The genealogy of some recently evolved vertebrate proteins. R. F. Doolittle (1985) *Trends Biochem. Sci.* **10**, 233–237.

Evolution of a biosynthetic pathway: the tryptophan paradigm. I. P. Crawford (1989) *Ann. Rev. Microbiol.* **43**, 567–600.

7.6.C. Protein Engineering

Protein engineering refers to the design and production of new proteins with predetermined properties. There is a practical motivation to be able to construct 'tailor-made' proteins with improved properties that will be useful in industry, medicine, agriculture, etc. It is also useful in that it increases our knowledge of the fundamental structure–function properties of natural proteins. It can involve the re-engineering of existing proteins or *de novo* attempts to design entirely new proteins (Section 9.4.D). Altering the gene for a protein has the advantage of being extremely specific, and the gene can be made to order (Section 6.6.E). The gene for the desired protein is altered or constructed, and the gene is expressed to produce the protein with the desired amino acid sequence. The properties of the protein are then characterized.

A prerequisite is that an efficient expression vector is available with which to produce the protein in large quantities (Section 6.3.A). For the cloned gene to be translated correctly, it must be inserted in such a way that the correct triplet reading frame is in phase with the start and stop codons (Figure 7-21). The polypeptide produced might also be extended with extra residues to facilitate identification or purification of the protein. There might be six extra His residues, which bind nickel ions and cause the polypeptide chain to adsorb to solid supports containing these ions. The sequence might be recognized by an antibody molecule, which enables the protein to be identified and purified readily. Entire protein domains that are specific for certain ligands can also be used. Popular examples include the enzyme glutathione-S-transferase, which will bind to affinity columns containing attached glutathione. The sequence might also be engineered so that it is easy to cleave the polypeptide chain, remove the additional residues and liberate just the desired polypeptide chain. Such sequences are usually the sites for cleavage by proteinases. They make it possible to attach the desired protein to the C-terminus of a model protein that is synthesized efficiently in an expression system. It is then synthesized efficiently along with its host protein and can be released subsequently.

It is usually desirable to maximize the amount of protein produced by an expression vector, and there are various stratagems. The secondary structure at the 5′-end of a messenger RNA or within a coding portion will usually inhibit translation but it can be minimized by making judicious changes to the sequence. On the other hand, the secondary structure at the 3′-end of the messenger RNA can stabilize it against degradation. Systems that produce large quantities of a foreign protein can overwhelm the protein biosynthesis apparatus of the cell and are often detrimental to the host organism, so it is desirable to be able to control the expression levels so that they are maximized only at an appropriate time.

The choice of host is very important. Cells use the various genetic codons (Figure 7-21) with different efficiencies, so the nucleotide sequence of the gene can be fine-tuned to best suit the host. In general, eukaryotic proteins are best produced in eukaryotic hosts, but these are the most difficult technically to use; yeasts are the least difficult, but the **baculovirus** system in insect cells is widely used. Prokaryotic systems such as *Escherichia coli* are the easiest to use, but their proteinases often degrade the polypeptide chains produced and they do not have the apparatus for making the usual post-translational modifications to eukaryotic polypeptide chains, such as disulfide bond formation and glycosylation. They often produce foreign proteins in an insoluble form, usually described as **inclusion bodies**. This can be an advantage in that the protein is protected from proteinases and easy to isolate, but it needs to be solubilized and folded before it can serve a useful purpose. Secreting the protein outside the cell can be beneficial, but the outcome is very unpredictable. If no suitable expression vector is found, small amounts of protein can be produced in cell-free systems. Protein expression remains an empirical subject.

Where the gene is expressed in an appropriate organism, genetic approaches can be very useful for studying proteins. The gene coding for a protein is altered and the effect on the organism is determined, so the function of the protein product of the gene is being studied in its natural *in vivo* environment. Specific site-directed mutagenesis (Section 6.6.E) is ideally suited for testing hypotheses about genes and proteins that are relatively well-characterized. Alternatively, mutational alterations of the gene and protein that produce a desired physiological effect can often be selected after random mutagenesis. Such an approach is ideally suited for initially characterizing the functional properties of a new gene and protein. No prior knowledge of the gene or protein is required, and there is little possibility of missing or overlooking important residues of the protein, so long as the mutagenesis and selection procedures are sufficiently rigorous.

Protein engineering 20 years on. J. A. Brannigan & A. J. Wilkinson (2002) *Nature Rev. Mol. Cell Biol.* **3**, 964–970.

Incorporation of nonnatural amino acids into proteins. T. L. Hendrickson *et al.* (2004) *Ann. Rev. Biochem.* **73**, 147–176.

A recurring theme in protein engineering: the design, stability and folding of repeat proteins. E. R. Main *et al.* (2005) *Curr. Opin. Struct. Biol.* **15**, 464–471.

Tn5 transposition: a molecular tool for studying protein structure-function. W. S. Reznikoff (2006) *Biochem. Soc. Trans.* **34**, 320–323.

Connecting protein structure with predictions of regulatory sites. A. V. Morozov & E. D. Siggia (2007) *Proc. Natl. Acad. Sci. USA* **104**, 7068–7073.

Protein production and purification. S. Graslund *et al.* (2008) *Nature Methods* **5**, 135–146.

~ CHAPTER 8 ~

POLYPEPTIDE CONFORMATION

The structures of all molecules extend to three dimensions. Two-dimensional (2-D) chemical representations usually suffice for small molecules because their three-dimensional (3-D) structures are defined reasonably well by the fixed bond lengths and bond angles of their covalent structures. These representations are not sufficient for larger molecules, however, because rotations about the many covalent bonds dramatically alter the relative positions of all the atoms. 3-D aspects of structure are especially important for polymers such as nucleic acids and proteins, in which many bonds can rotate. Polymers have an intrinsic tendency to be very flexible, when no one structure, or **conformation** (Section 1.2), is sufficiently more stable than all the others to predominate, but nevertheless many biological macromolecules, especially the proteins and nucleic acids (Chapters 2, 4 and 9), tend to adopt single, stable conformations.

The abilities of polypeptide chains to adopt a wide variety of 3-D conformations is crucial so that they can carry out their many different functions (Chapters 12–15). These conformational properties of polypeptide chains are controlled to a great extent by the flexibility of the polypeptide backbone and by local interactions with amino acid side-chains, which are described in this chapter.

Conformation of polypeptides and proteins. G. N. Ramachandran & V. Sasisekharan (1968) *Adv. Protein Chem.* 23, 293–437.

Principles of Protein Structure. G. E. Schulz & R. H. Schirmer (1979) Springer, NY.

Protein Structure. N. J. Darby & T. E. Creighton (1993) IRL Press, Oxford.

Introduction to Protein Structure, 2nd edn. C. Branden & J. Tooze (1999) Garland Publishing, NY.

8.1. LOCAL FLEXIBILITY OF THE POLYPEPTIDE BACKBONE: THE RAMACHANDRAN PLOT

A portion of the backbone of a polypeptide chain is illustrated in Figure 8-1, showing the conventions used in describing polypeptide conformation. Rotations about covalent bonds are described as **torsion** or **dihedral** angles, which are usually taken to lie in the range –180° to +180° (Section 1.2).

Rotation about the N–C$^\alpha$ bond of the peptide backbone is denoted by the torsion angle *phi*, ϕ, rotation about the C$^\alpha$–C' bond by *psi*, ψ, and that about the peptide bond (C'–N) by *omega*, ω. The maximum value of 180° (which is the same as –180°) is given to each of the torsion angles in the maximally extended chain, as shown in Figure 8-1, when the N, C$^\alpha$ and C' atoms are all *trans* to each other. At the other extreme, the *cis* conformation, the rotation angles have a value of zero. Rotation about a bond from this position so that the atoms behind the rotated bond move clockwise, when viewed along the bond (Equation 1.23), are given positive values; counterclockwise rotations are given negative values. Rotations about single bonds are intrinsically equivalent, and the relative preference for each particular torsion angle is determined by the energetics of the noncovalent interactions between the atoms and their environment.

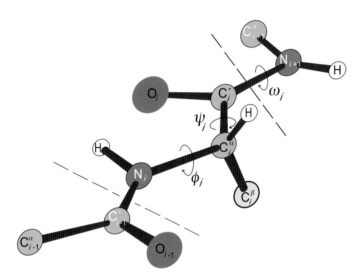

Figure 8-1. Perspective drawing of a segment of polypeptide chain comprising two peptide units. Only the C$^\beta$ atom of each side-chain is shown. The limits of a single residue (number *i* of the chain) are indicated by the *dashed lines*. The recommended notations for atoms and torsion angles are indicated. The polypeptide chain is shown in the fully extended conformation, where $\varphi = \psi = \omega = \pm 180°$.

Torsion angles of the side-chain are designated by χ_j, where *j* is the number of the bond working outwards from the C$^\alpha$ atom of the main chain. The commonly used designations of the side-chain atoms are given in Figure 7-2. Bond angles are usually denoted by the symbol τ, with a subscript *i* that gives the number of the residue in the chain (counting starting from the *N*-terminus), followed by symbols, in brackets, of the atoms that define the bond angle. For example, $\tau_i[\text{NC}^\alpha\text{C}']$ designates the angle formed by the N–C$^\alpha$ and C$^\alpha$–C' bonds at the C$^\alpha$ atom of the *i*th residue.

The peptide bond (Figure 8-2) has a tendency to be double-bonded, due to resonance:

$$\text{C}^\alpha\text{–C(=O)–N(H)–C}^\alpha \rightleftharpoons \text{C}^\alpha\text{–C(–O}^-\text{)=N}^+\text{(H)–C}^\alpha \quad (8.1)$$

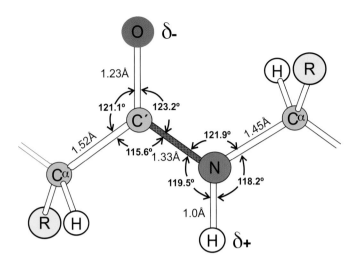

Figure 8-2. Section of a polypeptide structure showing the *trans* peptide bond (*orange*) connecting two amino acid residues. The dimensions given are the averages observed crystallographically in amino acids and small peptides. The partial charges arise from the partial double-bond character of the peptide bond (Equation 8.1).

The partial double-bond character of the peptide bond causes it and the two neighboring C^α atoms to act as a rigid planar unit. For this reason, this group is often designated as a **peptide unit**. The unit more commonly used, however, is the **residue**, in which all the atoms originate from the same amino acid. The resonance also causes charge separation and gives the peptide bond a substantial **dipole moment** of about 3.5 Debye:

$$\text{(8.2)}$$

The partial double-bond character of the peptide bond also means that it cannot rotate freely, and only two conformations are usually available, defined by the angle ω (Figure 8-1), the *cis* and *trans* conformations (Section 1.1.A.5):

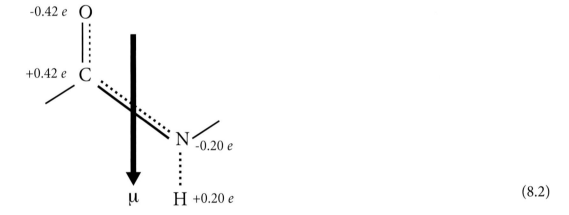

trans
ω=±180°

cis
ω=0°

$$\text{(8.3)}$$

The angle ω has the value 0° when the peptide bond is *cis*, and ±180° when *trans*; these angles are observed in polypeptide chains to vary somewhat from the ideal values. In the *cis* form, the C^α atoms and any side-chains of neighboring residues are in too-close proximity, as indicated by the wavy line in Equation 8.3, so the *trans* form (ω = 180°) is favored approximately 10^3-fold. When residue $i + 1$ is Pro, however, there is very little difference between the two:

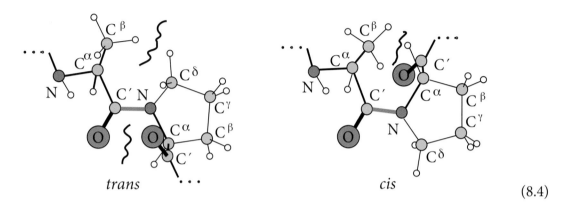

(8.4)

The *trans* form in this case is only slightly favored, generally by a ratio of about 4:1. The peptide bond preceding a Pro residue probably does not have the double-bond character that specifies the planar form (Equation 8.1) and small deviations from planarity of either the *cis* or *trans* form, with Δω = −20° to +10°, are believed to be only slightly unfavorable energetically.

The possible values of ϕ and ψ are also constrained geometrically, due to steric clashes between nonneighboring atoms. The permitted values of ϕ and ψ were first determined by Ramachandran and colleagues, using hard-sphere models of the atoms and fixed geometry of the bonds. The permitted values of ϕ and ψ are generally indicated on a 2-D map of the ϕ–ψ plane, known as a **Ramachandran plot**. An example for an Ala residue is illustrated in Figure 8-3-A. The normally allowed values, where there are no steric overlaps, are indicated by the darkest colors; the extreme limits with some unfavorable contacts are indicated by the lighter colors. The connecting region, indicated by the light color, becomes permitted if slight alterations of bond angles are allowed. Only 7.5% of the total area of the Ramachandran map is fully allowed, 22.5% partially allowed, which gives a quantitative measure of the limitations on flexibility of the polypeptide chain.

Gly residues have no C^β atom, and the restrictions on allowed conformations are much less severe (Figure 8-3-B): 45% of the total area is fully allowed, 61% within the extreme limits. In this case, **the Ramachandran plot is symmetric because the Gly residue is not chiral**. This extra flexibility tends to decrease the overall average dimensions of unfolded polypeptide chains, as the Gly residues permit the chain to reverse direction more readily. The extra flexibility of the Gly residue in an unfolded polypeptide chain is also believed to increase the conformational entropy and decrease the tendency to adopt fixed conformations.

With longer, larger side-chains of the other amino acids, additional restrictions arise, and the allowed region of the Ramachandran plot is accordingly smaller. The **Pro** residue is a special case, because its relatively rigid five-membered ring drastically limits the value of ϕ to be close to −60°.

More precise Ramachandran plots are produced from calculations of the relative energies of each

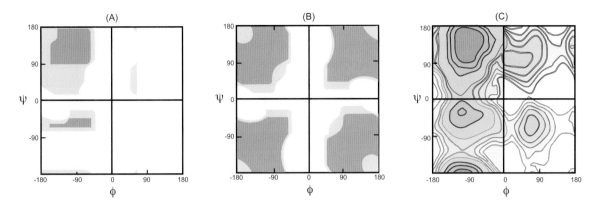

Figure 8-3. Ramachandran plots of the permitted values of φ and ψ for different residues. The original plots that considered only repulsions between hard-sphere atoms are shown in (A) and (B) for Ala and Gly residues, respectively. The fully allowed regions are *darkest*, the partially allowed regions *less colored*. The connecting regions in the *lightest color* are permissible with slight flexibility of bond angles. Each 2-D plot is continuous at the edges, because a rotation of −180° is the same as one of +180°. The much greater flexibility of the Gly residue compared with Ala is apparent in (B), as is the symmetry of the plot for Gly residues resulting from the absence of a chiral side-chain. Data from G. N. Ramachandran & V. Saisekharan (1968) *Adv. Protein Chem.* **23**, 283–437. (C). Ramachandran plot computed for the dipeptide N-acetyl-Ala-Ala-amide using molecular dynamics simulations and including water as the solvent. The apparent free energies for the various values of φ and ψ are given as contours of 2 kJ/mol (0.5 kcal/mol) relative to the lowest free energy in the upper left-hand corner that is colored *darkest*. Note that the differences between allowed and disallowed regions are much less distinct than in (A) and (B). Data from a figure kindly provided by J. Hermans in T. E. Creighton (1993) *Proteins: structures and molecular properties*, 2nd edn, W. H. Freeman, NY, p. 173.

conformation, permitting appropriate flexibility of bond lengths and angles and evaluating all favorable and unfavorable interactions, including those with the solvent (Figure 8-3-C). The more detailed calculations indicate much smaller energy differences between the so-called 'allowed' and 'disallowed' regions than might be expected from just steric considerations using rigid models. Flexibilities of bond lengths and angles permit conformations that are not possible with hard-sphere atoms and rigid bonds, and the variety of interactions that occur between atoms can cause the energies to vary substantially.

The Ramachandran plot demonstrates that **the restrictions on the flexibility of a polypeptide chain are considerable**, as only 30% of the combinations of local φ/ψ angles are possible with most residues. Nevertheless, there is still considerable scope for each residue in a polypeptide chain to adopt a range of conformations, and a random-coil polypeptide chain adopts a wide variety of conformations (Section 8.2).

A non-planar peptide bond in L-seryl-L-valine. A. Moen *et al.* (2004) *Acta Crystallogr. C* **60**, 564–565.

Exploiting the right side of the Ramachandran plot: substitution of glycines by D-alanine can significantly increase protein stability. B. Anil *et al.* (2004) *J. Am. Chem. Soc.* **126**, 13194–13195.

Correlation between omega and psi dihedral angles in protein structures. L. Esposito *et al.* (2005) *J. Mol. Biol.* **347**, 483–487.

The Ramachandran plots of glycine and pre-proline. B. K. Ho & R. Brasseur (2005) *BMC Struct. Biol.* **5**, 14.

8.2. RANDOM-COIL POLYPEPTIDE CHAINS

The substantial flexibility of the polypeptide chain gives it significant conformational entropy (Section 1.3) and makes the random coil the most favored conformation under many conditions. Most studies of random polypeptide chains have been done with polyamino acids, in which one type of amino acid is polymerized into a homopolypeptide chain. Yet it is not clear that any polypeptide chain adopts a completely random-coil conformation, as various interactions can usually be observed, especially between chemical groups close in the covalent structure. Many unfolded proteins that previously were thought to approximate random coils are now thought to contain significant amounts of nonrandom conformation, like that observed in poly(Pro) II (Section 8.3.D), but this might just be the most favorable stereochemistry for the polypeptide chain.

8.2.A. Statistical Properties

As with other disordered polymers (Section 1.3), the properties expected of random-coil polypeptides can only be estimated and are averages over time and over all the molecules of the population. In fact, there are so many different conformations possible with a long polypeptide chain that each molecule of a practical population is likely to have a unique conformation at any point in time. For example, a 1-g sample of a protein with a molecular weight of 10,000 (with fewer than 100 residues) contains only 6×10^{19} molecules. If an average residue were able to adopt four different conformations, such a random-coil chain would be able to adopt 10^{60} ($= 4^{100}$) different conformations. Even just two conformations per residue would produce 10^{30} different conformations. This is considerably larger than the number of molecules present in a practical sample.

Most calculations of polypeptide random coils simplify the detailed chemical architecture of the polypeptide backbone and use a single **virtual bond** for each residue, which is a vector joining adjacent C^α atoms. With a planar *trans* peptide bond, it has a length of 3.8 Å. For long polypeptide chains of n residues of amino acids other than Gly, the r.m.s. **end-to-end distance** is provided approximately by:

$$\langle r^2 \rangle_0^{1/2} = (130\,n)^{1/2} \tag{8.5}$$

This corresponds to a **characteristic ratio** (C_∞; Section 1.3.A.3) of about 9.0. This relatively large value of C_∞ is indicative of a relatively stiff polymer and is not only due to the steric restrictions on rotation about angles ϕ and ψ of the polypeptide backbone (Figure 8-3) but also the presence of only L-amino acid residues. A polymer of alternating D- and L-Ala residues has a C_∞ of only 0.9, because the D-amino acid residues cause the direction of the chain to be effectively reversed from that favored by L-amino acids. The increased flexibility possible with Gly residues and the symmetry of the allowed torsion angles causes poly(Gly) to have a C_∞ of only 2.0. A small percentage of Gly residues distributed randomly in a polypeptide chain greatly decreases the dimensions of random polypeptides. The presence of Pro residues also decreases the value of C_∞ because the accompanying *cis* form of the peptide bond (Equation 8.4) causes the polypeptide chain to change its direction.

The radius of gyration (Section 1.3.A.2) of random-coil polypeptide chains of various lengths is illustrated in Figure 8-4 and compared with other possible conformations. Note that the dimensions of the random coil are not directly proportional to the length of the polypeptide chain.

The distance between two residues in a random-coil polypeptide chain has been measured by

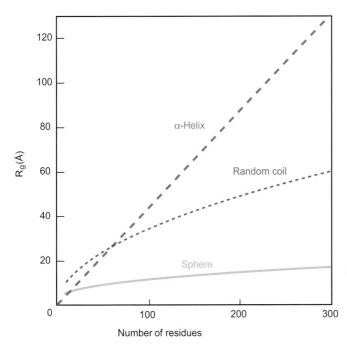

Figure 8-4. The radius of gyration, R_g, for polypeptide chains of various lengths in α-helical, random coil and compact spherical conformations. Adapted from T. E. Creighton (1993) *Proteins: structures and molecular properties*, 2nd edn, W. H. Freeman, NY, p. 177.

incorporating a fluorescence donor at one end and a fluorescence acceptor at the other and measuring the **fluorescence energy transfer** between them. This yields the radial distribution function of the distances between the two groups (Figure 8-5).

The proximities of two groups in a random polypeptide chain can also be expressed as their concentration relative to each other, the **effective concentration**. The tendencies of Cys residues up to 50 residues apart in unfolded proteins to form disulfide bonds (Section 11.4.C) indicate values of 1–70 mM, depending upon the number of residues and the sequence between the Cys residues.

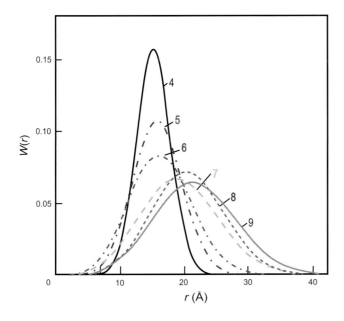

Figure 8-5. Radial distribution function of the distances between naphthalene and dansyl groups attached to the ends of peptides of 4–9 residues of *N*-hydroxyethyl-Gln, measured by fluorescence energy transfer. Data from E. Haas *et al.* (1975) *Proc. Natl. Acad. Sci. USA* **72**, 1808–1811.

Dimensions of protein random coils. W. G. Miller & C. V. Goebel (1968) *Biochemistry* **7**, 3925–3935.

How alike are the shapes of two random coils? A. D. McLachlan (1984) *Biopolymers* **23**, 1325–1331.

8.2.B. Rates of Conformational Change

The free-energy barriers separating all the ϕ and ψ bond rotations of the polypeptide backbone are believed to be only about 0.5–1.5 kcal/mol (2–6 kJ/mol), so **each of these two bonds would be expected to rotate at rates of the order of 10^{12} s^{-1}**. The movements of individual bonds in polymers are complex, however, and not easily described, because no internal bonds in a polymer can change independently of all the others: if only one bond near the middle of a chain were to rotate by 180°, the ends of the chain would need to undergo extremely large movements. This is implausible in a viscous solvent, including water. It seems intuitively obvious, therefore, **that the rotations of bonds throughout the polymer backbone must be coordinated in such a way as to produce more plausible types of movements**; a complete description of how this might occur is possible only by computer simulations.

The average rates at which individual bonds in disordered polypeptides change conformations have been measured from their **relaxation times** from ^{13}C nuclear magnetic resonance. The relaxation time is the average time taken by a population of molecules to change in some way by e^{-1} of their equilibrium positions. The relaxation times of the C$^{\alpha}$ atoms of the backbone of a disordered polypeptide chain have been measured to be 1.4–2.6 ns. These values are longer than they would be in a small molecule, as expected if the atoms cannot move independently but must be coordinated with rotations of other bonds in the polypeptide backbone; **the relaxation times of the side-chains are shorter and decrease for atoms further from the backbone** (Table 8-1).

Table 8-1. Rotational relaxation times in a random polypeptide chain

Residue	**Carbon atom**	**Relaxation time (10^{-9} s)**
Ala	C$^{\beta}$	0.21
Thr	C$^{\beta}$	1.56
	C$^{\gamma}$	0.18
Lys	C$^{\beta}$	0.81
	C$^{\gamma}$	0.54
	C$^{\delta}$	0.60
	C$^{\varepsilon}$	0.27
Peptide	C$^{\alpha}$	1.4 to 2.6

The values were measured at 45°C on performic acid-oxidized ribonuclease A using ^{13}C nuclear magnetic resonance by V. Glushko *et al.* (1972) *J. Biol. Chem.* **247**, 3176–3195.

The rates at which the ends of a random-coil polypeptide chain are moving relative to each other by diffusion have been measured using fluorescence energy transfer between donor and acceptor groups, as in Figure 8-5. The rates of motion were found to be an order of magnitude lower than when the fluorescent donor and acceptor were not tethered to the polypeptide chain, indicating that **the polypeptide chain possesses appreciable internal friction that resists motion**. Nevertheless, residues separated by 50–100 residues in a random-coil polypeptide chain tend to move through distances comparable to their average separation in 10^{-5}–10^{-6} s. Therefore **two groups on a short random-coil polypeptide chain should come into proximity roughly 10^5–10^6 times per second**.

Similar studies using fluorescence correlation spectroscopy found the dependence of the rate of end-to-end contact on polypeptide chain length to be that expected from Gaussian chain theory. The effects of solvent viscosity and temperature on the rates of contact indicated that the movements of the chain occur by an entropy-controlled process of simple diffusion. Intrachain diffusion is fast even in the presence of high concentrations of other macromolecules, in spite of the diffusion being hindered by repulsive interactions with the other macromolecules. This may due to the excluded volume effect reducing the disorder of the chain and therefore accelerating the loop search process. Even with diffusion being hindered by the high concentrations of other macromolecules in a cellular environment, unfolded proteins can still sample their available conformations quite efficiently.

Segmental relaxation in macromolecules. A. Perico (1989) *Acc. Chem. Res.* **22**, 336–342.

Dynamics of unfolded polypeptide chains in crowded environment studied by fluorescence correlation spectroscopy. E. Neuweiler *et al.* (2007) *J. Mol. Biol.* **365**, 856–869.

---

1. X–Pro Peptide Bond cis/trans *Isomerization*

The only bond in disordered polypeptide chains known to rotate relatively slowly is that about the peptide bond, interconverting between *cis* ($\omega = 0°$) and *trans* ($\omega = \pm180°$). Both the normal amide –CO–NH– peptide group and the imide –CO–N< formed by a Pro residue (Equations 8.3 and 8.4) can undergo *cis/trans* isomerization by rotation about the torsion angle ω of the peptide bond.

Cis/trans isomerization of the –CO–NH– moiety occurs with a half-time of less than 1 s at room temperature and leads to a very small percentage, much less than 1%, of *cis* isomer at equilibrium. For the X–Pro bond, however, the *cis* isomer is more prevalent and there is a relatively high rotational barrier $\Delta G^{\ddagger}$ of about 20 kcal/mol (83 kJ/mol). **The half-time for X–Pro bond isomerization can range from seconds to hours** depending on the polypeptide structure and conditions. In the absence of structural constraints, it occurs with a half-time of ~20 min at 0°C. The rate is very temperature-dependent, and within the usual temperature range the rate increases by a factor of 3.3 for each 10°C rise in temperature.

The normal rate constants are independent of pH in the physiological range, unless ionizable groups are located adjacent to the Pro residue. O-protonation of the peptide bond in acidic solution, with $pK_a = -1$, decreases the rate of isomerization. The backbone N atom can also become protonated in very strong acids ($pK_a = -7$), which accelerates the rate, as do most organic solvents, micelles and phospholipid vesicles. Large aromatic amino acids preceding the Pro residue decrease the rate of isomerization and increase the *cis* population up to 40%, whereas small aliphatic side-chains in the same position have the opposite effects, increasing the rate and lowering the *cis* content to 5–10%. The rates of isomerization of other peptide bonds not adjacent to Pro are not known, as the *cis* isomer is not populated significantly.

The combination of a substantial population of both isomers and their slow interconversion causes *cis* X–Pro bonds to have considerable influence on conformational changes of the polypeptide backbone. In the absence of a folded conformation, polypeptide chains can theoretically form 2^n *cis/trans* isomers, where *n* is the number of X–Pro bonds in the molecule. In simple cases, with only a few Pro residues in a polypeptide chain, all the possible isomers have been detected and measured in solution. Nonrandom structure in the peptide chain can, however, reduce the number of isomers substantially and alter the rate of isomerization.

8.3. Regular Structures

The random coil might be considered the natural state of a polymer, favored by its conformational entropy and interactions with solvent. Certainly a single stabilizing hydrogen bond, salt bridge or van der Waals interaction that might be possible in a random polypeptide chain is unlikely to be stable in most instances. The values of the intrinsic association constants (K_{AB}) for these interactions between independent molecules in water range from 0.01 to 0.9 M^{-1}, and the effective concentrations (C_{eff}) of pairs of groups in random coils are generally no greater than 0.1 M (Section 8.2.A). The value for the equilibrium constant (K_{eq}) for such interactions in random polypeptide chains expected from these considerations is:

$$K_{eq} = K_{AB} C_{eff} \tag{8.6}$$

The maximum value of K_{eq} expected in an unfolded polypeptide chain is then about 0.09; **any one such interaction might be present in an unfolded polypeptide chain at most about 8% of the time**.

On the other hand, multiple interactions present simultaneously in any particular conformation would be expected to stabilize each other, and that conformation might be sufficiently stable to predominate over all others that are possible. Many synthetic **polyamino acids**, where one or a few amino acids are polymerized in a regular sequence, have been found to adopt a few such regular conformations. The regularity of the conformation is a result of the regularity of the amino acid sequence in these cases, for each residue will tend to adopt the same conformation. That conformation can be specified by just a few dihedral angles for a single repetitive unit. If all residues have the same conformation, a helical conformation will result, which can be characterized by the number of residues per turn of helix and the distance traversed along the helix axis by each residue (Figure 1-1). The values for the regular conformations described here are presented in Table 8-2.

In spite of the special properties of polyamino acids, the regular conformations observed with them are also found in natural proteins, where they are known as the **secondary structure** (Section 9.1.B). Which conformation, if any, is adopted by any particular polypeptide chain depends upon its amino acid sequence and the conditions.

The discovery of the α-helix and β-sheet, the principal structural features of proteins. D. Eisenberg (2003) *Proc. Natl. Acad. Sci. USA* **100**, 11207–11210.

Table 8-2. Parameters for classical regular polypeptide conformations

Conformation	Torsion angle (°)			Residues per turn (n)	Translation per residue (Å)
	ϕ	ψ	ω		
Antiparallel β-sheet	−139	+135	−178	2.0	3.4
Parallel β-sheet	−119	+113	180	2.0	3.2
Right-handed α-helix	−57	−47	180	3.6	1.50
3_{10}-helix	−49	−26	180	3.0	2.0
Π-helix	−57	−70	180	4.4	1.15
Polyproline I	−83	+158	0	3.33	1.9
Polyproline II	−78	+149	180	3.0	3.12
Polyglycine II	−80	+150	180	3.0	3.1

Data from G. N. Ramachandran & V. Sasisekharan (1968) *Adv. Protein Chem.* **23**, 283–437; IUPAC-IUB Commission on Biochemical Nomenclature (1970) *Biochemistry* **9**, 3471–3479.

8.3.A. α-Helix

α-Helices are the simplest regular conformation adopted by polypeptides and the most prevalent form of secondary structure in folded proteins. They are formed from consecutive stretches of residues of the polypeptide chain adopting a right-handed helical backbone and a regular repeating pattern of backbone hydrogen bonds (Figure 8-6). **Each turn of an α-helix contains 3.6 residues and translates the backbone by 5.4 Å (or 1.5 Å per residue)**. The backbone angle ϕ is approximately −60°, and ψ −50° (Table 8-2). These values are in the fully allowed region in the lower left region of the Ramachandran plot (Figure 8-7). The side-chains of each residue project outwards from the axis of the helix and tend to point towards its N-terminus.

The α-helix is most noted for its regular hydrogen bond pattern: the backbone carbonyl oxygen of residue (i) hydrogen bonds to the backbone amide of residue ($i + 4$) in the polypeptide chain, with a length of 2.86 Å between the O and N atoms. Each acceptor and donor pair is separated by 13 atoms. All the peptide bonds in an α-helix are involved in very favorable hydrogen bonds, apart from the first three amide groups and last three carbonyl O atoms. Besides having a favorable backbone conformation and hydrogen bonds with favorable geometry, the atoms of the polypeptide chain pack closely together, making favorable van der Waals interactions.

All the hydrogen bonds and dipoles of each of the peptide bonds (Equation 8.2) are oriented in the same direction. Consequently, the helix itself has a significant **dipole moment**, with a partial positive charge at the N-terminus and a partial negative charge at the C-terminus.

The side-chains project out into solution from the α-helix and need not interfere with the peptide backbone, but there are various restrictions on the conformations of the side-chain. In particular, the g^+ rotamer (Equation 1.18) of angle χ_1 between the C^α and C^β atoms is almost forbidden because any side-chain will overlap with atoms of the previous turn of the helix. Side-chains with branched C^β atoms (Val, Ile and Thr) are most restricted in their conformations. Polar groups on some side-chains, such as those of Ser, Thr, Asp and Asn, can hydrogen bond to the backbone peptide groups and interfere with the hydrogen bonding of the α-helix. Such interactions are often found at the ends of α-helices in folded proteins, where normally there would be four carbonyl oxygens or peptide –NH groups that are not involved in helix hydrogen bonding. These hydrogen bonds can be thought of as terminating or 'capping' the helix.

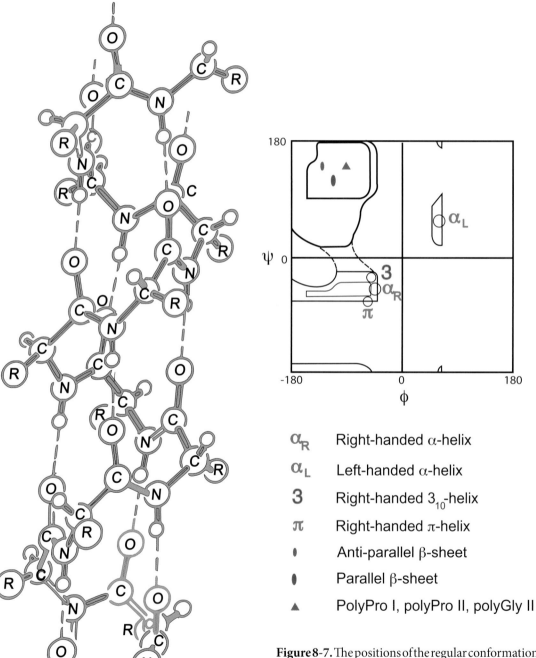

Figure 8-6. A classical α-helix.

α_R	Right-handed α-helix
α_L	Left-handed α-helix
3	Right-handed 3_{10}-helix
π	Right-handed π-helix
●	Anti-parallel β-sheet
●	Parallel β-sheet
▲	PolyPro I, polyPro II, polyGly II

Figure 8-7. The positions of the regular conformations of polypeptides on a Ramachandran plot. Data from G. N. Ramachandran & V. Sasisekharan (1968) *Adv. Protein Chem.* **23**, 283–437.

The **left-handed α-helix**, having the same hydrogen-bonding pattern as right-handed α-helices but backbone ϕ and ψ angles of +60° and +50° (with opposite signs to the right-handed α-helix) is feasible but not usually observed, as this conformation results in steric clashes between atoms of the backbone and side-chains. A Gly residue, however, has only an H atom as its side-chain and can adopt this conformation (Figures 8-3 and 8-7).

Helices in folded proteins are often observed to be **amphipathic**, having polar or charged side-chains arranged on one side of the helix and hydrophobic side-chains on the opposite side. This pattern permits an α-helix to bind on one side to a nonpolar surface, such as a membrane or another part of a folded protein, while the other side interacts with the aqueous solvent. An amphipathic helix can be predicted from just the amino acid sequence, using a **helical wheel** (Figure 8-8), in which the staggered arrangement of side-chains around the helix is visualized in two dimensions in a projection down the axis of the helix. Polar side-chains are on one side of an amphipathic helix and nonpolar side-chains on the other. Such a segregation of polar and nonpolar residues in a protein structure is often expressed as a **hydrophobic moment**. It is analogous to a dipole moment of electrical charge (Equation 8.2) but it measures the spatial separation of polar and nonpolar groups, rather than positive and negative charges.

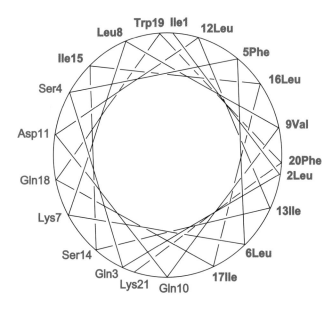

Figure 8-8. Helical wheel representation of an α-helix. The positions of the side-chains are shown in projection down the helix axis. In an ideal α-helix, there are 3.6 residues per complete turn, or a rotation of 100° per residue. The helical wheel consequently repeats after five turns of 18 residues; the 19th to 21st residues would overlap with the 1st to 3rd residues here, but are offset slightly to make them visible. With the amphipathic helix shown, the hydrophobic residues are indicated in *blue* and they lie solely on one side of the helix; the opposite side is composed solely of polar residues (*red*). This peptide has the sequence Ile–Leu–Gln–Ser–Phe–Leu–Lys–Leu–Val–Gln–Asp–Leu–Ile–Ser–Ile–Leu–Ile–Gln–Trp–Phe–Lys.

Hydrogen-bond lengths in polypeptide helices: no evidence for short hydrogen bonds. S. Aravinda *et al.* (2004) *Angew. Chem. Int. Ed. Engl.* **43**, 6728–6731.

Helices in biomolecules. K. Cahill (2005) *Phys. Rev. E* **72**, 062901.

Peptide helices based on α-amino acids. M. Crisma *et al.* (2006) *Biopolymers* **84**, 3–12.

Helices. N. Chouaieb *et al.* (2006) *Proc. Natl. Acad. Sci. USA* **103**, 9398–9403.

Structure and stability of the α-helix: lessons for design. N. Errington *et al.* (2006) *Methods Mol. Biol.* **340**, 3–26.

Origin of the pK_a perturbation of N-terminal cysteine in α- and 3_{10}-helices: a computational DFT study. G. Roos *et al.* (2006) *J. Phys. Chem.* **110**, 557–562.

1. 3_{10}- and Π-helices

Two possible variations of the α-helix conformation are defined by their hydrogen-bonding patterns. The 3_{10}-**helix** has backbone hydrogen bonds between residues (*i*) and (*i* + 3), so it is more tightly wound than the α-helix. In contrast, the Π-**helix** would be more loosely wound, with hydrogen bonds between the (*i*) and (*i* + 5) residues. The name of the 3_{10}-helix arises because it has three residues per turn and 10 atoms between hydrogen bond donor and acceptor. (According to this nomenclature, the α-helix is a 3.6_{13}-helix.) The 3_{10}-helix is not stable on its own, as the backbone atoms are packed too closely, and it is usually observed only as a local distortion at the ends of α-helices. The Π-helix is not expected to be stable, as there would be an empty space down the interior of the helix, and it is not normally observed.

Different spectral signatures of octapeptide 3_{10}- and α-helices revealed by two-dimensional infrared spectroscopy. H. Maekawa *et al.* (2006) *J. Phys. Chem. B* **110**, 5834–5837.

8.3.B. β-Sheet

A β-**sheet** consists of two or more adjacent β-**strands** linked by hydrogen bonds (Figure 8-9). A β-strand has an extended backbone conformation, corresponding to the favorable upper left region of the Ramachandran plot (Figures 8-3 and 8-7). It can, however, also be considered a helix with a helical repeat of exactly 2 (Figure 1-1-C). A single β-strand is not stable in isolation, as there are few interactions to stabilize it. β-**strands are stable only when incorporated into a β-sheet**, in which the β-strands lie side-by-side, linked together by hydrogen bonds between the backbones of adjacent strands. The side-chains of consecutive residues on a β-strand lie alternately above and below the sheet. This gives a pleated appearance to the β-sheet backbone and the β-sheet is sometimes referred to as a β-**pleated sheet**.

The β-strands in a β-sheet may be oriented in the same direction, forming a **parallel** β-**sheet**, or in alternating directions, in an **antiparallel** β-**sheet**. The parallel and antiparallel β-sheet structures have different hydrogen-bonding patterns (Figure 8-9). The parallel β-sheet has the hydrogen bonds evenly spaced along the β-strand and at an angle to the other strand. The hydrogen bonds are more perpendicular to the β-strands in an antiparallel β-sheet, and the spacings between adjacent hydrogen bonds alternate between wide and narrow. The backbone conformations of parallel and antiparallel β-sheets also differ: parallel sheets have $\phi = -119°$ and $\psi = 113°$, while antiparallel have $\phi = -139°$ and $\psi = 135°$. Classical β-sheets were believed to be planar and flat, but most of those observed experimentally have a right-handed twist of about 10°, with slightly more positive values of ϕ and ψ (Table 8-2). This conformation is believed to be more stable because it buries more nonpolar surface area than a flat sheet does.

The various amino acids appear to have different propensities to form β-strands and β-sheets, but attempts to measure this quantitatively have produced markedly different values; the degree to which a particular residue stabilizes the β-strand conformation depends substantially upon its neighbors in the β-sheet.

The formation of β-sheets has not been studied extensively, for most polyamino acids that form β-sheets do so intermolecularly, so the sheets grow without limit and precipitate. A solely intramolecular β-sheet requires segments of polypeptide chain at each end of the β-strands to reverse

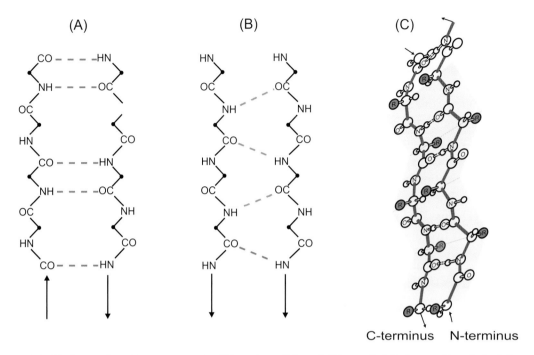

Figure 8-9. Hydrogen-bonding patterns in (A) antiparallel and (B) parallel β-sheets. The *dashed lines* are hydrogen bonds between the backbone carbonyl and NH groups of the peptide bonds. (C) Two strands of an antiparallel β-sheet, with the planes of the peptide bonds highlighted to emphasize the pleated appearance. Note that the side-chains of residues adjacent in the sequence are on opposite sides of the sheets, but those on adjacent strands project in the same direction.

the polypeptide chain, so another type of structure is involved. The most simple β-sheet would consist of two antiparallel β-strands connected by a reverse turn; such structures can be designed to be stable in water. They are observed to be formed within a few microseconds. In each case, however, the turn linking the individual strands is also part of the structure. **The stability of the antiparallel β-sheet appears to increase with the lengths of the strands, but only up to about seven residues, and with the number of strands**, as would be expected. The side-chains of residues on adjacent strands can interact, which will affect the stability of the sheet. The hydrogen bonds between strands contribute to stability, with those in the interior of the sheet contributing most.

The impetus for understanding β-sheet formation has increased with the realization that **amyloid proteins** that are associated with diseases such as Alzheimer's form aggregates of proteins in cells that comprise β-sheets.

Natural polypeptide scaffolds: β-sheets, β-turns, and β-hairpins. K. S. Rotondi & L. M. Gierasch (2006) *Biopolymers* **8**, 13–22.

Model systems for β-hairpins and β-sheets. R. M. Hughes & M. L. Waters (2006) *Curr. Opinion Struct. Biol.* **16**, 514–524.

De novo design of monomeric β-hairpin and β-sheet peptides. D. Pantoja-Uceda *et al.* (2006) *Methods Mol. Biol.* **340**, 27–51.

Hydrophobic surface burial is the major stability determinant of a flat, single-layer β-sheet. S. Yan *et al.* (2007) *J. Mol. Biol.* **368**, 230–243.

8.3.C. Polyglycine

Poly(Gly) has much greater conformational flexibility than other amino acid polymers, due to the absence of a side-chain (Figure 8-3-B), yet it is claimed to adopt an extended conformation in solution and it does adopt a fixed conformation under some conditions. Poly(Gly) with more than nine residues is insoluble in aqueous solution, and in the solid state its backbone is observed to adopt either of two alternative conformations, poly(Gly) I and poly(Gly) II, depending upon the conditions. Poly(Gly) I has long been believed to be an antiparallel rippled (rather than pleated) β-sheet, but it has more recently been suggested that poly(Gly) I adopts two different conformations, depending on the length of the polypeptide chain. In crystals of the short chains, the molecules have an unusual extended conformation generated by alternation of two mirror-symmetrical residual conformations along the chain. The molecules are parallel, and each chain forms interpeptide hydrogen bonds with four adjacent chains. The longer chain structures consist of antiparallel chains that are enantiomorphs and are linked by hydrogen bonds to form rippled sheets.

Poly(Gly) II is a helical conformation with three residues per turn (Table 8-2) and no intrachain hydrogen bonds; it is similar to the conformation of poly(Pro) II (Section 8.3.D).

Interest in the properties of poly(Gly) has increased with the observation that insertion of a few consecutive Gly residues into certain proteins causes medical abnormalities in humans.

Dimorphism of polyglycine I: structural models for crystal modifications. A. V. Kajava (1999) *Acta Crystallogr. D* **55**, 436–442.

Conformational preference of polyglycine in solution to elongated structure. S. Ohnishi *et al.* (2006) *J. Am. Chem. Soc.* **128**, 16338–16344.

8.3.D. Polyproline

A Pro residue has no H atom on the backbone nitrogen and so is unable to form the usual hydrogen bonds of secondary structure. Furthermore, the cyclic side-chain of Pro restricts the backbone conformation, so that ϕ must be close to –80°. Consequently, poly(Pro) does not adopt α-helix or β-strand conformations. Instead, the poly(Pro) backbone adopts one of two possible conformations, depending upon the conditions, called poly(Pro) I and poly(Pro) II. Their backbone conformations are very similar (ϕ angles of –83° and –78°, ψ angles of 158° and 149°, for forms I and II, respectively; Table 8-2) but form I is a right-handed helix, with 3.3 residues per turn, whereas form II is a left-handed helix, with 3.0 residues per turn (Figure 8-10). The primary difference between the two is that **poly(Pro) I has all its peptide bonds in the *cis* conformation ($\omega = 0°$) and poly(Pro) II has them all *trans* ($\omega = 180°$)**.

The poly(Pro) II conformation is unusual in that, due to steric constraints, its polypeptide backbone cannot readily form hydrogen bonds. It is unable to make local hydrogen bonds like those of α-helices, nor can it easily satisfy the hydrogen-bonding potential of neighboring residues, analogous to those in β-strands. **The major determinant of the poly(Pro) II backbone conformation is believed to be that the backbone polar groups are highly solvated**. Although Pro is the residue that most readily adopts the poly(Pro) II conformation, other residues can as well, and each residue has its own propensity to adopt the poly(Pro) II conformation: short, bulky side-chains occlude backbone from

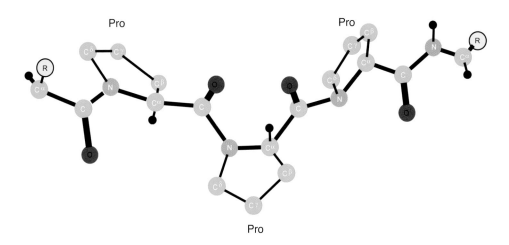

Figure 8-10. Schematic representation of the poly(Pro) II-helix for three sequential Pro residues. The small *black* spheres are H atoms on C^α atoms; the others are labeled.

solvent and thus disfavor the poly(Pro) II conformation, while the lack of a side-chain or long, flexible side-chains, especially those that can form hydrogen bonds to the backbone, tend to favor it. Some polar residues may also 'cap' the ends of the helical segment, as occurs in α-helices. The determinants of the poly(Pro) II conformation are strictly local and involve minimal interactions between residues, so unfolding of the helical conformation is not cooperative, in contrast to that of α-helices (Section 8.4).

The poly(Pro) I and II forms can be interconverted by changing the solvent. The interconversion occurs by a 'zipper' mechanism in which the intrinsically slow *cis–trans* isomerization of each of the peptide bonds (Section 8.2.B.1) starts at one end and progresses sequentially along the polypeptide chain. The I ⟶ II interconversion starts at the amino end of the chain, whereas the reverse II ⟶ I interconversion starts at the other end.

Study of poly(Pro) II has accelerated recently, as this local conformation within folded proteins has been found to be involved in a number of biological phenomena. Also, the polypeptide backbone is now believed to possess a significant propensity to adopt the poly(Pro) II-helical conformation and play a major role in the conformations of unfolded proteins (Section 11.2).

Properties of polyproline II, a secondary structure element implicated in protein–protein interactions. M. V. Cubellis *et al.* (2005) *Proteins* **58**, 880–892.

Single peptide bonds exhibit poly(pro)II ('random coil') circular dichroism spectra. I. Gokce *et al.* (2005) *J. Am. Chem. Soc.* **127**, 9700–9701.

Unusual compactness of a polyproline type II structure. B. Zagrovic *et al.* (2005) *Proc. Natl. Acad. Sci. USA* **102**, 11698–11703.

Stereoelectronic effects on polyproline conformation. J. C. Horng & R. T. Raines (2006) *Protein Sci.* **15**, 74–83.

8.4. α-HELIX FORMATION FROM A RANDOM COIL

The α-helix is a relatively simple structure and involves only a single, continuous segment of polypeptide chain. Therefore it is a good model system in which to study protein conformational changes, and considerable effort has gone into understanding the mechanism of forming a monomolecular α-helix from a random-coil polypeptide chain.

Initial studies of helix formation used polyamino acids in which all their residues were the same. Although such systems are simple in theory, studying the helix–coil transition experimentally in water is not. Polyamino acids that are soluble in water do not form the α-helix, and those that do form an α-helix are not water-soluble. The most informative studies require the use of short peptides with distinct amino acid sequences that are designed to be water-soluble and adopt the α-helical conformation to a substantial, measurable extent.

8.4.A. Factors Stabilizing the α-Helix

All peptide α-helices unfold with increasing temperature, so helix formation is an enthalpically favorable process and proceeds with the release of heat. The hydrogen bonds formed between atoms in the peptide backbone (>C=O...H–N<) probably provide the main enthalpic stabilization to the helical conformation, and each is believed to contribute 0.5–1.5 kcal/mol in stability. Opposing this favorable enthalpy change is the conformational entropy lost upon freezing the backbone conformation, plus, to a more limited extent, that of the side-chains. This unfavorable entropy change is about the same magnitude as the favorable enthalpy change, meaning that in isolation α-helices are at best marginally stable.

Different amino acid residues have different intrinsic tendencies to form helices (Table 8-3), known as the **helix propensity**. If this were an intrinsic property, it should be characteristic of a given amino acid and independent of its context. Measurements of helix propensities in different systems have, however, provided markedly different results, because **interactions between side-chains can also affect the stability of helices**. Side-chains three and four residues apart in the amino acid sequence will be close in space in a helical conformation (Figure 8-11). Consequently, favorable electrostatic interactions, hydrogen bonds or hydrophobic contacts between these side-chains can contribute up to 1 kcal/mol to helix stability. Conversely, any favorable interactions that are possible in the nonhelical form of the peptide or protein will decrease the amount of α-helix. The special amino acid residues Pro and Gly are exceptional. Gly has no side-chain and a much more flexible backbone that disfavors adopting any fixed conformation, including the α-helix. In the case of Pro, the amide nitrogen that would be hydrogen bonded is absent, cyclized with the Pro side-chain. Single Pro residues can be accommodated within long α-helices by distorting the helical geometry locally, but the α-helix stability is decreased by about 3 kcal/mol in model peptides. On the other hand, Pro is a preferred residue at the N-terminus of the helix, where other residues would have an unpaired backbone amide.

Many other interactions involving the amino acid side-chains affect the stability of an isolated α-helix. The first four >N–H and the last four >C=O groups of the helix will not be hydrogen bonded to a backbone partner. **Polar groups without hydrogen bonds to water or other protein groups are destabilizing**, so it is important that all polar groups be hydrated or have other intramolecular partners. The side-chain of the first residue of the helix (the **N-cap**) can fill this role if it is able to

Table 8-3. Relative helical propensities of amino acid residues

Amino acid residue	Relative stabilization of α-helical conformation (kcal/mol)
Ala	0.77
Arg	0.68
Lys	0.65
Leu	0.62
Met	0.50
Trp	0.45
Phe	0.41
Ser	0.35
Gln	0.33
Glu	0.27
Cys	0.23
Ile	0.23
Tyr	0.17
Asp	0.15
Val	0.14
Thr	0.11
Asn	0.07
His	0.06
Gly	0
Pro	≈ −3

Measured by substituting each amino acid residue into a solvent-accessible position in a synthetic peptide that forms an α-helical dimer that is in equilibrium with a randomly coiled monomeric state. The equilibrium constants for the monomer–dimer equilibrium were determined with the various peptides to provide a measure of the differences in stabilities of the α-helical conformation. The value for Gly was arbitrarily chosen as zero. Data from K. T. O'Neil & W. F. DeGrado (1990) *Science* **250**, 646–651.

hydrogen bond to one or two otherwise unsatisfied backbone >N–H groups at that end of the helix. Analogous **C-cap** interactions can occur at the C-terminus. These 'capping' hydrogen bonds have been proposed as important factors in determining where helices begin and end in proteins, and each is believed to contribute 1–2 kcal/mol to stability of the α-helix.

The helical conformation has the peptide groups and their associated dipoles (Equation 8.2) aligned in the same direction, whereas in random coil forms the dipoles are distributed randomly. Alignment of the individual peptide group dipoles in a single direction is usually considered to give rise to a **helix macrodipole** that is simply the additive effects of the individual dipoles: the N-terminus of the helix has a partial positive charge, while the C-terminus carries a partial negative charge (Figure 8-11). Residues with charged side-chains at the ends of a helix can interact favorably or unfavorably with this dipole, or with the microdipoles of the individual N–H and C=O groups (Equation 8.2); the free energy contribution of each such interaction is believed to be as much as 1 kcal/mol. Such an interaction has an equal effect on the ionization of the residue; for example, a Cys residue at the N-terminus of an isolated α-helix can stabilize the helix and have the pK_a of its thiol group lowered by as much as 1.6 pH units. The most substantial interactions appear to be due to hydrogen bonding or electrostatic interactions with the microdipoles of the individual peptide bonds (Equation 8.2), rather than with a macrodipole charge at the entire end of the helix, as is frequently assumed.

Thermodynamics of α-helix formation. G. I. Makhatadze (2005) *Adv. Protein Chem.* **72**, 199–226.

Hydrogen bonding is the prime determinant of carboxyl pK_a values at the N-termini of α-helices. M. A. Porter *et al.* (2006) *Proteins* **63**, 621–635.

Alpha-helix stabilization by alanine relative to glycine: roles of polar and apolar solvent exposures and of backbone entropy. J. Lopez-Llano *et al.* (2006) *Proteins* **64**, 769–778.

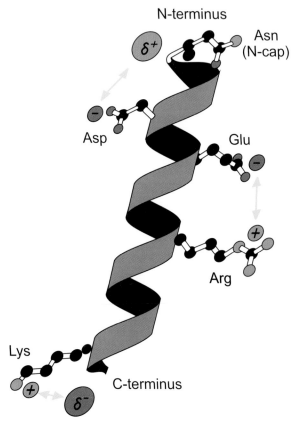

Figure 8-11. Interactions involving amino acid side-chains that increase the stability of isolated α-helices. A ribbon model of an α-helix is illustrated, with stabilizing interactions involving hydrogen bonding of the N-cap Asn side-chain at the N-terminus, favorable electrostatic interactions of ionized Asp and Lys side-chains with the helix dipole, and electrostatic interactions between oppositely charged Glu and Arg side-chains on adjacent turns of the helix. Of course, interactions between side-chains with the same charge will destabilize the helix, but hydrophobic interactions between nonpolar side-chains can also be significant.

8.4.B. Helix–coil Transitions

Although appearing to be very stable structurally (Figure 8-6), **isolated α-helices are generally only marginally stable**, if at all, in aqueous solution under normal circumstances. **They are usually dynamic systems in aqueous solution, being rapidly formed and unfolded some 10^3–10^7 times per second**. Formation of a helix has been believed to occur very rapidly, within 10^{-5}–10^{-7} s, but unraveling of the helix is usually just as rapid. Some recent measurements, however, indicate that the time-scale is about 10^{-3} s. The rate of generating the helix is generally independent of the length of the polypeptide chain, but the rate of unraveling is strongly size-dependent, and **the overall stabilities of α-helices depend upon the length of the polypeptide chain**.

Initiation of the helix in a random coil is the slowest step, whereas subsequent growth of the helix is rapid. Initiation of a helix can occur anywhere in the random coil, but growth and unraveling occur only at the ends of helical segments. For a polyamino acid of moderate length with a single helical region, this may be expressed simply by interconversion of the two conformations of each residue: coil, C, and helical, H_i, where i is the number of hydrogen bonds involved in the helical conformation (Figure 8-12-A). The rate constant for adding a residue to the end of a helix is k_f, and k_b is that for removing a residue. Both have very similar values, usually believed to occur in the region of 10^8–10^{11} s^{-1}. The value of the equilibrium constant for each residue added to the helix, s ($= k_f/k_b$),

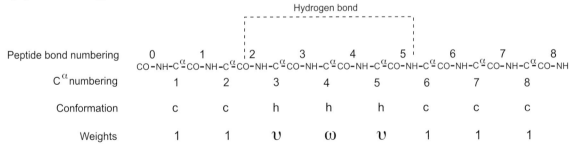

Figure 8-12. Two models for the α-helix–coil transition. (A) The relatively simple model of Zimm and Bragg treats the transition of a polypeptide chain, or a long segment of one, as simply being the equilibrium between one state, C, in which there is no helix, and other states, in which there is a helical segment of increasing lengths, H_i. The rate constants for adding and subtracting helical residues from the ends of the helical segment are assumed to be constant, except that the rate of forming the first turn of helix is lower by the nucleation parameter σ. The helix propensity of a residue is given by the equilibrium constant s. (B) The Lifson–Roig theory denotes the conformation of each residue as being either random coil (c) or helical (h). The statistical weights are defined using the coil state as a reference (weight = 1), with helix nucleation sites having weight v and internal helical residues w.

has a value not far from unity. In contrast, the rate constant for forming the first turn of helix is σk_f; the **nucleation factor** σ reflects the difficulty in initiating a helix, and the most accurately measured values are between 2×10^{-3} and 3.5×10^{-3}.

Initiating a helix is so much more difficult than adding an additional residue because an α-helix is defined by hydrogen bonds between residues four apart in the polypeptide chain. Consequently, four residues that will make up the first turn must be fixed in space simultaneously before the first turn of the helix and the first hydrogen bond can be formed. After that, adding an additional hydrogen bond and lengthening the helix by one residue requires that only one residue be fixed; this residue would already be in reasonable proximity to the end of the helix owing to its position as the next residue along the chain. In other words, **the entropic cost of forming the first turn and hydrogen bond of the helix is much greater than that of adding additional residues**; the effective concentrations of two residues four apart in the random coil are much lower than when one is part of a helix and the other is the next residue in the polypeptide chain. An additional factor is that the dipoles of neighboring residues in one turn of the α-helix are aligned parallel, which is unfavorable energetically in isolation. Once the helix is formed, however, this is compensated for by favorable head-to-tail interactions between the dipoles three and four residues apart. The unfavorable dipole interactions would predominate while forming the first turn of helix, but the subsequent favorable dipole interactions would assist helix propagation.

Owing to the difficulty of nucleation, **the α-helix $\rightarrow$ coil transition is cooperative**. The equilibrium constant between the two conformations is given by:

$$K_n = \frac{[H_n]}{[C]} = \sigma s^n \tag{8.7}$$

With σ-values of about 2×10^{-3} and s-values not much greater than unity, large values of n are required to give values of K_n greater than unity, i.e. a stable α-helix, and only long polypeptides are expected to be helical under such conditions. The total average helix content, however, will be given by the sum of all the possible helices of various lengths. For example, with $\sigma = 2 \times 10^{-3}$ and $s = 1.2$, a 13-residue peptide will be totally helical only 2.1% of the time, but the partial helices will increase the average helix content to nearly 20%.

At the midpoint of the helix–coil transition in very long chains, the average length of a helix is given by $\sigma^{-1/2}$. With $\sigma = 2 \times 10^{-3}$, therefore, the average helical segment at the midpoint of the transition should be 22 residues long, followed by an average of 22 coil residues. When $s > 1$, the observed rate constants for helix formation (k_{+1}) and unraveling (k_{-1}) are given by:

$$\begin{aligned} k_{+1} &= \sigma k_f \frac{s-1}{s} \\ k_{-1} &= k_b \frac{s-1}{s} \left(\frac{1}{s}\right)^{n-1} \end{aligned} \tag{8.8}$$

The term $(s - 1)/s$ gives the probability that helix formation will be completed once a nucleus has been formed. The value of σ is essentially independent of temperature, but the value of s generally decreases with increasing temperature. Consequently, the α-helix is most stable at low temperatures and may usually be 'melted out' by heating.

This simple theory assumes that adding a residue to either end of the helix is equally probable. This is not the case with mixed sequences or with ionized amino acid residues. Adding a positively charged residue to the carboxyl, negative end of a helix is more favorable, by a factor of about four, than to the amino end; the opposite holds for negatively charged side-chains or at the other end of the helix. Other polar side-chains may also interact with the peptide backbone to varying extents at the two ends of the helix, and hydrophobic or ionic interactions between side-chains can also have significant effects (Figure 8-11).

The different amino acids are believed to have different tendencies to form α-helices, but it has not been possible to measure their intrinsic values of σ and s using the various polyamino acids because most are not soluble or they preferentially form other conformations. Values for s were estimated by incorporating the amino acids to varying extents as a 'guest' in a water-soluble 'host' polypeptide, but this method is susceptible to interactions of the guest side-chains with the peptide dipole, each other and those of the host. The best currently available values of the relative intrinsic helical tendencies of different amino acids have been measured using short peptides of defined sequence (Table 8-3). These measurements indicate that the helical propensities of the different amino acids vary more than was thought previously. For example, Ala has an s-value of approximately 1.56, while that of Gly is only 0.15 due to its greater flexibility. One factor that limits the helical propensities of residues with branched side-chains is that only certain conformations are compatible with the α-helical conformation.

8.4.C. Helix–coil Models

The two most popular models for the helix–coil transition in peptides were developed independently by Zimm and Bragg and by Lifson and Roig. They share similar features, with a few important differences (Figure 8-12). For each model, helix formation is a function of three basic parameters: (1) the chain length or number of residues in the polypeptide, (2) a helix propagation parameter (s- or w-value) and (3) a helix nucleation parameter (σ or v^2). In each model, a single unit or residue can exist in only one of two possible conformations, helix or coil, h or c. Enumeration of all possible combinations of h and c residues for a polypeptide chain of a given length, along with the assigned statistical weights, provides the **partition function** for the system. The partition function can then be used to calculate the average properties of the population, notably the fraction of helical residues, the average length and number of helical stretches and other characteristics of the population.

Helix–coil theories: a comparative study for finite length polypeptides. H. Qian & J. A. Schellman (1992) *J. Phys. Chem.* **96**, 3987–3994.

Chain-length dependence of α-helix to β-sheet transition in polylysine: model of protein aggregation studied by temperature-tuned FTIR spectroscopy. W. Dzwolak *et al.* (2004) *Biopolymers* **73**, 463–469.

Electron density redistribution accounts for half the cooperativity of α-helix formation. A. V. Morozov *et al.* (2006) *J. Phys. Chem. B* **110**, 4503–4505.

Infrared temperature-jump study of the folding dynamics of α-helices and β-hairpins. F. Gai *et al.* (2007) *Methods Mol. Biol.* **350**, 1–20.

1. Zimm–Bragg Model

The Zimm–Bragg model has three key parameters: n, the number of peptide units in the chain; s, the helix propagation parameter; and σ, the helix nucleation parameter (Figure 8-12-A). This model defines n as the number of peptide units (–CONH–) in the chain, which causes some problems in ascribing s-values to particular side-chains, since a peptide unit encompasses parts of two amino acid residues.

In the full treatment of the Zimm–Bragg theory, a 4×4 correlation matrix is needed to assign statistical weights to helical or coil residues (Figure 8-12) and calculate the partition function for the peptide system. Residue-specific values of s and σ can be used. A somewhat simplified version of the model uses a 2×2 correlation matrix that can include nearest-neighbor interactions. A useful version of the Zimm–Bragg model for short peptides treats the peptide as a homopolymer and uses the single-sequence approximation, in which it is assumed that only one stretch of helical conformation exists at one time for a given short peptide chain. Since helix nucleation is an unfavorable event, it can be treated as occurring only once for a given peptide, and propagation of the helix 'zips up' the chain. The fraction of helical residues can be expressed as:

$$f_H = \frac{\sigma s}{(s-1)^3 n} \left\{ \frac{ns^{n+2} - (n+2)s^{n+1} + (n+2)s - n}{1 + \left[\sigma s/(s-1)^2\right]\left[s^{n+1} + n - (n+1)s\right]} \right\} \tag{8.9}$$

This version of the Zimm–Bragg model is used most frequently with simple repeating homopolymers for analyzing the cooperative dependence of helix formation on the peptide length.

Parameters of helix–coil transition theory for alanine-based peptides of varying chain lengths in water. J. M. Scholtz et al. (1991) *Biopolymers* **31**, 1463–1470.

---

2. Lifson–Roig Model

The Lifson–Roig model has some important differences. Most important is that the helical unit is defined as an amino acid residue, centered on the C^α atom and the attached side-chain, including the adjacent NH and CO groups. A residue is classified as being either helical, h, or coil, c, based on its (ϕ, ψ) angles. A residue that lies in helical (ϕ, ψ) space is classified as being helical, h, and all other conformations are considered nonhelical, or coil, c. The statistical weights (v or w) reflect the microscopic equilibrium constant for the transition between h and c. Helical residues flanked on both sides by other h residues are assigned a weight w, whereas h residues at the ends of a stretch of h residues are assigned a weight of v. Figure 8-12-B illustrates this for the simple case of a helical peptide with one hydrogen bond stabilizing the helical conformation. The parameter w is the helix propagation parameter, similar to the s-value in the Zimm–Bragg model, while v is a nucleation parameter for each end of the helical stretch of residues; v^2 is related to the Zimm–Bragg σ parameter. To assign the appropriate statistical weights to the partition function, the full Lifson–Roig model uses a 3×3 correlation matrix between adjacent residues.

The Lifson–Roig model is used more frequently because the values of w and v are readily assigned to specific residues (Figure 8-12-B); it can also be modified to include other interactions involving specific side-chains. The Lifson–Roig parameters make it possible to predict the amount of helix present in a peptide of any sequence.

Addition of side chain interactions to modified Lifson–Roig helix–coil theory: application to energetics of phenylalanine–methionine interactions. B. J. Stapley *et al.* (1995) *Protein Sci.* **4**, 2383–2391.

A model for the coupling of α-helix and tertiary contact formation. A. C. Hausrath (2006) *Protein Sci.* **15**, 2051–2061.

8.4.D. Trifluoroethanol (TFE)

2,2,2-Trifluoroethanol (CF_3CH_2OH; TFE) as a co-solvent stabilizes the α-helical conformation in peptides, at least those with some intrinsic tendency to adopt that conformation. Maximum stabilization requires relatively high concentrations, roughly 40–50% (v/v) TFE. In certain instances, TFE stabilizes the β-structure as well. How and why TFE stabilizes the polypeptide structure is unclear. It might interfere with the ability of water to solvate peptide groups of the polypeptide backbone; this is most significant with the disordered conformation, which would be destabilized. It has also been suggested that the structure-stabilizing effect of TFE is caused by the preferential aggregation of TFE molecules around the peptides. This coating would displace water, thereby removing alternative hydrogen-bonding partners and providing a low dielectric environment that favors the formation of intrapeptide hydrogen bonds. An unstructured peptide, whose backbone peptide groups are exposed to water, would be destabilized by the presence of TFE, and structures in which these groups are hydrogen bonded would be favored.

TFE interacts only weakly with nonpolar residues, so low concentrations do not disrupt hydrophobic interactions within peptides. At high concentrations, however, TFE weakens hydrophobic interactions and can be used as a protein denaturant to unfold proteins (Section 11.1.C); the unfolded protein is not a random coil. For example, the denatured state of hen egg-white lysozyme in 70% TFE consists of regions of high helical content that correspond to the regions that are helical in the native protein but lacking other interactions normally present in the folded structure.

Lifson–Roig nucleation for α-helices in trifluoroethanol: context has a strong effect on the helical propensity of amino acids. J. R. Lawrence & W. C. Johnson (2002) *Biophys. Chem.* **101**, 375–385.

Comparison of the effects of 2,2,2-trifluoroethanol on peptide and protein structure and function. J. F. Povey *et al.* (2007) *J. Struct. Biol.* **157**, 329–338.

8.5. FIBROUS PROTEINS

Natural fibrous proteins are similar to the above regular structures in that they have extended structures. Most play structural roles and have regular structures that represent a level of complexity that is intermediate between pure secondary structure as described above and the tertiary structures of globular proteins to be described in Chapter 9. The basis for their regular conformations is usually apparent in their amino acid sequences. On the other hand, the large sizes of fibrous proteins and their general insolubility make them more difficult than soluble proteins to characterize experimentally. Described here are the classical fibrous proteins that are best understood structurally. Their structures are diverse but they can be grouped conveniently into three classes: (1) α-fibrous proteins, (2) β-fibrous structures and (3) collagen proteins. Proteins that assemble into filamentous arrays but are individually globular in form are often considered to be fibrous; they will be described in Section 9.1.H.5.

All fibrous proteins have repeating elements in their amino acid sequences. Often they are in the form of tandem sequence motifs that specify the secondary structure as well as its assembly to higher levels of structure.

Fibrous proteins: new structural and functional aspects revealed. D. A. Parry & J. M. Squire (2005) *Adv. Protein Chem.* **70**, 1–10.

Structural and functional implications of sequence repeats in fibrous proteins. D. A. Parry (2005) *Adv. Protein Chem.* **70**, 11–35.

8.5.A. α-Fibrous Structures: Coiled Coils

α-Fibrous proteins occur naturally as individual elongated molecules, such as laminin in basement membranes and fibrinogen in blood plasma, or as filamentous assemblies, such as myosin in muscle thick filaments, desmin in type III intermediate filaments and fibrin molecules in blood clots. **All members of this family adopt α-helical conformations and their protein sequences have high helical propensities** (Section 8.4). The individual α-helices aggregate into rope-like structures in which two to five α-helices are intertwined. This assembly is stabilized and dictated by nonpolar interactions between the helices and electrostatic interactions involving certain amino acid sidechains. Precisely what structure is formed depends upon the amino acid sequence, which dictates both types of interaction.

The sequences of α-fibrous proteins are characterized by **heptad repeats** in which nonpolar residues alternate three and four residues apart. The repeat can be represented as $(a - b - c - d - e - f - g)_n$. About 75% of the *a* and *d* positions are occupied by nonpolar residues such as Leu, Ile and Val (Table 8-4 and Figure 8-13). These sequences adopt a right-handed α-helical conformation with about 3.6 residues per turn. The apolar residues in the heptad repeat are 3.5 residues apart, on average, so they form an apolar stripe on the surface of the α-helix that winds around the helix in a left-handed manner. Two or more such α-helices can come together, optimize the packing of the apolar residues along their interface and wind around one another to generate a left-handed coiled-coil, rope-like structure (Figure 8-14). The individual right-handed α-helices are distorted from their normal straight geometry into the left-handed superhelical conformation, but only slightly.

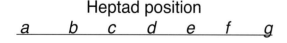

```
          Heptad position
  a    b    c    d    e    f    g
                                Arg-
Met- Lys- Gln- Leu- Glu- Asp- Lys-
Val- Glu- Glu- Leu- Leu- Ser- Lys-
Asn- Tyr- His- Leu- Glu- Asn- Glu-
Val- Ala- Arg- Leu- Lys- Lys- Leu-
Val- Gly- Glu- Arg
```

Figure 8-13. Heptad distribution of residues in the amino acid sequence of the GCN4 leucine zipper. The nonpolar residues are *blue* whereas the ionized residues are *red*.

Table 8-4. Frequency of occurrence of amino acid residues in the heptad repeats of the coil–coil proteins myosin, tropomyosin and paramyosin and intermediate filaments

Residue	% Average occurrence at position						
	a	b	c	d	e	f	g
Ala	**10.2**	**12.3**	8.1	**22.2**	4.4	**11.1**	9.3
Cys	0.9	0.0	0.4	0.3	0.2	0.5	0.1
Asp	0.1	**13.4**	**13.0**	1.0	4.0	9.8	8.0
Glu	0.8	**21.2**	**19.3**	5.5	**31.5**	**14.7**	**20.1**
Phe	2.0	0.4	1.4	2.1	0.4	0.4	0.0
Gly	0.6	1.7	3.5	1.0	1.3	4.2	1.2
His	1.2	2.7	1.6	1.1	0.8	2.6	0.7
Ile	**13.2**	0.9	2.2	6.3	2.4	2.2	2.2
Lys	7.7	**15.3**	**11.5**	0.6	9.0	**10.5**	**14.9**
Leu	**32.2**	1.6	3.9	**34.7**	6.4	3.9	5.6
Met	4.9	0.9	0.9	2.3	0.9	1.0	0.4
Asn	3.6	4.3	4.6	1.1	5.7	4.8	2.7
Pro	0.0	0.2	0.1	0.0	0.0	0.0	0.0
Gln	0.8	8.7	7.5	4.1	**14.0**	6.1	**13.2**
Arg	5.5	6.2	8.2	0.9	6.6	**13.2**	**10.7**
Ser	2.1	3.6	8.1	2.2	5.2	8.1	4.5
Thr	1.2	3.8	3.7	2.2	5.1	3.6	3.5
Val	8.9	1.9	1.9	6.0	1.9	3.1	2.7
Trp	0.1	0.0	0.0	0.7	0.0	0.1	0.0
Tyr	4.1	0.9	0.3	5.7	0.3	0.2	0.2

Data from C. Cohen & D. A. D. Parry (1990) *Proteins* 7, 1–15.

In many such coiled coils, the nonpolar residues at positions *a* and *d* are predominantly Leu residues, thus initially they had the alternative name **leucine zippers** (Table 8-4). Leu residues in position *d* point directly at the interface, whereas the Val residues that dominate at position *a* point outwards from the interface. This is the explanation for the observed predominance at position *a* of β-branched apolar residues, such as Val and Ile, where part of their side-chains can be redirected back into the interface region and away from the aqueous environment.

Favorable electrostatic interactions can also be made between the chains, which help to specify both the relative chain direction and the axial displacement between the chains. Their amino acid sequences

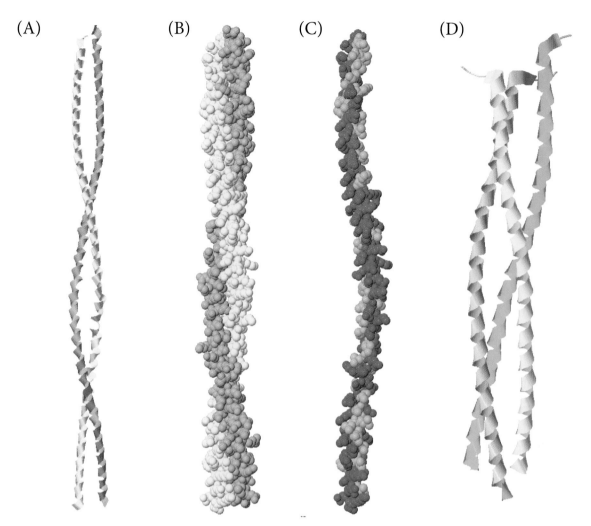

Figure 8-14. The structure of α-helical coiled coils. (A) A parallel dimeric coiled coil from the C-terminus of rabbit skeletal α-tropomyosin (PDB file *2d3e*). The coiled ribbon represents the α-helical course of the polypeptide backbone of each polypeptide chain. (B) A space-filling model of the same structure, with the two polypeptide chains differentiated. (C) One polypeptide chain of the structure has been removed, to expose the nonpolar surface that had been buried in the other chain. Nonpolar groups are colored *grey*, polar groups *blue*. (D) Example of a trimeric parallel coiled coil (PDB file *1wt6*). Figure generated using the program Jmol.

contain a very regular linear disposition of acidic and basic residues. The periods are usually 180° out of phase, which generates a simple rod structure with alternating bands of positive and negative charge. Assembly into the filamentous form is specified in large part by interactions between the charged groups between the adjacent helices, in what has been called an **ionic zipper**. The interchain ionic interactions occur predominantly between oppositely charged residues in positions *e* and *g* of different chains (Figure 8-15). These residues are often Glu or Gln, with Arg and Lys residues prominent at position *g*.

Coiled coils are helical structures that have an axis that is itself helical. A general feature of all such coiled-coil conformations is that the handedness of coiling alternates at successive levels of structure: the individual helices are right-handed, but the coiling of the helices together is left-handed. This is similar to the principle by which ropes are made: it maximizes the interactions between strands and minimizes their slippage.

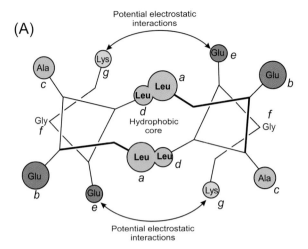

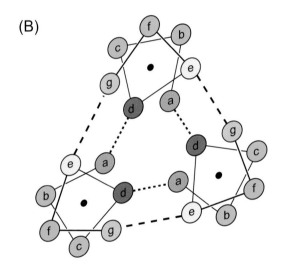

Figure 8-15. Helical wheel representations of one heptad repeat of the individual α-helices of a coiled coil, illustrating how the side-chains of residues pack together and interact in coiled coils with (A) two or (B) three strands.

The types of apolar residues in positions *a* and *d* are important in specifying, in subtle ways, the number of chains in the coiled-coil molecule, as are the electrostatic interactions between the residues at positions *e* and *g* (Figure 8-15). Two-stranded α-helical coiled coils have the strands parallel (rather than antiparallel) and in axial alignment, which is believed to maximize interactions between the two chains. The polypeptide chains can be identical or different. Three-stranded structures (Figure 8-15) occur in many proteins and contain the same or different polypeptide chains. Four-stranded ropes are found in the silks of bees, wasps and ants, as well as in globular proteins (Section 9.1.D.6). Five-stranded coiled coils have also been described. Which structure is adopted depends upon the amino acid sequences of the polypeptide chains, on the way that the apolar stripes achieve the most efficient packing, and on the electrostatic interactions.

There are exceptions to the repeating heptad, and Pro residues are occasionally found within heptads. Many sequences show discontinuities in their heptad substructures. All such short discontinuities can be classified as either *stutters* or *stammers*, which correspond to deletions of three or four residues, respectively, from an otherwise continuous repeat. Structurally, a stutter produces some local underwinding of the coiled coil that increases its supercoil pitch length, whereas a stammer causes local overwinding that shortens the length of the supercoil pitch. These discontinuities are assumed to produce functional variations of this regular structure.

Many of the proteins with coiled-coil structures have segments of polypeptide chain with different conformations at one or both ends. For example, **myosin** of muscle is a long, two-stranded coiled coil that has globular heads at each of the two carboxyl ends that are responsible for the catalytic and movement functions. **Intermediate filaments** have central double-helical rods 310 residues long, but other structures at both ends. Many coiled-coil proteins also participate in higher order structures, associating reversibly with identical or different molecules.

The α-fibrous proteins without periodicities in charged residues appear not to form regular filamentous assemblies. Instead, the molecules exist separately or assemble in a less regular manner, using nonhelical domains elsewhere in the molecule, usually at their N- and C-termini.

A parallel coiled-coil tetramer with offset helices. J. Liu *et al.* (2006) *Biochemistry* **45**, 15224–15231.

A seven-helix coiled coil. J. Liu *et al.* (2006) *Proc. Natl. Acad. Sci. USA* **103**, 15457–15462.

Self-assembly of coiled-coil tetramers in the 1.40 A structure of a leucine-zipper mutant. Y. Deng *et al.* (2007) *Protein Sci.* **16**, 323–328.

Crystal structures of tropomyosin: flexible coiled-coil. Y. Nitanai *et al.* (2007) *Adv. Exp. Med. Biol.* **592**, 137–151.

Considerations in the design and optimization of coiled coil structures. J. M. Mason *et al.* (2007) *Methods Mol. Biol.* **352**, 35–70.

Molecular basis of coiled-coil formation. M. O. Steinmetz *et al.* (2007) *Proc. Natl. Acad. Sci. USA* **104**, 7062–7067.

8.5.B. β-Fibrous Structures

Silk proteins are the best-known β-fibrous structures. The basic structure is an extended array of β-strands assembled into an antiparallel β-sheet. Two or more such sheets then aggregate. The sequences of many silks have repetitive amino acid sequences, and their amino acid compositions are generally very simple, with a predominance of only a small number of amino acids, especially Gly, Ala, Ser and Gln. **The sequences of many silks contain one or more type of dipeptide repeats**. The side-chains of the two residues of each repeat would be situated on opposite sides of a β-sheet, so dipeptide-repeat sequences will generate a β-sheet with one face consisting of the amino acid side-chains of one of the two residues, the other face by the second residue. For example, a face comprising Ala residues would be nonpolar and could readily interact hydrophobically with the same face in a second sheet. If the second residue were a polar amino acid, such as Ser, the opposite faces would be polar and would interact well with water. The β-sheets are probably not planar in these silk structures, as those in both feather and scale **keratins** are twisted, probably in a right-handed manner.

The crystalline β-sheet makes up only part of silk structures; they usually have regions believed to be sequential β-turns linked together into β-**spirals** plus 3_{10}-helices and linker and spacer segments. Certain silks almost certainly have three- and seven-residue repeats and form other structures.

β-structures in fibrous proteins. A. V. Kajava *et al.* (2006) *Adv. Protein Chem.* **73**, 1–15.

Molecular mechanisms of spider silk. X. Hu *et al.* (2006) *Cell. Mol. Life Sci.* **63**, 1986–1999.

β-silks: enhancing and controlling aggregation. C. Dicko *et al.* (2006) *Adv. Protein Chem.* **73**, 17–53.

Natural triple β-stranded fibrous folds. A. Mitraki *et al.* (2006) *Adv. Protein Chem.* **73**, 97–124.

8.5.C. Collagens

Collagen is the main constituent of extracellular matrices, including bones, tendons, skin, ligament, blood vessels and supporting membranous tissues of animals. In spite of this great diversity of roles, an individual usually has only about a dozen distinct types of collagen polypeptide chains, and they are closely related. Collagen polypeptide chains typically have just over 1000 residues, but there are exceptions.

Collagen polypeptides have distinctive repetitive sequences in which every third residue is Gly, i.e. (-Gly-Xaa-Yaa-)$_n$, with a preponderance of Pro residues as Xaa and Yaa. Many of the Pro residues (and also Lys residues) at position Yaa are hydroxylated during and after their biosynthesis. Pro residues are hydroxylated on the *gamma* carbon when in the sequence –Xaa–Pro–Gly– but on the *beta* carbon when in the sequence –Gly–Pro–:

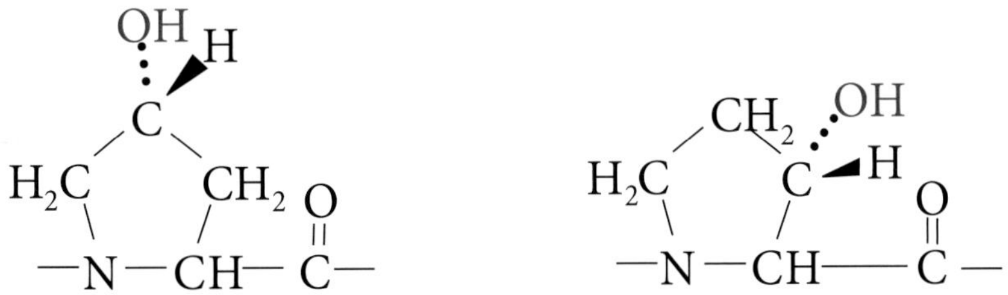

$$\text{γ-OH-Pro (4-hydroxy-Pro)} \qquad \text{β-OH-Pro (3-hydroxy-Pro)} \tag{8.10}$$

Lys residues in the sequence –Xaa–Lys–Gly are hydroxylated on the *delta* carbon:

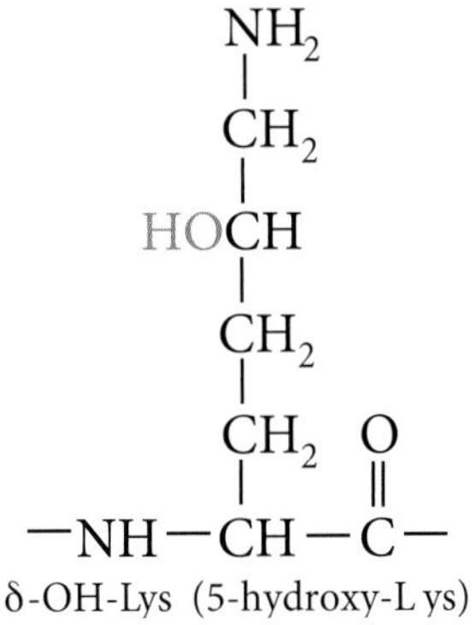

$$\text{δ-OH-Lys (5-hydroxy-Lys)} \tag{8.11}$$

The reason for the repeating sequence of the collagen chains is apparent from its 3-D structure (Figure 8-16). Three polypeptide chains are coiled about each other, each with a slightly twisted, left-handed three-fold helical conformation like that of poly(Pro) II and poly(Gly) II (Table 8-2 and Figure 8-10). There are 10 residues per three turns (108°/residue) or seven residues per two turns (103°/residue), with a pitch of 29 Å. The Gly residues at every third position come into close proximity with the other two chains in the triple helix, too close to permit a side-chain. The chains are linked together

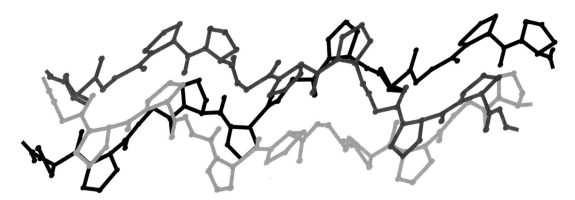

Figure 8-16. The structure of the triple helix of (Pro–Pro–Gly)$_{10}$, which is an idealized version of collagen. Adapted from K. Okuyama *et al.* (1981) *J. Mol. Biol.* **152**, 427.

by hydrogen bonds between the backbone –NH of the Gly residues and the backbone carbonyl group of residue Xaa of another chain, which are nearly perpendicular to the helical axis. The side-chains of residues Xaa and Yaa are exposed on the surface of the triple helix. The importance of these features is apparent from the observations that interruptions of the Gly–Xaa–Yaa repeats decrease the stability of the triple helix, and substitution of the Gly residue destabilizes and disrupts the triple helix. The Pro residues at positions Xaa and Yaa probably impart rigidity and stability to the structure, because their conformation is one of the few accessible to Pro residues.

The hydroxyl groups on hydroxyl–Pro residues (often abbreviated Hyp) are involved in hydrogen bonding between chains that stabilizes the triple helix. Hydroxylation of Pro residues in Pro–Gly or Pro–Yaa sequences increases the thermal stability markedly, although it is decreased if in the sequence –Gly–Pro–. Whether the hydroxyl group is at the *beta* or *gamma* position also affects the stability. A further role of hydroxylation in collagen is to stabilize the triple helix by adjusting to the right pucker of the Pro ring (and thus the appropriate ϕ angle) in position Yaa. There are repulsions between the Pro residues at position X and γ-hydroxy-Pro residues at position Y.

Another general feature of collagens is a low content of hydrophobic residues; the interior of the triple helix is not stabilized by hydrophobic interactions. Interactions between multiple triple helices are, however, stabilized to varying extents by hydrophobic interactions between nonpolar side-chains on their surfaces.

The three polypeptides within a collagen molecule are typically not identical. The most common collagen, collagen 1, contains two identical polypeptides (α_1) and a third (α_2) that is very similar. Polypeptide chains of 1000 residues produce a collagen triple helix that is 14 Å in diameter and 3000 Å in length, sufficiently large to be seen with an electron microscope. Collagen triple-helix molecules up to 28,000 Å (2.8 µm) in length have been observed.

The collagen triple helix can be unfolded by heating to above its physiological temperature, when it is converted to **gelatin**; the individual polypeptide chains are disaggregated, unraveled and disordered. The native conformation is not regenerated upon cooling; triple helical conformations are regenerated for short stretches by the polypeptide chains combining randomly, but not in the proper axial register. Collagens are assembled correctly *in vivo* because the polypeptide chains are synthesized as procollagen, with nonhelical, globular extensions of just over 100 and 300 residues at the amino and carboxyl ends, respectively. The C-terminal globular extensions associate specifically and serve to align the three polypeptide chains and to nucleate assembly of the triple helix. Folding occurs from the C-terminus to the N-terminus in a zipper-like fashion. The rate-limiting step is propagation of *cis* → *trans* isomerization of the peptide bonds preceding the many Pro residues; they can adopt the *cis* isomer in the unfolded molecule but need to be *trans* in natural collagen. Posttranslational modifications, especially hydroxylation and glycosylation of Pro and Lys residues at position Xaa, occur before the polypeptide chains are assembled into the triple helix. After completion of folding of the procollagen, the propeptide extensions at both ends of the chains are removed proteolytically.

A number of mutant collagens have been identified in which the amino acid replacement has varying effects on the stability of the collagen triple helix. Many of them involve replacement of the essential Gly residues by other amino acids. Besides destabilizing the triple helix, they also decrease the rate of folding of the triple helix at the point of the mutation; consequently, residues on the N-terminal side of the mutation are often subject to much more posttranslational modification than normal, with further adverse effects on stability of the structure.

Collagen functions *in vivo* by aggregating the triple helices side-by-side into microfibrils and assembling into larger supramolecular arrays. The dimensions of these arrays vary widely with different types of collagen and their site of assembly. The surface of the collagen triple helix is defined primarily by the side-chains of residues Xaa and Yaa, and interactions between these side-chains are believed to control their side-by-side interaction in the microfibril. The neighboring molecules are assembled in a parallel array, but mutually staggered, i.e. displaced axially with respect to one another, by approximately one-quarter of their length. This is known as the **quarter-staggered array** and accounts for the cross-striated periodic structure of collagen that is apparent in electron micrographs. This periodicity is due to each collagen monomer having eight highly charged regions 670 Å apart, corresponding to 234 residues, which can be stained specifically.

Once assembled, the collagen aggregates are stabilized by a variety of covalent cross-links between the triple helices, involving especially the hydroxy–Lys side-chains. The various collagens differ primarily in the nonhelical parts of their structures, which generally occur at the ends of the polypeptide chains. The fibrils are assembled into many different types of connective tissue, often by combining with other types of molecules. Tendon is almost pure collagen, and the parallel fibrils provide great tensile strength. The collagen fibrils in skin are cross-woven into sheets that can be stretched. The fibrils of cartilage are embedded in a matrix of proteoglycans. Collagen also comprises the matrix of bone, where it is cemented into a rigid structure by deposits of inorganic crystals similar to calcium hydroxyapatite.

Collagen structure: the Madras triple helix and the current scenario. A. Bhattacharjee & M. Bansal (2005) *IUBMB Life* **57**, 161–172.

Collagen fibril form and function. T. J. Wess (2005) *Adv. Protein Chem.* **70**, 341–374.

Molecular structure of the collagen triple helix. B. Brodsky & A. V. Persikov (2005) *Adv. Protein Chem.* **70**, 301–339.

Revision of collagen molecular structure. K. Okuyama *et al.* (2006) *Biopolymers* **84**, 181–191.

~ CHAPTER 9 ~

PROTEIN STRUCTURE

Most dynamic activities in cells and organisms are caused by proteins, and the key to understanding these phenomena is their structures. Protein structures, however, consist of many atoms, often thousands, and can be extremely complex. Natural proteins are also folded into precise three-dimensional (3-D) structures and are very different from random-coil forms of the polypeptide chain (Section 8.2), in that most of their covalent bonds adopt a single dihedral angle (Section 1.2.B) and Cys residues pair in specific disulfide bonds (Section 7.2.I.5), rather than the spectrum possible with disordered polypeptide chains (Table 7-3); this generally results in a folded conformation that can be considered unique. It is also very compact. **No changes in covalent structure need occur when these folded conformations are adopted**, except for the disulfide bonds that might be formed between Cys residues (Section 7.2.I.5); therefore the folded structure represents just one conformation of the many that are possible with a disordered polypeptide chain. With 20 different amino acid residues having diverse physical and chemical properties, a huge variety of protein sequences and conformations are possible. Some proteins are globular and water-soluble, others exist buried within membranes, while others adopt elongated and extended structures that serve primarily structural roles (Section 8.5). Protein structure is most impressive in its diversity.

Each protein is identified uniquely by its amino acid sequence, which is determined by the sequence of the gene from which the protein is produced, using the genetic code (Section 7.5.F). The amino acid sequence determines all further aspects of the structure of the protein and, consequently, its functions. It has become customary to dissect protein structure into four levels (Figure 9-1). The **primary structure** is the amino acid sequence. The **secondary structure** is any regular local structure of a linear segment of polypeptide chain, such as a helix or an extended strand (Section 8.3). The **tertiary structure** is the overall topology of the folded polypeptide chain. The **quaternary structure** is the aggregation of the individual polypeptides by specific interactions between them. As more protein structures have been determined, two intermediate levels of structure have become apparent; one comprises several elements of secondary structure packed together and is known as the **supersecondary structure** or **motif**; the other is the **domain**, which can be one self-contained part of a tertiary structure.

Details of virtually all the known structures of proteins can be obtained from the Protein Data Bank (PDB; www.rcsb.org). They can be viewed most readily online using the program Jmol at http://firstglance.jmol.org.

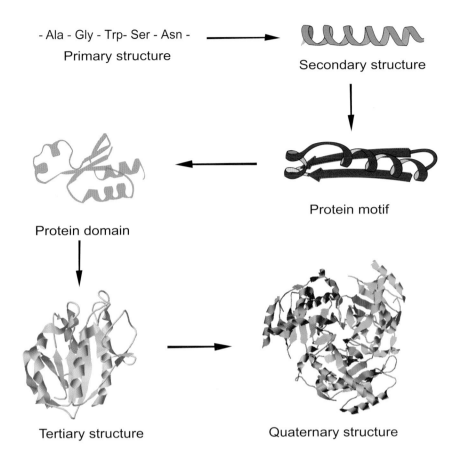

Figure 9-1. Different levels of protein structure, from primary to quaternary. Helices are depicted as *coils* and β-strands as *arrows*. The primary structure is the amino acid sequence of the polypeptide chain. The secondary structure example shown is an α-helix; the protein motif is a βαβ motif, where β is a β-strand, α an α-helix and the motif produces a parallel β-sheet. The protein domain is a folded structure that is stable in isolation. The tertiary structure is the folded structure of a complete polypeptide chain, in this case that of flavodoxin. The quaternary structure is produced by the aggregation of multiple polypeptide chains; in this case it is the homotrimer of chloramphenicol acetyltransferase.

Proteins: structures and molecular properties, 2nd edn, T. E. Creighton (1993) W. H. Freeman, NY.

Introduction to Protein Structure, 2nd edn, C. Branden & J. Tooze (1999) Garland Publishing, NY.

Introduction to Protein Architecture, A. M. Lesk (2000) Oxford University Press, Oxford.

Protein Structure and Function, G. Petsko & D. Ringe (2003) Blackwell, London.

Proteins: structures and functions, E. Whitford (2005) J. Wiley, NY.

9.1. 3-D STRUCTURES OF GLOBULAR PROTEINS: MOLECULAR COMPLEXITY

Most proteins of biological origin differ dramatically from the synthetic polypeptides with random or simple repetitive conformations and from the structural proteins described in Chapters 7 and 8. They have much smaller dimensions in solution and are of roughly spherical shape; hence they are generally referred to as **globular**. Moreover, these physical properties do not change in a continuous manner as the environment is altered, for example by changing the temperature, pH or pressure, as do the properties of random polypeptides. Instead, these physical properties generally exhibit little or no change until a point is reached where there is a sudden drastic change and, invariably, a loss of biological function: a phenomenon known as **denaturation** (Section 11.1). The denatured protein is much more like a random polypeptide than the original protein. The original protein possesses a folded 3-D conformation that is disrupted by denaturation, and upon which its biological properties are critically dependent. The detailed structures of a large number of globular proteins have been determined to atomic resolution, primarily by X-ray crystallography but also by nuclear magnetic resonance (NMR) and neutron and electron crystallography.

It is now generally accepted, much to the collective relief of protein crystallographers, that **the conformation of a protein is not substantially altered by its inclusion in a crystal lattice**, except possibly for the positions of intrinsically flexible side-chains and loops on the surface of the protein. A protein crystallized in a number of different crystal lattices, often by very different crystallization procedures, invariably has essentially the same structure in each. The specific interactions between adjacent molecules that determine the crystal lattice appear to be relatively weak and unable to perturb the general structure of a stable folded conformation. The intermolecular interactions in a crystal lattice are often similar to the intramolecular forces that specify the folded conformation; crystallization conditions that favor the crystal lattice therefore also often tend to stabilize the folded structure. **Related proteins are almost invariably found to have very similar folded conformations**; indeed, there have been many instances of surprising similarities in protein crystal structures (Section 9.3) whereas different structures have not been found when they were reasonably expected to be similar.

X-ray crystallography yields the electron density of the protein molecule, which must then be assigned to its appropriate atoms and interpreted in terms of its covalent structure. The result is a table of the 3-D coordinates of each atom in space. More immediately useful, however, is a description of the structure of the molecule in terms of the conformational angles of the polypeptide chain, as defined in Figure 8-1. An example for a very small protein of only 58 amino acid residues, bovine pancreatic trypsin inhibitor (BPTI), is given in Table 9-1.

Table 9-1. Conformational angles (°) of bovine pancreatic trypsin inhibitor

Residue	ϕ	ψ	ω	χ^1	χ^2	χ^3	χ^4	χ^5
Arg1		137	178	82	176	57	76	5
Pro2	−60	149	180	177	171	−165		
Asp3	−58	−33	179	−80	−10			
Phe4	−63	−16	177	66	75			
Cys5	−70	−19	177	−65	−84	−81		
Leu6	−89	−3	178	−53	−179			
Glu7	−78	150	177	−51	168	162		
Pro8	−72	158	−177	−174	−172	172		
Pro9	−67	143	−176	180	−174	168		
Tyr10	−121	115	−177	177	82			
Thr11	−77	−36	175	−71				
Gly12	90	176	−176					
Pro13	−83	−10	−177	−162	−172	165		
Cys14	−92	159	−170	−68	95	96		

Table 9-1 (cont.).   Conformational angles (°) of bovine pancreatic trypsin inhibitor

Residue	φ	ψ	ω	χ^1	χ^2	χ^3	χ^4	χ^5
Lys15	−119	32	169	−59	−74	174	−178	
Ala16	−77	173	−175					
Arg17	−130	81	−175	−66	179	−69	−94	−9
Ile18	−109	111	−178	−61	178			
Ile19	−78	120	176	−52	−58			
Arg20	−122	179	172	−68	−72	180	80	8
Tyr21	−114	147	168	−74	81			
Phe22	−130	160	173	69	101			
Tyr23	−87	127	176	180	−107			
Asn24	−100	97	−172	177	−6			
Ala25	−59	−28	173					
Gly28	79	13	−178					
Leu29	−156	171	−173	51	65			
Cys30	−95	146	−174	−66	−119	−88		
Gln31	−131	161	179	−65	175	160		
Thr32	−89	156	175	50				
Phe33	−151	164	178	75	114			
Val34	−97	113	174	179				
Tyr35	−86	135	180	174	44			
Gly36	−79	−9	171					
Gly37	99	7	172					
Cys38	−146	157	172	71	−117	96		
Arg39	62	36	−176	−47	−55	−171	−171	−3
Ala40	−56	156	164					
Lys41	−104	172	−168	−79	−170	171	163	
Arg42	−73	−26	179	−76	−159	88	172	7
Asn43	−82	74	−168	−169	8			
Asn44	−163	102	−169	−177	−44			
Phe45	−123	155	179	−56	87			
Lys46	−86	−8	177	−72	169	−150	−176	
Ser47	−151	158	174	73				
Ala48	−65	−30	180					
Glu49	−71	−39	175	−69	90	170		
Asp50	−70	−38	179	−85	3			
Cys51	−62	−45	179	176	−91	−88		
Met52	−72	−35	−179	−74	61	−71		
Arg53	−62	−38	180	178	149	39	−123	
Thr54	−82	−42	−166	−55				
Cys55	−111	−8	−167	−71	−65	−81		
Gly56	−76	−6	164					
Gly57	84	168	179					
Ala58	−65							

The most striking feature of the folded conformations of globular proteins is their complexity, and various ways of representing them are used (Figure 9-2). Depicting all the covalent bonds produces a very complicated structure (Figure 9-2-A) that is difficult to comprehend. Such detailed illustrations require stereodiagrams. If all the atoms are displayed with their van der Waals radii, one can perceive only the surface of the molecule (Figure 9-2-B, C). Consequently, less realistic models are necessary to convey the main features of the overall structure. Very often only the path of the polypeptide backbone is depicted, in a schematic fashion, and the amino acid side-chains are omitted (Figure 9-2-D–F). Protein structures are almost invariably constructed virtually, within a computer, but to build a physical 3-D model is the best way truly to understand the subtleties of protein structure.

The general properties of proteins of known structure presented here give a 'consensus picture' of protein structure, to which there are many exceptions. The common properties illustrate the general rules of protein architecture, but each protein is unique and generally attains its functional properties by incorporating specific exceptions to these generalities.

Multipolar representation of protein structure. A. Gramada & P. E. Bourne (2006) *BMC Bioinformatics* **7**, 242.

9.1.A. Tertiary Structure: the Overall Fold

The overall folded structures of small globular proteins are remarkably compact and roughly spherical in overall shape, but with a very irregular surface (Figure 9-2-B, C). Where a protein consists of more than about 200 residues, the structure usually appears to consist of two or more distinct structural units, generally referred to as **domains** (Figure 9-3). The individual domains are associated to varying extents, but they interact less extensively than do structural elements within the interiors of domains. Often a single segment of polypeptide chain links the domains, and each domain consists of a single stretch of polypeptide chain. But not always: in phosphofructokinase, pyruvate kinase and arabinose-binding protein, for example, there are two or three polypeptide connections between domains (Figure 9-3). In some cases, the ends of the polypeptide chain also interact with the other domain, appearing to serve as 'straps' holding the domains together. The definition of a domain is not rigorous, and the division of a structure into domains is a very subjective process that often is done in very different ways by different people. Other terms and subdivisions, such as **subdomain** and **folding unit**, are also often encountered in the literature. Nevertheless, there are many instances in which the presence of domains is clear to all observers (Figure 9-3). **Individual domains are most evident from their compactness**. This can be expressed quantitatively as the ratio of the accessible surface area of the domain to the surface area of a sphere with the same volume: values for obvious single domains are 1.64 ± 0.08.

Within a domain, the course of the polypeptide backbone is irregular but it generally pursues a moderately straight course across the entire domain and then turns on the surface and continues in a more-or-less direct path to the other side (Figure 9-4). The impression is of segments of somewhat stiff polypeptide chain interspersed with relatively tight turns or bends, which are almost always on the surface of the protein. This general type of structure has been compared to the behavior of a fire hose. It may be contrasted with other possible limiting situations: one more irregular, such as that obtained upon collapsing a flexible string, and the other more curved, as illustrated by a ball of string.

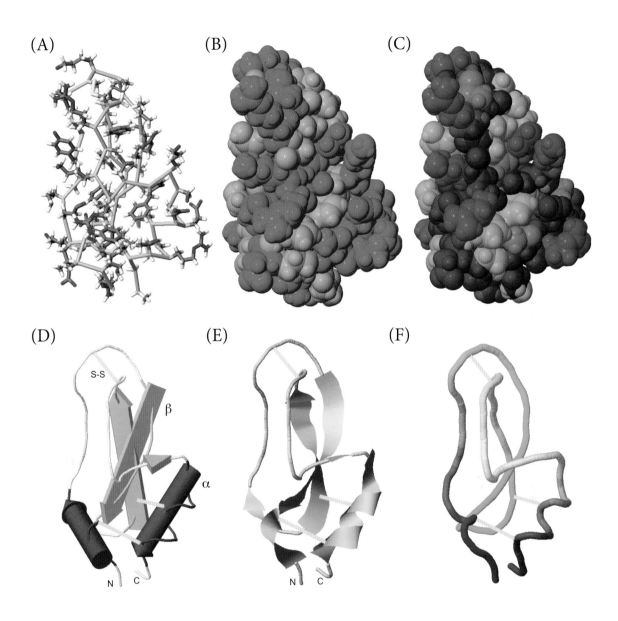

Figure 9-2. Various ways to represent the 3-D structures of globular proteins. The protein shown is one of the smallest known, bovine pancreatic trypsin inhibitor (BPTI), consisting of 58 residues, including three disulfide bonds. (A) Detailed structure showing the covalent bonds linking all the atoms, including the hydrogens, except that a single virtual bond links adjacent C^α atoms in place of the peptide bond. (B) The van der Waals surface of the atoms. Hydrophobic groups are *gray*, polar groups *colored*. (C) The ionized groups of the protein; acidic groups are *red*, basic groups *blue* and the polypeptide backbone *mauve*. (D) Schematic illustration of the secondary structure. Only the polypeptide backbone is shown, with *rockets* for α-helices and *slabs* for β-strands. The *arrowheads* point towards the C-terminus. The three *yellow rods* are the three disulfide bonds. (E) Cartoon version of the polypeptide backbone, using *ribbons* for β-strands and *coils* for α-helices. (F) Rainbow representation of the polypeptide backbone. The N-terminus is colored *blue*, the C-terminus *red*, progressing through the normal spectrum in between. Generated from PDB file *5pti* using the program Jmol.

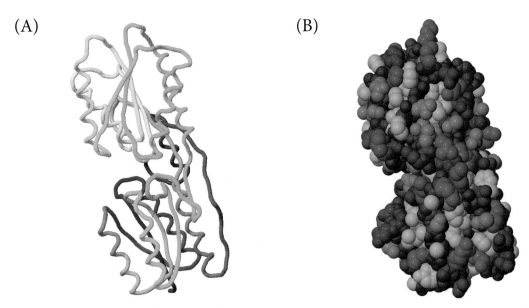

Figure 9-3. The structure of arabinose-binding protein, illustrating its separation into two structural domains. (A) The course of the polypeptide chain; the N-terminus is *blue*, the C-terminus *red*, in between varying with the colors of the rainbow. Note that the polypeptide chain crosses between the two domains three times. (B) A space-filling model of the van der Waals surface, with acidic groups *red*, basic groups *blue* and the polypeptide backbone *mauve*. Generated from PDB file *1abe* using the program Jmol.

The polypeptide backbone has never been observed to form a well-defined knot; i.e. if the polypeptide chain were grasped at each end and pulled out straight, a linear chain would always result. Minor exceptions occur where the two ends of the chain are somewhat entwined, but they could be pulled in a way that would avoid a knot.

Rotations about the individual bonds of both the backbone and side-chain are generally close to one of the conformations favored in the isolated structural unit. Consequently, the peptide bonds of the backbone are nearly always planar and the favored *trans* isomer ($\omega = 180°$), unless the next residue is Pro, when *cis* peptide bonds ($\omega = 0$) are not energetically unfavorable (Section 8.2). *Cis* peptide bonds occur in folded proteins in about 5% of the bonds that precede Pro residues, primarily at tight bends of the polypeptide backbone. Very few peptide bonds not involving Pro residues, no more than 0.05% of them, have been found to be *cis*. *Cis* peptide bonds may, however, have been missed in the past in protein structure determinations at low and moderate resolutions, and more *cis* peptide bonds may be observed as more refined, high-resolution structures appear. In carboxypeptidase A, for example, three non-Pro *cis* peptide bonds are present within a polypeptide chain of 307 residues; they were not noticed in the structure determined at 2 Å resolution but became apparent in the refined structures at 1.75 Å and 1.54 Å resolution.

The dihedral angles ϕ and ψ of the polypeptide backbone generally lie within the limits deduced for the isolated peptide unit (Figure 9-5). Similarly, rotations about the bonds of the side-chains are generally close to one of the three configurations in which the adjacent atoms are staggered, giving the greatest separation of the bulkiest groups (Figure 9-6). Unfavorable stereochemistry appears to be used in proteins only when required for their functional properties.

Many proteins also contain various ligands as intimate parts of the structure: prosthetic groups, coenzymes, metal ions and so forth (Chapter 12).

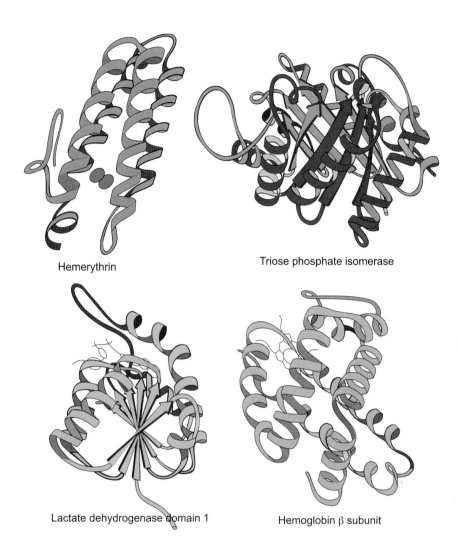

Figure 9-4. Artistic drawings of the polypeptide backbones of four proteins. α-Helices are depicted as helical *ribbons*, β-strands as *arrows*. The hemoglobin β-subunit has a heme group bound; hemerythrin has two iron atoms. Lactate dehydrogenase domain 1 has a molecule of the coenzyme NAD illustrated schematically. Adapted from drawings by J. S. Richardson.

Occurrence and role of *cis* peptide bonds in protein structures. D. E. Stewart *et al.* (1990) *J. Mol. Biol.* **214**, 253–260.

Partitioning protein structures into domains: why is it so difficult? T. A. Holland *et al.* (2006) *J. Mol. Biol.* **361**, 562–590.

9.1.B. Secondary Structure: Regular Local Structures

The extended segments of the polypeptide chain that traverse folded domains very often have regular conformations like those observed in model polypeptides (Section 8.3), especially **right-handed α-helices** and **extended β-strands associated into β-sheets**. Approximately 31% of the residues in known globular proteins occur in α-helices; 28% are in β-strands. Other regular conformations are

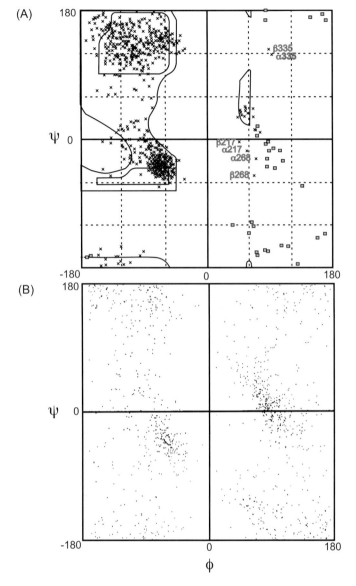

Figure 9-5. Ramachandran plots of the observed values of φ and ψ for all residues in creatine amidinohydrolase refined at 1.9 Å with an R-value of 17.7% (A) and for Gly residues in many proteins (B). Gly residues in (A) are indicated with *squares*. The only other residues to lie well outside the areas considered normally allowed are Asp217, Asp268 and Arg335, which are labeled. The α and β preceding the residue numbers refer to the two crystallographically independent subunits. Data from H. W. Hoeffken *et al.* (1988) *J. Mol. Biol.* **204**, 417–433 and C. Ramakrishnan *et al.* (1987) *Int. J. Peptide Protein Res.* **29**, 629–637.

much less frequent. A poly(Pro)-helix (Section 8.3.D) is frequently observed when there are two or more Pro residues close in the sequence. Short segments of left-handed, collagen-like helix, with Pro as every third residue (Section 8.5.C), are found occasionally, as are short three-stranded coiled coils (Section 8.5.A).

Hydrogen bonds involving the C=O and N–H groups of the polypeptide backbone usually define the secondary structure. The H⋯O distance is usually 1.9–2.0 Å, shorter than their combined van der Waals radii of about 2.6 Å. The covalent N–H distance is 1.03 ± 0.02 Å, so a typical N–H⋯O=C hydrogen bond has the N and O atoms 3.0 Å apart. The H atom is generally not observed directly in protein structures determined by X-ray crystallography, however, so hydrogen bonds in proteins are often inferred when the donor and acceptor atoms are closer than otherwise expected, and the hydrogen bond length is usually expressed as the distance between them. The positions of H atoms are determined most directly by neutron diffraction. Hydrogen bonds can also be detected by vibrational spectroscopy, because their absorbance is shifted to longer wavelengths, and by NMR, because the H atom is less shielded and consequently resonates at a lower magnetic field.

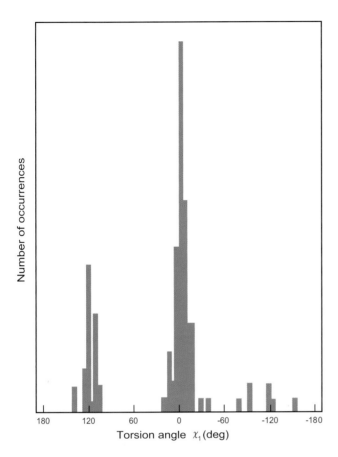

Figure 9-6. Histogram of the values of the C^α–C^β torsion angle, χ_1, observed for 151 Val residues in highly refined protein structures. Similarly sharp histograms are observed for other residues, but Val is special in having the greatest predominance of just one rotamer. Data from J. W. Ponder & F. M. Richards (1987) *J. Mol. Biol.* **193**, 775–791.

A regular secondary structure is a very useful focal point in comprehending a complex protein structure, so many schematic drawings of protein emphasize it, often in a very idealized manner (Figure 9-4). Helices are often indicated by cylinders or coiled ribbons, and extended strands of β-sheets by broad arrows, indicating the amino-to-carboxyl polarity of the polypeptide backbone.

The segments of α-helices and β-strands in globular proteins are generally relatively short, being limited to the diameter of the protein globule. The length of an α-helix is usually 10–15 residues, while that of a β-strand is 3–10 residues. Exceptions include the α-helix of 50 residues in influenza hemagglutinin and the 31-residue α-helices of calmodulin and troponin C, which dominate these structures. Secondary structures in proteins tend to be somewhat distorted, but the hydrogen bond lengths are most constant. Secondary structure is most apparent in larger protein domains, where it comprises most of the interior. At least one important property of secondary structure is that it provides an efficient means of pairing within hydrogen bonds the polar groups of the polypeptide backbone, which must occur within the protein interior and is a prerequisite for stability of the folded conformation (Section 11.3).

1. Helices

With a typical α-helix of 12 residues, only the central four have both of the hydrogen bonds that characterize this structure (Figure 8-6). The four residues at each end have only one hydrogen bond each, and the conformations of these residues are often irregular, frequently forming a local 3_{10}-helix. Which residues should be counted as part of the helix, therefore, is often not obvious. Various criteria have been used for defining secondary structure, such as hydrogen bonding, conformational angles ϕ and ψ, and distances between C^α atoms.

In solvent-exposed α-helices, the plane of the peptide bond is often rotated so that the carbonyl group is pointing outwards from the helix axis towards the solvent. The helix axis is often curved, with the surface on the outside somewhat extended, possibly because the hydrogen bonds there are weaker and slightly longer, due to interactions with water. The values of ϕ and ψ in α-helices in folded proteins average −62° and −41°, instead of −48° and −57° as in the standard α-helix (Table 8-2). Pro residues do not occur frequently in α-helices; when they do, they cause irregularities such as kinks and bends in the middle of the helix. The 3_{10}-helix (Section 8.3.A.1) is usually observed only at the ends of α-helices or in short stretches of four to five residues, and its geometry is usually distorted. The 3_{10}-helices become thinner and longer and less regular with increasing helix length, consistent with these helices being intrinsically unstable (Section 8.3.A.1). The Π-helix is not observed in protein structures.

Left-handed helices appear to be used only when important structurally or functionally, and they usually comprise only four residues, i.e. one turn. Poly(Pro) II-helices (Figure 8-10) are not common in globular proteins, but most contain at least one; they are usually shorter than five residues, although as many as 12 are known. Pro residues predominate in these structures, but Gln and positively charged residues are also favored. Gln's prevalence appears to be due to the ability of its side-chain to hydrogen bond with the backbone carbonyl oxygen of the following residue; this helps to fix the ψ angle of the Gln residue and the ϕ and ψ angles of the proceeding residue in poly(Pro) II-helical conformations, and it explains why Gln is favored at the first position.

A survey of left-handed helices in protein structures. M. Novotny & G. J. Kleywegt (2005) *J. Mol. Biol.* **347**, 231–241.

Amphipathic α-helices in proteins: results from analysis of protein structures. A. Sharadadevi *et al.* (2005) *Proteins* **59**, 791–801.

Polyproline helices in protein structures: a statistical survey. R. Berisio *et al.* (2006) *Protein Peptide Lett.* **13**, 847–854.

3_{10}-helices in proteins are parahelices. P. Enkhbayar *et al.* (2006) *Proteins* **64**, 691–699.

---

2. β-*Structure*

β-Sheets in folded proteins are generally twisted, rather than planar, with a right-handed twist of from 0° to 30° between strands. The conformational parameters can also deviate considerably from ideality. Somewhat more positive values of both ϕ and ψ than the standard values (Table 8-2) are generally observed with twisted sheets. Further distortions occur in β-sheets that consist of both parallel and antiparallel strands, because the ideal backbone conformations for the two types of sheet differ (Table 8-2). An extra residue is often present in a β-strand at the edge of a sheet, interrupting the hydrogen bond pattern and producing a β-bulge (Section 9.1.B.3).

β-Sheets can consist of entirely parallel or antiparallel strands or have a mixture of the two. Purely parallel sheets are least frequent, purely antiparallel sheets most common. Antiparallel sheets often consist of just two or three strands, whereas parallel sheets always have at least four. Mixed sheets generally contain 3–15 β-strands. The strands adjacent in a sheet tend to be segments that are also adjacent in the primary structure. This correlation is greatest for antiparallel strands and least for parallel strands.

Purely parallel or antiparallel sheets of six or eight strands are often said to curve around to close up the sheet into a continuous **β-barrel**, although in some cases they are probably described more accurately as a 'sandwich' of two β-sheets packed against each other. The most spectacular examples are the barrels of eight parallel β-strands, with an α-helix on the outside of the barrel connecting each pair of β-strands. This type of structure has been found in many different proteins but was first encountered in the enzyme triose phosphate isomerase (TIM) (Figure 9-4); it is often called the **TIM**, **(βα)$_8$** or **(αβ)$_8$** barrel (Section 9.1.I.3.a). In each case, the eight parallel strands slope at an angle of about 35° to the barrel axis, and adjacent strands are offset, or sheared, by the same amount relative to each other (Figure 9-7).

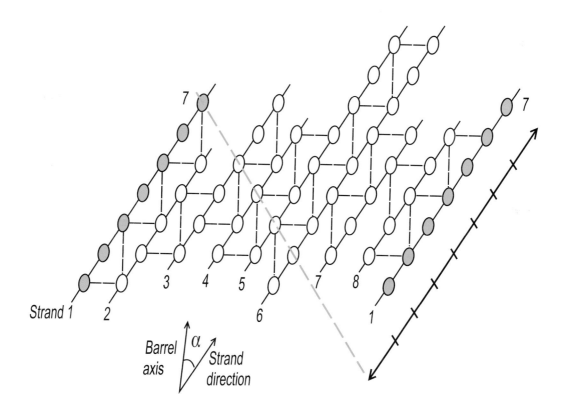

Figure 9-7. The geometry of β-sheet barrels, illustrated with the structure of TIM, (βα)$_8$ or (αβ) barrels. The β-sheet barrel of triosephosphate isomerase (TIM) has been unrolled schematically; the positions of the residues along the backbone are depicted by *circles*, hydrogen bonds are indicated by *dashed lines*. The N-terminal β-strand (number 1) is shown twice, at both sides of the sheet. The *thick dashed line* connects residues that are opposite each other in the β-sheet, starting on the left with residue number 7 of strand 1. At the other side of the sheet, it would intersect a residue that would be number –1 of strand 1, eight positions away from residue 7. The shear number is therefore 8. If the strands were vertical, with no shear, the dashed line would connect residue 7 in strand 1 at both sides of the sheet. The angle α gives the tilt of the β-strands from the vertical; its value in this case is 36°. Adapted from C. Chothia (1988) *Nature* **333**, 598–599.

Minimal surface as a model of β-sheets. E. Koh & T. Kim (2005) *Proteins* **61**, 559–569.

A new catalog of protein β-sheets. M. Parisien & F. Major (2005) *Proteins* **61**, 545–558.

Common evolutionary origin of swapped-hairpin and double-ϕ β-barrels. M. Coles *et al.* (2006) *Structure* **14**, 1489–1498.

3. β-*Bulge*

A β-bulge occurs when the regular hydrogen bond interactions and backbone conformation of a β-strand are disrupted by the presence of an extra residue. The additional residue is usually in a β-strand at one edge of the β-sheet, where the bulge can be accommodated most easily within the structure (Figure 9-8). Most proteins have two or more β-bulges. They can be conserved in the structures of related proteins, where they may play a functional role. The more general structural role of a β-bulge is to alter somewhat the direction of the β-strand; it also accentuates the normal twist of the β-sheet.

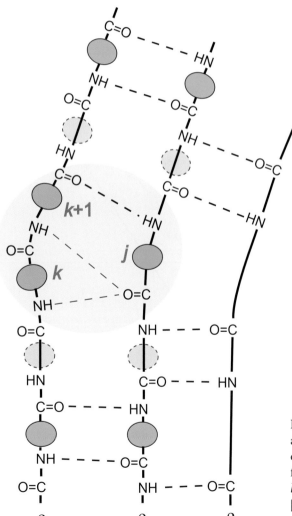

Figure 9-8. Schematic of a classic β-bulge caused by an additional residue in the β-strand on the left edge of the β-sheet. Residue j from the middle β-strand forms backbone hydrogen bonds (shown as *dashed lines*) with two residues (k and $k + 1$) from the left β-strand. For clarity, side-chains are indicated only as *ellipses*, *dashed* when below the plane of the paper.

About 90% of β-bulges occur between antiparallel β-strands. The two most common β-bulges are the classic and G1, which account for ~ 80% of all β-bulges. In the **classic β-bulge**, the residue at position 1 has backbone dihedral angles of $\phi = -100°$ and $\psi = -25°$, which are closer to those of an α-helical than a β-strand conformation, but residue two has angles closer to the β-strand conformation: $\phi = -180°$ and $\psi = 160°$. The name of the **G1 β-bulge** arises because the residue at position 1 is almost always Gly, because its conformational angles of $\phi = 85°$ and $\psi = 0°$ are not compatible with the other amino acid residues (Figure 8-3). The residue at position 2 of the G1 class has more normal β-strand dihedral angles of $\phi = -90$ and $\psi = 150$. The G1 β-bulge often occurs along with a type II β-turn, which requires a Gly at position 3 (Section 9.1.C.1). A β-bulge increases the right-handed twist of the β-strand from the usual 10° to between 35° and 45°, and it disrupts the alternating side-chain pattern of the usual β-sheet structure.

A β-bulge disrupts the regular β-sheet architecture, so it might be considered a destabilizing factor, but certain β-bulges are necessary for the structure and stabilize it due to other interactions. The same conformation can be induced by incorporating D-amino acid residues at those positions.

The role of a β-bulge in the folding of the β-hairpin structure in ubiquitin. P. Y. Chen *et al.* (2001) *Protein Sci.* **10**, 2063–2074.

Reconstruction of the conserved β-bulge in mammalian defensins using D-amino acids. C. Xie *et al.* (2005) *J. Biol. Chem.* **280**, 32921–32929.

9.1.C. Reverse Turns: Changing Direction

Proteins have roughly spherical structures, in spite of being composed of relatively straight segments of polypeptide chain, because **the polypeptide chain generally makes tight bends at the surface**, thereby reversing the direction of the chain (Figure 9-9). Gly or Pro residues are often found in turns, especially when unusual backbone conformations are required. Reverse turns include **β-turns, γ-turns** and **omega loops**. They occur most frequently in **β-hairpins**, which link two adjacent β-strands in an antiparallel β-sheet (Section 9.1.D.1). Turns are usually found at the surfaces of proteins, where the peptide bonds that do not interact with protein atoms can form hydrogen bonds with the solvent. It is these reverse turns or loops that give proteins their globularity, and nearly one-third of the residues of globular proteins are involved in them.

Because of their prevalence, reverse turns are frequently included as a third type of secondary structure. Turns are distinguished from the α-helix and β-sheet types of secondary structure in that they do not have repetitive backbone conformations and hydrogen-bonding patterns. Most residues in turns, however, adopt backbone conformations close to either the α-helical or β-strand conformations. The difference is that the backbone conformations of successive residues in a turn vary, whereas they are similar in β-strands and α-helices.

Role of local sequence in the folding of cellular retinoic a binding protein I: structural propensities of reverse turns. K. S. Rotondi & L. M. Gierasch (2003) *Biochemistry* **42**, 7976–7985.

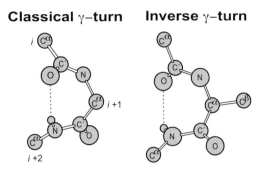
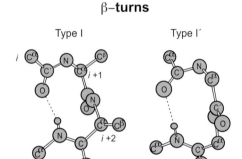
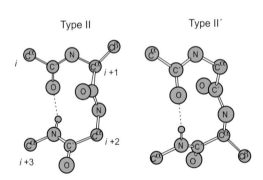

Figure 9-9. The most common γ- and β-turns connecting adjacent strands of an antiparallel β-sheet. The three and four residues, respectively, that are considered to define the turns are shown, with the first residue designated *i*. C^β atoms are shown only in positions where nonGly residues occur frequently. The last hydrogen bond of the β-sheet is shown as a *dashed line*. Only the H atoms involved in hydrogen bonds are shown. Adapted from G. D. Rose *et al.* (1985) *Adv. Protein Chem.* **37**, 1–109.

1. β-Turns

A β-turn has a hydrogen bond between the backbone carbonyl oxygen of one residue (i) and the backbone amide NH of the residue three positions further along the chain (i + 3) (Figure 9-9). There are several different types of β-turns, which are classified according to the backbone dihedral angles of the two middle residues, (*i* + 1) and (*i* + 2) (Table 9-2). Type I and II turns are the most common and account for about two-thirds of all β-turns. Type I turns prefer Asn, Asp or Ser as residue (*i*), Asp, Ser, Thr or Pro as (*i* + 1), Asp, Ser, Asn or Arg as (*i* + 2) and Gly, Trp or Met as (*i* + 3). Type II turns have very different preferences: Pro as residue (*i* + 1), Asn or Gly as residue (*i* + 2) and Gln or Arg as (*i* + 3). Type I′ and II′ β-turns differ by inversion of the sign of the ϕ and ψ angles; they are found almost exclusively in β-hairpins. The type IV β-turn is variable and includes any turn in a protein structure with an (*i*) to (*i* + 3) hydrogen bond where the ϕ and ψ angles differ by >40° from the values that define the other types of β-turn. Types VIa and VIb have a *cis* peptide bond before Pro residue (*i* + 2). The Pro residue in type VIa adopts a backbone conformation close to that of α-helical residues, whereas that in type VIb has a β-strand-like conformation.

Table 9-2. Structural features of γ- and β-turns

Bend type	Dihedral angles of central residues (°)[a]			
	ϕ_{i+1}	ψ_{i+1}	ϕ_{i+2}	ψ_{i+2}
γ				
Classical	70 to 85	−60 to −70		
Inverse	−70 to −85	60 to 70		
β				
I	−60	−30	−90	0
I'	60	30	90	0
II	−60	120	80	0
II'	60	−120	−80	0
III	−60	−30	−60	−30
III'	60	30	60	30
IV	Any bend with two or more angles differing by >40° from those given here			
V	−80	80	80	−80
V'	80	−80	−80	80
VIa[b]	−60	120	−90	0
VIb[b]	−120	120	−60	0
VII	Kink in chain created by ψ_{i+1} ~180°, $|\phi_{i+2}|<60°$, or $|\psi_{i+1}|<60°$, $|\phi_{i+2}|$~180°			
VIII	−60	−30	−120	120

[a] The central residue of a γ-turn is numbered $i + 1$; the two central residues of a β-turn are $i + 1$ and $i + 2$.

[b] The peptide bond between residues $i + 1$ and $i + 2$ is cis, and residue $i + 2$ is Pro.

Engineering diverse changes in β-turn propensities in the N-terminal β-hairpin of ubiquitin reveals significant effects on stability and kinetics but a robust folding transition state. E. R. Simpson et al. (2006) Biochemistry 45, 4220–4230.

Tuning the β-turn segment in designed peptide β-hairpins: construction of a stable type I' β-turn nucleus and hairpin–helix transition promoting segments. R. Rai et al. (2006) Biopolymers 88, 350–361.

Non-stereogenic α-aminoisobutyryl-glycyl (Aib-Gly) dipeptidyl unit nucleates type I' β-turn in linear peptides in aqueous solution. L. R. Masterson et al. (2007) Biopolymers 88, 746–753.

2. γ-Turns

A γ-turn has a hydrogen bond between the backbone carbonyl oxygen of one residue (i) and the backbone amide NH of residue (i + 2), two residues further along the chain (Figure 9-9). The classic and inverse types of γ-turns have inverted signs for the backbone ϕ and ψ dihedral angles for the ($i + 1$) intervening residue (Table 9-2). Consequently, their backbone conformations are mirror images of one another, analogous to the differences between the type I and I' and type II and II' β-turns. The classic γ-turn has somewhat unfavorable ϕ and ψ angles and is less common than the inverse type; it is found primarily connecting two adjacent antiparallel β-strands in a hairpin.

β- and γ-turns in proteins revisited: a new set of amino acid turn-type dependent positional preferences and potentials. K. Guruprasad & S. Rajkumar (2000) *J. Biosci.* **25**, 143–166.

First observation of two consecutive γ-turns in a crystalline linear dipeptide. A. I. Jimenez *et al.* (2005) *Angew. Chem. Int. Ed. Engl.* **44**, 396–399.

---

3. Omega Loops

The omega loop has a backbone conformation resembling the Greek letter omega (Ω). It comprises six or more amino acid residues that cause a reversal in the direction of the polypeptide chain of the protein, bringing the two ends of the loop into close proximity. They have no other common features and no regular hydrogen bonding or secondary structures. Nevertheless, their structures are compact and globular, with the side-chains often packing into the center of the loop. In a few cases, such as the phosphate-binding loop (P loop) of nucleotide-binding proteins (Section 12.4) and the hypervariable regions of antibodies that interact with antigens, omega loops have specific functions and are involved in the binding of substrates, catalysis and molecular recognition.

The nature of the turn in omega loops of proteins. M. Pal & S. Dasgupta (2003) *Proteins* **51**, 591–606.

9.1.D. Supersecondary Structures: Common Motifs

Certain assemblies of a number of secondary structure elements, including the connecting segments of polypeptide chains, have been observed a sufficient number of times that they are now recognized as a further level of structure, termed **super-secondary structures** or **motifs**. These structures are a higher level of structure than the secondary structure, but do not constitute entire structural domains.

The architectures of most proteins appear to be made up of segments of secondary structure packed together, so it is important to understand the basis for the interactions between helices, between sheets, and between helices and sheets. The general nature of their basic architectures causes the interactions between elements of secondary structure to be governed primarily by the amino acid side-chains on their surfaces. For example, interdigitation of the side-chains of two ideal α-helices would be expected only when their axes cross at angles of −82°, −60° or +19° (a rotation is positive when the lower helix is rotated clockwise relative to the top one). Helices pack onto β-sheets with their axes nearly parallel to those of the β-strands, because the twist of the β-sheet then matches the surface of the helix. Two normal twisted β-sheets should pack together face to face with the top sheet rotated clockwise between 20° and 50°, whereas two sheets folded over onto themselves should be at angles of about 90°. Proteins are found to observe these rules to a first approximation, but substantial variation is produced by the nonideality of their secondary structures and the variability of the amino acid side-chains involved in the contacts. In the great majority of cases, only nonpolar amino acid side-chains are involved in these interactions.

Structural classification of αββ and ββα supersecondary structure units in proteins. N. S. Boutennet *et al.* (1998) *Proteins* **30**, 193–212.

A supersecondary structure library and search algorithm for modeling loops in protein structures. N. Fernandez-Fuentes *et al.* (2006) *Nucleic Acids Res.* **34**, 2085–2097.

---

1. β-*Hairpin and* β-*Meander*

A β-hairpin consists of two adjacent antiparallel β-strands linked by a β-turn. A β-meander has a particular type of antiparallel β-sheet structure, with a very simple topology in which two or more β-strands that are consecutive in the primary structure are also adjacent to one another in the 3-D structure (Figure 9-10-C). Multiple linked β-hairpins produce a β-meander.

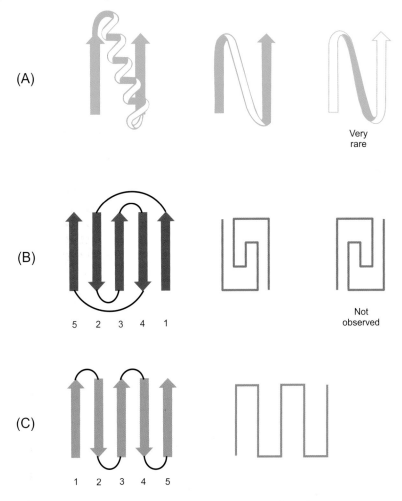

Figure 9-10. Super-secondary structures observed in proteins. (A) A β–α–β unit; the segment joining the two β-strands is almost always above their plane, not below. (B) The Greek key motif of antiparallel β-sheets, which also is always of one 'handedness'. (C), A so-called β-meander. Adapted from T. E. Creighton (1993) *Proteins: structures and molecular properties*, 2nd edn, W. H. Freeman, NY.

Engineering enhanced protein stability through β-turn optimization: insights for the design of stable peptide β-hairpin systems. E. R. Simpson *et al.* (2005) *Angew. Chem. Int. Ed. Engl.* **44**, 4939–4944.

Minimization and optimization of designed β-hairpin folds. N. H. Andersen *et al.* (2006) *J. Am. Chem. Soc.* **128**, 6101–6110.

De novo design of monomeric β-hairpin and β-sheet peptides. D. Pantoja-Uceda *et al.* (2006) *Methods Mol. Biol.* **340**, 27–51.

2. β-*Helix and* β-*Roll:* β-*Solenoids*

In contrast to the antiparallel β-sheets of other proteins comprising β-structures, the β-helix is characterized by **an unusual parallel β-sheet topology formed from three parallel β-sheets that are wound together into a right-handed helical structure** (Figure 9-11). The individual β-strands are short, consisting of only two to five residues. Each coil of the β-helix has the same 3-D arrangement of secondary structure elements, so the structure is repetitive. The side-chains of repeating residues are packed into the center of the helix and interact with one another to form structures given unusual names, such as 'asparagine ladders', 'serine stacks' and 'aromatic stacks'. These structures are now known collectively as 'β-solenoids'.

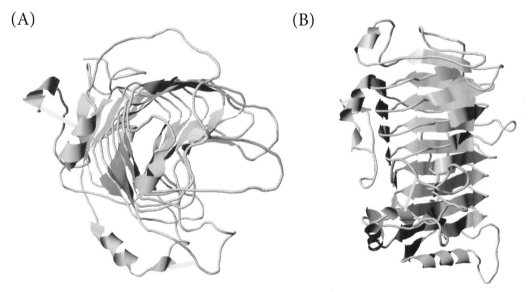

Figure 9-11. The β-helix of pectate lyase C from *Erwinia chrysanthemi*, illustrated in two orthogonal views. Each turn of the helix contains three β-strands, which associate into three parallel β-sheets that make up the sides of the resulting triangular prism shape. Generated from PDB file *2pec* using the program Jmol.

Sequence and structure analysis of parallel β-helices: implication for constructing amyloid structural models. H. H. Tsai *et al.* (2006) *Structure* **14**, 1059–1072.

Conformational and sequence signatures in β-helix proteins. P. Iengar *et al.* (2006) *Structure* **14**, 529–542.

β-rolls, β-helices, and other β-solenoid proteins. A. V. Kajava & A. C. Steven (2006) *Adv. Protein Chem.* **73**, 55–96.

3. Cystine Knot

The cystine knot consists of an antiparallel β-sheet and is characterized by three disulfide bonds, one of which passes through a ring formed by the other two (Figure 9-12). The six Cys residues are usually designated CI through CVI and are paired in disulfide bonds as CI–IV, CII–V and CIII–VI. Two families of cystine knot motifs are known.

Figure 9-12. The backbone structure of the cystine knot protein cycloviolacin. β-strands are shown as *arrows* and the three disulfide bonds are shown as *straight cylinders* connecting the two Cys residues. Generated from PDB file *1df6* using the program Jmol.

The first family is the growth-factor cystine knot family. Disulfide CI–IV passes through the ring formed by the CII–V and CIII–VI disulfide bonds, which consists of 8–14 residues. Four antiparallel β-strands make up the β-sheet. All these cystine knot growth factors are dimeric, but with different dimer interfaces.

The second family is that of the inhibitor cystine knots. In this case the knot is formed by the CIII–VI disulfide bond passing through the ring formed by disulfide bonds of CI–IV and CII–V. Furthermore, the antiparallel β-sheet is triple-stranded.

Evolution and classification of cystine knot-containing hormones and related extracellular signaling molecules. U. A. Vitt *et al.* (2001) *Mol. Endocrinol.* **15**, 681–694.

Oxidative folding of the cystine knot motif in cyclotide proteins. D. J. Craik & N. L. Daly (2005) *Protein Peptide Lett.* **12**, 147–152.

---

4. Greek Key Motif

The Greek key motif gets its name from being similar to a decorative pattern used in ancient Greece. It is one particular topology for arranging four β-strands into an antiparallel β-sheet (Figure 9-13). All protein Greek key motifs have the same chirality (Figure 9-10-B). The Greek key β-barrel is a type of antiparallel β-barrel, in which two Greek key β-sheets fold together to form an eight-stranded antiparallel β-barrel.

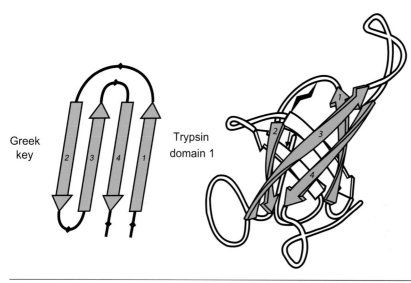

Figure 9-13. The Greek key motif (*left*) and its occurrence in domain 1 of trypsin (*right*). The four β-strands are numbered according to their occurrence in the primary structure. Adapted from drawings by J. S. Richardson.

A comprehensive analysis of the Greek key motifs in protein β-barrels and β-sandwiches. C. Zhang & S. H. Kim (2000) *Proteins* **40**, 409–419.

5. Jelly Roll Motif

The jelly roll motif is one order of eight β-strands in an antiparallel β-sheet that is frequently found in protein structures (Figure 9-14). The name comes from its similarity to a slice of rolled cake known as a jelly roll or Swiss roll. The jelly roll motif frequently forms a type of antiparallel β-barrel that is known as a jelly-roll β-barrel.

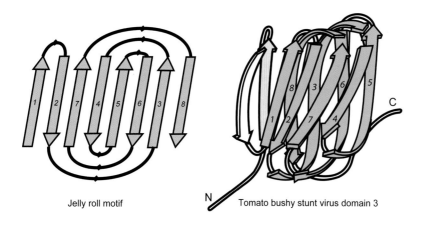

Figure 9-14. The jelly roll motif (*left*) and its occurrence in domain 3 of the tomato bushy stunt virus (*right*). The eight β-strands are numbered according to their order in the primary structure. Adapted from drawings by J. S. Richardson.

Sequence relationships in the legume lectin fold and other jelly rolls. A. Williams & D. R. Westhead (2002) *Protein Eng.* **15**, 771–774.

Crystal structure of the YML079w protein from *Saccharomyces cerevisiae* reveals a new sequence family of the jelly-roll fold. C. Z. Zhou *et al.* (2005) *Protein Sci.* **14**, 209–215.

6. Four-helix Bundle

The four-helix bundle is one of the simplest protein folds and can occur as an isolated 3-D fold or as a domain within a much larger and more complex protein structure. The four α-helices of a four-helix bundle are packed together lengthwise and antiparallel (Figure 9-15). The antiparallel packing may be caused by electrostatic interactions between the dipoles of the helices (Figure 8-11). The α-helices can be from four different polypeptide chains or from the same one. The α-helices of four-helix bundles tend to be about 20 residues long, longer than the usual average of 10. They are amphipathic (Figure 8-8), with the hydrophobic face of the helix buried and interacting with the other three helices.

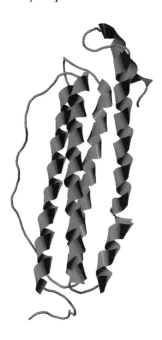

Figure 9-15. The four-helix bundle motif, as found in a single subunit of bullfrog ferritin. Only the polypeptide backbone is depicted, with the helices as *coils*. The topology is 'up-down-down-up', with a long loop connecting the two 'down' helices. There is an additional short helix at the C-terminus. Generated from PDB file *1mfr* using the program Jmol.

There are several different types of four-helix bundle: (1) the simple 'up-down-up-down' topology, where consecutive helices have short connections and alternating directions, (2) the 'up-down-down-up' topology, where there is a long loop or cross-over connection between helices 2 and 3 (Figure 9-15), and (3) the 'up-up-down-down' topology, with two long cross-over connections between helices 1 and 2 and helices 3 and 4. In most cases, the helices are aligned so that each adjacent pair is antiparallel.

The four-helix bundle is used by a wide range of proteins with diverse functions. A binding pocket for cofactors and metal ions can be generated by splaying apart the four helices (Section 12.2). Alternatively, the four helices may coil around each other to form a coiled coil (Section 8.5.A).

Sequence determinants of the energetics of folding of a transmembrane four-helix-bundle protein. K. P. Howard *et al.* (2002) *Proc. Natl. Acad. Sci. USA* **99**, 8568–8572.

A designed well-folded monomeric four-helix bundle protein prepared by Fmoc solid-phase peptide synthesis and native chemical ligation. G. T. Dolphin (2006) *Chemistry* **12**, 1436–1447.

Four-helix bundle cavitein reveals middle leucine as linchpin. J. Freeman *et al.* (2007) *Biopolymers* **88**, 725–732.

7. Epidermal Growth Factor (EGF) Motif

The EGF motif is a protein module of about 45 residues, six of which are Cys. The six conserved Cys residues are designated CI through CVI and pair up to form three disulfide bonds: CI–III, CII–IV and CV–VI (Figure 9-16). The motif gained its name because it was first found in epidermal growth factor (EGF). It occurs just once in EGF and in transforming growth factor-α but it is often present as multiple copies in much larger proteins (Figure 7-24). The protein fibrillin is an extreme case, with over 30 EGF domains. They are frequently present in combination with multiple copies of other modules, such as the Kringle domain (Section 9.1.D.9).

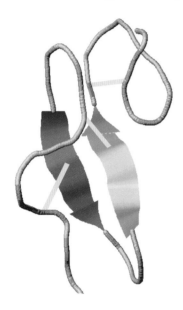

Figure 9-16. An EGF domain, that of epidermal growth factor. The two β-strands are depicted as *arrows* and the three disulfide bonds are indicated by *straight lines* connecting the pairs of Cys residues.

The amino acid sequences of EGF motifs are well conserved. The consensus sequence is X–X–X–X–Cys–$X_{(2-7)}$–Cys–$X_{(1-4)}$–(Gly/Ala)–X–Cys–$X_{(1-13)}$–t–t–a–X–Cys–X–Cys–X–X–Gly–a–$X_{(1-6)}$–Gly–X–X–Cys–X, where a is an aromatic residue, t is a nonhydrophobic residue and X is any amino acid. The numbers in parentheses indicate the variable number of X residues. Consequently, the EGF motif is often recognizable from just the amino acid sequence.

The 3-D structure of the EGF motif is also highly conserved (Figure 9-16). It consists of a two-stranded antiparallel β-sheet upon which the three disulfide bonds link the N- and C-terminal regions to the core. Some classes of EGF domains further stabilize the N-terminal region by binding calcium ions, which may produce long helical arrangements of multiple EGF motifs.

Evolution of distinct EGF domains with specific functions. M. A. Wouters *et al.* (2005) *Protein Sci.* **14**, 1091–1103.

---

8. Interleukin-1 Motif: Trefoils

The interleukin-1 motif has an unusual pseudothree-fold symmetry in which 12 β-strands form six hairpins. Three of the hairpins form a six-stranded antiparallel β-barrel with a hydrophobic core. The other three hairpins form a triplet structure that caps the barrel (Figure 9-17).

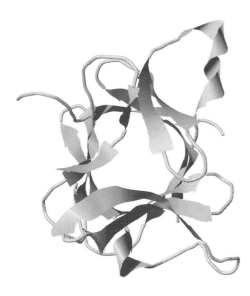

Figure 9-17. The interleukin-1 3-D structure. Generated from PDB file *1I2h* using the program Jmol.

The interleukin-1 motif is also known as a β-**trefoil** because it can be dissected into three equivalent Y-shaped (or trefoil) β-structural units. However, the β-trefoil terminology should not be confused with the 'trefoil peptide domain' that represents a very different structural motif of gastrointestinal protective peptides. Also be aware that all interleukins other than interleukin-1 are four-helix bundles.

There are few similarities in the sequences of proteins with this 3-D structure, other than a preference for the large hydrophobic residues that make up the core of the molecule.

A common sequence-associated physicochemical feature for proteins of β-trefoil family. R. Xu & Y. Xiao (2005) *Comput. Biol. Chem.* **29**, 79–82.

Structure of mammalian trefoil factors and functional insights. L. Thim & F. E. May (2005) *Cell. Mol. Life Sci.* **62**, 2956–2973.

---

9. Kringle Domain

The kringle domain is usually found in proteinases associated with blood clotting and fibrinolysis, such as thrombin, plasminogen and plasminogen activators (Figure 7-24), so it is both a structural and a functional unit. The kringle domain structure is a module of about 80 residues having six conserved Cys residues, which are designated as CI through to CVI and form three disulfide bonds with the pairings I–VI, II–IV and III–V (Figure 9-18). Usually one to five copies are present in a polypeptide chain, although apolipoprotein *a* has 38. It is often accompanied by multiple copies of other protein modules, such as the EGF motif (Figure 7-24).

The tertiary structure core of a kringle domain consists of a two-stranded antiparallel β-sheet and the close packing of two of the three disulfide bonds. There are also β-turns and several additional regions of short β-strands comprising only two residues. The amino acid sequences of the kringle domains are highly conserved and usually can be recognized from just the amino acid sequence of a protein.

The function of kringle domains appears to be molecular recognition, especially of proteins with an N-terminal Lys residue.

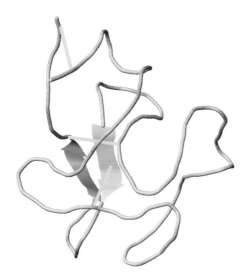

Figure 9-18. A kringle domain. The two β-strands are depicted as *arrows* and the three disulfide bonds by *straight lines* connecting the pairs of Cys residues. Generated from PDB file *1krn* using the program Jmol.

The kringle domains of human plasminogen. F. J. Castellino & S. G. McCance (1997) *Ciba Found. Symp.* **212**, 46–65.

Modes of evolution in the protease and kringle domains of the plasminogen–prothrombin family. A. L. Hughes (2000) *Mol. Phylogenet. Evol.* **14**, 469–478.

9.1.E. Contact Maps

Contact maps are 'fingerprints' of a protein structure, two-dimensional (2-D) representations of 3-D protein tertiary structures. Each protein can be identified from its contact map, which can reveal the secondary structure, fold topology and side-chain packing patterns. A protein structure composed of N structural units is expressed as an $N \times N$ array of the distances between each pair of units in the structure. The units might be all the individual atoms, selected types of atoms (e.g. C^α atoms), groups of atoms (e.g. the centers of mass of side-chains) or entire amino acid residues. For example, an $N \times N$ matrix of the distances between N protein C^α atoms can be transformed into a C^α-based contact map. Contact maps are often generated from such matrices by using a certain cut-off value for the pairwise distances: those C^α atoms closer to each other in the protein structure than the chosen cut-off distance are considered to be 'in contact' and marked by a symbol. This produces a 'black & white' contact map based on a binary $N \times N$ matrix. An example for the small protein BPTI (Figure 9-2) is illustrated in Figure 9-19. Alternatively, a set of several critical values for the distances between C^α atoms (or for other atoms or groups of atoms) will generate a 'gray-scale' contact map (Figure 9-20). A single cut-off distance is always inadequate in some way: too small a value misses some helix-to-helix contacts, while too large a value creates problems with identification of the secondary structure patterns. Gray-scale or colored maps with several cut-off ranges provide much more detailed structural information.

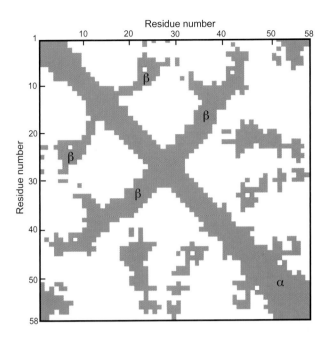

Figure 9-19. Contact map for bovine pancreatic trypsin inhibitor, BPTI. Both axes represent the amino acid sequence of the polypeptide chain; shading is present whenever two residues are within 10 Å of each other. The single α-helix at the C-terminus of the polypeptide chain is apparent from the broadening of the diagonal. The three-stranded antiparallel β-sheet generates the two features perpendicular to the diagonal. Adapted from a figure kindly provided by M. Levitt.

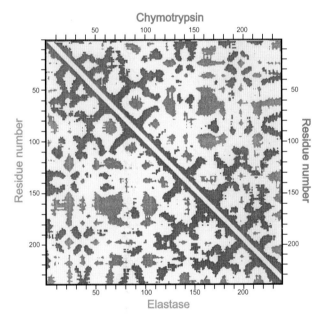

Figure 9-20. The similar contact maps of the homologous proteinases chymotrypsin (*top right*) and elastase (*bottom left*). Distances greater than 30 Å are *light gray*, whereas distances less than 15 Å are *dark gray*. Data from L. Sawyer *et al.* (1978) *J. Mol. Biol.* **118**, 137–208.

Contact maps based on C^α atoms (Figure 9-19) depict α-helices as a spread of close contacts along the diagonal, i.e. between residues nearby in the amino acid sequence, since C^α_i is in close proximity to C^α_{i-4}, C^α_{i-3}, C^α_{i+3} and C^α_{i+4}, where the subscript is the number of the residue in the polypeptide chain. Within a parallel β-sheet, in which the first two residues of two adjacent extended strands are i and j, C^α_i is next to C^α_j, C^α_{i+1} is adjacent to C^α_{j+1} and so on; this gives rise to a series of close contacts on a diagonal line *parallel* to the main diagonal but offset from it by $(i - j)$ residues. With two strands of antiparallel β-sheet, where residues i and j are, respectively, the first and last residues hydrogen-bonded, C^α_i is next to C^α_j, C^α_{i+1} is next to C^α_{j-1} and so forth. This gives rise to a series of contacts that define a diagonal line *perpendicular* to the main diagonal (Figure 9-19).

Structural similarity between a pair of proteins is immediately apparent by similarity of their contact maps (Figure 9-20); the contact map is independent of the orientation of the protein molecule, and there is no need to search all possible relative orientations of the two proteins to see if there is any match. The reconstruction of a protein structure from its contact map is possible, although not straightforward, and low-to-moderate-resolution 3-D models can be generated. Structural domains segregated along the polypeptide backbone are often apparent as segregated areas of contacts on the distance plots. Figure 9-20 illustrates this for the homologous two-domain proteins chymotrypsin and elastase.

Contact maps based on C^α atoms (Figure 9-19) reflect well the overall topology of the protein fold but do not provide many other structural details. Side-chain contact maps contain much more information about the topology of a protein fold and its secondary structure, and also about many fine details of the packing of the protein side-chains. A side-chain contact map can be constructed using either individual atoms or the centers of mass of the side-chains. In the first procedure, two residues are assumed to be in contact when any two heavy atoms (i.e. not including hydrogens) are closer than some cut-off distance. The various groups that make up the side-chains, for example CH_2, CH and NH_2, have similar sizes, so a good cut-off distance is 4.5–5.0 Å. The number of contacts detected in this range is relatively constant, and the packing of the side-chains is usually evident. An important and useful feature of these contact maps is the characteristic patterns of contacts between elements of secondary structure. The second procedure for generating a side-chain contact map uses the centers of mass of the side-chains. A larger cut-off distance needs to be used, and it can be made specific for certain amino acid pairs due to the different sizes of the side-chains. The two approaches lead to very similar protein representations, although those of atom-based contact maps are somewhat more informative.

A map of main-chain hydrogen bonds is a variant of a protein contact map.

Evolution and similarity evaluation of protein structures in contact map space. N. Gupta *et al.* (2005) *Proteins* **59**, 196–204.

A contact map matching approach to protein structure similarity analysis. R. C. de Melo *et al.* (2006) *Genet. Mol. Res.* **5**, 284–308.

Fast molecular shape matching using contact maps. P. K. Agarwal *et al.* (2007) *J. Comput. Biol.* **14**, 131–143.

9.1.F. Interiors and Exteriors

The surfaces of macromolecules are usually treated as the **accessible surface area**, which is that surface in contact with solvent water molecules. **The total accessible surface areas of proteins are approximately proportional to the two-thirds power of their molecular weights**, as would be expected for objects of similar, roughly spherical shapes. The accessible surface area of a protein is, however, **nearly twice that expected for a sphere of the same size**, which is some measure of the roughness of the surface. The accessible surface area (A_s, in units of Å^2) of a typical small monomeric protein is usually related to its molecular weight (M_r) by the approximate relationship:

$$A_s = 4.7 \, M_r^{0.76} \tag{9.1}$$

This is only 23–45% of the surface area of the unfolded polypeptide chain. The van der Waals volume, V, of a typical monomeric protein is directly proportional to its molecular weight:

$$V = 1.27\, M_r\, \text{Å}^3/\text{Dalton} \tag{9.2}$$

The interiors of globular proteins are densely packed, with adjacent atoms frequently in van der Waals contact (Figure 9-21). About 72–78% of the interior volume is filled with atoms, as defined by their van der Waals radii. This is close to the value of 74% possible with close packing of identical spheres and is within the range of 70–78% found with crystals of small organic molecules. The average volumes occupied by residues in folded proteins (Table 9-3) are virtually the same as those they occupy in crystals of amino acids (Figure 9-22). The close packing in protein interiors contrasts with the lower values observed with liquids, such as water (58%) and cyclohexane (44%). These packing densities of proteins are somewhat exaggerated, however, because many of the atoms in proteins are close because they are covalently bonded. The packing density does vary somewhat throughout the interior, generally being highest in areas where the polypeptide topology is most regular. In relatively few instances are there unfilled cavities of sufficient size to accommodate other molecules. Transfer measurements of small molecules from liquid water indicate that the tendency of single water molecules to enter small nonpolar cavities in proteins should be very small. The general dense packing of protein interiors is especially impressive when it is remembered that it must be compatible with the covalent bonds of the polypeptide chain.

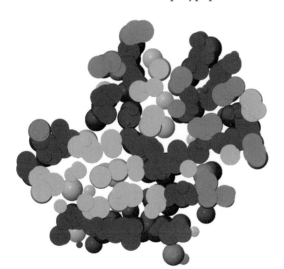

Figure 9-21. The dense packing of atoms in the interior of a globular protein, demonstrated by a thin slice through the interior of flavodoxin. Atoms of the protein are drawn with their van der Waals radii. Generated from PDB file *1czn* using the program Jmol.

Within the protein interior, **virtually all polar groups are paired in hydrogen bonds**; most of these polar groups are of the polypeptide backbone, and they tend to be hydrogen-bonded in secondary structure. **Water molecules are generally excluded from protein interiors**. When present, they appear to be integral parts of the protein structure (Figure 9-23): they are well-fixed in internal cavities of the protein that are isolated from the bulk solvent, they invariably form hydrogen bonds with polar groups of the protein, and they tend to be conserved in homologous structures. Some water molecules occur singly, some in pairs, and others in extended hydrogen-bond networks of up to eight molecules. Most make two or three hydrogen bonds with polar groups or other water molecules, although nearly half make the ideal four, involving the O atom as hydrogen acceptor in two and the two H atoms as donors in two further hydrogen bonds. Many bind to the NH and CO groups of the main-chain backbone, although some are attached to polar side-chains. Such buried water molecules appear to be important for filling holes and, probably the more important, pairing with internal polar groups of the protein in hydrogen bonds. There can be as many as 24 buried water molecules in a protein structure, as in the case of carboxypeptidase with 307 residues.

Table 9-3. The packing of residues in the interiors of proteins

Residue	Average volume of buried residues (Å^3)[a]	Fraction of residues at least 95% buried[b]	Relative free energy[c] of residue in interior to that on surface (kcal/mol)
Gly	66	0.36	0
Ala	92	0.38	−0.14
Val	142	0.54	−0.55
Leu	168	0.45	−0.59
Ile	169	0.60	−0.68
Ser	99	0.22	0.40
Thr	122	0.23	0.32
Asp	125	0.15	0.78
Asn	125	0.12	0.75
Glu	155	0.18	1.15
Gln	161	0.07	0.80
Lys	171	0.03	2.06
Arg	202	0.01	1.40
His	167	0.17	0.02
Phe	203	0.50	−0.61
Tyr	204	0.15	0.28
Trp	238	0.27	−0.39
Cys–SS–	106	0.40	−0.61
Cys–SH	118	0.50	
Met	171	0.40	−0.65
Pro	129	0.18	0.50

[a] From C. Chothia (1975) *Nature* **254**, 304–308.

[b] Average for 12 proteins; from C. Chothia (1976) *J. Mol. Biol.* **105**, 1–14.

[c] Calculated as −RT log$_e$ f, where f is the ratio of the occurrence of this amino acid residue in the interior to that on the surface, relative to the value for Gly, which was set to zero. From S. Miller *et al.* (1987) *J. Mol. Biol.* **105**, 641–656.

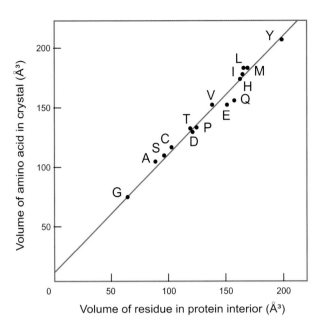

Figure 9-22. The amino acid residues are as tightly packed in protein interiors as in crystals of the amino acids. The mean volumes of the various amino acid residues buried in protein interiors are plotted against the volume of the corresponding amino acid in crystals. The line has a slope of 1 and an intercept of 11 Å²: the volume of the water molecule lost by an amino acid upon becoming a residue. Data from C. Chothia (1984) *Ann. Rev. Biochem.* **53**, 537–572.

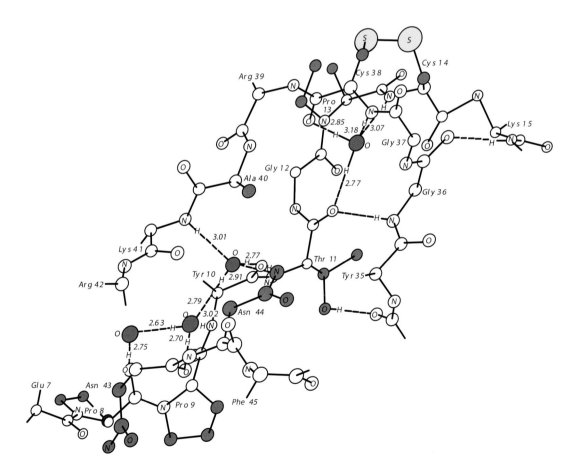

Figure 9-23. The four internal water molecules of bovine pancreatic trypsin inhibitor, BPTI. The O atoms observed crystallographically are shown in *red*, with the presumed positions of the H atoms indicated. The lengths in Ångstroms of the hydrogen bonds between N and O atoms are indicated. Adapted from T. E. Creighton (1978) *Prog. Biophys. Mol. Biol.* **33**, 231–297.

Water-soluble proteins have virtually all ionized groups on the surface, exposed to the solvent. On average, Asp, Glu, Lys and Arg residues comprise 27% of the protein surface and only 4% of the interior. Most proteins generally have one to two charged groups per 100 Å² of surface, but the charge density varies between 0.5 and 25 charges per 100 Å². The average contributions of charged, polar and nonpolar atomic groups to the accessible surface area are 14 ± 4%, 33 ± 7% and 53 ± 5%, respectively. Oppositely charged groups tend to be near each other on the surface, where they could form salt bridges, but they are rarely observed to do so in protein crystal structures, unless the side-chains are held in position by the rest of the protein. **Pairs of ionized acidic and basic groups hardly ever occur in the interiors of proteins**; although such a pair in close proximity might appear to have no net charge, it is energetically unfavorable to remove it from water. The surfaces of proteins have undoubtedly been selected in order to not bind to other proteins or substances that they encounter naturally, especially in the case of intracellular proteins that exist in very crowded environments.

Nonpolar side-chains predominate in the protein interior: Val, Leu, Ile, Phe, Ala and Gly residues comprise 63% of the interior residues. There is a reasonable correlation between the hydrophobicities of the residues and their tendency to occur in the protein interior (Figure 9-24). The size and complexity of many of the amino acid side-chains, plus the relatively small sizes of folded proteins, often make it difficult to classify residues as being simply buried or exposed. For example, the long side-chains of Lys and Arg residues have their ionized terminal groups almost invariably exposed to the solvent, but the remaining hydrophobic methylene carbons are often buried in the interior; consequently, the C^α atoms of these residues may be relatively far from the surface, and most of the side-chain may be buried. Small proteins have only one or two completely buried residues; even in large proteins, only 15% of the residues are totally inaccessible to solvent. A residue is commonly considered to be buried if greater than 95% of its surface area is inaccessible to solvent. For example, 54% of Val residues and 60% of Ile are at least 95% buried, but only 1% of Arg and 3% of Lys residues (Table 9-3). The hydrophobic residues are primarily involved in packing together the elements of secondary structure.

There are no conspicuous tendencies for the 20 different amino acid side-chains to be adjacent to each other or to the peptide backbone within protein interiors, other than the general tendencies

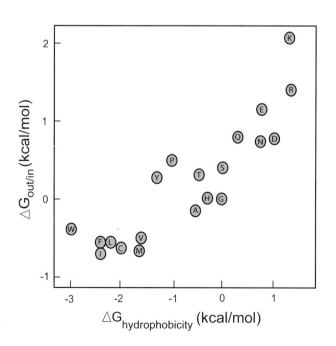

Figure 9-24. Correlation between the hydrophobicities of amino acid residues and their tendency to be in the interiors of folded globular proteins. The hydrophobicity was measured by the free energy of transfer of the N-acetyl amide forms of individual amino acids from water to nonpolar solvent (Table 7-6), and the tendency to occur in the interior of proteins was measured similarly as a free energy of transfer from the protein surface to the interior (Table 9-3). The amino acids are designated by their one-letter abbreviations (Figure 7-2). Data from T. E. Creighton (1993) *Proteins: structures and molecular properties*, 2nd ed., W. H. Freeman, NY, p. 254.

just described for hydrophobic groups to cluster, polar groups to be paired in hydrogen bonds, and oppositely charged groups to be near each other. The only other tendencies noted are for pairs of Cys residues to occur in positions where they can form disulfide bonds, at least in extracellular proteins where they tend to do so, and for aromatic residues to undergo favorable electrostatic dipole interactions with each other and with sulfur and oxygen atoms and amino groups. There are apparently no simple structural rules to relate conformation to the amino acid sequence. This point will become even more obvious from the variety of amino acid sequences that result in the same folded conformation (Section 9.3).

Description of atomic burials in compact globular proteins by Fermi-Dirac probability distributions. A. L. Gomes *et al.* (2007) *Proteins* **66**, 304–320.

Cubic equation governing the outer-region dielectric constant of globular proteins. H. Park & Y. H. Jeon (2007) *Phys. Rev. E* **75**, 021916.

A statistical approach to finding biologically relevant features on protein surfaces. F. K. Pettit *et al.* (2007) *J. Mol. Biol.* **369**, 863–879.

Detection of pockets on protein surfaces using small and large probe spheres to find putative ligand binding sites. T. Kawabata & N. Go (2007) *Proteins* **68**, 516–529.

Prediction of side-chain conformations on protein surfaces. Z. Xiang *et al.* (2007) *Proteins* **66**, 814–823.

9.1.G. The Solvent: Interactions with Water

The aqueous solvent surrounding protein molecules is an important aspect of protein structure but one of the most difficult to characterize. The structure determined crystallographically is averaged over both the extensive time required to collect the crystallographic data and over all the many molecules of the crystal lattice. The molecules of the liquid solvent are particularly mobile and by themselves would give only a uniform, average electron density throughout the nonprotein areas of the crystal.

Nonetheless, **distinct electron density peaks corresponding to relatively fixed solvent molecules are observed crystallographically near the surfaces of proteins**, especially in crystals at low temperatures. In well-refined and well-ordered crystal structures, an average of two fixed water molecules is observed for each amino acid residue. Of course, the electron density map does not identify the molecules, and crystal solvents usually contain high concentrations of salts or other agents that are added to induce crystallization, but most such fixed molecules appear from their electron densities and shapes to be water molecules; this is generally assumed in the absence of indications to the contrary.

Fixed water molecules occur primarily in positions where they can hydrogen bond to polar groups, and the degree of order of such solvent molecules is generally proportional to their proximity to the protein surface and to the extent of their participation in hydrogen bonding. Water molecules are most highly ordered, with crystallographic temperature factor (B) values as low as 13 Å^2 when extensively hydrogen-bonded in crevices on the protein surface or when bridging between different molecules in the crystal lattice. Almost invariably, the fixed water molecules are anchored by hydrogen bonding to fixed polar groups of the protein surface; in addition, one or more firmly anchored water molecules can apparently serve to fix adjacent water molecules in a hydrogen-bonded lattice. Hydrogen-bond acceptors, such as the carbonyl O atoms, interact about twice as frequently with water molecules as do hydrogen-bond donors, such as –NH– groups, probably because of a stronger interaction: the positive dipole of the water molecule is on the surface, where it can come closer to another atom, whereas the negative dipole is near the center of the water molecule. Beyond this, there is a continuum of degree of order of solvent molecules, with B-values increasing to the point where the molecules can be considered part of the bulk solvent and it is no longer deemed worthwhile to include them in the structure. All except the most ordered water molecules have only partial occupancies, indicating considerable flexibility in the solvent structure.

Ordered water molecules around nonpolar surfaces, as might be expected from the hydrophobic effect, have been observed only in exceptional situations, such as the exceptionally hydrophobic protein crambin (where all the water molecules within the crystal lattice can be localized). In other instances, such ordered networks must normally be only transient, and they might readily 'slip' along the nonpolar surface and thus be smeared out in the electron density map.

Properties of spanning water networks at protein surfaces. N. Smolin *et al.* (2005) *J. Phys. Chem. B* **109**, 10995–11005.

Water structure and interactions with protein surfaces. T. M. Raschke (2006) *Curr. Opinion Struct. Biol.* **16**, 152–159.

Exploring structurally conserved solvent sites in protein families. C. A. Bottoms *et al.* (2006) *Proteins* **64**, 404–421.

Protein surface hydration mapped by site-specific mutations. W. Qiu *et al.* (2006) *Proc. Natl. Acad. Sci. USA* **103**, 13979–13984.

Water structure and interactions with protein surfaces. T. M. Raschke (2006) *Curr. Opinion Struct. Biol.* **16**, 152–159.

Multiple solvent crystal structures: probing binding sites, plasticity and hydration. C. Mattos *et al.* (2006) *J. Mol. Biol.* **357**, 1471–1482.

9.1.H. Quaternary Structure: Initiating Macromolecular Assemblies

Many proteins exist naturally as aggregates of two or more polypeptide chains, either identical or different. Different polypeptide chains are usually designated by letters, for example normal adult hemoglobin is $\alpha_2\beta_2$ (Section 12.5.B) and aspartate transcarbamoylase is r_6c_6 (Section 15.1.C). Many proteins are dimers, trimers, tetramers or even high-order aggregates of identical polypeptide chains. The association of two molecules requires only spatial and physical complementarity of the interacting surfaces. The surfaces of proteins have probably been selected by evolution not to interact with the many other proteins that they normally come into contact with but should not bind to.

The quaternary structure is invariably observed in the structure determined crystallographically, but there may be uncertainty about whether the interactions between protein molecules in the crystal lattice are due to the quaternary structure or simply induced by the crystallization process and responsible for the crystal lattice. Fortunately, protein crystals generally contain 40–60% solvent and the crystal lattice interactions are not strong. In general, the more extensive the contacts, the stronger the interaction. The more extensive the direct interactions between protein molecules, the greater the probability that the interactions will survive outside the crystal lattice (Figure 9-25).

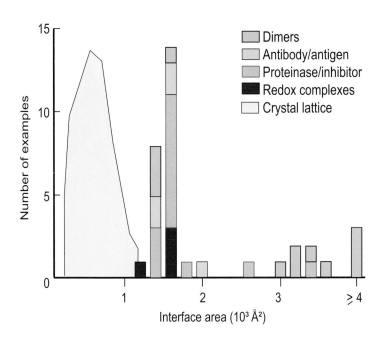

Figure 9-25. Histogram of the areas of interactions observed in protein crystal structures. The *line* indicates the interfaces between monomeric proteins that produce the crystal lattice. There is virtually no overlap with the surface areas that interact in protein dimers and three different types of protein complexes. Data from J. Janin & F. Rodier (1995) *Proteins* **23**, 580–587.

Crystal-packing interfaces tend to be more hydrophilic than those involved in quaternary structure, having an average of 15 water molecules per 1000 Å^2 of interface area, whereas the specific interfaces of homodimers and complexes have 10–11. Water molecules may form a ring around interfaces that remain 'dry' or they may permeate 'wet' interfaces. Most of the specific interfaces of quaternary structure are dry and most of the crystal-packing interfaces are wet, but there are exceptions.

Each subunit of a quaternary structure is generally folded into an apparently independent globular structure that then interacts with other subunits (Figure 9-26-A). There are spectacular exceptions, however, such as the dimeric *trp* repressor in which the two identical polypeptide chains are intimately entwined (Figure 9-26-C). The interfaces between different subunits are generally similar to the interiors of the individual molecules, in that they are closely packed and involve primarily hydrophobic interactions between nonpolar side-chains. The most abundant residue at subunit interfaces is Leu. Remarkably, however, the second most abundant residue is Arg; its very polar guanido group is almost always ionized and a surface group in monomeric proteins. This and other polar groups that are buried in subunit contacts contribute to the stability of the assembly by forming intersubunit hydrogen bonds and salt bridges, with a frequency of about one for each 150 Å^2 of interface area. The periphery of the interface is usually similar to the exteriors of monomeric proteins. Some interfaces involve interactions between elements of secondary structure, whereas others involve interactions between loops on the surfaces of the individual monomers. **The only common aspect of the interacting surfaces of the individual molecules is that they are highly complementary**, in both shape and pairing of polar groups.

Two fundamentally different types of interactions between identical subunits are possible, designated as **isologous** and **heterologous** (Figure 9-27). An isologous association involves the same surfaces on both monomers, which associate to produce a dimer with a two-fold symmetry axis. The two monomers are equivalent, which requires that the two halves of the interface of the monomer be complementary about a 'mirror plane' that coincides with the symmetry axis in the dimer. Nonequivalent association of the two surfaces, to form a nonsymmetrical dimmer, is unlikely because some of the complementarity

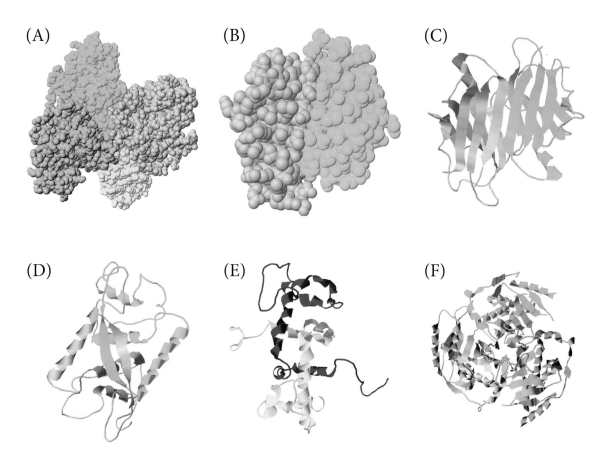
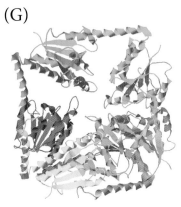

Figure 9-26. Examples of protein quaternary structures. (A) The isologous tetrameric structure of rabbit skeletal muscle aldolase, with three two-fold symmetry axes. The four polypeptide chains are depicted in different colors. The primary contacts between subunits involve hydrophobic side-chains and water molecules between them. Generated from PDB file *1zah*. (B) The prealbumin dimer, in space-filling form; the two monomers are depicted with *different colors*. (C) Cartoon description of the prealbumin dimer in the same orientation as in (B), in which *arrows* depict the β-strands. The two monomers associate to extend the two β-sheets, by two strands of one monomer hydrogen bonding to the corresponding strands from the other molecule. Two such dimers can further associate at nearly right angles to each other to form a tetramer by isologous interactions between the side-chains protruding outwards from the β-sheets. Generated from PDB file *2pab*. (D) The *met* repressor dimer, showing the course of the polypeptide chain schematically. The two polypeptide chains are depicted with *different colors*. Generated from PDB file *1mjk*. (E) Unusual quaternary structure of the *trp* repressor dimer, with the two polypeptide chains depicted with *different colors*. Instead of two independently folded polypeptide chains, the two polypeptide chains of the *trp* repressor are entwined. Generated from PDB file *1wrs*. (F) The heterologous trimeric structure of chloramphenicol acetyltransferase, in which each subunit interacts with its neighbor by extending its β-sheet by one strand, near the center of the molecule. Generated from PDB file *2id9*. (G) The octameric structure of bacterial phosphoribosylaminoimidazole carboxylase catalytic subunits; two of the subunits at the top of the top layer have been removed to illustrate the lower layer. Each layer is a heterologous tetramer, with four-fold symmetry. The two layers are joined together by isologous interactions. Generated from PDB file *1d7a*. Figure generated using the program Jmol.

sites would not be paired. Further association to produce tetramers requires another type of binding surface on each monomer, and most tetramers are formed by two sets of isologous interactions, to give three two-fold symmetry axes. A few proteins, including prealbumin and concanavalin A, associate isologously to dimers by extending the β-sheets of different molecules (Figure 9-26-C); then two such dimers pack together in another isologous manner to produce a tetramer. Isologous interfaces are those with two-fold symmetry; others are heterologous.

A heterologous association involves association of two different sites on each monomer that are complementary and do not overlap. The two monomers are not equivalent in a heterologous dimer because each has a different binding site unpaired. Consequently, each monomer can bind to another monomer; indefinite polymerization will result, unless the geometry of the association leads to a closed ring (Figure 9-27-C, D). **Any oligomeric protein with a fixed number of identical subunits that is not a power of two is likely to result from heterologous association**. Some elongated structural proteins, such as actin, are formed by indefinite heterologous polymerization of individual globular monomers (Section 9.1.H.5). Trimers (bacteriochlorophyll protein and KDPG aldolase), some tetramers (manganese superoxide dismutase and neuraminidase), pentamers (muconolactone isomerase) and a 17-mer (tobacco mosaic virus disk) are produced by closed heterologous association.

Morphological aspects of oligomeric protein structures. H. Ponstingl *et al.* (2005) *Prog. Biophys. Mol. Biol.* **89**, 9–35.

Mass spectrometry of macromolecular assemblies: preservation and dissociation. J. L. Benesch & C. V. Robinson (2006) *Curr. Opinion Struct. Biol.* **16**, 245–251.

Protein interactions probed with mass spectrometry. S. Kaveti & J. R. Engen (2006) *Methods Mol. Biol.* **316**, 179–197.

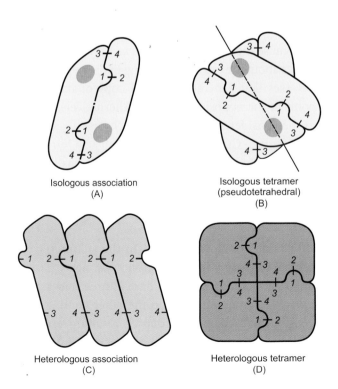

Figure 9-27. Schematic illustrations of isologous and heterologous association between protein subunits. (A) Isologous association to form a dimer with a two-fold symmetry axis perpendicular to the plane of the figure. (B) Forming a tetramer by isologous association, using two different bonding interactions, 1–2 and 3–4, and the cross-hatched circles of (A). (C) Heterologous association leading possibly to infinitely long polymers. (D) Heterologous association to form a closed, finite structure, in this case a tetramer, with a four-fold symmetry axis. Adapted from J. Monod *et al.* (1965) *J. Mol. Biol.* **13**, 88–118.

1. Symmetry

Molecular symmetry is crucial in quaternary structure. A polypeptide chain is chiral (Section 1.1.A) and cannot be symmetrical. The symmetry of oligomeric proteins therefore must arise from the presence of multiple identical polypeptide chains. There must not be inversion centers or mirror symmetry, which would invert the chirality of the L-amino acids (Section 1.1.A). The term **protomer** is used to designate the individual identical units in an oligomeric protein. A protomer may comprise more than one polypeptide chain if they are not identical in sequence or conformation. With different polypeptide chains in hetero-oligomers, the protomer must contain each type of chain; for example, the protomer of an immunoglobulin G molecule composed of two light chains and two heavy chains consists of one heavy and one light chain.

In a symmetrical homo-oligomer, a protomer must be related to another by a rotation, translation or screw rotation; the latter is a combination of a rotation and a translation. Applying these symmetry operations repeatedly to the protomer should reconstitute the whole object. Sets of symmetry operations that generate symmetrical objects of finite size are known as **point groups**. Each is characterized by the number of asymmetric units it contains, called its multiplicity (m). **Cyclic C_n symmetry has $m = n$.** The n protomers are related by rotations of $360/n$ degrees about a single symmetry axis c, and any integral value of n is feasible:

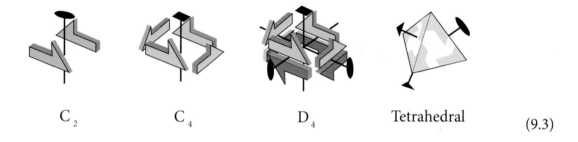

$$C_2 \qquad C_4 \qquad D_4 \qquad \text{Tetrahedral} \qquad (9.3)$$

Dihedral D_n symmetry is also described as $n2$ because the oligomer contains $m = 2n$ protomers. The protomers are related by rotations of $360/n$ degrees about the c axis, all pointing in the same direction. In addition, there are perpendicular two-fold axes that generate a second copy of the structure below the first, in which all the protomers point in the opposite direction. Alternatively, each pair of protomers related by the two-fold axes can be considered a dimer, in which case D_n symmetry simply describes the rotation symmetry of the n dimers. **Cubic symmetries** occur in a tetrahedron, an octahedron or an icosahedron; all have $m = $ some multiple of 12. A tetrahedron has both two-fold and three-fold symmetry axes that are not orthogonal. There must be three protomers around each of the four three-fold symmetry axes, so $m = 12$. Octahedral symmetry has four-fold axes, and icosahedral symmetry has five-fold axes.

The numbers of protomers and polypeptide chains are closely related to the symmetry of an oligomeric protein (Table 9-4). A homodimer can have only the point group symmetry C_2, which has a single two-fold axis and $m = 2$. A symmetrical homotrimer must have a three-fold axis with 120° rotations and cyclic C_3 symmetry, with $m=3$. A homotetramer, on the other hand, can have two different symmetries: either a four-fold axis in the cyclic point group C_4, resulting from cyclic heterologous association, or three orthogonal two-fold axes in the dihedral point group D_2 (also described as 222) resulting from two types of isologous association. Both point groups have $m = 4$ yet they yield very different quaternary structures, and D_2 is observed much more frequently than C_4.

Table 9-4. Symmetries of oligomeric proteins

Structure	Protein	Quaternary structure	Point group
Cyclic			
	HIV proteinase	α_2	C_2
	Porin	α_3	C_3
	Neuraminidase (viral)	α_4	C_4
	Pentraxins	α_5	C_5
	α-Hemolysin	α_7	C_7
	Hemoglobin	$\alpha_2\beta_2$	C_2 [a]
	Light harvesting complex II	$\alpha_9\beta_9$	C_9
Dihedral			
	Phosphofructokinase	α_4	D_2
	Hemerythrin	α_8	D_4
	GTP cyclohydrolase	α_{10}	D_5
	GroEL chaperonin	α_{14}	D_7
	Aspartate transcarbamoylase	$\alpha_6\beta_6$	D_3
Cubic			
	Phaseolin	α_{12}	Tetrahedral
	Apoferritin	α_{24}	Octahedral
	Virus coats	α_{60}, α_{180}	Icosahedral

[a] Hemoglobin also has approximate D_2 symmetry because the α- and β-chains are homologous and have very similar structures (see Figure 9-34).

Data from T. L. Blundell & N. Srinivasan (1996) *Proc. Natl. Acad. Sci. USA* **93**, 14243–14248.

Dihedral symmetry requires the number of subunits to be even ($m = 2n$) and is very common in soluble globular proteins. Homohexamers generally have D_3 symmetry. Octamers have D_4 symmetry, as illustrated by the bacterial phosphoribosylaminoimidazole carboxylase catalytic subunit (Figure 9-26-G). C_2 dimers and D_2 tetramers have only isologous interfaces; in other systems, at least one set of interfaces must be heterologous. Membrane proteins often have odd numbers of subunits and cyclic symmetry as a result of heterologous association. For example, porins of the bacterial outer membrane and the bacteriorhodopsin of *Halobacterium halobium* are homotrimers with C_3 symmetry, while the α-hemolysin of *Staphylococcus* is a heptamer with C_7 symmetry. Cubic symmetry is less common but it is found in ferritin and large assemblies such as the pyruvate decarboxylase complex and icosahedral viruses.

Structural symmetry and protein function. D. S. Goodsell & A. J. Olson (2000) *Ann. Rev. Biophys. Biomol. Struct.* **29**, 105–153.

Extent of protein–protein interactions and quasi-equivalence in viral capsids. C. M. Shepherd & V. S. Reddy (2005) *Proteins* **58**, 472–477.

Diversity and identity of mechanical properties of icosahedral viral capsids studied with elastic network normal mode analysis. F. Tama & C. L. Brooks (2005) *J. Mol. Biol.* **345**, 299–318.

Structures of T=1 and T=3 particles of cucumber necrosis virus: evidence of internal scaffolding. U. Katpally *et al.* (2007) *J. Mol. Biol.* **365**, 502–512.

2. Asymmetry and Approximate Symmetry

Symmetry is the norm in oligomeric structures, and **any asymmetry found is remarkable and usually involved in the function**. For example, the F1 fragment of the ATP synthase of mitochondria and chloroplasts has the formula $\alpha_3\beta_3\gamma$. The three-fold symmetry of the assembly of α- and β-chains is broken by the presence of a single γ-chain in the middle. Contacts with γ cause the three active sites carried by the three β-chains to be nonequivalent, which is an essential feature of the catalytic mechanism, in which the γ-chain rotates by steps of 120° relative to the remainder of the protein.

Approximate symmetry is common in assemblies that do not have exactly identical chemical units. In mammalian hemoglobins, the $\alpha_2\beta_2$ oligomer has the exact symmetry of point group C_2 with the two-fold axis relating the two $\alpha\beta$ units; in addition, there is an approximate D_2 symmetry because the tertiary structures of the homologous α- and β-chains are very similar and can be considered virtually equivalent (see Figure 9-34). Approximate symmetry between structural domains of a single-chain protein is also well documented, and it is indicative that the two halves arose by duplication of the sequence. The similarity in the two halves of ferredoxin in Figure 9-28 is reflected in the similarity in their amino acid sequences (Figure 7-33). In this case, **each of the two symmetry-related halves of the structure involves both halves of the primary structure**. It seems only plausible that this structure was originally a dimer of two identical polypeptide chains, which then became fused by duplication and extension of its gene. The two halves of the polypeptide chain, originally identical, were then free to acquire differences in their amino acid sequences while retaining the overall conformation.

Oligomeric proteins occasionally have symmetries that do not belong to one of the point groups mentioned above and cannot be exact. This is possible primarily because the protein molecule is composed of several domains that can move relative to each other. For example, the tetrameric *Escherichia coli lac* repressor (Figure 13-5) is assembled from two dimers, each of which has C_2 symmetry. The two dimers are related by an approximate two-fold axis that is not orthogonal to those of the dimers as should be required by D_2 symmetry. The purpose of this unorthodox structure is to have the DNA-binding headpieces of all four subunits on the same face of the tetramer, which would not be possible in a D_2 tetramer. The C-terminal α-helices of each subunit form most of the dimmer–dimer contacts on the opposite face. The four C-terminal α-helices assemble with D_2 symmetry, whereas the rest of the protein does not. This is made possible by a long connecting peptide adopting different conformations in two of the subunits.

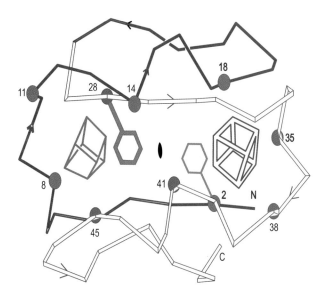

Figure 9-28. Internal symmetry in the structure of ferredoxin, reflecting the internal homology in its primary structure (Figure 7-33). The amino-terminal half of the polypeptide chain is in *red*, the other half *blue*. Note that each of the two symmetry-related halves of the structure involves both halves of the primary structure. The *distorted cubes* are the iron–sulfur clusters that bind electrons reversibly in the function of ferredoxin. *Circles* indicate Cys residues that are attached to the iron–sulfur clusters. Data from A. D. McLachlan (1979) *J. Mol. Biol.* **128**, 49–79.

An **icosahedron** has 60 monomers in identical symmetrical positions and is the largest symmetrical closed structure that is possible with identical monomers. Yet the shells of many icosahedral viruses comprise more than 60 monomers: tomato bushy stunt and southern bean mosaic viruses each have 180 identical monomers in their coats, and they are related by two-, three- and five-fold symmetry axes. All 180 monomers cannot all have equivalent positions, but this structure is possible because the monomers, although identical, occupy three different types of positions in the symmetrical structure. This structural **quasiequivalence** is possible because the virus coat protein molecule can adapt to these three different environments by moving several domains relative to each other in order to be able to adopt the three different packing arrangements that are required.

Two dimeric proteins from the human immunodeficiency virus (HIV), the HIV proteinase and the HIV reverse transcriptase, depart from two-fold symmetry. Alone, the **HIV proteinase dimer** displays exact C_2 symmetry but this is incompatible with its binding of a single asymmetric peptide substrate and its catalytic mechanism. The mechanism of hydrolysis requires one of two intrinsically equivalent active-site Asp residues to be protonated, the other deprotonated. The HIV proteinase symmetry is broken upon binding the substrate, due to minor structural changes for the protein to fit the asymmetric ligand; the complex retains approximate symmetry. Similar asymmetry is observed in dimeric threonine synthase from plants (Figure 15-9).

The **HIV reverse transcriptase** displays a much greater asymmetry. Its two polypeptide chains originated from the same gene and were initially identical, but one has undergone proteolytic processing that changes its primary structure. The original chain and the processed chain adopt very different folds in the dimer (Figure 9-29) and have very different functional roles. In this case, both the primary and 3-D structures are asymmetric.

HIV reverse transcriptase structures: designing new inhibitors and understanding mechanisms of drug resistance. J. Ren & D. K. Stammers (2005) *Trends Pharmacol. Sci.* **26**, 4–7.

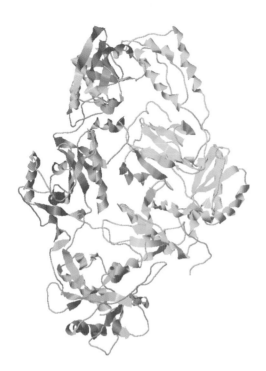

Figure 9-29. The asymmetric dimer of HIV reverse transcriptase. The two polypeptide chains were originally identical but the green one has had part of the polypeptide chain removed. The dimer that results is very asymmetric. Generated from PDB file *1bqm* using the program Jmol.

3. Interfaces

The contacts between subunits determine the stabilities of quaternary structures. The interfaces are generally very similar in character to the interiors of the individual subunits or domains. The extent of the contact between subunits is measured by the **interface area**, B, the area of the protein surface that is buried by the interaction and made inaccessible to the solvent. The interface areas observed in some oligomeric proteins are given in Table 9-5. The total area of the surface buried in the quaternary structure is B_{tot}. In all cases, it is at least 1400 Å^2, comparable to the surface area that is buried in contacts between an antibody and its protein antigen, or a proteinase and a proteinase inhibitor (Figure 9-25); this may be the minimum required for stable association. Most oligomeric proteins have much larger subunit interfaces. The tetrameric quaternary structure of bovine catalase with D_2 symmetry buries 70% of the protein-accessible surface. The six large interfaces are isologous, so only three are unique (Table 9-5).

The buried surface area B_{tot} is usually distributed among several pairwise interfaces, up to $n_{pair} = n(n-1)/2$ for an assembly of *n* subunits when each one is in contact with all others. In a dimer $n_{pair} = 1$, while in a tetramer $n_{pair} = 6$, and most D_2 tetramers have six pairwise interfaces like catalase. The four identical subunits of phosphofructokinase (Figure 15-7) can be labeled A, B, C and D. The AB and CD pairs are equivalent and make extensive contacts burying 4500 Å^2 each. As the equivalent AC and BD pairs bury significantly less, and the AD and BC pairs very little, phosphofructokinase appears, based on the size of the pairwise interfaces, to be a dimer of AB-like dimers.

Dividing large oligomers into smaller assemblies on the basis of their 3-D structure is often an arbitrary decision. For example, each bacterial phosphoribosylaminoimidazole carboxylase catalytic subunit in the octameric structure (Figure 9-26-G) has two symmetry-equivalent heterologous interfaces with neighbors in its tetramer and two isologous interfaces with subunits of the other tetramer. Both are relatively large, making it impossible to decide whether it is a dimer of cyclic tetramers or a cyclic tetramer of dimers.

Table 9-5. Interface areas between subunits of oligomeric proteins (Å²)

Protein	Total accessible surface area	Interface areas			
		B_{tot}	B_{pair}		
Dimers					
Avian pancreatic peptide	5300	1400			
Uteroglobin	7500	3000			
Superoxide dismutase	13,800	1350			
Triosephosphate isomerase	20,300	3180			
Alcohol dehydrogenase	29,000	3260			
Citrate synthase	28,500	9800			
Tetramers					
Mellitin	6300	4160	880	820	520
Glutathione peroxidase	28,600	6280	1520	1520	180
Phosphofructokinase	40,600	14,400	4500	2520	180
Catalase	60,900	42,300	9260	9140	4120
Hexamer					
Insulin	13,100	8600	1280	1400	180[a]
Octamer					
Hemerythrin	35,900	13,600	1780	260	1420[a]

Dimers have C_2 symmetry, while tetramers have D_2 symmetry, with isologous interfaces. Tetramers and larger oligomers contain several interfaces, with areas given per pair of subunits. Larger oligomers also contain heterologous interfaces, which are marked[a].

Data from J. Janin *et al.* (1988) *J. Mol. Biol.* **204**, 155–164.

Interface sliding as illustrated by the multiple quaternary structures of liganded hemoglobin. T. C. Mueser *et al.* (2000) *Biochemistry* **39**, 15353–15368.

Hydration of protein–protein interfaces. F. Rodier *et al.* (2005) *Proteins* **60**, 36–45.

Targeting protein–protein interactions by rational design: mimicry of protein surfaces. S. Fletcher & A. D. Hamilton (2006) *J. Roy. Soc. Interface* **3**, 215–233.

Protein–protein binding-sites prediction by protein surface structure conservation. J. Konc & D. Janezic (2007) *J. Chem. Inf. Model.* **47**, 940–948.

---

4. Oligomerization and Domain Swapping

To convert a monomeric protein to a dimer or higher oligomer requires only that the surface residues present a surface that is appropriately complementary. Presumably in this way, the globin polypeptide chains have evolved from the monomeric myoglobin to the tetrameric $\alpha_2\beta_2$ forms of hemoglobins (Figure 9-34).

Complementary surfaces already exist within the monomeric structure, however, between the various elements of secondary structure that interact with each other to make up the fold of the monomer. Another way of oligomerizing would be for these complementary surfaces to interact intermolecularly, rather than intramolecularly (Figure 9-30). This phenomenon is known as **domain swapping**: one domain of a multi-domain monomeric protein is replaced by the same domain from an identical protein chain. The result is an intertwined dimer or higher oligomer, with one domain of each subunit replaced by the identical domain from another subunit. The domain that is swapped can be as large as an entire globular domain or as small as an α-helix or β-strand. What controls whether such complementary surfaces interact intra- or intermolecularly is not known but it is probably due to differences in the effective concentrations of the two structural motifs during folding and assembly (Section 11.3.A).

Figure 9-30. An example of a dimeric protein generated by domain swapping, the dimeric bovine seminal ribonuclease. The two identical polypeptide chains are colored as *blue* and *green*. The two protomers are identical but both are made up of *blue* and *green* portions of polypeptide chain, because each N-terminal α-helix is situated in the other monomer. An unswapped monomeric structure occurs in bovine ribonuclease A. Generated from PDB file *1y92* using the program Jmol.

Evolution of protein function by domain swapping. M. Ostermeier & S. J. Benkovic (2000) *Adv. Protein Chem.* **55**, 29–77.

Protein folding and three-dimensional domain swapping: a strained relationship? M. E. Newcomer (2002) *Curr. Opinion Struct. Biol.* **12**, 48–53.

Snapshot of protein structure evolution reveals conservation of functional dimerization through intertwined folding. D. Y. Chirgadze *et al.* (2004) *Structure* **12**, 1489–1494.

Open interface and large quaternary structure movements in 3-D domain swapped proteins: insights from molecular dynamics simulations of the C-terminal swapped dimer of ribonuclease A. A. Merlino *et al.* (2005) *Biophys. J.* **88**, 2003–2012.

5. Filamentous Arrays

Filamentous arrays comprise protein molecules that are globular in shape but assemble either helically or as a long string to form, respectively, a regular or semiregular filamentous structure. A good example of the former system is **actin**, in which G-actin (globular) monomers assemble by heterologous interactions into F-actin filaments (fibrous). These, together with other proteins such as tropomyosin and troponin, form the thin filaments of muscle. **Titin**, an extremely long protein from muscle, consists of a series of globular domains strung together by the polypeptide chain and is thus an example of the second type of filament-forming fibrous protein. The short arms in laminin are likewise composed of globular regions strung together in a nearly linear manner. A great many types of filamentous arrays may be formed, depending upon the protein subunit.

9.1.I. Classification of Tertiary Structures: Order out of Diversity

On the basis of their secondary structures, protein structures have been classified into four classes: **α**, having only α-helices; **β**, primarily a β-sheet structure; **α + β**, having both helices and sheets but in separate parts of the structure; and **αβ**, where both helices and sheets interact and often alternate along the polypeptide chain. In α-proteins, about 60% of the residues are in α-helices, and these helices are usually in contact with each other. In β-proteins there are always two β-sheets, both usually antiparallel, that pack against each other. In the α + β-proteins, there may be a single β-sheet, which is usually antiparallel; the helices often cluster together at one or both ends of the β-sheet. The αβ-proteins have one major β-sheet of primarily parallel strands; a helix usually occurs in each of the segments of polypeptide chain connecting the β-strands, probably owing to the necessarily long lengths of these connections. The helices pack on both sides of the sheet, unless the sheet is closed into a barrel, when the α-helices pack around the outside of the barrel, as in the common TIM barrel first discovered in triose phosphate isomerase (Figure 9-4).

Since this classification system was proposed, the many more protein structures that have been determined subsequently have smudged the distinctions between them. **Protein structures seem to occupy a continuum in 'fold space'.** Nevertheless, this classification system is still used widely to describe protein structures, and it describes some of the major themes of protein structure.

Protein structure comparison: implications for the nature of 'fold space', and structure and function prediction. R. Kolodny *et al.* (2006) *Curr. Opinion Struct. Biol.* **16**, 393–398.

A framework for protein structure classification and identification of novel protein structures. Y. J. Kim & J. M. Patel (2006) *BMC Bioinformatics* **7**, 456.

---

1. α-*Structures*

The cores of these protein structures are formed by **short α-helices packed together and connected by loop regions**. The internal packing is primarily hydrophobic and holds the helices together in a stable structure, while the hydrophilic groups on the surface make the protein water-soluble. The simplest and most frequent α-helical domain consists of four α-helices arranged in a four-helix bundle (Section 9.1.D.6). Another example is the globin fold found in myoglobin and hemoglobin (Figures 9-4 and 9-34). Several exceptional cases are known: for example, the 450 N-terminal residues of a bacterial muramidase comprise a single α-helical domain containing 27 α-helices arranged in a two-layered ring. The ring has a right-handed superhelical twist and a large doughnut-like central hole ~ 30 Å in diameter.

α-Structures are dominated by the interactions between adjacent α-helices. In four-helix bundles, the α-helices lie almost parallel or antiparallel to each other, with an angle of only ~ 20° between their axes. In the globin fold, the angles are larger, usually ~ 50°. The packing angles of helices are probably determined by maximizing the interactions between their irregular surfaces.

A new perspective on analysis of helix–helix packing preferences in globular proteins. A. Trovato & F. Seno (2004) *Proteins* **55**, 1014–1022.

Novel approach for α-helical topology prediction in globular proteins: generation of interhelical restraints. S. R. McAllister *et al.* (2006) *Proteins* **65**, 930–952.

---

2. β-Structures

Proteins of the β-class are composed primarily of β-strands, varying in number from four to more than 10. The β-strands are usually arranged in an antiparallel orientation, in such a way that they form two β-sheets that are joined together and packed against each other. The β-sheets have the usual twist, and two packed together tend to form a barrel-like structure stabilized by a core of hydrophobic side-chains from the β-sheet inside the barrel.

The simplest topology is that of the up-and-down β-barrels, comprising eight or 10 β-strands. All connections between the β-strands are hairpins, so that β-strands adjacent in sequence are also adjacent in the structure. This topology is equivalent to a β-meander (Figure 9-10-C) but the barrel is closed by interactions between the first and last β-strands. Hydrophobic side-chains fill the interior of the barrel. Proteins having this type of structure often function as transporters or solubilizers of hydrophobic ligands, such as retinoids, fatty acids and bile salts, which bind in the interior of the barrel.

Other examples include the Greek key barrel (formed from two Greek key motifs; Figure 9-13) and the jelly roll EGF motif barrel (Figure 9-14). The various barrel motifs have different connections between the strands in the barrel. Other types of antiparallel β-barrels include the interleukin-1 motif (Figure 9-17). All these protein barrels comprise only antiparallel β-strands and are therefore classified as all-β structural domains. They differ significantly from the TIM barrel (Section 9.1.I.3.a and Figure 9-4), which is formed from a parallel β-barrel surrounded on the exterior by α-helices and is classified as an αβ-domain.

The β-strands of β-barrels are usually inclined relative to the barrel axis, and a characteristic feature of β-barrels is the shear number, a measure of the extent to which the β-strands are offset in their hydrogen-bonding pattern (Figure 9-7). It is defined as the change of residue numbers on a β-strand when a point moves in the left hydrogen-bond direction around the barrel and back to the initiating strand. All β-barrels have positive shear numbers, meaning that they are all twisted in the right-hand direction. Most have even shear numbers but exceptions exist; many contain β-bulges (Section 9.1.B.3).

Up-and-down β-sheets that do not fold into barrels are also known, such as that of the neuraminidase from influenza virus. Each of the four subunits of the tetramer is folded into a single domain comprising six similar motifs of up-and-down antiparallel β-sheets of four β-strands each. The strands are highly twisted so that the first and fourth of each sheet differ in their directions by nearly 90°. To a very rough approximation, the six motifs are arranged within each subunit with an approximate six-fold symmetry axis, like six blades of a propeller.

The up-and-down β-barrel proteins. J. LaLonde (1994) *FASEB J.* **8**, 1240–1247.

Shear numbers of protein β-barrels: definition refinements and statistics. W.-M. Liu (1998) *J. Mol. Biol.* **275**, 541–545.

---

3. αβ-Structures

The αβ-structures are made from combinations of β–α–β motifs that form a predominantly parallel β-sheet surrounded by α-helices. Crevices for binding ligands are generally formed by the loop regions. The α-helix connecting the two parallel β-strands always lies on one side of the sheet (Figure 9-10-A), designated right-handed because it has the same hand as a right-handed α-helix; the reason for this specificity is not clear but it may be due to twisting of the β-sheet. There are two main types of αβ-structures.

(1) The first has a core of twisted parallel β-strands arranged close together like the staves of a barrel. The α-helices connect the parallel strands and lie on the outside of the barrel. This is usually called the TIM barrel, from its first known example, the enzyme triosephosphate isomerase (Figure 9-4).

(2) The second class has an open twisted β-sheet surround by α-helices on both sides. These structures are much more variable in terms of the topology of the β-sheet and the number of β-strands. A typical example is the nucleotide-binding domain (Section 12.4).

Steric restrictions in protein folding: an α-helix cannot be followed by a contiguous β-strand. N. C. Fitzkee & G. D. Rose (2004) *Protein Sci.* **13**, 633–639.

Role of topology, nonadditivity, and water-mediated interactions in predicting the structures of α/β proteins. C. Zong *et al.* (2006) *J. Am. Chem. Soc.* **128**, 5168–5176.

Contact patterns between helices and strands of sheet define protein folding patterns. A. P. Kamat & A. M. Lesk (2007) *Proteins* **66**, 869–876.

---

a. TIM/$(\alpha\beta)_8$ Barrels

The TIM barrel is named after the enzyme **triosephosphate isomerase**, frequently abbreviated **TIM**, in which it was first found. It is also known as the $(\beta\alpha)_8$- or $(\alpha\beta)_8$- barrel. It has a central cylinder or barrel of β-sheet formed from eight parallel β-strands; the barrel is closed by hydrogen bonding between the first and last strands in the sequence (Figure 9-4). Adjacent β-strands are connected by an α-helix, and the eight α-helices form an external coat for the central β-barrel. Both the strands and helices of the TIM barrel have a pronounced right-hand twist (Figure 9-7).

This fold is very common and found in many proteins with diverse functions and without detectable sequence similarities; consequently, their similar structures might result from convergent evolution. The activities and sequences of these proteins vary, but their active sites are all formed by loop regions at the carboxyl ends of the β-strands that connect to the α-helices. The enzymatic activity of any particular enzyme is determined by the amino acid residues in these regions. Half-barrels, $(\beta\alpha)_4$, adopt folded conformations, of uncertain structure, that are nearly as stable as the complete barrel, so the possibility of duplicating, mixing and fusing together pairs of such half-barrels may be why this structure has arisen so frequently.

Mimicking enzyme evolution by generating new $(\beta/\alpha)_8$-barrels from $(\beta/\alpha)_4$-half-barrels. B. Hocker *et al.* (2004) *Proc. Natl. Acad. Sci. USA* **101**, 16448–16453.

Exploring the environmental preference of weak interactions in $(\alpha/\beta)_8$ barrel proteins. S. Chakkaravarthi *et al.* (2006) *Proteins* **65**, 75–86.

Mapping the structure of folding cores in TIM barrel proteins by hydrogen exchange mass spectrometry. Z. Gu *et al.* (2007) *J. Mol. Biol.* **368**, 582–594.

---

4. α + β and Other Structures

The α + β-proteins are built up from **a combination of discrete α- and β-motifs**; they usually have one small antiparallel β-sheet in one part of the domain packed against a number of α-helices clustered together at one or both ends of the β-sheet.

Other protein structures are irregular and have little secondary structure; they are predominantly very small proteins.

5. Protein Structure Classification Databases

There are now so many different protein structures and folds known that the above classification systems are no longer adequate. Structural classifications are now made and maintained in databases. At the beginning of 2007 the structures of more than 93×10^3 domains and more than 63×10^3 polypeptide chains were contained in these databases.

The **SCOP** database (**S**tructural **C**lassification **o**f **P**roteins; http://scop.mrc-lmb.cam.ac.uk/scop/) attempts to classify proteins in terms of their structural and evolutionary relatedness. Proteins that are clearly homologous because of >30% sequence identity and/or similar structures and functions (Section 9.3) are classified as a family. Proteins with low sequence identities but structures and functions suggesting that they might have a common evolutionary origin are placed together in a superfamily. Protein structures share a common fold if they have the same major secondary structures in the same arrangement and with the same topological connections. They often have peripheral regions that differ in size and conformation, which might comprise half the structure. Proteins with the same fold may have a common evolutionary origin, or the structural similarities could arise just from the physical principles of protein structure.

The **CATH** protein structure classification (www.cathdb.info) clusters proteins on the basis of their **C**lass (Section 9.1.I), **A**rchitecture, **T**opology and **H**omology. The architecture is defined by the orientations of the secondary structure elements but ignores the connectivities between them, while the topology includes the connectivities. Homology implies that the proteins arose from a common evolutionary ancestor and is based on degree of sequence identity and similarities in structure and function.

9.2. MEMBRANE PROTEINS: AVOIDING WATER

Proteins that penetrate into and usually traverse the lipid bilayer of a biological membrane are considered to be **intrinsic** or **integral membrane proteins**. They partition into the lipid interior of a membrane, rather than remain in aqueous solution, and are fundamentally different from proteins that are anchored to the membrane only via a fatty acid or prenyl group attached to one of their polypeptide chain termini. Some 20–25% of genes in the genomes of microorganisms and multicellular organisms are predicted to encode integral membrane proteins, which reflects the importance of membranes in the life of any cell. Membrane proteins function in membrane transport, secretion, bioenergetic processes and cell–cell communication.

Integral membrane proteins have characteristic chemical and structural properties. They are rich in exposed hydrophobic amino acids and restricted in their patterns of secondary structure, probably due to the need to pair the peptide groups in hydrogen bonds within the membrane environment. A consequence of the hydrophobic external surface means that an integral membrane protein can be solubilized in aqueous solution only in the presence of a detergent, which forms an interface between the protein and water and is essential for stability and biological activity. Membrane proteins can often be purified using the same types of methods used for soluble proteins, so long as all solutions contain a detergent at a concentration greater than its critical micelle concentration.

In spite of their great importance, 3-D protein structures at atomic resolution have been determined for relatively few membrane proteins. This is due primarily to the great difficulty in crystallizing membrane proteins, which is caused by three factors. First, the detergent has its own phase behavior, and its presence significantly complicates the search for the right crystallization conditions. Second, the polar parts of membrane proteins are often small, and they may also be masked by the polar headgroups of the detergent. It is these groups that usually participate in the protein–protein contacts that are critical for a crystal lattice, so there are many fewer possibilities with membrane proteins. Third, membrane proteins often assemble into large complexes. For example, photosynthetic reaction centers in plants and bacteria, ATP synthase and respiratory complexes contain a large number of subunits, pigments and bound metals. Similarly, molecular machines involved in protein secretion and ATP-driven active transport usually have several protein components, and ion channels are typically oligomeric.

Most of the structures of membrane proteins have been determined using 3-D crystals and X-ray diffraction, although some have been determined using electron crystallography on 2-D (single-layered) crystals of the protein reconstituted with lipid. Electron crystallography is becoming increasingly important in determining membrane protein structures.

The known membrane protein structures fall into four categories: **α-helical**, **β-barrel**, **monotopic** and **bitopic** (Figure 9-31). They differ from globular proteins primarily in that their external surfaces in contact with the nonpolar portion of the membrane bilayer are also nonpolar. Otherwise, the interiors of these proteins are remarkably similar to those of water-soluble proteins. An early suggestion that membrane proteins might be 'inside-out' turned out not to be correct. The elements of secondary structure that make up the protein interiors tend to be longer than those in water-soluble proteins, to be able to traverse the membrane. Arg and Lys residues are much more common on the side of the membrane that is facing the cell interior, leading to the positive-inside rule for defining the orientation of transmembrane segments in the membrane. The aromatic amino acids, especially Tyr and Trp, are found much more frequently at interfacial positions between the water and the membrane interior. The energetic cost of partitioning peptide bonds into membrane bilayers is prohibitive unless the peptide bonds participate in hydrogen bonds. However, even then there is a significant free energy penalty for dehydrating the peptide bonds, which can only be overcome by favorable hydrophobic interactions. **Membrane protein structure formation is thus dominated by hydrogen-bonding interactions**.

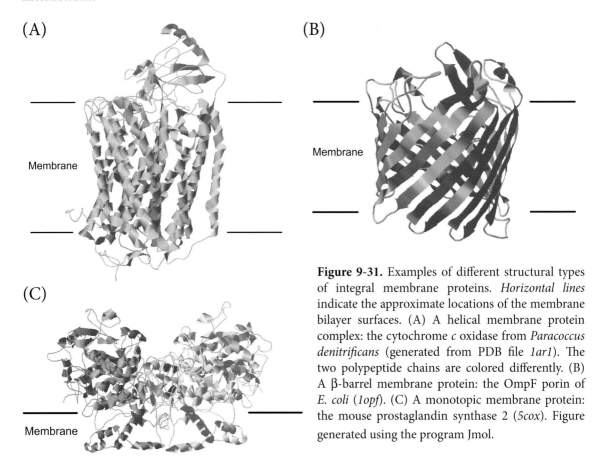

Figure 9-31. Examples of different structural types of integral membrane proteins. *Horizontal lines* indicate the approximate locations of the membrane bilayer surfaces. (A) A helical membrane protein complex: the cytochrome *c* oxidase from *Paracoccus denitrificans* (generated from PDB file *1ar1*). The two polypeptide chains are colored differently. (B) A β-barrel membrane protein: the OmpF porin of *E. coli* (*1opf*). (C) A monotopic membrane protein: the mouse prostaglandin synthase 2 (*5cox*). Figure generated using the program Jmol.

Solving the membrane protein folding problem. J. U. Bowie (2005) *Nature* **438**, 581–589.

How hydrogen bonds shape membrane protein structure. S. H. White (2005) *Adv. Protein Chem.* **72**, 157–172.

A limited universe of membrane protein families and folds. A. Oberai *et al.* (2006) *Protein Sci.* **15**, 1723–1734.

Membrane-protein topology. G. von Heijne (2006) *Nature Rev. Mol. Cell Biol.* **7**, 909–918.

Top-down mass spectrometry of integral membrane proteins. J. Whitelegge *et al.* (2006) *Expert Rev. Proteomics* **3**, 585–596.

From interactions of single transmembrane helices to folding of α-helical membrane proteins: analyzing transmembrane helix–helix interactions in bacteria. D. Schneider *et al.* (2007) *Curr. Protein Peptide Sci.* **8**, 45–61.

9.2.A. Helical Superfolds

Helical membrane proteins are assembled from **transmembrane α-helices that traverse the membrane** (Figure 9-31-A). Each transmembrane helix typically consists of 20–30 residues, sufficient to traverse the thickness of a membrane. Usually all these residues are hydrophobic, although single polar or potentially charged residues are occasionally inserted into the membrane-spanning sequence. The helical secondary structure is a natural way of embedding a polypeptide chain into lipid, because hydrogen bonding of the peptide groups is satisfied within the helix.

The structures of the photosynthetic reaction center (Figure 9-32) and the mitochondrial cytochrome *c* oxidase contain a total of 39 transmembrane α-helices, with the following properties that may be general features of transmembrane α-helices.

(1) The helices need to be 40 Å long in order to traverse the bilayer as straight rods. In most cases, however, they are slightly tilted relative to the membrane normal and consequently need to be slightly longer.

(2) Polar residues occur at the ends of the α-helices.

(3) The helices have a hydrophobic core that spans ~ 20 Å and is composed primarily of hydrophobic residues, such as Phe, Val, Leu, Met and Ile.

(4) On both sides of the core are found the polar aromatic residues Tyr and Trp. These buried Tyr and Trp residues are slightly further from the core than the aromatic residues closer to the hydrophobic core, which are exposed to the lipid headgroup region.

(5) The region outside the hydrophobic core is also populated by Asn and Gln residues, which are either exposed to the surface or buried in the protein structure.

(6) Pro residues occur most frequently around 25 Å from the core, just beyond the ends of the helices. Pro and Gly residues that occur in the hydrophobic core are buried in the structure.

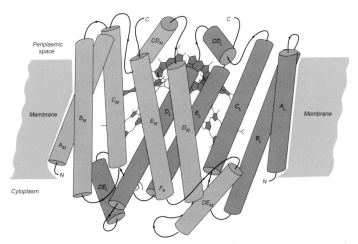

Figure 9-32. The photosynthetic reaction center. The two main subunits, L and M, are illustrated schematically, along with cofactors and prosthetic groups involved in the function. The major α-helices of each subunit are labeled A–E; minor helices are labeled by the major helices they connect. Note the approximate two-fold symmetry axis; the L and M subunits probably arose from a common ancestral protein that functioned as a dimer of identical subunits. The functional protein also contains an H subunit (not shown) that is located on the cytoplasmic face of the membrane but has one transmembrane helix. A cytochrome subunit (not shown) also binds to the L and M subunits on the cytoplasmic face. Adapted from H. Michel & H. Deisenhofer (1988) *Biochemistry* **27**, 1.

In the case of channels involved in the transport of solutes through the membrane, α-helices line the channel through which the solutes are transported. These helices tend to be tilted steeply relative to the membrane normal and are arranged consistently in a right-handed bundle. The helical framework also provides a supporting scaffold for structures associated with channel selectivity that do not span the membrane.

Various superfolds are characterized by the number of membrane-spanning segments and their inside/outside topology. In particular, there is a large family of proteins with seven transmembrane helices, the '7TM' superfamily, that is exemplified by bacteriorhodopsin as shown in Figure 9-33. All members of this family have seven α-helices that span the membrane, and the N-terminus is on the external side of the membrane. Many are G-protein coupled receptors that use hydrolysis of GTP to mediate signaling from the outside to the cellular interior, and have a sensory function in processes such as hearing, seeing, smelling and tasting, but the 7TM proteins are functionally very diverse. The seven-helix fold appears to be able to adapt to different functions by accommodating diverse ligands and prosthetic groups.

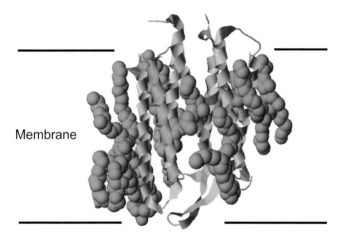

Figure 9-33. The structure of bacteriorhodopsin with lipid molecules bound. The seven transmembrane α-helices are indicated schematically as *coils*. The atoms of the bound lipid molecules are shown as *spheres* with their van der Waals radii. The bilayer will extend horizontally across the structure, encompassing the bound lipid molecules. The only parts of the protein exposed to the aqueous solvent are those at the top and bottom. Generated from PDB file *1c3w* using the program Jmol.

Some proteins are predicted to have 12 transmembrane helices, and they function in the transport of solutes and ions across membranes. Dimers of monomers containing six α-helices are a subclass of the 12-helix proteins. Another important example is the photosynthetic reaction center that plays a central role in photosynthesis (Figure 9-32). Other abundant superfolds are formed by four-helix bundles.

It is difficult to surmise the function of a novel integral membrane protein from its structure. For example, membrane transport is catalyzed by proteins with diverse structures. Even the tetrameric potassium channels can be divided into two subclasses. In one class, each monomer has two transmembrane helices, so the tetramer has eight. In the second case, the monomer has six membrane-spanning segments and the tetramer has 24. In both cases, however, the ion-conducting channel is lined by four α-helices, one from each of the four monomers.

Construction of helix-bundle membrane proteins. A. K. Chamberlain *et al.* (2003) *Adv. Protein Chem.* **63**, 19–46.

Sequence motifs, polar interactions and conformational changes in helical membrane proteins. A. R. Curran & D. M. Engelman (2003) *Curr. Opin. Struct. Biol.* **13**, 412–417.

Folding of helical membrane proteins: the role of polar, GxxxG-like and proline motifs. A. Senes *et al.* (2004) *Curr. Opin. Struct. Biol.* **14**, 465–479.

Computational analysis of α-helical membrane protein structure: implications for the prediction of 3-D structural models. T. A. Eyre *et al.* (2004) *Protein Eng. Des. Sel.* **17**, 613–624.

Transmembrane helices of membrane proteins may flex to satisfy hydrophobic mismatch. P. L. Yeagle *et al.* (2007) *Biochim. Biophys. Acta* **1768**, 530–537.

9.2.B. β-Barrel Membrane Proteins

The transmembrane portion of a few membrane proteins consists of an **up-and-down β-barrel** (Section 9.1.B.2) that is made up of eight, 14, 16 or 18 antiparallel β-strands (Figure 9-31-B). This structure occurs in the porins and membrane-bound forms of some toxins, such as α-hemolysin, which is a heptameric toxin from *Staphylococcus aureus*. These proteins facilitate the diffusion of small molecules through the barrel and across the membrane, and **the interior of the barrel is hydrophilic**. No more than 10 residues are needed to traverse the bilayer, due to the extended conformations of β-strands. The β-strands of the porin β-barrel range in length from six to 17 residues. Only every second residue of each strand faces the lipid core of the membrane, and only it is consistently hydrophobic. Only the hydrophobic nature of the lipid-facing residues needs to be conserved, not their identity, due to the fluid nature of the bilayer core. The rest of the residues in the barrel can be either hydrophilic or hydrophobic, depending on whether they are exposed to the aqueous pore or buried in the protein structure. As with helical and monotopic (Section 9.2.C) membrane proteins, the strands are rich in aromatic side-chains that interact with the lipid headgroups.

The β-barrel membrane proteins have a number of peculiar properties. First, their sequences are at least as hydrophilic as those of soluble proteins. Yet the lipid-exposed surface of the β-barrel is highly hydrophobic, and the porins are soluble only in the presence of a detergent. In contrast, the monomeric α-hemolysin is a soluble protein, which rearranges upon interaction with lipids or detergent to form the heptameric transmembrane pore. Second, the primary structures of porins form at least 10 families that show no clear sequence homology with each other, indicating that highly diverse primary structures can fold into these very similar 3-D structures. Third, porins are unusually stable, forming trimers that are resistant towards denaturation, even at elevated temperatures.

Folding and assembly of β-barrel membrane proteins. L. K. Tamm *et al.* (2004) *Biochim. Biophys. Acta* **1666**, 250–263.

β-Barrel transmembrane proteins: geometric modelling, detection of transmembrane region, and structural properties. I. K. Valavanis *et al.* (2006) *Comput. Biol. Chem.* **30**, 416–424.

A minimal transmembrane β-barrel platform protein studied by nuclear magnetic resonance. M. U. Johansson *et al.* (2007) *Biochemistry* **46**, 1128–1140.

β-Barrel membrane bacterial proteins: structure, function, assembly and interaction with lipids. S. Galdiero *et al.* (2007) *Curr. Protein Peptide Sci.* **8**, 63–82.

9.2.C. Monotopic and Bitopic Membrane Proteins

Some membrane proteins require detergents for solubilization but have no transmembrane α-helices or β-strands (Figure 9-31-C). They are known as **monotopic** membrane proteins. They associate with the surface of the membrane using four short α-helices that run approximately parallel to the plane of the membrane. The α-helices are amphipathic and contain a number of aromatic side-chains, which are probably exposed to the lipid headgroup region of the membrane, but the four-helix motif is only immersed into one layer of the bilayer. Although these proteins are genuine membrane proteins, they are not transmembrane proteins. The membrane-binding domains of cyclo-oxygenases are the least conserved part of the enzymes (33% sequence identity) but the 3-D structure of the membrane anchor is well preserved.

A **bitopic** membrane protein contains a single transmembrane α-helix that connects two domains on opposite sides of the membrane. These transmembrane helices were initially considered to be merely hydrophobic anchors to the membrane, but they are now known to play more important and functional roles.

Prediction of amphipathic in-plane membrane anchors in monotopic proteins using a SVM classifier. N. Sapay et al. (2006) *BMC Boinformatics* **7**, 255.

Surface-active helices in transmembrane proteins. J. P. Orgel (2006) *Curr. Protein Peptide Sci.* **7**, 553–560.

How important are transmembrane helices of bitopic membrane proteins? M. Zviling et al. (2007) *Biochim. Biophys. Acta* **1768**, 387–392.

9.2.D. Interactions with the Membrane

Proteins embedded in membranes differ from water-soluble proteins primarily in their surfaces, where they interact with lipids rather than water. One important parameter is the hydrophobic thickness of the lipid bilayer, defined by the lengths of the lipid fatty acyl chains. Binding to lipids is important for the vertical positioning and tight integration of proteins in the membrane and for assembly and stabilization of oligomeric and multi-subunit complexes and supercomplexes, as well as for their functional roles. Many membrane proteins selectively bind defined lipid species. Binding of lipids results from multiple noncovalent interactions between protein residues and the headgroups and hydrophobic tails of the lipids. Analysis of lipids with specific headgroups in membrane protein structures demonstrates distinct motifs for the interactions between the phosphodiester moieties and the side-chains of amino acid residues. There are differences in binding at the electropositive and electronegative sides of the membrane, plus preferential binding to the latter. Charged residues are not always buried, and few are paired with another charge; instead they often interact with phospholipid headgroups or with other types of residue.

On average, transmembrane residues that are accessible to the lipid tails are significantly more hydrophobic than the buried residues, less conserved and contain different types of residues. Membrane proteins are not rigid entities but tend to deform to ensure good hydrophobic matching to the surrounding lipid bilayer. A significant proportion of charged and polar residues in contact with the hydrophobic lipid tails form hydrogen bonds only with residues one turn away in the same helix. Residues that line a pore through the protein can be either hydrophilic or hydrophobic, and it is difficult to distinguish them in terms of either their type of residue or evolutionary conservation.

Lipids in membrane protein structures. H. Palsdottir & C. Hunte (2004) *Biochim. Biophys. Acta* **1666**, 2–18.

Specific protein–lipid interactions in membrane proteins. C. Hunte (2005) *Biochem. Soc. Trans.* **33**, 938–942.

Roles of bilayer material properties in function and distribution of membrane proteins. T. J. McIntosh & S. A. Simon (2006) *Ann. Rev. Biophys. Biomol. Structure* **35**, 177–198.

9.3. PROTEINS WITH SIMILAR FOLDED CONFORMATIONS: EVOLUTION IN 3-D

With the very wide diversity in amino acid sequences observed in proteins (Section 7.6), it might be expected that there is a corresponding diversity in folded conformations. On the contrary, **similarities between protein structures are observed remarkably often**. In particular, the tertiary structures of proteins have been much more conserved during evolution than have their primary structures. The extent of this evolutionary conservation is a matter of intense debate and speculation, because similarities in tertiary structures are being found in cases where there is no detectable sequence homology and no evolutionary or functional reasons to expect them. A further question subject to much debate is how many different protein folds are possible and found in nature.

Protein families and their evolution: a structural perspective. C. A. Orengo & J. M. Thornton (2005) *Ann. Rev. Biochem.* **74**, 867–900.

Fold usage on genomes and protein fold evolution. S. Abeln & C. M. Deane (2005) *Proteins* **60**, 690–700.

A structure-centric view of protein evolution, design, and adaptation. E. J. Deeds & E. I. Shakhnovich (2007) *Adv. Enzymol.* **75**, 133–191.

9.3.A. Homologous Proteins: Protein Families

Proteins with substantial similarities in their primary structures and functions are almost certain to have arisen from a common ancestor during evolution, and they **invariably have very similar folded conformations**. For example, the 3-D structures of horse and human hemoglobin are virtually identical in detail, even though they differ in 43 of the 287 residues of the α- and β-chains. Similarly, the cytochromes *c* of horse and tuna have extremely similar structures, even though differing at 17 of the 104 amino acid residues. Such conservation of conformation is understandable, in that these related proteins have virtually identical properties and probably serve essentially the same function in the different species, with very similar structural requirements.

That more distantly related and paralogous proteins also have very similar conformations was shown by the first two proteins whose structures were determined crystallographically, sperm whale myoglobin and horse hemoglobin. The single polypeptide chain of myoglobin and the α-, β- and γ-chains of hemoglobin are remarkably similar in the general topologies of their polypeptide backbones (Figure 9-34). These proteins have amino acid sequences that are sufficiently similar to indicate that they are related evolutionarily (Figure 7-31) and they have similar functions in reversibly binding oxygen at a bound heme group (Section 12.5.B). Other oxygen-binding, heme-containing proteins from a wide variety of vertebrates, a marine annelid worm, a larval insect and lupine root nodules also have

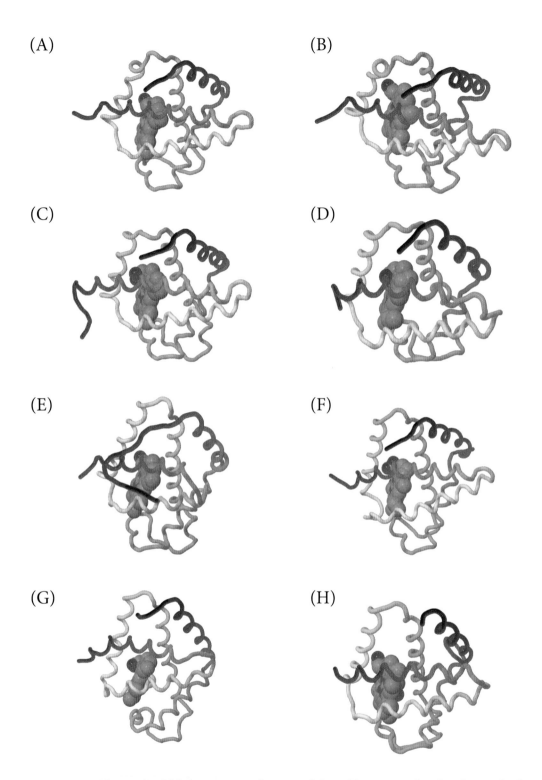

Figure 9-34. The similar folded structures of various globins. The course of each polypeptide chain is shown, with the N-terminus *blue*, the C-terminus *red* and the colors of the rainbow in between. The van der Waals surface of each heme group is shown. (A) Human hemoglobin α-chain (generated from PDB file *1a3n*). (B) Human hemoglobin β-chain (*1a3n*). (C) Sperm whale myoglobin (*1vxg*). (D) Erythrocruorin (*1eca*). (E) Lamprey hemoglobin (*1f5o*). (F) Glycera hemoglobin (*1jf3*). (G) Clam ferric hemoglobin (*1moh*). (H) *Ascaris* hemoglobin (*1ash*). Some of these proteins have minimal similarities in their amino acid sequences. Figure generated with the program Jmol.

very similar conformations (Figure 9-34). These proteins are usually grouped together as the **globin family**. In some of these instances, the primary structures are not detectably similar. In all, the known globins have only two amino acid residues in common. Nevertheless, intermediate sequences and their closely similar conformations indicate that they have all diverged from a common evolutionary ancestor.

Another family of proteins with similar conformations is that of the **cytochrome c-type** proteins. The conformations of the closely similar cytochromes c found in the mitochondria of vertebrates (Figure 7-29) resemble those in a variety of functionally related cytochromes from bacteria. The bacterial cytochromes can be recognized from their properties as similar to cytochrome c but are sufficiently different in their primary structures and physical and functional properties to be given distinct names such as c_2, c_{550}, c_{551} and c_{555}. The polypeptide chains of these proteins range in length from 82 to 135 residues and have only five residues in common, yet their 3-D structures are remarkably similar.

Another well-characterized family of homologous proteins is that of the trypsin-like **serine proteinases** (Section 7.5.B.1), so named because they all have an important Ser residue at the active site and similar catalytic mechanisms in hydrolyzing polypeptide chains (Section 15.2.D.1). This family contains the paralogous proteins of higher organisms, trypsin, chymotrypsin, elastase and thrombin, as well as similar enzymes from microbial sources. In the case of bovine α-chymotrypsin and elastase, only 39% of their residues are identical, yet their conformations can be superimposed so that on average their polypeptide backbone atoms differ in their relative positions by only 1.80 Å. This similarity in topology is immediately obvious in their contact maps (Figure 9-20). Trypsin and α-chymotrypsin may similarly be juxtaposed to within 0.75 Å average difference, even though only 44% of their residues are identical. The bacterial serine proteinases have diverged more in both sequence and overall conformation; fewer than 20% of their residues are identical with those of any of the mammalian proteins, with numerous insertions and deletions, not much more similar than random sequences would be expected to be. Yet their 3-D conformations are more obviously similar than are their primary structures; 55–64% of the residues of the mammalian and bacterial proteins are topologically equivalent, in that their C^α atoms occupy the same relative overall position. With the greater divergence of amino acid sequence, there is a somewhat greater change in structure (Figure 9-35), although **the structure is the more conserved**.

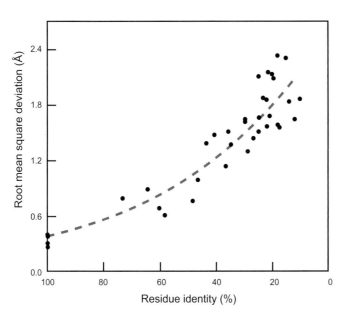

Figure 9-35. Changes in the tertiary structure in response to evolutionary variation of the primary structure of 32 pairs of homologous proteins. The percentage of residues that are identical in each pair of proteins is plotted versus the root mean square deviation of the backbone atoms of their common core. The common core is that part of the related proteins that retains the same fold. When only 20% of the residues are identical, the common core is generally only about half the residues of each protein. Data from C. Chothia & A. Lesk (1986) *EMBO J.* **5**, 823–826.

The first question that arises about these observations is how such wide variation in amino acid sequences is compatible with such similar folded conformations. The greatest variation in amino acid residues occurs at positions in the polypeptide chain where the side-chain is on the surface. Here the chemical nature of the side-chain is often changed; there is a tendency, however, for the residues at reverse turns to have one of the short, polar side-chains, or to be Gly or Pro, those residues found most frequently at these positions (Table 9-6). The conformational flexibility in these regions can also lead to surprising changes in local conformation. For example, bovine and porcine phospholipase A_2 have homologous sequences and very similar 3-D structures, except for a segment of 12 residues on the surface; these segments have very different conformations in the two proteins, even though they differ in sequence at only one position.

The fewest changes occur to interior residues, and the nonpolar nature of the side-chains is most conserved. The side-chain torsion angles are also conserved to a remarkable degree, as are interior polar groups involved in hydrogen bonds. There is generally a common core of residues, comprising the center of the molecule and the major elements of the secondary structure, that is most highly conserved. Structural changes occur in this common core only very slightly as the primary structure of the protein diverges, and there is a good correlation between the extent of the sequence and structural differences (Figure 9-35). Note that the changes in conformation are not directly proportional to the changes in sequence, but of a higher order: this would suggest that structural changes occur primarily after multiple amino acid substitutions.

The most highly conserved residues are those involved directly in the functional properties of the protein. Examples in cytochrome *c* are the invariant Cys14 and Cys17, to which the heme group is attached, and the ligands to the heme iron atoms, His18 and Met80 (Figure 7-29). Likewise, the His side-chains of hemoglobin that bind the heme iron (Figure 12-20) are conserved (Figure 7-31). As would be expected, changing residues that are conserved in many homologous sequences generally has substantial effects on the function and stability of the protein, although not always the dramatic effects that might be predicted.

Insertions and deletions within the polypeptide chain occur most frequently at reverse turns on the surface of the folded protein, usually with little perturbation of the interior. Disulfide bonds are often added or deleted. For example, in the serine proteinase family, trypsin has six, chymotrypsin five, elastase four, α-lytic proteinase three and proteinases A and B of *Streptomyces griseus* two each. The Cys residues usually do not change their pairings, however. If a disulfide bond is deleted, both Cys residues are generally changed to other amino acid residues, but their relative 3-D positions in the folded structure remain the same. Disulfide bonds obviously do not determine the conformation, nor are they absolutely necessary (Section 11.4.C).

Elements of the secondary structure can move relative to each other, change length or even disappear altogether, but a helix is not usually replaced by a β-strand or vice versa. In particular, neither the order nor the orientation of individual strands in a β-sheet, i.e. parallel or antiparallel, usually differ in proteins with homologous primary structures. The relative positions of α-helices are retained, but often within relatively wide limits. For example, in some of the widely differing globins, corresponding pairs of α-helices may differ in position by up to 7 Å and in relative orientation by up to 30°. Individual mutations that are acceptable seem to produce very small changes in structure but the cumulative effects of many such sequence changes can produce substantial differences (Figure 9-35).

Table 9-6. Conformational preferences of the amino acid residues[a]

Amino acid	Conformation			α-Helix[b]			Turn		
	α-Helix (P_α)	β-strand (P_β)	Reverse turn (P_t)	N-term	Middle	C-term	Type I	Type II	Other
Glu	**1.59**	0.52	1.01	**2.12**	1.18	1.21	1.12	0.84	1.06
Ala	**1.41**	0.72	0.82	1.33	**1.60**	**1.46**	0.74	0.94	0.58
Leu	**1.34**	1.22	0.57	1.03	1.50	**1.46**	0.61	0.53	0.75
Met	**1.30**	1.14	0.52	0.75	1.44	**1.92**	0.66	0.73	0.96
Gln	**1.27**	0.98	0.84	**1.39**	1.22	1.24	0.79	1.45	1.02
Lys	**1.23**	0.69	1.07	0.98	1.05	**1.68**	0.70	0.73	1.04
Arg	**1.21**	0.84	0.90	1.26	1.25	1.23	0.88	1.22	0.84
His	**1.05**	0.80	0.81	0.68	0.97	**1.57**	0.78	0.64	1.00
Val	0.90	**1.87**	0.41	1.00	1.09	1.08	0.39	0.61	0.48
Ile	1.09	**1.67**	0.47	0.96	1.31	0.99	0.39	0.43	0.93
Tyr	0.74	**1.45**	0.76	0.63	0.61	1.00	0.71	0.91	0.97
Cys	0.66	**1.40**	0.54	0.78	0.66	0.56	1.38	0.99	0.78
Trp	1.02	**1.35**	0.65	1.20	1.34	0.78	1.35	0.15	0.52
Phe	1.16	**1.33**	0.59	0.94	1.45	1.20	0.77	0.76	0.53
Thr	0.76	**1.17**	0.90	0.75	0.87	0.80	1.25	0.67	0.93
Gly	0.43	0.58	**1.77**	0.60	0.47	0.31	1.14	**2.61**	1.38
Asn	0.76	0.48	**1.34**	0.80	0.80	0.75	**1.79**	0.99	1.37
Pro	0.34	0.31	**1.32**	0.90	0.19	0.06	0.95	**1.80**	**1.51**
Ser	0.57	0.96	**1.22**	0.67	0.44	0.73	**1.47**	0.76	**1.49**
Asp	0.99	0.39	**1.24**	1.35	1.03	0.67	**1.98**	0.71	1.28

[a] The normalized frequencies for each conformation (e.g. P_α, P_β, P_t) were calculated from the fraction of residues of each amino acid that occurred in that conformation, divided by this fraction for all residues. Random occurrence in a conformation would give a value of unity.

[b] N-terminal and C-terminal include the four helical residues at the ends of a helical segment eight or more residues long, and three residues at the ends of segments six or seven long. Middle includes all helical residues between N- and C-terminals.

Data from R. W. Williams *et al. Biochim Biophys. Acta* (1987) **916**, 200–204; C. M. Wilmot & J. M. Thornton (1988) *J. Mol. Biol.* **203**, 221–232.

In summary, the 3-D structures of evolutionarily related proteins appear to have been remarkably conserved during evolution, suggesting that the 3-D structure was well-defined at the early stage of evolution in the common ancestor and that it has been conserved because it is crucial for the function of the protein. Changes in the primary structure due to genetic mutations have accumulated during evolutionary divergence, but only those that are compatible with the folded conformation and its function have been retained. These are the factors that are apparently responsible for the constraints on evolutionary change of the primary structure (Section 7.6.A.6). Nevertheless, the great variety of amino acid sequences that produce the same folded conformation indicates considerable flexibility in the rules of protein structure.

Knowledge of the 3-D structures of evolutionarily related proteins is very useful for inferring the genetic basis of the evolutionary divergence, in that it identifies which residues of the primary structure should be aligned: structurally equivalent residues are assumed to be those that are evolutionarily related. Alignment of the sequences of distantly related proteins is usually uncertain, especially when there have been numerous deletions and insertions. The usual approach of minimizing the number of genetic mutations has often given alignments that are inconsistent with the structural homology.

Analysis of the sequences of proteins defined by the genomes of various organisms has found that about 2/3 of these sequences can be assigned to as few as 1400 domain families for which structures are known. About 200 of these domain families are common to all kingdoms of life and account for nearly half of those identified in the various genomes. Some of these domain families have been very extensively duplicated within a genome and combined with other domains to generate various multi-domain proteins. These combinations tend to differ in the various organisms, so fewer than 15% of the protein families found within a genome appear to be common to all kingdoms of life.

With many more sequences and structures available, contrary examples have been encountered of apparently homologous proteins with globally distinct structures. Significant sequence conservation, local structural resemblance and functional similarity strongly indicate evolutionary relationships between these proteins despite pronounced structural differences at the level of the overall conformation. For example, the lysozymes of chicken and bacteriophage T4 have similar functions but their structures seemed initially to be quite different. Closer comparison, however, reveals significant structural similarities that probably indicate they diverged long ago from a common ancestor; this has been confirmed by finding lysozymes with intermediate structures. In this counter-example, the same function has been retained even though the structure has diverged substantially.

In another study, the sequence of a helical protein was deliberately altered in one step at half of the residues: the fold of the protein was converted from totally α-helical to totally β-sheet.

Protein families and their evolution: a structural perspective. C. A. Orengo & J. M. Thornton (2005) *Ann. Rev. Biochem.* **74**, 867–900.

Exploring the common dynamics of homologous proteins. Application to the globin family. S. Maguid *et al.* (2005) *Biophys. J.* **89**, 3–13.

Domain rearrangements in protein evolution. A. K. Bjorklund *et al.* (2005) *J. Mol. Biol.* **353**, 911–923.

Evolutionary plasticity of protein families: coupling between sequence and structure variation. A. R. Panchenko *et al.* (2005) *Proteins* **61**, 535–544.

Structural divergence and distant relationships in proteins: evolution of the globins. J. T. Lecomte *et al.* (2005) *Curr. Opinion Struct. Biol.* **15**, 290–301.

Protein family expansions and biological complexity. C. Vogel & C. Chothia (2006) *PLoS Comput. Biol.* **2**, e48.

Testing for spatial clustering of amino acid replacements within protein tertiary structure. J. Yu & J. L. Thorne (2006) *J. Mol. Evol.* **62**, 682–692.

1. Structural Homology within a Polypeptide Chain

Some proteins have two or more domains or motifs with very similar structures. Very often this is accompanied by homology in the sequence of the different segments of polypeptide chain, suggesting that **the repeated sequence probably arose originally by duplication and fusion of the gene for one such segment** (Figure 9-28). At an intermediate stage of evolution, the protein probably had several single-segment polypeptide chains aggregated into a symmetrical oligomer. In the example of Figure 9-28, each of the two symmetry-related halves of the structure involves both halves of the primary structure, so this conformation must have pre-dated the polypeptide elongation by gene duplication that produced a single polypeptide chain. The gene fusion might have occurred at the same time or after the gene duplication. Even when no amino acid sequence homology is detectable, similar motifs within the protein structure are found to be related by symmetry axes, very close to those expected if the similar structures were originally identical subunits in an oligomeric protein. **It is likely that these proteins also arose by oligomerization followed by gene duplication and fusion**, but the divergence of the amino acid sequences has been sufficient to mask their homology.

Structure function and evolution of multi-domain proteins. C. Vogel *et al.* (2004). *Curr. Opinion Struct. Biol.* **14**, 208–216.

9.3.B. Structural Similarity without Apparent Sequence Homology: Surprises

Has the folded conformation of a protein been so conserved during evolution that a similar conformation in another protein is indicative that both had a common evolutionary ancestor, even though there is no detectable homology in their primary structures? In this case, what degree of structural similarity is significant? At what point do similarities in 3-D structure reflect simply the general principles of protein structure? Does a common backbone topology and a similar function require a common ancestor? Alternatively, can a similar conformation originate by evolutionary convergence as a result of selection for a common function and common structural requirements for such a function?

These are just a few of the questions raised by observations of similar folded conformations where none was expected because the primary structures of the proteins are not apparently homologous. No well-documented examples of evolutionary convergence have been found for protein structures, but this may well be because convergence is difficult to prove at the molecular level. The best evidence for convergence would be differences in the order in the primary structure of elements of the secondary structure that have the same position within similar 3-D structures; rearrangements of the gene are considered unlikely to be compatible with retention of the 3-D structure, although it is not impossible.

The most plausible candidates for evolutionary convergence are the $(\beta\alpha)_8$ barrels (Section 9.1.I.3.a) that were first observed in triose phosphate isomerase (Figure 9-4). This structure has been observed in more than 16 different proteins, with no detectable amino acid homology among many of them. All are very similar, except that one protein structure has the order in the primary structure of the first helix and second β-strand reversed. The regularity of the $(\alpha\beta)_8$ structure makes it plausible that it is an example of convergent evolution for structural reasons. The proteins with this structure also have some functional similarities, however, in that all are enzymes and all have their active sites in the same position at the C-terminal end of the β-barrel. This might indicate that they evolved by drastic divergence from a common ancestor.

Other examples of structural similarities without detectable primary structure homology also involve functional similarities. The mononucleotide-binding domain occurs in many proteins that bind nucleotides; it consists of three parallel β-strands with two intervening α-helices (Section 12.4). Generally no amino acid sequence homology is detectable between many of them, only that necessary for structural and functional reasons. Almost all of these supersecondary structures bind nucleotides in a similar way, however, and they are generally considered to have arisen by divergence from a common ancestor.

The best examples of evolutionary convergence in proteins are of similar active sites and catalytic mechanisms, but without apparent sequence or conformational homology. Very similar geometries of residues at their active sites are found in the bacterial proteinase subtilisin, in members of the mammalian trypsin proteinase family and in a serine carboxypeptidase, but no other structural similarities; remarkably, the serine carboxypeptidase is structurally homologous to a different class of proteinases, that including carboxypeptidase A. Thermolysin and bovine carboxypeptidase A have similar catalytic mechanisms with similar active sites, but no other detectable similarities. The possibility that these similarities are the vestige of a distant evolutionary origin obscured by extreme divergence is believed to be ruled out by the different positions in the primary structure occupied by the equivalent catalytic site residues. The genetic rearrangements that would be necessary during divergence from a common ancestor are considered to be very improbable.

Understanding the importance of protein structure to nature's routes for divergent evolution in TIM barrel enzymes. E. L. Wise & I. Rayment (2004) *Acc. Chem. Res.* **37**, 149–158.

Catalytic versatility, stability, and evolution of the $(\beta/\alpha)_8$-barrel enzyme fold. R. Sterner & B. Hocker (2005) *Chem. Rev.* **105**, 4038–4055.

Comparison of sequence-based and structure-based phylogenetic trees of homologous proteins: inferences on protein evolution. S. Balaji & N. Srinivasan (2007) *J. Biosci.* **32**, 83–96.

9.3.C. Sequence Similarity without Structural Homology: New Folds?

The presence of significant sequence similarity generally implies that the 3-D structure will also be homologous. The possibility of sequence similarity without a common evolutionary ancestry seems so remote as to be impossible, and the 3-D structure is generally conserved more than is the sequence. Nevertheless, one case in which there is sequence similarity without structural homology has been found in the proteins with **Leu-rich repeats**. These LRR proteins are a diverse group characterized by distinctive repeating sequences in their primary structure that include several Leu residues or other aliphatic residues. For example, a typical repeating module in an LRR has the sequence XLXXLXLXXNX*a*XX*a*XXXX*a*XXLX, where L is leucine, *a* an aliphatic residue, often Leu, and X any residue. The number of residues in the repeat varies from 20 to 29, and the number of repeats in a protein varies from one to 41. LRR proteins are associated with functions such as signal transduction, cell adhesion and protein–protein interactions.

The structures of two LRR proteins, ribonuclease inhibitor and leucine-rich variant (LRV) protein, both exhibit repeating 3-D structural protein motifs that form a supercoil, but the motifs differ (Figure 9-36). In the case of ribonuclease inhibitor, each LRR unit forms a short β-strand followed by a loop and an α-helix, arranged roughly parallel to one another. The repeating α–loop–β units generate a super helix, with the β-strands forming a parallel β-sheet that lines the interior of a horseshoe-shaped structure. In the LRV protein, in contrast, there is no β-sheet structure; instead the repeating units form alternating α-helix and 3_{10}-helix motifs arranged in a right-handed superhelix of a comma-shaped structure. In this case, it is the parallel α-helices that line the inner surface of the protein. In both ribonuclease inhibitor and LRV protein, however, the conserved leucine residues form the core of the protein. In this case, therefore, the sequence similarities are seen to arise for structural reasons and probably did not originate by divergent evolution.

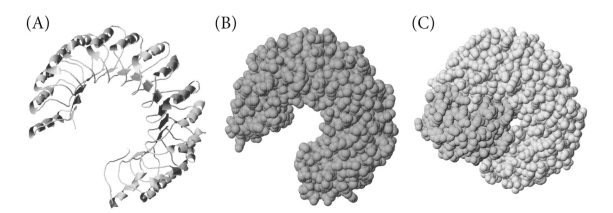

Figure 9-36. The structure of the leucine-rich repeat (LRR) protein ribonuclease inhibitor. (A) The cartoon depiction of the course of the polypeptide backbone shows β-strands as *arrows* and α-helices as *coils*. (B) Space-filling model, with the atoms depicted as their van der Waals volumes. (C) The complex with the enzyme ribonuclease A. Generated from PDB files *2bnh* and *1dfj* using the program Jmol.

Other repeating substructures could be used in LRR architecture. For example, the repeating three β-strands of pectate lyase that form a β-helix structure (Figure 9-11) have been suggested as a model for some LRRs. The molecular recognition and protein interaction roles shared by LRRs are thought to be a result of the nonglobular nature of these proteins and their conformational flexibility.

Protein engineering studies are increasingly observing significant changes in the fold of a structure upon mutation in just a few residues. Consequently, protein structures may not be as fixed as described above but liable to change and evolve gradually.

Directed evolution of highly homologous proteins with different folds by phage display: implications for the protein folding code. P. A. Alexander *et al.* (2005) *Biochemistry* **44**, 14045–14054.

Sequence determinants of a conformational switch in a protein structure. T. A. Anderson *et al.* (2005) *Proc. Natl. Acad. Sci. USA* **102**, 18344–18349.

Sequential reorganization of β-sheet topology by insertion of a single strand. M. Sagermann *et al.* (2006) *Protein Sci.* **15**, 1085–1092.

Protein structure comparison: implications for the nature of 'fold space', and structure and function prediction. R. Kolodny *et al.* (2006) *Curr. Opinion Struct. Biol.* **16**, 393–398.

Protein structure: evolutionary bridges to new folds. T. O. Yeates (2007) *Curr. Biol.* **17**, R48–50.

9.4. PROTEIN STRUCTURE PREDICTION

It should be possible to infer the unique 3-D protein structure of any protein from simply its amino acid sequence, because many proteins are known to unfold and then refold *in vitro* and no other factors need be involved (Section 11.4); therefore, **the 3-D structure must be determined by the amino acid sequence**. Sufficient protein structures and sequences are known to be able to devise and test prediction methods, but the problem has not been solved, in spite of many approaches attempted over the past 40 years.

The solution is usually thought to be just a matter of finding the folded conformation with the lowest apparent free energy. The apparent free energy of a conformation can be calculated, taking into consideration the various kinds of interactions within a protein molecule and between the protein and surrounding medium, i.e. water. The conformation can then be altered gradually to lower the calculated free energy. The problem is difficult for two reasons. The first is **the extremely large number of conformations that a polypeptide chain may adopt** (Section 8.2). Computer simulations are effective only for short polypeptide chains. The difference in magnitudes of the calculations for a 10-residue peptide and a 100-residue protein is not a factor of 10, because the magnitude of the calculation increases exponentially: making the very conservative assumption that every residue has on average two different choices of conformation, the number of conformations that would need to be considered with 10 residues would be 2^{10} ($\sim 10^3$), but 2^{100} ($\sim 10^{30}$) with 100. The magnitude of the latter number makes this approach impractical, and it is a very conservative estimate. Furthermore, in a few cases it is apparent that the native conformation of a protein is not the one with the lowest free energy, the most intrinsically stable, but is determined by the folding pathway (Section 11.4.B); this implies that finding the final conformation would require simulation of the entire folding process.

The second difficulty arises because **protein folding is a cooperative transition between the unfolded and folded states** (Section 11.1.B). Individual protein molecules frequently remain as either one of these two states, without adopting intermediate conformations to any substantial extent or for significant periods of time. In a cooperative folded structure, each part of the protein molecule is influenced by the entire structure and vice versa. Due to this cooperativity, there is no simple one-to-one correspondence between the sequence and the structure. A segment of polypeptide with a certain sequence might adopt a β-strand in one protein but an α-helix in another, which has been observed with pentapeptides in natural proteins. Therefore the relationship between sequence and structure is only probabilistic, rather than deterministic, which makes prediction difficult.

Several other approaches to protein structure prediction have been developed. One uses empirical rules relating sequence and structure that are derived from the known protein structures. This approach has had some success with secondary structure prediction, but not with tertiary structures. Another approach is known as 'threading protein sequences'. It assumes that the folded conformation is already known and present in the structure databases as some other protein, so it simply examines whether the new sequence is likely to fit any of these known structures. It has revealed the 3-D structures of some proteins but it cannot predict any truly novel structures.

Practical lessons from protein structure prediction. K. Ginalski *et al.* (2005) *Nucleic Acids Res.* **33**, 1874–1891.

Progress in modeling of protein structures and interactions. O. Schueler-Furman *et al.* (2005) *Science* **310**, 638–642.

Protein structure prediction: inroads to biology. D. Petrey & B. Honig (2005) *Mol. Cell* **20**, 811–819.

Sequence comparison and protein structure prediction. R. L. Dunbrack (2006) *Curr. Opinion Struct. Biol.* **16**, 374–384.

Into the fold. Advances in technology and algorithms facilitate great strides in protein structure prediction. P. Hunter (2006) *EMBO Rep.* **7**, 249–252.

Large-scale prediction of protein structure and function from sequence. S. C. Tosatto & S. Toppo (2006) *Curr. Pharm. Des.* **12**, 2067–2086.

9.4.A. *Ab Initio* Predictions: the Ultimate Goal

The term *ab initio* is derived from the Latin 'from the beginning', or 'from the sequence' in this case. The term **ab initio predictions** lumps together all the approaches of structure prediction that do not require a known protein structure. Such methods would be applicable to any protein and would also explain the mechanism of protein folding. However, the huge number of conformations accessible to a polypeptide chain makes predicting a single conformation extremely difficult. For example, the chance of randomly predicting the native backbone structure of a 150-amino acid residue protein to low resolution, say within 6 Å, is only 10^{-14}. To predict the structure to high resolution, say to 2 Å, randomly would be immensely more difficult, with only 10^{-57} probability.

Attempts to minimize the free energy of a protein conformation frequently reduce the computation time by simplifying both the geometry of the polypeptide chain and the atomic force fields employed to estimate the energy. Another line of approach is first to predict the secondary structure elements, the α-helices and β-sheets, and then to attempt to pack them into a globule using empirical rules or imposing general constraints. If an algorithm could be devised, the combination with secondary structure prediction (Section 9.4.B) could provide a complete prediction scheme starting from simply the amino acid sequence.

More simple folding simulations that use a lattice have also been developed. Lattice models permit each residue to occupy only a point intersection on the lattice, and adjacent residues or atoms in the sequence must be on adjacent lattice points. They have the advantage of reducing dramatically the number of possible conformations, actually digitizing them into countable objects. Lattice models of the folded structure are able to mimic quite well the overall shape of a protein but they are poor for local structure; just imagine the difficulty of representing an α-helix on a cubic lattice. More sophisticated lattices have been devised but such models are based on the unstated assumption that the folding process occurs by building up the structure in a stepwise manner.

To compare the effectiveness of different methods, a world-wide prediction contest, named **CASP** (**C**ritical **A**ssessment of techniques for protein **S**tructure **P**rediction), is held regularly. A completely blind test is arranged in which every predictor makes their own predictions for target proteins whose 3-D structures are not yet known but are in the process of being determined. An appointed assessor then evaluates all the predictions by comparing them with the structures that are determined experimentally subsequently and presents the results at a joint meeting of all the predictors. The results have been promising and suggest that it should be possible to infer the correct folds of target proteins from their sequences alone.

A decade of CASP: progress, bottlenecks and prognosis in protein structure prediction. J. Moult (2005) *Curr. Opinion Struct. Biol.* **15**, 285–289.

Computer-based design of novel protein structures. G. L. Butterfoss & B. Kuhlman (2006) *Ann. Rev. Biophys. Biomolec. Structure* **35**, 49–65.

Worth the effort. An account of the Seventh Meeting of the Worldwide Critical Assessment of Techniques for Protein Structure Prediction. A. Tramontano (2007) *FEBS J.* **274**, 1651–1654.

9.4.B. Secondary Structure Prediction: A One-dimensional Problem

The goal of secondary structure prediction is to infer the location of α-helices and β-strands along the polypeptide chain, without necessarily determining how the latter associate into a β-sheet. The prediction problem is therefore only one-dimensional (1-D), relating two linear strings: the amino acid sequence and the linear sequence of elements of secondary structure. The secondary structure of each residue is often denoted by letters: A, α-helical; B, β-stranded; C, coil, irregular conformations; and, sometimes T, for turns. A prediction algorithm produces a string of As, Bs, Cs or Ts along the sequence being analyzed. When the structure is known already, the success rate of the prediction is measured simply as the percentage of residues that are predicted correctly, counted on a residue-by-residue basis. The random level of a three-state prediction accuracy might be expected to be close to 33%, but it is actually ~ 40% due to an uneven distribution of the three states in known structures; on average, irregular is the most abundant and β-strand the rarest.

The basic logic of most of the many prediction methods that have been developed to date is similar to that developed initially by **Chou and Fasman**; individual amino acid residues are assumed to have their own intrinsic propensities to form α-helix, β-strand, irregular and turn conformations. For example, every Ser residue has a certain propensity for A, B, C and T conformations, while every other type of residue has different propensities. The propensities for the various amino acid residues can be estimated from the known protein structures (Table 9-6). The propensity of each segment of a sequence to adopt a particular conformation is assumed to be the sum of the individual propensities of the various residues. This approach can be extended by incorporating the influence of near-neighbor residues or considering pairs or triplets of residues. Sophisticated computing algorithms of informatics theory can be useful, including neural networks, genetic algorithms and the 'hidden Markov' method.

The simple method of Chou and Fasman to predict helix, strand or irregular conformations requires only 60 parameters: 20 amino acid residues × three conformational states each. More detailed methods can require about 1000 parameters, for example 20 amino acids × 17 positions of neighboring residues

× 3 states. More than 10,000 parameters are required for methods that use the propensities of pairs of residues: 20 × 20 amino acid pairs × 10 positions × 3 states. Chou and Fasman developed their method when only 29 protein structures, with a total of roughly 6000 residues, were known. This was sufficient to derive the 60 parameters for their method, but not for more complex methods. The number of known structures has increased dramatically since then and made it possible to derive reliable parameters for more complex methods. The prediction accuracy has also increased: the initial prediction accuracy of the Chou–Fasman technique was no greater than 55%, which was significant, but the accuracy has now reached 70% or more. Predictions can also be improved by including homologous sequences that have been aligned accurately: they will adopt basically the same fold, so using a number of such sequences increases the information available.

It is almost certain that 100% accuracy of secondary structure prediction is not attainable. The main reason is that the secondary structure adopted by a segment of polypeptide chain in a folded protein is not dictated solely by its sequence, but also by the context of the 3-D structure in which it occurs. The various amino acid residues have different propensities to adopt the α-helical conformation (Table 8-3) but β-sheets seem to be determined more by the tertiary structure than by the local amino acid sequence.

Amino acid propensities for secondary structures are influenced by the protein structural class. S. Costantini *et al.* (2006) *Biochem. Biophys. Res. Commun.* **342**, 441–451.

Analysis of an optimal hidden Markov model for secondary structure prediction. J. Martin *et al.* (2006) *BMC Struct. Biol.* **6**, 25.

Improved Chou–Fasman method for protein secondary structure prediction. H. Chen *et al.* (2006) *BMC Bioinformatics* **7**, S14.

Amino acid pairing preferences in parallel β-sheets in proteins. H. M. Fooks *et al.* (2006) *J. Mol. Biol.* **356**, 32–44.

---

1. Identifying Transmembrane Helices: Hydropathy

Transmembrane α-helices comprise a number of primarily hydrophobic amino acid residues and consequently can be identified in the primary structure of a membrane protein with relatively high confidence using **hydropathy analysis**. Hydropathy is a term that refers to the spectrum of amino acid side-chains in the hydrophilicity–hydrophobicity scale (Table 7-6) and each residue is assigned a characteristic hydrophobicity value. The term hydropathy was originally coined by Kyte and Doolittle when describing an algorithm that identifies helical transmembrane spans and was sufficiently successful to be widely used (although it is not to be confused with a dubious medical treatment of the same name). Their scale is based on the transfer free energies from water to vapor of model compounds for the amino acid side-chains and on the exterior–interior distribution of amino acids in protein structures (Table 9-3); in addition, the values for some residues were adjusted subjectively. A more realistic hydrophobicity scale (Table 7-6) is based on the free energies of transfer from water to oil of amino acid residues that are parts of a helix, corrected for unfavorable hydrophilic contributions due to polar atoms and protonation/deprotonation events of charged residues upon transfer to the nonpolar phase.

To calculate a hydropathy profile, the hydrophobicities of a chosen number of neighboring residues within a window of the sequence are summed and divided by the number of residues in the window; the resulting average score is assigned to the residue in the middle of the window. Then the window is moved on in the sequence by one residue and the procedure repeated. An example is shown in Figure 9-37. In the cases of bacteriorhodopsin, cytochrome oxidase and the bacterial photosynthetic reaction center, this simple analysis correctly identified the approximate positions of all the transmembrane helices, with no false positives.

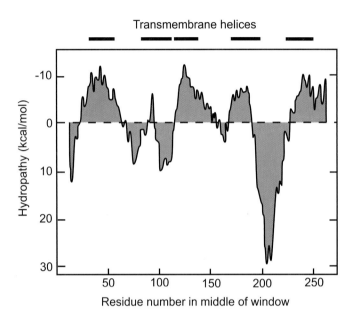

Figure 9-37. Hydropathy analysis of the L subunit of the photosynthetic reaction center of *Rhodobacter sphaeroides*. A window length of 19 residues was used. The known locations of the transmembrane helices are indicated above, demonstrating that the method identifies them at least approximately.

More advanced methods make use of the observation that deletions and insertions occur in homologous sequences only outside the transmembrane sequences. Multiple homologous sequences improve the prediction significantly. Likewise, modern algorithms use statistical analysis of amino acid preferences within actual known transmembrane helices and their flanking loops.

The **"positive inside"** rule has been incorporated into the prediction protocols to determine the orientation of the polypeptide in the membrane. This rule is based on the empirical observation that Arg and Lys residues in the loops between transmembrane helices occur more frequently on the inner side of the membrane than on the outer surface, so long as the loops are shorter than 60 residues.

Transmembrane protein structures without X-rays. S. J. Fleishman *et al.* (2006) *Trends Biochem. Sci.* **31**, 106–113.

Prediction of buried helices in multispan α-helical membrane proteins. L. Adamian & J. Liang (2006) *Proteins* **63**, 1–5.

Study and prediction of secondary structure for membrane proteins. S. R. Amirova *et al.* (2007) *J. Biomol. Struct. Dyn.* **24**, 421–428.

Membrane protein structure: prediction vs reality. A. Elofsson & G. von Heijne (2007) *Ann. Rev. Biochem.* **76**, 125–140.

Formation of transmembrane helices *in vivo*: is hydrophobicity all that matters? G. von Heijne (2007) *J. Gen. Physiol.* **129**, 353–356.

9.4.C. Homology Modeling

Two proteins with sequences identical at 30% or more of the amino acid residues, over a span of more than 100 residues, are almost certainly homologous, belong to the same family and will have essentially the same structure. If the structure of one is known, the structure of a homologous protein can be modeled on the first one. It is possible that examples of virtually all the folded structures are already known, in which case this method should be able to predict the structure of any protein of known sequence.

The first step in homology modeling is to align the new sequence against the sequence of a protein homolog of known structure. The two aligned sequences may contain insertions or deletions (indels) here and there and produce a gap in one of the sequences, which introduces uncertainty into the alignment. As the sequence similarity decreases, the number of gaps increases, and the entire alignment becomes increasingly less certain. If a structure is modeled based on an incorrect alignment, the resulting model will also be incorrect. A sequence identity of at least 40–50% of the residues was required initially for accurate homology modeling but now 30% identity is often sufficient to predict the structure with an accuracy equivalent to a low-resolution X-ray structure.

Using the sequence alignment, the new amino acid sequence is mounted onto a template backbone from the known structure, and the necessary amino acid side-chains are replaced. The first problem is to place the new side-chains into the correct orientation. The orientations of the new side-chains of interior residues will be determined by the packing of all the atoms within the protein interior, so it is simplest to adjust the conformations of the new side-chains to accommodate the fixed conformations of the polypeptide backbone and those side-chains that are not changed. The orientations can also be selected from those observed most frequently in known protein structures, known as **rotamer libraries**, and those calculated to have the most favorable energies.

The second problem is to account for insertions and deletions. There is no template for residues that have been inserted; they generally occur at the protein surface, so there are few interactions and steric hindrance to guide the structure design. The only approach is to generate an appropriate loop structure with an energetically favorable conformation and join its two termini smoothly to the template structure. Once an initial model has been devised, the **simulated annealing** method used to generate structures from NMR data can be used with the entire model structure; this allows all of the structure to vary while searching for the optimum, energetically most stable, conformation as a whole.

Computer programs for homology modeling are commercially available.

On the origin and highly likely completeness of single-domain protein structures. Y. Zhang *et al.* (2006) *Proc. Natl. Acad. Sci. USA* **103**, 2605–2610.

Multiple mapping method: a novel approach to the sequence-to-structure alignment problem in comparative protein structure modeling. B. K. Rai & A. Fiser (2006) *Proteins* **63**, 644–661.

Sequence comparison and protein structure prediction. R. L. Dunbrack (2006) *Curr. Opinion Struct. Biol.* **16**, 374–384.

Advances in homology protein structure modeling. Z. Xiang (2006) *Curr. Protein Peptide Sci.* **7**, 217–227.

1. Inverse Folding Problem

This procedure is opposite to protein structure prediction, where one starts with the sequence and tries to find the 3-D structure. Instead, one starts from a known 3-D structure and works towards a sequence that is compatible with this structure but has the desired new properties.

The suitability of each type of amino acid residue for each position in the protein structure is calculated using an interaction energy potential known as the **mean force energy potential**. This energy depends upon the amino acid types and the distances between the C^β atoms of pairs of side-chains in the structure. It is calculated as the sum of the interaction energies calculated between the side-chains. The procedure can identify the one sequence that should be best suited to adopt any known protein structure. Other sequences can be ranked in terms of their suitability, and the one that best suits the desired criteria can be selected.

At each stage of the procedure, the optimal packing of the atoms within the protein interior needs to be determined. The polypeptide backbone of the known structure can be kept constant but a major problem is how to deal with the amino acid side-chains of the new sequence. One approach is to use a simplified treatment of side-chains in which each of the 20 side-chains of the normal amino acids is represented by one point with the appropriate bulkiness and other physicochemical properties. The C^β atom is often chosen as the point for all residues except Gly. Its position is determined by the dihedral angles of the backbone alone, so it is not necessary to specify any side-chain conformation.

The veracity of the inverse folding procedure can be examined by mounting the native amino acid sequence (*a*) of one known protein structure (A) onto another structure (B) that is larger than A. Mounting sequence *a* onto structure B can be done in many ways, with various alignments, shifting residues one-by-one without introducing any gaps. Each time, the total energy of the structure is estimated. Upon completion of this calculation, structure B is replaced by structure A. If the procedure is valid, the native combination of sequence and structure should give the lowest energy.

In silico protein design: fitting sequence onto structure. B. I. Dahiyat (2006) *Methods Mol. Biol.* **316**, 359–374.

Knowledge-based potentials in protein design. A. M. Poole & R. Ranganathan (2006) *Curr. Opinion Struct. Biol.* **16**, 5080513.

Potential energy functions for protein design. F. E. Boas & P. B. Harbury (2007) *Curr. Opinion Struct. Biol.* **17**, 199–204.

---

2. Threading Protein Sequences

The inverse folding problem just described suggested the threading method to predict protein structures. Inverse folding considers a certain protein structure relative to all the possible sequences, whereas the threading method considers a certain amino acid sequence relative to all known protein structures. The amino acid sequence (*a*) of a protein of unknown structure (A) can be tested one-by-one for its fitness for each of the known protein structures. All feasible alignments of sequence to structure are examined. This procedure should identify the known structure that is most compatible with sequence *a*. It is likely to be similar to the structure A that needs to be predicted. This procedure is known as the "**3D-1D compatibility**" method, "**fold recognition**" or "**prediction by threading**". It extends the conventional comparison of 1-D sequences alone but should be more effective in detecting similarities between proteins because the 3-D structure is more conserved than the sequence during evolution.

Threading a given primary sequence through a given tertiary structure involves a large number of possibilities, known as the **alignment problem**. To reduce the magnitude of the computational problem, the **3-D profile table** was introduced (Figure 9-38). This is a (20 × *n*) table constructed from the known protein structure, where columns of the 20 amino acid residues are arrayed along the *n* residues of the structure. The table describes the fitness of each type of amino acid residue for each given residue site, which depends upon its secondary structure, hydrophilic/hydrophobic environment, etc. It is straightforward to compare such a 3-D profile table with any sequence to obtain the best 3-D/1-D alignment, including gaps, plus the alignment score. A set of 3-D profiles generated from a library of structures is scanned with a query sequence, to find the model structure and alignment that produces the highest score or lowest energy. The computation time is no greater than that for the usual sequence homology search methods.

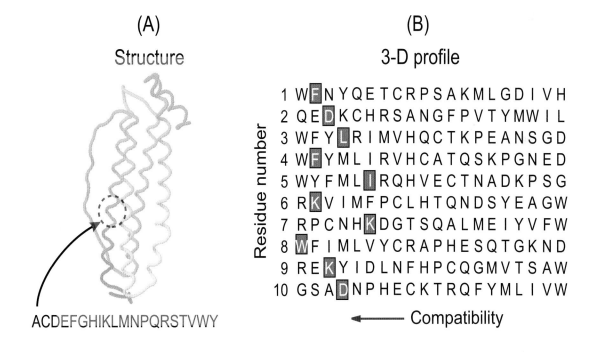

Figure 9-38. Testing the compatibility of a protein sequence with a known 3-D structure by threading the sequence through the structure. (A) Each of the 20 amino acid residues is placed one at a time at each of the residues of a known 3-D protein structure to evaluate its compatibility with each position of that structure. The 20 residues are then sorted in decreasing order of their compatibility with each position. (B) A '3-D profile table' is assembled, using the one-letter abbreviation for each amino acid residue. The residues in the native original protein are highlighted; not surprisingly, all are on the left side of the table, implying that they are compatible with the 3-D structure. Other sequences might be even more compatible, and the 3-D profile table will indicate the relative compatibility of any other amino acid sequence with that structure.

The fitness between a sequence and a structure is determined by calculating the energies of the interactions of each residue with all its neighboring residues. The problem is that the surrounding residues will not be known until after the alignment is fixed. Until then, the native amino acid residues of the model structure are used in the surrounding sites.

Other types of methods can be envisaged for directly threading the entire sequence through a structure, without using the 3-D profile, and they could use many types of energy potential functions. The problem, however, is the large number of alignments that are possible; the computations for direct threading approaches are very time consuming. The fundamental idea of threading might be converted into a quasi-*ab initio* method, if the backbone conformation were no longer fixed but allowed to vary.

Protein sequence threading: averaging over structures. A. J. Russell & A. E. Torda (2002) *Proteins* **47**, 496–505.

Optimal protein threading by linear programming. J. Xu *et al.* (2003) *J. Bioinform. Comput. Biol.* **1**, 95–117.

9.4.D. *De Novo* Protein Design

It is not yet routine to predict the 3-D structure of a protein from just its amino acid sequence, but the principles of protein structure are now understood sufficiently well to make it feasible to design protein sequences that adopt at least a few folded conformations. **The amino acid sequence must be designed not only to be capable of adopting the desired conformation but also to avoid forming incorrect structures**.

A successful procedure is well illustrated (Figure 9-39) by the initial design of a four α-helix bundle conformation, which is found in several natural proteins (Section 9.1.D.6 and Figure 9-15). In the first step, a 16-residue peptide was designed to adopt an α-helical conformation and form a tetramer of four such peptides. Only Leu residues were used to form the hydrophobic interface between the desired helices, because this is also a helix-favoring residue (Table 8-3). Gly residues were incorporated at each end of the helix, to terminate the helix and anticipate reverse turns to be incorporated there. The other residues, on the solvent-facing side of each helix, were made either Lys or Glu, in positions to interact electrostatically with each other and with the helix dipole favorably (Figure 8-11). After the peptide sequence had been optimized for forming the tetrameric helix bundle, a loop with the sequence –Gly–Pro–Arg–Arg–Gly– was designed to connect pairs of helices in an antiparallel orientation. This two-helix peptide formed stable dimers. Finally, a similar loop was inserted to link two such peptides. The final 74-residue peptide adopts a very stable monomeric conformation that is at least approximately like that anticipated. A similar approach has produced a synthetic $(\beta\alpha)_8$ barrel, designated octarellin.

In practice, however, it has been found more feasible to obtain a new protein with a desired function from an existing one by changing the amino acid sequence of the existing protein randomly and selecting for those sequences that give the desired function. Our understanding of protein structure and function is still insufficient for designing many new proteins.

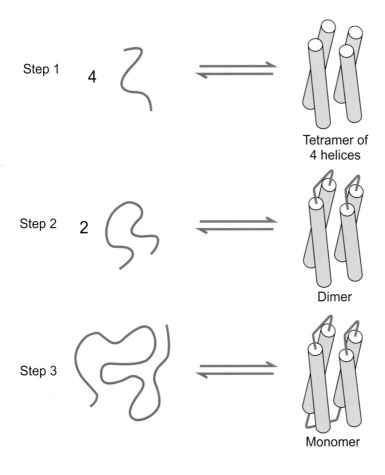

Figure 9-39. Schematic illustration of the incremental approach to the design of a four-helix bundle protein. Kindly provided by W. F. DeGrado in T. E. Creighton (1993) *Proteins: structures and molecular properties*, 2nd edn, W. H. Freeman, NY, p. 258.

Computer-based design of novel protein structures. G. L. Butterfoss & B. Kuhlman (2006) *Ann. Rev. Biophys. Biomol. Struct.* **35**, 49–65.

Sequence search methods and scoring functions for the design of protein structures. H. Madaoui *et al.* (2006) *Methods Mol. Biol.* **340**, 183–206.

Emulating membrane protein evolution by rational design. M. Rapp *et al.* (2007) *Science* **315**, 1282–1284.

~ CHAPTER 10 ~

PHYSICAL PROPERTIES OF FOLDED PROTEINS

Natural proteins in their biological environments are usually in solution or embedded within membranes, where they interact in a variety of ways with their environments. These interactions can have numerous effects on the physical and biological properties of the protein, so it is important to understand how the protein structure responds to its environment. **The folded conformations of proteins give them physical, chemical and biological properties quite unlike those of unfolded polypeptides or of the sum of their constituent amino acids**. The compactness of their folded conformations causes native proteins to diffuse and rotate relatively rapidly. The individual domains of proteins are relatively resistant to proteinases when they are folded (Section 7.5.B) and multi-domain proteins can often be cleaved specifically between the domains. Some individual domains can be cleaved by proteinases at peptide bonds in mobile surface loops, but even then the folded structure generally remains otherwise intact and folded. If dissociated, the polypeptide fragments can often recombine spontaneously under the appropriate conditions and regenerate the folded structure.

The folded conformation places atoms of the protein in unique environments, which can have marked effects on their physical and chemical properties. Two or more functional groups are often held in proximity, so that their effective concentrations relative to each other are extremely high; consequently, reactions may occur between them that would be negligible between the same two groups on separate molecules.

The tertiary structure of a compact folded protein structure can be considered unique except for local aspects of its surface. A native protein does not generally adopt two or more substantially different folded conformations without a change in covalent structure. Nevertheless, the literature is full of references to conformational changes in proteins; protein chemists automatically tend to suggest changes in conformation to rationalize any unexpected behavior of a protein. Many of these so-called conformational changes involve strictly localized alterations in conformation or changes in the degree of flexibility. In nearly all cases where substantial conformational changes within folded proteins have been characterized, they involve primarily quaternary rearrangements of subunits or structural domains relative to each other. The individual subunits or domains maintain their overall conformation, and it is these individual structural units for which the architecture is apparently unique.

Very small proteins appear to have the most mobile conformations. Some are unable to maintain a single conformation in solution and are probably more appropriately considered to be peptides. The 29-residue hormone glucagon is a notable example; it approximates a random coil in dilute solution, and only in concentrated solution does it tend to acquire a trimeric helical structure like that observed in the crystal lattice. There appears to be a lower limit (roughly 50) to the number of amino acid residues that must comprise a protein for it to maintain a single conformation in solution. In the case of smaller proteins (which may turn out simply to be large peptides) it is necessary to determine the relevance of their crystal structures by examining their conformational properties in solution.

Proteins in solution are not the rigid molecules that might be inferred from most representations of their structures. **It is a thermodynamic requirement that molecules the size of proteins have substantial transient fluctuations**. The mean-square fluctuation in energy (δE) of any object as it exchanges thermal energy to-and-fro with its surroundings is given by:

$$\langle \delta E^2 \rangle = k_B T^2 C \tag{10.1}$$

where k_B is Boltzmann's constant, T temperature and C heat capacity. For globular proteins at room temperature, these fluctuations in energy can be comparable to the net free energy stabilizing the folded conformation. The most common and best-understood movements of atoms in molecules are the small-scale vibrations of bond lengths and angles that occur in all molecules at room temperature and are detectable by infrared and Raman spectroscopy techniques. These vibrations in proteins are similar to those observed in small molecules, and they occur at frequencies of between 6×10^{12} and 10^{14}/s. Longer time scales permit larger movements, such as those of the individual domains of large proteins, linked together by relatively flexible 'hinge' regions. For example, the individual domains of antibody molecules are moderately independent and rotate relative to each other on a time scale of 10^{-8}–10^{-7} s. At the other extreme of time scale, protein folded conformations undergo complete unfolding spontaneously with a frequency of somewhere between 10^{-4} and 10^{-12}/s. Protein folded conformations are only marginally stable (Section 11.3), so even under conditions optimal for stability they must spontaneously undergo unfolding that is transient but complete. If the folded conformation of a protein has an optimal net stability of 5 kcal/mol, which corresponds under the same conditions to an equilibrium constant of 2×10^{-4} for the unfolded protein, each protein molecule will have that probability of being unfolded at any instant of time. Under normal circumstances, it will rapidly refold and will have been unfolded for only a very short period of time (Section 11.1). **Protein flexibility therefore involves movements of widely varying magnitudes on a time scale spanning perhaps 26 orders of magnitude**. Such flexibility is often claimed to be crucial for protein function but there are few direct demonstrations of that. Nevertheless, protein flexibility is real and must be taken into account in all considerations of protein structure and function.

Describing protein flexibility is not straightforward, except for that of the side-chains on the protein surface, which usually can move like those in small molecules and unfolded proteins. Within the protein interior, however, the close packing of atoms (Figure 9-21) means that movements of neighboring atoms must usually be coordinated. The complexity of such movements means that they can best be described by computer simulation.

The rates at which conformational changes occur is only one aspect of protein flexibility; the other is the energies of the various conformations. In some cases, such as the rotation of a symmetrical side-chain, one conformation has the same energy as another and each is equally likely to occur. In others, the perturbed conformation has a much higher free energy and is encountered only infrequently and briefly. Only the low-energy conformations are normally present to a substantial extent, and here it is apparent that there are severe constraints on the extent to which a folded protein conformation normally varies. These constraints are indicated by the slight variations that are produced by crystallization in different crystal lattices, by varying the temperature and pressure, and upon binding of ligands (Chapter 12). The results indicate that **proteins can be thought of as existing normally as a range of distinct, but closely related, microconformations that are usually interconverted rapidly at room temperature**; they collectively make up the average **macroconformation** that is generally observed. At very low temperatures, however, different molecules may become trapped in different **microstates**.

What vibrations tell about proteins. A. Barth & C. Zscherp (2002) *Quart. Revs. Biophys.* **35**, 369–430.

Protein dynamics studied by neutron scattering. F. Gabel *et al.* (2002) *Quart. Rev. Biophys.* **35**, 327–367.

'Four-dimensional' protein structures: examples from metalloproteins. M. Fragai *et al.* (2006) *Acc. Chem. Res.* **39**, 909–917.

10.1. SOLUBILITIES AND VOLUMES OF PROTEINS IN WATER

Water must be included in all attempts to understand the properties of proteins. Water should not be treated as an inert environment but rather as an integral and active component of biomolecular systems, where it has both dynamic and structural roles.

The interactions of a protein with solvent are determined primarily by its surface. **The most favorable interactions of proteins with aqueous solvent are provided by the charged and polar groups of the polypeptide chain and hydrophilic side-chains**. The surfaces of most water-soluble globular proteins are covered uniformly by charged and polar groups, although with some nonpolar patches, and their solubilities are governed primarily by the interactions of the polar groups with water. In contrast, membrane proteins and others that are relatively insoluble in water, such as crambin, have largely nonpolar surfaces that interact with the nonpolar interior of the membrane. Varying one position on the surface of one globular protein between all 20 amino acids revealed a wide range of contributions to protein solubility even among the hydrophilic amino acids. Asp, Glu and Ser residues contributed to the solubility significantly more than the other hydrophilic amino acids, especially when the protein had a large net charge. It seems likely that such results will depend upon the detailed nature of the site that is altered.

The solubility of a molecule is also governed by its relative free energy in an amorphous or solid state, which is determined by its interactions with all the other molecules that might be present. For example, many structural proteins have polar surfaces and would be expected to be soluble, but they are complementary, both physically and chemically, with surfaces on other molecules, to which they can bind tightly and produce large, insoluble complexes. The importance of the solid-phase makes this a very complex situation and **no quantitative general explanation of protein solubility is available**.

The solubility of a globular protein in water is generally at a minimum at its **isoelectric point**, the pH where it has zero net charge, and the solubility generally increases as the pH is moved further away (Figure 10-1). **The greater the net charge on the protein and the lower the ionic strength,**

the greater the electrostatic repulsions between different molecules, which tend to keep them in solution. The effect is not primarily on the free energy of the protein in solution but when it is precipitated, aggregated or crystallized in the solid-phase. Most proteins unfold at some extreme pH value, however, due to polar groups that cannot ionize within the folded conformation (Section 11.3.A). Unfolding usually has drastic consequences for the solubility of the protein, due to the exposure of its nonpolar surfaces.

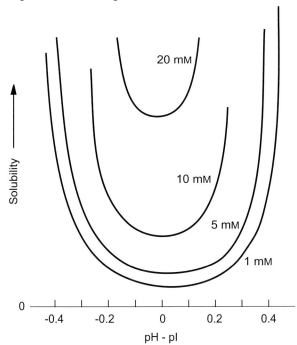

Figure 10-1. The variation with pH of the solubility of a typical globular protein at several different NaCl concentrations. The pH is expressed relative to the isoelectric point (pI) of the protein.

Water mediation in protein folding and molecular recognition. Y. Levy & J. N. Onuchic (2006) *Ann. Rev. Biophys. Biomol. Struct.* **35**, 389–415.

Amino acid contribution to protein solubility: Asp, Glu, and Ser contribute more favorably than the other hydrophilic amino acids in RNase Sa. S. R. Trevino *et al.* (2007) *J. Mol. Biol.* **366**, 449–460.

10.1.A. Hydration Layer

The hydration layer bound to proteins is usually more ordered and less mobile than bulk water and has a 10% greater density and a 15% greater heat capacity. This hydration layer generally contains 0.3 g of water per gram of protein, which is equivalent to an average of nearly two H_2O molecules per amino acid residue of the total protein. The hydration layer increases the volume of the protein by about 43% and its radius of about 12%. The interactions of water with protein surfaces have been investigated by gradually adding water to a dry protein, in many cases hen egg-white lysozyme, which contains 129 residues. The first 60 molecules of water interact primarily with the charged groups on the lysozyme surface, which can then ionize normally. The next 160 water molecules bind primarily to the remaining polar groups on the lysozyme surface. These two phenomena involve approximately one water molecule per charged or polar atom. The next 80 water molecules complete a monolayer around the protein molecule, which now has virtually all of the properties of a lysozyme molecule in aqueous solution. It is even active as an enzyme (Chapter 14) and water is one of its substrates. Although the water molecules that are interacting strongly with the ionized and polar groups on the surface are held tightly, they are kinetically labile and exchange with other water molecules of the bulk solvent on the 10^{-9} s time scale. Even those water molecules that are buried and integral parts of the protein structure (Figure 9-23) exchange with the solvent, although somewhat more slowly, with residence times as long as 10^{-2} s.

The interactions between proteins and water molecules in solution are very short-lived, 0.3 ns or less. The vast majority of water molecules in the protein hydration layer suffer a mere two-fold decrease in their mobilities compared with bulk water. **The water molecules in contact with the protein move nearly as rapidly as those in the bulk liquid.**

Protein surface hydration mapped by site-specific mutations. W. Qiu *et al.* (2006) *Proc. Natl. Acad. Sci. USA* **103**, 13979–13984.

H/D isotope effects on protein hydration and interaction in solution. A. S. Goryunov (2006) *Gen. Physiol. Biophys.* **25**, 303–311.

Preferential hydration and solubility of proteins in aqueous solutions of polyethylene glycol. I. L. Shulgin & E. Ruckenstein (2006) *Biophys. Chem.* **120**, 188–198.

Preferential hydration of lysozyme in water/glycerol mixtures: a small-angle neutron scattering study. R. Sinibaldi *et al.* (2007) *J. Chem. Phys.* **126**, 235101.

10.1.B. Partial Volumes

The partial specific volumes of folded proteins in normal aqueous solutions fall in the range of 0.70–0.75 cm³ g⁻¹, depending on their structure and amino acid composition (Table 10-1). The majority exhibit values in the range of 0.72–0.75 cm³ g⁻¹, within the range of the partial specific volumes of amino acids, which range from 0.57 cm³ g⁻¹ for aspartic acid to 0.88 cm³ g⁻¹ for leucine and isoleucine (Table 7-1). The **volumes occupied by amino acid residues in folded proteins are virtually identical with their volumes in solution** (Figure 9-22). The partial volumes in solution of the amino acid residues are 18–39% greater than their volumes defined by the van der Waals radii of their atoms; the largest increases are observed with nonpolar side-chains, the smallest with polar side-chains, because the polar atoms interact more tightly with the water of hydration, known as **electrostriction**.

The apparent partial specific volume of a protein, φ_2, may usually be estimated accurately from just its amino acid composition:

$$\varphi_2 = \Sigma \varphi_i \, w_i / \Sigma w_i \qquad (10.2)$$

where φ_i is the apparent specific volume of the ith amino acid residue (Table 10-1) and w_i its weight percentage. There is no *a priori* reason why this relationship should hold but the interiors of proteins are nearly as close packed as possible (Section 9.1.F) and the volume occupied by a residue is virtually identical to its partial molar volume in solution (Figure 10-2). Consequently, Equation 10.2 yields relatively accurate results, but changes in the environment (solvent composition, temperature, pressure, etc.) may cause significant changes in this volume.

Table 10-1. Partial specific volumes ($\bar{v}_2^0$), adiabatic and isothermal compressibilities (β_s and β_T) and volume fluctuations (δV_{rms}) of proteins in water at 20 to 25°C

Protein	$\bar{v}_2^0$ (cm³ g⁻¹)	β_s (10⁻¹² cm² dyn⁻¹)	β_T (10⁻¹² cm² dyn⁻¹)[a]	δV_{rms} (cm³ mol⁻¹)[b]
Globular proteins				
Peroxidase	0.702	2.36	6.70	68.3 (0.24)
Ribonuclease A	0.704	1.12	5.48 (5)	36.2 (0.38)
Lysozyme	0.712	4.67	7.73 (12.3)	44.2 (0.43)
Soybean trypsin inhibitor	0.713	0.17	4.44	41.1 (0.20)
α-Chymotrypsin	0.717	4.15	8.32	62.2 (0.33)
Trypsin	0.719	0.92	5.16	46.0 (0.28)
Cytochrome c	0.725	0.07	4.27	30.8 (0.34)
Serum albumin	0.735	10.5	14.6 (13.4)	135.0 (0.27)
α-Lactalbumin	0.736	8.27	12.4	56.9 (0.54)
Carbonic anhydrase	0.742	6.37	10.5	76.0 (0.34)
Ovalbumin	0.746	9.18	12.1	101.0 (0.30)
Myoglobin	0.747	8.98	13.1	64.1 (0.51)
β-Lactoglobulin	0.751	8.45	11.8	63.6 (0.46)
Fibrous proteins				
Gelatin	0.689	−2.5		
F-actin	0.720	−6.3		
Myosin	0.724	−18.0		
Tropomyosin	0.733	−41.0		

[a] The values in parentheses represent values measured experimentally; other values were estimated from β_s.

[b] The values in parentheses express δV_{rms} as the percentage of the total protein volume.

Data from K. Gekko (1999) in *Encyclopedia of Molecular Biology* (T. E. Creighton, ed.), Wiley-Interscience, NY, pp. 553–555.

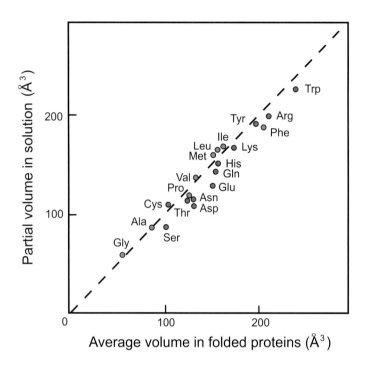

Figure 10-2. Correspondence between the average volume occupied by each amino acid residue in solution and in folded proteins. The line has a slope of unity. Data from T. E. Creighton (1993) *Proteins: structures and molecular properties*, 2nd edn, W. H. Freeman, NY.

The partial specific volume of a protein in water at infinite dilution, $\bar{v}_2^0$, is the sum of three contributions: (1) the constitutive volume, the sum of the van der Waals volumes of the constitutive atoms ($\bar{v}_c$); (2) the volume of the cavities in the molecule due to imperfect atomic packing ($\bar{v}_{cav}$); (3) the volume change due to solvation or hydration ($\bar{v}_{cav}$):

$$\bar{v}_2^0 = \bar{v}_c + \bar{v}_{cav} + \Delta\bar{v}_{sol} \tag{10.3}$$

The value of $\bar{v}_{sol}$ is usually negative because hydration involves the contraction of solvent water. The change in volume upon transferring a protein from one solvent to another is the difference in $\bar{v}_2^0$ between the two solvents, which will include any change in volume resulting from a change in the conformation. The value of $\bar{v}_c$ is constant, so any observed volume change is ascribed primarily to differences in $\bar{v}_{cav}$ and $\bar{v}_{sol}$. Any process that increases the solvent accessible surface area, such as unfolding or dissociation into subunits, is generally expected to produce negative volume changes due to the increased hydration. This is generally observed, but some measurements have found no change or even slightly positive volume changes. It should be noted that the volume change observed at a significant protein concentration may differ, even in sign, from that measured at infinite dilution.

Volumetric properties of proteins. T. V. Chalikian (2003) *Ann. Rev. Biophys. Biomolec. Structure* **32**, 207–235.

Partial molar volume of proteins studied by the three-dimensional reference interaction site model theory. T. Imai *et al.* (2005) *J. Phys. Chem. B* **109**, 6658–6665.

10.2. CHEMICAL REACTIVITIES

The reactivities of groups on proteins to various chemical reagents (Section 7.2) have long been used to probe the structure, function and dynamics of proteins. Reaction with a chemical reagent requires that the reactive group be accessible to it; therefore, the protein group must be on the surface, or the reagent must diffuse through the structure to the reactive group. Even totally buried groups in proteins are usually observed to react at a finite rate with an appropriate reagent. The small extent to which a buried residue does react must reflect the flexibility of the protein structure and the transient occurrence of the buried group in a structure where it is accessible to the reagent. Groups on the surface of the molecule are expected to react as readily as the same groups on small molecules, although this may be perturbed by their particular environments. Groups involved in protein function are very often on the surface of the protein, in an active site where they are accessible to their substrates. Their chemical modification will usually inactivate that protein function, whereas the reaction of other accessible groups will have little or no effect. Particularly important groups often have physical properties that are markedly altered from the norm, which can be detected from their rates of chemical modification. Residues that exhibit different reactivities in the presence and absence of a ligand are usually important for the function.

Studies of the chemical reactivities of groups in proteins have uncovered a bewildering range of reactivities, from nearly total unreactivity to hyperreactivity, where the reaction is much more rapid than with a normal group. Two factors were believed to be usually involved: (1) the effect of the environment on the electronic state of a group, i.e. its intrinsic reactivity, and (2) steric effects on accessibility to the modifying reagent. Rationalization of such results in the light of a protein's crystal structure was not, however, always convincing. Groups buried in the interior can be much more reactive than those at the surface; an example is cytochrome *c*, where the two internal Tyr residues (48 and 67) are much more reactive towards tetranitromethane (Section 7.2.L) than the two at the surface of the molecule (residues 74 and 97). In contrast, iodination modifies primarily residues 67 and 74.

When the relative reactivity of a residue depends upon the nature of the reagent, the crucial factor is usually the **local concentration of the reagent**, determined by its interaction with the neighboring parts of the protein. Steric or electrostatic repulsions may reduce the local concentration of reagent to far below that of the bulk solvent, leading to unreactivity with a nearby group. In contrast, binding of a reagent to the protein can produce extremely high local concentrations, perhaps up to the equivalent of about 10^{10} M. This could lead to apparent hyperreactivity of a group that is nearby and in appropriate proximity to react.

One of the classic examples of **hyperreactivity** involves the reactive Ser residue characteristic of the so-called serine proteinases (e.g. trypsin, chymotrypsin and elastase; Section 15.2.D.1). A unique Ser residue (number 195 in the usual numbering system based upon chymotrypsinogen) of these related enzymes reacts rapidly with acylating reagents, such as diisopropyl fluorophosphate (DFP):

$$[(CH_3)_2CH\text{-}O\text{-}]_2 \overset{\overset{O}{\|}}{P}\text{—}F \;+\; HOCH_2\text{—} \;\longrightarrow\; [(CH_3)_2CH\text{-}O\text{-}]_2 \overset{\overset{O}{\|}}{P}\text{—}OCH_2\text{—} \;+\; F^- + H^+ \quad (10.4)$$

DFP$$Ser 195

A Ser hydroxyl normally is not very reactive chemically and does not react with such acylating reagents (Section 7.2.C); Ser195 of the serine proteinases does not react when the native conformation is disrupted. Therefore, the occurrence of this reaction in the native protein was attributed to a greatly enhanced nucleophilicity of this Ser hydroxyl group by the folded conformation. It is now clear, however, that reagents such as DFP bind in the active sites of serine proteinases much like the substrates of these enzymes and produce very high local concentrations and very great apparent reactivities (Section 14.4.A).

Another classic example of hyperreactivity of specific residues is the reaction of His12 and His119 of ribonuclease A with iodoacetate, iodoacetamide and other alkyl halides (Table 10-2). These reagents are normally most reactive with thiol groups (Section 7.2.I.1) but normal His residues will also react with these reagents, although much less rapidly. The rates of reaction of both His12 and His119 of ribonuclease A are considerably enhanced. The rate also depends upon the nature of the reagent, indicating that interactions between the reagent and surrounding protein are important. The orientation of the His residues in the folded conformation is also important for determining their accessibility to reagent, as His12 invariably is alkylated on the $N^{\varepsilon 2}$ atom, His119 on the $N^{\delta 1}$ atom.

Table 10-2. Rates of alkylation of histidine and two His residues of ribonuclease A

	Second-order rate constant $(10^{-4} \text{ s}^{-1} \text{ M}^{-1})$[a]		
		Ribonuclease A[b]	
Alkylating reagent	L-Histidine	His12	His119
Iodoacetate		7.3	51.1
Iodoacetamide	0.012	1.1	0
Bromoacetate	0.086	20.5	184.5
L-α-Bromopropionate	0.0027	0.19	0.66
D-α-Bromopropionate	0.0028	4.16	1.84
D-α-Bromo-n-butyrate		3.60	1.11
β-Bromopyruvate		0	911
β-Bromopropionate	0.0229	0	6.33

[a] At 25°C and pH 5.3–5.5.

[b] His12 is always alkylated at atom $N^{\varepsilon 2}$ atom, whereas His119 reacts at $N^{\delta 1}$; reaction of one N atom inhibits reaction at the other.

Data from R. L. Heinrickson *et al.* (1965) *J. Biol. Chem.* **240**, 2921–2934 and R. G. Fruchter & A. M. Crestfield (1967) *J. Biol. Chem.* **242**, 5807–5812.

A method for quantitatively characterizing the chemical properties of specific groups in proteins is **competitive labeling.** A small amount of a radioactively labeled reactive reagent is incubated with the protein under conditions where only a small fraction of the protein molecules reacts. This minimizes complications of the modification altering the properties of the protein that would become apparent if molecules reacted more than once. The relative extents to which the various groups are modified should reflect their relative reactivities in the original protein. The modification of the protein is subsequently completed using a different isotopic form of the same reagent. The modified residues are identified by chemical procedures, and the incorporation of radioactivity from the first reagent is measured. The reactivity of each group is compared to that of an internal standard of a suitable model compound.

For example, only the unionized form of amino groups reacts with most reagents (Section 7.2.E); therefore, the variation of reactivity with pH can be used to determine the apparent pK_a value of the amino group in a protein and the relative reactivity of its unionized form relative to the standard. The reactivities of various residues of a particular type often vary uniformly with their pK_a values because the affinity of a group for protons can also tend to reflect its affinity for other reagents. This is usually illustrated with a **Brønsted plot** (Figure 10-3) constructed with data from model compounds. Departures of protein groups from this relationship reflect the effect of the protein structure on the reactivity. When the three α-amino groups of the three polypeptide chains of α-chymotrypsin were reacted with acetic anhydride (Section 7.2.E), the pK_a and reactivity of Cys1 were found to be normal (Figure 10-3), consistent with its exposed situation in the protein's crystal structure. The α-amino group of Ile16 has a relatively high pK_a value and low reactivity, which is consistent with its buried position in the folded protein, where it interacts in a salt bridge with Asp194. The α-amino group of Ala149 has a somewhat elevated pK_a value and diminished reactivity; it is partly buried in the folded protein. Its reactivity increases above pH 9, which is attributed to the effect of ionization of the nearby Ile16 α-amino group.

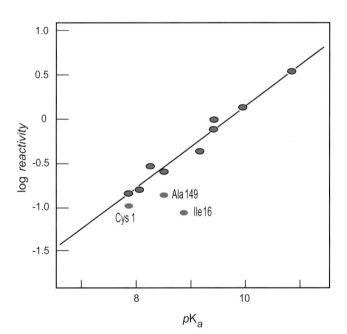

Figure 10-3. Brønsted plot of the reactivities of the three α-amino groups of α-chymotrypsin (Cys1, Ile16 and Ala149, *red circles*) with acetic anhydride compared with those of model compounds (*blue circles*). The extent to which the reactivities of the protein α-amino groups differ from those of the model compounds is due to the effects of the protein conformation. Data from H. Kaplan (1972) *J. Mol. Biol.* **72**, 153–162.

The close proximities of other groups in a folded protein can also lead to unexpected intramolecular reactions. In hen lysozyme, iodine unexpectedly attacks preferentially one of the least accessible Trp residues (Trp108). Other reagents tend to react with other Trp residues, so presumably the reactive iodine species, the I^+ ion, binds to the protein near Trp108. In solution, oxidation of an indole ring is believed to occur by attack by the I⁺ ion, followed by displacement of HI by a hydroxide ion or water molecule, to yield a residue of oxindole alanine:

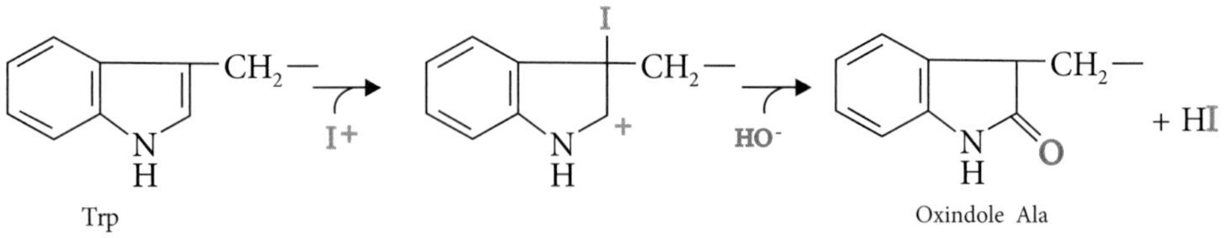

$$\text{Trp} \xrightarrow{I^+} \text{intermediate} \xrightarrow{HO^-} \text{Oxindole Ala} + HI \tag{10.5}$$

In the case of lysozyme Trp108, however, the indole ring is buried in the folded structure and a hydroxyl group or water molecule is not available. The carboxyl group of the neighboring Glu35 is nearby, and it reacts with the intermediate to yield an internal ester cross-link (Figure 10-4).

In general, groups held in close proximity and appropriate orientation by the folded conformation of a protein can undergo reactions that are not significant between separate molecules in solution.

Chemical characterization of functional groups in proteins by competitive labelling. N. M. Young & H. Kaplan (1989) in *Protein Function: a practical approach* (T. E. Creighton, ed.), pp. 225–245, IRL Press, Oxford.

Chemical modification. T. Imoto & H. Yamada (1989) in *Protein Function: a practical approach* (T. E. Creighton, ed.), pp. 247–277, IRL Press, Oxford.

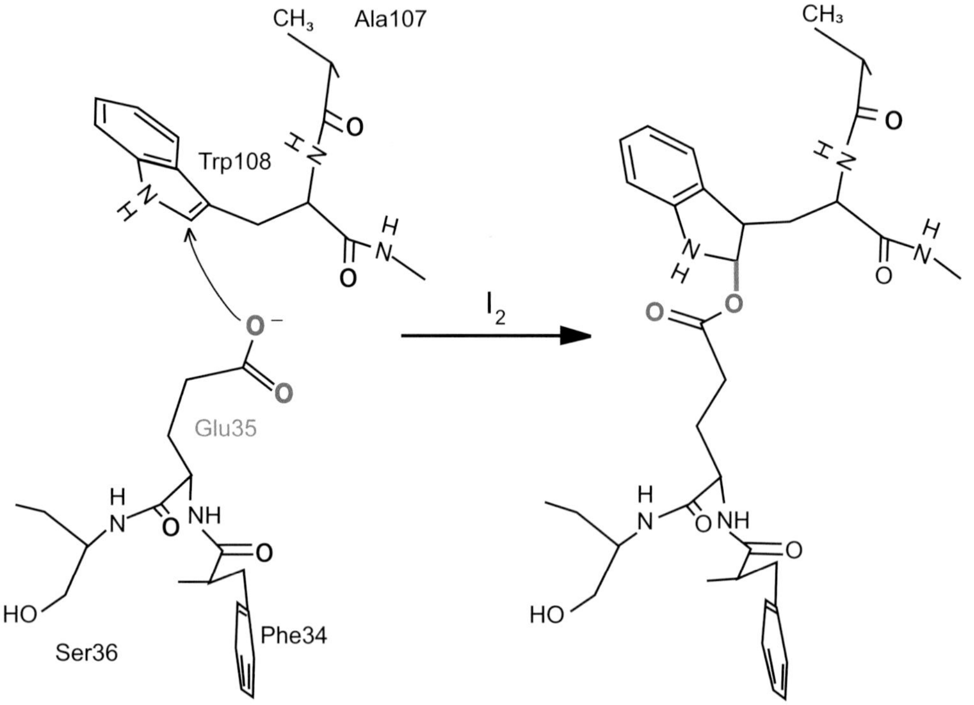

Figure 10-4. Covalent cross-link between Glu35 and Trp108 of hen lysozyme produced by treatment with iodine. The positions of these residues in native lysozyme are shown on the *left*. Iodine presumably reacts initially with Trp108 but then the adduct reacts preferentially with Glu35, rather than with water, owing to the proximity of the Glu side-chain. The structure of the cross-linked protein is shown on the *right*. Data from C. R. Beddell *et al.* (1975) *J. Mol. Biol.* 97, 643–654.

10.2.A. Ionization: Electrostatic Effects

The folded conformations of proteins have a wide variety of effects on the ionization of individual groups in proteins. Many charged groups are brought into close proximity on the surface of a folded protein, so ionization of groups that would increase the net charge can be suppressed. This general electrostatic effect can affect the ionization of all the groups. Specific interactions also occur, such as hydrogen bonding or salt bridges, which affect primarily the ionization of particular groups. The pK_a values of individual groups can be affected by many environmental and electrostatic factors, even in small molecules. The variety of environments encountered in folded proteins can produce very unique ionization properties of various groups. The pK_a values of residues of the same type consequently can vary widely within a protein due to their different environments, often over a range of 3–4 pH units. For example, the observed pK_a values of the His residues in sperm whale myoglobin range from 5.5 to 8.1, compared with the values of 6.0–7.0 observed in isolated His residues (Section 7.2.F). In many cases, buried His residues cannot ionize unless the protein unfolds first (Section 11.1.F).

Understanding and simulating electrostatic effects in the very heterogeneous environments within a folded protein immersed in water or a membrane is much more complex than in a homogeneous liquid, where a simple dielectric constant can describe the effect of the environment. Detailed modeling of electrostatic effects in proteins requires consideration of all the atoms and the charges of both the protein and solvent, plus their atomic polarizabilities. Electrostatic effects in proteins are also complicated by the presence of counterions in the aqueous solvent and by possible binding of ions by the protein. The ionization of each group on a protein is affected by its environment, the protein and the solvent, plus the ionization of other groups on the protein. Consequently, it is not easy to predict the ionization behavior of any one group or the titration curve of the total protein.

The effects of other charged groups on the ionization of any particular residue can be determined by mutating the other changed groups individually. The results of such a series of mutations on the ionization of residue His64 at the active site of subtilisin are given in Table 10-3. Removing a positive charge on another residue increases the pK_a of His64 by 0.1 pH unit, but removing a single negative charge decreases its pK_a by up to 0.4 pH units. Replacing a negatively charged group by one positively charged decreases the pK_a of His64 by 0.6 pH units, while doing the same simultaneously for two groups decreases it by 1 pH unit. Using the known distances between His64 and the mutated charged groups in the crystal structure of the normal protein, the effective dielectric constant for each of these electrostatic interactions were calculated to range between 45 and 173. These relatively high values result from the groups being at the surface of the protein, accessible to water. That some of the effective dielectric constants are greater than that of water (80) is probably a result of the polar groups on the protein being in appropriate positions and orientations to 'solvate' the charges more effectively than does water. Probably for that reason, the occasional isolated charged groups found within protein interiors are generally surrounded by appropriate protein dipoles: the dipoles of C=O groups for positive charges and of N–H for negative charges. The pK_a values of many internal groups imply apparent internal protein dielectric constants of 10 or higher, substantially higher than the dielectric constants of 2–4 measured experimentally with dry proteins. For example, a Glu residue was buried in the hydrophobic core of staphylococcal nuclease by replacement of Val66. Its pK_a was measured to be 4.3 units higher than the pK_a of a normal Glu residue in water. A similar increase of 4.9 units was measured for the pK_a of a Lys residue buried at the same location. These ΔpK_a values are equivalent energetically to the transfer of a charged group from water to a medium of dielectric constant about 12. When such buried residues do ionize, there are local changes to the protein structure to accommodate the buried charge.

There are many unanswered questions about electrostatic effects in proteins.

On the use of different dielectric constants for computing individual and pairwise terms in Poisson–Boltzmann studies of protein ionization equilibrium. V. H. Teixeira *et al.* (2005) *J. Phys. Chem. B* **109**, 14691–14706.

High apparent dielectric constant inside a protein reflects structural reorganization coupled to the ionization of an internal Asp. D. A. Karp *et al.* (2007) *Biophys. J.* **92**, 2041–2053.

Table 10-3. Measurement of the effective dielectric constant of subtilisin by the effects of various replacements of ionized residues on the apparent pK_a value of His64 at low ionic strength

Replacement	Measured ΔpK_a [a]	Mean distance from charge to His64 N atoms (Å) [b]	Effective dielectric constant (D^{eff}) [c]
Glu156 → Ser	−0.38	14.4	45
Asp99 → Ser	−0.40	12.6	48
Glu156 → Lys	−0.63	15.5	50
Asp99 → Lys	−0.64	13.8	55
Ser156 → Lys	−0.25	16.5	59
Ser99 → Lys	−0.25	15.0	65
Asp36 → Gln	−0.18	15.1	90
Lys213 → Thr	+0.08	17.6	173
Asp99 → Ser and Glu156 → Ser	−0.63		
Asp99 → Lys and Glu156 → Lys	−1.00		

[a] Values for the pK_a shifts are the mean values of the two His imidazole N atoms; the normal pK_a value for His64 is 7.0–7.1, depending upon the ionic strength.

[b] Mean distances are the average from the side-chain N or O atoms of the ionized side-chain to the two His imidazole N atoms. Mean distances for the replacement side-chain were obtained assuming the side-chain to be fully extended.

[c] The effective dielectric constant was calculated using the equation:

$$D_{eff} = \frac{244}{(\Delta q)\, r\, (\Delta pK_a)}$$

where Δq is the change in the number of charges and r is the distance in Ångstroms.

Data from M. J. E. Sternberg *et al.* (1987) *Nature* **330**, 86–88.

10.3. Isotope (Hydrogen) Exchange

The best established evidence for extensive flexibility of protein structure is that groups that are internal and inaccessible to the solvent do react at a finite rate with appropriate reagents in solution, even if only very slowly. Either the normally buried protein group must occasionally be at the surface, accessible to the reagent, or the reagent must permeate the protein interior; both would require disruption of the normal protein conformation. Such reactions are, however, often complicated by the tendencies of reagents to be repelled from the protein or to bind to it and react rapidly with nearby groups, perhaps also perturbing the conformation. The most useful reagent, therefore, is one that is normally present, namely water. It may be used in its isotopic forms (^1H_2O, ^2H_2O, ^3H_2O) to measure the tendencies of the various H atoms of the protein to exchange with the solvent.

A protein with H atoms of one isotope is transferred to water of a different isotope, and the exchange between the two is measured. H atoms attached to various atoms exchange with solvent at varying intrinsic rates, depending upon the tendency of the group to ionize. H atoms on O, N or S atoms exchange rapidly, whereas those attached to C atoms do so at much lower rates that are usually not significant. Most measurements of exchange in proteins use the –NH– groups of the polypeptide backbone. These exchange at convenient rates, and they also have the advantage that exchange can be monitored at specific sites by nuclear magnetic resonance (NMR) because ^1H gives an NMR signal whereas ^2H does not.

Designating the two isotopes as H and H*, the remainder of the macromolecule as M, and O as the oxygen atom of water, the reaction can be represented as:

$$MH + -OH^* \leftrightarrow MH^* + -OH \tag{10.6}$$

Separate samples can be exposed to a labeling pulse for varying time periods, then the exchange quenched to trap the label by adjusting the pH; the subsequent time–course of the loss of label from each site can be monitored. The pulse conditions can be chosen to label selectively the sites of interest for the particular application. Characteristic patterns of protection factors are associated with H atoms in α-helices or β-sheets. When hydrogen exchange involves a cooperative unfolding of an entire structural domain, all the hydrogens in that domain show the same protection factor, so the observed rates provide information about the stability of the native structure and how it responds to changes in the environment.

10.3.A. Exchange in Macromolecules

A rapidly exchanging, presumably flexible, protein (insulin) is compared with one that exchanges slowly and is relatively inflexible (BPTI) in Figure 10-5. In BPTI, the most slowly exchanging H atoms do so at about 10^{-8} times the rate observed in model amides. Note that the rates are measured over eight orders of magnitude by manipulating the pH from 3.0 to 9.2, which alters the intrinsic rate of exchange in a predictable manner due to catalysis by H^+ and ^-OH. That the measurements generally follow the same curve (except at the most acidic pH value) indicates that varying the pH does not alter the structures of the proteins or the mechanisms of exchange.

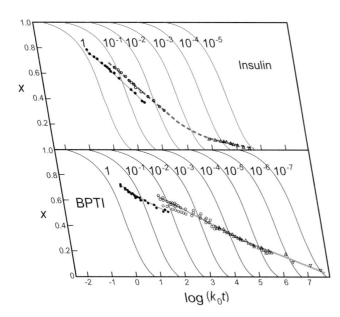

Figure 10-5. Exchange-rate curves of insulin and bovine pancreatic trypsin inhibitor (BPTI) at 25° C. x is the fraction of original isotope remaining, k_0 is the exchange rate of an average solvent-exposed peptide group, and t is time. The measurements were made in the time interval from about 5 min to 10 h after dissolution of the protein and at the following values of pH: insulin 3.0 (*closed circles*), 4.1 (*open circles*), 7.7 (*triangles*); bovine pancreatic trypsin inhibitor 3.4 (*closed circles*), 5.1 (*open circles*), 7.3 (*triangles*), 9.2 (*inverted triangles*). The *solid curves* are calculated for the various probabilities of exposure to solvent of the peptide groups indicated. Data from A. Hvidt & E. J. Pedersen (1974) *Eur. J. Biochem.* **48**, 333–338.

An NH group at the surface of a protein and hydrogen bonded only to water usually is observed to exchange at a rate like that of an appropriate model small molecule. H atoms that are protected in the interior of a macromolecule and by internal hydrogen bonding exchange more slowly. The degree of this retardation provides information about the stability of the macromolecular structure and the flexibility that permits exchange, but the mechanism of exchange must be understood. Slow exchange of groups in biological macromolecules is widely believed to be retarded because the H atom is involved in a hydrogen bond to some adjacent group. Certainly that hydrogen bond must generally be broken to expose the NH to solvent and permit it to exchange. Simple interpretation is confounded, however, because inaccessibility usually accompanies hydrogen bonding in folded macromolecules (Section 9.1.F).

In spite of many extensive investigations, the available evidence is not sufficient to decide how hydrogen exchange occurs in any particular protein. There are, however, two limiting and extreme models: (1) **local unfolding** (or breathing) of the protein structure and (2) **solvent penetration** of the structure. The evidence most indicative of the classical local unfolding process is the observation that the rate of exchange is correlated with the solvent accessibility of the residue and its neighbors but not with the distance of its –NH from the surface. Also, substantial effects on the rate of exchange of *all* slowly exchanging atoms in the well-studied small protein BPTI (Figure 9-2) are produced by localized covalent modifications of the protein. The magnitudes of these effects are roughly proportional to the extent to which the stability of the folded conformation is decreased by the modification. Also, binding of a ligand at one site on a folded protein almost invariably diminishes the rate of exchange of all the interior groups throughout the protein. Observations more consistent with the solvent permeation mechanism are that (1) exchange occurs within the crystalline state of at least some proteins at rates very similar to those in solution; unfolding would be expected to be greatly diminished by the crystal lattice; (2) the rate of exchange in solution is not increased by low concentrations of denaturing agents, such as urea, that would be expected to favor unfolding, until total unfolding becomes significant and all the sites tend to exchange at the same rate (Figure 10-6); and (3) the rate is decreased by increased pressure, suggesting a need to create channels through the protein.

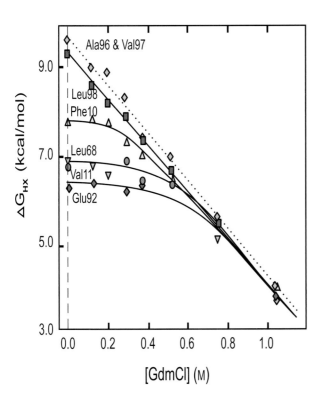

Figure 10-6. Dependence of the rate of exchange of backbone NH groups of cytochrome *c* as a function of guanidinium chloride (GdmCl) concentration. The rate of exchange is inversely proportional to the free energy change required (corrected for the various intrinsic exchange rates), so the fastest-exchanging rates are at the bottom. The rates of exchange of the rapidly exchanging residues Glu92, Val11 and Leu68 barely change at low concentrations of denaturant, until complete unfolding becomes significant, when all the residues exchange at the same rate. Residues Leu98, Val97 and Ala96 appear to exchange only when the protein unfolds completely. Data from Y. Bai *et al.* (1994) *Proteins* **20**, 4–14.

It seems most likely that hydrogen exchange occurs at different sites in folded proteins by different processes that depend upon the site, the protein and the conditions.

Mechanisms and uses of hydrogen exchange. S. W. Englander *et al.* (1996) *Curr. Opinion Struct. Biol.* **6**, 18–23.

Protein analysis by hydrogen exchange mass spectrometry. A. N. Hoofnagle *et al.* (2003) *Ann. Rev. Biophys. Biomolec. Structure* **32**, 1–25.

Native state hydrogen-exchange analysis of protein folding and protein motional domains. C. Woodward *et al.* (2004) *Methods Enzymol.* **380**, 379–400.

10.3.B. Solvent-penetration Model

According to this model of hydrogen exchange, water and catalytic H$^+$ or OH$^-$ gain access to the interior of the macromolecule through channels, perhaps generated transiently by conformational fluctuations. The rate of exchange is then governed by the extent to which those buried H atoms come into contact with the solvent atoms and the environment within the altered protein structure; the latter factor is unknown, so the rates of exchange cannot be interpreted quantitatively and this model is not considered very useful, although it more readily accounts for the experimental observations with proteins.

10.3.C. Local-unfolding Model

According to the other model, called the local-unfolding model, flexibility of a segment of the molecule breaks a set of hydrogen bonds and other interactions that hold the macromolecule in its stable conformation. The NH groups of that segment are thereby exposed to solvent, where they can undergo exchange. This model accords nicely with the dynamic nature of macromolecular structure but it does not explain why exchange is not decreased by incorporating the macromolecule into a crystal lattice nor increased by low concentrations of denaturants but is decreased by increased pressure. The last observation suggests a need to open a protein structure and create channels. It is probably an oversimplification to assume a single locally unfolded state from which exchange at a site occurs; proteins can probably achieve a variety of locally unfolded states, from ones involving only minor local fluctuations to those involving global unfolding.

The local-unfolding model has the great advantage that it can be analyzed quantitatively. The basic mechanism is postulated to be:

$$\text{Closed} \underset{k_{\text{fold}}}{\overset{k_{\text{unfold}}}{\rightleftharpoons}} \text{Open} \xrightarrow{k_{\text{ex}}} \text{Exchanged product} \quad (10.7)$$

Here k_{unfold} is the rate constant for the motion that converts the closed form to the open one, k_{fold} is the rate constant for reversion back to the closed form, and k_{ex} is the intrinsic rate constant for hydrogen exchange. The general expression for the observed rate constant is:

$$k_{\text{obs}} = \frac{k_{\text{unfold}} k_{\text{ex}}}{k_{\text{fold}} + k_{\text{ex}}} \quad (10.8)$$

Two limiting mechanisms can be distinguished.

1. EX1 Mechanism

If $k_{\text{ex}} \gg k_{\text{fold}}$, Equation 10.8 simplifies to the EX1 mechanism:

$$k_{\text{obs}} = k_{\text{unfold}} \quad (10.9)$$

The rate-limiting step is the opening of the protein structure to produce the locally unfolded state, from which exchange is rapid. Under these conditions, the rate of exchange should be independent of the pH, except insofar as k_{unfold} is pH-dependent. These conditions can be induced at high pH, where k_{ex} becomes sufficiently great, but in practice it is rare to achieve these conditions for a protein without denaturation becoming significant (Section 11.1). Then exchange generally occurs from the totally unfolded conformation.

2. EX2 Mechanism

If $k_{\text{fold}} \gg k_{\text{ex}}$, Equation 10.8 simplifies to the EX2 mechanism:

$$k_{\text{obs}} = K_{\text{op}} k_{\text{ex}} \quad (10.10)$$

where K_{op} is the equilibrium constant for forming the locally unfolded state:

$$K_{\text{op}} = k_{\text{unfold}}/k_{\text{fold}} = [\text{open}]/[\text{closed}] \quad (10.11)$$

Exchange occurs only during that fraction of the time when the macromolecule is unfolded at least locally. The value of k_{ex} can be assumed to be the same as the measured value for an appropriate small molecule model under the same conditions, so the experimental value of k_{obs} provides an estimate of K_{op}. This equilibrium constant can then be converted to the free energy change involved in the local unfolding:

$$\Delta G° = - RT \log_e K_{op} \qquad (10.12)$$

This value is an estimate of the energy that normally stabilizes the closed native state of the macromolecule, or of the energy required to allow the local unfolding. Alternatively, the degree to which the exchange is retarded by the protein structure is often expressed as a **protection factor**, P:

$$P = k_{ex}/k_{obs} = 1/K_{op} \qquad (10.13)$$

Protein hydrogen exchange mechanism: local fluctuations. H. Maity *et al.* (2003) *Protein Sci.* **12**, 153–160.

EX1 hydrogen exchange and protein folding. D. M. Ferrarro *et al.* (2004) *Biochemistry* **43**, 587–594.

10.4. FLEXIBILITY DETECTED CRYSTALLOGRAPHICALLY

The protein models resulting from crystallographic structure determinations are usually depicted as being static, but this should not be taken to imply that the molecule is equally as static, even within the crystal lattice. The electron density observed crystallographically is averaged over all the molecules within the crystal and over the time of the measurements, so the electron density is smeared out if an atom of the protein is not in exactly the same position within each unit cell of the crystal. This will reflect flexibility of the protein molecule in the lattice, but unfortunately it can also result from disorder in the crystal lattice.

Indirect evidence for protein flexibility within the crystal comes from chemical exchange of buried H atoms of the proteins with those of the solvent, which can be observed by neutron diffraction using the isotopes ¹H and ²H; they have very different neutron scattering properties.

Mapping protein dynamics by X-ray crystallography. D. Ringe & G. A. Petsko (1985) *Prog. Biophys. Mol. Biol.* **45**, 197–235.

Atomic motions in molecular crystals from diffraction measurements. J. D. Dunitz *et al.* (1988) *Angew. Chem. Int. Ed. Engl.* **27**, 880–895.

10.4.A. Effects of Different Crystal Lattices

The crystal lattice must put some pressure on a protein molecule to adopt a conformation compatible with a feasible crystal lattice, so crystallizing the same protein in different crystal lattices should give some idea of the flexibility of that protein. A flexible protein would be expected to adopt somewhat different conformations in different crystal lattices. In contrast, the general observation is that identical protein molecules crystallized in different ways, often using very different solvents, or, when observed in different environments within the same crystal, have very similar conformations. Significant differences in conformation are observed primarily only with the smallest proteins, which appear to be more intrinsically flexible, with alternative conformations that have very similar free energies. Otherwise, the folded conformations of larger proteins do not normally adapt much to different crystal lattices; the greatest conformational differences are usually observed to be local changes in the conformations of flexible side-chains and polypeptide loops on the surface of the molecule, which are known to be flexible. Close inspection is required to find significant differences in the overall conformation. For example, the structure of a small, relatively rigid, protein, BPTI (Figures 9-2 and 10-5), is altered in three different crystal lattices by an average of 0.4–0.5 Å in the relative positions of the backbone C^α atoms. This is not much greater than the differences observed in independent crystal structure determinations but is probably the best measure of the intrinsic flexibility of the protein molecule.

A mutant T4 lysozyme displays five different crystal conformations. H. R. Faber & B. W. Matthews (1990) *Nature* **348**, 263–266.

10.4.B. The Temperature Factor: Mobility or Disorder?

The extent to which the electron density is smeared out in an electron density map is usually expressed as the temperature factor, B, although this should not be taken literally. Proteins have considerably larger B-values than those determined for crystals of small molecules, where the value of B is generally 3–4 Å^2 and seldom greater than 10 Å^2. For interior atoms of well-ordered protein crystals, the B-values are usually in the range of 10–20 Å^2. Such B-values imply average amplitudes of vibrations of 0.3–0.5 Å, and instantaneous deviations of up to 1.0–1.5 Å would be expected frequently.

The value of B generally varies throughout a protein structure, with those atoms on the surface of the molecule usually having the greatest values. The average atomic fluctuation is linearly related to the square of the atomic distance from the center of mass of the protein. In a few instances, surface side-chains or small portions of the polypeptide chain may be invisible, owing to a large possible number of different conformations. In the case of the proteins trypsinogen and IgG(Kol), entire domains of the proteins are not visible, which has been taken to indicate that these parts have extreme flexibility.

Interpretation of B-values is not straightforward because the temperature factor is affected by errors in the crystallographic phases and disorder within the crystal lattice, as well as by flexibility of the protein molecule. Lattice disorder and protein flexibility can be distinguished in theory by comparing different parts of the electron density map to determine which variation can be explained by slightly different packing of rigid molecules within the unit cell, by comparing maps of the same molecule in different crystal lattices or in different environments within the same lattice (e.g. when there is noncrystallographic symmetry) or by varying the temperature, to see if the flexible portions of the molecule can be frozen in fixed conformations at low temperatures. None of these approaches have led to definitive conclusions.

Flexibility analysis of enzyme active sites by crystallographic temperature factors. Z. Yuan *et al.* (2003) *Protein Eng.* **16**, 109–114.

Evolutionary conservation of protein backbone flexibility. S. Maguid *et al.* (2006) *J. Mol. Evol.* **63**, 448–457.

Protein dynamics from X-ray crystallography: anisotropic, global motion in diffuse scattering patterns. L. Meinhold & J. C. Smith (2007) *Proteins* **66**, 941–953.

10.5. FLEXIBILITY DETECTED BY NMR

NMR provides much dynamic information about the flexibilities of macromolecules. In particular, whether equivalent groups, such as the three H atoms of a methyl group, give individual or averaged signals depends upon the rate at which they are interconverted on the NMR time scale.

Simultaneous determination of protein structure and dynamics. K. Lindorff-Larsen *et al.* (2005) *Nature* **433**, 128–132.

New tools provide new insights in NMR studies of protein dynamics. A. Mittermaier & L. E. Kay (2006) *Science* **312**, 224–228.

NMR: prediction of protein flexibility. M. Berjanskii & D. S. Wishart (2006) *Nature Protocol.* **1**, 683–688.

10.5.A. NMR Time Scales

Protein dynamics and thermodynamics can be characterized through NMR measurements of the relaxation rates of side-chain ^{1}H and ^{13}C and backbone ^{15}N nuclei. The rates reflect protein motions on time scales from picoseconds to milliseconds. **Information about the rates of motions comes from the line shapes of NMR spectra**. With rapid interconversions of two conformations, only one averaged signal is observed. Individual signals for the two conformations are observed when the two conformations are very slowly interconverted. At intermediate rates, the peak can be so severely broadened that it becomes undetectable. If the individual signals are known, the line shape can provide a very accurate estimate of the rate constants for the interconversion of the two states. The line shape itself can be analyzed, or the transverse relaxation time (T_2), which is another measure of the line width, can be measured. The time scale for these phenomena, with which the rates of interconversion must be compared, is given by the difference in the resonance frequency of the spins in the two states. With NMR the practical time scale ranges from tens of seconds to 100 μs, depending upon the type and chemical environment of the nucleus being observed.

Side-chains on the surfaces of proteins, as well as terminal methyl groups of side-chains within the interior, are observed in this way to be comparable to those in unfolded proteins or in small-molecule analogs, rotating on time scales of 10^{-11}–10^{-8} s. The rotational correlation times measured by ^{13}C NMR in a disordered polypeptide chain are given in Table 8-1. The slower motions undergone by internal groups, however, are masked in many techniques by the overall rotation of the entire protein molecule.

NMR data on the flexibility of proteins is most readily interpreted if both ^{1}H and ^{15}N nuclei are used. The data can then be interpreted in terms of the order parameter (S) of each nucleus. A nucleus with no freedom within the folded protein conformation will have an order parameter approaching unity, whereas totally flexible nuclei have values approaching zero. **In folded proteins, the order parameters of most nuclei are greater than 0.9, indicating only modest degrees of flexibility**; only atoms in very mobile loops or at the ends of the polypeptide chain have smaller values. Examples of the order parameters of a partly folded intermediate in folding are illustrated in Figure 10-7.

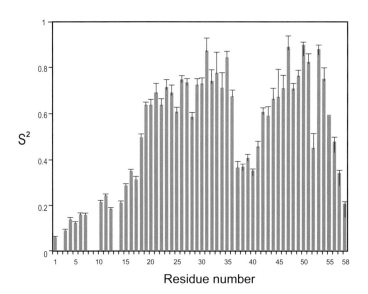

Figure 10-7. The flexibility of a partly folded protein molecule expressed as the order parameter S. The protein is the intermediate with only the Cys30–Cys51 disulfide bond in the disulfide folding pathway of BPTI (Figure 11-26). Residues 19–36 and 42–54 are largely folded; they comprise the main elements of secondary structure of the native protein. In contrast, the other residues, especially the N-terminal 17 residues, are much more flexible. Residues without values indicated are Pro residues, for which the procedure is not appropriate. Data from C. P. M. van Mierlo et al. (1993) *J. Mol. Biol.* **229**, 1125–1146.

NMR probes of molecular dynamics: overview and comparison with other techniques. A. G. Palmer (2001) *Ann. Rev. Biophys. Biomol. Structure* **30**, 129–155.

Thermodynamic interpretation of protein dynamics from NMR relaxation measurements. L. Spyracopoulos (2005) *Protein Peptide Lett.* **12**, 235–240.

10.5.B. Aromatic Ring Flipping

The two *ortho* ($C^{\delta 1}$ and $C^{\delta 2}$) and *meta* ($C^{\varepsilon 1}$ and $C^{\varepsilon 2}$) H atoms of Tyr and Phe residues are in unique environments in the interiors of folded proteins and would be expected to give individual resonances in ^{1}H NMR spectra. If, however, the rings are flipping by 180° rotations about the C^β–C^γ bond rapidly on the NMR time scale, the two atoms of each spend equal amounts of time in both environments and will then give only a single, averaged resonance.

Most proteins give averaged spectra for Phe and Tyr residues, suggesting that the rings are rotating by 180° flips at least 10^4 times per second, even when the rings are fully buried. Of course, such apparently averaged ^{1}H NMR spectra could arise by accidental coincidence of the individual resonances, but this is unlikely in general. It has been shown not to be the case in **BPTI, where the two Tyr and three Phe residues that are most buried in the protein interior give immobilized spectra at low temperatures** (Table 10-4); the other aromatic rings give averaged spectra. As the temperature is increased, averaged spectra are obtained, indicating that rotations are now occurring more frequently. **Even rapidly rotating rings appear fixed on the ^{13}C NMR time scale, indicating that the flips occur with frequencies no greater than 5×10^7/s.**

Table 10-4. Flipping of aromatic rings in BPTI

Residue	Frequency of 180° rotations (s⁻¹) at temperature			Activation parameters		
	4°C	40°C	80°C	Enthalpy $\Delta H^\ddagger$ (kcal/mol)	Entropy $\Delta S^\ddagger$ (cal/mol K)	Volume $\Delta V^\ddagger$ (Å³)
Tyr10	Rapid	Rapid	Rapid			
Tyr21	Rapid	Rapid	Rapid			
Tyr23	<5	3×10^2	5×10^4	26	35	
Tyr35	<1	16	5×10^4	37	68	60
Phe4	Rapid	Rapid	Rapid			
Phe22	Rapid	Rapid	Rapid			
Phe33	Rapid	Rapid	Rapid			
Phe45	30	1.7×10^3	5×10^4	17	11	50

Data from G. Wagner *et al.* (1976) *Biophys. Struct. Mech.* **2**, 139–159 and (1987) *J. Mol. Biol.* **196**, 227–231.

Flipping of aromatic rings within the close-packed interior of a protein requires movement of the surrounding atoms. The usual representation of Tyr and Phe side-chains as flat planar rings would seem to imply that their flipping requires substantial flexibility of the protein structure. The movements required are, however, not as great as might be expected because these rings are actually more like flattened spheres, or oblate ellipsoids, with thickness 3.4 Å and diameter 6.8 Å. The perturbations of the structure required for flipping of the aromatic rings have been studied extensively with BPTI; computer simulations indicate that **adjacent atoms need move by small bond rotations of no greater than 17°.**

Ring flipping is probably an infrequent process that occurs rapidly once initiated, because the half-rotated intermediate state should be very unstable. The frequencies of flipping of Phe45 and Tyr35 of BPTI are decreased by elevated pressures, indicating that **the transition states for ring flipping have substantially increased volumes.**

In contrast to Tyr and Phe residues, **buried Trp and His side-chains of folded proteins do not detectably undergo such ring flipping.** This is undoubtedly due to the much greater size of the indole ring and the absence of symmetry in both His and Trp side-chains. Full 360° rotations would consequently be required to occur in both cases, with much greater conformational adjustments. The side-chains of buried Trp residues generally do undergo much more restricted vibrations about their mean positions on the 10^{-8} s time scale.

Side-chain rotational isomerization in proteins: a mechanism involving gating and transient packing defects. J. A. McCammon *et al.* (1983) *J. Am. Chem. Soc.* **105**, 2232–2237.

Rates and energetics of tyrosine ring flips in yeast iso-2-cytochrome *c*. B. T. Nall & E. H. Zuniga (1990) *Biochemistry* **29**, 7576–7584.

10.6. Varying the Temperature

The expansion of a macromolecule with increasing temperature reflects its atomic packing and flexibility. The apparent and partial specific volumes ($\bar{v}_2^0$) of proteins in aqueous solution (Section 10.1.B) vary linearly with changes in the temperature in the range 4–45°C. The values for $d\bar{v}_2^0/dT$ are positive and in the range of 2.5–10 ($\times 10^{-4}$ cm^3 g^{-1} K^{-1}); the majority lie in the range 3.5–5 ($\times 10^{-4}$ cm^3 g^{-1} K^{-1}). These effects are more pronounced than observed with amino acids, small peptides or denatured proteins, so they are probably caused by expansion of cavities and dehydration of the protein surface.

Crystallographic studies have shown that **increasing the temperature causes both the crystal lattice and the protein structure to expand**. Increasing the temperature of crystals of myoglobin from 80 K (–193°C) to 255 K (–18°C) or 300 K (+27°C) caused volume expansions of 5% for the crystal lattice and 3% for the protein molecule. Greater expansions are exhibited by liquid water and hydrocarbons. The coefficients of thermal expansion of the structures of three model proteins were measured to vary between 0.4 and 1.4 ($\times 10^{-4}$ K^{-1}). Most of the expansion of the proteins occurred in the hydrophobic packing between helices, which individually did not change much. Proteins composed primarily of α-helix expanded more with temperature than did those composed primarily of β-sheet. The largest movement in hen lysozyme was undergone by the two subdomains of the structure, which moved further apart as the temperature was raised. Temperature factors for the room-temperature structure averaged 15.2 Å^2, but only 8.1 Å^2 for the low-temperature structure.

These changes in the structure of the protein are considerably smaller than those measured with the volume in solution, so the effects on the solvation shell must be greater than those on the protein molecules themselves. The hydration shell around proteins is believed to expand about 40% more than does bulk water.

All proteins appear to undergo a dramatic change in their dynamic properties at a temperature of approximately 220 K (–53°C). At higher temperatures, their dynamic behavior is dominated by large-scale collective motions of bonded and nonbonded groups of atoms. At lower temperatures, simple harmonic vibrations predominate. The transition has been described as a **glass transition** to emphasize certain similarities between the change in dynamic behavior of individual protein molecules and the changes in viscosity and other properties of liquids when they form a glass. The glass transition of proteins is frequently claimed to reflect the intrinsic temperature-dependence of the motions of atoms in the protein itself, and the function of the protein is altered significantly below this transition temperature. All these observations were necessarily made in cryosolvents, however, and the protein glass transition is similar to what occurs in the glass transition of aqueous solutions of glycerol. Also, the protein dynamics are said to be suppressed mainly by inertia of the bulk glass and to a lesser extent by specific interactions at the protein–solvent interface. Consequently, the observed protein internal dynamics as a whole appear to have been dictated by the cryosolvent rather than being a reflection of the intrinsic properties of the protein molecule itself.

Observation of fragile-to-strong dynamic crossover in protein hydration water. S. H. Chen *et al.* (2006) *Proc. Natl. Acad. Sci. USA* **103**, 9012–9016.

Glass transition in biomolecules and the liquid–liquid critical point of water. P. Kumar *et al.* (2006) *Phys. Rev. Lett.* **97**, 177802.

Inertial suppression of protein dynamics in a binary glycerol–trehalose glass. J. E. Curtis *et al.* (2006) *J. Phys. Chem.* **110**, 22953–22956.

10.7. EFFECTS OF HIGH PRESSURE

The effect of increasing the pressure is to favor structures that occupy smaller volumes, including the solvent. The degree to which the volume decreases is measured by the **compressibility**.

The intrinsic compressibilities of various globular proteins in solution are similar, with an average value of 25×10^{-6} bar^{-1} and a range of 10 –30 ($\times 10^{-6}$ bar^{-1}) (Table 10-1). For comparison, the compressibilities of liquid water, benzene and hexane are greater at 45, 96 and 165 ($\times 10^{-6}$ bar^{-1}). **Proteins are less compressible than liquids, and nearly as incompressible as crystalline solids**; for example, the corresponding value for ice is 13×10^{-6} bar^{-1}.

The measured value of the compressibility of a protein reflects the effect of pressure on the packing of the interior, by reducing the sizes of cavities, as well as on the hydration of the molecule. The compressibility effect on hydration is usually negative, whereas that on cavities is positive. Two types of compressibility can be measured: the adiabatic compressibility is measured at constant entropy, whereas the isothermal compressibility is measured at constant temperature.

Compressibility of protein transitions. N. Taulier & T. V. Chalikian (2002) *Biochim. Biophys. Acta* **1595**, 48–70.

Decomposition of protein experimental compressibility into intrinsic and hydration shell contributions. V. M. Dadarlat & C. B. Post (2006) *Biophys. J.* **91**, 4544–4554.

Evaluation of intrinsic compressibility of proteins by molecular dynamics simulation. K. Mori *et al.* (2006) *J. Chem. Phys.* **125**, 054903.

10.7.A. Adiabatic Compressibility

Most compressibility studies of proteins have been of adiabatic compressibility because of its accuracy and technical convenience. The value of β_s for various proteins (Table 10-1) varies over a wide range and is sensitive to the structural characteristics of the individual proteins. For example, fibrous proteins show negative β_s, indicating that the dominant effect is on hydration of the surface, rather than on the volumes of internal cavities. In contrast, most globular proteins have positive β_s, which can be ascribed to a large cavity effect overcoming the opposite hydration effect.

The effect of solvation would be expected to be proportional to the surface area of the protein, whereas the effect from cavities should be proportional to the volume. In that case, β_s would be expected to increase with increasing molecular weight of the protein, but no such correlation is evident. Instead, a good positive correlation is found between the value of β_s and the partial specific volume. Hydrophobic proteins show large positive β_s, probably due to enhanced imperfect packing of nonpolar residues localized in the interior of the molecule. Typical α-helical proteins, such as myoglobin and bovine serum albumin, also have very large β_s, while essentially nonhelical proteins, such as trypsin and soybean trypsin inhibitor, have small values. The α-helix seems to be a relatively rigid structure that limits the efficiency with which it can be packed together with other helices in an irregular protein structure, leaving cavities. Four amino acid residues (Leu, Glu, Phe and His) show statistically a strong ability to increase β_s, while another four (Asn, Gly, Ser and Thr) decrease it, although the basis for this is obscure. A single amino acid substitution can bring about a noticeable change in β_s, presumably due to changes in the atomic packing, although it is rare to observe any visible changes in the tertiary structures of such mutant proteins by X-ray crystallography. Lysozyme has a considerably smaller β_s than does α-lactalbumin, even though they have very similar primary and tertiary structures. Such observations are not readily explained.

10.7.B. Isothermal Compressibility

The volume fluctuations and pressure-dependent properties of proteins are theoretically related to the partial specific isothermal compressibility, β_T, rather than the adiabatic one, β_s. Determining β_T, however, requires that the solution density or partial specific volume be measured at constant temperature as a function of pressure, using either hydrostatic pressure or centrifugal force. Such measurements are difficult, and high pressure can cause protein denaturation and modify the preferential solvent interaction, so few β_T data are available for proteins. Approximate values of β_T may be estimated from β_s if the thermal expansion coefficient α and the heat capacity at constant pressure, C_p, are known or can be reasonably inferred. Such calculated β_T-values are listed in Table 10-1, along with a few experimental data. The calculated and experimental values of β_T are not very different: the value of β_T is greater than β_s by (3–4) ($\times 10^{-12}$ cm^2 dyn^{-1}); these differences are comparable to those measured for the amino acids.

The average fluctuation of the partial molar volume, δV_{rms}, estimated using β_s instead of β_T, is listed in the last column of Table 10-1. The volume fluctuation is only about 0.3% of the overall dimensions of the protein. The solvation factor is still included in the value of β_T that is used, so the values estimated by this method are smaller than the probable fluctuations. A volume fluctuation up to 50% greater might be expected for a protein without hydration if the intrinsic isothermal compressibility of a protein itself is used instead of β_T. If the volume fluctuation were concentrated in one part of a protein at any particular moment, the fluctuation in volume could produce sufficient cavities or channels to allow the entry of solvent or probe molecules to explain phenomena such as the hydrogen exchange (Section 10.3) and quenching by small molecules of the fluorescence due to buried Trp residues of folded proteins.

10.7.C. Structural Effects of High Pressure

What happens to the protein architecture at high pressures has been determined crystallographically in several instances. **Increasing the hydrostatic pressure on lysozyme crystals to 1000 atmospheres decreases the volume of the protein molecule by only 0.3%**, corresponding to $\beta_T = 4.7 \times 10^{-12}$ cm^2 dyn^{-1}. Contraction of the molecule is distributed nonuniformly and occurs predominantly in only one of the two domains. Least change occurs to the elements of secondary structure, the β-sheet and individual α-helices. The greatest compression is accomplished by increasing the packing density of side-chain atoms. Increasing the pressure, as with decreasing the temperature, causes many of the crystallographic B-values to decrease, suggesting they reflect primarily inherent flexibility of the protein molecule. In the case of myoglobin at 155 atmospheres of N_2, molecules from the gas appeared to fill small cavities within the protein interior, with slight readjustments of the internal packing. A mutant protein containing a cavity due to deletion of an internal Leu side-chain had the same response to high pressures as the normal protein without the cavity. Surprisingly, the cavity was most rigid while other regions deformed substantially. The cavity filled with four water molecules at high pressures. These observations argue against simplistic structural interpretations of compressibility data.

Computer simulations can also give estimates of compressibilities; normal mode analysis of myoglobin indicated $\beta_T = 9.37 \times 10^{-12}$ cm^2 dyn^{-1}. The compressibility of the hydration shell is estimated to be very negative and to change only slightly with temperature. This is similar to that expected for multiple polar groups, but more extreme, which would suggest that the hydration shells around proteins are large cooperative networks of water molecules.

NMR snapshots of a fluctuating protein structure: ubiquitin at 30 bar-3 kbar. R. Kitahara *et al.* (2005) *J. Mol. Biol.* **347**, 277–285.

Conformational fluctuations of proteins revealed by variable pressure NMR. H. Li & K. Akasaka (2006) *Biochim. Biophys. Acta* **1764**, 331–345.

Protein stability and dynamics in the pressure-temperature plane. F. Meersman *et al.* (2006) *Biochim. Biophys. Acta* **1764**, 346–354.

The use of high-pressure nuclear magnetic resonance to study protein folding. M. W. Lassalle & K. Akasaka (2007) *Methods Mol. Biol.* **350**, 21–38.

Structural rigidity of a large cavity-containing protein revealed by high-pressure crystallography. M. D. Collins *et al.* (2007) *J. Mol. Biol.* **367**, 752–763.

10.8. SPECTRAL PROPERTIES

The chromophores in folded proteins exist in a wide variety of environments that affect their absorption and emission of light in various ways.

The aromatic residues Phe, Tyr and Trp are responsible for most of the absorbance properties of proteins in the near-ultraviolet (UV) wavelength region of 250–300 nm and for the effects of the folded conformation. Their absorbance spectra are shifted to somewhat longer wavelengths (red-shifted) in a nonpolar environment like the interior of a protein. For example, the spectrum of a Tyr side-chain is shifted by 3 nm when the solvent is changed from water to CCl_4. The extent of the shift in wavelength in a folded protein can be used to estimate the exposure to solvent of these aromatic groups in the folded structure. A complementary method, known as solvent perturbation, measures the extent to which the spectrum is altered by changing the solvent. Both types of measurements yield primarily an average over all the aromatic residues of the protein.

Fluorescence by the aromatic side-chains is much more sensitive to their environments, but in an unpredictable way. The fluorescence intensity can be increased or decreased by the folded conformation, depending primarily upon what fluorescence-quenching groups are nearby. The magnitude of the fluorescence intensity is not very informative in itself but it can be a very sensitive monitor of changes in the structure of the protein. The wavelength at which fluorescent light emitted is more informative. For example, Trp residues that are exposed to water fluoresce maximally at a wavelength of 350 nm, whereas totally buried side-chains emit at about 330 nm.

The fluorescence of a folded protein is especially complex in the usual situation in which there is more than one aromatic side-chain. The folded structure brings all these groups into relatively close proximity, so the phenomenon of **fluorescence resonance energy transfer (FRET)** between them is very efficient. In this process, the light energy absorbed by one chromophore is transferred to another that absorbs at a longer wavelength and it can then be emitted as fluorescence by the second residue. The absorbance wavelengths of the aromatic residues are in the order Phe < Tyr < Trp, so proteins containing all three types of residues generally emit fluorescence typical of Trp residues. Fluorescence by Tyr residues is observed only in the absence of Trp, while that of Phe residues is observed only when it is the sole aromatic amino acid.

The circular dichroism spectra of proteins are very informative. The spectrum in the near-UV region of 250–300 nm is dominated by the aromatic residues and reflects the degree of asymmetry of their environments. The spectrum at shorter wavelengths is due primarily to the chiral nature of the polypeptide backbone and depends primarily upon the secondary structure present in the protein. By comparing a protein spectrum to that of standard conformations, the amounts of the various types of secondary structure in a protein can be estimated.

First-principles calculations of protein circular dichroism in the far-ultraviolet and beyond. M. T. Oakley *et al.* (2006) *Chirality* **18**, 340–347.

Protein modification for single molecule fluorescence microscopy. M. S. Dilllingham & M. I. Wallace (2008) *Org Biomol. Chem.* **6**, 3031–3037.

Concentration-independent estimation of protein secondary structure by circular dichroism: a comparison of methods. P. McPhie (2008) *Anal. Biochem.* **375**, 379–381.

Protein secondary structure analyses from circular dichroism spectroscopy: methods and reference databases. L. Whitmore & B. A. Wallace (2008) *Biopolymers* **89**, 392–400.

10.9. INTEGRAL MEMBRANE PROTEINS

Natural membranes are usually very concentrated mixtures of proteins embedded in the membrane bilayer. They usually contain about 50% of their mass as proteins, although this can vary from as low as 25% to as great as 75%. Individual proteins generally diffuse rapidly in the two-dimensional plane of the membrane, with diffusion coefficients of about 10^{-10} cm²/s, although this can be decreased substantially if they are interacting with other molecules inside or outside the membrane. The embedded proteins usually retain their orientations with respect to the inside and outside surfaces of the membrane and do not flip between the two surfaces. The orientation in the membrane of a protein can be determined by its covalent modification from reagents on each side of the membrane. The membrane lipids diffuse about 100-fold more rapidly in the plane of the membrane, which corresponds to a mean velocity of approximately 2 μm/s, although this is decreased in the proximity of the membrane proteins. Embedded proteins usually induce disorder in the adjacent lipid bilayer and restrict the diffusion of neighboring lipid molecules. The restricted lipid molecules exchange rapidly with those in the bulk membrane, so there are no strong interactions between them and the protein molecules.

Proteins in membranes tend to interact with each other much more than do proteins in solution. This is due to their high concentrations and to being restricted to two dimensions rather than three. There is a large **excluded volume effect** because there is not much empty space in the bilayer for the protein to move into. The proteins are restricted in their orientation and to a two-dimensional plane, so fewer degrees of freedom need to be lost for them to interact with another molecule. Perhaps for this reason many proteins in membranes are oligomeric.

Lipids in the assembly of membrane proteins and organization of protein supercomplexes: implications for lipid-linked disorders. M. Bodanov *et al.* (2008) *Subcell. Biochem.* **49**, 197–239.

The assembly of membrane proteins into complexes. D. O. Daley (2008) *Curr. Opinion Struct. Biol.* **18**, 420–424.

Vertebrate membrane proteins: structure, function, and insights from biophysical approaches. D. J. Müller *et al.* (2008) *Pharmacol. Rev.* **60**, 43–78.

~ CHAPTER 11 ~

PROTEIN DENATURATION: UNFOLDING AND REFOLDING

The native structures of proteins are usually delicate and readily disrupted, a process known as denaturation. It is vividly evident whenever one boils an egg: the many proteins of the egg white precipitate. In another everyday case, heating collagen separates its three strands (Section 8.5.C) to produce gelatin. In these cases denaturation is irreversible, but there are many instances, particularly with small proteins, where it is readily reversible and involves only unfolding of the protein conformation; the denatured protein can then be referred to as being unfolded. This type of denaturation provides insight into the forces that maintain the folded state and need to be disrupted upon denaturation; the reverse of denaturation is the folding of the polypeptide chain; both are fundamental aspects of protein structure and function.

The electrostatic, van der Waals and hydrogen bond interactions that are possible between atoms of proteins are individually weak or nonexistent in the presence of water. Ionized and polar groups interact with water almost as favorably as they interact with suitable other ionized and polar groups, and it is energetically unfavorable to remove them from aqueous solution. Nonpolar groups prefer to interact with each other, rather than with water, but even the resulting hydrophobic interaction is not very strong when measured between pairs of nonpolar molecules. Yet such interactions produce stable folded conformations of proteins and nucleic acids. How can they do this? It is primarily because **there are many such interactions within macromolecules, and they are intramolecular and cooperate**, in contrast to the intermolecular interactions that have been considered in previous chapters. The main difference between these two types of interactions is entropic in nature.

Some factors in the interpretation of protein denaturation. W. Kauzman (1959) *Adv. Protein Chem.* **14**, 1–63.

Protein denaturation. C. Tanford (1968, 1970) *Adv. Protein Chem.* **23**, 121–275; **24**, 1–95.

Protein Folding. T. E. Creighton, ed. (1992) W. H. Freeman, NY.

The experimental survey of protein-folding energy landscapes. M. Oliveberg & P. G. Wolynes (2005) *Quart. Rev. Biophys.* **38**, 245–288.

11.1. REVERSIBLE UNFOLDING AT EQUILIBRIUM

Protein denaturation is the process of disappearance under nonphysiological conditions of the specific properties of a native, functionally active protein; usually it involves unfolding of the polypeptide chain, but it can also be caused by simple precipitation or chemical modification. Denaturation can be caused by heating, cooling, high pressure, extremes of pH or the addition of chemical reagents known as denaturants (Figure 11-1). Protein denaturation is of greatest interest when it involves only unfolding of the polypeptide chain and is reversible. The following discussion assumes that the protein being studied is in a pure form, with no covalently modified variants present in substantial quantity, and that the observations are made at equilibrium.

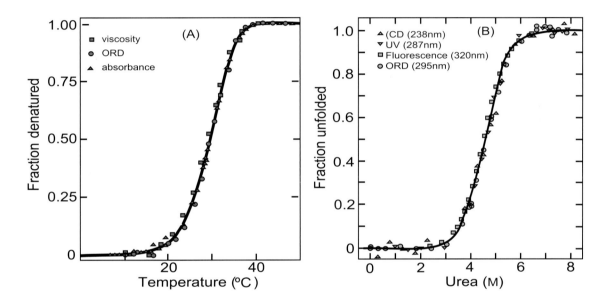

Figure 11-1. Typical denaturation of a small globular protein caused by heating (A) or the addition of the denaturant urea (B). (A) Thermal denaturation of ribonuclease A was followed by changes in the intrinsic viscosity, optical rotatory dispersion (ORD) at 365 nm and ultraviolet (UV) absorbance. (B) Urea denaturation of ribonuclease T_1 was measured by changes in the circular dichroism (CD) at 238 nm, UV absorbance at 287 nm, ORD at 295 nm and fluorescence emission at 320 nm. Data from A. Ginsburg & W. R. Carroll (1965) *Biochemistry* **4**, 2159, and J. A. Thomson *et al.* (1989) *J. Biol. Chem.* **264**, 11614.

The unfolding and refolding of proteins can be followed experimentally by any technique that is sensitive to the conformational properties of the polypeptide chain (Figure 11-1). They are too numerous to specify but include most of the techniques used with proteins.

11.1.A. Reversibility of Denaturation

The native conformation of a protein can often be restored by removing the denaturing conditions. The change of conformation of a protein upon denaturation, even complete unfolding of the polypeptide chain, can be a reversible process that is determined solely by the environmental conditions and the amino acid sequence of the protein. In other words, **a denatured protein can usually refold into its native conformation in a self-assembly process that is directed solely by its amino acid sequence**. The greatest exception to this is when the polypeptide chain has been chemically modified after its folding, such as the conversion of proinsulin to insulin by proteolytic cleavage of the precursor polypeptide chain after it has reached its folded conformation. Chemical modifications can also occur during denaturation. Some of the groups of proteins that are exposed upon denaturation are highly reactive, especially Cys thiol groups (Section 7.2.I), and they can easily be modified covalently: thiol groups can react with small molecules, with each other to form disulfide bonds, and with the groups of other molecules to form aggregates. In addition, exposure of the nonpolar groups of a protein upon unfolding decreases the protein's solubility in water (Section 10.1) and aggregation of the denatured protein is a common occurrence. Consequently, that protein denaturation is reversible is not always apparent, especially in the case of large proteins. **Proteins with multiple domains tend to interact intermolecularly under conditions where denaturation is not complete, to produce protein aggregates**, in place of the normal intramolecular association (Figure 11-2); this phenomenon is similar to the domain swapping that is observed with some native proteins (Section 9.1.H.4).

(A) Independent monomers (intermolecular)

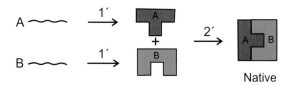

(B) Part of the same polypeptide chain (intramolecular)

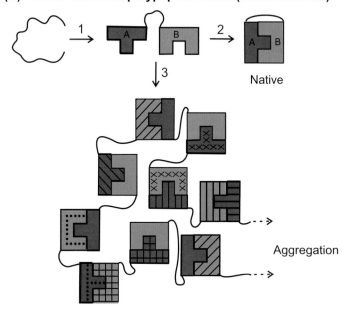

Figure 11-2. Schematic illustration of the possibility of intermolecular interactions between the folded domains of a two-domain protein. (A) The folding (step 1′) and assembly (step 2′) of the two domains when they are independent polypeptide chains. (B) Intermolecular interactions between the folded domains of a two-domain protein. The interactions are the same as those in (A) but they should occur intramolecularly in this case. Intermolecular interactions can happen at the midpoint of the unfolding transition, where about half the domains are unfolded, or if the association step 2 is slower than the rate of folding of the monomers, step 1. This is believed to be one reason why multi-domain proteins often precipitate upon refolding. It is similar to the phenomenon of domain swapping (Figure 9-30).

For denaturation to be reversible often requires special conditions: (1) low concentrations of protein to minimize the tendency of the unfolded protein molecules to aggregate; (2) increased electrostatic repulsions between the protein molecules with an appropriate pH and ionic strength of the solution; (3) avoidance of anything that might react irreversibly with the protein; and (4) the inclusion of co-solutes that can increase the solubility of the unfolded proteins, such as the denaturants urea or guanidinium chloride (GdmCl) (Section 11.1.C). Unfolding tends to become increasingly irreversible at higher temperatures, especially above 70°C, and with prolonged incubation, especially at alkaline pH, due to temperature-induced chemical changes to the protein, for example rearrangements of disulfide bonds (Section 7.2.I.6) and deamidation of Asn and Gln residues (Section 7.2.H).

11.1.B. Cooperativity of Unfolding

In the case of small globular proteins, usually containing no more than 200 amino acid residues, all parameters sensitive to the protein structure usually change abruptly and simultaneously at some temperature, pH, pressure or denaturant concentration (Figure 11-1). In such cases, **denaturation appears to be a single-step process in which the entire protein undergoes a single cooperative transition from the unique native state, N, to the denatured unfolded state, D**. No intermediate partly folded states are present to any significant extent, so denaturation of such proteins is approximated at equilibrium by a two-state transition:

$$N \leftrightarrow D \tag{11.1}$$

In other words, the native structure breaks down and refolds in an all-or-none manner (Figure 11-3). **Any partly folded intermediate states are unstable and populated at only very low levels at equilibrium**. All the amino acid residues of such proteins change their conformation simultaneously and cooperatively. Such proteins usually consist of a single structural domain (Section 9.1.A).

The Gibbs free energy difference ($\Delta G°$) between N and D depends on all the parameters that specify the external conditions (e.g. temperature, pressure and concentration of denaturant). One can measure $\Delta G°$ by thermodynamic analysis of the process of protein denaturation upon varying one of the parameters, keeping the others fixed. Its value can be determined quantitatively, however, only if the unfolding process is complete, reversible and highly cooperative, close to a two-state transition. In this case, the relative concentrations of the N and D states determine the equilibrium constant for the reaction, K_{eq}, which gives the value of $\Delta G°$ in the usual way ($\Delta G° = -RT \log_e K_{eq}$). Under conditions where both N and D are populated significantly, the value of K_{eq} can be determined directly using any observable parameter sensitive to the state of the protein, $Y(x)$, while varying parameter x of the environment:

$$\Delta G°(x) = -RT \log_e K_{eq} = -RT \log_e [\frac{Y(x) - Y_n}{Y_d - Y(x)}] \tag{11.2}$$

where Y_n and Y_d are values of the parameter characteristic of the pure native and pure denatured states, respectively.

In contrast, **multi-stage denaturation is normal for large proteins and frequently reflects the presence of multiple domains**. In some cases, the domains are relatively independent and can be excised from the protein and retain their folded structure and denaturation properties (Figure 11-4). The various domains of a protein usually have different stabilities, so they tend to unfold under different denaturing conditions; the denaturation profiles of proteins in which the domains unfold independently can be very complex. In other cases, however, the domains interact to varying extents in the folded protein, usually stabilizing each other, so individually they are less stable and denature more readily. The unfolding transition will then be dominated by the initial breakdown of the multi-domain structure, and individual folded domains will not be stable under these conditions, so a single unfolding transition will be observed. A single unfolding transition also might result from individual independent domains denaturing similarly.

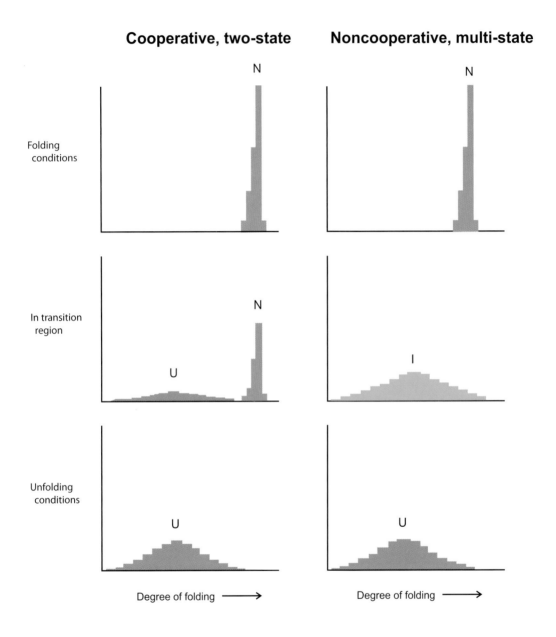

Figure 11-3. Illustration of a cooperative two-state folding transition (*left*), as normally observed with single-domain proteins, compared with a noncooperative multi-state transition (*right*). Histograms of the numbers of molecules (*vertical axis*) with different degrees of folding (*horizontal axis*) are illustrated for conditions favoring folding (*top*) and unfolding (*bottom*) and the transition region (*middle*). The folded state, N, is represented as a narrow distribution of folded conformations, the unfolded state, U, as a broad distribution of less compact conformations. Within the transition region, a two-state transition will have two distinct populations in equilibrium, similar to those at the two limiting conditions of folding and unfolding. In contrast, a noncooperative transition will have the distribution of the entire population shifted so that most molecules have intermediate partially folded conformations (I) within the transition region.

A severe test of the cooperativity of the denaturation process consists of comparing the effective thermodynamic parameters derived from the equilibrium analysis with the real thermodynamic parameters measured by direct experimental methods. The thermal unfolding of a protein can be measured in a differential scanning calorimeter, which produces fundamental information about the thermodynamics of the unfolding/refolding transition at equilibrium. Calorimetric observations of the effects of heating solutions of a typical small globular protein at three different pH values are illustrated in Figure 11-5. The quantity measured by the calorimeter is the energy required to raise the temperature of the entire solution, which is its heat capacity at constant pressure, C_p. The contribution of the protein is determined by subtracting the corresponding measurements of just the aqueous solvent, which gives the **partial C_p** of the protein. The partial C_p is not just that of the protein but includes any effects it has on the surrounding solvent. The partial C_p of the folded protein initially changes only very slightly as the temperature is increased but unfolding is apparent by a large absorption of heat, giving a peak in the heat capacity curve (Figure 11-5). When unfolding is complete, at higher temperatures, the partial C_p of the protein becomes more constant and it is usually greater than its original value.

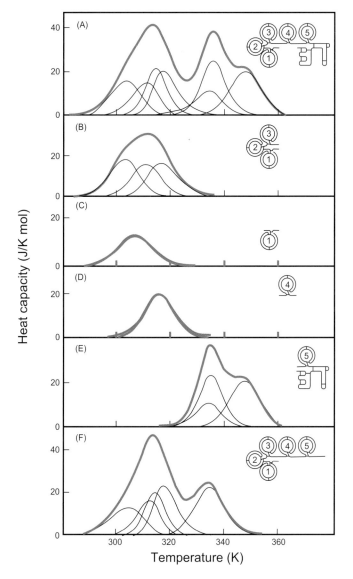

Figure 11-4. Calorimetric characterization of the thermal unfolding of plasminogen and its proteolytically liberated domains. The *thin lines* show the deconvolution of the excess heat absorption profiles into individual two-state transitions for the various domains of each of the molecules. The *inserts* show schematically the domain structures of the protein molecule used; complete plasminogen is at the *top*. Starting from the N-terminus, there are five homologous regions, labeled 1 to 5, which are known as kringle domains from their disulfide bond connectivities (Section 9.1.D.9). At the other end is a serine proteinase domain that appears to unfold as two structural domains. The first four kringle domains unfold independently, but the stabilities of the fifth kringle and of the serine proteinase domain are mutually dependent, indicating that they interact and stabilize each other. Data from V. V. Novokhatny et al. (1984) *J. Mol. Biol.* **179**, 215–232.

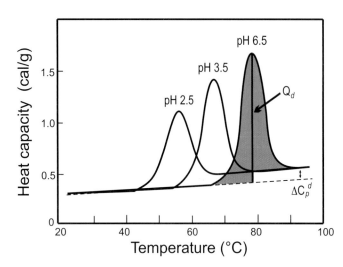

Figure 11-5. Typical unfolding of a small globular protein as a function of temperature measured calorimetrically at three different pH values. The parameter measured is the heat capacity at constant pressure (C_p) of the sample. The simplicity of each curve, with a single peak, and its shape are consistent with the protein unfolding reversibly in a single-state cooperative transition. The *shaded region* denotes the heat of the transition Q_d at a temperature of 78°C, the midpoint of the unfolding transition at pH 6.5. The difference in heat capacities of the folded and unfolded states is indicated by the *double-headed arrow* at the far right.

If a reversible thermal unfolding transition is truly two-state, the enthalpy change measured calorimetrically should be the same as the effective or van't Hoff enthalpy calculated from the sharpness of the transition, from the temperature-dependence of the equilibrium constant in a van't Hoff plot. The results with single-domain proteins generally show good correspondence between the two measurements of enthalpy, with ratios of about 1.05 confirming that these proteins unfold cooperatively, with few or no partially folded species present at significant levels at equilibrium (Figure 11-3).

Nuclear magnetic resonance (NMR) is a powerful method for monitoring the conformations of proteins. Using the chemical shifts of individual proton resonances, the states of individual amino acid residues of a protein and their rate of hydrogen exchange (Section 10.3) can be monitored as the environmental conditions are changed. Such analysis confirms that small proteins unfold cooperatively, although some reveal indications of subdomain organization, where a part of the polypeptide chain remains at least partially folded under some conditions.

The cooperativity of unfolding demonstrates that the folded conformation is stable only if all its stabilizing interactions are present simultaneously. Disrupting any part of the structure requires, and causes, unfolding of the entire polypeptide chain.

The linkage between protein folding and functional cooperativity: two sides of the same coin? I. Luque *et al.* (2002) *Ann. Rev. Biophys. Biomol. Structure* **31**, 235–256.

Cooperativity principles in protein folding. H. S. Chan *et al.* (2004) *Methods Enzymol.* **380**, 350–379.

Protein folding cooperativity: basic insights from minimalist models. A. F. de Araujo (2005) *Protein Peptide Lett.* **12**, 223–228.

11.1.C. Denaturants

Denaturants are various co-solutes that, when added to aqueous solutions, denature proteins. The most commonly used denaturants are urea and GdmCl:

$$\underset{\text{Urea}}{H_2N-\overset{\overset{O}{\|}}{C}-NH_2} \quad \underset{\text{Guanidinium ion}}{H_2N-\overset{\overset{NH_2}{|}}{C}=NH_2^+} \tag{11.3}$$

Most proteins will denature in concentrated solutions (e.g. up to 8 M) of these compounds and remain soluble. The dependency of the protein conformation on the concentration of these denaturants is the simplest and most popular method of quantitatively measuring protein stability. It is usually also the most reversible type of denaturation, probably because the high concentrations of the denaturant tend to keep the unfolded protein soluble. Furthermore, the relatively large changes in the hydrodynamic and optical properties of proteins in concentrated solutions of these denaturants suggest that they cause the most complete unfolding of polypeptide chains (Section 11.2).

Unfolding of small single-domain proteins usually occurs very sharply, over a small change in the denaturant concentration, and usually is a cooperative two-state transition (Figure 11-1). In this case, the relative free energies of the folded and unfolded states can be measured throughout the transition region (Equation 11.2). Although there is no obvious reason why it should, the apparent value of $\Delta G°$ usually varies linearly with the denaturant concentration, [D]:

$$\Delta G^{app} = \Delta G° - m\,[D] \tag{11.4}$$

where $\Delta G°$ is the value of ΔG^{app} at zero denaturant concentration, which is the best measure of the stability of the folded conformation. The dependence of ΔG^{app} on the denaturant concentration is given by the parameter m. Its value is usually proportional to the amount of protein surface exposed to solvent upon unfolding.

Urea is a destabilizing agent only while GdmCl has a dual role: it destabilizes the protein native state at high concentrations but it stabilizes it at low concentrations. The reason is that GdmCl is a salt, in contrast to urea, so an increase in its concentration also increases the ionic strength of the solution, which reduces the electrostatic interactions within proteins that destabilize them. If the ionic strength of the solution is kept constant by addition of neutral salts, while varying the GdmCl concentration, the stability of the protein is found to be directly proportional to the concentration of GdmCl over an extended range of concentrations. The stability of the protein in the absence of denaturants can then be estimated using Equation 11.4 and it is usually found to be in good agreement with the estimates of $\Delta G°$ measured in other ways (Section 11.3.A).

The mechanism of action of these denaturants is not fully understood but two mechanisms appear to predominate. First, **denaturants increase the aqueous solubility of small nonpolar molecules and the nonpolar side-chains that form the compact cores of globular proteins** (Figure 11-6). Consequently, they decrease the magnitude of the hydrophobic effect by up to one-third. Second, these denaturants are highly polar, with **multiple groups capable of hydrogen bonding** (Equation 11.3). They appear to disrupt the structural properties of liquid water, which may account for their effect on the hydrophobic interaction, and they also **form hydrogen bonds with polar groups of proteins, particularly the peptide groups**, increasing the probability of their exposure. To have these effects, the hydrogen bonds involving the denaturant would have to be stronger than those formed with water. A strong interaction of urea with polar groups is apparent from its influence on

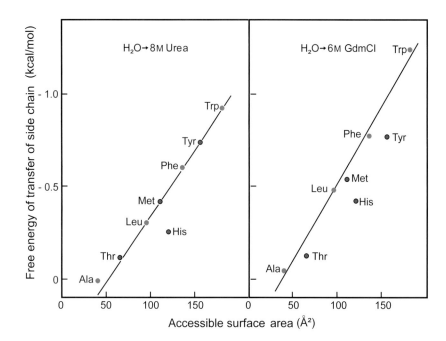

Figure 11-6. Correlation between the accessible surface areas of the amino acid side-chains and their free energy of transfer from water to either 8 M urea or 6 M GdmCl. The *solid lines* have slopes of 7.1 and 8.3 cal/mol/Å², respectively, but note that they do not pass through the origin. Residues indicated by *blue circles* have polar groups on their side-chains. Data from T. E. Creighton (1979) *J. Mol. Biol.* **129**, 235–264.

the solubilities of model compounds. Calorimetry shows that urea and GdmCl interact with proteins in both the folded and unfolded states, with a strong heat effect that depends on the temperature and is proportional to the exposed surface area. This effect may be sufficient to explain their denaturing properties. Increasing the temperature causes the bound denaturants to dissociate from the protein, with a gradual absorption of heat that causes an increase in the apparent heat capacity of the protein.

Any co-solvent, L, that binds more to one reactant than another will shift the equilibrium between the reactants (A ↔ B) towards the one to which it binds most. If L binds more tightly to A than to B, the reaction can be written as:

$$A \leftrightarrow B + \Delta N_L \, L \tag{11.5}$$

This indicates that the difference in the number of binding sites on the two reactants, ΔN_L, can be inferred from the effect of the co-solvent on the apparent equilibrium constant between A and B, K^{app}:

$$\Delta N_L = \frac{\partial \log_e K^{app}}{\partial [L]} \tag{11.6}$$

The partial derivative is used because all other parameters must be kept constant. The apparent number of binding sites and the binding constant can be obtained by analyzing calorimetric data

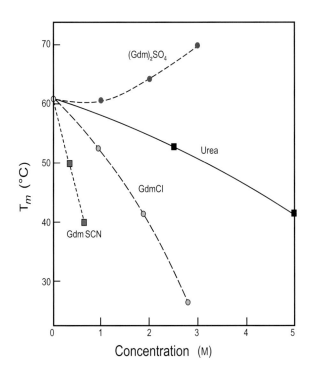

Figure 11-7. Thermal stability of ribonuclease A as a function of the concentration of urea and various guanidinium (Gdm$^+$) salts. The stability is expressed as the temperature at the midpoint of the thermal unfolding transition, T_m. Data from P. H. Von Hippel & K.-Y. Wong (1965) *J. Biol. Chem.* **240**, 3909–3923.

using the standard model for multi-site ligand binding, but it must take into account the competition between water and high concentrations of the solvent for the protein surface. Data on the effects of denaturants indicate that the overall interaction of denaturants with proteins is a consequence of the strong favorable interactions of denaturants with the proteins at exchangeable sites, which is compensated for by unfavorable interactions that are almost as strong at sites on the protein surface at which the denaturants cannot displace water.

Specific interactions of urea with polar groups on the surfaces of proteins have been observed by X-ray crystallography. The numbers of binding sites for urea and GdmCl observed with two proteins are close to the numbers estimated from calorimetric data. Binding of the denaturant significantly decreases the crystallographic temperature factor of the protein, i.e. the mobility of its native state. The reason is that each molecule of urea or GdmCl interacts simultaneously with several groups of the protein by forming multiple hydrogen bonds.

The denaturation potencies of guanidinium salts are affected by the nature of their anions, according to the Hofmeister series. Guanidinium thiocyanate is more potent than the chloride, whereas guanidinium sulfate actually stabilizes proteins (Figure 11-7). **The stabilizing effect of the sulfate ion overcomes the destabilizing effect of the guanidinium ion**. Other salts affect the stability of folded proteins according to their ranking in the Hofmeister series (Figure 11-8).

Preferential interactions of urea with lysozyme and their linkage to protein denaturation. S. N. Timasheff & G. Xie (2003) *Biophys. Chem.* **105**, 421–448.

Competitive model on denaturant-mediated protein unfolding. R. Murugan (2003) *Biophys. J.* **84**, 770–774.

Thermodynamics of denaturant-induced unfolding of a protein that exhibits variable two-state denaturation. A. C. Ferreon & D. W. Bolen (2004) *Biochemistry* **43**, 13357–13369.

Molecular crowding effects on protein stability. F. Despa *et al.* (2005) *Ann. NY Acad. Sci.* **1066**, 54–66.

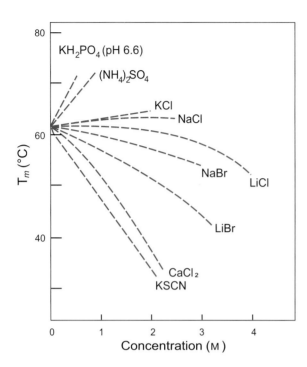

Figure 11-8. The effects of salts of the Hofmeister series on the thermal stability of ribonuclease A. Data from P. H. Von Hippel & K.-Y. Wong (1965) *J. Biol. Chem.* **240**, 3909–3923.

11.1.D. Heat Denaturation

Increased thermal motion at higher temperatures can cause disruption of the native conformation of a protein. In the case of small globular proteins, breakdown of the native structure takes place over a relatively narrow temperature range and appears to be a cooperative two-state transition (Figure 11-1). If unfolding is induced by increased temperature, it must be associated with an increase in the enthalpy and absorption of heat. The thermodynamics of heat denaturation can be measured directly by scanning microcalorimetry as an intense peak of heat absorption (Figure 11-5).

For a two-state transition, there are equal concentrations of the folded and unfolded states at the melting temperature, T_m, the equilibrium constant has the value of unity and the free energy change is zero:

$$-R\, T_m \log_e K_{eq} = \Delta G(T_m) = \Delta H(T_m) - T_m \Delta S(T_m) = 0 \tag{11.7}$$

Under these conditions the entropy of the transition is given by:

$$\Delta S(T_m) = \Delta H(T_m)/T_m \tag{11.8}$$

A notable feature of the thermal denaturation of proteins is that **the partial heat capacity of the denatured protein is 25–50% greater than that of the native state**. The increase in the heat capacity depends upon the protein and the temperature of unfolding, decreasing at higher temperatures; this decrease in the denaturation heat capacity could be due to several reasons: (1) any residual structure in the unfolded protein might melt gradually with continued increase in temperature; (2) the conformational freedom of the polypeptide chain might increase after disruption of the rigid

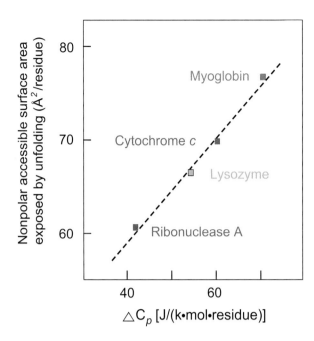

Figure 11-9. Relationship between the measured change in heat capacity per residue upon unfolding of several proteins and the nonpolar surface area buried within the interior of the protein and assumed to be exposed to solvent upon unfolding. Note that the relationship is not one of direct proportionality, in that it does not extrapolate to the origin. Data from P. L. Privalov & G. I. Makhatadze (1990) *J. Mol. Biol.* **213**, 385–391.

compact native structure; or (3) groups that are buried in the folded protein become exposed to water upon unfolding. The first suggested reason does not seem to be significant because the heat capacity change observed upon thermal denaturation is almost the same as with denaturants, so long as the effects of denaturant solvation are taken into account properly, and proteins denatured by urea and GdmCl do not generally have residual cooperative structures (Section 11.2). The contribution of the increased conformational freedom is believed not to be large. Consequently, **the higher partial heat capacity of the protein denatured state is believed to be due primarily to more groups, especially the nonpolar groups, being exposed to water**.

Transfer of nonpolar groups into water results in an increase in the heat capacity, while exposure of polar groups has the opposite effect, and aromatic groups are intermediate. These three classes of atoms contribute to the heat capacity in proportion to their water-accessible surface areas exposed upon unfolding: ΔA_{np}, ΔA_{arom} and ΔA_{pol}. The change in heat capacity upon protein unfolding can be calculated using the equation:

$$C_p^{hyd}(25°C) = -(2.14\,\Delta A_{np} + 1.55\,\Delta A_{arom} - 1.27\,\Delta A_{pol})\ \text{Å}^{-2}\text{JK}^{-1}\text{mol}^{-1} \tag{11.9}$$

The coefficients in this equation depend on the temperature, and those given are for 25°C. The surface area exposed upon unfolding of the nonpolar and aromatic groups is greater than that of the polar groups, so unfolding of a protein produces an overall increase of the partial heat capacity. **The magnitude of the increase in heat capacity is roughly proportional to the nonpolar surface area that is buried in the native conformation** and believed to be exposed to solvent upon unfolding (Figure 11-9).

A most important consequence of the change in the heat capacity is that **the enthalpy and entropy of denaturation depend upon the temperature**. Indeed, since:

$$\frac{\partial \Delta H}{\partial T} = \Delta C_p \quad \text{and} \quad \frac{\partial \Delta S}{\partial T} = \frac{\Delta C_p}{T} \qquad (11.10)$$

the enthalpy and entropy are given by:

$$\Delta H(T)_{pH} = \Delta H(T_t)_{pH} + \int_{T_t}^{T} \Delta C_p(T) dT \qquad (11.11)$$

$$\Delta S(T)_{pH} = \frac{\Delta H(T_t)_{pH}}{T_t} + \int_{T_t}^{T} \frac{\Delta C_p(T)}{T} dT \qquad (11.12)$$

Since ΔC_p is positive and decreases with increasing temperature, the enthalpy and entropy changes upon protein denaturation increase asymptotically to some constant level at very high temperatures.

These temperature-dependent changes in the enthalpy and entropy of denaturation cause the thermodynamics of the unfolding transition to be complex and can produce denaturation at low temperatures (Section 11.1.E) as well as high.

Relationships between the temperature dependence of solvent denaturation and the denaturant dependence of protein stability curves. M. E. Zweifel & D. Barrick (2002) *Biophys. Chem.* **101**, 221–237.

Heat capacity in proteins. N. V. Prabhu & K. A. Sharp (2005) *Ann. Rev. Phys. Chem.* **56**, 521–548.

Using circular dichroism collected as a function of temperature to determine the thermodynamics of protein unfolding and binding interactions. N. J. Greenfield (2006) *Nature Protoc.* **1**, 2527–2535.

11.1.E. Cold Denaturation

Some enzymes in aqueous solution, in the absence of urea and GdmCl, are inactivated at low temperatures (about 0°C) but their activities are restored at room temperature. Most of these proteins are oligomeric, and their cold inactivation is probably due to their dissociation into subunits. The quaternary structure of proteins is usually less stable than the tertiary, and its stability generally decreases at lower temperatures. But **even small model proteins can be observed to unfold reversibly at low temperatures**. Low-temperature unfolding has the opposite thermodynamic characteristics of that at high-temperatures, in that heat is released upon unfolding at low temperatures but is taken up upon unfolding at high temperatures (Figure 11-10). The heat capacity change is the same, however, and the unfolded states are indistinguishable thermodynamically. On the other hand, one cold-denatured state that has been characterized physically appears to be relatively compact, although its dimensions increase at decreasing temperatures.

The large ΔC_p of protein unfolding is responsible for cold denaturation. It causes there to be a temperature at which stability of the folded state is at a maximum (Figure 11-11) and the net stability decreases at both higher and lower temperatures. All such proteins with a large ΔC_p of unfolding would be expected to unfold at both high and low temperatures, but unfolding at low temperatures is observed only under circumstances where it occurs within an accessible temperature range, above the freezing point of water (Figure 11-12). **The temperature of cold denaturation is highest for proteins with the greatest change in heat capacity upon unfolding**, which are those that expose the

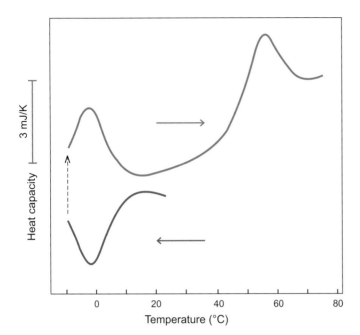

Figure 11-10. Unfolding of apomyoglobin at both high and low temperatures measured calorimetrically. In the *lower trace*, folded apomyoglobin at room temperature was cooled to −10°C. The trough in the heat capacity is caused by the release of heat upon unfolding at −6°C. The cooled solution was then warmed, to produce the *upper trace*. The peak at −6°C corresponds to the uptake of heat as the apomyoglobin refolds. This is followed by a second peak of heat uptake as the protein unfolds above 50°C. Data from Y. Griko *et al.* (1988) *J. Mol. Biol.* **202**, 127–138.

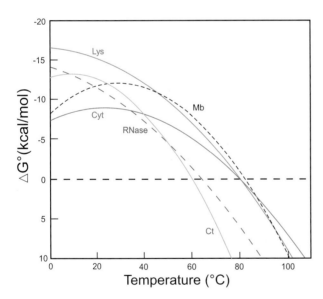

Figure 11-11. Temperature dependence of the difference in free energy between the folded and unfolded states of several proteins, expressed per mole of protein. Lys, hen lysozyme; RNase, ribonuclease A; Mb, metmyoglobin; Ct, α-chymotrypsin; Cyt, cytochrome *c*. The pH of each solution was that at which the protein is most stable. Data from P. L. Privalov & N. N. Khechinashvili (1974) *J. Mol. Biol.* **86**, 665–684.

greatest amount of nonpolar surface. The temperature of cold denaturation can be raised significantly by an appropriate choice of pH, ionic strength, buffer and presence of denaturants. Only in a few such instances does cold unfolding occur at temperatures above 0°C.

The opposite values of ΔH and ΔS upon unfolding at high and low temperatures are simply a consequence of the large heat capacity change. Physically, **cold unfolding can be thought of as being a result of the increasing propensity at low temperatures of ordered water to solvate the nonpolar surface areas of the protein that are exposed upon unfolding**, which diminishes the magnitude of the hydrophobic effect at low temperatures.

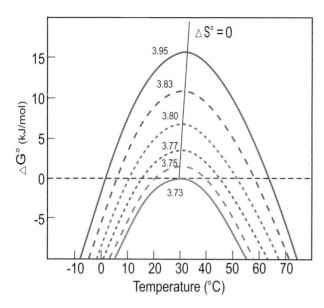

Figure 11-12. The Gibbs free energy difference between the denatured and native states of metmyoglobin in acetate buffer solution at the pH values indicated on the curves. Under these conditions, the protein cold denatures within the temperature range 0–30°C. Each curve is at a maximum when the entropy of unfolding is zero. Adapted from P. L. Privalov *et al.* (1986) *J. Mol. Biol.* **190**, 487–498.

Direct access to the cooperative substructure of proteins and the protein ensemble via cold denaturation. C. R. Babu *et al.* (2004) *Nature Struct. Biol.* **11**, 352–257.

Combined NMR-observation of cold denaturation in supercooled water and heat denaturation enables accurate measurement of ΔC_p of protein unfolding. T. Szyperski *et al.* (2006) *Eur. Biophys. J.* **35**, 363–366.

The cold denatured state is compact but expands at low temperatures: hydrodynamic properties of the cold denatured state of the C-terminal domain of L9. Y. Li *et al.* (2007) *J. Mol. Biol.* **368**, 256–262.

11.1.F. pH Denaturation

Many proteins denature at pH values less than about 5 or greater than 10. In the case of small single-domain proteins, reversible unfolding usually takes place over an extremely narrow range of pH. In the case illustrated in Figure 11-13, **staphylococcal nuclease changes from being totally folded to being totally unfolded if the pH is changed by only 0.3 pH unit**. Such an abrupt transition cannot be caused by the titration of a single ionizable group, which should require 2 pH units to go from 9% ionization to 91%. The abruptness of the unfolding transition is indicative of a very cooperative transition involving multiple ionizing groups. Denaturation induced by pH is accompanied by either the release or uptake of protons. Titration of protein groups having approximately normal pK_a values produces minimal and gradual changes in their properties, while the gross conformational changes associated with pH denaturation are accompanied primarily by the simultaneous unmasking of several buried groups with abnormal pK_a values.

Unfolding at extremes of pH usually occurs because the folded protein has several groups buried in nonionized form that can ionize only after unfolding. Most prevalent are His and Tyr residues, which can cause unfolding at acid and alkaline pH values, respectively. The simultaneous ionization of several such groups after unfolding accounts for the cooperativity of the unfolding transition (Figure 11-13). The general electrostatic repulsion between the ionized groups on the surface of a protein might also tend to cause unfolding when the protein has a substantial net charge, because such repulsions would be diminished in the unfolded state. Although proteins tend to be most stable near their isoelectric point, **there is little evidence for repulsions between groups with the same**

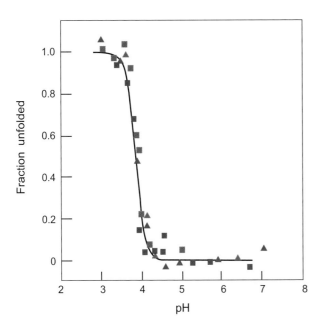

Figure 11-13. The acid-induced unfolding of staphylococcal nuclease A, measured by viscosity (*squares*) and CD at 220 nm (*triangles*). The *red* symbols were measurements made during acidification, the *blue* symbols upon raising the pH. Data from C. B. Anfinsen (1972) *Biochem. J.* **128**, 737–749.

electrostatic charge being sufficient to produce unfolding: in particular, reversing the net charge on basic groups on the surface of a protein often has little effect on the stability of the protein.

Specific salt bridges between ionizing groups can provide substantial stabilizing contributions to the folded state in some cases and can be disrupted at extremes of pH, where one of the groups is no longer ionized. The stability of this type of interaction can be estimated from the extent to which the pK_a values of the two groups are perturbed. An exceptional example is the salt bridge formed between the side-chains of Asp70 and His31 in T4 lysozyme, which perturbs the apparent pK_a values of these two residues from their normal respective values of 3.5–4.0 and 6.8 in the unfolded state to 0.5 and 9.1, respectively, in the folded state where the salt bridge exists. These pK_a values imply that the salt bridge is stable by 3–5 kcal/mol, and the salt bridge must contribute the same to the stability of the folded state, because ionization and conformational stability are linked functions (Figure 11-21). The greater the influence of a stabilizing salt bridge, the less susceptible it is to disruption at extremes of pH.

The unfolding induced at extremes of pH depends on temperature and, therefore, because they are linked functions (see Figure 11-21), **temperature-induced denaturation must depend on the pH**. Because the temperature-induced denaturation of a small single-domain protein appears to be a cooperative two-state transition, the transition induced by variation of pH at a fixed temperature should also be two-state, and the initial and final characteristics of the protein should be independent of the sequence of pH or temperature variation. Consequently, the Gibbs free energy difference between the native and denatured states can be determined from the pH-dependence of the protonation difference between these states, $\Delta v(pH)$:

$$\Delta G°(pH) = 2.3\,RT \int_{pH_t}^{pH} \Delta v(pH)\,dpH \qquad (11.13)$$

where pH_t is the transition pH at which half of the molecules are in the denatured state. The Gibbs free energy values of pH- and temperature-induced denaturation are in very good agreement with small model proteins; in these cases, **pH-induced unfolding can also be regarded as a two-state transition**.

The pH denaturation of proteins is associated with significant heat effects that depend on the temperature (Figure 11-14). After correction for the heats of protonation, the enthalpy of pH-induced protein transitions appears to be identical to the enthalpy of the temperature-induced transition and depends in the same way on temperature. Thus the pH-induced denaturation of protein results in a heat capacity increase that is similar to that observed by temperature induced-denaturation.

The origin of pH-dependent changes in *m*-values for the denaturant-induced unfolding of proteins. S. T. Whitten *et al.* (2001) *J. Mol. Biol.* **309**, 1165–1175.

Molecular mechanisms of pH-driven conformational transitions of proteins: insights from continuum electrostatics calculations of acid unfolding. C. A. Fitch *et al.* (2006) *Proteins* **63**, 113–126.

The pH-dependent unfolding mechanism of P2 myelin protein: an experimental and computational study. E. Polverini *et al.* (2006) *J. Struct. Biol.* **153**, 253–263.

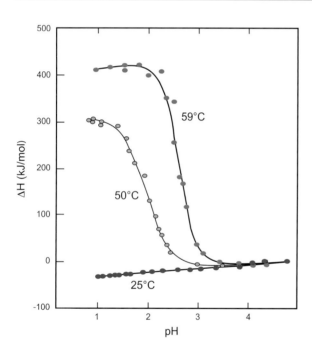

Figure 11-14. Enthalpy change of lysozyme upon acid-induced unfolding. The native protein was transferred from pH 4.8 to the indicated acidic pH values, at constant temperature. At 298 K (25°C) the protein remains folded at all pH values, but it unfolds at low pH at the higher temperatures. Data from W. Pfeil & P. L. Privalov (1976) *Biophys. Chem.* **4**, 23–32.

11.1.G. Denaturation by High Pressures

Proteins are not very sensitive to pressure, being very incompressible (Section 10.7), and only at extremely high pressures (usually above 2 kbar) do they appear to unfold. The pressure-induced denaturation of proteins takes place over a relatively narrow pressure range, depending on the temperature and solvent conditions, particularly the pH (Figure 11-15). At the same time, the temperature and pH at which denaturation takes place also depend on the pressure. These three parameters are interdependent, and varying any one of them at fixed values of the others can produce unfolding of the native protein structure. Pressure-induced denaturation is also believed to be a cooperative two-state transition, although there are observations to the contrary and pressure-denatured proteins seem to retain more residual structure than when denatured by heat, pH or denaturants. In certain multi-domain proteins, the individual domains seem to undergo independent structural changes at specific well-defined pressures. In the case of oligomeric proteins, the first effect of high pressure is to dissociate them into subunits, followed by their unfolding.

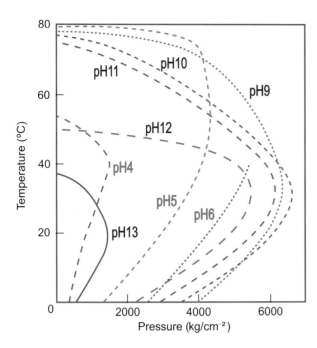

Figure 11-15. Pressure-denaturation of myoglobin at various temperatures and pH values. Isobars of its half-conversion to the denatured state are shown in the pressure–temperature plane at various pH values. At low pressures, to the left of each curve, the protein is folded, whereas it is predominantly unfolded at higher pressures. The protein is more easily denatured by high pressure at pH values more acidic or more basic than neutrality. Data from A. Zipp & W. Kauzmann (1973) *Biochemistry* **12**, 4217–4228.

A shift in equilibrium caused by increased pressure should be associated with a decrease in volume. Assuming that it is a two-state transition, the effective change of molecular volume in any transition, ΔV, can be calculated from the effect of pressure on the equilibrium constant, K_{eq}:

$$\Delta V = -RT \frac{\partial \log_e K_{eq}}{\partial P} \tag{11.14}$$

where R is the gas constant, T the temperature and P the pressure. The inefficiency of pressure in inducing protein denaturation implies that the volume change upon denaturation, ΔV_d, is very small at room temperature. It is usually estimated to be between 10 and 100 ml/mol at room temperature, less than 0.5% of the total protein volume, and even smaller at higher temperatures. Exposure of nonpolar groups to water on unfolding should cause a significant decrease in volume upon unfolding at low pressures; with increasing pressure, the volume reduction should become less negative and change sign at about 2 kbar. This is not observed with protein denaturation, so the interior of a globular protein cannot be compared with an oil drop or a solid hydrocarbon. The difference is that, in contrast to liquid and solid hydrocarbons, the native protein interior is densely packed but also has some cavities and loose parts (Figure 9-21). Increasing hydrostatic pressure may then force water molecules into the protein interior, gradually filling cavities and eventually breaking up the protein structure.

Revisiting volume changes in pressure-induced protein unfolding. C. A. Royer (2002) *Biochim. Biophys. Acta* **1595**, 201–209.

Compressibility of protein transitions. N. Taulier & T. V. Chalikian (2002) *Biochim. Biophys. Acta* **1595**, 48–70.

Role of protein cavities on unfolding volume change and on internal dynamics under pressure. P. Cioni (2006) *Biophys. J.* **91**, 3390–3396.

Protein stability and dynamics in the pressure-temperature plane. F. Meersman *et al.* (2006) *Biochim. Biophys. Acta* **1764**, 346–354.

11.1.H. Breakage of Disulfide Bonds

Many proteins that have disulfide bonds between their Cys residues (Section 7.2.I.5) require them for stability of their folded conformation; **the protein will unfold upon simply breaking the disulfide bonds** by either reduction or oxidation. This type of unfolding upon reducing the disulfide bonds has been studied extensively as the reverse of disulfide-coupled protein refolding (Section 11.4.C). The effects of breaking the various disulfide bonds of a protein depend upon their position within the folded protein structure and their stabilities. Disulfide bonds on the surface of a protein are generally the weakest and most accessible to a reducing agent, so they tend to be reduced first. In many cases the folded conformation is still sufficiently stable to be retained, although weakened. Breaking further disulfide bonds within the interior of the protein requires large-scale disruption of the folded structure, if only transiently, which makes other disulfide bonds also accessible; consequently, the remaining disulfide bonds tend to be reduced nearly simultaneously, and the fully reduced and unfolded protein results. Consequently, **reducing the disulfide bonds that are necessary for the folded conformation is usually a cooperative process**.

One advantage of using the disulfide bonds is that the unfolding process can be studied with the same conditions under which the disulfide-bonded protein is stable, with no need for denaturants; the only difference is the presence of small amounts of a reducing agent, so no other factors can complicate interpretation of the results. **The stabilities of the various disulfide bonds of the protein at each stage of folding can be measured directly**, relative to the disulfide bond of a reference compound, by the equilibrium constant for interchange of the disulfide bonds of the protein and the reference (Equation 7.34).

Unfolding probably results from the loss of the covalent disulfide cross-links, plus replacement of the S atoms of the disulfide bond by more polar and larger thiol groups, which will tend to ionize above pH 8. Any covalent cross-link will stabilize any folded conformation in which it is stable, because the cross-link destabilizes the unfolded conformation by decreasing its conformational entropy (Section 1.3.B.1).

Disulphide bonds and protein stability. T. E. Creighton (1988) *BioEssays* 8, 57–63.

11.2. UNFOLDED PROTEINS

The folded conformations of proteins are known in great detail (Chapter 9) but the same cannot be said of the unfolded state that is produced by various types of denaturation. The completeness of unfolding of the polypeptide chain is one of the most debated subjects in protein science. The ideal fully unfolded polypeptide chain would be a **random coil**, which is best defined as a polymer in which the conformational properties of each residue are independent of those of other residues not close in the covalent structure (Section 1.3.A). Characterizing an unfolded polypeptide chain is difficult because it is rapidly interconverting between a very large number of different conformations, so the most powerful methods of protein structure determination are inapplicable, especially crystallography. The varying microenvironments of each nucleus in a folded conformation that are responsible for their widely varying chemical shifts in NMR disappear upon unfolding, and the amino acid residues revert to their intrinsic NMR patterns, where the residues of the various types are virtually indistinguishable (Figure 11-16). With other techniques, there is no definitive reference unfolded polypeptide chain to indicate what results would be expected for a random coil. In fact, the ideal random-coil state can probably not be achieved in practice, because of the diversity of chemical groups that make up a protein (Section 7.2). Extreme conditions can change the balance of forces maintaining the native protein structure and destroy specific interactions between clusters of protein groups, but they certainly cannot eliminate completely all the interactions between the great variety of groups present. In other words, there probably is no theta solvent for polypeptide chains containing 20 different amino acid residues (Section 1.3.B). Some local interactions between groups close in the primary structure are often detected to small extents in unfolded proteins. Spectroscopic observations have suggested that unfolded proteins tend to adopt a poly(Pro) II conformation (Section 8.4.D) but the strong propensity of Pro residues with *trans* peptide bonds to adopt this conformation results from local interactions between the pyrrolidine ring and backbone; it is not due to any long-range interactions, and the polypeptide backbone has a slight intrinsic preference for this conformation. In spite of uncertainty about the details, it is clear that **unfolded proteins have much greater molecular dimensions on average than the folded form of the protein**.

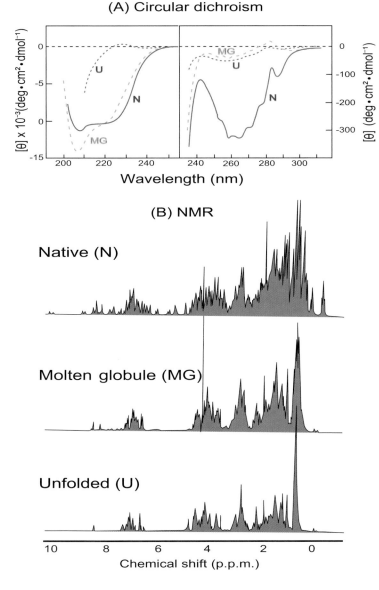

Figure 11-16. Spectral properties of the native, molten globule, and fully unfolded states of α-lactalbumin. (A) CD spectra at 4.5°C in the far-UV and near-UV regions of bovine α-lactalbumin in the native state at pH 7.0 (N), the molten globule at pH 2.0 (MG) and the fully unfolded state in 6 M GdmCl at pH 2.0 (U). Data from K. Kuwajima *et al.* (1985) *Biochemistry* **24**, 874–881. (B) ^{1}H NMR spectra at 52°C of guinea-pig α-lactalbumin in the following states: native at pH 5.4, molten globule at pH 2.0 and fully unfolded at pH 2.0 in 9 M urea. Data from J. Baum *et al.* (1989) *Biochemistry* **28**, 7–13.

It is generally believed that proteins are most unfolded in concentrated solutions of urea or GdmCl because these co-solutes increase significantly the intrinsic viscosities of proteins at room temperature and decrease the circular dichroism (CD) ellipticity at 220 nm almost to zero, as expected for a random coil. The viscosity is a measure of the hydrodynamic volume of the polypeptide chain, while the CD at 220 nm reflects the content of helical structure. The average dimensions of the large majority of chemically denatured proteins are effectively indistinguishable from those of a random-coil ensemble, in that they increase with the 0.598 power of the length of the polypeptide chain, very close to the 0.588 expected for an ideal random coil. The least unfolded seem to be thermally denatured proteins at high temperatures. The intrinsic viscosities of unfolded proteins at high temperatures are significantly smaller than in concentrated solutions of denaturants at room temperature, and thermally unfolded proteins show some residual CD signal. Furthermore, the observed temperature-dependence of the apparent equilibrium constant for unfolding has been thought to indicate that denaturation by denaturants produces a much larger change in heat capacity than denaturation by heating. The change in heat capacity is believed to be an indicator of the degree of exposure of nonpolar groups to water (Figure 11-9), so this observation could be considered evidence of a more complete unfolding of proteins by denaturants than by temperature. The heat capacity change could also have positive contributions from noncovalent interactions within the unfolded protein, because these interactions would be expected to change with temperature more in nonnative conformations than they do in the native state.

On the other hand, **the intrinsic viscosities of unfolded polypeptide chains decrease significantly with increasing temperature** (Figure 11-17). This appears to be the primary reason why the intrinsic viscosity of a heat-denatured protein measured at high temperatures is significantly lower than that in concentrated solutions of GdmCl or urea measured at room temperature. The CD ellipticity of proteins in concentrated solutions of denaturants also depends on temperature, and at high temperatures is similar to that of heat-denatured proteins in water. In addition, the increase in heat capacity observed upon protein denaturation by urea or GdmCl appears to be greater than that on thermal denaturation simply because denaturants interact strongly with proteins and this interaction is enthalpic. With increasing temperature, denaturant molecules tend to dissociate from proteins, absorbing heat, which increases the value of the apparent change in heat capacity. If this effect is taken into account, very similar changes in heat capacity are produced by denaturants and temperature. Increasing the temperature causes polypeptide chains in solutions of denaturants to approach the random-coil conformation because of the increase in thermal motion and the dissociation of bound denaturant molecules. Denaturants do not affect much the conformation of a polypeptide chain at high temperatures, and it approaches that of a heat-denatured protein in the absence of denaturants. Some nonspecific, random interactions between the groups of the polypeptide chain may persist, especially attractive hydrophobic interactions between nonpolar groups, which increase with increasing temperature and decrease the dimensions of the polypeptide chain, but they do not form specific structures.

The enthalpies of unfolding of small globular proteins by GdmCl, acid and temperature measured calorimetrically suggest that the enthalpy of unfolding is a universal function of temperature, so long as the heats of ionization of groups upon unfolding and of solvation by denaturants are taken into account. Any interactions between the various parts of denatured proteins appear not to be significant energetically, so long as the proteins do not aggregate. Therefore, all these observations

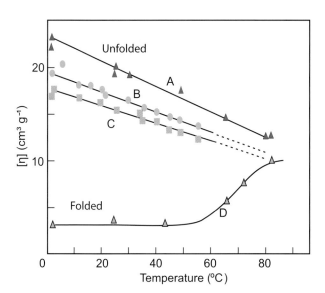

Figure 11-17. Temperature-dependence of the intrinsic viscosity of proteins that are folded at low temperatures and undergo thermal unfolding transitions (D) compared with those that are unfolded at all temperatures (A-C). (A) apomyoglobin and myoglobin at pH 3.0; (B) lysozyme without disulfide bonds at pH 2.2; (C) ribonuclease A without disulfide bonds at pH 2.2; (D) myoglobin at pH 5.0. Data from P. L. Privalov et al. (1989) *J. Mol. Biol.* **205**, 737–750.

suggest that **the denatured state of a small single-domain protein can usually be considered the reference unfolded state and to approach the ideal random-coil state**, although this conclusion is not universally accepted.

The situation is less clear with large polypeptide chains, those of proteins consisting of more than one structural unit or subdomain. In these cases, the denatured state might not always be fully unfolded but have one or more subdomains tending to preserve their native structure. A comprehensive study of the effects of amino acid replacements on the unfolding of staphylococcal nuclease suggested that some amino acid replacements were affecting the stability of the unfolded state; this would not be expected if the unfolded protein were a truly random coil, when changing one residue for another should affect only the stability of the folded state. It is now clear, however, that these observations result from the unfolding transition of this protein not being truly two-state; a partly folded intermediate state can also be populated and is affected by some of the mutations.

Polypeptide chains that have been unfolded simply by breaking the disulfide bonds that are required for stability of their folded conformation are the simplest to study, because there is no need to use denaturants and the folded and unfolded states can be studied under the same conditions. The bulk of these studies indicate that **the reduced forms of small model proteins, such as BPTI and ribonuclease A, approximate random coils**, although each amino acid residue probably has a somewhat different conformational tendency, and there are weak interactions apparent between groups no more than two residues apart in the primary structure, involving especially aromatic and nonpolar side-chains. The NMR spectra of reduced BPTI can be accounted for by the combined spectra of short peptides corresponding to the entire sequence, so there are no apparent interactions between groups distant in the primary structure. In spite of this, there is still much controversy regarding the compactness of these polypeptide chains, with some arguing that they are nearly as compact as the folded protein.

Many studies of protein folding and stability use proteins that contain covalent cross-links, especially disulfide bonds, in their folded structures. When these cross-links are not broken upon unfolding, remember that they persist in the unfolded protein, so it is then not fully unfolded but cross-linked, often extensively.

A few proteins are known as **natively unfolded proteins** because they are unfolded even under physiological conditions. They adopt folded conformations, and become functional, only after interacting with other molecules. In these cases, folding is coupled to binding of a ligand. The reasons are not clear, even though some of these proteins are very important physiologically; examples are the nucleoporins, which are part of the nuclear pores, and *tau*, which regulates the organization of neuronal microtubules and is also found in high concentrations in neurofibrillary tangles of Alzheimer's disease and other neurodegenerative disorders. Many natively unfolded proteins are involved in critical steps of the cell cycle and regulatory processes, and it is possible that adopting a folded structure only upon interacting with a ligand enables them to interact with multiple targets.

Conformational entropy of alanine versus glycine in protein denatured states. K. A. Scott *et al.* (2007) *Proc. Natl. Acad. Sci. USA* **104**, 2661–2666.

Characterization of denatured proteins using residual dipolar couplings. E. B. Gebel & D. Shortle (2007) *Methods Mol. Biol.* **350**, 39–48.

Highly populated turn conformations in natively unfolded tau protein identified from residual dipolar couplings and molecular simulation. M. D. Mukrasch *et al.* (2007) *J. Am. Chem. Soc.* **129**, 5235–5243.

Small-angle X-ray scattering of reduced ribonuclease A: effects of solution conditions and comparisons with a computational model of unfolded proteins. Y. Wang *et al.* (2008) *J. Mol. Biol.* **377**, 1576–1592.

11.2.A. Molten Globule

The molten globule state is a conformational state that is intermediate between the native and fully unfolded states of certain globular proteins. It is most frequently observed under mild denaturation conditions, at slightly acid pH, or as a transient intermediate kinetic species that is formed rapidly upon transfer of the unfolded state to refolding conditions. The characteristics of the molten globule state are (1) a native-like content of secondary structure; (2) absence of a normal tertiary structure produced by the tight packing of amino acid side-chains; (3) a compact overall shape of the protein molecule, with a radius only 10–30% greater than that of the native state; (4) a loosely packed hydrophobic core that is accessible to solvent; (5) an enhanced compressibility relative to the folded state, in contrast to fully unfolded states, which are less compressible. These observations suggest that **the molten globule is a compact globule with a 'molten' side-chain structure that is stabilized primarily by nonspecific hydrophobic interactions**. Little beyond this is certain.

The molten globule state is characterized experimentally by having a native-like circular dichroism peptide spectrum below 250 nm and an unfolded-like CD spectrum between 250 and 320 nm (Figure 11-16-A). The former suggests the presence of a native-like content of secondary structure, while the latter indicates a lack of specific packing of the aromatic side-chains. Some H atoms of the peptide backbone are protected from hydrogen exchange with the solvent (Section 10.3), perhaps due to the secondary structure, but are much less protected than in the native state. The NMR spectrum of the molten globule is closer to that of the unfolded protein, and confirms the absence of a specific tertiary structure in that there is little, if any, chemical shift dispersion in the spectrum (Figure 11-16-B). The individual resonances in the NMR spectrum are broader than those in the unfolded state, which may be due to conformational fluctuations in the molten globule state. Hydrophobic fluorescent dyes that bind to solvent-accessible hydrophobic surfaces of a protein molecule, such as 8-anilinonaphthalene-

1-sulfonate (ANS), bind much more strongly to a molten globule than to the fully folded or unfolded forms of the protein.

In the case of some proteins, such as carbonic anhydrase and α-lactalbumin, the molten globule state can be observed at equilibrium in intermediate concentrations of a strong denaturant, such as 2 M GdmCl. The acidic or alkaline unfolding transitions of these proteins also demonstrate the molten globule state. In contrast, many globular proteins undergo cooperative two-state unfolding transitions caused by denaturants or extremes of pH without the intermediate being apparent. The acidic or alkaline unfolding transitions of proteins such as cytochrome *c*, apomyoglobin and β-lactamase produce a more extensively unfolded state, and the addition of salt refolds the protein molecule from the unfolded to the molten globule state. The salt-induced refolding to the molten globule state is caused by counterion binding of the salt to the protein molecule, which eliminates the electrostatic repulsion between the charged groups. The molten globule state can also be produced by hydrostatic pressures, alcohols, removal of the bound metal ion from a metal-ion binding protein, as in the case of apo-α-lactalbumin produced by removal of the bound Ca^{2+}, and covalent modification. Whether or not the molten globule state is detectable as a stable intermediate is believed to depend upon its stability relative to that of the native and unfolded states.

The molten globule state can also be observed as a kinetic intermediate during the refolding of some proteins. In this case, the molten globule intermediate is in rapid equilibrium with the unfolded state, and the two are converted at the same rate to the fully folded conformation. This makes it impossible to determine unequivocally the kinetic role of the intermediate, whether or not it is on or off the folding pathway.

Whether the molten globule is a thermodynamic state, comparable to the fully unfolded and folded states, is an important but unanswered question. In some cases it appears to be, whereas in most cases it appears to be just a variant of the unfolded protein and in rapid equilibrium with it. Another unresolved question is how close the topology of the polypeptide chain of a molten globule approximates that in the native conformation. A native-like topology would seem necessary if one believes, as some do, the molten globule to be an important intermediate in protein folding. Indications of the presence of a few native-like contacts have been observed in some molten globules, but many of these molten globule proteins retain covalent aspects of the native structure, such as four native disulfide bonds or ligation to prosthetic groups. When permitted to interchange, the four disulfide bonds of molten globule α-lactalbumin do so spontaneously to many different disulfide pairings, which must have lower free energies, so the native disulfide bonds do not stabilize the molten globule conformation, and the native topology is not preferred in this molten globule. Moreover, the species with rearranged disulfide bonds tend to retain the molten globule conformation; the existence of molten globules with non-native disulfide bonds would seem to demonstrate that they cannot have native topologies.

Energetic basis of structural stability in the molten globule state: α-lactalbumin. Y. V. Griko (2000) *J. Mol. Biol.* **297**, 1259–1268.

Local and global cooperativity in the human α-lactalbumin molten globule. C. M. Quezada *et al.* (2004) *J. Mol. Biol.* **338**, 149–158.

Using nuclear magnetic resonance spectroscopy to study molten globule states of proteins. C. Redfield (2004) *Methods* **34**, 121–132.

11.2.B. Conformational Equilibria in Polypeptide Fragments

Any nonrandom conformation that is present in fragments of a polypeptide chain is also likely to be present in the unfolded protein unless the other parts of the intact protein actively interfere with it. A polypeptide chain will appear from its average properties to be a random coil simply if no conformation is populated by a substantial fraction of the molecules. Consequently, a seemingly random-coil polypeptide could have some particular conformation populated by, say, 10^{-2} of its molecules, even though this value would be expected in a truly random coil to be populated much less, perhaps 10^{-n}, where n is the number of amino acid residues. The actual value is important for understanding the energetics of polypeptide conformation. Such conformational equilibria are usually expressed by K_{conf}, the equilibrium constant between the particular folded conformation of interest, F, and all other conformations, U:

$$K_{conf} = \frac{[F]}{[U]} \tag{11.15}$$

Proteins that appear to be fully unfolded are unlikely to have any values of K_{conf} greater than about 10^{-2}, which would make it unlikely that the nonrandom conformation could be detected directly.

Fortunately, such small values of K_{conf} can be measured immunochemically, using the conformational specificity of antibodies against proteins. Antibodies generated against the native protein tend to be specific for the native conformation, at least for the unique arrangement of groups from several adjacent residues on its surface that is normally recognized by an antibody. Antibodies directed against unfolded proteins are quite different and recognize many linear segments of the polypeptide chain in very many different conformations. Because the unfolded conformation is constantly changing, the affinities of antibodies directed against unfolded proteins are much lower than the affinities of anti-native antibodies for the native conformation. **The cross-reaction between the two sets can be used to measure the probability that an unfolded polypeptide is in the native-like conformation, and therefore binds to antibodies against that conformation, or that a folded protein is unfolded transiently and binds to antibodies against the unfolded protein.**

Consider a polypeptide chain that possesses all the parts recognized by antibodies against the folded protein, N, but that is in this conformation only infrequently and in equilibrium with other conformations, for example U. **A reasonable assumption is that when this polypeptide is in the N-like conformation, it will bind to the anti-N antibodies with the same affinity, K_N, as the folded protein:**

$$U \underset{K_{conf}}{\rightleftharpoons} N \underset{K_N}{\overset{Ab}{\rightleftharpoons}} Ab\cdot N$$

$$K_N = \frac{[Ab\cdot N]}{[Ab][N]} \tag{11.16}$$

where Ab is an antibody-combining site. The observed affinity for the polypeptide involved in the conformational equilibrium will be given by:

$$K_{app} = \frac{[Ab \cdot N]}{([U]+[N])[Ab]} = \frac{[Ab \cdot N]}{\left(1 + \dfrac{1}{K_{conf}}\right)[N][Ab]} = \frac{K_N}{\left(1 + \dfrac{1}{K_{conf}}\right)} \quad (11.17)$$

Therefore, the affinity of the polypeptide for the anti-N antibodies will be lower by the factor $[1 + (1/K_{conf})]$. If K_{conf} is very small, this factor becomes $1/K_{conf}$ (Figure 11-18).

Values of K_{conf} measured with a few unfolded proteins or fragments are in the range of 10^{-3}–10^{-4}. These values are consistent with the generally unfolded state of the protein fragments but are considerably larger than might be expected for random occurrence of a unique conformation in a fully unfolded polypeptide chain.

This technique has not been widely used, probably because it is difficult to know exactly how much of the conformation the antibodies are recognizing, but the few data available are very valuable.

An immunological approach to the conformation equilibrium of staphylococcal nuclease. B. Furie *et al.* (1975) *J. Mol. Biol.* **92**, 497–506.

Understanding protein folding through peptide models. J. J. Osterhout (2005) *Protein Peptide Lett.* **12**, 159–164.

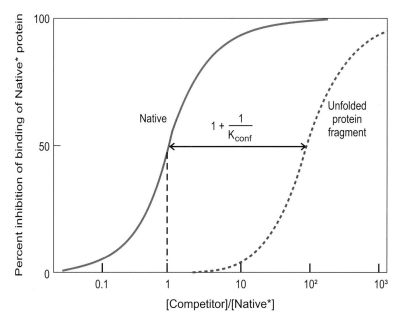

Figure 11-18. Immunochemical measurement of the equilibrium between U and N conformations in a protein fragment using antibodies that recognize only the N conformation. A constant amount of radioactive folded protein, designated Native*, is mixed with varying amounts of nonradioactive protein or fragment to be tested for their ability to bind to the antibody and compete with the labeled protein. This mixture is added to a limiting amount of antibodies recognizing the N conformation. Binding of radioactivity to the antibodies is measured. Binding of the unlabeled competing protein to the antibodies is reflected in an inhibition of binding of radioactive protein. The *solid curve* is that expected with competitor unlabeled protein in the folded conformation, indistinguishable from the radioactive protein; an equivalent amount of radioactive and nonradioactive proteins should produce 50% inhibition. The *dashed curve* is that expected for a polypeptide that is in the N conformation only 1% of the time, i.e. $K_{conf} = 10^{-2}$. In general, such a curve is offset to the right by the factor $(1 + 1/K_{conf})$. Adapted from T. E. Creighton (1993) *Proteins: structures and molecular properties*, 2nd edn, W. H. Freeman, NY, p. 315.

11.3. PROTEIN STABILITY

How do proteins manage to adopt and maintain a specific folded conformation, rather than the multitude of conformations that comprise the unfolded state? One approach is to measure the difference in Gibbs free energy between the unfolded (denatured) and native states of the protein as some condition (x) is varied to produce unfolding (Equation 11.2). The value of $\Delta G°(x)$ can be measured directly only in the denaturation transition zone, where the concentrations of both N and U can be measured, and its value under more pertinent physiological conditions requires extrapolation. Only in the cases of thermal and pH denaturation can extrapolation to standard conditions be done exactly, because the functional dependence of $\Delta G°$ values on temperature and pH can be determined experimentally (Equations 11.13 and 11.18). There is also the empirical observation that the value of $\Delta G°$ is linearly related to the concentrations of urea and GdmCl (Equation 11.4).

The folded conformations of most proteins are not very stable, even under optimal conditions. **The free energy of the folded state is at most only 5–15 kcal/mol (20–60 kJ/mol) less than that of the unfolded state**. Consequently, even under optimal conditions, there is a significant probability (roughly 10^{-4}–10^{-11}) of molecules being totally unfolded. **Larger proteins are not necessarily more stable than small ones**. A small protein might contain only 50–100 amino acid residues, so it appears that each amino acid residue contributes on average at most 0.1 kcal/mol (0.5 kJ/mol) to the stabilization of the native state. This value is five times smaller than the energy of thermal motion at room temperature, RT = 0.6 kcal/mol (2.5 kJ/mol). The stability of the native state usually decreases in magnitude at acidic and alkaline pH values and at both high and low temperatures. The temperatures of maximum stability for proteins are usually close to or slightly lower than the temperature at which they normally function.

The complete thermodynamic characterization of the protein lysozyme (Figure 11-19) illustrates the complexity of the physical basis of protein structure. **The observed net stability of the folded state is a very small difference between very large, but compensating, individual factors**. For example, both the enthalpies and entropies of both the folded and unfolded states are very temperature-dependent, as reflected in their substantial heat capacities. Their contributions to the free energy vary by up to 700 kcal/mol over the range 0°–100°C (Figure 11-19-A,B). But the enthalpy and entropy vary in similar fashion and compensate for each other, an example of **enthalpy–entropy compensation**, so the free energy is a relatively small difference between the two and varies only one-tenth as much (Figure 11-19-C). The free energies of the folded and unfolded states also vary similarly; thus the difference between them, the observed net stability, is no greater than 16 kcal/mol (Figure 11-19-D). Consequently, accounting for the net stabilities of proteins in terms of the primary interactions stabilizing them is a hazardous accounting procedure, in which the net result is miniscule relative to the individual terms. An error of greater than 2% in a term of 700 kcal/mol would obliterate the net difference of only 16 kcal/mol. In spite of this, the small single-domain proteins studied have similar marginal net stabilities of their folded structures (Figure 11-11).

The stabilities of different small single-domain proteins can be compared by normalizing their thermodynamic quantities to correct for their different sizes. When this is done, **the measured entropy and enthalpy changes of unfolding of the various proteins differ somewhat, but upon extrapolation they converge to a common value at a high temperature**, known as T* (Figure 11-20). When ΔC_p was believed to be independent of temperature, this temperature of convergence was thought to be about 110°C. ΔC_p is now known to decrease at higher temperatures, however, which causes the curves of Figure 11-20 to flatten at high temperatures, and the convergence temperature is

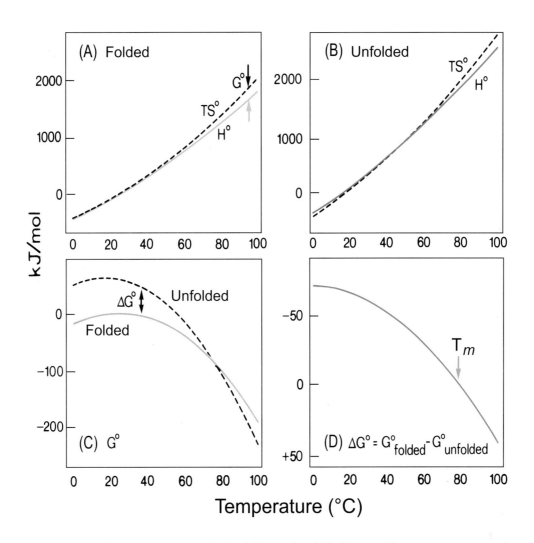

Figure 11-19. Thermodynamic parameters for the folded and unfolded forms of lysozyme at pH 7.0 and various temperatures. (A, B) The enthalpic (H) and entropic (T S) contributions to the free energies of the folded and unfolded states are plotted as a function of temperature. (C) The Gibbs free energy (G°) of each state is plotted as a function of temperature for both the folded and unfolded states; it is the difference between the enthalpic and entropic contributions, as indicated in (A). The net stability of the folded state, ΔG°, is plotted in (D) as a function of temperature; it is the difference between the free energies of the two states, as indicated in (C). Note the change in energy scale from (A) and (B) to (C) and then to (D); the final ΔG° illustrated in (D) is a very small difference between the individual enthalpy and entropy contributions of (A) and (B). Data from W. Pfeil and P. L. Privalov, *Biophys. Chem.* **4**, 41–50 (1976). Adapted from T. E. Creighton (1993) *Proteins: structures and molecular properties*, 2nd edn, W. H. Freeman, NY, p. 298.

generally taken to be approximately 140°C. On the other hand, it is a reasonable approximation, and much more convenient, to take ΔC_p as constant at lower temperatures. In that case, **the stabilities of many different small proteins can be described quantitatively at most temperatures, T, by a simple equation in which the only variable is the ΔC_p of unfolding**:

$$\Delta G_{unfold} = \Delta H^* - T\,\Delta S^* + \Delta C_p\,[(T - T^*) - T \log_e (T/T^*)] \tag{11.18}$$

where $\Delta H^* = 1.54$ kcal (mol residue)$^{-1}$ or 6.4 kJ (mol residue)$^{-1}$ and $\Delta S^* = 4.35$ cal (mol residue)$^{-1}$

or 18.1 J (mol residue)$^{-1}$, which are the common values at the convergence temperature T*. The value of T* is taken to be 112°C with the approximation that the ΔC_p of unfolding is independent of temperature.

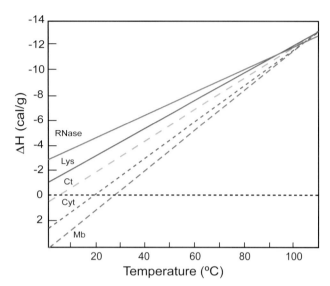

Figure 11-20. The specific enthalpy difference, ΔH per gram of protein, between the folded and unfolded states of five proteins: RNase, ribonuclease A; Lys, hen lysozyme; Ct, bovine α-chymotrypsin; Cyt, cytochrome *c*; Mb, metmyoglobin. The pH of each solution was that at which the protein is most stable. Data from P. L. Privalov and N. N. Khechinashvili (1974) *J. Mol. Biol.* **86**, 665–684.

It is remarkable that the convergence temperature T* is virtually the same as the temperature at which the entropy of transfer of nonpolar molecules from liquid to water is zero and at which the enthalpy of transfer from gas to water is zero. It is tempting to conclude that all proteins have the same enthalpy and entropy of unfolding at T* because the nonpolar surfaces that are exposed to water upon unfolding have no net interactions with the water. In this case, their common enthalpies and entropies would reflect the other common features of protein structure. The thermodynamics of unfolding of proteins share many features of the hydrophobic interaction measured in model systems, including the large change in heat capacity, the large temperature-dependence of the changes in enthalpy and entropy, the occurrence of a temperature at which the free energy change is a maximum, and the convergence of thermodynamic parameters. These similarities are undoubtedly due to the exposure of nonpolar surfaces to water that occur in both.

There are substantial differences, however, in that there are additional enthalpic factors that stabilize the folded state and entropic factors that stabilize the unfolded state. The thermodynamics of unfolding of proteins are much more similar to the dissolution of crystals of cyclic dipeptides of nonpolar amino acids. This is not unexpected, in view of the close packing of the protein interior, which approximates a crystal much more than a liquid. What is most surprising is that Equation 11.18 and experimental measurements indicate that the more hydrophobic the interior of a protein, i.e. the greater the value of ΔC_p of unfolding (Figure 11-9), the lower its net stability. This is not what would be expected if the hydrophobic interaction were the major force stabilizing the folded conformation.

Thermodynamic problems of protein structure. P. L. Privalov (1989). *Ann. Rev. Biophys. Biophys. Chem.* **18**, 47–69.

Physical basis of the stability of folded conformations of proteins. P. L. Privalov (1992) in *Protein Folding* (T. E. Creighton, ed.), W. H. Freeman, NY, pp. 83–126.

Energetics of protein structure. G. I. Makhatadze & P. L. Privalov (1995) *Adv. Protein Chem.* **47**, 307–425.

Stability of proteins: temperature, pressure and the role of the solvent. C. Scharnagl *et al.* (2005) *Biochim. Biophys. Acta* **1749**, 187–213.

11.3.A. Physical Basis of Protein Stability

The physical basis for the stability of the folded conformations of proteins has been one of the most contentious subjects, and there is still no consensus. In part this is due to the complexity of protein structures, both folded and unfolded, the variety of interactions that they participate in, and the various accounting procedures that are possible. For example, the excluded volume effect due to atoms not being able to occupy the same volume can be very substantial in unfolded proteins. The change in excluded volume on unfolding can be greater than the molar volume of the protein itself. Excluded volume contributes to stabilization of the native structure, but it is difficult to take it into account quantitatively.

The obvious importance of hydrogen bonds for the architecture of α-helices, β-sheets and folded proteins led initially to the assumption that the hydrogen bonds were also responsible for their stability. It was realized subsequently, however, that proteins when unfolded would make at least the same number of hydrogen bonds to water, so the hydrogen bonds were thought to provide no stabilization, only architectural specificity; the hydrophobic interaction was assumed to be the driving force for folding, and there were some similarities in the properties of the two phenomena. The thermodynamics of unfolding, however, indicate that this is probably not correct. Protein stability does not have the same thermodynamic properties as the hydrophobic interaction. The ΔC_p term in Equation 11.18 is proportional to the nonpolar surface area exposed to water upon unfolding, to the hydrophobicity of the protein interior (Figure 11-9). The other term by which it is multiplied, involving only the temperature, is positive for all temperatures less than 100°C. The term reflecting the hydrophobicity of the protein therefore *decreases* the stability of the folded protein, i.e. **the more hydrophobic the protein, the lower its net stability at lower temperatures**. This observation is the opposite of what would be expected if hydrophobicity were the predominant factor in protein stability. Current analysis, although not universally accepted, indicates that **the main contributors to stabilizing folded proteins are, in order of decreasing importance, (1) hydrogen bonding, (2) van der Waals interactions and (3) the hydration of nonpolar groups**. In this accounting procedure, the hydrophobic effect has been split into van der Waals interactions and hydration effects.

The only way that hydrogen bonds could contribute to the net stability of the folded state of a protein would be if those within a folded protein were much stronger than those between the unfolded protein and water. This is possible, as the hydrogen bonds within a protein are intramolecular, whereas those between the unfolded protein and water are intermolecular. The strengths of intramolecular interactions depend upon the **effective concentrations** of the interacting groups, and the very large values (up to 10^{10} M) that can be measured in intramolecular interactions in small molecules and in macromolecules have very important consequences for understanding the stabilities of the folded structures of macromolecules. If the hydrogen-bonding groups have effective molarities in the folded state that are greater than the solvent concentration, 55 M for water (or 110 M for H atoms), the hydrogen bonds within the folded state will be more stable than those between the unfolded state and water, and they will stabilize the folded conformation. Another way of thinking about this is to consider the increased entropy of the water molecules that are liberated when the protein folds and forms intramolecular hydrogen bonds between groups that were hydrogen-bonded to water. Such intramolecular hydrogen bonds will be especially stable if a number of them within a folded protein comprise a cooperative system. In this case, the simultaneous presence of numerous intrinsically weak interactions gives a much greater contribution to net stability than is possible with individual interactions.

The cooperativity of protein folded structures arises because partially folded conformations are unstable relative to the fully folded and fully unfolded conformations. Therefore, weakening one or a few interactions upon partial unfolding weakens the other interactions so that their contributions to stability are decreased, and the free energy rises substantially. The unfolding transition is much more abrupt than expected from the disruption of a single interaction (Figure 11-13).

There are probably two major reasons for the cooperativity of folding transitions. The first concerns **unfavorable interactions in the partially folded states that are not present in either the fully folded or unfolded states**. Two plausible examples would be the increase in free energy produced by (1) breaking an internal hydrogen bond without supplying comparable hydrogen-bonding partners to the acceptor or donor and (2) pulling apart two nonpolar surfaces sufficiently far that the van der Waals interactions are greatly diminished, but not so far that other atoms can intervene, so there is no gain of comparable interactions with other surfaces or the solvent. Any such conformational strain present in partially folded structures, but not in U or N, should contribute to the cooperativity of folding, but not to the net stability of the folded state.

The second reason for cooperativity is the simultaneous presence of many interactions within a single conformation, which should contribute to the net stability of the folded state. For two groups to interact requires a decrease in their freedom, an energetic cost in loss of entropy. Any further interactions realized in the same conformation do not incur any such cost; even if further changes in conformation are necessary for additional groups to interact, the cost in entropy loss will be lower due to the presence of the first interaction. Although basically entropic, this effect can also contribute to the enthalpy if the intramolecular interactions are as a consequence also more favorable enthalpically. For example, most hydrogen bonds within water and between a protein and water are usually present only a fraction of the time, whereas those within folded proteins are present essentially all of the time; the latter should consequently have the more negative enthalpy.

Disulfide bonds are the only interaction within proteins whose free energy can be measured directly (relative to a reference disulfide bond) and produce measurements of the effective concentrations possible in folded proteins. The equilibrium constant for a thiol–disulfide exchange with an appropriate disulfide reagent, such as glutathione:

$$\text{Protein}_{SH}^{SH} + \text{GSSG} \rightleftharpoons \text{Protein}_{S}^{S} + 2\,\text{GSH} \tag{11.19}$$

has units of molarity and gives directly the effective concentration of the two thiol groups in the protein lacking the disulfide bond, because GSSG can be considered to have an 'intermolecular' disulfide bond because it produces two independent GSH molecules when reduced.

In an unfolded protein, the disulfide bonds formed are only weak, with effective concentrations in the millimolar range; however, **the disulfide bonds are 10^5–10^8 times more stable in the native conformation**, with effective concentrations in the range of 100–10^5 M. The largest values are for the disulfide bonds in the central core of the molecule, where the interacting S atoms would be expected to be held in position to interact, whether or not they are interacting; this would generate a high effective concentration, and such disulfide bonds would be expected to be the most stable. These measurements, however, probably overestimate the disulfide bond stability, as the large values arise in part because in these instances the protein without the disulfide bond has two thiol groups buried in the interior. This is energetically unfavorable, because two thiol groups are somewhat polar, and they are larger than two S atoms that are linked covalently. Nevertheless, even the least stable disulfide bond, on the surface of the protein, where the S atoms have considerable mobility when not interacting in a disulfide bond, has a high effective concentration of 100 M.

Such measurements illustrate directly how a disulfide bond stabilizes a folded conformation and how the folded conformation stabilizes the disulfide bond. The two are linked functions, and **whatever effect a disulfide bond has on the stability of the protein conformation, that conformation must have same effect on the stability of the disulfide bond** (Figure 11-21). The same considerations apply to other interactions within the folded protein, such as hydrogen bonds, not just disulfide bonds. The disulfide bonds, at least in the instances that have been well studied, merely stabilize the folded conformation; the very same folded conformation is observed with BPTI when any one of its three disulfide bonds is missing, and in one case when only one of the three is present. **The disulfide bonds do not determine the folded structure**, even though this is stated frequently.

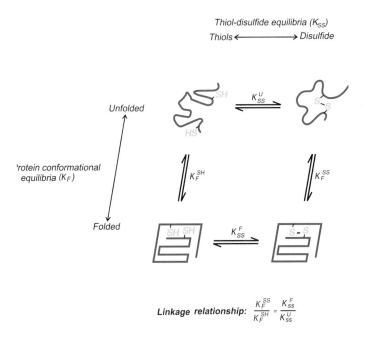

Figure 11-21. An example of linked functions, in this case of protein disulfide bond stability and conformational stability. A protein with two Cys residues that can form a disulfide bond is illustrated in the unfolded and folded conformations. The folded conformation has the two Cys residues in proximity for forming a disulfide bond. The indicated equilibrium constants represent the stabilities of the disulfide bonds, K_{SS}, and of the folded conformation, K_F. The thiol and disulfide reagent that is used to interconvert the thiols and disulfide bond of the protein (Equation 7.34) are omitted. The linkage relationship results from the general requirement that the free energy change around any cycle be zero. In this case, it states that whatever effect the folded conformation has on the stability of the disulfide bond, the disulfide bond must have the same effect on the stability of the folded conformation. Any two phenomena that affect each other may be represented in this way, and they are subject to the same linkage relationship. In particular, comparable linkage relationships pertain to all interactions within the folded conformation, not just disulfide bonds.

The most important aspect of protein structures is their cooperativity. Only by having a cooperative network in which individually weak interactions occur simultaneously and stabilize each other will a stable structure result. For example, the disulfide bonds of BPTI are about 10^5–10^8 times more stable in the folded conformation than in the unfolded protein. They are stabilized by the remainder of the protein structure, and they stabilize it (Figure 11-21). This suggests how a cooperative folded structure can occur. A single weak interaction in the unfolded protein will be weak, due to the low effective concentration of the interacting groups in the unfolded protein, and the equilibrium constant for any one interaction will be considerably less than unity. Forming any interaction, even transiently, however, will bring other groups into closer proximity and increase somewhat their effective concentrations. Such second interactions will consequently be more stable than in the absence of the first interaction. Further interactions will likewise be stabilized, with increased effective concentrations, until the equilibrium constant for generating a further interaction is greater than unity. Up until this point adding further interactions has been increasingly improbable, i.e. the apparent free energy has increased, but with sufficient interactions and folded conformation to give effective concentrations greater than unity, the addition of further interactions is favored energetically, and the free energy of the various species will decrease until the folded conformation becomes more stable than the initial unfolded protein, with an overall equilibrium constant greater than unity. At this point, the individual weak interactions will have cooperated to build up a cooperative stable structure.

It seems reasonable, therefore, to conclude that most of the polar interactions within a folded protein, especially the hydrogen bonds, are more favorable energetically, in both enthalpy and free energy, than the corresponding interactions of the unfolded state with itself and water. The stabilities of all the polar interactions within the fully folded state are consequently greater than in any other conformation, making the fully folded state a unique conformation.

The various contributions to stability of a folded protein can be dissected on the basis of Equation 11.18. If it is assumed that the significance of the convergence temperature, T^*, is that there are then no net interactions between the exposed nonpolar surfaces of the protein and the solvent, the observed value of ΔH^* would represent all the net stabilizing interactions within the folded state, such as the van der Waals interactions, hydrogen bonds, electrostatic interactions, etc., that are disrupted upon unfolding. Table 11-1 gives the magnitudes of these values for a protein the size of hen egg white lysozyme, consisting of 129 residues (Figure 11-19). The value of ΔS^* would represent primarily the greater conformational entropy of the unfolded state; its value corresponds to a reasonable eight-fold increase in the average number of conformations per residue upon unfolding. According to this analysis, a protein folded conformation like that of lysozyme is held together by the –198 kcal/mol contribution of the stabilizing interactions, but this is balanced by the increased entropy of the unfolded polypeptide chain and the solvation of the nonpolar surfaces by water at lower temperatures. The contribution of the entropy of unfolding is +167 kcal/mol at 25°C. This destabilizing factor increases in magnitude at higher temperatures and is responsible for thermal unfolding. The contribution to the free energy of the solvation of nonpolar surfaces is only +17 kcal/mol at 25°C but this destabilizing contribution increases in magnitude at lower temperatures and is responsible for cold unfolding.

If hydrophobicity is not the primary driving force stabilizing folded conformations, the contributions of the van der Waals interactions between nonpolar atoms in the protein interior must be considerably less than those that are present in a liquid. Otherwise, the surface area buried in the protein interior could provide a free energy of stabilization from the hydrophobic effect that would be adequate to account for all the interactions believed to stabilize the folded state. Nevertheless, the thermodynamic analysis given above indicates that the hydrophobic contribution is less than half of this (Table 11-1). Therefore, the van der Waals interactions between the atoms within a protein interior must be less favorable than those in a nonpolar liquid, even though the interiors of proteins appear to be densely packed. The dependence of the van der Waals attraction on the sixth power of the distance implies that there could be a substantial decrease in the strength of the interaction, with only a slight decrease in packing interactions. Although proteins are densely packed in their interiors, they are not closely packed, with all adjacent atoms in van der Waals contact simultaneously (Figure 9-21). Imperfect van der Waals interactions imply that there must be constraints that are preventing them from improving, which are most likely to be the constraints of the covalent structure. There may then be a dynamic tension in the folded state between the tendency to maximize the van der Waals interactions and the energetic cost of the conformational strain that would be necessary due to the covalent constraints.

The concept of hydrophobicity now appears to have less relevance for the stabilities of folded protein structures than had previously been thought. Although interactions between nonpolar atoms are important, it is perhaps more meaningful to consider separately the van der Waals interactions involved in the folded state and the interactions of the same surfaces with water in the unfolded state. Van der Waals interactions within proteins appear to be much less than optimal, and considerably poorer than those that occur within a liquid. Hydrogen bonds and other specific interactions appear to provide the greatest contribution to stability of protein folded conformations, probably as a result of the entropic cooperativity between them. Interactions with directional requirements, like hydrogen bonding, should gain most from entropic effects. The role of the aqueous solvent is primarily to interact favorably with the charged and polar groups on the protein surface, which it can do equally well in both the folded and unfolded states. The interaction of water with nonpolar groups, especially in the unfolded state, is increasingly favorable at low temperatures, which decreases the stability of the folded conformation and can lead to cold-induced unfolding.

Table 11-1. Contributions to the free energy of folding at 25°C of a typical protein the size of hen lysozyme, with 129 residues, estimated on the basis of Equation 11.18

Contribution	$G°_N - G°_U$ (kcal/mol)
Greater conformational entropy of U[a]	+167
Net stabilizing interactions[b]	−198
Solvation of increased nonpolar surfaces in U[c]	+17
Net stability	−14

[a] $T \Delta S°_{conf}$; $\Delta S°_{conf}$ = 4.35 cal/(residue mol K).

[b] Sum of van der Waals interactions in N and net greater stability of hydrogen bonds and other polar interactions in N relative to U, minus any conformational strain. Calculated from $\Delta H^* = 1.54$ kcal/mol residue.

[c] Favorable interactions of nonpolar surface with water at 25°C, calculated from $\Delta C_p [T - T^* - T \log(T/T^*)]$, where $T^* = 112°C$ and $\Delta C_p = 12.5$ cal/(residue mol K), the measured value for hen lysozyme.

Data from T. E. Creighton (1993) *Proteins: structures and molecular properties*, 2nd edn, W. H. Freeman, NY.

Even though the folded state is stabilized by many interactions, it has only marginal net stability because of the large conformational entropy of the unfolded state. Consequently, the net stability of the folded state, $\Delta G°_N$, is a relatively small difference between the substantial compensating interactions stabilizing the folded and unfolded states (Figure 11-19).

Energetics of protein structure. G. I. Makhatadze & P. L. Privalov (1995) *Adv. Protein Chemistry* **47**, 307–425.

Stability and stabilization of globular proteins in solution. R. Jaenicke (2000) *J. Biotechnol.* **79**, 193–203.

Polar group burial contributes more to protein stability than nonpolar group burial. C. N. Pace (2001) *Biochemistry* **40**, 310–313.

---

1. Water and Co-solvents

The effect of a co-solvent on the stability of the folded conformation of a macromolecule is determined strictly by the difference between the interactions of the co-solvent with the macromolecule in the unfolded and native states. This depends upon the preferential binding (or hydration) and the transfer free energy, i.e. the free energy of interaction of the co-solvent with, or binding to, the macromolecule.

The unfolding process can be represented by a simple equilibrium between the native (N) and denatured (D) states, N ↔ D, that is affected by a ligand, in this case the co-solvent. The effect of the co-solvent is determined simply by whether it binds more tightly to the N or D states. **Greater binding to the D state enhances the unfolding reaction, while greater binding to N stabilizes the native structure**. It can also be considered in terms of the relative transfer free energies of the folded and unfolded forms of the macromolecule from water to the solution with the co-solvent.

Salts affect protein stability according to their ranking in the Hofmeister series (Figure 11-8).

Interpreting the effects of small uncharged solutes on protein-folding equilibria. P. R. Davis-Searles *et al.* (2001) *Ann. Rev. Biophys. Biomolec. Structure* **30**, 271–306.

Protein stability in mixed solvents: a balance of contact interaction and excluded volume. J. A. Schellman (2003) *Biophys. J.* **85**, 108–125.

Destabilization and stabilization of proteins. J. A. Schellman (2006) *Quart. Rev. Biophys.* **38**, 351–361.

---

a. Stabilizers

The agents that are most frequently used to stabilize macromolecules are (1) 1 M solutions of sucrose and other sugars, such as trehalose and glucose, (2) 3–4 M glycerol, (3) neutral amino acids, such as glycine, proline and alanine, at concentrations of about 1 M, (4) methyl amines, including betaine, sarcosine and trimethylamine-N-oxide, (5) polyols, such as sorbitol, inositol and mannitol, and (6) sulfate salts, such as sodium and ammonium sulfates. **All of these stabilizers have the property of being preferentially excluded from the surfaces of proteins**, which imparts stabilization and precludes their contact with the surface that could affect the protein structure. They can also be considered to stabilize proteins by increasing the chemical potential of the denatured state. Most stabilizers used in the laboratory are also osmolytes and cryoprotectants, and many have been used in nature to maintain high osmotic pressures in organisms, such as amphibians, to protect them from adverse temperatures, desiccation, salinity, etc., and to protect against freezing.

The osmophobic effect: natural selection of a thermodynamic force in protein folding. D. W. Bolen & I. V. Baskakov (2001) *J. Mol. Biol.* **310**, 955–963.

Predicting the energetics of osmolyte-induced protein folding/unfolding. M. Auton & D. W. Bolen (2005) *Proc. Natl. Acad. Sci. USA* **102**, 15065–15068.

Influence of the Hofmeister anions on protein stability as studied by thermal denaturation and chemical shift perturbation. X. Tadeo *et al.* (2007) *Biochemistry* **46**, 917–923.

---

b. Destabilizers

The agents that destabilize macromolecular structures are generally considered to be **denaturants** (Section 11.1.C). **They tend to bind preferentially to proteins, especially to the unfolded state**. The most common denaturants of proteins are urea, GdmCl (Equation 11.3), sodium dodecyl sulfate (SDS), trichloracetic acid, $CaCl_2$ and LiBr. The effects of urea and GdmCl on proteins have been described in Section 11.1.C. SDS is a detergent that binds tightly to unfolded polypeptide chains, trichloroacetic acid is a strong acid (although its denaturing action is not understood), and $CaCl_2$ and LiBr are salts in the Hofmeister series.

Effect of alcohols on protein hydration: crystallographic analysis of hen egg-white lysozyme in the presence of alcohols. A. Deshpande *et al.* (2005) *Acta Crystallogr. D* **61**, 1005–1008.

Preferential solvation in urea solutions at different concentrations: properties from simulation studies. H. Kokubo & B. M. Pettitt (2007) *J. Phys. Chem. B* **111**, 5233–5242.

---

c. The Role of Water

The preceding discussion of the importance of hydration for the stability of the folded state may have given the impression that water is essential for the stability of the folded state, and this would also be expected if the hydrophobic interaction were the most important factor, as it results from nonpolar atoms avoiding contact with water. On the contrary, **folded protein structures can be more stable in the absence of water, and water destabilizes folded structures**. The reason for this is most apparent if one considers hydrogen bonding in the folded state: hydrogen bonds stabilize folded structures only if they are more stable than those formed between water and the polar groups on the unfolded molecule. In the absence of water, this stabilizing effect is enhanced.

Electrospray ionization mass spectrometry is one of the few techniques available to study macromolecules *in vacuo*. Isolated α-helices, which are barely populated in water, are extremely stable in its absence. Peptides as short as eight residues adopt this structure *in vacuo*, but only if the molecules have little or no net charge. Electrostatic interactions predominate in the gas-phase, where they are no longer shielded by the high dielectric constant of water. Consequently protein molecules with substantial net charges unfold in the gas-phase, probably due to the electrostatic repulsions between groups held in close proximity in the folded conformation. On the other hand, polar groups are less likely to ionize in the gas-phase: individual amino acids exist as zwitterions in solution but not in a vacuum. If intramolecular salt bridges or ion pairs can be formed, however, the ionized forms of the groups are much more stable.

Peptides and proteins in the vapor phase. M. F. Jarrold (2000) *Ann. Rev. Phys. Chem.* **51**, 179–207.

Extreme stability of an unsolvated α-helix. M. Kohtani *et al.* (2004) *J. Am. Chem. Soc.* **126**, 7420–7421.

Helices and sheets *in vacuo*. M. F. Jarrold (2007) *Phys. Chem. Chem. Phys.* **9**, 1659–1671.

11.3.B. Effects of Varying the Primary Structure

The stability of a protein depends upon its primary structure but in a complex manner. The amino acid sequences of proteins change during evolutionary divergence, but their folded conformations change relatively little (Section 9.3), as do their net stabilities (Figure 11-11). Yet single amino acid replacements can alter the stability of a protein quite drastically. Selective pressures probably act during protein evolution to maintain the stability of a protein at about its optimum level. Net stability appears not to be maximized in nature, as proteins are generally only marginally stable under the conditions where they function, even though that stability could be increased. Both the natural variation that exists in proteins and that introduced by chemical modification, genetic mutations or site-directed mutagenesis, provide a plethora of data concerning the physical basis of protein stability.

1. Natural Proteins of Exceptional Stability

Proteins that exist naturally under so-called 'physiological' conditions (37°C, 0.15 M salt, neutral pH, 1 atmosphere pressure) tend to have similar stabilities to unfolding (Equation 11.18), even though they have different folds and different amino acid sequences. These proteins generally come from organisms known as **mesophiles**. Many organisms are adapted to other types of environments, such as low or high temperature (**psychrophiles** and **thermophiles**, respectively), high salt concentrations (**halophiles**), etc. Normal mesophilic proteins would be unfolded or insoluble under many of these other conditions, so it is not surprising that the proteins from these other organisms have adapted to their environments. How they have done so continues to intrigue protein chemists.

Thermophilic proteins usually are intrinsically more resistant to high temperatures than are their mesophilic counterparts. Even when purified, they generally resist unfolding at the elevated temperatures where they normally function (up to 90°C) and they are often impervious to denaturants. Most thermophilic proteins have increased their melting temperatures simply by increasing their stability at all temperatures. Nevertheless, **it is very difficult to distinguish between thermophilic and mesophilic proteins on the basis of just their structures**. The most significant structural differences between them seem to be that **thermophilic proteins have more salt bridges on their surfaces or incorporate ions as part of their folded structures**. They appear to rely upon additional polar interactions for their greater stability, although it is plausible that the charged groups are more important for maintaining the solubility of the protein. It might have been thought that thermophilic proteins would be more hydrophobic, because the hydrophobic effect in model systems increases energetically with increasing temperature, but that would be inconsistent with the observed tendency of more hydrophobic proteins to be less stable (Equation 11.18).

Proteins from halophiles often exist in ionic environments equivalent to saturated KCl solutions. When isolated, they tend to require high ionic strengths for their folded conformations to be stable. The few halophilic proteins that have been studied bind anomalously large amounts of water and salt. They appear to have incorporated ions into their folded conformations, although the structural details remain to be elucidated.

The above observations have come from a few studies of a few proteins, so their generality needs to be established. Even some proteins from ordinary sources can have exceptional stabilities. For example, the small protein BPTI (Figure 9-2) comes from that well-known mesophile the cow, but it does not unfold at temperatures as high as 100°C, at extremes of pH or in concentrated denaturants such as 6 M GdmCl and 8 M urea.

Electrostatic and hydrophobic interactions play a major role in the stability and refolding of halophilic proteins. T. Arakawa & M. Tokunaga (2004) *Protein Pept. Lett.* **11**, 125–132.

Crowding in extremophiles: linkage between solvation and weak protein–protein interactions, stability and dynamics, provides insight into molecular adaptation. C. Ebel & G. Zaccai (2004) *J. Mol. Recognit.* **17**, 382–389.

Lessons in stability from thermophilic proteins. A. Razvi & J. M. Scholtz (2006) *Protein Sci.* **15**, 1569–1578.

Mechanisms for stabilisation and the maintenance of solubility in proteins from thermophiles. R. B. Greaves & J. Warwicker (2007) *BMC Struct. Biol.* **7**, 18.

---

2. Mutagenic Studies

The principles of protein stability have been examined extensively by measuring the effects of altering the primary structure, either chemically or mutationally. The latter approach uses site-directed mutagenesis of the gene to make specific alterations (Section 7.6.C). Alternatively, mutations can be introduced randomly, either to the entire gene or to just a segment of it, and those mutations that produce the desired effect can be selected by screening a population of microorganisms making the variant proteins. For example, mutations that decrease a protein's stability sufficiently to inhibit significantly its biological function *in vivo* can often be selected on the basis of their disruption of its biological function in the mutant form. The method of selection and the changes in stability that are detectable in this way will depend upon the particular protein, but the principles are general. Once mutants that are inactive have been isolated, further mutagenesis and selection can identify amino acid replacements that reverse the particular destabilization or increase stability generally. If random mutagenesis is targeted to a certain part of the gene and protein, and is sufficiently drastic to have produced all possible amino acid replacements, those mutations that are not identified by the selection process can be assumed to have passed the selection criterion. Negative observations thereby become significant.

The effects of any specific mutation on stability depend upon the role that the original residue played in the folded structure and the role of the introduced residue, plus any alterations in conformation caused by the replacement. The quantitative effect of two mutations on a protein can be additive, partially additive, synergistic, antagonistic or absent. Each of these five possible types of interactions has its own mechanistic explanation. **Additive effects indicate that the two altered residues function independently**. Departures from additivity reflect interactions between them: **partial additivity** indicates cooperativity, **synergy** indicates anticooperativity and **antagonism** indicates opposing structural effects of the two mutations. If there is no additional effect, there is either partial additivity or antagonism between the two residues. A significant conceptual simplification is achieved by applying inverse thinking, namely by using the parameters of the double mutant rather than those of the wild-type enzyme as the reference point. For example, to explain partially additive effects, inverse thinking starts with the double mutant. Restoring only one residue increases the stability by factor A, while restoring only the other residue has a factor B effect. If restoring both residues has an effect greater than A + B (in terms of energy), the excess is a measure of the cooperativity between the two residues in the structure.

The folded structures of proteins are generally not altered substantially by a single amino acid replacement, which probably explains why evolutionary change of the structure of a protein requires multiple amino acid replacements (Figure 9-35). One of the largest structural changes observed upon mutagenesis involves extension of a helix upon replacing a Pro residue; the original Pro residue presumably terminates the α-helix prematurely. The structures of proteins seem to be sufficiently plastic to accommodate small replacements, but this plasticity varies throughout the protein structure. The packing of atoms is usually observed to change locally around a mutation, but the extent of the change depends upon both the magnitude of the covalent change and the site in the protein. Partly for these reasons, it is difficult to predict or rationalize the structural and energetic consequences of any particular mutation.

The effects of individual mutations on stability of the folded structure vary widely, both within and between proteins. Only a few generalities may be made. Replacements on the surface of a protein generally have little or no effect on the stability, unless either the original or the introduced side-chains have a specific role in the structure. Charged residues on the protein surface can be interchanged or replaced by nonpolar residues, and vice versa, with only very small effects on stability, unless the charged groups are involved in specific salt bridges. **The interior of the protein is generally most sensitive energetically to mutation**. Large decreases in net stability are usually produced by replacing buried hydrophobic side-chains by even larger ones that do not fit, by smaller ones that leave cavities or cause substantial repacking of the interior, and by polar or charged ones that are in unfavorable environments. The net stability of the folded state can be decreased by up to 9 kcal/mol by a single mutation. This is comparable to the overall net stability of the folded state, so just one such mutation can cause a protein to unfold. The greatest energetic effects generally occur at sites within the best-packed, least flexible part of the protein structure. Other parts of the protein appear more able to accommodate mutations by altering the local packing of adjacent atoms. Inserting extra residues within α-helices or β-strands generally causes a substantially greater decrease in stability than insertions within less regular parts of the polypeptide chain. The effects of insertion and deletion are more variable, because of the number of ways in which the change can be accommodated by the protein structure.

Within the protein interior, there is often a rough correlation between the hydrophobicity of the side-chain introduced and the net stability of the protein. The effect on stability of deleting a nonpolar group can be up to two to three times greater than that expected from the free energy of transfer from water to a nonpolar liquid (Table 7-6). The protein interior is not comparable to a liquid, and the correlation breaks down when the side-chain introduced is significantly larger than the original one. There appears to be an energetic cost to a folded protein rearranging its interior packing substantially. Multiple replacements that maintain the hydrophobicities and volumes of interior residues can still have substantial effects on net stability, indicating that it is not just hydrophobicity that is important. Any correlations with hydrophobicity may be misleading and may actually reflect the more important aspects of the extent of the van der Waals interactions present between nonpolar atoms.

Altering specific interactions or aspects of the structure can also affect protein stability. Replacing a charged residue involved in a particularly stabilizing salt bridge can decrease the stability by up to 5 kcal/mol. The special backbone conformational properties of Gly and Pro residues are often important, so replacing or introducing them can have substantial effects on stability. Substituting residues within an α-helix that increase the helical propensity, either due to the intrinsic helical propensity of the residue (Table 8-3) or by interacting electrostatically with the helix dipole (Figure 8-11), can increase protein stability so long as any other consequences of the mutation do not predominate.

Most amino acid replacements are found to have little effect on the stability of the natural folded state, or to decrease it, and **few replacements increase protein stability**. Those that do seem to introduce new stabilizing interactions (van der Waals interactions, salt bridges, disulfide bonds, etc.), to improve the packing of internal side-chains and relieve slightly unfavorable aspects of the native structure. In almost all instances, however, the additional stability is less than might be expected from the stabilizing interaction introduced, so there is generally an energetic cost to be paid upon altering a protein structure, possibly due to unfavorable interactions that are introduced inadvertently. The proteins that occur naturally do not have the greatest possible stability, but alterations to the covalent structure are most likely to be destabilizing. Nevertheless, protein stability has been improved by natural selection to produce thermophilic proteins (Section 11.3.B.1).

Disulfide bonds and Gly and Pro residues alter the conformational entropy of the unfolded state (Section 1.3.B.1), and this is one of the ways in which their addition or replacement can alter protein stability.

The consequences of an amino acid replacement are often attributed to its effect on either the folded or unfolded state. This not straightforward, however, for the free energies of proteins with different covalent structures cannot be compared directly. The best that can be done is to consider unfolding (N ⟶ U) and the mutational event (A ⟶ B) as linked functions (see also Figure 11-21):

$$N_A \underset{}{\overset{K^A_{unfold}}{\rightleftharpoons}} U_A$$
$$K^N_{mut} \updownarrow \qquad \updownarrow K^U_{mut} \qquad (11.20)$$
$$N_B \underset{}{\overset{K^B_{unfold}}{\rightleftharpoons}} U_B$$

$$\frac{K^A_{unfold}}{K^B_{unfold}} = \frac{K^N_{mut}}{K^U_{mut}} \qquad (11.21)$$

$$\Delta\Delta G°_{unfold} = \Delta G°_{mut} \qquad (11.22)$$

The difference in stability caused by the replacement A → B should therefore be the same as the difference in free energies in making the mutation in the folded and unfolded states. The latter factor, $\Delta G°_{mut}$, cannot be measured experimentally but it can be simulated computationally by **free energy perturbation** methods. In this computer procedure, the covalent model of the protein is gradually changed from A to B. Its free energy can be calculated if there is no substantial change in the structure. This provides a feasible method of computing the effects of mutations on conformational stability. It would be impractical to compute the free energies of the unfolding process directly, due to the enormous conformational changes that occur. Instead the equal value of $\Delta G°_{mut}$ is calculated. The linkage relationships of Equations 11.21 and 11.22 are also useful for rationalizing the effects of mutations on stability, by considering the difference in the energetic consequences of making the mutation in the folded and unfolded states. **Amino acid replacements are generally observed to affect primarily the folded state**, as would be expected.

The simplest way to stabilize the folded conformation of a protein is to add a ligand (L) that binds specifically and only to the folded conformation (N) (Chapter 12):

$$U \underset{K_f}{\rightleftharpoons} N \underset{K_L}{\overset{L}{\rightleftharpoons}} N \cdot L \tag{11.23}$$

$$K_{f,L} = \frac{[N]+[N \cdot L]}{U} = \frac{[N]}{[U]}(1 + K_L[L]) = K_f(1 + K_L[L]) \tag{11.24}$$

The increase in stability as measured by the equilibrium constant should be directly proportional to the free ligand concentration and its affinity for the folded conformation.

Genetic analysis of protein stability and function. A. A. Pakula & R. T. Sauer (1989) *Ann. Rev. Genetics* **23**, 289–310.

Selection of mutations for increased protein stability. B. van den Burg & V. G. Eijsink (2002) *Curr. Opinion Biotechnol.* **13**, 333–337.

Combinatorial approaches to protein stability and structure. T. J. Magliery & L. Regan (2004) *Eur. J. Biochem.* **271**, 1595–1608.

Ligand effects on protein thermodynamic stability. J. M. Sanchez-Ruiz (2007) *Biophys. Chem.* **126**, 43–49.

11.3.C. Structural Stability of Membrane Proteins

Experimentally studying the folding and stability of membrane proteins (Section 9.2) is much more complicated than for water-soluble proteins, because they are embedded within a membrane. There the protein is interacting with the hydrocarbon core of the lipid bilayer, the polar groups of the bilayer interface and the aqueous solvent at the surface of the protein. The proteins, whether folded or unfolded, are generally insoluble in aqueous solution, and most agents to unfold them will also disrupt the membrane. During their biosynthesis *in vivo*, membrane proteins are extruded into the appropriate membrane, where they fold. Some toxins that are initially soluble but function by being inserted into membranes are synthesized as water-soluble proteins, with normal polar surfaces, but upon encountering an appropriate membrane they undergo a drastic structural rearrangement to convert them into a protein capable of existing within the membrane bilayer. All these complex interactions must be taken into account. The folding of membrane proteins is frequently analyzed in terms of a four-step model, involving (1) partitioning of the polypeptide chain into the membrane, (2) folding into secondary structure elements (usually α-helices), (3) insertion into the membrane and (4) association of the elements of secondary structure.

In contrast to soluble proteins, membrane proteins fold and remain in an environment from which water is largely excluded. The absence of water forces the protein to form hydrogen bonds between peptide groups within the molecule itself, driving the formation of transmembrane helices (Section 9.2.A) and β-barrels (Section 9.2.B). **The hydrogen bonds in α-helices and β-sheets within the membrane are probably much stronger than the average such bonds in soluble proteins** because there is no competition to form hydrogen bonds with water. As a consequence, each transmembrane helix or β-sheet is a stable and rigid structure that can be considered an independent folding unit. In water-soluble proteins, the tertiary fold at least partially results from the action of the hydrophobic effect, which brings about formation of the hydrophobic core. In membrane proteins, in contrast, the association of the transmembrane helices within the bilayer in the absence of water cannot be driven by the hydrophobic effect. Mutagenesis studies and analysis of known structures suggest that **optimizing the van der Waals interactions is responsible for the packing of protein α-helices within the bilayer**. Residues buried between helices are more conserved during evolution, but also more polar, than those exposed to lipid, suggesting that a subtle stereochemical fit plays a key role in the formation of the tertiary structure.

Probably as a consequence of this greater stability of membrane-spanning helices, they are less susceptible to denaturation by temperature or commonly used denaturants, such as SDS, urea and GdmCl. Upon gel electrophoresis in the presence of SDS, integral membrane proteins with multiple transmembrane helices often migrate faster than expected from their true molecular weight. This may be due to (1) the maintenance of secondary and/or tertiary structures under conditions in which soluble proteins are totally denatured, and/or (2) increased binding of the anionic detergent.

Reversible unfolding of β-sheets in membranes: a calorimetric study. W. C. Wimley & S. H. White (2004) *J. Mol. Biol.* **342**, 703–711.

Polytopic membrane protein folding and assembly *in vitro* and *in vivo*. P. J. Booth & S. High (2004) *Mol. Membr. Biol.* **21**, 163–170.

Energetics of membrane protein folding and stability. C. A. Minetti & D. P. Remeta (2006) *Arch. Biochem. Biophys.* **453**, 32–53.

Deciphering molecular interactions of native membrane proteins by single-molecule force spectroscopy. A. Kedrov *et al.* (2007) *Ann. Rev. Biophys. Biomol. Structure* **36**, 233–260.

11.4. PROTEIN REFOLDING *IN VITRO*

Natural proteins can often be denatured and then renatured to their original conformation. When they cannot, it is usually because the unfolded protein has precipitated or aggregated, or has been subjected to covalent modification. Folded proteins can often be cleaved by proteinases at specific flexible loops on their surfaces, and the folded conformation is maintained. Such cleaved proteins can be unfolded and dissociated into two or three polypeptide fragments; very often these fragments can recombine and regenerate the folded conformation. Therefore, the information for the secondary, tertiary and quaternary structures of a protein resides in its primary structure. How this occurs is one of the major questions of molecular biology; if it were understood, it should be possible to predict the three-dimensional structure of a protein from just its amino acid sequence. The amino acid sequences of proteins are much, much easier to determine, usually from their gene sequences (Section 7.5.F), than are their three-dimensional structures (Chapter 9). The immense discrepancy in the number of known primary and tertiary structures continues to grow with the elucidation of the nucleotide sequences of entire genomes, with all their genes specifying protein sequences. There is an immense need to understand the basis of protein folding, and much effort has been expended, both theoretically and experimentally, but the answer is still not known. Nor is there a consensus regarding how to interpret what is known.

Protein folding is simply a conformational change, an isomerization unless disulfide bonds are formed, but its unusual aspect is the enormous number of conformations that an unfolded protein can adopt. If an average amino acid residue is able to adopt j conformations, a polypeptide chain with $N + 1$ residues (N peptide units between them to define the conformation) would be able to adopt up to j^N different conformations. The value of j is believed to be approximately 8, so a relatively small polypeptide chain of 100 amino acid residues should be able to sample some 10^{89} different conformations. This is an immense number, greater than the number of atoms in the universe. If the rate constant for one unfolded conformation to change to another is k_v, the average time to sample all these conformations will be given by:

$$\tau = (N k_v)^{-1} j^N \tag{11.25}$$

Unfolded conformations cannot change more rapidly than 10^{13} times per second (and probably do so some 10^9–10^{10} times per second), so it would require, on average, more than 10^{66} years to sample 10^{89} conformations. Some of these conformations would be impossible because atoms would overlap, but there is plenty of scope to be more conservative and still end up with very large numbers. Even if there were an average of only two conformations possible per residue, there would still be 10^{30} conformations for $N = 100$ and a requirement of 10^7 years for random searching. Nevertheless, many proteins are observed to refold *in vitro* within seconds or minutes, some within microseconds. Clearly, protein folding does not occur by a random searching of all possible conformations to find the unique native conformation, and there are likely to be pathways of folding, although not everyone would agree.

A further important aspect is that experimental observation of protein folding by most methods requires some 10^{15}–10^{18} molecules. Unfolded proteins under denaturing conditions (e.g. 8 M urea or 6 M GdmCl) at least approximate random coils (Section 11.2), so each unfolded molecule would be expected to have a different conformation at each instant of time (and a slightly different one some 10^{-10} s later); therefore, each molecule of a large population is initiating folding from a different starting point. Yet in many cases all the molecules fold with the same rate constant, indicating that they all have the same probability of folding in any instant of time.

Role of cofactors in metalloprotein folding. C. J. Wilson *et al.* (2004) *Quart. Revs. Biophys.,* **37**, 285–314.

Protein folding and the organization of the protein topology universe. K. Lindorff-Larsen *et al.* (2005) *Trends Biochem. Sci.* **30**, 13–19.

Conformational dynamics and ensembles in protein folding. V. Munoz (2007) *Ann. Rev. Biophys. Biomol. Struct.* **36**, 395–412.

11.4.A. Refolding of Single-domain Proteins

Experimental studies *in vitro* of the kinetics of protein refolding generally start with the fully unfolded protein under unfolding conditions, and refolding is initiated by changing the conditions abruptly to favor the folded state. Usually this involves diluting a denaturant or changing the pH. Changes in the conformational properties of the population of protein molecules are then monitored as a function of time. Observations with a number of small, model proteins have given the following general picture of how proteins fold (although there are frequent exceptions).

The fully folded state N generally begins to appear immediately, without a detectable lag period, so there is no sequence of obligatory intermediates that accumulate to substantial levels. The rate of folding depends upon the identity of the protein and the refolding conditions. The rate of refolding is greater the more physiological the conditions, but it is independent of the conditions that were used to unfold the protein. Consequently, unfolded proteins adapt very rapidly to changes in their environments:

$$\begin{matrix} U\ (80°C) \\ U\ (pH\ 2) \end{matrix} \longrightarrow U(25°C,\ pH\ 7) \xrightarrow{\text{slow}} N \tag{11.26}$$

Unfolding is observed by the reverse process of abruptly changing the conditions to those favoring unfolding, for example by rapidly adding a denaturant or changing the pH or temperature. The more denaturing the conditions, the greater the rate of unfolding.

In the cases of both unfolding and refolding, all the molecules have the same probability of undergoing the unfolding or refolding transition, and a single rate constant is observed for the population unless the molecules differ covalently or conformationally in a way that is only slowly interconverted. The latter usually occurs when there are both *cis* and *trans* isomers of peptide bonds (Section 8.2). The *trans* form of the peptide bond is intrinsically the more stable, and the *cis* form is usually not present substantially, unless the following residue is Pro, when the *cis* peptide bond is nearly as stable as the *trans*. Within a folded protein, the *cis* form can occur, usually preceding Pro residues, and this peptide bond is normally *cis* in all the molecules. In the unfolded state, however, there is an equilibrium mixture of both isomers of each peptide bond, according to their intrinsic stabilities, and the *cis* isomer is usually present at about 20% of the peptide bonds preceding Pro residues. As a consequence, a fraction of the unfolded molecules has one or more isomers that are not appropriate for the folded state: **the greater the number of Pro residues and cis isomers in the folded state, the greater the**

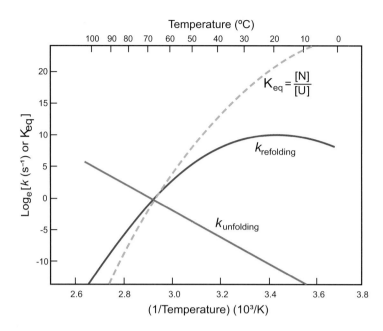

Figure 11-22. Typical temperature-dependence of the rates and equilibria of protein folding transitions not involving intrinsically slow isomerizations. The natural logarithms of the rate constants for unfolding and refolding are plotted as a function of (temperature)$^{-1}$ in an Arrhenius plot. A similar plot of the equilibrium constant (K_{eq}) between the folding (N) and unfolded (U) states is a van't Hoff plot. The curvature of the van't Hoff plot is due to the greater apparent heat capacity of U than N. The linear Arrhenius plot for the rate of unfolding indicates that the folding transition state has the same heat capacity as N. The greater heat capacity of U is reflected entirely in the curvature of the Arrhenius plot for the rate of refolding, because $\log_e K_{eq} = \log_e k_{refolding} - \log_e k_{unfolding}$. As a consequence of this curvature, the rate of refolding decreases with increasing temperature, in contrast to most chemical reactions. The data used to construct this diagram are for hen egg white lysozyme at pH 3, extrapolated to the absence of GdmCl. Although $k_{refolding} = k_{unfolding}$ at $K_{eq} = 1$, it is a coincidence that the rate constants also had the value 1 s^{-1} at this temperature, so that all three curves intersect at a common point. The experimental data are from S.-I. Segawa & M. Sugihara (1984) *Biopolymers* **23**, 2473–2488, and W. Pfeil & P. L. Privalov (1976) *Biophys. Chem.* **4**, 41.

fraction of unfolded molecules with non-native *cis/trans* isomers. Such *cis/trans* isomerization is intrinsically slow (Section 8.2.B.1) and the folding of the molecules containing incorrect peptide bonds can be slowed or prevented by the incorrect isomer; the rate of folding can be determined by the isomerization.

Cis/trans isomerization is a complicating factor that obscures the actual folding process, so the following discussion will be limited to those few cases of small model proteins without such an intrinsically slow process in folding. In the absence of Pro peptide bond isomerization, **both unfolding and refolding are usually observed to follow first-order kinetics** with a single rate constant for each and **no lag period**. Typical observations of the rates of unfolding and refolding are illustrated in Figure 11-22. Note that **protein refolding is a very unusual reaction in that the rate decreases with increasing temperature**, except at very low temperatures. The ratio of the rate constants for unfolding and refolding is generally consistent with the measured equilibrium constant for the transition, so microscopic reversibility appears to apply and there seems to be a classical **transition state** for the unfolding/refolding transition. That all the many conformationally diverse molecules of a population of unfolded protein refold with the same rate constant, independent of how the protein was unfolded, suggests that all the unfolded molecules equilibrate rapidly and reversibly before undergoing the same rate-limiting step, which depends only upon the final folding conditions (Equation 11.26).

Some unfolded proteins adopt partly folded or molten globule conformations very rapidly upon being transferred to refolding conditions, and the native conformation is generated more slowly. Such partly folded species are often considered responsible for the rapidity of folding, and much effort has gone into characterizing them; this is difficult, because they are populated only transiently before being converted to N. These intermediate species are usually in rapid equilibrium with the unfolded state, and it is generally not possible to determine whether they are on- or off-pathway intermediates, I:

$$U \longrightarrow I \longrightarrow N \tag{11.27}$$

$$\begin{array}{c} U \longrightarrow N \\ \updownarrow \\ I \end{array} \tag{11.28}$$

Not all unfolded proteins adopt partly folded species but remain unfolded until being converted to N in an apparently all-or-nothing transition. These tend to be the smaller proteins, and they usually refold more rapidly than those that adopt partly folded conformations. Therefore **the presence of stable partly folded intermediates is not necessary for rapid protein folding** and, in general, the presence of an intermediate at substantial levels seems more likely to slow refolding.

Protein unfolding is almost always an all-or-nothing transition, and partly folded species are generally not detected as intermediates during unfolding, even with proteins that adopt partly folded intermediates in refolding. Some exceptions have been reported, but the significance of any intermediates in unfolding is even more uncertain than of those observed in refolding.

Using protein folding rates to test protein folding theories. B. Gillespie & K. W. Plaxco (2004) *Ann. Rev. Biochem.* **73**, 837–859.

Protein folding studied by real-time NMR spectroscopy. M. Zeeb & J. Balbach (2004) *Methods* **34**, 65–74.

Distinguishing between cooperative and unimodal downhill protein folding. F. Huang et al. (2007) *Proc. Natl. Acad. Sci. USA* **104**, 123–127.

---

1. Characterizing the Transition State for Folding

Most important for the folding process is its transition state. It can be characterized only indirectly by measuring the rate of unfolding or refolding as a function of altering the covalent structure of the protein or changing the conditions. The relative sensitivities of the rates of the unfolding and refolding to the denaturant concentration indicate how exposed the transition state is to the denaturant, relative to the starting form of the protein. Similarly, the temperature dependence of the rates indicates the relative enthalpy, entropy and heat capacity of the transition state. Plots of the logarithm of the rate constant for unfolding versus the denaturant concentration are generally closely linear, comparable to what is observed with the equilibrium stability of the protein (Equation 11.4). This indicates that the nature of the transition state, and the pathway, remains constant and that the transition state has about the same exposure to denaturant as the folded state. The linearity of the Arrhenius plot for the rate of unfolding of Figure 11-22, in contrast to that for the rate of refolding, indicates that the transition state has the same heat capacity as the native state, different from that of the unfolded state. The slope of the Arrhenius plot indicates that the transition state has a somewhat lower enthalpy than the folded state. Similar observations are commonly made with other proteins, usually by varying the denaturant concentration rather than the temperature. **The thermodynamics of protein stability, especially the greater heat capacity of the unfolded state, account for the unusual phenomenon of the rate of refolding decreasing at higher temperatures**, in that the transition state has lower enthalpy than the native state.

The role of each amino acid residue in folding can be determined by using protein engineering to alter the structure of the protein systematically and then measuring the effects on the rates of unfolding and refolding. Changes in the rates are expressed as changes in the free energy of the transition state and are compared with the effect of the replacement on the stability of the folded state. The ratio of the two is usually expressed as a **phi-value**: a value of 0 indicates that the changed amino acid residue has no effect on the transition state, whereas a value of 1 indicates it has the same effect in the transition state as in the fully folded protein. Numerous such studies have indicated that many, if not all, of the stabilizing interactions of the native state have been disrupted in the transition state for unfolding and refolding. A very small phi-value is usually interpreted as indicating that that residue is unfolded in the transition state, but interpreting changes in free energy in terms of structure is not straightforward; an alternative explanation, with some experimental support, is that the cooperativity that produces strong interactions in the folded state has been disrupted. The residue may be in a nearly native conformation, but without its cooperativity. These types of studies generally suggest that the transition state is an expanded form of the native state, with a weakened but poorly hydrated core and a loosened periphery. The surface residues in such an expanded conformation are, on average, farther apart than are those in the center of the molecule. **The transition state appears to be close to the final conformation structurally but to lack the cooperativity that gives net stability to the fully folded conformation**.

Ligands that bind tightly to the native protein generally do not increase the rate of refolding of small proteins; instead, they usually decrease the rate of unfolding, and this is why they increase the stability of the folded conformation (Equation 11.23). These observations indicate that **the transition state does not bind its ligands tightly**, which raises the interesting and unresolved question as to whether a transient transition state would have time to bind a ligand, even if it had a binding site. Exceptions are some metalloproteins where the metal ions can bind specifically to their appropriate amino acid residues in the unfolded state if several of them are situated appropriately, usually close, in the unfolded polypeptide chain. In these cases, the metal ion can bind to the unfolded protein and increase the rate of folding.

Φ-Values for BPTI folding intermediates and implications for transition states. G. Bulaj & D. P. Goldenberg (2001) *Nature Struct. Biol.* **8**, 326–330.

The protein folding transition state: what are Φ-values really telling us? D. P. Raleigh & K. W. Plaxco (2005) *Protein Peptide Lett.* **12**, 117–122.

Φ-analysis at the experimental limits: mechanism of β-hairpin formation. M. Petrovich *et al.* (2006) *J. Mol. Biol.* **360**, 865–881.

2. Kinetic Schemes for Folding

Three types of folding mechanisms that are frequently discussed are illustrated in Figure 11-23. The general experimental observations are inconsistent with a scheme in which a nucleation event takes place in the unfolded protein followed by rapid folding (Figure 11-23-A); in this case, one would expect to see partially folded intermediates in unfolding, and not in refolding, the opposite of what is observed. The possibility that each unfolded molecule folds by a distinct pathway (Figure 11-23-B) is inconsistent with the observation of first-order kinetics (which implies that all molecules

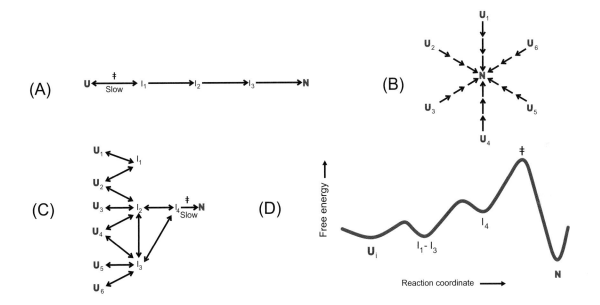

Figure 11-23. Examples of kinetic schemes for protein folding, in the absence of intrinsically slow isomerizations. U_i are various unfolded molecules with different conformations at the start of folding, I_i are partially folded molecules and N is the fully folded protein. All kinetic steps indicated by *arrows* are rapid, except for those labeled 'slow'; ‡ indicates the occurrence of the overall transition state. *Single-headed arrows* indicate steps that effectively occur only in the indicated direction under conditions strongly favoring folding. (A) The nucleation, rapid-growth model, in which a nucleation event in the unfolded protein, here indicated as formation of I_1, is the rate-limiting step. The nucleation event is very local and occurs randomly, so it could occur in all the unfolded molecules at a reasonable rate. Subsequent steps through various intermediates are rapid and essentially irreversible under strongly folding conditions. (B) The 'jig-saw puzzle' model in which each unfolded molecule refolds by a unique sequence of events. The different pathways only converge at the fully folded conformation. Each pathway will occur at a unique rate. (C) The general model indicated by the experimental data. All the unfolded molecules rapidly equilibrate under folding conditions with a few partially folded, marginally stable intermediates, which are also in rapid equilibrium. All the molecules pass through a common slow step, which involves going through a transition state that is a distorted form of the native-like conformation. Any intermediates that occur after the rate-limiting step are probably very unstable relative to N and rapidly convert to it. (D) Extremely simplified schematic diagram of the energy landscape for folding of a single-domain protein by the scheme in (C) under conditions where the folded structure N is maximally stable. The one-dimensional reaction coordinate at the bottom is some measure of the degree of folding of the polypeptide chain, such as its conformational entropy or compactness. The highest free energy barrier (‡) is the kinetic barrier to unfolding and maintains the folded protein N in its cooperative and stable folded structure. It is also the rate-determining energy barrier to the reverse process of refolding. The partially folded intermediates, I, have free energies greater than that of U and consequently will not be populated to significant levels during folding or unfolding, even transiently. The rapid equilibration between U and I indicated by the low energy barriers is the reason why all the molecules fold with the same rate constant and without a detectable lag period.

have exactly the same probability of folding at any instant of time) and the other indications that there is a single transition state (or a family of closely related transition states). The general kinetic scheme that is suggested by the experimental observations for folding of a single protein domain is illustrated in Figure 11-23-C: **all the unfolded molecules equilibrate rapidly prior to refolding, probably through a limited number of partially folded but very unstable intermediates, before they complete folding through a single transition state or a family of closely related transition states**.

This scheme is also consistent with, and could be imagined to be the result of, the cooperativity of protein folding (Figure 11-23-D). The highest free energy barrier (i.e. the transition state) would occur late in the folding process. In unfolding, it would represent the energy required to disrupt the strong cooperative interactions within the folded state. A scheme illustrating how a cooperative structure would be attained could be an idealized picture of the energy profile of a protein folding transition. The transition state would be that of highest free energy, where sufficient entropy has been lost during refolding that incorporating additional interactions is now energetically favorable (proceeding with a decrease in free energy). On unfolding, the transition state represents the point where the cooperativity has been disrupted sufficiently that breaking further interactions is energetically favorable. All the intermediates in this hypothetical scheme are unstable relative to the unfolded and fully folded state, and their exact identities need not be known. Of course, a less idealized scheme would allow for different intermediates to have varying free energies, determined by their conformational properties and not just their number of stabilizing interactions.

On the other hand, some believe that folding of single-domain protein structures should be downhill energetically, described as sliding down a funnel-shaped energy landscape, with a constantly decreasing free energy as the reaction proceeds and no energy barrier. If this were to occur, however, there would also be virtually no energy barrier to unfolding, so the folded structure would not be closely defined and would not be cooperative. A cooperative folded structure demands a free energy barrier to both unfolding and refolding, and virtually all proteins fold with free energy barriers that represent the transition state.

Cooperativity is the key to understanding protein structure and folding.

The present view of the mechanism of protein folding. V. Daggett & A. Fersht (2003) *Nature Rev. Mol. Cell Biol.* **4**, 497–502.

All-atom simulations of protein folding and unfolding. R. Day & V. Daggett (2003) *Adv. Protein Chem.* **66**, 373–403.

A unified mechanism for protein folding: predetermined pathways with optional errors. M. M. Krishna & S. W. Englander (2007) *Protein Sci.* **16**, 449–464.

11.4.B. Kinetic Determination of Folding

If nonrandom folding pathways are crucial because proteins cannot fold on a reasonable time scale by randomly sampling all conformations, the folded state that results might not be the most stable conformation possible but instead could be the form that is most accessible kinetically. A protein might normally fold for solely kinetic reasons to a state that is only metastable and not the most stable thermodynamically. Also, if a kinetic pathway of folding is so vital, interfering with that pathway should block folding. Relatively few examples are known.

1. Bacterial Proteinases

Bacterial proteinases such as **α-lytic protease** are synthesized as inactive precursors with amino-terminal extensions of the polypeptide chain that, after folding, are removed proteolytically to generate the active, native proteinase (Section 15.2.D). The pro-proteinase unfolds and refolds *in vitro* but the mature form does not. Instead, the unfolded proteinase adopts molten globule-like conformations; it refolds only when the pro segment is folded or added (it need not be part of the covalent structure). The pro-region of α-lytic protease is a 166-residue folded globular domain, almost as large as the 189-residue mature proteinase, that inhibits the folded proteinase by binding tightly ($K_i = 10^{-10}$ M) to its active site. The pro-region is then cleaved by the proteinase to release the pro-region and generate the active proteinase. Remarkably, the final folded proteinase is somewhat less stable than its unfolded form, and it should unfold spontaneously; it does, but only very slowly, with a half-time of 1.2 years, because of the very large kinetic barrier, with a transition state about 30 kcal/mol higher in free energy. Binding of the pro-region stabilizes the fully folded state. It stabilizes the folding transition state even more, so that both unfolding and refolding are increased in rate by a factor of 3×10^9. Consequently, the complete pro-precursor folds with a half-time of 23 s, rather than the 1800 years that are required in the absence of the pro-region. The transient presence of the pro-region effectively decouples folding and unfolding. The unfolding transition of mature α-lytic protease is also atypical in being primarily entropic in nature and having a remarkably large change in heat capacity. The folded α-lytic protease is also unusually rigid, which minimizes its degradation by other proteinase molecules.

It might be expected that this protein would have a very unusual structure, but α-lytic protease is homologous with other serine proteinases, such as bovine trypsin and chymotrypsin (Section 15.2.D), that unfold and refold more normally. They all have closely related tertiary structures that exhibit no obvious differences, except for some indications of locally strained conformations in α-lytic protease that might account for its remarkable properties. These properties appear to have evolved to extend the lifetimes of the proteinase, by decreasing its flexibility and suppressing local unfolding. This has been accomplished by increasing the cooperativity of the folded structures (i.e. by maximizing the effective concentrations of the interactions stabilizing the structure) but this increased stability is used to incorporate some conformational strain that physically restricts the mobility of the protein structure. This example also illustrates how it is the cooperativity of the fully folded state that determines the rate of folding, as the transition state is a distorted or incomplete high-energy form of the final conformation (Figure 11-23); this is best understood by considering the process of unfolding. The rate of unfolding is decreased by the high energy of the transition state, but so is the rate of refolding. The result is a requirement for a catalyst for folding. The penalty for evolving kinetic stability is remarkably large, in that each factor of 2.4–8 in proteinase resistance is accompanied by a cost of about 10^5 in the spontaneous folding rate and 5–9 kcal/mol in thermodynamic stability.

These remarkable proteinases also illustrate how evolution can select for proteins with distinctive structural features, which suggests that the similar and typical properties of 'normal' proteins (e.g. Equation 11.18) are likewise the result of selection for such 'normal' properties and not the result of some intrinsic properties of folded proteins.

2. Serpin Proteinase Inhibitors

The **serpin proteinase inhibitor** family includes some, such as plasminogen activator inhibitor-1 and α1-antitrypsin, that are synthesized in a form that is active as an inhibitor but relatively unstable. After biosynthesis, the active form slowly converts to a form that is more stable but inactive as an inhibitor, known as the latent form. If this latent form is unfolded and then refolded, the active metastable form is regenerated, before again undergoing the slow conversion to the inactive but stable form. Therefore **folding does not produce the most stable folded conformation directly but occurs only through a metastable intermediate**. The difference between the two forms involves a remarkably large change in a β-sheet of the protein (Figure 11-24). That this final and drastic change in conformation is slow is not surprising, but there is no current explanation as to why these serpins fold first to the metastable conformation.

Nevertheless, these proteins demonstrate that protein folding need not produce the most stable folded structure. Some claim to be able to predict the rate of folding of a polypeptide chain from its known three-dimensional structure, but such studies invariably omit the cases described here.

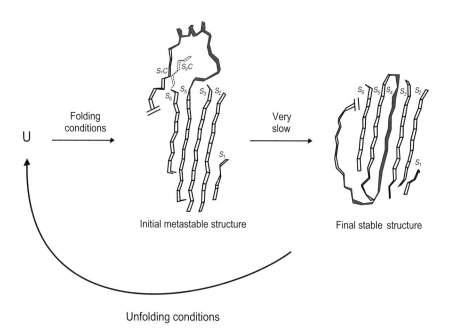

Figure 11-24. Kinetic determination of the folding pathway of α1-antitrysin. The unfolded protein folds initially to a metastable structure that slowly converts to the final stable structure. Although the final structure is the more stable, it is not generated directly from the unfolded protein without going through the metastable intermediate. The final slow step involves insertion of a new β-strand into the middle of a β-sheet. The rest of the protein structure is barely changed and is not shown.

Comprehensive analysis of protein folding activation thermodynamics reveals a universal behavior violated by kinetically stable proteases. S. S. Jaswal *et al.* (2005) *J. Mol Biol.* **347**, 355–366.

How well can simulation predict protein folding kinetics and thermodynamics? C. D. Snow *et al.* (2005) *Ann. Rev. Biophys. Biomol. Structure* **34**, 43–69.

A protein family under 'stress': serpin stability, folding and misfolding. G. L. Devlin & S. P. Bottomley (2005) *Front. Biosci.* **10**, 288–299.

11.4.C. Folding Coupled to Disulfide Formation

Folded proteins that contain disulfide bonds between the thiol groups of Cys residues often unfold if these bonds are broken or reduced. No denaturants are required and the reduced protein remains unfolded even under physiological conditions. When the disulfide bonds are permitted to reform, the protein can regenerate the native disulfide bonds and conformation. **Folding is then coupled to disulfide formation**. The great advantages of this type of folding are that (1) disulfide bond formation and breakage can be controlled experimentally using thiol–disulfide exchange with a disulfide reagent, (Section 7.2.I.6); (2) intermediates with incomplete or incorrect disulfide bonds can be forced to accumulate to substantial levels; (3) any disulfide bonds present in a protein molecule can be trapped in a form that can be stable indefinitely (Figure 11-25); (4) the pathway of disulfide formation can be dissected by replacing Cys residues or blocking their thiol groups irreversibly, decreasing the number of disulfide possibilities; (5) all the disulfide intermediates can be identified and characterized, and their roles in the folding process can be determined, usually unambiguously, to establish experimentally the folding pathway in terms of the disulfide bonds; and (6) the conformations of the trapped intermediates reveal the pathway in terms of the protein conformation.

The disulfide folding pathway that is best-characterized is that of BPTI (Figures 9-2 and 11-26). It is cooperative and has the general folding properties observed with proteins not containing disulfide

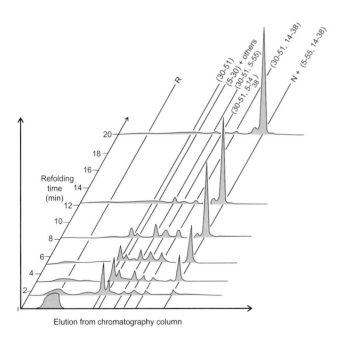

Figure 11-25. Isolation of intermediates trapped during the kinetics of refolding of reduced BPTI. Refolding and disulfide formation in the reduced protein were initiated by addition of glutathione disulfide, when all the disulfide bonds are formed at about the same rate. Disulfide formation was stopped at the indicated times by addition to 0.1 M iodoacetate, which rapidly reacts with all free thiol groups, converting free Cys residues to acidic carboxymethyl–Cys residues. Protein molecules with different numbers of disulfide bonds then differ in their net charges and are readily separated by ion-exchange chromatography. BPTI molecules are eluted roughly in order of their content of disulfide bonds, although molecules with the same number, but different pairings, of disulfides are also resolved, as the charge distribution on the protein molecule is also important in its binding to the resin. Fully reduced BPTI, R, is present initially and largely disappears after 1.5 minutes, when one-disulfide intermediates are near their maximum levels. Two-disulfide intermediates accumulate more slowly, reaching their maximum levels after about 5 minutes. The major intermediates are identified by the disulfides they contain. The peak labeled N eventually predominates; it contains fully refolded BPTI plus a quasi-native species lacking the 30–51 disulfide bond, in which the thiol groups are buried and unreactive. The small peak preceding N contains incorrectly folded molecules, with nonnative disulfide bonds. Data from T. E. Creighton (1984) *Methods Enzymol.* **107**, 305–329.

bonds. Reduced BPTI is a very unfolded polypeptide chain that approximates a random coil, but with a few weak local interactions between residues close in the primary structure. Of the four such interactions that have been well-characterized, only two contribute to the folding process and are retained in the final conformation; the other two are disrupted by the folding process. **The initial formation of disulfide bonds is approximately random** after any differences in reactivities of the thiol groups have been taken into account. The one-disulfide intermediates that are generated are not random, however; that with the Cys30–Cys51 disulfide bond prredominates because it adopts a stable partly folded conformation (Figure 10-7) in which the two major elements of secondary structure of the folded protein, the α-helix and β-sheet, are stable and interacting. The structure of this intermediate restricts and determines which disulfide bonds can be formed subsequently. There is a kinetic block of a large energy barrier in forming either the 30–51 or 5–55 disulfide bonds if the

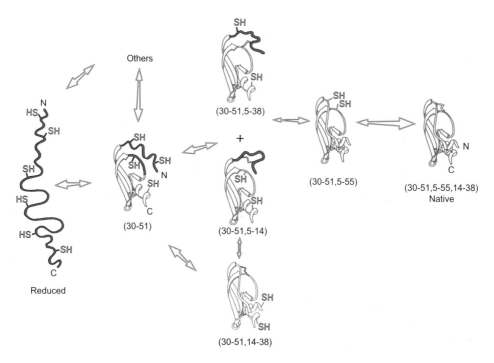

Figure 11-26. The disulfide folding pathway of BPTI. The polypeptide backbone of the protein (Figure 9-2) is depicted by a *smooth open line* when its conformation is not regular or well-defined, by *arrows* for β-strand conformation, and by a *coil* for α-helix. The approximate conformations of the intermediates are indicated. Disordered conformations are indicated in *red*, native-like conformations in *blue*. The thiol groups and disulfide bonds of the six Cys residues are depicted in *yellow*. The intermediates are designated by the disulfide bonds they contain in brackets. The relative rates of the intramolecular steps at alkaline pH are indicated semiquantitatively by the thickness of the appropriate arrowhead; the wider the arrowhead, the greater the rate in that direction. The fully reduced protein is unfolded under the normal conditions of folding. Consequently, formation of the initial disulfide bonds is nearly random. The (30–51) one-disulfide intermediate predominates. It is partly-folded (Figure 10-7) and in rapid equilibrium with the other one-disulfide intermediates, which seem to be largely unfolded. Three different second disulfide bonds, 14–38, 5–14 and 5–38, are formed readily in intermediate (30–51), 10^5-fold more rapidly than is disulfide 5–55. These three second disulfide bonds are rearranged intramolecularly to the native-like intermediate (30–51, 5–55), which readily forms the 14–38 disulfide bond to complete disulfide formation and refolding. The '+' between intermediates (30–51, 5–14) and (30–51, 5–38) indicates that they have comparable kinetic roles. The nonproductive quasi-native species (5–55) and (5–55,14–38) are omitted. Unfolding and disulfide breakage occur most readily by the reverse of this pathway. Data from T. E. Creighton (1990) *Biochem J.* **270**, 1–16.

other is already present, even though, but because, this would generate the native-like (30–51, 5–55) intermediate. This large energy barrier is the one that separates the fully folded conformation from those with partial or no fixed conformations. Native-like intermediates are normally formed most readily by intramolecular disulfide rearrangements of two intermediates with non-native second disulfide bonds (Figure 11-26). Non-native conformation does not guide this disulfide rearrangement pathway but this pathway predominates because the rearrangements are intramolecular, while disulfide formation involves bimolecular reactions with a disulfide reagent.

The disulfide pathways elucidated to varying extents with a few other small proteins show similar properties, but with variations. Initial disulfide formation in the reduced protein is approximately random until a stable folded conformation is adopted, which can either favor or disfavor formation of further disulfide bonds. Some proteins, such as BPTI (Figure 11-26), adopt partly folded conformations; some, such as α-lactalbumin, adopt the molten globule conformation; whereas others, such as ribonuclease A, remain unfolded until the entire folded conformation appears simultaneously. In the absence of a folded conformation, adding additional disulfide bonds becomes increasingly slow, and the disulfide bonds formed are less stable because of constraints on the flexibility of the unfolded polypeptide chain due to the existing disulfide bonds. There is a kinetic block in forming a disulfide bond if that bond will become buried in a stable folded conformation that results. Such kinetic blocks are due to the high energy barrier that is most apparent in the reverse direction, upon reducing a buried disulfide bond, which requires distorting the folded conformation to make the disulfide bond accessible to the reducing reagent. This kinetic barrier is usually overcome most readily by intramolecular rearrangements of protein disulfide bonds, instead of the intermolecular reactions that are required for direct protein disulfide formation or reduction by a disulfide reagent (Equation 7.34). The same energy barrier and the same conformational transitions appear to be involved in the two processes. The overall transition state occurs just before adopting the final native conformation and is a distorted high-energy form of the native state. It reflects the energy required to disrupt the cooperativity of the fully folded state and is the energy barrier limiting the conformational flexibility of that state.

When a protein has all but one or two of its native disulfide bonds, it can adopt a stable quasi-native conformation: the normal native conformation of BPTI is adopted if any one of the three disulfide bonds is missing, or if only the 5–55 disulfide bond is present. The stability of this quasi-native conformation depends upon which disulfide bonds are missing. The quasi-native conformation inhibits formation of native disulfide bonds that will be buried, but favors formation between those Cys residues on the surface that are brought into proximity by the native-like conformation. These quasi-native conformations illustrate that the disulfide bonds merely stabilize the native conformation and do not determine it.

Two-state folding of lysozyme versus multiple-state folding of α-lactalbumin illustrated by the technique of disulfide scrambling. L. Li & J. Y. Chang (2004) *Protein J.* **23**, 3–10.

Alteration of the disulfide-coupled folding pathway of BPTI by circular permutation. G. Bulaj *et al.* (2004) *Protein Sci.* **13**, 1182–1196.

Folding of small disulfide-rich proteins: clarifying the puzzle. J. L. Arolas *et al.* (2006) *Trends Biochem. Sci.* **31**, 292–301.

11.4.D. Proteins with Multiple Domains

Large proteins are comprised of two or more domains (Section 9.1.A). Each is usually comparable to a small single-domain protein. Such multi-domain proteins differ in the extent to which the domains interact with and stabilize each other. If the domains do not interact, or not very much, a single domain can be excised from the protein and it will usually maintain the same conformation. After unfolding, it can refold (Figure 11-4). **The folding of a single domain within a multi-domain protein is thought to involve the same principles as with single-domain proteins**, and multi-domain proteins are believed to fold in a modular fashion by their individual domains folding and then associating. The final step, in which the domains interact, often appears to be rate-limiting. Very often, the isolated domain folds more rapidly than when in the intact protein, suggesting that the unfolded domains interact and interfere with each other during folding, or that their rate of association is slow.

The folded domains and subunits that are present during refolding of these complex proteins often can recognize and bind their specific ligands. Consequently, the presence of such a ligand can increase the rate of refolding and assembly in the case of multi-domain proteins.

Multi-domain proteins are especially susceptible to aggregation during refolding. This is thought to be due to specific complementary interactions between the domains of different molecules, similar to those that should occur between the domains of the same polypeptide chain (Figure 11-2) or in domain swapping (Figure 9-30).

Stability and folding of domain proteins. R. Jaenicke (1999) *Prog. Biophys. Mol. Biol.* **71**, 155–241.

Autonomous protein folding units. Z. Y. Peng & L. C. Wu (2000) *Adv. Protein Chem.* **53**, 1–47.

Stepwise unfolding of ankyrin repeats in a single protein revealed by atomic force microscopy. L. Li *et al.* (2006) *Biophys. J.* **90**, L30–32.

Apparent cooperativity in the folding of multidomain proteins depends on the relative rates of folding of the constituent domains. S. Batey & J. Clarke (2006) *Proc. Natl. Acad. Sci. USA* **103**, 18113–18118.

11.4.E. Proteins with Multiple Subunits

Quaternary structure consists of multiple polypeptide chains that can be dissociated and unfolded under denaturing conditions; often they can then refold and reassemble to regenerate the original tertiary and quaternary structures. Folding of the individual domains and their reassembly can usually be observed as separate transitions (Figure 11-4). The assembly process can be monitored by covalent cross-linking at different times. **Whether folding or assembly is rate-limiting depends upon the protein concentration**, as the rate of assembly is second-order, or greater, and thus more rapid at higher protein concentrations, whereas folding is unimolecular and first-order. Kinetic results with a number of proteins are consistent with the individual subunits folding and then reassembling. The assembly process is often slow, however, so it may involve some changes in the tertiary structures of the individual subunits. Elucidating the details has been difficult due to the large sizes and complexities of multi-subunit proteins. The proteins also usually have a tendency to aggregate during refolding, especially at high protein concentrations. Other proteins do not usually have an effect on aggregation, unless very closely related, so the interactions responsible for aggregation appear to be specific (Figure 11-2).

Some dimeric proteins have the two polypeptide chains very entwined in their native conformations (e.g. Figure 9-26-C). It would not seem possible for the same conformation to be stable with a single chain, in the absence of the other subunit, but such proteins can refold and reassemble. In the case of the intertwined dimer of the *trp* repressor (Figure 9-26-C), a rapid dimerization reaction between partially folded monomers is followed by isomerization of the dimeric intermediates to yield the native dimer. The individual monomers exist in two distinct, slowly interconverting, conformations. The dimer is formed by random association of the two different monomeric forms, which involves alternative packings of the core dimerization domain and the DNA-binding domain. One, perhaps both, of these packing modes involves non-native contacts.

Folding and association of oligomeric and multimeric proteins. R. Jaenicke & H. Lilie (2000) *Adv. Protein Chem.* **53**, 329–401.

How the protein concentration affects unfolding curves of oligomers. R. Ragone (2000) *Biopolymers* **53**, 221–225.

Recognition between flexible protein molecules: induced and assisted folding. A. P. Demchenko (2001) *J. Mol. Recognit.* **14**, 42–61.

Rough energy landscapes in protein folding: dimeric *E. coli* Trp repressor folds through three parallel channels. L. M. Gloss *et al.* (2001) *J. Mol. Biol.* **312**, 1121–1134.

11.4.F. Competition with Aggregation and Precipitation

Unfolded proteins are notoriously insoluble and frequently precipitate when unfolding has been caused by thermal denaturation or extremes of pH. Denaturants such as urea and GdmCl act by solubilizing proteins, so the unfolded protein remains soluble, but it frequently precipitates as soon as the denaturant is diluted to permit refolding. This precipitation is usually specific to the protein, so specific interactions are believed to occur between molecules due to specific conformations adopted by the unfolded protein, like those thought to occur between protein domains or subunits (Figure 11-2). Precipitation can often be avoided by lowering the concentration of the protein, but this requires working with large quantities of solution and recovery of the dilute protein can be problematic.

Precipitation is a major problem in the production of useful proteins; many of those produced by genetic engineering are synthesized in very high quantities within cells that are not their normal hosts, and the newly synthesized polypeptide chains tend to aggregate with other components in the cell. They end up in **inclusion bodies**, dense bodies of aggregated proteins that can be dissolved only under very extreme conditions. The protein released is invariably unfolded and must be renatured before it can be useful. One way to avoid the problems of working at high dilution is to adsorb the unfolded protein while in urea to a solid matrix, such as an ion-exchange resin, on which it can refold while remaining adsorbed. The urea is removed gradually by washing the matrix with an inverse gradient of decreasing urea concentrations to permit the protein to refold. The protein can then be eluted in a folded form, which can provide a purification step as in chromatography.

Single-step recovery and solid-phase refolding of inclusion body proteins using a polycationic purification tag. M. Hedhammar *et al.* (2006) *Biotechnol. J.* **1**, 187–196.

On-column refolding of recombinant chemokines for NMR studies and biological assays. C. T. Veldkamp *et al.* (2007) *Protein Expr. Purif.* **52**, 202–209.

Current status of technical protein refolding. A. Junbauer & W. Kaar (2007) *J. Biotechnol.* **128**, 587–596.

11.4.G. Differences with Folding *in Vivo*

Protein folding *in vivo*, during or after biosynthesis of the polypeptide chain, differs fundamentally from refolding *in vitro*: (1) domains can fold sequentially as the polypeptide chain is synthesized from its N-terminus; (2) folding involves the concerted action of **chaperone** proteins, which bind to the polypeptide chain before folding and thereby minimize illicit interactions like those in Figure 11-2; (3) the slow steps, such as *cis–trans* peptide bond isomerization and disulfide bond formation, breakage and rearrangement, are catalyzed by proteins such as cyclophilins and protein disulfide isomerase. These intriguing and complex phenomena have been characterized extensively but are beyond the scope of this volume.

Molecular chaperones and protein quality control. B. Bukau *et al.* (2006) *Cell* **125**, 443–451.

Pharmacological targeting of catalyzed protein folding: the example of peptide bond *cis/trans* isomerases. F. Edlich & G. Fischer (2006) *Handb. Exp. Pharmacol.* **172**, 359–404.

Molecular chaperones: assisting assembly in addition to folding. R. J. Ellis (2006) *Trends Biochem. Sci.* **31**, 395–401.

Protein misassembly: macromolecular crowding and molecular chaperones. R. J. Ellis (2007) *Adv. Exp. Med. Biol.* **594**, 1–13.

~ CHAPTER 12 ~

LIGAND BINDING BY PROTEINS

The biological functions of macromolecules almost invariably depend upon their direct physical interaction with other molecules. All organisms and cells survive only because they interact effectively with their environments. Extraneous molecules must be distinguished as either useful or dangerous and dealt with accordingly. Organisms have sophisticated sensory organs for being aware of their environments. At a lower level, all cells have a variety of receptors for different molecules for this purpose. Virtually every small molecule in a cell was first bound specifically by the enzyme that produced it or by the receptor on the cell that enabled it to enter that cell. Every aspect of the structure, growth and replication of an organism depends upon proteins binding small molecules, other proteins, nucleic acids, polysaccharides or lipids. Of crucial importance is the specificity of such interactions. In the crowded interior of a cell, each molecule must interact only with the appropriate molecules and not with any of the others that are present, often in extremely high concentrations.

The following discussion will be general but will apply primarily to proteins binding smaller ligands, such as their substrates for enzymatic reactions, although the catalytic events that take place after binding on the enzyme will be saved for Chapter 14. The interactions between nucleic acids and proteins are described in Chapter 13; base-pairing interactions between complementary strands of DNA and RNA are described in Chapter 5, and their interactions with small molecules are described in Sections 2.5 and 14.6.

Only binding to specific sites on a protein will be considered here, and phenomena such as the electrostatic binding of counterions to a polyanion or polycation (Section 2.3) will not be considered here as ligand binding. Likewise, little attention will be given to interactions with normal components of the solvent, such as water molecules and co-solvents in the case of water-soluble proteins and lipids in the case of membrane proteins. These interactions are not fundamentally different but they do not occur at specific sites on the protein and are generally weak, occurring only because the solvent molecules are present at high concentrations.

Binding and Linkage: functional chemistry of biological macromolecules. J. Wyman & S. J. Gill (1990) University Science Books, CA.

Ligand binding. A. Levitzki (1997) in *Protein Function: a practical approach*, 2nd edn (T. E. Creighton, ed.), IRL Press, Oxford, pp. 101–129.

Structural and mechanistic determinants of affinity and specificity of ligands discovered or engineered by phage display. B. A. Katz (1997) *Ann. Rev. Biophys. Biomol. Structure* **26**, 27–45.

Differences between non-specific and bio-specific, and between equilibrium and non-equilibrium, interactions in biological systems. J. Israelachvili (2005) *Quart. Revs. Biophys.* **38**, 331–337.

12.1. GENERAL PROPERTIES OF PROTEIN–LIGAND INTERACTIONS

Proteins usually bind only very specific ligands and can discriminate between closely related molecules. This specificity is usually crucial for their biological functions. Proteins are generally classified according to the purpose and consequences of their binding; examples are structural proteins, enzymes, repressors, lectins, toxins, immunoglobulins, hormones, receptors, membrane transport proteins and proteins of motility. The physical principles of the interactions are similar in all these cases. The following discussion focuses on the protein; whatever molecule it interacts with, even if another protein, is designated the **ligand**. A protein with its ligand bound is known as the **holo** form; that without the ligand is the **apo** form. Some examples of specific complexes are illustrated in Figure 12-1.

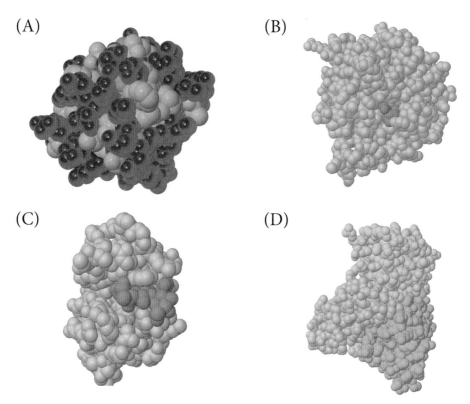

Figure 12-1. Examples of the structures of protein–ligand complexes. (A) Parvalbumin binding two Ca^{2+} ions (Section 12.3), which are barely visible as the two *green spheres*. All the atoms are shown as the appropriate van der Waals spheres; hydrophobic residues are *gray*, polar residues *mauve*. Generated from PDB file *2pas*. (B) The enzyme carboxypeptidase A binding the inhibitor D-((N-hydroxyamino)carbonyl)phenylalanine. A zinc ion is also bound beneath the inhibitor as part of the enzyme active site. All the protein atoms are *blue*, whereas the inhibitor is *green*. Generated from PDB file *1hdq*. (C) Hen lysozyme binding a trisaccharide in the substrate-binding cleft. Generated from PDB file *1uib*. (D) Hen lysozyme bound to the variable domain of an antilysozyme antibody molecule; the lysozyme is *rose* colored, while the two polypeptide chains of the variable domain are *blue* and *green*. Generated from PDB file *2eks*. Figure generated using the program Jmol.

Specific ligands for proteins, such as the substrates of enzymes, usually have one specific binding site on each polypeptide chain or domain, and multiple binding sites for the same ligand occur primarily when the protein has multiple subunits. One structural domain can bind more than one ligand at separate binding sites, but it is unusual for there to be binding sites for more than two or three different ligands on any one domain. A protein that binds a number of different ligands often binds them on separate domains (Figure 12-2). Nevertheless, there are always exceptions to these generalities. For example, cytochrome c_3 is a single polypeptide chain of only 118 residues yet it binds four identical heme groups in different environments. The bacteriochlorophyll protein from green photosynthetic bacteria binds seven chlorophyll molecules, each in a different position within a 'string bag' of a 15-stranded β-barrel.

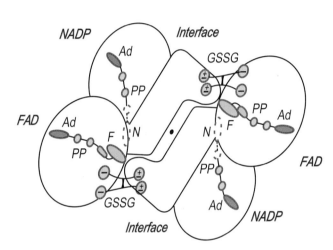

Figure 12-2. The domain structure and ligand binding sites of the dimeric enzyme glutathione reductase. The two-fold axis relating the two polypeptide chains is indicated by the dot in the center. The names of the domains are in *red italics*; the other letters refer to the ligands. The two interface domains bind to each other and determine the dimeric structure. The NADP-binding domain has the adenine (Ad), ribose (circles), diphosphate (PP) and nicotinamide (N) moieties indicated; only the nicotinamide portion is fully buried. The FAD-binding domain has the coenzyme indicated similarly, except that F is the flavin moiety; only the adenine group is exposed to the solvent. The substrate, oxidized glutathione (GSSG), binds between the two subunits. From a diagram by G. E. Schulz.

Small ligands tend to bind to depressions on or near the surfaces of proteins, or in the clefts between domains, presumably so that they can readily bind and then dissociate (Figure 12-1). The size of the depression depends upon the size of the ligand. A very small ligand, such as a metal ion (Section 12.2), tends to be bound inside a relatively spherical protein molecule, where it is surrounded by the protein, probably to ensure sufficient interactions so that only the appropriate ligand is bound; flexibility of the structure permits small ligands to diffuse through the interior rapidly. Extremely small ligands, such as electrons and oxygen molecules, usually bind to a special bound cofactor, rather than directly to the protein; some examples are given in Figure 12-3. Long, linear ligands, such as polysaccharides, tend to be bound in clefts on the surface. If protein and ligand are of similar size, their interface tends to be flat and extensive (Section 9.1.H). With very large ligands, such as nucleic acids, the protein tends to bind to depressions on the surface of the ligand (Section 13.3).

There is generally both steric and physical complementarity between the ligand and its protein binding site. These interactions are similar to those observed within protein interiors; indeed, deleting a group of atoms from a protein structure introduces a site for binding an external ligand that can mimic the deleted atoms. The interface between ligand and protein is closely packed, and all polar groups are paired in hydrogen bonds, often with intervening water molecules. Electrostatic charges are generally neutralized, often by the dipoles at the ends of α-helices (Section 8.4.A), or very effectively 'solvated' by multiple hydrogen bonds to a number of groups (Figure 12-4). Such hydrogen bond arrays appear to disperse the formal buried charge.

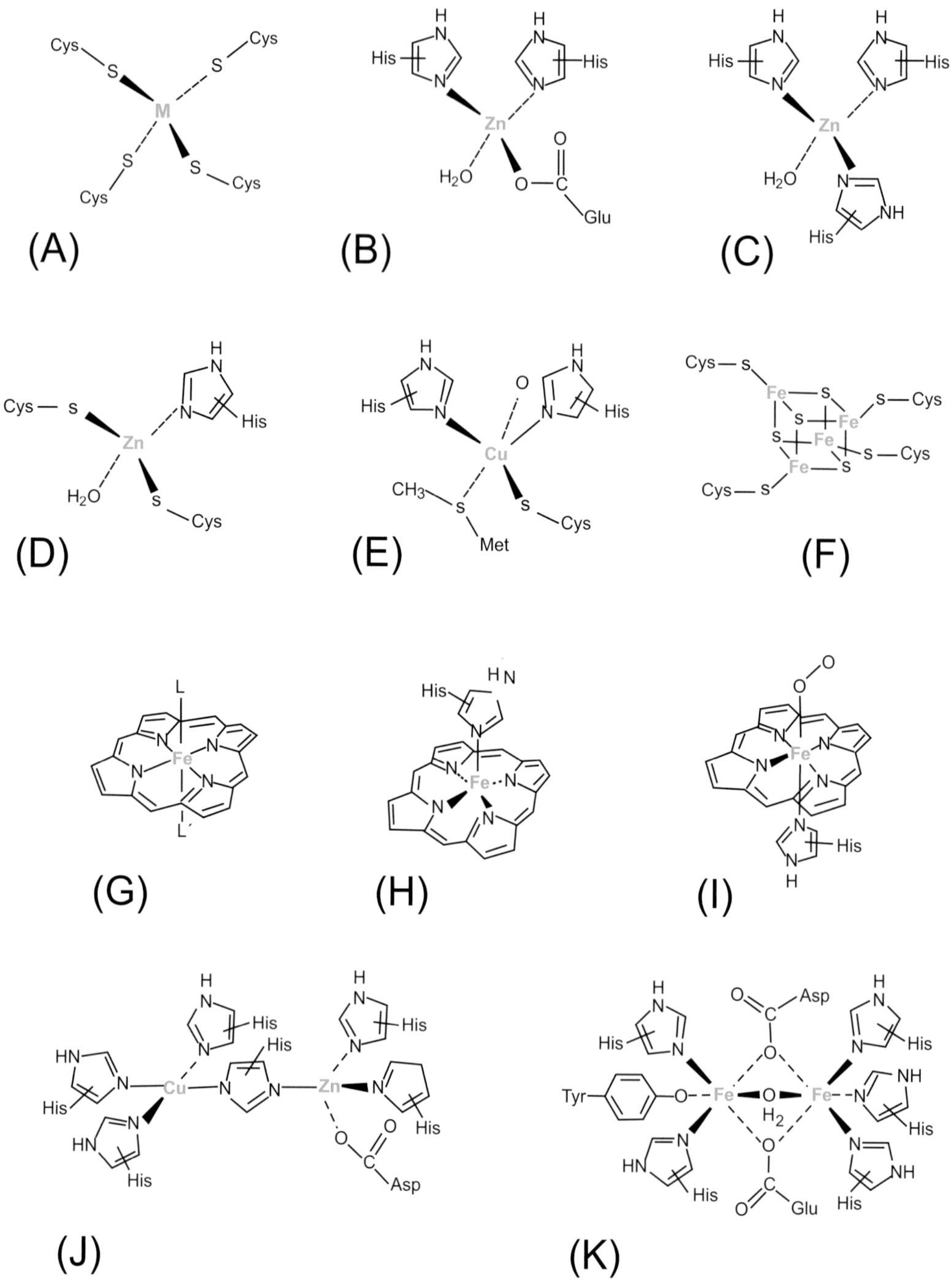

Figure 12-3. Examples of some metal ions bound tightly to proteins, with the partial structures of the amino acid side-chains to which they are attached. (A) M is either Fe, as in rubredoxin, or Zn, as in aspartate transcarbamoylase and liver alcohol dehydrogenase. (B) Carboxypeptidase A. (C) Carbonic anhydrase and insulin. (D) Liver alcohol dehydrogenase. (E) Azurin and plastocyanin. (F) Some ferredoxins and high-potential iron protein; planar (2Fe–2S) and (3Fe–3S) complexes are also found in other ferredoxins. (G) Heme group of cytochromes; L is usually His and L' is His in cytochrome b_5 or Met in cytochrome c. (H) Deoxy heme group in myoglobin and hemoglobin. (I) Oxy form of (H). (J) Methemerythrin. (K) Superoxide dismutase.

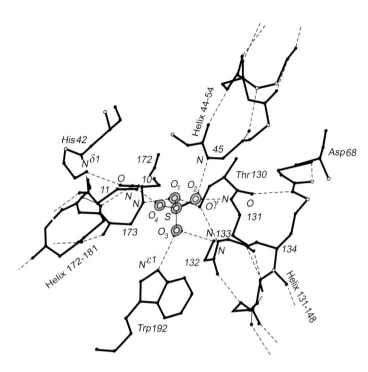

Figure 12-4. Binding of the SO_4^{2-} dianion to the sulfate-binding protein involved in bacterial active transport. The SO_4^{2-} molecule (depicted with *yellow* atoms) is entirely inaccessible to solvent yet there is no positively charged group nearby, only the N-termini of three α-helices. The crystal structure is consistent with seven hydrogen bonds between the ligand and the protein; all hydrogen bonds in the vicinity of the anion are indicated by the *dashed lines*. From a drawing by F. A. Quiocho.

The structure of a protein domain generally does not change substantially when it binds a small ligand; the apparent movements are usually similar to those expected from the known flexibility of protein structures. The most extreme changes within domains generally involve movements of intrinsically flexible loops on the protein structure; often they close over a bound ligand, especially if it is to be altered by enzyme catalysis. Some flexibility is probably required for ligands to associate and dissociate, but this is limited in scope. Exceptions are usually of functional importance, as in allostery (Section 12.5).

An analogy is often made with a key (the ligand) fitting into a lock (the protein). The rigidity implied by this analogy is too extreme, but it does indicate that the complementarity between ligand and protein on the whole usually pre-exists in the folded protein structure. This is undoubtedly important for the specificity of binding (Figure 12-5). A very malleable protein would adapt its shape to match that of many ligands and would bind many of them with similar affinities. The difference in affinity for two ligands is limited by the energy that would be required to distort the normal conformation to one that is complementary to the low-affinity ligand. This energy appears to be substantial and the low affinities of certain ligands simply reflect their noncomplementarity to the pre-existing binding site.

Ligands that bind between two or more domains, or subunits, often do cause changes in the relative positions of the domains or subunits. These changes can produce functional changes at other sites on the protein, as in allosteric proteins (Section 12.5). Usually, the domains or subunits move together to engulf the ligand (Figure 12-6). This movement maximizes the interactions between the protein and the ligand and minimizes interactions with other components of the solvent, while permitting the ligand to associate and dissociate.

There are exceptions to all the above generalities, especially when required for the function of the protein.

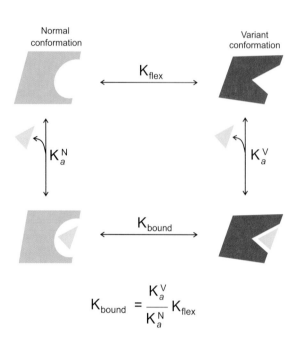

Figure 12-5. Relation between flexibility of a protein structure and its ability to bind ligands specifically. The normal conformation of a protein is imagined not to be complementary to a potential ligand (designated by the *triangle*) and consequently has poor affinity for that ligand. A variant conformation of the protein resulting from its flexibility might be complementary to the ligand and bind it tightly, so $K_a^V \gg K_a^N$, but this conformation would not normally be populated substantially, i.e. $K_{flex} \ll 1$. When the ligand is bound, however, the variant conformation should be more stable by the ratio of the two binding affinities, so it might be populated in the liganded state. For example, if $K_{flex} = 10^{-3}$ and the variant conformation binds the ligand 10^4 times more tightly than the normal conformation, K_{bound} will be 10, so the variant conformation should be present in the complex 91% of the time. Adapted from T. E. Creighton (1993) *Proteins: structures and molecular properties*, 2nd edn, W. H. Freeman, NY, p. 337.

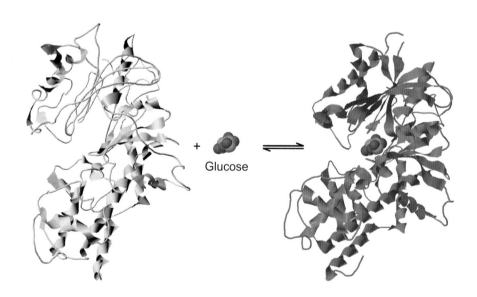

Figure 12-6. The conformational change that occurs upon glucose binding to yeast hexokinase. The two lobes of the enzyme engulf the glucose molecule. The hexokinase is illustrated schematically, while the glucose is illustrated with the van der Waals spheres for its atoms. Figure generated from PDB files *1hkg* and *3b8a* using the program Jmol.

Investigation of molecular interaction within biological macromolecular complexes by mass spectrometry. S. Akashi (2006) *Med. Res. Rev.* **26**, 339–368.

On the nature of cavities on protein surfaces: application to the identification of drug-binding sites. M. Nayal & B. Honig (2006) *Proteins* **63**, 892–906.

Water mediation in protein folding and molecular recognition. Y. Levy & J. N. Onuchic (2006) *Ann. Rev. Biophys. Biomolec. Structure* **35**, 389–415.

How different are structurally flexible and rigid binding sites? Sequence and structural features discriminating proteins that do and do not undergo conformational change upon ligand binding. K. Gunasekaran & R. Nussinov (2007) *J. Mol. Biol.* **365**, 257–273.

12.2. METALLOPROTEINS

More than 30% of cellular proteins bind metal ions, either permanently as prosthetic groups or more reversibly as ligands. These metal ions play a variety of roles: transfer of electrons, maintenance of the protein structure, binding oxygen, generation of coordinated hydroxide radicals, binding of substrates and electrophilic catalysis. Metal ions are especially simple ligands, simple spheres, and their binding and discrimination requires some special chemistry; in these cases, a number of groups from the protein interact with the spherical metal ion ligand and are designated as the ligands to it. Some examples are given in Figure 12-3.

The most important aspect of a metal ion to govern its interaction with proteins is its size (Table 12-1). This is often even more important than its charge. The radius of any metal ion increases with the number of its ligands (the **coordination number**) because greater numbers of bonds to ligands weaken the strength of any one bond. Ca^{2+} and Na^+ often interchange in biological systems, in spite of having different charges, because they are of similar size. Ba^{2+} is poisonous because it competes with K^+, which has a similar radius, and not with Ca^{2+}, which is of identical charge but smaller radius.

The incorporation of a particular trace metal ion into an apoprotein is influenced by its ionic radius, thermodynamic stability, ligand-substitution kinetics and charge (Table 12-2). The amino acid residues that regularly act as metal ligands in proteins are the thiol groups of Cys residues, imidazole groups of His, carboxyl groups of Asp and Glu, and the phenolic hydroxyl groups of Tyr (Section 7.2). Each metal is considered a Lewis acid and favors different sets of protein ligands, which are frequently dictated by the 'hard–soft' theory of acids and bases. The coordination number and geometry of each metal site is determined by the metal's oxidation state, although substantial distortions from the idealized structures occur in metalloproteins. Three to four ligands, usually the side-chains of Cys and His residues, most typically complete the coordination sphere about a bound metal ion. The metal stoichiometry is usually measured by titration or atomic absorption spectrometry. The nature of the metal coordination sphere can be probed by electronic absorption spectroscopy in cases involving iron or copper coordination.

X-ray absorption spectroscopy is particularly useful for the investigation of the local environment within 5 Å of metals in proteins. When X-rays are absorbed by metal atoms, they liberate electrons that are backscattered from neighboring atoms and produce an interference phenomenon that provides very precise measurements of the distances to neighboring atoms and provides information about their identities, as well as about the valence state and coordination geometry of the metal. The extended X-ray absorption fine structure (EXAFS) region of the spectrum contains information about the number and average distances of neighboring atoms and their relative disorder.

Table 12-1. Effective ionic radii (Å) of some cations

	Coordination number[a]					
Ion	4	5	6	7	8	9
Ca^{2+}			1.00	1.06	**1.12**	1.18
Cd^{2+}	0.78	0.87	**0.95**	1.03	1.10	
Co^{2+}	0.58	0.67	**0.74**		0.90	
Cu^{2+}	0.57	0.65	0.73			
Fe^{2+}	0.63		**0.78**[b]		0.90	
Fe^{3+}	0.49	0.58	**0.64**[c]		0.78	
K^+	1.37		1.38	1.46	**1.51**	1.55
Mg^{2+}	0.57	0.66	**0.72**		0.89	
Mn^{2+}	0.66	0.75	**0.83**	0.90	0.96	
Na^+	0.99	1.00	**1.02**	1.12	1.18	1.24
Ni^{2+}	0.55	0.63	**0.69**			
Pb^{2+}	0.98		1.19	1.23	1.29	1.35
Zn^{2+}	0.60	0.68	0.74		0.90	

[a] The radius of the most common coordination number is in bold.

[b] High spin; the low-spin radius is 0.61 Å.

[c] High spin; the low-spin radius is 0.55 Å.

Data from R. B. Martin (1996) in *Encyclopedia of Molecular Biology and Molecular Medicine* (R. A. Meyers, ed.), VCH, Weinheim, p. 128.

Table 12-2. Coordination environments preferred by common metal ions found in proteins

Metal ion	Coordination number	Geometry	Interacting groups
Hg^{2+}	3	Trigonal	Thiol
Cd^{2+}	4	Tetrahedral	Thiol
Cu^+	4	Tetrahedral	Thiol
Fe^{2+}	4	Tetrahedral	Thiol, imidazole, carboxyl
Zn^{2+}	4	Tetrahedral	Thiol, imidazole

Data from W. R. Ellis & L. S. Powers (1999) in *Encyclopedia of Molecular Biology* (T. E. Creighton, ed.), Wiley-Interscience, NY, p. 1477.

Metalloproteins differ in the role of the metal ion in the folding and structure of the protein. Some metal ions bind only after the apoprotein has completed folding and presents an intact binding site; in this case, the metal ion can be removed without altering the protein's structure. In other cases, the metal ion binds to the protein before it has completed folding, and subsequently removing the ion often disrupts the folded structure.

Ligand–metal ion binding to proteins: investigation by ESI mass spectrometry. N. Potier *et al.* (2005) *Methods Enzymol.* **402**, 361–389.

Investigation of structure, dynamics and function of metalloproteins with electrospray ionization mass spectrometry. I. A. Kaltashov *et al.* (2006) *Anal. Bioanal. Chem.* **386**, 472–481.

'Four-dimensional' protein structures: examples from metalloproteins. M. Fragai *et al.* (2006) *Acc. Chem. Res.* **39**, 909–917.

Stereochemistry of guanidine–metal interactions: implications for L-arginine–metal interactions in protein structure and function. L. DiCostanzo *et al.* (2006) *Proteins* **65**, 637–642.

12.2.A. Chelation: Synergy Between Ligands

Metal ions have intrinsic affinities for certain atoms but they are relatively weak. Tight binding is obtained, however, by combining several such groups on the same molecule in appropriate positions. For example, Ni^{2+} ions bind weakly to individual NH_3 molecules, but 4×10^4 times more tightly when two amino groups are in reasonable proximity on one molecule (Table 12-3). With additional amino groups, the binding affinity increases, until with five the association constant has increased 6×10^{14} times! Similarly, Zn^{2+} ions bind weakly to acetate groups, with $K_d = 0.1$ M, but they bind to the four carboxyl groups of EDTA (Figure 12-7) simultaneously with a dissociation constant of 4×10^{-17} M (Table 12-4). **Simultaneous binding by multiple groups held in proximity is the basis of the chelate effect**.

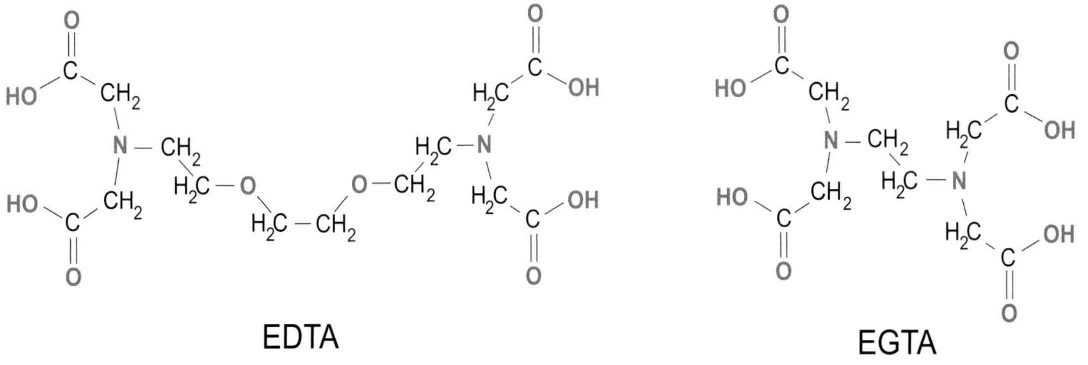

Figure 12-7. The chemical structures of EDTA (ethylenediamine-*N*,*N*,*N'*,*N'*-tetracetic acid) and EGTA (ethyleneglycol bis(β-aminoethyl ether) *N*,*N*,*N'*,*N'*-tetracetic acid). The atoms involved in binding directly to metal ions are in *red*.

Table 12-3. The chelate effect on the binding of amines to Ni^{2+}

Ligand	Dissociation constants (M)		
	K_1 [a]	K_2 [b]	K_3 [c]
NH_3 [d]	2.1×10^{-3}	7.6×10^{-3}	0.024
$H_2N(CH_2)_2NH_2$	5×10^{-8}	1×10^{-6}	5×10^{-5}
$H_2N(CH_2)_2HN(CH_2)_2NH_2$	1.7×10^{-11}	1×10^{-8}	
$H_2N(CH_2)_2HN(CH_2)_2NH(CH_2)_2NH_2$	5×10^{-15}		
$H_2N(CH_2)_2HN(CH_2)_2NH(CH_2)_2NH(CH_2)_2NH_2$	3.3×10^{-18}		

[a] For first ligand.

[b] For second ligand molecule.

[c] For third ligand molecule.

[d] Up to six NH_3 molecules can bind simultaneously to one Ni^{2+} ion.

Data from the Chemical Society (London) Special Publications **17** (1964) and **25** (1971).

Table 12-4. Affinities of EDTA and EGTA for various cations

	Dissociation constant (M)	
Metal ion	EDTA	EGTA
Mg^{2+}	2.3×10^{-9}	5.0×10^{-6}
Ca^{2+}	1.0×10^{-11}	1.3×10^{-11}
Zn^{2+}	4.0×10^{-17}	2.5×10^{-13}
Cd^{2+}	4.0×10^{-17}	3.2×10^{-17}
Mn^{2+}	1.6×10^{-14}	
Cu^{2+}	2.0×10^{-19}	
Fe^{2+}	4.7×10^{-15}	
Fe^{3+}	5.9×10^{-25}	
Co^{2+}	5.0×10^{-17}	
Co^{3+}	1.0×10^{-36}	
Pb^{2+}	5.0×10^{-19}	

At pH 7.0, 25°C, and 0.1 M ionic strength. Data from A. E. Martell & R. M. Smith (1974) *Critical Stability Constants*, Plenum Press, NY.

EDTA is used widely to control the concentrations of divalent and trivalent cations in aqueous solution. The flexibility of the molecule, and the variety of potential liganding sites within it, permit EDTA to accommodate and bind tightly a wide variety of cation types, with various preferred ligand geometries and bond lengths (Table 12-4). In contrast to many other chelating agents, EDTA forms only a 1:1 complex with either divalent or trivalent cations, in which a single ion is chelated by all four carboxyl groups. The binding affinity is dependent upon the pH, because protonation of the carboxyl groups at low pH lowers the affinity for cations.

A related agent, **EGTA** (Figure 12-7), is often used to control Ca^{2+} levels in the presence of physiological concentrations of Mg^{2+} ions, because it binds Ca^{2+} ions 3×10^5-fold more tightly. The much greater affinity for Ca^{2+} is explained because EGTA wraps around the ion in a very stable octadentate complex (Figure 12-8). Such a complex is not possible with the smaller Mg^{2+} ion because of its more stringent binding requirements. The preference of Mg^{2+} for oxygen ligands over nitrogen is so great that it binds to EGTA via the carboxylate groups only.

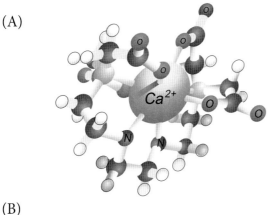
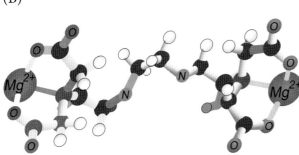

Figure 12-8. Structures of the complexes of EGTA with (A) Ca^{2+} and (B) Mg^{2+} ions. The *small white spheres* are H atoms, the *darkest spheres* C. The other atoms are labeled. EGTA binds a single Ca^{2+} ion very tightly, using its four carboxyl groups and two N atoms, but it binds Mg^{2+} ions much less tightly, because this ion binds to only two carboxyl groups.

Energetics of Ca^{2+} EDTA interactions: calorimetric study. Y. V. Griko (1999) *Biophys. Chem.* **79**, 117–127.

12.2.B. Zinc-binding Proteins

Zinc is, after iron, the most plentiful of the trace metal ions in humans. More importantly, zinc is soluble, forms a stable Zn^{2+} ion and does not undergo biological reduction–oxidation chemistry. Zinc is a strong Lewis acid (having empty *d* electron orbitals that attract electrons), exchanges ligands rapidly and is content in several different coordination geometries. It is important in some enzymes, using its electrophilic character, or it can play solely a structural role. Zinc-binding proteins are most conspicuous as transcription factors (Section 13.3.C), such as the zinc fingers, where the zinc ions normally play a structural role. In general, structural zinc ions adopt a tetrahedral arrangement formed by atoms from the side-chains of Cys or His residues, occasionally Asp or Glu. **Of the protein ligands to structural Zn^{2+} ions, about half are His residues, a quarter Cys, about 15% Asp and about 5% Glu.** The amino acid sequences characteristic of the zinc-binding motifs are described in Table 12-5.

Table 12-5. Major families of zinc metalloregulatory proteins

Family	Consensus sequence	Ligand
GAL4	$CX_2CX_{9-27}CX_2CX_6C$	DNA
Retroviral nucleocapsid	$CX_2CX_4HX_4C$	RNA
Nuclear hormone receptor	$CX_2CX_{13}CX_2CX_{15}CX_5CX_{12}CX_4C$	DNA
Zinc finger	$CX_{2-4}CX_{12}HX_{3-5}H$	DNA and RNA

C, cysteine residue; H, histidine; X, any residue.

Data from W. R. Ellis & L. S. Powers (1999) in *Encyclopedia of Molecular Biology* (T. E. Creighton, ed.), Wiley-Interscience, NY, p. 2815.

Coordination numbers of 4, 5 and 6 are observed with Zn^{2+}, which are designated as T_4, T_5 and T_6 (Figure 12-9). The ideal geometries minimize repulsions between the electron pairs of the interacting ligands. Such ideal geometries are not possible if two atoms that are closely bonded in a molecule, such as the two oxygens of a carboxyl group, interact simultaneously in a bidentate manner with the Zn^{2+} ion. Sulfur-containing ligands display a marked preference for Zn^{2+} ions, so it is not surprising that Cys residues play such a prominent role in binding zinc in biology.

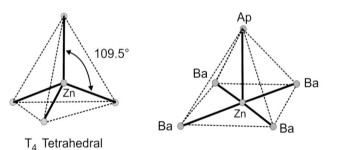

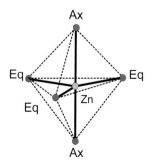

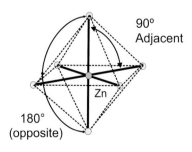

Figure 12-9. The ideal geometries adopted by Zn^{2+} ions in complexes. The ligands are labeled as Ba (basal), Ap (apical), Eq (equatorial) and Ax (axial). Adapted from I. L. Alberts *et al.* (1998) *Protein Sci.* **7**, 1700–1716.

The majority of structural Zn^{2+} sites in proteins have tetrahedral geometry, although all four of the geometries of Figure 12-9 are found. Those in nuclear hormone receptors possess tetrahedral environments of four Cys residues. Fungal transcription factors contain a $(Cys)_3$-Zn- (Cys_2) -Zn- $(Cys)_3$ dimer, with two Cys residues bridging the two zinc centers. Nucleocapsids of retroviruses (e.g. HIV-1) contain large amounts of Zn^{2+}, probably ligated by three Cys residues and one His.

Zn^{2+} is colorless and consequently not useful for spectroscopic studies, but it can be replaced with Co^{2+}, a chromophoric probe that possesses a comparable ionic radius, is kinetically labile and prefers similar coordination environments. Co^{2+} ions have 10^3-fold lower affinities, however, for the tetrahedral environments usually occupied by Zn^{2+} ions.

Analysis of zinc binding sites in protein crystal structures. I. L. Alberts *et al.* (1998) *Protein Sci.* **7**, 1700–1716.

Structural investigations of calcium and zinc binding in proteins. D. E. Brodersen & M. Kjeldgaard (1999) *Sci. Prog.* **82**, 295–312.

A new zinc binding fold underlines the versatility of zinc binding modules in protein evolution. B. K. Sharpe *et al.* (2002) *Structure* **10**, 639–648.

12.2.C. Metallothioneins

Metallothioneins are synthesized by organisms in response to heavy metal ions, especially Cd^{2+}, Cu^{1+} and Zn^{2+} ions, and they sequester these ions in protein-bound metal clusters (Figure 12-10). **Metallothioneins generally consist of 61–68 residues, of which 20 are Cys, with no aromatic residues** (Phe, Tyr, Trp or His). Humans produce four different metallothioneins, which differ somewhat in the residues other than Cys. The amino acid sequence of metallothionein-2 from humans is (using the one-letter code):

$$\overset{1}{\text{Acetyl-M}}\text{-D-P-N-C-S-C-A-A-G-D-S-C-T-C-A-G-S-C-K-C-K-E-C-K-C-T-S-}\overset{30}{\text{C-K-}}$$

$$\overset{31}{\text{K-S-}}\text{C-C-S-C-C-P-V-G-C-A-K-C-A-Q-G-C-I-C-K-G-A-S-D-K-C-S-C-}\overset{61}{\text{C-A}}$$

(12.1)

The 20 conserved Cys residues are in yellow. The thiol groups of these Cys residues are the ligands for up to seven divalent metal ions (Zn^{2+} or Cd^{2+}), which are bound in two clusters of three and four metal atoms in the two domains of the protein. A flexible hinge region containing a conserved Lys–Lys segment connects the two domains. The Zn^{2+} or Cd^{2+} ions in both clusters are tetrahedrally coordinated by individual Cys thiol groups and eight bridging S atoms that bind to two different metal atoms. The metallothioneins occur naturally with ions bound in this way.

Alternatively, up to 12 Cu^{1+} ions can be bound by one molecule. Six are bound to each domain, and each Cu^{1+} ion is bound by two or three Cys residues.

Although the metal ion clusters appear to be very stable, the ions exchange rapidly with others in solution. In the absence of metal ions, the thionein proteins are largely unfolded.

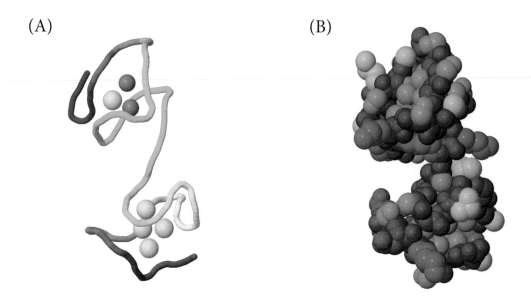

Figure 12-10. The structure of metallothionein-2 from rat liver with five Cd^{2+} and two Zn^{2+} ions bound. (A) Rainbow representation of the polypeptide chain, with the N-terminus *blue* and the C-terminus *red*. The Cd^{2+} ions are the *lighter colored spheres*. Residues 1–30 comprise the N-terminal domain (at the top), with one Cd^{2+} and two Zn^{2+} ions bound to nine Cys residues. Residues 31–61 comprise the C-terminal domain, with four Cd^{2+} ions bound to 11 Cys residues. (B) Space-filling representation of the same view. The metal ions are not visible or accessible to the solvent. Figure generated from PDB file *4mt2* using the program Jmol.

Metallothionein: the multipurpose protein. P. Coyle *et al.* (2002) *Cell. Mol. Life Sci.* **59**, 627–647.

Advances in metallothionein structure and functions. M. Vasak (2005) *J. Trace Elem. Med. Biol.* **19**, 13–17.

12.2.D. Iron Transport and Storage Proteins

Iron is the most abundant transition metal in the earth's crust but it is poorly available biologically. The stable ionic form of iron at neutral pH in an oxidizing environment is Fe^{3+}, and it forms insoluble ferric hydroxides in the presence of O_2 at neutral pH. The maximum solubility of Fe^{3+} ions at neutral pH is no greater than about 10^{-20} M, when a typical cell would contain only about 0.01 Fe^{3+} ion on average. Organisms have evolved various strategies to take up iron by chelating it. **Once inside a cell, free iron is quite toxic** because Fe^{2+} and Fe^{3+} ions react with superoxide ion, O_2^-, and peroxide, H_2O_2, to generate hydroxyl radicals that will oxidize the cell indiscriminately. Hence the levels of iron need to be regulated closely; important proteins bind and store iron, and their synthesis is often controlled by the levels of iron.

Ferritin is a large protein (~ 500 kDa) that consists of an aggregate of 24 polypeptide chains of two functionally distinct subunits, ferritin H and L, which form a hollow shell in the apo form. Each apoferritin structure can take up approximately 4500 Fe^{3+} atoms and prevent them from reacting with peroxide and superoxide. The protein shell encloses a microcrystalline inorganic matrix with the approximate composition $(FeOOH)_{8n}(FeO{:}OPO_3H_2)_n$. The protein does not provide specific binding sites for each stored ferric ion; instead, the H subunit has the ferroxidase activity that is necessary for uptake and oxidation of ferrous iron, while the L subunit is involved in nucleation of the iron core.

The protein shell has eight pores for transporting Fe, 12 mineral nucleation sites and up to 24 oxidase sites.

Transferrins are much smaller proteins that bind to iron (Figure 12-11) and then to the transferrin receptor at the surface of a cell. The complex is internalized by endocytosis, and the iron then dissociates into the cytoplasm, where it is free to be taken up by ferritin and other cellular proteins. The transferrin polypeptide chain is folded into two lobes, representing its N- and C-terminal halves. The two lobes are homologous and have similar structures that comprise two $\alpha\beta$-domains with a deep cleft between them. A single Fe^{3+} ion is bound very tightly ($K_d = 10^{-22}$ M) to each lobe in this cleft, using the side-chains of two Tyr residues, one Asp and one His. In addition, a CO_3^{2-} ion bridges between the metal ion and an anion-binding site on one domain. Because of their direct interaction,

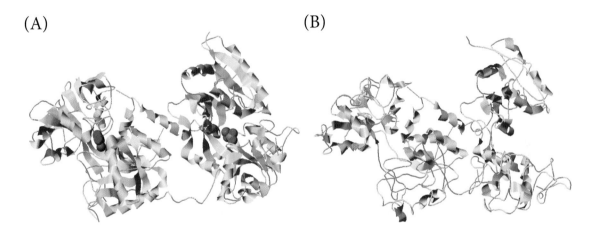

Figure 12-11. The structure of chicken transferrin with (A) and without (B) an Fe atom and a carbonate group bound to each of its two lobes. The atoms of iron and carbonate are depicted as *spheres*, while only the polypeptide backbone of the protein is depicted schematically; *arrows* depict β-strands and *coils* are α-helices. The *yellow rods* are disulfide bonds. Figure generated from PDB files *1n04* and *1ryx* using the program Jmol.

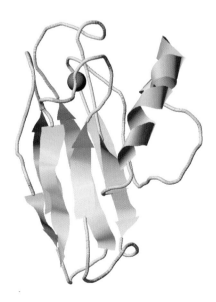

Figure 12-12. The three-dimensional structure of the azurin from *Pseudomonas aeruginosa*. The sphere is the redox-active copper ion. *Arrows* indicate β-strands, whereas the α-helix is depicted by the *coil*. Figure generated from PDB file *1azu* using the program Jmol.

the binding of Fe^{3+} and CO_3^{2-} is highly cooperative. In their absence, the cleft is opened by a large rotation of 63° by the two domains that comprise each half of the molecule (Figure 12-11).

Lactoferrin and transferrin: functional variations on a common structural framework. E. N. Baker *et al.* (2002) *Biochem. Cell. Biol.* **80**, 27–34.

Ferritin: at the crossroads of iron and oxygen metabolism. E. C. Theil (2003) *J. Nutr.* **133**, 1549S–1553S.

12.2.E. Blue-copper Proteins

Blue-copper proteins get their name from their intense color, due to absorption of light with a wavelength of around 630 nm, which arises from the unique coordination of the single bound copper ion. They are also called **cupredoxins**, and the best-characterized members of this family are **azurin** (from bacteria) and **plastocyanin** (from plants). They are believed to function in electron transport, and their copper ions oscillate between Cu^{1+} and Cu^{2+} redox states. *In vitro*, Zn^{2+} can bind in place of copper.

The best-studied proteins of this family are the **azurins**. They are small proteins of about 128 amino acid residues, consisting of one α-helix and eight β-strands that fold into a β-barrel structure (Figure 12-12) with a double-wound Greek key topology (Section 9.1.D.4). In the azurin from *Pseudomonas aeruginosa*, the copper ion is bound about 7 Å below the surface, coordinated by three equatorial ligands (the imidazole groups of His46 and His117, the thiol group of Cys112) and two weaker ligands in axial positions (the S atom of Met121 and the backbone carbonyl of Gly45) (Figure 12-3-E). All except the last ligand are conserved in azurins from other species. The distorted trigonal bipyramidal geometry (Figure 12-9) is intermediate between the preferred geometries for Cu^{1+} (tetrahedral) and Cu^{2+} (tetragonal planar) and stabilizes the Cu^{1+} form more than Cu^{2+}. **The different redox states of a metal ion usually have different preferred ligation states and proteins often adjust the redox potential of the ion by binding it in a structure that favors the desired redox state.** The blue color is due to a ligand-to-metal charge transfer involving the thiolate ligand as the electron donor; the color is more than 100 times more intense than in simple Cu^{2+} complexes. The hyperfine splitting in the electron paramagnetic resonance spectrum is unusually narrow (about 0.008 cm^{-1}), about half that observed in inorganic copper complexes, which is attributed to delocalization of the unpaired copper electron onto the electrons of the Cys S atom reducing the nuclear–electron interaction. These properties of the Cu atom are altered markedly by replacing Met121 with other amino acids, depending on the hydrophobicity of the residue.

Removing the copper ion does not alter the structure of the protein but it does decrease the net stability of its folded state by about 3–5 kcal/mol. The folded conformation of the Cu^{1+} form is somewhat less stable than the Cu^{2+}, implying that the folded conformation stabilizes Cu^{1+} relative to Cu^{2+}. Remarkably, the copper ion remains bound to unfolded azurin, but is then ligated by only three residues (Cys112, His117 and Met121) that are in close proximity in the primary structure. This should favor their simultaneous binding to the copper ion, and even just a short 13-residue peptide including these residues binds the ion tightly, with altered spectroscopic properties. Folding of the polypeptide chain is much faster when the copper is bound already to the unfolded protein, but this is simply because the metal ion can enter the folded binding site only very slowly, with a half-time of about 20 minutes.

Role of cofactors in folding of the blue-copper protein azurin. P. Wittung-Stafshede (2004) *Inorg. Chem.* **43**, 7926–7933.

Active site structures and the redox properties of blue copper proteins: atomic resolution structure of azurin II and electronic structure calculations of azurin, plastocyanin and stellacyanin. K. Paraskevopoulos *et al.* (2006) *Dalton Trans.* 3067–3076.

Reduction potential tuning of the blue copper center in *Pseudomonas aeruginosa* azurin by the axial methionine as probed by unnatural amino acids. D. K. Garner *et al.* (2006) *J. Am. Chem. Soc.* **128**, 15608–15617.

12.3. CALCIUM-BINDING PROTEINS

Ca^{2+} is very important biologically in acting as a 'second messenger' in a wide range of key intracellular and extracellular systems, and calcium-binding proteins play important roles in mediating each of these effects. The initial stimulus to the system alters the Ca^{2+} concentration, which then produces the desired physiological response by binding to the appropriate proteins. **A variety of structural motifs bind Ca^{2+} but all involve O atoms of the protein backbone or side-chains, reflecting the intrinsic affinity of Ca^{2+} for O atoms**.

Many enzymes require bound Ca^{2+} ions for stability. The ions are bound in a variety of ways. Members of both the trypsin and subtilisin families of serine proteinases bind one to three Ca^{2+} ions, which are important for structural stability but do not directly participate in catalysis. In most cases, the Ca^{2+} ions are coordinated by ligands dispersed throughout the structure, although trypsin binds one Ca^{2+} ion using a single 12-residue surface loop.

In other enzymes, **the Ca^{2+} ions are bound at the active site and involved directly in catalysis**. Staphylococcal nuclease hydrolyzes both DNA and RNA and uses Ca^{2+} bound at the active site to polarize the phosphate at the phosphoester bond to be hydrolyzed. More important physiologically are the proteins involved in sensing Ca^{2+} concentrations as a second messenger and in blood clotting.

Ca^{2+} and Mg^{2+} ions seem superficially to be very similar, and it is often considered surprising that they do not compete more effectively for protein binding sites, but there are a number of differences between them. Ca^{2+} favors a larger and more diverse coordination number (six or eight) than does Mg^{2+} (six) due to its larger size. Mg^{2+} tends to form well-defined octahedral complexes with six ligands and precise bond lengths, whereas Ca^{2+} prefers looser complexes of higher and more variable coordination number, without directionality and with variable bond lengths (Figure 12-8). Both ions favor oxygen ligands, but Mg^{2+} binds more strongly than Ca^{2+} to N atoms. The Ca^{2+} ion has greater affinity for large multi-dentate anionic ligands. Being smaller, Mg^{2+} does not form a structure like the most insoluble of calcium phosphates, hydroxyapatite.

Calcium-binding sites in proteins: a structural perspective. C. A. McPhalen *et al.* (1991) *Adv. Protein Chem.* **42**, 77–144.

Guidebook to the Calcium-Binding Proteins. M. Celio, ed. (1996) Oxford University Press, Oxford.

Extracellular calcium-binding proteins. P. Maurer *et al.* (1996) *Curr. Opin. Cell Biol.* **8**, 609–617.

Calcium as a Cellular Regulator. C. B. Klee & E. Carafoli, eds (1997) Oxford University Press, Oxford.

12.3.A. EF-hand Calcium-binding Proteins

EF-hand calcium-binding proteins have important roles in sensing the concentration of Ca^{2+} ions in biological systems. They are characterized by a highly conserved helix–loop–helix motif known as the **EF-hand motif**, which consists of a 12-residue Ca^{2+}-binding loop flanked by two α-helices (Figure 12-13). The name 'EF-hand' was introduced with the first three-dimensional structure determined, that of parvalbumin (Figure 12-1-A), made up of α-helices designated as E and F.

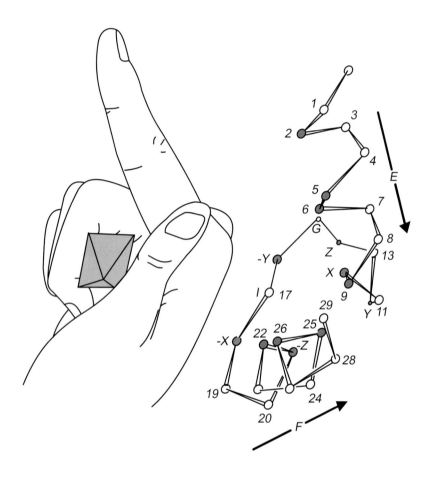

Figure 12-13. The EF-hand that binds one Ca^{2+} ion in parvalbumin. The hand shown on the *left* illustrates the octahedral arrangement of amino acid residues that provide the O atoms chelating the central Ca^{2+} ion. The forefinger of the hand represents the E-helix, the thumb the F-helix. The positions of the C^α atoms in parvalbumin are illustrated on the *right*. The Ca^{2+} ion is bound within the turn connecting the E and F α-helixes. O atoms that chelate the ion at the vertices of the octahedron come from the residues designated X, Y, Z, –X, –Y, and –Z by their octahedral positions. The Ca^{2+} ion is seven-coordinated; the Glu residue at position –Z is a bidentate ligand that uses both O atoms of its side-chain. The Gly and Ile residues at positions 15 and 17, respectively, are highly conserved in different Ca^{2+} binding proteins. Helices E and F have hydrophobic side-chains at residues 2, 5, 6, 9, 22, 25, 26 and 29 that pack against another EF-hand related by an approximate two-fold axis of rotation. The structure of the protein with Ca^{2+} ions bound is illustrated in Figure 12-1-A. Kindly provided by R. H. Kretsinger in T. E. Creighton (1993) *Proteins: structures and molecular properties*, 2nd edn, W. H. Freeman, NY, p. 363.

The canonical EF-hand comprises a 30-residue contiguous segment of polypeptide chain that contains an α-helix of residues 1–10 that was designated as E in parvalbumin, a loop of residues 10–21 around the calcium ion, and a second α-helix of residues 19–29 designated as F. The consensus sequence is:

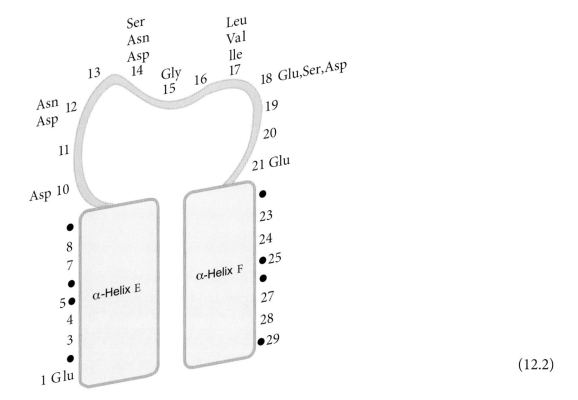

(12.2)

Any amino acids can occur at residues marked with only the number. The residues of the E- and F-helices marked '•' must be nonpolar for packing of the two helices. A central hydrophobic core is formed from the side-chains of residues 2, 5, 6 and 9 of helix E, 22, 25, 26 and 29 of helix F, and 17 from the binding loop. The side-chain carboxyl groups of residues 10, 12 and 14 provide monodentate oxygen ligands to the Ca^{2+} ion, and residue 21 provides a bidentate oxygen ligand. Residue 16 coordinates Ca^{2+} directly using its main-chain O atom, while residue 18 binds Ca^{2+} indirectly through a bound water molecule.

Almost all EF-hand proteins are composed of pairs of EF-hands. This pairing of Ca^{2+}-binding sites is believed to stabilize the protein conformation, increase the Ca^{2+} affinity and make Ca^{2+} binding cooperative. The cooperativity between these two sites allows for an 'all-or-nothing' response to Ca^{2+} binding that is crucial for the function of these proteins as intracellular Ca^{2+} sensors. The conformations of the EF-hands often change when Ca^{2+} is not bound (Figure 12-14).

Thermodynamic analyses of calcium binding to troponin C, calmodulin and parvalbumins by using microcalorimetry. K. Yamada (1999) *Mol. Cell. Biochem.* **190**, 39–45.

EF-hand protein dynamics and evolution of calcium signal transduction: an NMR view. F. Capozzi *et al.* (2006) *J. Biol. Inorg. Chem.* **11**, 949–962.

1. Calmodulin and Troponin C

Calmodulin and **troponin C** are the best-known members of the EF-hand family. Each has two largely independent domains connected by a flexible linker (Figure 12-14). Each domain contains two EF-hands, so **these proteins each bind four Ca^{2+} ions**. In the resting cell, with little or no Ca^{2+} present, they exist in an inactive state, with either Mg^{2+} or no ion bound. When the intracellular Ca^{2+} concentration rises in response to a signal, the proteins bind Ca^{2+} at both ends. Significant conformational changes occur within the individual domains upon Ca^{2+} binding (Figure 12-14). In the apo state, each domain occupies a 'closed' conformation, in which the four helices in the domain are nearly antiparallel, with interhelical angles near 180°. When Ca^{2+} is bound, each domain adopts an 'open' conformation, in which the four α-helices are nearly perpendicular to each other and exposes a hydrophobic surface of about 125 Å² that can interact with target proteins.

Figure 12-14. The crystal structure of troponin C, with Ca^{2+} ions bound at two of the EF-hand binding sites at the C-terminus of the molecule (*bottom*) but not at the two sites at the N-terminus (*top*). The *spheres* are the calcium ions; the polypeptide chain is depicted with *coils* for the α-helices. Each of the four Ca^{2+} binding sites is similar to that of the EF-hand of parvalbumin (Figure 12-13). The three-dimensional structure of the top domain, without Ca^{2+}, is substantially different from that of the lower domain. The long helix linking the two domains is flexible in solution (Figure 12-15). Figure generated from PDB file *5tnc* using the program Jmol.

Calmodulin with calcium bound binds to peptides from its target proteins very tightly, with dissociation constants ranging from 10^{-7} to 10^{-11} M. The target peptides have an intrinsic tendency to adopt an amphiphilic α-helical conformation and tend to have a bulky hydrophobic residue at either end, often (but not always) spaced 12 residues apart, but their sequences are not highly similar. Calmodulin is able to bind so tightly to such a wide array of targets due to its own plasticity (Figure 12-15). Binding of a target peptide alters the relative disposition of the two domains, but there is little change within the Ca^{2+}-activated domains themselves. The calmodulin–peptide complex forms a well-packed ellipsoid, which contrasts sharply with the dumbbell shape of calmodulin observed in the absence of target. The flexible central α-helix connecting its two domains can function as an 'expansion joint', allowing calmodulin to bind to peptides with different numbers of residues between the two bulky hydrophobic anchors. Furthermore, relatively nonspecific van der Waals interactions tend to dominate the interaction between calmodulin and the peptides, and hydrogen bonds are not as important, so the binding interactions are not very specific. Met residues are also commonly involved in the interaction; their flexible and polarizable side-chains probably help to mold the interacting surfaces.

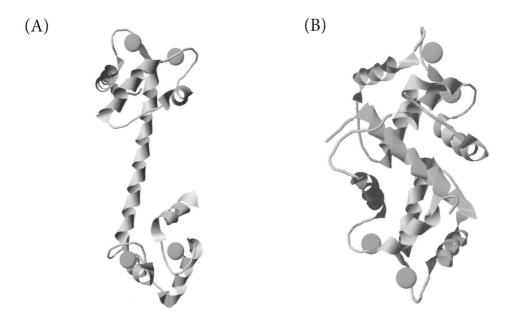

Figure 12-15. The three-dimensional conformations of calmodulin with four Ca^{2+} ions bound alone (A) and bound to a target peptide (B). The *spheres* are the Ca^{2+} ions, while the course of the polypeptide backbone is depicted schematically, with α-helices as *coils*. The target peptide is from myosin light chain kinase and is colored *green*; it is bound in the center of the molecule in an α-helical conformation. The two Ca^{2+}-binding domains of calmodulin do not change upon binding the target peptide, but the long α-helix linking them collapses to bring them much closer together. Figure generated from PDB files *1exy* and *1qtx* using the program Jmol.

Normally, these calmodulin-binding peptides are part of a larger protein, whose properties are altered by the binding of calmodulin. At least 50 different structures in complex with fragments derived from calmodulin-regulated proteins are known.

A closed compact structure of native Ca^{2+}-calmodulin. J. L. Fallon & F.A. Quiocho (2003) *Structure* **11**, 1303–1307.

Regulation of muscle contraction by tropomyosin and troponin: how structure illuminates function. J. H. Brown & C. Cohen (2005) *Adv. Protein Chem.* **71**, 121–159.

Prediction of three dimensional structure of calmodulin. K. Chen *et al.* (2006) *Protein J.* **25**, 57–70.

12.3.B. Carboxylation and Hydroxylation of Asp, Asn and Glu Residues

The carboxyl groups of Asp and Glu residues are frequently used in binding Ca^{2+} ions, but in some cases **the affinity for Ca^{2+} is augmented by adding another carboxyl or hydroxyl group**. Many proteins involved in blood clotting contain **γ-carboxyglutamic acid** residues (Gla):

$$\begin{array}{c} {}^{-}O_2C \diagdown \quad \diagup CO_2^{-} \\ {}^{\gamma}CH \\ | \\ {}^{\beta}CH_2 \\ | \end{array} \qquad (12.3)$$

which are glutamic acid residues that have been carboxylated using HCO_3^- in a vitamin K-dependent process. The additional carboxyl group markedly increases the intrinsic affinity of these side-chains for Ca^{2+} ions. The majority of Gla-containing proteins are zymogen forms of serine proteinases involved in the blood coagulation cascade: prothrombin, factor VII, factor IX, factor X, protein C, protein S and protein Z. Binding Ca^{2+} is necessary for them to bind to phospholipids, which is required for the function of the proteins. Ca^{2+} binding also appears to be necessary for the formation of the native conformation of the Gla domain in some instances.

The Gla domain of **prothrombin** involves nine to 10 turns of α-helix, divided amongst three separate helices. Seven Ca^{2+} ions interact with 24 O atoms from 16 of the 18 carboxyl groups of the nine Gla residues that are ordered in the structure. A 10th Gla residue is disordered and does not participate in Ca^{2+} binding. The coordination geometries of the Ca^{2+} ions do not correspond to any idealized polyhedra. Five of the Ca^{2+} ions are involved in a polymeric array with 18 of the liganding O atoms. Four of these Ca^{2+} ions are completely buried in the protein; this complex structure has no significant net charge and is thought to facilitate the folding of the Gla domain. The complexity and irregularity of this structure explains the selectivity for Ca^{2+} ions: Ca^{2+} is able to adopt different and distorted coordination geometries, but Mg^{2+} is fairly rigid in its requirement for six ligands and cannot accommodate the unusual network of ligands in the Gla domain. The remaining two metal ion sites in the Gla domain of prothrombin are accessible to solvent and have a net charge of about +0.5 each. Because of the positive charge, these sites are thought to be involved in neutralizing negatively charged phospholipids, allowing the protein to associate with membranes.

Certain **EGF motifs** (Section 9.1.D.7), especially those present in proteins involved in the blood clotting cascade, contain β-**hydroxy-aspartic acid** residues. The additional hydroxyl group increases the affinity of these residues for Ca^{2+} ions. Asn residues are similarly hydroxylated in other proteins. In each case, the hydroxyl group is added post-translationally in a specific orientation to the β-carbon of the side-chain. The hydroxyl group introduces a new center of chirality and is always the *erytho* isomer:

$$\begin{array}{cc} HO\diagdown \overset{H}{\underset{|}{C}} \diagup CO_2^- & HO\diagdown \overset{H}{\underset{|}{C}} \diagup CONH_2 \\ | & | \\ e\text{-}\beta\text{-Hydroxy-Asp} & e\text{-}\beta\text{-Hydroxy-Asn} \end{array} \quad (12.4)$$

Vitamin K-dependent biosynthesis of γ-carboxyglutamic acid. B. Furie *et al.* (1999) *Blood* **93**, 1798–1808.

Ca^{2+} binding to proteins containing γ-carboxyglutamic acid residues. E. Persson (2002) *Methods Mol. Biol.* **172**, 81–95.

γ-Glutamate and β-hydroxyaspartate in proteins. F. J. Castellino *et al.* (2002) *Methods Mol. Biol.* **194**, 259–268.

12.4. NAD- AND NUCLEOTIDE-BINDING PROTEINS

The nucleotide-binding motif is frequently observed in protein structures that bind nucleotides, and it has a characteristic sequence that can usually be detected in the primary structure. There are two classes of the motif, the **dinucleotide-binding fold** and the **mononucleotide-binding fold**. The former binds dinucleotides such as nicotinamide adenine dinucleotide (NAD) and flavin adenine dinucleotide (FAD), the latter binds mononucleotides such as adenosine triphosphate (ATP) and is also called the **ATP-binding motif**.

12.4.A. Dinucleotide-binding Motif

NAD-binding proteins bind the dinucleotide NAD:

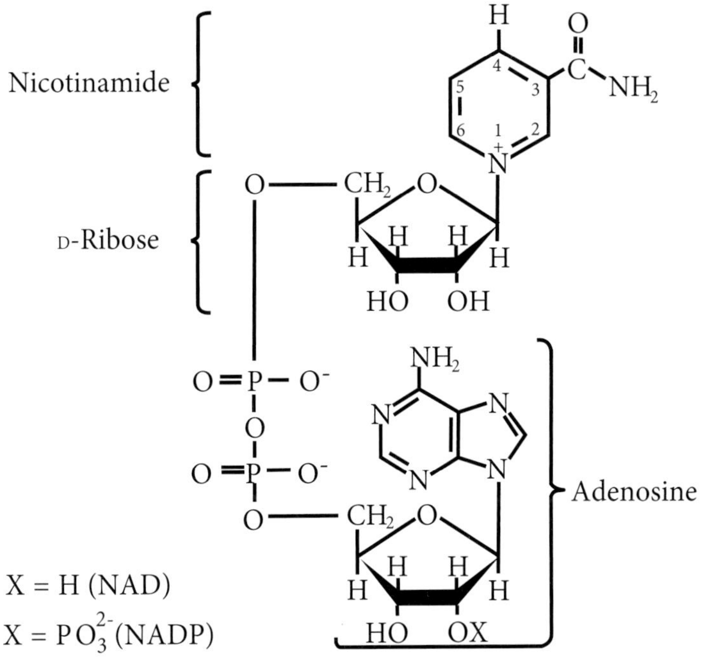

$$X = H \text{ (NAD)}$$
$$X = PO_3^{2-} \text{ (NADP)}$$

(12.5)

Generally they are enzymes that catalyze a redox reaction in which a H atom is transferred from the carbon of a substrate alcohol group to NAD, thereby oxidizing the substrate and reducing NAD to NADH. Examples of NAD-utilizing enzymes include lactate dehydrogenase, alcohol dehydrogenase and malate dehydrogenase; they catalyze the same basic reaction, but on different substrates, simply as a result of having differing specificities for that substrate.

The structures of these dehydrogenases are composed of two domains; one binds the substrate and the other binds NAD. The substrate-binding domains of the different proteins are unrelated but their **NAD-binding domains** have a common structural fold (Figure 12-16). The NAD-binding domain is a symmetrical six-stranded β-sheet with α-helices on both sides. It is formed from two β–α–β–α–β protein motifs, called **Rossmann folds**. Each binds one half of the dinucleotide; the adenine moiety of NAD or FAD binds to the first of these structural motifs, and nicotinamide binds to the second.

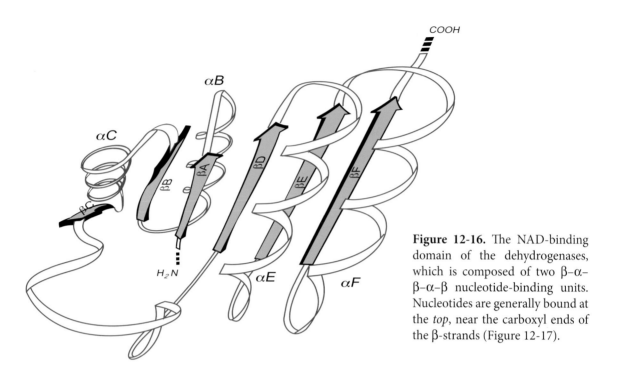

Figure 12-16. The NAD-binding domain of the dehydrogenases, which is composed of two β–α–β–α–β nucleotide-binding units. Nucleotides are generally bound at the *top*, near the carboxyl ends of the β-strands (Figure 12-17).

The first three elements of secondary structure in the common adenine-binding motif, β1–αA–β2, have similar sequences amongst the NAD-binding proteins (Figure 12-17). This fingerprint sequence has small hydrophobic residues conserved in β1, followed by a Gly-rich region (–Gly–X–Gly–X–X–Gly, where X is any residue), additional small hydrophobic residues in the αA-helix, and an Asp or Glu at the end of helix αA. The Gly residues adopt conformations forbidden to other residues and allow close packing of the strands and helix, plus a close approach between the adenine pyrophosphate and the N-terminus of the αA-helix. The Asp/Glu residue interacts with the ribose 3′-hydroxyl group. **This signature sequence of ~30 residues can be used to predict NAD-binding proteins simply from their primary structure.**

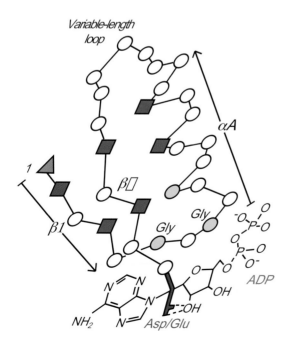

Figure 12-17. A typical β–α–β nucleotide-binding domain with an ADP molecule bound at *lower right*. Residues in the α-helix and β-strands indicated with *blue squares* are usually small and hydrophobic (Ala, Ile, Leu, Val, Met or Cys) and form the hydrophobic core between the β-strands and the α-helix. *Filled green circles* indicate Gly residues that permit a sharp turn between the first β-strand and the helix. The first residue of the polypeptide chain shown, indicated by an *orange triangle* and numbered, is usually basic or hydrophilic (Lys, Arg, His, Ser, Thr, Gln or Asn). The last residue of the chain illustrated is either Asp or Glu, with the side-chain hydrogen-bonded to the ligand. The loop between the helix and the second β-strand can be variable in length. Proteins containing this structure can be identified from their sequences on the basis of this pattern of residues.

The fingerprint sequence for binding the dinucleotide nicotinamide adenine dinucleotide phosphate (NADP), in which the 2′-ribose hydroxyl of NAD is phosphorylated (Equation 12.5), is slightly different. In this case, the third Gly residue (in the αA-helix) is replaced by Ala, which prevents the close approach of helix αA to the β-strands. This provides additional space where the 2′-phosphate group binds. In addition, the 2′-phosphate group interacts electrostatically with the Arg residue that replaces the conserved Asp/Glu of the NAD/FAD sequence.

12.4.B. Mononucleotide-binding Motif

The classical mononucleotide- or **ATP-binding motif** is typified by **adenylate kinase** (Figure 14-20). It is also an αβ-protein fold, but a five-stranded parallel β-sheet with a connectivity that differs from the dinucleotide fold. Similarly, the bound nucleotide binds to a β1–αA–β2 region with a characteristic sequence: Gly–X–X–Gly–X–Gly–Lys. The Gly-rich sequence is located entirely in the loop between β1 and αA, so all three Gly residues are part of the loop and they adopt conformations that are forbidden for other residues. This contrasts with the dinucleotide sequence, where the third Gly residue is part of helix αA (Figure 12-17). The loop forms a large pocket to which the phosphate of the mononucleotide binds; consequently, it is also called the **P-loop**. The backbone amides and the conserved Lys residue interact with the phosphate of the nucleotide.

The P-loop fingerprint sequence can be used to identify ATP- or GTP-binding proteins from just their primary structure. It has also been used to classify a broad group of protein structures, the **nucleoside triphosphate hydrolase** fold, which has a P-loop structure within an αβ-domain comprising a central β-sheet surrounded by α-helices. Different families within this classification have varying numbers and connectivities of β-strands.

The P-loop: a common motif in ATP- and GTP-binding proteins. M. Saraste *et al.* (1990) *Trends Biochem. Sci.* **15**, 430–434.

Binding of nucleotides by proteins. G. E. Schulz (1992) *Curr. Opin. Struct. Biol.* **2**, 61–67

12.5. ALLOSTERY: INTERACTIONS BETWEEN DIFFERENT BINDING SITES

Many proteins bind ligands cooperatively, in that the binding of a ligand to one site on the protein can alter the affinities of other sites for the same or different types of ligands, either increasing or decreasing them. To distinguish this type of effect between distant sites from that where the ligands interact directly, either at adjacent sites or by competition for binding at the same site, the term 'allosteric' was coined. **In allosteric systems, the ligands bind to different sites and can be unrelated structurally, but the different sites interact**. Interactions between the same ligand binding to multiple sites are known as **homotropic**, whereas interactions between different ligands are known as **heterotropic**.

Cooperativity: a unified view. L. Acerenza & E. Mizraji (1997) *Biochim. Biophys. Acta* **1339**, 155–166.

Allostery in very large molecular assemblies. K. E. van Holde *et al.* (2000) *Biophys. Chem.* **86**, 165–172.

Quantitative analysis and interpretation of allosteric behavior. G. D. Reinhart (2004) *Methods Enzymol.* **380**, 187–203.

Allosteric mechanisms of signal transduction. J. P. Changeux & S. J. Edelstein (2005) *Science* **308**, 1424–1428.

12.5.A. Structural Models

Two fundamentally different types of structural models have been proposed to account for cooperativity of binding. They are compared in Figure 12-18.

1. Sequential Model: Direct Interactions

In the sequential model, structural changes upon ligand binding are simply transmitted from one site to another through the structure of the protein. Any type of interaction between the sites is possible in such a system, and it can readily explain both negative and positive cooperativity. In this view, cooperativity might be expected in virtually every protein molecule that binds more than one ligand

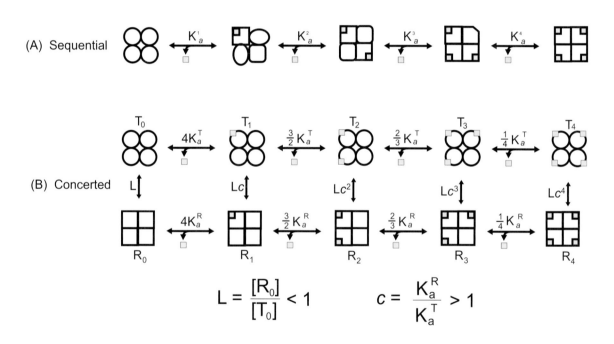

Figure 12-18. Comparison of the sequential and concerted models for positive allosteric cooperativity of ligand (*squares*) binding, using a hypothetical tetrameric protein. (A) In the sequential model, binding of each ligand induces a conformational change in the subunit to which it is bound and in the other subunits, thereby increasing their affinities for ligand, so that $K_a^1 < K_a^2 < K_a^3 < K_a^4$. The conformations depicted and the sequence of binding by the individual subunits are arbitrary. (B) In the concerted model, there are only two conformational states, T and R, with intrinsic affinities for ligand of K_a^T and K_a^T, respectively, which are not altered by ligand binding to other subunits. Instead, binding of each ligand shifts the equilibrium from T to R by the factor $1/c$, which is the ratio of the two intrinsic affinities, K_a^T/K_a^R. The association constants for binding of the first, second, etc., ligand molecules include statistical factors to take into account the numbers of free and occupied sites on the protein molecule. The intrinsic affinities of all sites in the T state are postulated to be identical, as are those in the R state.

molecule simultaneously. However, **cooperativity is the exception rather than the rule, and many proteins bind ligands at different sites without any interaction between them**. In nonallosteric systems, changes in the structure of the protein upon binding a ligand are usually minimal, but one of the largest movements detected is a 12 Å movement of a peptide loop upon binding of the cofactor NAD to each of the four sites on tetrameric lactate dehydrogenase; yet there is no effect of this on the NAD affinities of the other three subunits.

Is allostery an intrinsic property of all dynamic proteins? K. Gunasekaran *et al.* (2004) *Proteins* **57**, 433–443.

2. Concerted Model: Quaternary Structure Changes

The second allosteric model, known as the **symmetric, two-state, concerted** or **Monod–Wyman–Changeux (MWC) model** (Figure 12-18), was inspired by the observations that (1) nearly all regulatory allosteric proteins and enzymes are composed of multiple subunits, either the same or different; (2) each subunit has a binding site for one or more ligands, and binding of ligand to one subunit can affect the affinities of the other subunits; and (3) ligand binding is usually accompanied by a substantial change in quaternary structure. The archetypal allosteric protein is vertebrate hemoglobin, which binds four oxygen molecules with positive cooperativity, plus varying types of interactions with the binding of other ligands (Section 12.5.B). The two-state allosteric model of Monod, Wyman and Changeux postulated that:

(1) each molecule of the native enzyme contains multiple subunits, each with intrinsically identical binding sites for substrates and ligands;

(2) the regulatory behavior depends on structural interactions between the subunits, the effectors act indirectly on the other binding sites, and no direct interactions between substrate(s) or effectors are required;

(3) a pre-existing reversible conformational equilibrium in the quaternary structure of the protein is responsible for the actions of the effectors;

(4) only two conformations are possible for the subunits of an oligomer, and all of the subunits in a given oligomer have the same conformation;

(5) the two conformations have different affinities for a ligand;

(6) the effectors simply pull the equilibrium towards whichever state they bind to most tightly.

The name 'concerted' refers to the all-or-none nature of the conformational transition. The two conformations are usually denoted by the terms **R-state** (for 'relaxed') and **T-state** (for 'tense'). **The R-state normally has the higher affinity for the primary ligand, while the T state is imagined to be constrained in some manner**. The two forms with *i* ligand molecules bound are often referred to as T_i and R_i. According to the model (Figure 12-18-A), **these two conformations coexist even in the absence of ligand**, with an equilibrium constant L between T_0 and R_0. The T_0 form would normally be favored, L < 1, and the protein would have relatively low affinity for the first ligand molecule. **The R conformation has the higher affinity for ligand by a factor *c***, so ligand molecules will be bound to it preferentially. This will pull the conformational equilibrium towards the R state, because the conformational equilibrium and ligand binding are linked functions (Figure 11-14). The conformational equilibrium between the two conformations with one ligand molecule bound will be Lc. The other vacant sites on all the R_1 molecules will then be in the high-affinity form, so the average affinity of the vacant sites of the entire population will be increased. Consequently, **positive cooperativity in binding is explained by a large predominance of the unliganded protein in the low-affinity T-state, with a shift to the high-affinity R-state upon binding of ligand.**

The difference in affinities of the two states R and T for a ligand must be caused structurally by the difference in their quaternary structures. Consequently, the values of L and c are probably not independent, as has been observed in hemoglobins (Equation 12.12).

Heterotropic interactions involving other ligands arise because these ligands bind preferentially to either the T or R conformations and thus their binding tends to pull the equilibrium towards that conformation. Secondary ligands that bind preferentially to the T state will decrease the affinity for the first ligand, whereas the opposite effect will be caused by preferential binding to the R state. The model imagines that each type of ligand controls the apparent affinity of the protein for other ligands simply by shifting the equilibrium between the T and R states, which have different affinities for each ligand.

On the nature of allosteric transitions: a plausible model. J. Monod *et al.* (1965) *J. Mol. Biol.* **12**, 88–118.

3. Comparison of the Sequential and Concerted Models

The sequential and concerted models differ in terms of protein conformation primarily with respect to the conformations and ligand affinities of the partially liganded states. Both models imply that ligand binding has effects on the protein conformation. In the sequential model, such effects extend directly to the other binding sites and alter their affinities for ligands. In the concerted model, the conformational effects need extend only to the interface between the subunits to alter the conformational equilibrium between the T and R states. The two models also differ in that the concerted model envisages the two conformations to be present even in the absence of the ligand, R_0 and T_0, whereas in the sequential model the R-like conformation is induced only upon ligand binding. The sequential model predicts that the conformational change upon ligand binding should parallel the extent of ligand binding, whereas with the concerted model the two need not coincide because the conformational change should tend to occur at one particular stage of ligand binding, that when Lc^i becomes greater than unity.

The concerted model is much more restrictive than the sequential model. The only parameters that can be varied are L, the conformational equilibrium constant in the absence of any ligand, and c, the relative affinities of the two states for each ligand. Moreover, the two parameters L and c are observed not to be independent in the case of hemoglobin (Section 12.5.B), in that alteration of one also changes the other. **The greatest restriction of the concerted model is that it does not readily explain negative cooperativity** (Section 12.5.C); a ligand can pull the R $\leftrightarrow$ T conformational equilibrium only towards the form with higher affinity for it.

12.5.B. Hemoglobin and Myoglobin

Hemoglobin (Hb) is the best-understood allosteric protein. It is one of the heme-containing oxygen-binding proteins, with a globin polypeptide chain and a protoheme IX prosthetic group containing a ferrous iron atom (Fe^{2+}) that actually binds the O_2 molecule (Figure 12-3-I). It was the inspiration for the concerted allosteric model and still largely fits that model. It is instructive to contrast hemoglobin with the closely related **myoglobin** (Figure 9-34), which does not have any allosteric properties. The primary physiological function of hemoglobin is to transport oxygen throughout the bloodstream, whereas that of myoglobin is to store oxygen within muscle.

All globins have the **globin fold** (Figure 12-19), usually composed of eight α-helical segments, designated A through H, connected by loop segments AB, BC,... GH. The nonhelical segments at the amino and carboxyl termini are designated as NA and HC, respectively. The globin fold is maintained even when the sequence identity between pairs of globins is as low as 16% (Figure 9-34). In most vertebrates, individual globin polypeptide chains consist of 141–153 amino acid residues.

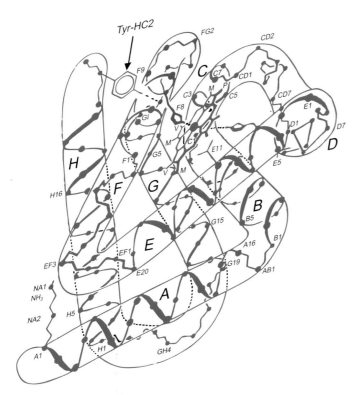

Figure 12-19. The globin fold, illustrated by that in vertebrate hemoglobins. The eight α-helices are indicated by the letters A through H, starting at the N-terminus (*lower left*). Individual residues are designated by their positions within the helices or the intervening segments. Some crucial residues are labeled. The heme group is wedged between the E- and F-helices. The methyl, vinyl and propionate side-chains of the heme group are labeled as M, V and P.

Proteins have no groups capable of binding O_2 tightly and reversibly, so they use the heme group, which is protoporphyrin IX with a ferrous iron atom (Fe^{2+}) at its center (Figure 12-3-G, H, I). The heme group is inserted into a cleft in the globin fold between the E- and F-helices, known as the **heme pocket** (Figure 12-20). The Fe atom is linked covalently to the imidazole N^ε of the *proximal* His residue at position F8 (the eight residue of the F-helix). There are also many van der Waals interactions between the heme group and the amino acid side-chains lining the heme pocket.

Under physiological conditions, the Fe atom is in the ferrous (Fe^{2+}) state, when it is capable of binding one molecule of O_2 on the *distal* side of the heme group (Figure 12-20). The side-chain of the distal His–E7 forms a hydrogen bond with one of the O atoms, stabilizing the bound oxygen. In this state, the ferrous iron is hexa-coordinated: to one of the O atoms, N^ε of His–F8, and the four pyrrole N atoms of protoporphyrin IX. When oxygen is not bound, the sixth coordination site for oxygen is left vacant and the Fe atom is only penta-coordinated, although a water molecule occupies the heme pocket.

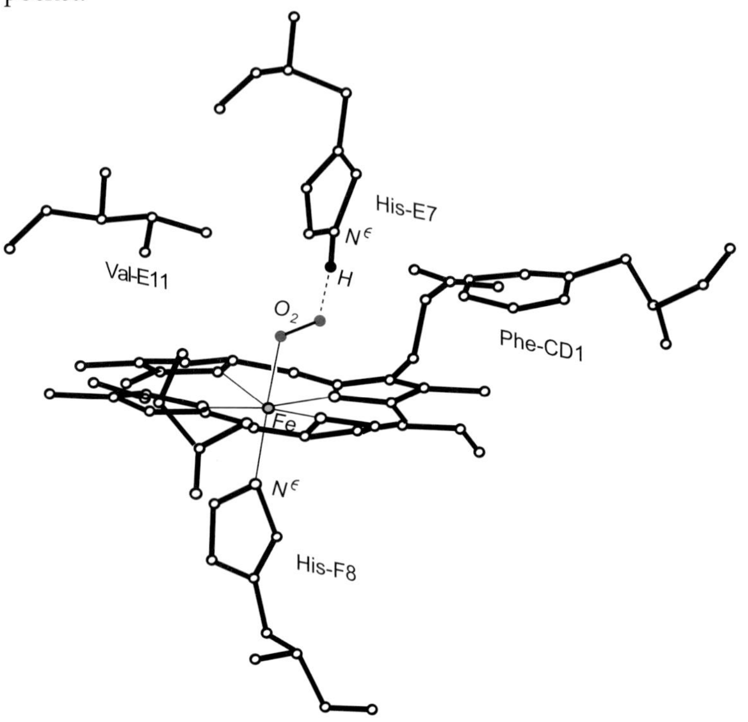

Figure 12-20. The heme group and surrounding residues of sperm whale myoglobin, with O_2 bound. Residue His–F8 is the proximal ligand to the ferrous Fe atom, and His–E7 is the distal His residue, forming a hydrogen bond to the bound oxygen molecule. Two other residues important in controlling the oxygen affinity are Val–E11 and Phe–CD1. This drawing depicts the individual atoms almost as points, but their van der Waals volumes would largely fill the space. From a drawing by M. F. Perutz.

Allosteric hemoglobin assembly: diversity and similarity. W. E. Royer *et al.* (2005) *J. Biol. Chem.* **280**, 27477–27480.

The allosteric properties of hemoglobin: insights from natural and site directed mutants. A. Bellelli *et al.* (2006) *Curr. Protein Peptide Sci.* **7**, 17–45.

Asymmetric cooperativity in a symmetric tetramer: human hemoglobin. G. K. Ackers & J. M. Holt (2006) *J. Biol. Chem.* **281**, 11441–11443.

---

1. Structure

The adult form of human hemoglobin, HbA, is a tetramer composed of two α-subunits and two β-subunits, each of which has one bound heme group (Figure 12-21). The α- and β-chains consist of 141 and 146 amino acid residues, respectively. Both adopt very similar globin folds, although α-chains lack α-helix D, and the tetramer approximates a homotetramer with identical polypeptide chains. There are relatively few contacts between the pair of α-subunits and the pair of β-subunits within the $\alpha_2\beta_2$ tetramer, whereas they are extensive between the unlike subunits. Hemoglobin is best considered a dimer of identical αβ-pairs (Figure 12-21). The two α-subunits are designated α^1 and α^2, and the β-subunits are numbered so that the distance between the α^1 and β^1 heme Fe atoms is greater than between α^1 and β^2. Symmetry dictates that the $\alpha^1\beta^1$ pair is identical to $\alpha^2\beta^2$, while $\alpha^1\beta^2$ is the same as $\alpha^2\beta^1$. Within the center of the hemoglobin tetramer is a sizeable aqueous cavity that is accessible to small molecules.

(A)   (B)

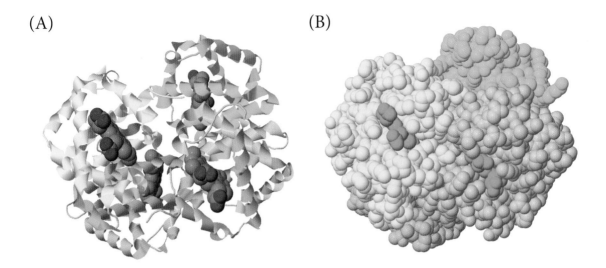

Figure 12-21. The tetrameric structure of vertebrate hemoglobins. (A) The protein is shown schematically, whereas the heme groups are shown with a van der Waals sphere for each atom. (B) Space-filling model of the same view, demonstrating that only the edges of the heme groups are accessible to the solvent. The tetrameric molecule consists of two pairs of α- and β-subunits, with an exact two-fold symmetry axis. The α- and β-chains are homologous, with very similar globin folds (Figure 9-34), so the tetramer has approximate four-fold symmetry. Each polypeptide chain has a bound heme group and is illustrated in a different color. Figure generated from PDB file *1xz2* using the program Jmol.

In contrast to hemoglobin, the **closely related myoglobin is a monomer**. It binds only a single molecule of molecular oxygen and demonstrates no allosteric properties. Myoglobin serves as a nonallosteric control to illustrate the allosteric properties of hemoglobin.

Quaternary structure of hemoglobin in solution. J. A. Lukin *et al.* (2003) *Proc. Natl. Acad. Sci. USA* **100**, 517–520.

Myoglobin: an essential hemoprotein in striated muscle. G. A. Ordway & D. J. Garry (2004) *J. Exp. Biol.* **207**, 3441–3446.

The enigma of the liganded hemoglobin end state: a novel quaternary structure of human carbonmonoxy hemoglobin. M. K. Safo & D. J. Abraham (2005) *Biochemistry* **44**, 8347–8359.

---

2. Oxygen Binding

The environment of the heme group is crucial for making it a site for the reversible binding of O_2.

Heme groups are used in other proteins to undergo or catalyze redox reactions in which the Fe atom alternates between ferrous (Fe^{2+}) and ferric (Fe^{3+}). The heme pocket in the globin fold provides a highly hydrophobic environment that maintains the heme Fe atom in the ferrous state; otherwise, binding of O_2 by the ferrous Fe atom would oxidize it to the Fe^{3+} form, producing **methemoglobin**, which is unable to bind oxygen. Several amino acid residues surrounding the heme group are important for stabilizing the linkage between the protein and heme group and for reversible binding of O_2 (Figure 12-20). Residue His–F8 is the ligand to the heme Fe atom on the proximal side. Residues His–F8 and Phe–CD1 are strictly conserved in all globins (Figure 7-31). Residue E7 stabilizes the bound oxygen by forming a hydrogen bond with one of its atoms (Figure 12-20); it is usually His, but is occasionally replaced by Gln, Val, Leu or Tyr in some globins. The residues at positions E11, B10 and CD4 are also known to be important for ligand binding.

The properties of the heme group in globins change significantly upon binding O_2. In the ferrous deoxy form, with no ligand, the Fe atom is somewhat too large to fit in the center of the heme group and is forced 0.5–0.6 Å out of the mean plane of the heme. Partly as a consequence, the heme group in the unliganded state is domed. When O_2 or CO binds, the Fe atom becomes smaller and lies in the heme plane, and the heme group is flat (Figure 12-22).

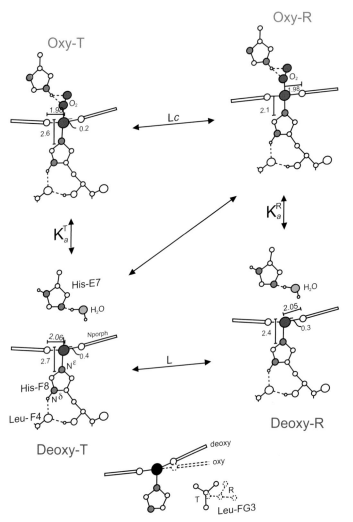

Figure 12-22. Changes in heme stereochemistry on binding of O_2 by the α-subunits in the R and T structures of hemoglobin. The *bottom* diagram shows the change in conformation of the heme that ideally should take place upon binding O_2. The heme tends to flatten, but this is prevented in the T structure by the side-chains of Leu–FG3 and Val–FG5 (Val–FG5 would be hidden behind Leu–FG3 in the diagram and is not depicted). Consequently, upon uptake of oxygen by the T structure, the heme remains domed and the iron remains displaced from the porphyrin plane in the oxy T state. Consequently, the T state has low O_2 affinity. This unfavorable stereochemistry is alleviated by the change to the R quaternary structure. The R structure does not constrain the heme group to be either flat or domed. The heme therefore can alternate between the oxy and deoxy conformations, and the R state has normal O_2 affinity. The numbers in the diagram indicate the distances (in Å) of N^ε of His–F8 from the mean plane of the porphyrin carbons and nitrogens, the mean distance between the iron and the porphyrin nitrogens (N_{porph}), and the displacement of the iron from the plane of the porphyrin nitrogens. Adapted from M. F. Perutz *et al.* (1987) *Acc. Chem. Res.* **20**, 309–321.

Myoglobin binds O_2 in the normal way, with a single binding constant (Figure 12-23). Binding of O_2 by hemoglobin differs in having a lower oxygen affinity for the first oxygen molecule to bind, but then increasing to a somewhat greater affinity for subsequent molecules. The sigmoidal shape of the direct binding curve (Figure 12-23-A) is ascribed to positive cooperativity of oxygen binding, with **the fourth O_2 molecule being bound some 500 times more tightly than the first**; the fourth molecule is bound with an affinity very similar to that of myoglobin. **The Hill coefficient is in the region of 2.9 for hemoglobin but 1.0 for myoglobin** (Figure 12-23-C). The sigmoidal shape of the oxygen-binding curve is important for hemoglobin's function, increasing the difference in oxygen saturation between arterial and vinous blood and enhancing the amount of oxygen bound in the lungs and released at the tissues.

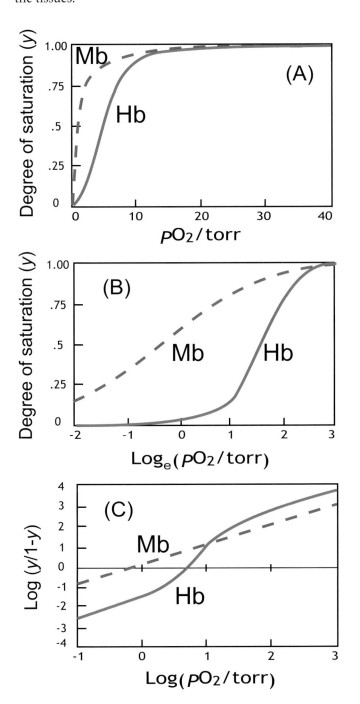

Figure 12-23. Oxygen binding curves for hemoglobin (Hb) and myoglobin (Mb). The oxygen concentration is specified by its partial pressure. The binding curves of myoglobin illustrate normal binding, but those for hemoglobin display positive cooperativity. (A) Direct plot of extent of binding (y) versus the oxygen concentration. (B) Extent of binding versus logarithm of the oxygen concentration. (C) Hill plot. Adapted from J. Wyman & S. J. Gill (1990) *Binding and Linkage*, University Science Books, CA, p. 3.

The oxygen affinity of hemoglobin is also affected by other factors, such as organic phosphates, CO_2, anions such as Cl^- and protons (the Bohr effect; Section 12.5.B.5). In contrast, no cofactor is known to affect the oxygen-binding properties of myoglobin and it has no significant Bohr effect.

The related ligand CO, **carbon monoxide**, has an intrinsic affinity for heme groups that is several thousand-fold greater than that of O_2. Such tight binding to myoglobin or hemoglobin would be very deleterious, for CO is a poison and is produced in cells. Consequently, **the affinity of heme groups for CO is lowered about 100-fold in both myoglobin and hemoglobin**. CO binds best when it is perpendicular to the plane of the porphyrin ring, but this is believed to be prevented by steric clashes with the distal His residue and a Val residue of the heme pocket. In contrast, O_2 normally binds at an angle to the heme plane, and this is accommodated by the heme pocket (Figure 12-20). CO is observed crystallographically to be forced to adopt a similar position, with deleterious consequences for its affinity.

The heme pocket is buried within the interiors of both myoglobin and hemoglobin, with no channel to the solvent apparent in their crystal structures (Figure 12-21). Yet O_2 and CO associate and dissociate rapidly from the proteins. Moreover, the unligated heme pocket is normally occupied by a water molecule, which must be displaced before O_2 or CO can enter. Fluctuations of the protein structures are presumed to produce a channel of sufficient size between the solvent and heme pocket. There is much debate as to how these small ligands enter and leave the heme pocket.

Binding of either O_2 or CO to the heme group produces substantial changes in its absorption spectrum, so the kinetics of binding and dissociation can be studied spectroscopically. Also, bound O_2 or CO can be released rapidly from the heme iron by a short flash of intense light. The released ligand can then rebind immediately or diffuse into the heme pocket before either rebinding or diffusing out of the protein. All these processes occur on the femtosecond to nanosecond time scales at room temperature and have been studied exhaustively, although usually at very low temperatures in a cryosolvent. Time-resolved crystallography has observed released CO to move about 2 Å from the heme iron very rapidly after photolysis; in addition, changes in the position of the Fe atom relative to the heme group occur very rapidly.

Structure of a ligand-binding intermediate in wild-type carbonmonoxy myoglobin. K. Chu *et al.* (2000) *Nature* **403**, 921–923.

Ligand binding and conformational motions in myoglobin. A. Ostermann *et al.* (2000) *Nature* **404**, 205–208.

A hierarchy of functionally important relaxations within myoglobin based on solvent effects, mutations and kinetic model. D. Dantsker *et al.* (2005) *Biochim. Biophys. Acta* **1749**, 234–251.

Nature of the FeO_2 bonding in myoglobin and hemoglobin: a new molecular paradigm. K. Shikama (2006) *Prog. Biophys. Mol. Biol.* **91**, 83–162.

---

3. Cooperativity of Oxygen Binding

Upon binding oxygen, the hemoglobin quaternary structure undergoes an extensive change in the relative orientations of the four subunits, due primarily to changes in the $\alpha^1\beta^2$ and $\alpha^2\beta^1$ interfaces (Figure 12-24). If crystals of either form of hemoglobin have oxygen added or removed, the crystals shatter. This switch between two alternative quaternary structures is the archetypal T ↔ R allosteric transition.

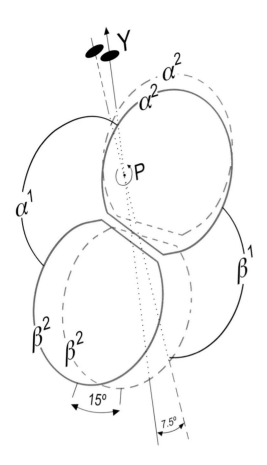

Figure 12-24. The quaternary structures of deoxy (T) and liganded (R, oxy) hemoglobins. The $\alpha^1\beta^1$ dimers of both are superimposed, and the relative positions of the $\alpha^2\beta^2$ dimer are depicted in *solid lines* for T and *dashed lines* for R. The $\alpha^2\beta^2$ unit rotates 15° relative to $\alpha^1\beta^1$ about the axis P perpendicular to the plane of the paper and to the two-fold symmetry axis Y. The only direct contacts between the subunits are between α^1 and β^1 and between α^1 and β^2 (plus the symmetrical contacts involving α^2). The binding site for the organic phosphate allosteric effectors in the T state is at the bottom of the molecule straddling the two-fold axis between the two β-chains. In the R state, the two chains move closer together, disrupting this binding site. Adapted from a drawing by J. M. Baldwin.

The main difference between the R and T structures of hemoglobin is that the two αβ-dimers are rotated and translated relative to each other (Figure 12-24); **the difference is primarily of the quaternary structure, and the structures of the individual α- and β-subunits are changed much less**. The two β-subunits are about 5 Å further apart in the T quaternary structure, due primarily to differences at the $\alpha^1\beta^2$ (and the corresponding $\alpha^2\beta^1$) interface. They involve alternative interdigitations of groups that must be either one way or the other, not intermediate (Figure 12-25). Consequently, intermediate quaternary structures would not be structurally feasible or stable. The T quaternary structure appears to be unique, but that of R is a mixture of two, known as R and R2. The dimer interface in R is less stable by 6.3 kcal/mol than that in T and acts as a 'molecular slide bearing' that permits the two αβ-pairs to slide back and forth between the two R forms. One can speak of a T ↔ R allosteric transition, but keep in mind that R is a mixture of two species.

Cooperativity of binding O_2 largely requires the change in quaternary structure. There is no evidence of any cooperativity within the R quaternary structure, although there is some within the T structure; in this case, binding of O_2 at one heme group does cause some changes in affinity at the others

The α- and β-subunits of hemoglobin are very similar, but not identical, and they have somewhat different oxygen affinities. This complicates greatly the quantitative analysis of oxygen binding by the tetramer. Instead of there being only five ligation states (with 0–4 ligand molecules bound) of a symmetrical tetramer, there are 10 distinguishable liganded forms of the $\alpha_2\beta_2$ structure. As the α- and β-subunits have somewhat different oxygen affinities, a negative Hill coefficient results in the absence of any cooperativity of oxygen binding. A further complication is that the hemoglobin tetramer, especially in the R state, tends to dissociate reversibly into αβ-dimers, which are not cooperative.

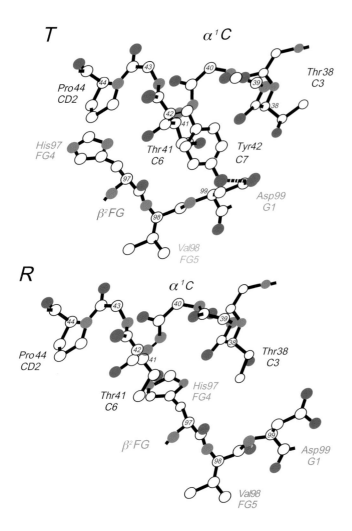

Figure 12-25. Part of the $\alpha^1\beta^2$ interface of hemoglobin, illustrating the large differences there between the R (oxy) and T (deoxy) states. Oxygen atoms are *red*, nitrogens are *blue* and carbons are *white*. At the top of each structure is the C-helix of the α^1 chain; at the bottom the irregular corner linking the F- and G-helices of the β^2 chain. The conformations of both subunits are virtually the same individually in the R and T structures, but the contacts differ markedly, owing to a shift of one subunit relative to the other. His97 of the β-chain is in contact with Thr41 of the α-chain in T, but with Thr38 in R. These two residues are on adjacent turns of the C-helix. Intermediate positions of the two subunits would be unstable because His97 and Thr41 would be too close together. Therefore, hemoglobin $\alpha_2\beta_2$ molecules must be in either the R or T quaternary structures. Adapted from a drawing by J. M. Baldwin.

Description of hemoglobin oxygenation under universal solution conditions by a global allostery model with a single adjustable parameter. K. Imai *et al.* (2002) *Biophys. Chem.* **98**, 79–91.

Crystallographic evidence for a new ensemble of ligand-induced allosteric transitions in hemoglobin: the T-to-T(high) quaternary transitions. J. S. Kavanaugh *et al.* (2005) *Biochemistry* **44**, 6101–6121.

---

4. Heterotropic Interactions

Several other classes of ligand are bound at other specific sites on the deoxy T quaternary structure of hemoglobin: protons (the basis of the Bohr effect), CO_2, chloride ions and organic phosphates, such as 2,3 bisphosphoglycerate (2,3-DPG), ATP, GTP and inositol phosphates (Table 12-6). Each of these effectors binds more avidly to the deoxy form of hemoglobin, so they shift the equilibrium to the T state and in effect (although not directly) compete with oxygen for binding. Because all these binding reactions are linked functions (Figure 11-14), the binding of oxygen must have the same effect on the affinity of hemoglobin for these nonheme ligands, and the presence of each of these ligands affects the binding of the others.

Under physiological conditions, the nonheme ligands lower the oxygen affinity and increase the cooperativity of oxygen binding by HbA. The shift of the oxygen binding curve to higher partial oxygen pressures, i.e. the lowering of the overall oxygen affinity, upon acidification or increasing the CO_2 concentration or temperature, has physiological consequences because this enhances the release of oxygen at the peripheral tissues, where such changes result from active metabolism.

The nonheme ligands bind more tightly to the T state than to the R state. The most dramatic are the organic phosphates, such as 2,3-DPG. They all bind at the same single site per $\alpha_2\beta_2$ tetramer; only one site is present because it straddles the β-subunits across the two-fold symmetry axis. This binding site is destroyed when the quaternary structure changes to the R state, because the β-chains move much closer together (Figure 12-24).

CO_2 binds as the carbamate to the α-amino groups of both α- and β-chains in the T state:

$$-NH_2 + CO_2 \leftrightarrow -NH-CO_2^- + H^+ \qquad (12.6)$$

Binding to the T state is favored because the ionized carbamate groups have electrostatic interactions with positively charged groups of the T state that are not in the same position in the R structure.

No binding sites for Cl^- ions have been identified. Cl^- might simply neutralize the excess of positive charges within the central cavity of $\alpha_2\beta_2$, which produces electrostatic repulsions, especially in the T state.

The chloride effect in human hemoglobin. A new kind of allosteric mechanism. M. F. Perutz *et al.* (1994) *J. Mol. Biol.* **239**, 555–560.

Heterotropic effectors control the hemoglobin function by interacting with its T and R states: a new view on the principle of allostery. A. Tsuneshige *et al.* (2002) *Biophys. Chem.* **98**, 49–63.

Allosteric effects of chloride ions at the intradimeric $\alpha^1\beta^1$ and $\alpha^2\beta^2$ interfaces of human hemoglobin. I. N. Rujan & I. M. Russu (2002) *Proteins* **49**, 413–419.

---

5. Bohr Effect

The effect of pH on oxygen affinity is known as the **Bohr effect**, and the change in oxygen affinity with pH defines the **Bohr coefficient**, δ, which is the number of protons (per heme group) bound to hemoglobin upon its full oxygenation:

$$Hb + 4O_2 + 4\delta H^+ \leftrightarrow Hb(O_2)_4(H^+)_{4\delta} \qquad (12.7)$$

When the shape of the oxygen dissociation curve is independent of pH, at least within the range of fractional occupancies (y) of $0.1 < y < 0.9$, δ is independent of the degree of O_2 binding and specifies the dependency of oxygen affinity on the pH:

$$\delta = \partial \log P_{50}/\partial pH \qquad (12.8)$$

where P_{50} is the oxygen affinity at half-saturation ($y = 50\%$).

Above pH 6.3, δ is negative and protons are released upon oxygenation of hemoglobin, whereas at lower pH values δ is positive and protons are taken up upon oxygenation. These two phenomena are known as the 'alkaline' and 'acid' Bohr effects. At physiological pH, 7.4, human HbA has δ = –0.6 in the presence of 0.1 M NaCl. This value is halved in the absence of Cl⁻ ion.

The Bohr effect arises from changes in the pK_a values of particular ionizable groups of hemoglobin that result from changes in their environments when the R and T states interconvert. The alkaline Bohr effect of human HbA is believed to be due primarily to the imidazole side-chain of C-terminal His146 of the β-chain. Its pK_a value changes from 8.0 in T_0 to 7.1 in R_4, contributing the greater part of the alkaline Bohr effect in the absence of Cl⁻. In addition, the α-amino groups of the α-chain change pK_a from about 7.8 in T_0 to 7.0 in R_4. These changes in pK_a are produced by involvement of the groups in salt bridges in the T structure, but not in R. In the presence of Cl⁻, binding of these ions to several ionizable groups lining the central cavity of the hemoglobin molecule, including the α-amino groups of the α-chains and the ε-amino groups of Lys82 of the β-chains, affects the protonation of those groups. Consequently, the release of Cl⁻ ions upon binding of oxygen produces an additional, Cl⁻-dependent Bohr effect. His143 of the β-chains is thought to be one of the groups responsible for the acid Bohr effect, but the others are still unidentified.

The Bohr effect is important physiologically. CO_2 produced in tissue cells diffuses into erythrocytes and produces HCO_3^- and H^+:

$$CO_2 + H_2O \leftrightarrow H_2CO_3 \leftrightarrow H^+ + HCO_3^- \tag{12.9}$$

The HCO_3^- ions move to the blood plasma, in exchange for chloride ions, and are transported to the lungs. Equation 12.9 is shifted to the right by hemoglobin absorbing the protons produced. Moreover, dissolution of the CO_2 is greater in the tissue capillaries than in the alveolar tissues because of the uptake of protons upon oxygen release by hemoglobin:

$$Hb\,O_2 + \delta H^+ \leftrightarrow Hb(H^+)_\delta + O_2 \tag{12.10}$$

This oxygen-linked proton binding is known as the **Haldane effect** and is the reciprocal expression of the Bohr effect; δ is the same as the Bohr coefficient. The Bohr effect and the Haldane effect are thermodynamically equivalent because they are linked functions (Figure 11-14); whatever effect changes in the pH have on the affinity of hemoglobin for oxygen, so oxygen binding must have the same effect on the affinity of hemoglobin for protons:

$$\left(\frac{\partial H^+}{\partial y}\right)_{pH} = \left(\frac{\partial \log P_{O_2}}{\partial pH}\right)_y \tag{12.11}$$

The release of oxygen in the tissue capillaries enhances the uptake of protons by hemoglobin, thereby causing a further shift to the right of Equation 12.9. Part of the CO_2 is transported directly to hemoglobin as a carbamino moiety (Equation 12.6). This CO_2 binding is also oxygen-linked. Dissociation of oxygen in tissue capillaries enhances the binding of CO_2 by hemoglobin, increasing the CO_2 content of the venous blood. Consequently, as a result of its allosteric properties, hemoglobin plays a key role in transport of CO_2 as well as of oxygen.

Interaction of hemoglobin with hydrogen ions, carbon dioxide, and organic phosphates. J. V. Kilmartin & L. Rossi-Bernardi (1973) *Physiol. Rev.* **53**, 836–890.

The Bohr effect and the Haldane effect in human hemoglobin. I. Tyuma (1984) *Jap. J. Physiol.* **34**, 205–216.

6. Allosteric Mechanism of Hemoglobin

Hemoglobin largely follows the concerted allosteric model, and was the inspiration for it, although some variations are apparent. The two quaternary structures R and T can be shown to coexist, in the absence of any ligand, by modifying the protein at the $\alpha^1\beta^2$ interface or by disrupting the salt bridges that preferentially stabilize the T state. In this way, the value of L, the equilibrium constant between R and T (Figure 12-18), can be increased from its normal value of approximately 10^{-4}, favoring the T state, to a value close to unity, so that R becomes populated substantially even in the deoxy form. Likewise, liganded hemoglobin can often be pulled into the T state by high concentrations of the allosteric effectors that bind preferentially to the T state (Table 12-6).

Table 12-6. Allosteric effectors of O_2 affinity of hemoglobin

	Major binding site on T state	
Decrease O_2 affinity	α-chain	β-chain
Organic phosphates		α-NH^{3+}
(e.g. diphosphoglycerate)		His2
		Lys82
		His143
CO_2	α-NH$_2$	α-NH$_2$
Anions (e.g. Cl$^-$)	α-NH$_3^+$	Lys82
	Arg141	
H$^+$ (Bohr effect)	α-NH$_2$	His146
Increase O_2 affinity	**Binding site on R state**	
	α-chain	β-chain
O_2 and other heme ligands	Heme Fe	Heme Fe

The O_2 affinity of the R state is not very sensitive to changes in the conditions, and its conformation is close to that of the individual α- and β-chains in isolation and of monomeric myoglobin. The T state is almost invariant structurally, but its oxygen affinity varies widely, depending upon which allosteric effectors are present. Mutations that affect the relative stability of R and T also affect the oxygen affinity of the T state. The greater the stability of T relative to R, i.e. the smaller the value of L, the lower is the O_2 affinity of T. Consequently, the two parameters L and c of the concerted model are not independent in hemoglobin but are related by:

$$\log c = A - 0.25 \log L \tag{12.12}$$

where A is a constant.

The T state is considered to be constrained at the heme group, and the degree of the constraint depends upon the degree of stabilization of the T state. Although constrained, the T state is more stable than the R state when all the hemes are deoxy. Its quaternary structure has somewhat more extensive interactions between the subunits and extra salt bridges, involving certain C-terminal residues of both the α- and β-chains, that are not possible in the R state. The T state is also much less liable to dissociate into dimers than is the R state, and it is further stabilized by the preferential binding of the other allosteric effectors.

The low oxygen affinity of the T state is believed to be due to residues surrounding the heme group preventing it from adopting the planar conformation, with the Fe atom in its plane, that normally occurs upon ligand binding. The greater the stability of the T state, the stronger these constraints and the lower the oxygen affinity (Figure 12-22). These constraints are largely absent in the R structure, so the heme can take up the geometry preferred by either the liganded or deoxy heme group.

The movement of the Fe atom upon ligand binding appears to be crucial for causing the change in quaternary structure from T to R. The Fe atom is attached to the proximal His side-chain, and it is here that the largest changes in tertiary structure of the individual chains are observed (Figure 12-26). Movement of the proximal His residue causes corresponding changes in the position of the F α-helix, of which it is part. This part of the protein in the T state appears to prevent the Fe atom from moving towards the heme plane upon oxygen binding and to be responsible for the low oxygen affinity of the T state. Indeed, when the very strong ligand nitrous oxide is bound tightly to the Fe atom, pulling the tetramer into the T state by the presence of high concentrations of allosteric effectors such as inositol hexaphosphate actually breaks the iron–His bond.

When O_2 is bound, the constraint is reversed, and the Fe atom and His residue are pulled towards the heme group, which also pulls the F-helix in that direction. This movement is transmitted to the ends of the relatively rigid helix, which is in contact at its carboxyl end with the other subunit across the $\alpha^1\beta^2$ interface (Figure 12-25). This is the interface that changes most markedly in the T ↔ R quaternary structure change. Another route linking ligand binding to the quaternary structure may result from the tendency of the liganded heme group to flatten and to push on the side-chains of Leu–FG3 and Val–FG5 at the turn between the F- and G-helices (Figure 12-25); these residues are also part of the $\alpha^1\beta^2$ interface.

When the first O_2 molecule binds to one subunit of the $\alpha_2\beta_2$ tetramer in the T state, the unfavorable interactions at the heme group are transmitted to the $\alpha^1\beta^2$ interface, which weakens the T state and favors its switch to the R state. This weakening of the T state may also cause the affinities of the other unoccupied heme groups in the T state tetramer to increase somewhat (Equation 12.12). Upon binding second and third O_2 molecules, the T state is further destabilized, and at this stage it becomes less stable than the R state, so tetramers adopt the R quaternary structure. All the heme groups of the R state have very high O_2 affinity, and **this shift to the R state is the main cause of the increased affinity for oxygen that produces the sigmoidal oxygen binding curve** (Figure 12-23). With four O_2 molecules bound, the R state predominates.

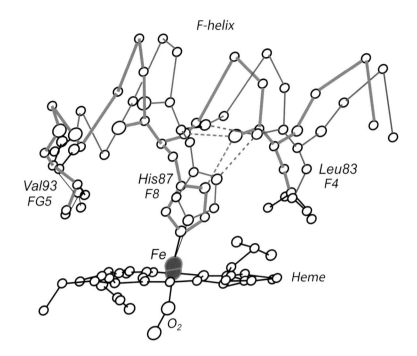

Figure 12-26. Changes at the proximal side of the heme group upon binding of O_2 to the α-chains of hemoglobin. Similar changes occur in the β-chains. The average positions of the heme groups of the deoxy (T) and oxy (R) structures were superimposed to illustrate the relative movements of the protein atoms. Some of the neighboring residues of the F-helix and the FG corner between the F- and G-helices are shown in *thick red lines* for oxy, *thin blue lines* for deoxy. Bound O_2 is shown *below* the heme group. Upon binding O_2 (or CO), the Fe atom moves about 0.6 Å from above to into the plane of the heme group. The Fe atom pulls with it the proximal His87 (also known as F8, i.e. the eight residue of the F-helix) and the F-helix of which it is a part. This produces changes at the FG corner, on the *left*, which is in contact with the other α-chain across the $\alpha^1\beta^2$ interface (Figure 12-25). It is this interface that changes in the T ↔ R quaternary structure change, and the above sequence of events describes a plausible, but unproved, mechanism by which O_2 binding can affect the quaternary structure and vice versa. The *dashed lines* illustrate hydrogen bonds that are believed to be important for orienting the His–F8 side-chain. Kindly provided by B. Shaanan in T. E. Creighton (1993) *Proteins: structures and molecular properties*, 2nd edn, W. H. Freeman, NY.

The allosteric properties of hemoglobin are generally those expected by the concerted allosteric model (Section 12.5.A) but some of the interactions that occur within the T state are most readily explained by the sequential model. Since both models seem physically possible, it is perhaps not surprising that both would have been used.

Deciphering the molecular code of hemoglobin allostery. G. K. Ackers (1998) *Adv. Protein Chem.* **51**, 185–253.

A tertiary two-state allosteric model for hemoglobin. E. R. Henry *et al.* (2002) *Biophys. Chem.* **98**, 149–164.

Allosteric action in real time: time-resolved crystallographic studies of a cooperative dimeric hemoglobin. J. E. Knapp *et al.* (2006) *Proc. Natl. Acad. Sci. USA* **103**, 7649–7654.

12.5.C. Negative Cooperativity

Homotropic allosteric interactions in which the binding of one molecule of ligand *decreases* the affinity of other sites for the same ligand are difficult to explain with the concerted model, in which binding of ligand can pull the T ↔ R transition only towards the higher affinity state; this can produce only positive homotropic cooperativity. Nevertheless, **negative cooperativity is observed with some proteins**. On the other hand, it is often found subsequently to be due to heterogeneity of the protein preparation, which produces binding curves like those expected with negative cooperativity. Well-established cases of negative cooperativity demonstrate Hill coefficients of 0.55–0.8 for dimers, 0.3–0.64 for tetramers and 0.16–0.45 for hexamers. In extreme cases, proteins even exhibit **half-of-the sites reactivity**, in which only half the expected sites of an oligomer bind a ligand or half the individual groups react with a modifying reagent; the other half of the sites bind weakly or react slowly, or not at all.

One possible explanation for apparent negative cooperativity is that the equivalent binding sites on a symmetric protein oligomer overlap, or are sufficiently close to interact sterically or electrostatically. This could occur because they include, or are near to, the symmetry axes of the oligomer. Binding of the first ligand molecule would block, or perhaps just inhibit, binding of a second molecule at the overlapping or adjacent site. For example, the binding site for organic phosphates on hemoglobin straddles the two-fold axis, and only one molecule is bound per $\alpha_2\beta_2$ hemoglobin molecule, whereas two would be expected for a symmetric dimer. Effects due to direct interactions between adjacent or overlapping sites, or due to nonidentical sites, are not allosteric in the original sense of the term.

An allosteric explanation for such behavior would be that initially identical and equivalent binding sites on an oligomeric protein are made nonidentical, and of lower affinity or reactivity, by binding of ligand or reagent to another site. Alternatively, the binding sites on the oligomeric protein may not be equivalent initially, even though the subunits are identical. In both cases, **the asymmetric structure would need to be more stable than that with either none or all of the sites liganded**.

All of these phenomena appear to occur in various instances of negative cooperativity. For example, the enzyme **tyrosyl tRNA synthetase** exhibits half-of-the-sites reactivity in solution by binding tightly only one molecule of tyrosine or tyrosine adenylate per dimer molecule. It was concluded to be asymmetric in solution, even though it crystallizes as a symmetrical dimer with each subunit having a complete binding site for tyrosine and tyrosine adenylate. The symmetry in solution of the molecule without ligand was investigated by constructing heterodimers in which one polypeptide chain was truncated slightly and mutating the binding site of one or the other polypeptide. Which site bound ligand initially was found to be random, but the same site was used to bind successive ligand molecules. This dimeric enzyme appears to be intrinsically asymmetric, even in the absence of ligands, but the structural basis is not known.

Glyceraldehyde-3-phosphate dehydrogenase (GPD) in some instances binds the cofactor NAD to the four sites on its tetramer with negative cooperativity. This has been concluded to be due to asymmetry of the tetramer induced by ligand binding. The crystal structures of both the apo and holo GPD (without and with bound NAD) show the four subunits to be equivalent. Local rearrangements occur upon NAD binding, and the crystal structure of the tetramer with one NAD molecule bound shows it to be asymmetric, with the liganded subunit in the holo conformation and the other three subunits in the apo form. Only very slight changes were observed in the unliganded subunits, however, so why the affinities of these sites change is not clear.

Asymmetry is most apparent in the structure of the bacterial **phosphopantetheine adenylyltransferase** enzyme, which binds some substrates and products under some conditions to only three of its six otherwise identical subunits. Its quaternary structure is two heterologous rings of three subunits interacting isologously, comparable to that of the catalytic subunits of aspartate transcarbamoylase (Figure 15-4). The crystal structure of the protein has the product bound to the three active sites in one of the two cyclic trimers, while the opposite trimer has its active sites empty (Figure 12-27). Superimposing the two subunits indicates only very small differences in the conformations of their polypeptide chains. Binding of the ligand causes small shifts in the local structure that affects the interface between the two trimers. The asymmetric structure appears to be more stable, having a greater buried surface area, than the symmetric structures with either no ligands or all six active sites liganded. The apo form appears to be symmetric, so the asymmetry must be induced by binding of ligand to the first sites.

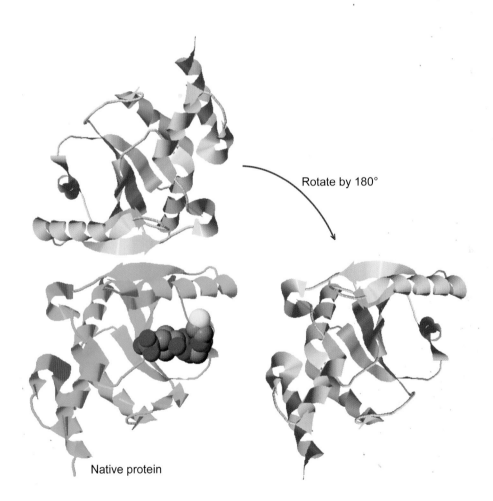

Figure 12-27. The asymmetry between the two halves of phosphopantetheine adenylyltransferase, one with the substrates 4-phosphopantetheine and an analog of ATP bound to only one half (at the *bottom*). One subunit from each of the two cyclic trimers is shown on the *left*; the view is approximately down the normal two-fold symmetry axis. The three-fold symmetry axis relating these two subunits to the other four is vertical in the plane of the page. The backbone of the polypeptide chain is indicated by *coils* for helices and *arrows* for β-strands. The upper subunit without the ligand has been rotated by 180° and translated to the *lower right*, so that it can be compared with the subunit containing the ligand; there is very little difference in the two subunit conformations. Figure generated from PDB file *1qjc* using the program Jmol.

The structural basis of negative cooperativity: receptors and enzymes. D. E. Koshland (1996) *Curr. Opinion Struct. Biol.* **6**, 757–761.

High-resolution structures reveal details of domain closure and 'half-of-sites-reactivity' in *Escherichia coli* aspartate beta-semialdehyde dehydrogenase. C. E. Nichols *et al.* (2004) *J. Mol. Biol.* **341**, 797–806.

Substrate-induced asymmetry and channel closure revealed by the apoenzyme structure of *Mycobacterium tuberculosis* phosphopantetheine adenylyltransferase. V. K. Morris & T. Izard (2004) *Protein Sci.* **13**, 2547–2552.

Phosphopantetheine adenylyltransferase from *Escherichia coli*: investigation of the kinetic mechanism and role in regulation of coenzyme A biosynthesis. J. R. Miller *et al.* (2007) *J. Bacteriol.* **189**, 8196–8205.

1. Negative Cooperativity or Heterogeneity of Sites?

One of the most difficult aspects of ligand binding studies is distinguishing between true negative cooperativity and simple heterogeneity of binding sites; the two give very similar binding curves, with the apparent affinity decreasing with the extent of binding. In the case of negative cooperativity, this occurs because binding of the initial ligand molecules decreases the affinity of the vacant sites for further ligand molecules. In the case of heterogeneity, it is simply because the high affinity sites are filled first, followed by the lower affinity sites.

Methods to distinguish between them would examine the effects of occupying some of the sites on the affinity of the remaining sites. A fraction of the sites might be occupied irreversibly by affinity ligands that react covalently at their binding site, but the test requires that the affinity ligand react equally with all sites present, even if some have high and others low affinity for the original ligand. If the affinity ligand reacted randomly and induced negative cooperativity, the affinities of the remaining sites would be lower than those of the original population, whereas the affinities of the heterogeneous population would be unchanged. This requirement that the affinity label induce negative cooperativity, but not discriminate between heterogeneous sites, is a severe one; it can be tested by characterizing the kinetics of the interaction of the affinity label with the original population of sites, to determine whether it reacts equally with all of them.

A related approach uses a second ligand, rather than an affinity label, to occupy some of the sites. As before, such a ligand must not bind with negative cooperativity and it must not distinguish between different classes of heterogeneous sites. **If the presence of this ligand alters the apparent cooperativity of binding of the first ligand, as measured by the Hill coefficient, it can be concluded that the system is truly cooperative negatively**. If no such change in cooperativity occurs, however, no firm conclusion can be drawn, as the second ligand may not have had the requisite properties.

Rather than using equilibrium affinities, the rates of dissociation of ligands can be measured. A small amount of radioactive ligand is added, so that only a fraction of the sites on the macromolecule are occupied. This equilibrium mixture is then diluted, so that some of the ligand dissociates, and the rate of dissociation is measured. The effect of excess nonlabeled ligand on this rate of dissociation is measured. **With heterogeneity of sites, the unlabeled ligand should have no effect on the rate of dissociation of the labeled ligand, whereas that rate should be increased by unlabeled ligand if there is negative cooperativity**. This method has given inconsistent results, however, and **the rates of dissociation of ligands appear to be affected by excess ligand even in the absence of negative cooperativity**. This should not happen if dissociation is an all-or-none event, but it can be explained if the ligands bind by attachment to multiple sites on the macromolecule: the ligand can dissociate temporarily from one or a few individual sites, without dissociating completely from the binding site, and a competing ligand can intervene and bind to those sites, thus accelerating the dissociation of the original ligand, which is now bound to only some of the multiple sites. If nothing else, this illustrates graphically the possible complexity of the binding of even single ligands to individual binding sites on macromolecules.

Ligand competition curves as a diagnostic tool for delineating the nature of site–site interactions: theory. Y. I. Henis & A. Levitzki (1979) *Eur. J. Biochem.* **102**, 449–465.

~ CHAPTER 13 ~

NUCLEIC ACID–PROTEIN INTERACTIONS

The interactions between proteins and nucleic acids are some of the most important in biology. RNA binds to proteins mainly for structural reasons, although in some cases the RNA is also involved in catalysis (Section 14.6). Binding to DNA serves two principal functions: to (1) organize and compact the DNA of chromosomes and (2) effect and regulate the processes of gene transcription, DNA replication and DNA recombination. These two functions are very different and have different structural and binding requirements. **Chromosomal DNA is organized by abundant proteins that can bind to many sites irrespective of the sequence** (Sections 13.3.F and 13.3.G). At the other extreme, **regulation of the enzymatic processes that manipulate DNA require precise targeting to particular DNA sequences**, which involves specific recognition of the nucleotide sequence by one or more specific proteins. Many such proteins act as repressors or activators of gene expression, either by themselves or in combination with corepressors or coactivators. A repressor usually acts simply by binding to a specific site on the DNA adjacent to a gene and blocking access of the proteins that would transcribe the gene into messenger RNA, which then gets translated into protein. Others do not prevent access but block further progression along the DNA of the RNA polymerase enzyme (Section 6.2). An activator usually acts by being part of the site next to the gene that is recognized and bound by the transcription apparatus; it assists in the binding of the other components of the transcription machinery. Most prominent are the **transcription factors**, which are part of the complex machinery for controlling the transcription of DNA into messenger RNA. In these cases, the specificity of binding is paramount.

It is not straightforward for a protein to distinguish between different nucleotide sequences in double-stranded DNA, for the nucleotides of the two antiparallel strands are base-paired within the interior of the double helix. The exterior surface of the double helix is almost independent of its nucleotide sequence, being comprised primarily of the constant phosphate–sugar backbone. Only the edges of the nucleotides are accessible to the solvent and a protein, primarily within the major groove of the DNA double helix (Figure 13-1). Nevertheless, this is sufficient for **the nucleotides to be discriminated by the different polar groups that are accessible**.

The interactions between proteins and nucleic acids can be very strong. They must often be sufficient for a protein to find a single specific DNA binding site in a cell. A typical eukaryotic cell has a diameter of 10 μm, which corresponds to a volume of 5×10^{-10} liters, so **one or two specific binding sites on a genome would be present at a concentration of only about 10^{-12} M**; the dissociation constant must be no greater than this for binding to be stable. **One consequence of such tight binding is that the complex usually dissociates very slowly**. A further consequence is that the association reaction can also be very slow at the very low reactant concentrations approaching the dissociation constant, because the rates of association and dissociation are equal at equilibrium.

Nucleic acids, especially DNA, can be very large molecules, with many sites at which proteins can bind. One must then distinguish between (1) relatively nonspecific binding, in that it occurs with many different nucleotide sequences and at many different sites on the DNA molecule, and (2) specific

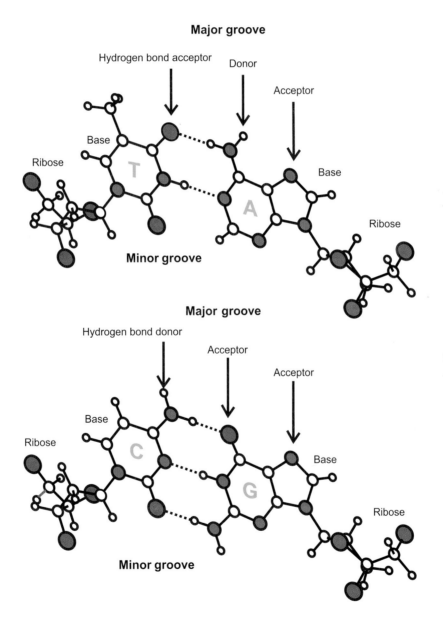

Figure 13-1. Hydrogen-bond interactions that can be used to distinguish between the various base pairs in double-helical B-form DNA. The upper edges of the bases are exposed in the major groove, and each has a unique pattern of hydrogen bond donors and acceptors. Each base pair can be reversed, leading to two more different patterns. The *red circles* represent O atoms, the *blue circles* are N, and the *smallest circles* are H atoms; the others are carbons.

binding to a certain sequence of nucleotides. **Proteins often bind to long DNA molecules by binding loosely and nonspecifically at the first site they encounter, then diffusing one-dimensionally along the double helix until they encounter their specific binding site**; they bind much more tightly to it, and consequently remain bound to this site. This is the basis for the seemingly paradoxical observation that a protein can find a specific binding site on a DNA molecule more rapidly if the DNA molecule is elongated with nonspecific sequences.

Finding a single site on genomic DNA is not a trivial matter; the human genome consists of 3×10^9 base pairs. The sequence of nucleotides that make up genomic DNA is not very regular, and to a first approximation the sequence can be taken to be random. It is then possible to estimate how many nucleotides must make up a binding site in order to have a reasonable likelihood of being unique and occurring only once in the genome. Each residue in DNA can be occupied by four different nucleotides (or base pairs), so a segment of N nucleotides can exist as 4^N different sequences. This number must be at least as large as the number of base pairs in the genome for it to have a reasonable chance of being unique. A 16-nucleotide segment can exist in 4.29×10^9 $(= 4^{16})$ different sequences, so **a binding site comprising 16 nucleotides has a reasonable chance of being unique in the human genome**. Human proteins that recognize a single site on the genome would be expected to recognize 16 base pairs specifically. The genomes of prokaryotes are smaller, that of *Escherichia coli* comprising 4×10^6 base pairs, but 11 nucleotides ($4^{11} = 4.19 \times 10^6$) are still probably needed to define a unique site.

Understanding the interactions between proteins and nucleic acids requires that the sequence specificity be determined, that the affinities of the protein for this and other sequences be measured, and that the structure of the complex be determined.

Protein–DNA recognition patterns and predictions. A. Sarai & H. Kono (2005) *Ann. Rev. Biophys. Biomolec. Structure* **34**, 379–398.

Recognizing DNA. R. Lavery (2005) *Quart. Revs. Biophys.* **38**, 339–344.

An extra dimension in nucleic acid sequence recognition. K. R. Fox & T. Brown (2005) *Quart. Revs. Biophys.* **38**, 311–320.

Identification and characterization of DNA-binding proteins by mass spectrometry. E. Nordhoff & E. Lehrach (2007) *Adv. Biochem. Eng. Biotechnol.* **104**, 111–195.

13.1. TECHNIQUES FOR MEASURING PROTEIN–DNA INTERACTIONS

Some of the techniques that are used to characterize the binding of ligands to macromolecules are also applicable to nucleic acids and proteins. Spectroscopic measurements, such as ultraviolet (UV) absorbance, fluorescence and circular dichroism, can usually detect the difference between free and associated nucleic acids or proteins, but the signals are often small and they do not reveal which binding sites are being used. Affinity chromatography can be very useful, as either the nucleic acid or the protein can be bound covalently to a column or blotting matrix.

The interactions between nucleic acids and proteins differ from those of other ligands in that both reactants are macromolecules and the complex frequently dissociates only slowly. Consequently, techniques such as ultracentrifugation can distinguish between free and bound protein and nucleic acid, and the stoichiometry of the interaction can often be inferred from the change in the sedimentation coefficient. Both free reactants and their complexes can often be observed by electron microscopy, or those in a complex can be cross-linked covalently. Some of the techniques for small ligands, such as equilibrium dialysis, are not applicable with two macromolecules, but other techniques take advantage of the special properties of nucleic acids and can be extremely informative.

Most important is that specific complexes with proteins dissociate only very slowly, because such complexes are very stable, usually having dissociation constants of 10^{-7}–10^{-14} M. In this case, if the rate constant for binding is 10^6 s^{-1} M^{-1}, the rate constant for dissociation will be between 10^{-1} and 10^{-8} s corresponding to half-times for dissociation of between 7 and 7×10^7 s. Consequently, tight complexes can be separated from the free components by many techniques, such as chromatography, filtration and electrophoresis.

Protein–DNA interactions. V. Dotsch (2001) *Methods Enzymol.* **339**, 343–357.

Circular dichroism for the analysis of protein × DNA interactions. M. L. Carpenter *et al.* (2001) *Methods Mol. Biol.* **148**, 503–510.

Ultraviolet crosslinking of DNA–protein complexes via 8-azidoadenine. R. Meffert *et al.* (2001) *Methods Mol. Biol.* **148**, 323–335.

Analysis of DNA–protein interactions by intrinsic fluorescence. M. L. Carpenter *et al.* (2001) *Methods Mol. Biol.* **148**, 491–502.

Surface plasmon resonance applied to DNA–protein complexes. M. Buckle (2001) *Methods Mol. Biol.* **148**, 535–546.

Scanning transmission electron microscopy of DNA–protein complexes. J. S. Wall & M. N. Simon (2001) *Method Mol. Biol.* **148**, 589–601.

Covalent trapping of protein–DNA complexes. G. L. Verdine & D. P. G. Norman (2003) *Ann. Rev. Biochem.* **72**, 337–366.

Crystallization of protein–DNA complexes. T. Hollis (2007) *Methods Mol. Biol.* **363**, 225–2237.

13.1.A. Filter-binding Assays

Complexes that dissociate slowly can be measured using filter-binding assays, which are simple, rapid, sensitive and versatile. They are based on the empirical observation that **most proteins, but not double-stranded DNA molecules, bind to the surfaces of nitrocellulose membrane filters**. Therefore, when a DNA-binding protein is incubated with a specific DNA sequence and then passed through a filter, the DNA•protein complex that was formed and any excess protein will bind to the membrane filter, while uncomplexed DNA will pass through. If the DNA used is radiolabeled, the amount of the DNA•protein complex bound to the membrane filter can be determined easily from the amount of radioactivity retained. Denatured single-stranded DNA, however, also binds to nitrocellulose and cannot be studied in this way.

The use of highly radioactive DNA makes it possible to use very low concentrations of DNA and protein, and the protein need not be pure. The lower limit for the concentration of radioactive DNA is usually about 10^{-11} M. A typical experiment involves titration of a constant amount of radiolabeled

DNA with the DNA-binding protein of interest. The amount of radioactivity retained on the membrane increases with increasing protein concentration until all the radioactive DNA that can bind to that DNA-binding protein is depleted from the incubation mixture. The binding affinity can be determined by measuring the level of radioactive complexes formed in the presence of a nonradioactive competitor DNA of known concentration. The kinetics of association of a specific DNA•protein complex can also be studied in this way, by mixing the components at time zero and then measuring the time–course of binding by filtering aliquots at various times; the binding reaction can be quenched by the addition of an excess of unlabeled DNA to the aliquots to bind to the free protein molecules. The kinetics of dissociation are measured by mixing radioactive DNA with an excess of the protein, to saturate all the DNA binding sites; all the radioactivity of an aliquot will bind to the filter. At time zero, an excess is added of the same DNA, but not radioactive. As protein dissociates from the radioactive complexes, it is effectively sequestered by binding to the nonradioactive DNA and the radioactivity of aliquots retained by the filter will decrease with time and reflect the rate of dissociation of the complex.

Filter-binding assays cannot differentiate between dissimilar DNA•protein complexes, so this assay is most useful when only a single DNA•protein complex is formed. Most proteins bind to nitrocellulose but their individual binding affinities vary. The optimum conditions for binding a DNA•protein complex to the nitrocellulose membrane will differ for each protein and must be determined for each case.

Filter-binding assays. P. G. Stockley (1994) in *Methods in Molecular Biology*, Vol. 31 (G. G. Kneate, ed.), Humana Press, NJ, pp. 251–262.

Is nitrocellulose filter binding really a universal assay for protein–DNA interactions? S. Oehler *et al.* (1999) *Anal. Biochem.* **268**, 330–336.

13.1.B. Gel Retardation Assay

The gel retardation assay is also known as the **bandshift assay** or **electrophoretic mobility shift assay**. It can be used to detect proteins that bind to DNA or RNA in crude cell extracts or to study the binding activities of purified proteins. The assay is based on the observations that protein•DNA complexes remain intact when separated gently by nondenaturing gel or capillary electrophoresis and that they migrate as distinct bands more slowly than does the free DNA fragment (Figure 13-2). The gel matrix can be polyacrylamide for DNA fragments shorter than 500 base pairs, agarose gels for fragments longer than 1000 base pairs, and mixtures of polyacrylamide and agarose for the intermediate lengths.

The assay is simple, quick and very sensitive when highly radioactive DNA is used. It is useful for both sequence-specific proteins and nonspecific proteins, such as histones, so long as the complexes have dissociation constants $<10^{-7}$ M and do not dissociate during the electrophoretic separation. Including unlabeled nonspecific DNA in the binding mixture to sequester any proteins that would bind nonspecifically ensures that only specific complexes are observed. Any complex observed should be shown to be specific by demonstrating that radiolabeled complex is not formed in the presence of an excess of the specific DNA that is not radioactive.

Gel retardation assays can be used quantitatively to estimate dissociation constants of protein•DNA complexes, so long as dissociation is negligible during the electrophoretic separation. Additionally,

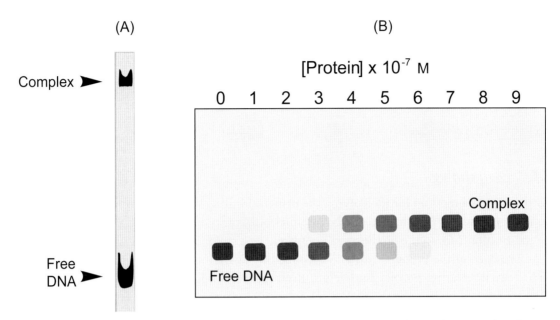

Figure 13-2. Idealized bandshift assays. Electrophoresis in 0.7% (w/v) agarose electrophoresis gels is from *top* to *bottom*. The DNA molecules are radioactive, and only they are detected using autoradiography. (A) Binding of a 150-kDa protein causes the radioactive 62-base pair DNA fragment containing the binding site to migrate as a distinct band, at the top of the gel, much more slowly than the unbound DNA at the bottom of the gel. (B) Titration of a 21-base pair oligonucleotide with increasing amounts of a binding protein. The DNA concentration was 1×10^{-8} M and the protein concentration was varied as shown. When the DNA concentration is less than the dissociation constant, the value of the dissociation constant can be estimated by the concentration of protein that produces 50% binding, in this case approximately 4×10^{-7} M.

such assays can detect interactions between one protein that binds DNA and another protein that does not. The binding of such a second protein to a protein•DNA complex to form a triple complex will be apparent by a further retardation of mobility that is called a **supershift**. Supershift experiments can also be used to assay the binding of a second DNA-binding protein to the same DNA molecule or the binding of a second molecule of DNA to the protein. What proteins are present in a complex can be determined in purified systems by excising the band of the complex and then separating the proteins by SDS–polyacrylamide gel electrophoresis or mass spectrometry.

The gel retardation assay can also detect bending of DNA induced by a DNA-binding protein. When binding of a protein to a DNA fragment of length 100–250 base pairs causes it to bend, the complex migrates more slowly through the pores of the gel. The degree of retardation depends on the site where the protein binds. The retardation is greatest when the binding site is in the middle of the DNA molecule. This method can be used to obtain an estimate of the degree of DNA bending induced by a protein by comparing it with proteins that induce known bends.

Disadvantages of this assay are that the electrophoretic separation is relatively slow, at least compared with the filter-binding assay, so some complexes might dissociate during the separation. Also, electrophoresis requires that the buffer have a low ionic strength, although this will probably stabilize most complexes by enhancing electrostatic interactions. Often protein–nucleic acid complexes are observed to persist within gels for much longer than would be expected on the basis of their free solution lifetimes. Their lifetimes in free solution may be limited by their frequency of encounter with other molecules of DNA or with protein•DNA complexes, which will be decreased in gels.

Electrophoretic mobility shift assays for the analysis of DNA–protein interactions. M. A. Laniel *et al.* (2001) *Methods Mol. Biol.* **148**, 13–30.

Capillary DNA–protein mobility shift assay. J. Xian (2001) *Methods Mol. Biol.* **163**, 355–367.

A modified quantitative EMSA and its application in the study of RNA–protein interactions. Y. Li *et al.* (2004) *J. Biochem. Biophys. Methods* **60**, 85–96.

13.1.C. Footprinting

Footprinting is a method for identifying the protein-binding sites on DNA *in vivo* and *in vitro*. A bound protein molecule can protect DNA from cleavage or modification by a nuclease or chemical reagent (Section 2.6), so the binding site is apparent as a region of the DNA with reduced reactivity in the presence of the protein. This is determined by analyzing the products of a partial cleavage reaction by denaturing polyacrylamide gel electrophoresis (Figure 13-3). The cleavage pattern is usually simplified by labeling the DNA molecule with radioactivity at only one end; only the DNA fragments containing this end are detected using autoradiography. Only partial cleavage will produce a collection of DNA fragments of different lengths that reflect the sites at which cleavage occurred. **The 'footprint' of the protein appears as a gap in the normal pattern of cleavage.** In some instances, the bound protein can cause intensification of a band, implying that the bound protein made that DNA site hyperreactive, usually by changing its local conformation.

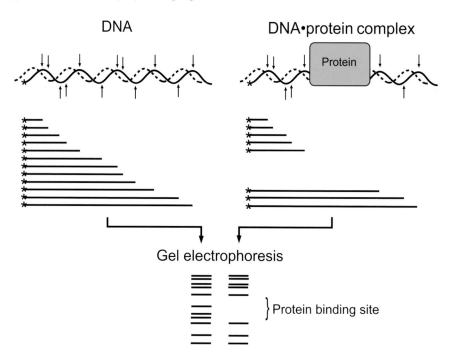

Figure 13-3. Schematic description of the footprinting assay. A naked DNA fragment and its fragmentation pattern produced by cleavage at the *arrows* is shown at the *left*. The DNA is labeled radioactively at one end, so only the fragments containing this end of the DNA are visualized by autoradiography after gel electrophoresis under denaturing conditions (*bottom*). With protein bound to a specific segment of DNA, potential cleavage sites within this segment are protected and cleavage does not occur. This results in a gap in the gel electrophoresis pattern corresponding to the binding site for the protein. Adapted from D. Rhodes & L. Fairall (1997) in *Protein Function: a practical approach*, 2nd edn (T. E. Creighton, ed.), IRL Press, Oxford, p. 228.

The extent of the cleavage reaction is usually chosen so that, on average, a given DNA molecule will have at most one cleavage. The nuclease DNase I is commonly used *in vitro* for footprinting, while dimethylsulfate and hydroxyl radicals are used for footprinting using chemical cleavage methods. **DNase I** cleaves preferentially in the minor groove of DNA, also depending on the local conformation of the bound DNA; cleavage is enhanced when the minor groove is on the outside of DNA wrapped around a protein. It is usually used to cleave only one of the two DNA strands, rather than both. **Dimethylsulfate** methylates the N7 position of guanine in the major groove and the N2 position of adenine in the minor groove, which sensitizes the sugar–phosphate backbone to mild alkaline hydrolysis and makes it possible to determine the sites of reaction. For **hydroxyl-radical** footprinting, the OH• radicals are generated by the **Fenton reaction**, in which hydrogen peroxide reacts with a complex of Fe^{2+} and EDTA:

$$[Fe\,(EDTA)]^{-2} + H_2O_2 \rightarrow [Fe\,(EDTA)]^- + OH^- + OH^{\bullet} \tag{13.1}$$

The OH• radicals react with deoxyribose moieties of the DNA backbone, which can cause cleavage of the backbone. This reagent attacks primarily the minor groove irrespective of the nucleotide sequence. It is, however, sensitive to the width of the groove and cleaves preferentially where the minor groove is wide. The opposite selectivity is exhibited by a similar type of reagent, **Mn-porphyrin**, which cleaves DNA preferentially where the minor groove is narrow.

The modification reaction need not cleave the DNA double helix but merely modify it so that it can no longer be copied by an enzyme like DNA polymerase. In this case, a radioactive oligonucleotide primer complementary to an appropriate sequence on the DNA molecule is used to initiate synthesis of the complementary strand, as in the chain-termination method of sequencing DNA (Section 6.4.B). The newly synthesized complementary strand will be extended until a site of modification is reached. The length of the strand synthesized indicates the site of the modification.

The various cleavage methods can provide complementary information. For example, the enzyme DNase I is a large molecule, whereas a hydroxyl radical is very small and thus can provide a more detailed picture. On the other hand, DNase I cleavage can provide information about the local DNA conformation, whereas cleavage by hydroxyl radicals tends to be too uniform over all the possible sites in the DNA molecule.

Footprinting techniques can also be used to probe the three-dimensional (3-D) structures of nucleic acids, especially RNA (Section 4.2).

Mapping nucleic acid structure by hydroxyl radical cleavage. T. D. Tullius & J. A. Greenbaum (2005) *Curr. Opinion Chem. Biol.* **9**, 127–134.

DNase I footprinting of small molecule binding sites on DNA. C. Bailly *et al.* (2005) *Methods Mol. Biol.* **288**, 319–342.

Visualising DNA: footprinting and 1-2D gels. A. R. Urbach & M. J. Waring (2005) *Mol. Biosyst.* **1**, 287–293.

Radiolytic footprinting with mass spectrometry to probe the structure of macromolecular complexes. K. Takamoto & M. R. Chance (2006) *Ann. Rev. Biophys. Biomol. Structure* **35**, 251–276.

13.2. PRINCIPLES OF PROTEIN–DNA RECOGNITION

Individual sequence-specific proteins usually recognize DNA sequences that are 3–15 nucleotides long. They use two types of interactions.

(1) The principal basis for sequence selectivity is direct contact between the polypeptide chain of the DNA-binding protein and the exposed edges of the DNA base pairs, primarily in the major groove of B-DNA. There the four different Watson–Crick base pairs can be distinguished by their distinctive pattern of hydrogen-bond donor and acceptor groups (Figure 13-1). **There is, however, no clear one-to-one correspondence between amino acid residues and the bases that they recognize**. The contacts may involve either hydrogen bonds (Figure 13-4) or van der Waals interactions, which usually detect the methyl group of thymine nucleotides. Small molecules, such as water molecules that are tightly bound to both proteins and DNA and integral components of the macromolecular structure, may also participate in these interactions and so provide binding specificity by proxy. The interfaces between DNA and protein are as tightly packed as are the interiors of proteins and DNA, so the interactions are probably as energetically favorable in all these cases. Cavities in the DNA–protein interfaces tend to be larger but are more likely to be occupied by water molecules.

(2) The specificity of the direct interactions can be supplemented by the bendability or deformability of the DNA molecule, which limits the energetically favorable conformations of a particular binding site. The conformational flexibility of the DNA double helix depends upon the nucleotide sequence, so it provides additional information about the sequence. Bending of DNA is achieved by a variety of mechanisms, notably the induction of bends by spatial constraints on a rigid protein surface, the insertion of hydrophobic residues between adjacent bases in the DNA duplex, and charge neutralization on one face of the double helix. **Compared with direct contacts to the base pairs, bending of the DNA is a second-order effect that modulates the affinity for the whole binding site**.

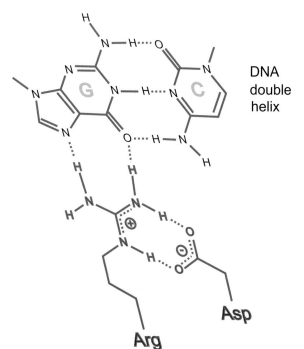

Figure 13-4. Hydrogen bonds involved in the recognition of guanine bases by zinc fingers. The Arg and Asp residues are involved in an ion pair, which may help to align the guanido group of the Arg residue to form two hydrogen bonds to the guanine base. Adapted from N. P. Pavletich & C. O. Pabo (1991) *Science* **252**, 809–817.

The binding energy available from direct specific interactions with the base pairs of binding sites of average length, although significant, is not by itself in general sufficient to produce a stable bound complex. The additional binding energy required results from other interactions between the protein and DNA that are independent of the nucleotide sequence, especially direct electrostatic interactions between basic amino acid residues and the negatively charged sugar–phosphate backbone. The spatial constraints imposed by these types of interaction may also constrain the conformation of the DNA when bound to a protein. Both components of the interaction contribute to the overall binding constant, but the first, sequence-specific, interactions determine the specificity of a protein for its binding of DNA.

Recognition of specific nucleotide sequences by proteins is a complex and subtle phenomenon that is not well understood. Consequently, **it is not yet possible to predict the specific nucleotide sequences that a given protein will recognize**.

Multiprotein–DNA complexes in transcriptional regulation. C. Wolberger (1999) *Ann. Rev. Biophys. Biomol. Structure* **28**, 29–56.

DNA mimicry by proteins and the control of enzymatic activity on DNA. D. T. Dryden (2006) *Trends Biotechnol.* **24**, 378–382.

13.2.A. Specificity of DNA–Protein Binding

The binding of proteins to specific DNA sequences is fundamentally different from the problem of how to discriminate between various small-molecule ligands, which can be based at least in part on their size and shape (Chapter 12). Binding specific sequences is unique in that the ligand is a short segment of DNA that is immersed in a sea of other DNA sequences that are part of the same molecule and very similar chemically and structurally to the specific ligand. **All proteins that bind to specific DNA sequences also have significant affinity for nonspecific sites**, binding that is equally probable at any particular point along the DNA and independent of the sequence. The two phenomena of specific and nonspecific binding are usually measured by the apparent dissociation constants of the complex of a protein (P) with DNA that contains a DNA site that is specific (DNA^S) or completely unrelated (DNA^N):

$$P + DNA^S \longrightarrow P{\bullet}DNA^S \quad K_d^S = [P][DNA^S]/[P{\bullet}DNA^S] \tag{13.2}$$

$$P + DNA^N \longrightarrow P{\bullet}DNA^N \quad K_d^N = [P][DNA^N]/[P{\bullet}DNA^N] \tag{13.3}$$

The dissociation constants for specific and nonspecific binding are compared in Table 13-1 for a number of different types of DNA-binding proteins. Each pair of specific and nonspecific binding constants was measured under the same conditions. Most pertinent is their ratio, because the absolute values depend upon a number of variables. The dissociation constants measured depend on the conditions, such as the temperature, pH and concentration and type of cations and anions. In addition, the stability of a specific complex can depend on the DNA sequences flanking the specific binding site, varying as much as 17-fold even if the flanking sequences do not contact the protein

Table 13-1. Specificities of DNA-binding proteins

Protein	Specific site	K_d^N (M)	K_d^S (M)	K_d^N/K_d^S	$-\Delta\Delta G°$ (kcal/mol)
Transcription factors					
CRP	lac promoter	1.4×10^{-6}	1.2×10^{-11}	1.2×10^5	6.8
E12	CAGGTG	1.9×10^{-7}	1.4×10^{-8}	1.4×10^1	1.5
GCN4	ATGACTCAT	1.1×10^{-6}	3.5×10^{-8}	31	2.0
	ATGACGTCAT	1.1×10^{-6}	2.7×10^{-8}	41	2.2
IHF	λ attP H'	2.9×10^{-6}	1.6×10^{-9}	1.8×10^3	4.4
MASH-1	CAGGTG	1.9×10^{-8}	5.9×10^{-9}	3.2	0.7
MEF-2C	TATAAATA	$>1.2 \times 10^{-5}$	1.1×10^{-7}	$>1.1 \times 10^2$	>2.7
MyoD/E12	CAGGTG	7.9×10^{-8}	7.5×10^{-9}	11	1.4
TBP	TATAAAAG	$>1.0 \times 10^{-5}$	3.7×10^{-9}	$>2.7 \times 10^3$	>4.6
Repressors					
lac	O^{sym}	2.4×10^{-4}	8.3×10^{-12}	2.9×10^7	10.1
λ CI	$O_R 1$	4.0×10^{-4}	1.2×10^{-10}	3.3×10^6	8.7
	$O_R 2$	4.0×10^{-4}	6.7×10^{-10}	6.0×10^5	7.7
	$O_R 3$	4.0×10^{-4}	6.3×10^{-9}	6.3×10^4	6.4
λ Cro	$O_R 1$	1.5×10^{-6}	8.3×10^{-12}	1.8×10^5	6.6
	$O_R 2$	1.5×10^{-6}	1.2×10^{-10}	1.3×10^4	5.1
	$O_R 3$	1.5×10^{-6}	2.0×10^{-12}	7.5×10^5	7.3
MetJ	Met Box	1.6×10^{-6}	5.0×10^{-4}	3.1×10^2	3.4
P22 Arc	Left half-site	4.7×10^{-8}	2.7×10^{-11}	1.7×10^3	4.3
trp	Wild-type op.	4.2×10^{-5}	2.2×10^{-9}	1.9×10^4	5.8
Restriction endonucleases					
EcoRI	GAATTC	2.8×10^{-4}	5.9×10^{-12}	4.7×10^7	10.3
EcoRV	GATATC	1.0×10^{-5}	3.8×10^{-10}	2.6×10^4	6.0

Data from R. Alleman (1999) in *Encyclopedia of Molecular Biology* (T.E. Creighton, ed.), Wiley-Interscience, NY, p. 744.

directly. Furthermore, the apparent dissociation constants of the protein complexes with nonspecific DNA depend on the length of the DNA probe, as the protein can bind to any of the nucleotides. Consequently, all the values for K_d^N in Table 13-1 were corrected for the length of the DNA probe by dividing the measured K_d^N by double the difference between the lengths of the specific and nonspecific DNA molecules. **The specificity of DNA binding is best expressed as the ratio K_d^N/K_d^S.**

The dissociation constants of the specific complexes vary over approximately six orders of magnitude and are as small as 10^{-12} M (Table 13-1). The complexes of bacterial proteins tend to be more stable than those of eukaryotic proteins, probably because the specific DNA binding sequences in the prokaryotes are longer. In contrast, the dissociation constants of the nonspecific complexes span only approximately four orders of magnitude, and the eukaryotic proteins bind more tightly. Therefore, prokaryotic proteins tend to bind DNA with greater specificity than do most eukaryotic transcription factors.

The longer specific sequences and the greater specificities in prokaryotes seem counterintuitive. The *E. coli* genome consists of only 4×10^6 base pairs, whereas those of eukaryotes tend to be much larger. Of the proteins listed in Table 13-1, the restriction enzyme *Eco*R displays the highest DNA-binding specificity, 4.7×10^7. This specificity is greater than the number of base pairs in the genome, so binding of the *Eco*R1 protein could occur predominantly to a single specific sequence on the chromosome, yet the specific DNA-binding site of *Eco*R1 has the sequence GAATTC, and such a hexamer sequence should occur statistically approximately 1000 times in the *E. coli* genome. Binding to a unique site in a human genome would require that the specificity be $>3 \times 10^9$, the number of base pairs in the human genome. This would require statistically a minimal length of 16 base pairs, but eukaryotic transcription factors that are involved in gene expression recognize much shorter sequences. For example, proteins that contain the basic helix–loop–helix motif (BHLH) (Section 13.3.D) bind to the sequence CAGGTG, which occurs approximately 7×10^5 times on a mammalian chromosome. **Most eukaryotic transcription factors bind to DNA sequences that are too short to be unique on the mammalian chromosome, and with insufficient specificity.** The reason is that transcription factors are designed to bind to a family of similar sequences within the genome, and they regulate multiple genes. Consequently, they need some flexibility in their specificity, so they usually bind significantly to several related DNA sequences. MyoD recognizes DNA through a BHLH domain and can activate myogenesis in a wide variety of cell types, including myoblasts and fibroblasts, while the BHLH protein MASH-1 promotes the differentiation of committed neuronal precursor cells. BHLH proteins need to bind to DNA with a specificity of approximately 4×10^3 in order to bind with equal probability to a nonspecific site and to one of the approximately 700,000 specific sites on the mammalian chromosome. But even then, MASH-1 would still activate transcription from MyoD target promoters and vice versa. Clearly, unique binding sites on chromosomes need to consist of many more than just a few sequential nucleotides. **A single DNA sequence recognition motif typically recognizes only three to four base pairs**, which is insufficient to allow highly selective discrimination between all the sequences in a genome.

In practice, the effective size of a recognition site can be increased by using a larger protein assembly or the conjunction of two sequence-specific DNA-binding motifs in the same polypeptide chain that recognize longer DNA sequences. Accordingly, **regulation of gene expression in higher organisms relies on multiprotein complexes with the potential for combinatorial interactions that involve a substantial segment of DNA**. The complexes often include a stable homo- or heterodimer; some dimerization surfaces, especially the coiled coils in bZIP proteins (Section 13.3.D), are sufficiently similar or pliable that two different polypeptide chains can interact to form a nonsymmetrical heterodimer. In other cases, however, cooperative interactions between proteins bound at contiguous or distant DNA sites are required. One example of such interactions is the cooperative binding of λ C_I repressor dimers to the leftward and rightward operators in lambda phage DNA, each of which contains three C_I binding sites. Stable binding to any two of these sites depends on interactions between separate dimers, so occupation is sensitive to small changes in the concentration of repressor molecules over a certain range (Section 13.3.A.2).

Interactions can also occur between distant binding sites on the DNA in that binding to both simultaneously generates a loop of intervening DNA. Loop formation of this type can involve only a single stable protein assembly, as in the case of the tetramer of the *lac* repressor containing four helix-loop–helix motifs (Figure 13-5). Alternatively, cooperative interactions between proteins bound at the separate sites may be required. **When bound at distant sites, the two proteins should be bound on the same face of the double helix**, because the torsional rigidity of DNA prevents bringing the two proteins into appropriate spatial register when they are bound initially to opposite faces of the double helix. An example is the binding of the *E. coli* AraC protein to two sites separated by about 231 base pairs on the *araBAD* promoter. Altering the separation of these sites by integral numbers of double-helical turns has little effect on loop formation. In contrast, altering the separation by 0.5, 1.5 or 2.5 turns severely impairs loop formation and regulatory function.

Thermodynamic and kinetic methods of analyses of protein–nucleic acid interactions. From simpler to more complex systems. W. Bujalowski (2006) *Chem. Rev.* **106**, 556–606.

---

1. Specific Interactions

As with other protein–ligand interactions (Section 12.1), the precise recognition of a defined DNA sequence by a DNA-binding protein requires optimal shape and chemical complementarity between the interacting species. Consequently, a large number of noncovalent interactions occur at the interface between them. **Individual interactions often contribute only a small amount to the overall stability of the complex, but all of these interactions are important for the preferred binding of the protein to the specific DNA site.**

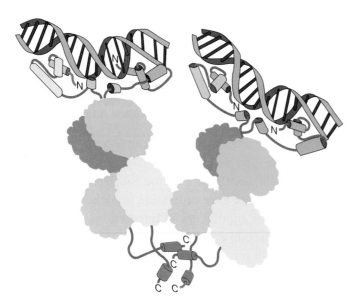

Figure 13-5. The tetrameric *lac* repressor bound to two DNA fragments, each with a palindromic sequence. Each arm of the V-shaped tetramer consists of a dimer containing a single DNA-binding site that has two-fold symmetry (i.e. is palindromic). The two helix–turn–helix motifs of each dimer bind in two successive major grooves, while shorter hinge helices bind adjacent to each other in the minor groove between the two major groove binding sites. The four subunits of the tetramer are held together by the C-terminal helices, which form a four-helix bundle. Adapted from M. Lewis *et al.* (1996) *Science* **271**, 12347–1254.

Initially the specificity of DNA•protein complexes was thought to be due to 'direct readout' of the interactions between the functional groups of the proteins and the DNA bases. The initial structures of DNA•protein complexes demonstrated such specific interactions to be due to hydrogen bonds, electrostatic interactions between charged groups and van der Waals interactions. Hydrogen bonds can be made between the polypeptide backbone or amino acid side-chains of the protein and the nucleobases of the DNA. Longer side-chains are more flexible, and the need to fix them should decrease their entropy and the stability of the interaction. However, they are often orientated in space by other side-chains or by making multiple contacts to the same base, adjacent bases or a base and phosphodiester group (Figure 13-4). In Figure 13-6, the carboxyl group of a Glu345 makes hydrogen bonds to both an adenine and adjacent cytosine, thereby specifying the first two bases of the binding site. A neighboring Arg residue orients the Glu side-chain in space through the formation of a 'clamp' between the phosphate backbone and peptide. If these groups are fixed in position before the DNA binds, they should contribute energetically to binding just as any other group. Further specificity arises from van der Waals interactions of the β- and γ-CH_2 groups of Glu345 with the methyl group of a nearby thymine base. Such van der Waals interactions appear to stabilize a complex by 0.5–2 kcal/mol.

The thermodynamics of the binding interactions between amino acid side-chains and DNA bases have been studied using site-directed mutation of both the protein and the DNA. The results from such experiments are difficult to interpret, as they appear to be complicated by subtle energetic changes that result from perturbation of the local structure of the DNA and/or the protein and changes in solvation. Nevertheless, **each hydrogen bond between the protein and the DNA appears to stabilize a complex by roughly 1.4 kcal/mol**, about the same value measured for hydrogen bonds in folded proteins (Section 11.3.A).

DNA-binding proteins make extensive contacts to the phosphodiester groups of the DNA backbone through charged and uncharged side-chains, but it is difficult to measure the contributions of these interactions to the overall stability of the complex. One approach is to determine the contribution of the polyelectrolyte effect from the dependence of the reaction free energy on the concentration of univalent salts (Section 13.2.A.5) and add a contribution for the interactions between the DNA

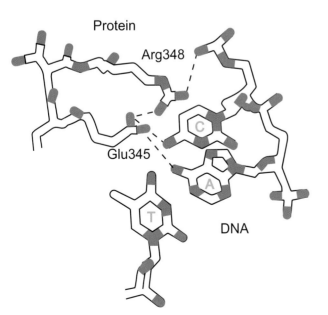

Figure 13-6. The network of hydrogen bonds between residues Glu345 and Arg348 of protein E47 and the CpA dinucleotide of the E-box DNA sequence. N atoms are *blue*, O atoms are *red*. The *dashed lines* are presumed hydrogen bonds. The carboxyl group of Glu345 is hydrogen bonded to N4 of cytosine and to N6 of adenine. Note the 'clamp' function of the side-chain of Arg348 connects the phosphate backbone to Glu345 and clamps the conformation of its side-chain. The β- and γ-CH_2 groups of Glu345 are in van der Waals contact with the methyl group of the thymine base. Adapted from a figure by R. Alleman.

phosphates and uncharged amino acid residues. The polyelectrolyte effect is, however, a consequence of the fact that DNA is a polyanion of high axial charge density (Section 2.3) and the association reaction is driven by the release of monovalent cations, at least at low salt concentrations. **The salt-dependence of binding to various specific and nonspecific DNA sequences is often different, indicating that protein–phosphate contacts are involved in determining the specificity of DNA binding.**

Anions also affect the dissociation constants of DNA•protein complexes, especially at high concentrations and most likely through the Hofmeister effect in which anions affect many types of interactions through their effects on the structure of water. The nature of the anion is important; for example, replacing chloride with glutamate can increase the affinity for a specific DNA sequence as much as 80-fold. At 25°C in a low salt buffer, the leucine zipper protein GCN4 (Section 13.3.D) binds with the same affinity to the sequences ATGAC**G**TCAT and ATGACTCAT. In a buffer containing 250 mM potassium glutamate, however, the affinity for the latter sequence is approximately one order of magnitude greater than for the other.

Is there a code for protein–DNA recognition? Probab(ilistical)ly. P. V. Benos *et al.* (2002) *BioEssays* **24**, 466–475.

Cation-π/H-bond stair motifs at protein–DNA interfaces. M. Rooman *et al.* (2002) *J. Mol. Biol.* **319**, 67–76.

Protein–DNA interactions: amino acid conservation and the effects of mutations on binding specificity. N. M. Luscombe & J. M. Thornton (2002) *J. Mol. Biol.* **320**, 991–1009.

A simple physical model for the prediction and design of protein–DNA interactions. J. J. Havranek *et al.* (2004) *J. Mol. Biol.* **344**, 59–70.

---

2. Nonspecific Complexes

Proteins that bind nonspecifically to DNA, or with little sequence specificity, recognize primarily the DNA backbone of phosphate and sugar groups. Electrostatic interactions with the ionized phosphate group of each nucleotide appear to provide the major driving force for binding and maintain a loose association between the protein and DNA. Both specific and nonspecific binding have been characterized with the ***lac* repressor**. It binds nonspecifically to any DNA with a K_d of 10^{-6} M at 0.1 M salt, but 10^8-fold more tightly to the correct sequence. Increased salt concentrations decrease the strength of the nonspecific binding, but specific binding is not so sensitive to the ionic strength. The interactions in the specific complex cause the *lac* repressor and DNA to bend by 25° and 36°, respectively, but the DNA maintains a canonical B-DNA conformation in the nonspecific complex, because the protein interacts mainly with its phosphate backbone. **The same set of residues of the *lac* repressor switch roles from a purely electrostatic interaction with the DNA backbone in the nonspecific complex to a highly specific binding mode with the base pairs of the specific sequence.**

Glucocorticoid receptor (Section 13.3.C.2) binds to DNA sequences containing inverted repeats separated by three base pairs. When the spacing is increased to four base pairs, one protein subunit binds specifically to one inverted repeat, while the other can not reach its specific binding site. Both subunits insert an α-helix into the major groove of the DNA double helix, but it is opened by approximately 2 Å in the specific complex, probably due to the greater number of interactions. The restriction enzyme ***EcoRV*** inserts a loop into the major groove of the DNA; in the nonspecific complex, it is found to be partially disordered and not so buried in the grooves; the surface area buried is 1570 Å^2 greater in the specific complex.

Nonspecific binding is very important for proteins that play a structural role for DNA irrespective of its nucleotide sequence, but it is also important for proteins that bind specifically to find those sites quickly. They bind nonspecifically and flexibly to the DNA molecule and diffuse along it one-dimensionally until they find their specific sequence.

Crystal packing interaction that blocks crystallization of a site-specific DNA binding protein–DNA complex. O. Littlefield & H. C. Nelson (2001) *Proteins* **45**, 219–228.

Structure and flexibility adaptation in nonspecific and specific protein–DNA complexes. C. G. Kalodimos *et al.* (2004) *Science* **305**, 386–389.

---

3. Water-mediated Contacts

Water molecules are observed at the interfaces of most protein•DNA complexes, where they appear to be involved in specific interactions between protein and DNA. They are believed to fill voids in the structure, screen electrostatic repulsions between like charges, act as linkers between complementary charges, and be intermediaries in hydrogen bonding.

Water-mediated contacts are the principal means for site-specific recognition between the *trp* repressor and its operator (Section 13.3.A.5). Of the 90 water molecules that are integral parts of the complex, 26 are located within the protein•DNA interface, with 13 in each half-site. Three of these 13 are involved in contacts between the nucleobases and the protein. These three water molecules are already in place in the free DNA, and two others are in place in the free protein. Such water molecules should therefore be considered as intrinsic parts of the DNA and protein structures.

In contrast, some DNA•protein interfaces are completely devoid of water, such as the perfectly complementary interface between TATA-binding protein and TATA box-containing DNA sequences (Section 13.3.B). The complex of the transcription factor GATA-1 with DNA (Figure 13-19-D) is devoid of water molecules that mediate contacts between the protein and the nucleobases, but several water molecules bridge the protein and the phosphate backbone.

Wet and dry interfaces: the role of solvent in protein–protein and protein–DNA recognition. J. Janin (1999) *Structure* **7**, R277–279.

The role of water in protein–DNA recognition. B. Jayaram & T. Jain (2004) *Ann. Rev. Biophys. Biomolec. Structure* **33**, 343–361.

---

4. Dehydration Effects

Sequence-specific DNA-binding reactions are characterized by relatively large negative changes in the heat capacity, ΔC_p. Consequently, the enthalpic ($\Delta H°$) and entropic ($T\Delta S°$) contributions to the free energy change (ΔG°) of the binding reactions vary with temperature in an almost parallel manner, making ΔG° nearly independent of temperature. Large negative changes of ΔC_p generally indicate that large amounts of nonpolar surface area are buried when the complex is formed, releasing water molecules. The areas buried in specific protein•DNA complexes vary between 1000 and 5500 Å^2, but in many cases these are too small to account for the observed values of ΔC_p. The discrepancies are thought to result from a series of temperature-dependent equilibria that are coupled to the binding equilibrium, such as ion binding, protonation and protein conformational changes.

Binding of proteins to DNA in a nonspecific fashion involves almost no changes in heat capacity, indicating that **the formation of nonspecific complexes does not involve major dehydration**. An interesting exception is provided by the DNA-binding domain of the transcription factor MASH-1, which appears to bind in only one manner. The association reaction between MASH-1 and an oligonucleotide containing the natural target of MASH-1 is characterized by a ΔC_p of –733 (±99) cal mol^{-1} K^{-1}, while the formation of a nonspecific complex with heterologous DNA results in a ΔC_p of –575 cal mol^{-1} K^{-1}. Unlike other DNA-binding proteins, all DNA complexes of MASH-1 demonstrate the thermodynamic characteristics of specific complexes, irrespective of the particular DNA sequence.

The number of water molecules released on formation of a complex between a protein and DNA has been estimated by varying the chemical activity of water, a_w. This can be accomplished through the addition of neutral salts and can be measured by the decrease in the vapor pressure of the solution. Varying a_w yields a value for the net number of water molecules released, i.e. the difference between the number of water molecules released and those taken up, which will include those resulting from changes in the exposure of surface area due to unfolding or refolding of the DNA or protein. For example, the interaction between a protein and DNA in the presence of a neutral solute (S) can be represented as:

$$\text{protein} + \text{DNA} \leftrightarrow \text{protein} \cdot \text{DNA} + \nu_w \, H_2O + \nu_s \, S \tag{13.4}$$

where ν_w and ν_s are the net numbers of water and solvent molecules released from the protein and DNA. If molecules are taken up by the complex, the values of ν_w and ν_s will be negative. In general, for such a situation:

$$\nu_i = \frac{\partial \ln K_{obs}}{\partial \ln a_i} \tag{13.5}$$

where ν_i is the difference in the number of molecule i associated with reactants and products of the reaction with equilibrium constant K_{obs} and a_i is the activity (the concentration corrected for nonideality effects) of i. Consequently, the situation for Equation 13.4 will be:

$$\frac{d \ln K_{obs}}{d \ln a_w} = \nu_w + \nu_s \frac{d \ln a_s}{d \ln a_w} \tag{13.6}$$

Many osmolytes are preferentially excluded from the immediate surfaces of macromolecules, decreasing ν_s, so ν_w will be significantly greater than ν_s, and the term ν_w will predominate on the right side of Equation 13.6. Therefore, measurement of the effect of varying the concentration of such excluded osmolytes on the binding free energy (the left side of Equation 13.6) will yield the number of water molecules involved in the interaction, ν_w.

This method indicates that formation of the complex between the cyclic AMP receptor protein (CRP) and C1 site in the *lac* promoter (Section 13.3.A.6) is accompanied by the release of 79 ±11 water molecules. When CRP is transferred from the C1 site to a nonspecific site, 56 ±10 of these water molecules are taken up again. When gal repressor binds to the O_I operator site, between 100 and 180 water molecules are released, depending upon the neutral salt used. A further 6 ± 3 water molecules are released when the repressor moves from O_I to the O_E operator, to which it binds with greater affinity. Changing the water activity affects the DNA binding by ultrabithorax and deformed homoeodomains differently, even though they have closely similar amino acid sequences (Section 13.3.A.3); DNA binding by ultrabithorax releases 22–27 water molecules but only five are released in the case of deformed.

Complexes of lambda Cro repressor (Section 13.3.A.2) with varying DNA sequences and dissociation constants demonstrate that the number of water molecules sequestered by nonspecific complexes varies linearly with the binding free energy. One extra bound water molecule corresponds to the loss of approximately 150 cal/mol of binding free energy. Conversely, every 10-fold decrease in binding constant at constant salt and temperature is associated with eight to nine additional water molecules sequestered in the noncognate complex. The water seems to be sterically sequestered, because the measured number of water molecules does not depend upon the nature of the osmolyte used.

Clearly, the thermodynamics of the interactions between DNA and proteins are complex and not well understood.

Calorimetry of protein–DNA complexes and their components. C. M. Read & I. Jelesarov (2001) *Methods Mol. Biol.* **148**, 511–533.

Heat capacity effects of water molecules and ions at a protein–DNA interface. S. Bergqvist *et al.* (2004) *J. Mol. Biol.* **336**, 829–842.

Sequestered water and binding energy are coupled in complexes of lambda Cro repressor with non-consensus binding sequences. D. C. Rau (2006) *J. Mol. Biol.* **361**, 352–361.

---

5. Release of Condensed Counterions

An entropic driving force for proteins to bind to DNA is believed to come from the release of the condensed counterions that are bound electrostatically to a polyelectrolyte-like DNA (Section 2.3). This factor can be estimated by measuring the affinity at various salt concentrations and plotting the logarithm of the dissociation constant against the logarithm of the salt concentration. A straight line should be obtained, with slope Z reflecting the number of counterions released upon forming the complex; this is commonly interpreted as the number of ionic interactions generated in the complex. A large slope is taken to imply that the electrostatic interactions predominate, whereas a small value for Z or a negative slope indicates that they do not.

The affinities of most proteins for DNA decrease with increasing salt concentration, indicating that Z is positive. On the other hand, a TATA-binding protein from a halophilic organism adapted to living in conditions of high salt and temperature exhibits the opposite behavior, which has been attributed to only three amino acid residues of the protein. The halophilic complex is believed to sequester cations that become part of it, so the stability of this complex is increased by increasing salt concentrations (Section 11.3.B.1).

Association of protein–DNA recognition complexes: electrostatic and nonelectrostatic effects. J. Norberg (2003) *Arch. Biochem. Biophys.* **410**, 48–68.

Halophilic adaptation of protein–DNA interactions. S. Bergqvist *et al.* (2003) *Biochem. Soc. Trans.* **31**, 677–680.

13.2.B. Changes in the Protein Conformation

The binding sites of many proteins that bind DNA specifically are remarkably flexible in the absence of the DNA, and binding to DNA induces specific local conformations. In other cases, the quaternary structure can alter. In general, complex formation does not result from a simple alignment of rigid, complementary surfaces, commonly known as the 'lock-and-key' model, but rather follows what is generally known as an **induced fit** mechanism (Section 14.4.E). For example, the DNA-binding pocket of the endonuclease *Eco*RV is not accessible in the free protein and the cleft between the two protein subunits is too narrow. The pocket and cleft open to accommodate the DNA through a combination of tertiary and quaternary structure changes. The relative orientations of the amino and carboxy terminal domains of CRP change significantly on DNA binding. In extreme cases, the portion of the protein that will bind the DNA is unfolded until the DNA binds. The observed changes in conformation of the protein upon binding are believed to account for the discrepancy between the measured values of ΔC_p and those expected on the basis of the nonpolar surface buried in the complex.

Most DNA-binding proteins have their specific affinities for DNA modulated by another ligand, a corepressor or coactivator. This ligand is frequently bound to a protein domain other than that which binds to the DNA. In the one case where the mechanism is clearly known, the corepressor ligand alters the affinity of the *trp* repressor for DNA by altering the dimeric protein structure so that the spacing of the two DNA binding sites does or does not coincide with the spacing of the palindromic sequences of the DNA (Figure 13-16).

Use of urea and glycine betaine to quantify coupled folding and probe the burial of DNA phosphates in *lac* repressor–lac operator binding. J. Hong *et al.* (2005) *Biochemistry* **44**, 16896–16911.

Molecular flexibility in protein–DNA interactions. S. Gunther *et al.* (2006) *Biosystems* **85**, 126–136.

13.2.C. Changes in the DNA Conformation

Many proteins recognize regular B-DNA through the formation of a number of hydrogen bonds and van der Waals interactions between amino acid side-chains and functional groups of the bases, but others recognize and generate bent conformations of the DNA. A dramatic example is provided by the structure of **integration host factor (IHF)** bound to the H′ site of phage λ DNA, in which the 34-base pair piece of DNA is literally wrapped around the protein, creating a buried protein•DNA interface of 4600 Å^2 (Figure 13-7) and bending the DNA by nearly 180°. This bending completely reverses the direction of the DNA within a small number of base pairs.

The DNA-binding affinities of the **434** and **Cro repressors** (Section 13.3.A.2) depend upon the flexibility of bases at the center of the binding site that are not in direct contact with the protein. Varying the nucleotide sequence of a binding site reveals a linear relationship between the free energy of the binding reaction with the proteins and the flexibilities of their central sequences. **The intrinsic sequence-dependent properties of the DNA are important for the formation of DNA•protein complexes.**

Double-stranded DNA often bends at the junctions between regions of G•C and A•T base pairs (Section 2.4). The transition from G•C to A•T base pairs causes this region of the DNA to be flexible and capable of potentially undergoing a bend. The crystal structure of the dodecamer d(CGCGAATTCGCG) has a bend of 18° at the GC/AT junction at one end of the helix, but not at the other. Although such sequences need not be bent, they usually are when proteins bind there. Changing the nucleotide sequence to disfavor bending destabilizes the complex, and there is a correlation between the affinity for a mutant site and its intrinsic bendability.

MEF-2C binds with maximal affinity to a DNA site with an alternating run of eight thymines and adenines, while the affinity for an intrinsically rigid sequence of only A nucleotides is reduced by more than one order of magnitude. Binding of DNA by MEF-2C is accompanied by DNA bending of approximately 70°, irrespective of the particular DNA sequence. When such a protein binds to a more rigid DNA sequence (with lower affinity), it often does not bend the DNA, or a substantial amount of the binding free energy must be used to bend the DNA target. Consequently, **proteins often target the most deformable base step within their target sequence.** The **purine repressor, PurR**, binds to runs of four consecutive adenines and thymines, interrupted by the dinucleotide CpG. PurR bends the DNA into the major groove by 45° through the insertion of a hinge into the minor groove and intercalation of the side-chains of two Leu residues into the CpG step. Purine–pyrimidine steps stack more stably than do purine–purine or pyrimidine–pyrimidine steps by approximately 0.52 kcal/mol, while the stacking energy of pyrimidine–purine steps is reduced by another 0.39 kcal/mol. Therefore, CpG is the step that can be unstacked most easily within the *pur* operator. Similarly, the bending of 22° of the PRDI site within the interferon-β promoter in the complex with interferon regulatory factor occurs at the CpA step, which is the most easily unstacked dinucleotide in the PRDI sequence ACTTT**CA**CTTCTC. The DNA complex of the chromosomal protein Sac7d from the hyperthermophile *Sulfolobus acidocaldarius* has the DNA kinked by 61° through the intercalation of two amino acid side-chains between the bases of the dinucleotide CpG in the sequence GC**CG**ATCGC. This is the base step that is most easily unstacked.

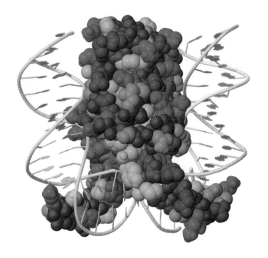

Figure 13-7. The IHF–DNA complex, illustrating both the large contact surface of ~ 4600 Å² between the protein and DNA and the massive, nearly180°, bend in the DNA induced by the protein. The DNA is represented as a skeletal model, while the protein atoms are represented as van der Waals spheres; acidic groups are *red*, basic groups *blue* and the polypeptide backbone *mauve*. Generated from PDB data file *1ihf* using the program Jmol.

DNA bending should be an energetically costly process, even for sequences of enhanced bendability. Some of the binding free energy must be used to bend a DNA molecule that, in the absence of protein, would not be bent. Therefore, **binding at the more readily bendable sites can contribute to the overall specificity of a DNA-binding protein**. Numerous observations indicate that **proteins recognize DNA not only through specific contacts with the nucleobases but also through the inherent conformational properties of the DNA that vary with the sequence**.

Proteins use primarily two mechanisms to bend DNA. In the first, **kinks are introduced through the intercalation of amino acid side-chains between adjacent base pairs in the minor groove**. In the IHF–DNA complex (Figure 13-7), two intercalating Pro residues projecting from the tip of each β-hairpin arm introduce large kinks at symmetrically displaced ApA nucleotide steps. Other proteins use nonaromatic, hydrophobic amino acids to intercalate into base steps from the minor groove side of the double helix, creating a widened minor groove and dramatically bent DNA. TATA box-binding protein induces strong bends of approximately 45° at either end of its specific sequence, the TATA box, through the intercalation of two Phe rings between two adjacent base pairs (Figure 13-8). The hyperthermophile chromosomal protein Sac7d bends DNA by 72°, primarily through the intercalation of Val26 and Met29 at a single CpA base step. This causes the DNA to bend into the major groove and away from the protein molecule.

The second mechanism for the introduction of bends into DNA involves **asymmetric neutralization of the negative charges of the phosphodiester groups by cationic amino acid residues**. Unbalanced electrostatic repulsions between the negative charges on DNA cause it to collapse towards the bound protein (Figure 13-8). Asymmetric charge neutralization is at least partly responsible for the DNA bend angle of 72° in the target DNA bound to serum response factor (SRF). Several positively charged amino acid residues of SRF bind the phosphate groups on only one side of the SRF site (Figure 13-9). The three positively charged amino acids interact with and pull the distal ends of the DNA. In addition, Arg143 lies in an extended conformation along the floor of the minor groove and stabilizes the relatively large propeller twists at the central base steps that are characteristic of AT-rich regions and facilitate bending of the DNA. In some of these cases, the DNA sequences have a propensity to bend.

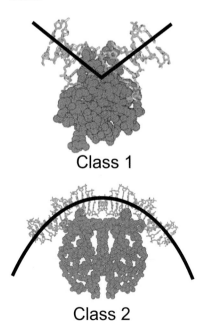

Figure 13-8. Two classes of DNA-bending proteins. Class 1-bending proteins, such as the TATA box-binding protein illustrated here, bind to the minor groove, unwind the DNA and induce bending away from the protein by intercalation of hydrophobic amino acid side-chains between base pairs. Class 2-bending proteins, such as CRP, form complexes in which the DNA bends towards the protein to maximize their interaction. Adapted from L. D. Williams & L. J. Maher (2000) *Ann. Rev. Biophys. Biomol. Structure* **29**, 497–521.

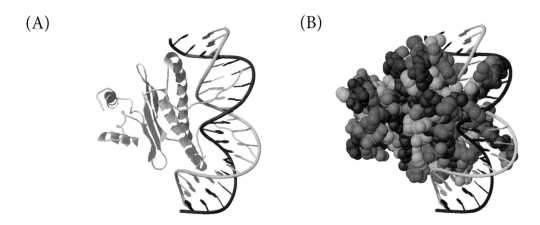

Figure 13-9. Interactions between positively charged amino acid side-chains and DNA phosphates in the serum response factor–DNA complex. (A) Skeletal model, illustrating the bend of the DNA and the polypeptide extensions that bind in the grooves of the DNA. (B) Space-filling model of the protein, emphasizing the polypeptide extensions. Basic groups are *blue*, acidic groups *red*. The contacts between the protein and DNA occur on only one side of the double helix, causing the DNA to bend around the protein. The His and Lys residues at either end serve as handles to bend the DNA further, although the DNA used is too short to allow formation of all the contacts involving the His residue. Generated from PDB file *1srs* using the program Jmol.

Bending is the most dramatic change observed upon DNA binding to proteins, but many more subtle conformational rearrangements of the DNA occur. A slight bending of the DNA of approximately 25° occurs in the 434 repressor–DNA complex (Figure 13-10) but more obvious is that the DNA in the region of the central four base pairs of the binding site is significantly overwound by approximately 20° (Section 3.1). These four base pairs are not in direct contact with the repressor, but sequences with A•T or T•A base pairs at these positions bind more strongly than do those with C•G or G•C. The intrinsic twist of a sequence affects its affinity; in this case, **sequences with lower affinity are more underwound**.

The role of intercalating residues in chromosomal high-mobility-group protein DNA binding, bending and specificity. J. Klass *et al.* (2003) *Nucleic Acids Res.* **31**, 2852–2864.

Integration host factor: putting a twist on protein–DNA recognition. T. W. Lynch *et al.* (2003) *J. Mol. Biol.* **330**, 493–502.

Predicting indirect readout effects in protein–DNA interactions. Y. Zhang *et al.* (2004) *Proc. Natl. Acad. Sci. USA* **101**, 8337–8341.

DNA twisting flexibility and the formation of sharply looped protein–DNA complexes. T. E. Cloutier & J. Widom (2005) *Proc. Natl. Acad. Sci. USA* **102**, 3645–3650.

13.3. DNA-BINDING STRUCTURAL MOTIFS

The structures of many complexes of protein and nucleic acid have been determined in great detail by X-ray crystallography and nuclear magnetic resonance (NMR). To promote crystallization of the complex and to keep it a practical size, the protein is bound to a nucleic acid of the smallest possible size, ideally containing only the specific sequence of nucleotides recognized.

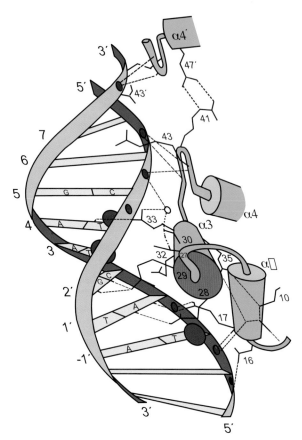

Figure 13-10. Interactions between one helix–turn–helix motif of the 434 repressor and its DNA binding site. Only part of the repressor protein is shown, with the helices depicted as *cylinders*. The DNA double helix is shown schematically, with the backbone of the individual chains depicted as *ribbons* and the positions of the phosphate groups by *circles*. The base pairs are indicated as the *flat segments* connecting the ribbons; the *dark spheres* on the thymidine nucleotides are the methyl groups. *Dashed lines* represent likely hydrogen bonds. Solvent molecules are required as bridges in some of these interactions, as illustrated by the interaction of the side-chain of residue 33 with a DNA backbone phosphate group and with the side-chain of residue 30. Adapted from A. K. Aggarwal *et al.* (1988) *Science* **242**, 899–907.

DNA binding has been found to be specified by a large number of disparate protein motifs: **many different protein structures can bind DNA specifically**. All these have one characteristic in common: **the protein needs to have interacting groups that protrude substantially from the protein surface**, to be able to contact the nucleotides at the base of the major groove and discriminate between different DNA base pairs by interacting with their edges (Figure 13-1). The most commonly encountered types of motif are listed in Table 13-2. The degree of sequence specificity can vary enormously within any particular class of motif. For example proteins with the helix–turn–helix–motif (Section 13.3.A) range from the *lac* repressor, with a high degree of sequence specificity, to the FIS protein, with little. Frequently more than one type of binding motif is found in a particular protein or even in a conserved sequence motif.

The best-studied protein structures will be described here briefly. A database of the structures and stabilities of DNA•protein complexes is available at www.rtc.riken.go.jp/jouhou/pronit/pronit.html.

Structural basis of DNA–protein recognition. R. G. Brennan & B. W. Matthews (1989) *Trends Biochem. Sci.* **14**, 286–290.

Structural studies of protein–nucleic acid interaction: the sources of sequence-specific binding. T. A. Steitz (1990) *Quart. Rev. Biophys.* **23**, 205–280.

A structural taxonomy of DNA-binding domains. S. C. Harrison (1991) *Nature* **353**, 715–719.

Transcription factors: structural families and principles for DNA recognition. C. O. Pabo & R. T. Sauer (1992) *Ann. Rev. Biochem.* **61**, 1053–1098.

13.3.A. Helix–turn–helix Motif

The best-characterized structural motif of DNA-binding proteins is **the helix–turn–helix**; it is present in proteins from both prokaryotes and eukaryotes. **It consists of two short α-helices separated in the sequence by a Gly residue** (Figure 13-11) that, in concert with its neighbors, acts as a flexible hinge allowing the polypeptide chain to bend between the two α-helices so that they can make hydrophobic contacts with each other. It is observed in a number of different proteins that have no other structural similarities (Figure 13-10). This motif **protrudes from the protein surface to allow one of the helices, the recognition helix, to enter the major groove of the DNA double helix** (Figure 13-11). The second helix stabilizes the structural motif by packing against the recognition helix, primarily through interactions between nonpolar side-chains. This is especially important for the structural integrity of the motif, as it can make few other stabilizing interactions with the rest of the protein. For this reason, the residues participating in this packing interaction are highly conserved in motifs that otherwise have little sequence or structural similarity. This conserved pattern of residues often makes it possible to detect the helix–turn–helix motif from just the primary structures of suspected DNA-binding proteins.

The distal recognition helix binds in the DNA major groove and recognizes the DNA sequence, primarily by hydrogen bonds from amino acid side-chains, especially those of Arg, Asn, Asp, Gln and Glu residues, which can make multiple hydrogen bonds. Water molecules are sometimes involved as bridges in those hydrogen bonds. Van der Waals contacts between protein and DNA are also involved.

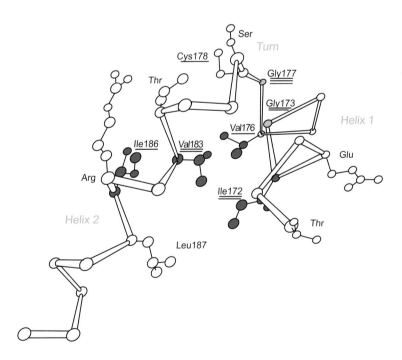

Figure 13-11. Structure of the helix–turn–helix motif of CRP. The first α-helix is to the *right*, the second is the recognition helix and is to the *left*. The degree of conservation of the various residues in other motifs is indicated by the number of *underlines*. The Gly177 residue is generally conserved in different proteins because it is important for the bend. Residues Ile172, Val176, Val183 and Ile186 are usually hydrophobic in other such motifs because they pack between the two helices. Data from T. A. Steitz (1990) *Quart. Rev. Biophys.* **23**, 205–280.

There is no simple code relating the amino acid sequence of the motif to the sequence of the DNA it recognizes; several side-chains can interact with each base pair, and more than one base pair can be in contact with a side-chain. The second, proximal helix interacts with the recognition helix and with the sugar–phosphate backbone of the DNA. The structure of this domain is highly conserved, but the orientation of the recognition helix within the major groove varies (Figure 13-10). The two DNA-binding helices are generally stabilized by interactions with other protein segments, such as two additional helices in the lambda CI repressor or a single additional helix in the homeobox domain.

Table 13-2. Protein motifs Involved in DNA binding

Motif	Target	Sequence specificity	Examples
β-strand	Major groove		Met repressor, Arc repressor
bZip	Major groove	High	GCN4, Fos-Jun
GATA	Major + minor grooves		GATA-1
GRP motif	Minor groove	Low preference	Hin recombinase, HMG14/17
Helix–loop–helix	Major groove	High	Myc, Achaete
Histone-fold	Backbone	None	Core histones, TAFs
HMG domain	Minor groove	Variable, but low	HMG1, SRY, TCF-1, UBF
HU class	Minor groove	Variable, but low	HU, IHF, TF1
Smad		Major groove	Smad MH-1
TBP domain	Minor groove	High	TBP
Helix–turn–helix	Major groove	Highly variable	FIS, C_I repressor, CRP
Homeodomain			Fushi tarazu
Winged helix			HNF3, histone H5
ETS domain			Fli-1, Elk-1
POU 'domain'			Oct1
Zinc-containing motifs			
Zinc finger	Major groove	Generally high	TFIIIA, Tramtrack, GAGA
Receptor DBD	Major groove		Glucocorticoid receptor
Gal4 DBD	Major groove		GAL4

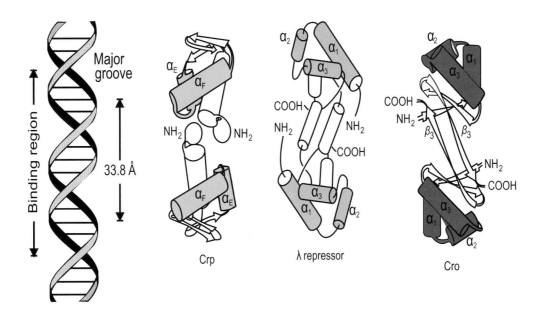

Figure 13-12. Comparisons of the dimeric structures of three DNA-binding proteins with the helix–turn–helix motif, relative to the structure of dsDNA (*left*). The three proteins are the Cro repressor, the amino-terminal domain of λ repressor and the carboxyl-terminal domain of CRP. The α-helices of each helix–turn–helix motif are *colored*. That in the foreground of each protein, and labeled as α_3 or α_F, is the recognition helix, which fits most closely into the major groove of the DNA. Note that the spacing between the two recognition helices of the dimer is very close to the 33.8 Å spacing between the DNA major grooves, but that the orientations of the helices relative to the major groove differ. Original figure kindly provided by B. W. Matthews (1993) in *Proteins: structures and molecular properties*, 2nd edn (T. E. Creighton, ed.), W. H. Freeman, NY, p. 356.

Eukaryotic helix–turn–helix proteins have essentially the same structure but contain longer turns or loops. These eukaryotic variants include the **homeodomain** fold, which contains both a helix–turn–helix element and a short extended minor groove-tracking sequence, the **POU**-specific 'domain', and the winged-helix fold. The last fold contains an extended β-sheet immediately following the recognition helix.

The eukaryotic proteins usually bind as monomers, but their prokaryotic counterparts normally bind as homodimers. The latter have both equivalent binding sites, making the same interactions to DNA with the same sequence, which requires that the sequence be **palindromic** or have **inverted repeats** (Section 2.2.C.6). For example, the binding site for the *lac* repressor has the following sequence of its two complementary DNA strands:

$$5'\text{-}\mathbf{TGTGTG}\text{G}\mathbf{AATTGT}X_9\mathbf{ACAATT}\text{T}\mathbf{CACACA}$$
$$3'\text{-}\mathbf{ACACAC}\text{C}\mathbf{TTAACA}Y_9\mathbf{TGTTAA}\text{A}\mathbf{GTGTGT} \quad (13.7)$$

All but one of 13 base pairs of each half obeys the palindromic nature of this binding site, indicated in bold. The nine nucleotides between the palindromic sequences are indicated by X_9 and Y_9. The palindromic sequences define a local two-fold symmetry axis at right angles to the DNA helix that relates the two halves of the DNA binding site. **The dimeric protein binds with its two-fold axis coinciding with that of the DNA sequence, and the distance between the two halves of the protein is the same as that of one turn of B-form DNA** (Figure 13-12). Having a double binding site increases markedly the strength and specificity of the binding interaction.

Although similar in many ways, the various helix–turn–helix motifs may not have evolved from a single evolutionary ancestor; instead, there may have been at least six to 11 different such domains in the last universal common ancestor of all life forms.

The many faces of the helix–turn–helix domain: transcription regulation and beyond. L. Aravind *et al.* (2005) *FEMS Microbiol. Rev.* **29**, 231–262.

Detecting DNA-binding helix–turn–helix structural motifs using sequence and structure information. M. Pellegrini-Calace & J. M. Thornton (2005) *Nucleic Acids Res.* **33**, 2129–2140.

---

1. Lac Repressor

The *lac* repressor is a member of a family of almost a dozen similar, mostly dimeric, repressors in *E. coli*. It represses the production of β-galactosidase in the absence of an inducer, such as lactose, which inactivates the *lac* repressor by binding to it.

The *lac* repressor is a tetramer of identical subunits of 360 amino acid residues each. The tetramer has a V-shape, in which each arm comprises two polypeptide chains; the two dimers are held together at the base of the V (Figure 13-5). Each polypeptide chain consists of five domains.

(1)   The C-terminal tails of the four subunits of one *lac* repressor molecule, residues 330–360, form a four-helix bundle composed of four coiled coils, each carrying two leucine heptad repeats (Section 8.5.A). This domain links the four polypeptide chains and is responsible for the tetrameric structure.

(2 & 3)   Two globular domains, residues 60–329, form the core of *lac* repressor and form the dimeric part of *lac* repressor that binds the inducer.

(4)   A helix–turn–helix motif at the N-terminus of each polypeptide chain binds directly to the DNA.

(5)   A hinge α-helix preceding the helix–turn–helix motif in the sequence binds to the minor groove of the DNA.

Recognition of the *lac* operator by the *lac* repressor involves specific interactions between residues in the repressor's recognition helix and bases in the DNA major groove. Each dimer arm binds a separate copy of a palindromic DNA sequence (Equation 13.7). The two recognition α-helices of each dimer bind to successive major grooves of the DNA, while the two hinge helices of each dimer are close together in the structure and interact with the DNA minor groove between the two major-groove binding sites; both types of binding are specific for the DNA sequence. The hinge helices distort the structure of the DNA and open the minor groove, bending the DNA away from the protein.

Tetrameric *lac* repressor can bind to two palindromic sequences (Equation 13.7) if they are on different pieces of DNA or the same piece at an appropriate distance, corresponding to integral turns of the DNA double helix with a turn every 10.5 base pairs. The intervening DNA sequence is looped out from the tetramer. A single symmetrical substitution in each 10-base pair half-site of a symmetric *lac* binding sequence reduces the observed equilibrium association constant by three to four orders of magnitude; double symmetrical substitutions of the neighboring nucleotide reduce the binding by nearly six orders of magnitude, to approach the affinity for nonspecific sequences. The free energy effects of multiple substitutions at adjacent sites are nonadditive: the first reduces the free energy of binding by 3–5 kcal/mol, approximately halfway to the nonspecific level, whereas the second is less deleterious, reducing it by less than 3 kcal/mol.

If the C-terminal residues are deleted, *lac* repressor cannot form the tetramer, only a dimer. The dimer binds only weakly to a single palindromic sequence and in an asymmetric manner, with a different pattern of specific contacts between the two halves of the palindromic sequence. This involves a combination of elongation and twist by 48° of one subunit relative to the other, significant rearrangement of many side-chains, as well as sequence-dependent deformation of the DNA.

Plasticity in protein–DNA recognition: *lac* repressor interacts with its natural operator 01 through alternative conformations of its DNA-binding domain. C. G. Kalodimos *et al.* (2002) *EMBO J.* **21**, 2866–2876.

Role of hydration in the binding of *lac* repressor to DNA. M. G. Fried *et al.* (2002) *J. Biol. Chem.* **277**, 50676–50682.

Toward an integrated model of protein–DNA recognition as inferred from NMR studies on the *lac* repressor system. C. G. Kalodimos *et al.* (2004) *Chem. Rev.* **104**, 3567–3586.

The *lac* repressor. M. Lewis (2005) *C. R. Biol.* **328**, 521–48.

---

2. Lambda CI and Cro Repressors

The lambda bacteriophages encode two helix–turn–helix repressor proteins, CI and Cro, containing helix–turn–helix motifs. **The structures of these two repressor proteins differ substantially** (Figure 13-13). Whereas the Cro protein contains 56 amino acid residues and exists in solution as a monomer or dimer, the CI repressor contains 236 residues and can exist as a monomer, dimer, tetramer or octamer in solution. Each CI monomer contains two domains: an N-terminal domain of some 90 residues that contains the helix–turn–helix motif and binds specifically to the operator sites, and a C-terminal domain that does not interact with DNA. Between these two domains is a flexible hinge. Dimers of Cro are required to recognize operator DNA, but dimerization is weak, with dissociation constants in the micromolar concentration range. At nanomolar concentrations of subunits, assembly to dimers occurs only upon binding to DNA. It is slow, and Cro binding to DNA *in vivo* may be under kinetic rather than thermodynamic control. Binding of the Cro repressor is accompanied by large changes in both the protein and DNA: the subunits of Cro rotate 53° with respect to each other, and 19 base pairs of the DNA are bent by 40°. Recognition is achieved almost entirely by direct hydrogen bonding and van der Waals contacts between the protein and the exposed bases within the major groove of the DNA.

Both of these repressors bind to a set of six similar, but nonidentical, sites or operators in the phage genome. These sites are grouped in sets of three on each side of the structural gene for the CI protein. The affinities of the CI and Cro repressors for these six sites differ. When bound to DNA, the CI repressor dimers have the potential to form tetrameric complexes involving their C-terminal domains. The organization of the tripartite leftward and rightward operators of phage λ allows cooperative interactions between dimers bound to two adjacent sites. Because the occupancy of adjacent binding sites by CI repressor dimers depends on an interaction between two molecules, the binding curve is

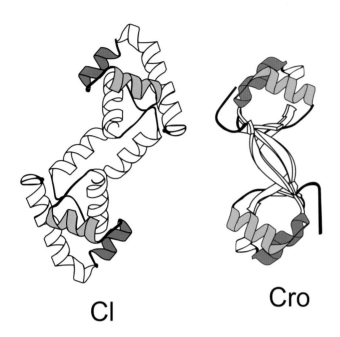

Figure 13-13. Structures of the CI and Cro repressors of lambda bacteriophage. Both have helix–turn–helix motifs (*colored*) with a spacing corresponding to that of successive turns of the major groove of B-DNA (Figure 13-12). In each case, a bound DNA double helix would lie with its helical axis approximately vertical. Adapted from a figure supplied by S. E. V. Phillips.

no longer hyperbolic, as it would be for independent binding, but instead is sigmoidal. An important consequence of this cooperativity is that occupancy changes rapidly over a small range of CI repressor concentrations and attains a level of >99% at concentrations significantly lower than those that would be required for the same level of independent binding. This cooperative protein•DNA binding implies that any modification of the protein–protein interactions required for cooperativity will result in a substantial change in occupancy.

Cooperativity in long-range gene regulation by the lambda CI repressor. I. B. Dodd *et al.* (2004) *Genes Dev.* **18**, 344–354.

Slow assembly and disassembly of lambda Cro repressor dimers. H. Jia *et al.* (2005) *J. Mol. Biol.* **350**, 919–929.

---

3. Homeodomains

Homeodomain-containing proteins are extremely important transcription regulators that control the coordinated expression of genes involved in development, differentiation and cellular transformation. The highly conserved 60-amino acid residue homeodomain allows these proteins to bind to DNA and modulate the expression of numerous target genes.

The homeodomain structure is built up from three α-helices, connected by short loop regions (Figure 13-14). α-Helices 2 and 3 and the loop region between them comprise the helix–turn–helix motif, but the recognition helix 3 is longer than normal. Homeodomains frequently bind to DNA as a monomer. The recognition helix 3 is positioned in the major groove of B-form DNA, and the ninth residue of this helix plays a crucial role in the specificity. The N-terminal residues, preceding helix 1, interact with the minor groove on the other side of the DNA double helix (Figure 13-14). Many other residues, including some from helix 2, interact nonspecifically with the DNA double helix.

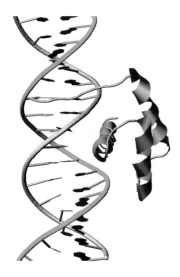

Figure 13-14. The structure of a homeodomain bound to dsDNA. The N-terminus of the polypeptide chain is at the *upper left*, whereas the C-terminus is behind, at the end of the α-helix. The three α-helices are depicted as *coils*. The helix at the C-terminus lies in the major groove of the DNA molecule. Generated from PDB file *1hdd* using the program Jmol.

The homeodomain binds specifically to the DNA sequence ATTA with a dissociation constant of about 1 nM. It also binds nonspecifically to other DNA, but with an affinity 10^2 lower. The homeodomain-containing proteins exhibit specificity *in vivo* but very similar DNA-binding affinities *in vitro*. Consequently, other mechanisms such as protein–protein interactions are probably involved in modulating their function.

Homeodomain revisited: a lesson from disease-causing mutations. Y. I. Chi (2005) *Human Gen.* **116**, 433–444.

NMR structural and kinetic characterization of a homeodomain diffusing and hopping on nonspecific DNA. J. Iwahara *et al.* (2006) *Proc. Natl. Acad. Sci. USA* **103**, 15062–15067.

---

4. POU Domains

POU proteins play a prominent role in the differentiation of the nervous system of several organisms, but also in more general cellular housekeeping. The POU region is a conserved sequence motif defining a family of eukaryotic transcription factors that contain **two structurally independent DNA-binding domains connected by a flexible linker of variable length and sequence** (Figure 13-15). The N-terminal POU-specific domain is a four-helix bundle containing the helix–turn–helix motif, while the C-terminal domain is structurally homologous to the homeobox domain. Both domains make sequence-specific contacts with DNA and together bind with high affinity to a conserved eight-base pair DNA binding site (ATGCAAAT) that is found in the promoter regions of a variety of genes. The N-terminal POU-specific domain binds to the ATGC part of the site, whereas the homeodomain recognizes the second half of the sequence.

The two domains are intrinsically flexible, which appears to confer functional diversity; **they encircle the DNA and are able to assume a variety of conformations that depend upon the DNA sequence**. In various members of this family, the orientations of the POU-specific domains relative to the POU homeodomain can be flipped, so that they interact with either the same or opposite sides of the DNA double helix. There is remarkable flexibility of the POU-specific domain in adapting to variations in sequence within the site.

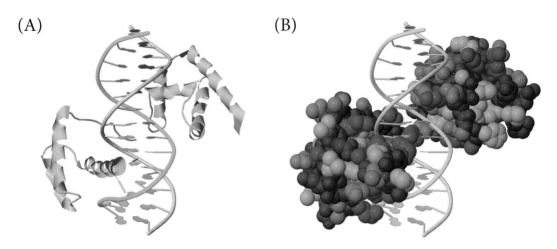

Figure 13-15. The structure of the POU domain and its binding to DNA. (A) Both the DNA and the protein are shown as simplified models; α-helices of the protein are depicted as *coils*. (B) The protein shown as a space-filling model. The domain like the homeodomain is at the *lower left*, while the four-helix bundle domain is at the *top right*. The polypeptide chain linking the two domains is flexible and only part is shown. Generated from PDB file *1oct* using the program Jmol.

The POU•protein complex provides various types of structures that can recruit specific regulator proteins to control transcription of the gene, providing a mechanism for assembling different combinations of proteins, as well as allosteric molecular recognition. One POU domain transcription factor, Oct4, interacts with another type of transcription factor, with which it controls gene expression, by dimerizing through interaction domains onto DNA in distinct conformational arrangements that are dictated by the nucleotide sequence of the binding site. Structures of the Oct1 transcription factor in the presence of two different DNA sites demonstrate how its POU DNA-binding domains can form two unrelated dimer arrangements, depending upon the DNA sequence. One arrangement allows binding of the Oct1-specific coactivator OBF-1, but binding of this coactivator is blocked in the second arrangement because the binding site is involved in its own dimer assembly. Conversely, the same overall assembly is maintained in the presence of two different DNA sequences by another POU transcription factor, Pit-1. The distances of the two Pit-1 half-binding sites on these two sequences differ by two base pairs, so the overall dimensions of the two complexes vary, allowing binding of a specific repressor in one conformation but not in the other.

Despite the various DNA-mediated molecular mechanisms, the net result is the same: conformation-dependent binding of further regulator proteins. **The DNA motif serves not only as a binding site for specific transcription factors, but also regulates their function by mediating specific transcription factor assemblies**, which determine binding to conformation-dependent coregulators.

Differential activity by DNA-induced quaternary structures of POU transcription factors. A. Remenyi *et al.* (2002) *Biochem. Pharmacol.* **64**, 979–984.

Crystal structure of a POU/HMG/DNA ternary complex suggests differential assembly of Oct4 and Sox2 on two enhancers. A. Remenyi *et al.* (2003) *Genes Dev.* **17**, 2048–2059.

5. Trp Repressor

The *trp* repressor from bacteria is a dimer of a 107-residue polypeptide chain. Each polypeptide chain comprises six α-helices, but three of these helices from each chain are entwined about each other (Figure 13-16); **rather than each chain making a globular structure and then dimerizing, the globular structure requires the dimer of these pairs of three N-terminal α-helices**. This central core supports structurally two 'heads' that interact with the palindromic binding site on DNA, one at each end of the molecule. Helices 4 and 5 of each polypeptide chain comprise a helix–turn–helix motif that interacts with the DNA. Recognition in this system is controlled by the binding of the corepressor ligand, L-tryptophan, as well as by the conformational and dynamic properties of the DNA target sequences, DNA sequence-dependent control of the oligomerization properties of the repressor, water-mediated interactions and specific interactions involving the peptide backbone and DNA phosphate moieties.

The *trp* repressor binds to DNA specifically only when two molecules of the amino acid L-tryptophan are bound, one to each half of the dimer (Figure 13-16). This binding causes the two 'heads' to tilt outwards and changes the orientation of the recognition helices. These two helices are now in the right position, 34 Å apart, to bind to the appropriate palindromic DNA sequence within the major groove. Without the bound tryptophan, the distance between the two heads is too short by 5–6 Å to permit such binding. **This is the best example of how binding of another ligand, a corepressor, can control binding of the protein to a DNA molecule**. The relative affinities for different sequences are conserved in the absence of the corepressor, and it is in all cases significantly higher than that observed for holo repressor binding to nonspecific DNA; therefore, the apo repressor does not participate only in nonspecific binding.

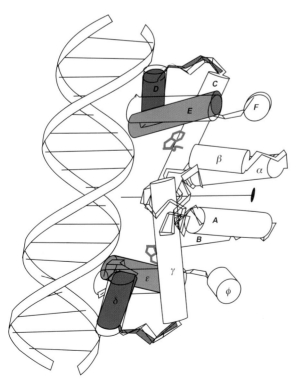

Figure 13-16. Structural change in the *trp* repressor induced by binding of the corepressor, L-tryptophan. Where the structure is altered by binding of L-tryptophan, the *shaded* model is that without the ligand. α-Helices A through F are from one polypeptide chain of the dimer, α through φ are from the other. The tryptophan corepressor bound to each of the two sites of the dimer is shown as a red skeletal model. The corepressor binds between helices B, C and E of each half of the dimeric protein. In doing so, it pushes helices D and E of the helix–turn–helix motif away from the center of the molecule and places the two equivalent recognition helices E and ε the correct distance apart to bind in adjacent major grooves of normal double helical DNA. The binding affinity for DNA is increased 10^3-fold by the presence of the corepressor. Adapted from R.-G. Zhange *et al.* (1987) *Nature* **327**, 591–597.

The *trp* repressor binds specific sequences even though there are no direct contacts between the protein and the bases. In the complex of the *trp* repressor with a symmetrical, palindromic, 18-base pair sequence, TGT**ACTAG**TT•AA**CTAGT**AC, the two ACTAG sequences centered five base pairs from the dyad (indicated by the dot) are those most important for binding. Remarkably, these sequences make no direct hydrogen bond or nonpolar contacts with the protein that might account for the specificity of binding; the only direct contact lies outside the specific sequence. Direct contacts between the protein and DNA are principally with the phosphate groups of the DNA backbone, which maintain the conformation of the backbone. The contacts between protein and DNA are mediated by three well-ordered water molecules in the major groove. Replacing the amino acid residues that interact with these water molecules reduces the affinity and, in some examples, alters the specificity of binding. The *trp* repressor illustrates how **some sequence specificity can require water molecules to mediate direct recognition of the operator sequence**.

Even on very small (25-base pair) targets, up to three molecules of repressor can bind simultaneously and slightly cooperatively. A 2:1 dimer–DNA complex forms first; at higher concentrations a third dimer binds with significantly lower affinity. There appears to be an inherent asymmetry in the normal target sequence and in interactions between neighboring repressor molecules. The *trp* repressor tends to aggregate even in the absence of DNA, but the interactions when on DNA depend upon the DNA sequence.

Probing the role of water in the tryptophan repressor–operator complex. M. P. Brown *et al.* (1999) *Protein Sci.* **8**, 1276–1285.

Probing the physical basis for trp repressor–operator recognition. A. O. Grillo *et al.* (1999) *J. Mol. Biol.* **287**, 539–554.

Water-mediated contacts in the trp-repressor operator complex recognition process. F. R. Wibowo *et al.* (2004) *Biopolymers* **73**, 668–681.

---

6. Cyclic AMP Receptor Protein/Catabolite Gene Activator Protein

The **cyclic AMP receptor protein** (CRP) of *E. coli* is also known as the **catabolite gene activator protein** (CAP). It binds cyclic AMP (cAMP) and stimulates expression of the *lac* operon. It is the **paradigm for gene activator proteins**. CRP, in conjunction with cAMP, participates as a global transcription factor in a wide regulatory network, both activating and repressing a large number of operons. One mechanism of **catabolite repression**, which is a mechanism for modulating the expression of specific genes in response to the overall nutritional state of the organisms, is due to the reduction in the intracellular concentrations of both cAMP and CRP.

The CRP protein is a dimer composed of two identical subunits of 209 amino acid residues each (Figure 13-17). It binds to DNA when complexed with cAMP, which behaves as an allosteric effector; there are two cyclic AMP-binding sites in each monomer. CRP binds to specific DNA sites through a helix–turn–helix motif. When bound, it interacts with RNA polymerase and other regulatory proteins to regulate transcription of the target operons.

Each subunit has two domains. The larger N-terminal domain is responsible for dimerization and binding of cAMP binding. The second domain contains the helix–turn–helix motif that is responsible for DNA binding. Without cAMP, CRP binds to DNA nonspecifically. The binding of cAMP causes a change in conformation, which may involve realignment of the subunits and rearrangement of the domains, to one that binds to specific sequences with a dyad symmetry. The DNA sequences of many CRP binding sites are variations of the palindromic sequence 5′-AAA**TGTGA**TCT•AGA**TCACA**TTT-3′. The two TGTGA motifs highlighted tend to be conserved in various promoters. The cAMP–CRP complex binds to target DNA by inserting the helix–turn–helix motifs of the two subunits into successive major grooves of the DNA. The DNA is bent by 90° (Figure 13-8) but the significance of this bending is not clear.

The basic role of the cAMP–CRP complex is to enhance binding of RNA polymerase and facilitate its transcription of the adjacent gene. The complex of cAMP–CRP and RNA polymerase with DNA is cooperative, the binding of one stimulating binding of the other. Both CRP and RNA polymerase must bind to the same face of the DNA, and cAMP–CRP directly contacts the RNA polymerase. CRP and RNA polymerase holo enzyme associate cooperatively to form a 2:2 complex *in vitro*, with or without cAMP. A surface-exposed loop in the C-terminal domain of CRP and the C-terminal part of the α-subunit of RNA polymerase are primarily responsible for the contact between two proteins. An additional contact between the N-terminal domain of CRP and the N-terminal domain of α-subunit is involved in some cases.

CRP and the *lac* repressor (Section 13.3.A.1) associate cooperatively to form a 2:1 complex *in vitro*. Forming the complex is stimulated by cAMP, with a net uptake of one cAMP per molecule of CRP. CRP probably binds in the cleft between dimeric units in the *lac* repressor tetramer (Figure 13-5). These CRP–*lac* repressor interactions may play important roles in regulatory events that take place at overlapping CRP and repressor binding sites in the lactose promoter.

The *Escherichia coli* cyclic AMP receptor protein forms a 2:2 complex with RNA polymerase holoenzyme, *in vitro*. D. Dyckman & M. G. Fried (2002) *J. Biol. Chem.* **277**, 19064–19070.

Catabolite activator protein: DNA binding and transcription activation. C. L. Lawson *et al.* (2004) *Curr. Opinion Struct. Biol.* **14**, 10–20.

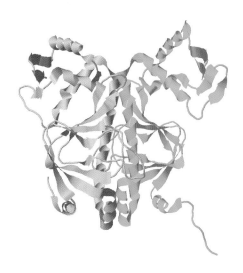

Figure 13-17. Structure of the CRP dimer. The helix–turn–helix motifs that interact with the major groove of DNA are at the top of the molecule; the bound DNA would lie horizontally along the top of the molecule. The spacing between the two sets in the dimer corresponds to the spacing between successive turns of the major groove of B-DNA. Generated from PDB file *1i5z* using the program Jmol.

13.3.B. TATA-binding Protein

The DNA sequence TATA occurs approximately 30 base pairs upstream of the start site of transcription of eukaryotic genes and helps to define the initiation site for transcription of the gene. The association of monomeric TATA-binding protein (TBP) with promoter DNA is an essential first step in many current models of eukaryotic transcription initiation. This step is followed by others in which additional transcription factors, and finally RNA polymerase, assemble at the promoter.

The C-terminal domain of TBP is composed of two direct repeats of about 80 amino acid residues; it resembles a saddle that binds with high affinity ($K_d \leq 5$ nM) and sits astride the DNA, in contact with its minor groove (Figure 13-18). Two Phe residues of TBP are inserted between the two last base pairs of the TATA element, causing the DNA structure to bend by about 80° and widening the minor groove without interfering with base pairing (Figure 13-8). Sequences at these sites that favor flexibility increase the affinity for TBP up to more than 100-fold. The rate of dissociation of the TBP–DNA complex is decreased, rather than increasing the rate of its formation. The importance of the DNA kinking varies, depending upon the nature of the TATA box and the surrounding sequences.

Little information is available concerning the structure of full-length TBP containing both the conserved C-terminal and the more variable N-terminal domains. The latter is involved in a monomer–dimer equilibrium with an apparent dissociation constant of approximately 8 μM.

Interactions with other components of the transcription machinery are vital for the function of TBP. Bending the DNA around the TATA box allows contacts between regulatory factors bound to their specific sequence, on one side, and the basal transcription machinery on the other side. One example of the mechanisms that regulate the interactions of TBP with promoter DNA is autorepression, where TBP sequesters its DNA-binding surface through dimerization. Once TBP is bound to DNA, some factors induce TBP to dissociate, while others convert the TBP–DNA complex into an inactive state. TBP *in vivo* is part of **transcription factor TFIID**, a multiprotein complex containing TBP and 14 TBP-associated factors (TAFs). The overall structure of this large complex resembles a molecular clamp formed by three major lobes connected by thin linking domains. The TAFs are dispensable for basal transcription *in vitro* but are required for the response to activator proteins and have selective effects on gene expression. The complexity of all these interactions of TBP with other proteins explains how TBP can be involved in the regulation of so many different genes, each with their own expression requirements.

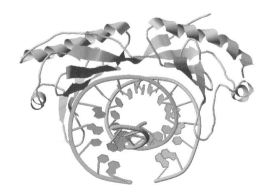

Figure 13-18. Structure of the complex of TATA-binding protein (TBP) with a TATA box. The protein binds the TATA sequence from the minor groove and bends the DNA into the major groove. The α-helices are represented as *coils* and the β-strands as *arrows*. Generated from PDB file *1cdw* using the program Jmol.

Self-association of the amino-terminal domain of the yeast TATA-binding protein. C. A. Adams *et al.* (2004) *J. Biol. Chem.* **279**, 1376–1382.

Comparison of the effect of water release on the interaction of the *Saccharomyces cerevisiae* TATA binding protein (TBP) with 'TATA Box' sequences composed of adenosine or inosine. S. Khrapunov & M. Brenowitz (2004) *Biophys. J.* **86**, 371–383.

13.3.C. Zinc-containing DNA-binding Motifs

A number of different DNA-binding proteins and motifs contain one or more Zn atoms that coordinate with amino acid residues, usually Cys or His, in the appropriate spatial orientation and thereby stabilize the structure of the domain. Six types of these domains have been identified, yet they are structurally diverse and the only common feature is the presence of zinc (Figure 13-19).

1. Zinc Fingers

Zinc fingers occur in many different versions in virtually every eukaryotic genome. They provide one more structural means for an α-helix to protrude from the protein surface sufficiently to interact with nucleotide bases in the major groove of DNA. **The most conspicuous feature of this structure is a Zn^{2+} ion chelated by two Cys and two His residues that occur in a characteristic sequence motif, $-X_3-Cys-X_{2-4}-Cys-X_{12}-His-X_{3-4}-His-X_4-$**, where X is any amino acid. Hundreds of such sequence motifs have been found in proteins involved in many aspects of gene regulation in eukaryotes. The four residues that chelate the zinc ion are held in the appropriate positions by the 3-D structure of the finger polypeptide chain, which consists of a two-stranded β-hairpin and a single α-helix (Figure 13-20). This fold is apparently stabilized by several hydrophobic side-chains that make up the core of the structure. **Each zinc finger seems to be a stable, autonomous, structural unit, even though it generally consists of only 28–31 residues.**

When bound to DNA, the proximal end of the α-helical region is directed into the major groove and makes contact with specific bases (Figure 13-4). Successive amino acid residues that make contact are separated by approximately one turn of the α-helix (Figure 13-20). One of the His residues that coordinate the zinc ion also contacts the sugar–phosphate backbone. A single zinc finger generally contacts three successive base pairs in DNA, usually in the same strand of the double helix (Figure 13-21).

Most zinc finger proteins contain multiple (up to 30) contiguous fingers separated by short linkers, frequently with the sequence –Thr–Gly–Glu–Lys– (Figure 13-20). An individual finger binds to DNA only weakly, but simultaneous interactions involving a number of zinc fingers increase the binding substantially. **Adjacent zinc fingers often individually specify four-base pair subsites that overlap by one base pair.** Different zinc fingers have different specificities for the triplet that they recognize. Proteins containing multiple zinc fingers are therefore of modular construction, and **nature seems to have devised a variety of such proteins with specificities for different DNA sequences by 'mixing and matching' various individual fingers with specificities for different triplets**. The simplicity of these direct interactions means that there is almost a direct code between the amino acids of the zinc finger and the bases of the DNA that they recognize, so it is possible to design zinc fingers to recognize specific DNA sequences. Adding domains with nuclease activity can be used to cleave specific DNA sequences.

There is considerable variation in the structures and functions of the zinc finger modules. The structure of the first finger is often especially variable. In the case of the SWI5 protein, for example, the first finger is preceded by a short α-helix, and an additional β-strand precedes the normal β-hairpin loop. Some of the multiple fingers may mediate interactions with other proteins and not be involved in DNA binding. In some proteins, zinc fingers are associated with conserved protein domains that help to define the functions of these regulators. Some zinc fingers bind specifically to RNA molecules.

In addition to the classical zinc fingers described here, there are at least 14 different classes of zinc fingers that differ in the nature and arrangement of their zinc-binding residues.

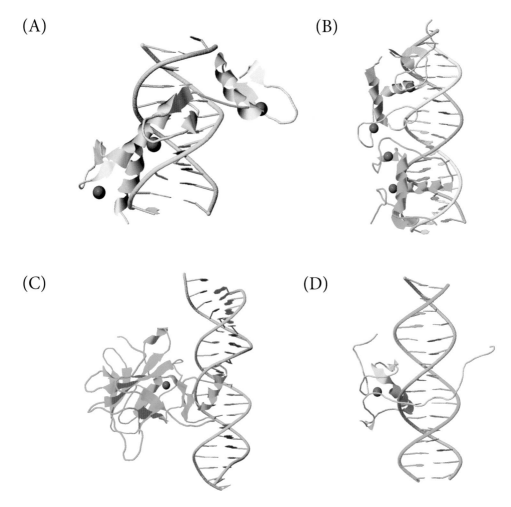

Figure 13-19. Four different zinc-containing transcription factors bound to their specific DNA sequences. The α-helices are depicted as *coils*, β-sheets as *arrows* and zinc ions as *spheres*. (A) Three zinc fingers of murine Zif268 (Section 13.3.C.1; generated from PDB file *1jk1*). (B) The dimeric DNA-binding domain of estrogen receptor (Section 13.3.C.2; generated from PDB file *1hcq*). (C) The core domain of the p53 tumor suppressor, a β-sandwich that serves as a scaffold to orient a loop–sheet–helix motif and two large loops. It is stabilized by a zinc ion that is coordinated by three Cys and one His residues. The zinc domain orients the side-chain of an Arg residue that contacts the minor groove of the DNA, while two Arg residues of the loop–sheet–loop motif contact the major groove (generated from PDB file *1tup*). (D) The DNA-binding domain of the chicken erythroid transcription factor GATA-1. A single zinc ion stabilizes the core of two irregular antiparallel β-strands and an α-helix. The α-helix and the loop connecting the two antiparallel strands interact with the major groove of the DNA, while the C-terminal tail wraps around the DNA to contact the minor groove (generated from PDB file *2gat*). Figure generated using the program Jmol.

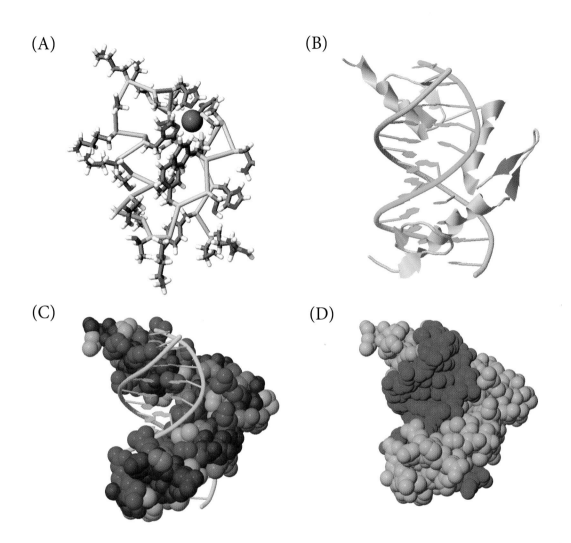

Figure 13-20. The structure of a single two-Cys, two-His zinc finger (A) and the interaction of a three-finger protein with DNA (B, C, D). (A) All the covalent bonds of the protein are depicted, except that a single bond links adjacent C^α atoms. The *sphere* is the Zn^{2+} ion, which is ligated by two His and two Cys side-chains. (B, C, D) The crystal structure of a polypeptide chain comprising three zinc fingers from protein Zif268. The protein is depicted in cartoon fashion in (B) and as space-filling in (C). In (D), both protein and DNA are depicted with their atoms as *spheres* of the appropriate van der Waals radii. Amino acid side-chains from residues at the amino end of each helix contact the edges of three sequential base pairs in the major groove of the DNA (Figure 13-21). Generated from PDB files *3znf* and *1a1f* using the program Jmol.

Designer zinc-finger proteins and their applications. M. Papworth *et al.* (2006) *Gene* **366**, 27–38.

Designer zinc finger proteins: tools for creating artificial DNA-binding functional proteins. M. Dhanesekaran *et al.* (2006) *Acc. Chem. Res.* **39**, 45–52.

Invariance of the zinc finger module: a comparison of the free structure with those in nucleic-acid complexes. D. Lu & A. Klug (2007) *Proteins* **67**, 508–512.

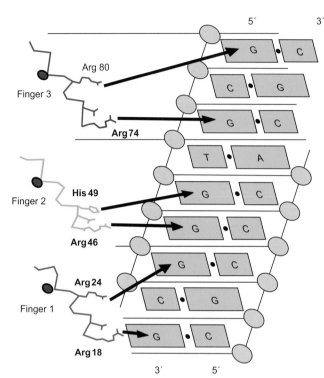

Figure 13-21. Hydrogen bonds from Arg and His residues of three zinc fingers to guanine bases in the DNA. The *arrows* indicate hydrogen bond interactions with the indicated bases of the DNA. The Zn atoms of the protein are *orange*; the phosphate groups of the DNA backbone are *blue*. Adapted from N. P. Pavletich & C. O. Pabo (1991) *Science* **252**, 809–817.

2. Steroid Hormone Receptors

The DNA-binding domains of the steroid hormone receptors, such as the **glucocorticoid** and **estrogen receptors**, consist of another type of DNA-binding motif that contains zinc and eight Cys residues that bind two zinc ions. **Each zinc ion is coordinated by four Cys residues in a loop that occurs in Cys–Cys … Cys–Cys sequences**. Unlike the zinc fingers, the distal Cys residues are followed immediately by an α-helical region (Figure 13-22). The second α-helix is in direct hydrophobic contact with the first, forming a single structural domain known as the **double-loop zinc helix**. The α-helix immediately distal to the first Cys tetrad sequence contains the residues responsible for recognizing specific DNA sequences.

The DNA-recognition domains of each receptor exist as dimers and bind to a palindromic sequence that contains three conserved bases on each side. They differ in the separation between the conserved sequences in such a way that the individual components of the dimer must have different relative rotational orientations, as a result of differences in their dimerization.

What sets steroid hormone receptors (and other nuclear hormone receptors) apart from other families of sequence-specific transcriptional activators is the presence of **a ligand-binding domain that acts as a molecular switch to turn on transcriptional activity when a hormonal ligand induces a conformational change in the receptor**. The ligand-binding domains possess a conserved structure with 12 α-helices surrounding a central hydrophobic core. Upon binding hormone, helix 12 changes position and generates a pocket with helix 3 and helix 5, where transcriptional coactivators bind. The novel coactivator protein complex has an essential role in receptor-mediated transcriptional activation. Coactivators function as adaptors in a signaling pathway that transmits transcriptional responses from the DNA-bound receptor to the basal transcription machinery. Hormone **agonists** induce a conformational change in the C-terminal transcriptional activation domain, AF-2, that creates a new protein interaction site on the surface of the ligand-binding domain that is recognized

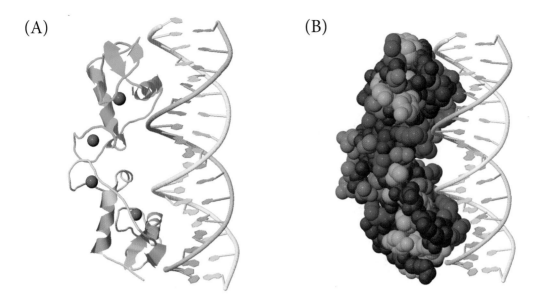

Figure 13-22. A dimer of the DNA-binding domain of the estrogen receptor binding to its specific DNA sequence. (A) The *spheres* are the four zinc ions, *coils* are α-helices and *arrows* β-strands. (B) The protein is shown in space-filling form. Generated from PDB file *1hcq* using the program Jmol.

by Leu–X–X–Leu–Leu sequence motifs in the p160 family of coactivators. In contrast, steroid **antagonists** for the estrogen receptor induce an alternate conformation in AF-2 that occludes the coactivator binding site and recruits corepressors that can actively silence steroid responsive genes. Thus **the cellular availability of coactivators and corepressors is an important determinant in the biological response to both steroid hormone agonists and antagonists**.

To affect transcription, the steroid hormone receptors must be able to gain access to their binding sites on the genomic DNA within the compacted chromosomes (Section 13.3.F). They do this by interacting directly with higher order chromatin structures and perturbing them using enzymatic multi-complexes that can either remodel or modify the chromosomal structure.

Structure and function of the glucocorticoid receptor ligand binding domain. R. K. Bledsoe *et al.* (2004) *Vitam. Horm.* **68**, 49–91.

Gene regulation by the glucocorticoid receptor: structure–function relationship. R. Kumar & E. B. Thompson (2005) *J. Steroid Biochem. Mol. Biol.* **94**, 383–394.

---

3. GAL4 Type

The GAL4 protein is typical of a third type of zinc-containing domain that binds DNA. **Two Zn atoms in each monomer are coordinated by six Cys residues; two of them interact with both Zn atoms**. This structure is connected by a flexible region of polypeptide chain to a short amphipathic α-helix (Figure 8-8) that interacts with another monomer to form a short coiled coil (Section 8.5.A) of two parallel α-helices. The GAL4 protein functions as a symmetrical dimer and binds to a palindromic DNA recognition site. Direct contact with the bases in the DNA major groove is made by amino acid residues from a short α-helical region (Figure 13-23). Most of the contacts to the sugar–phosphate backbone involve amino acid residues in the more extended polypeptide chain.

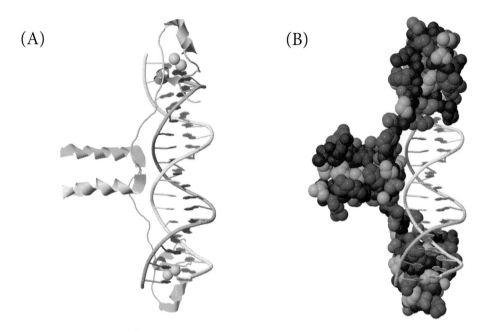

Figure 13-23. Structure of the GAL4 dimer–DNA complex. The protein is shown as a cartoon in (A) and with space-filling representation in (B). The two Zn atoms of each monomer are shown as *spheres*; they are coordinated by six Cys residues. Residues 8–40 of GAL4 comprise the α-helical DNA-recognition element, which interacts with the major groove of the DNA at each end of the elongated molecule. Residues 49–64 form the coiled-coil dimerization element, at the center, while residues 41–48 are in an extended conformation, interacting with the DNA backbone by its basic residues. Generated from PDB file *1d66* using the program Jmol.

Regions of GAL4 critical for binding to a promoter *in vivo* revealed by a visual DNA-binding analysis. A. Mizutani & M. Tanaka (2003) *EMBO J.* **22**, 2178–2187.

Yeast Gal4: a transcriptional paradigm revisited. A. Traven *et al.* (2006) *EMBO Rep.* **7**, 496–499.

13.3.D. bZip and Helix–Loop–Helix Domains

The **basic leucine zipper (bZip)** DNA-binding domain, and the closely related **helix–loop–helix (HLH)** domain, are found in many eukaryotic transcription factors (Figure 13-24). Both bind in the major groove of DNA. They also are able to form homo- or heterodimers with proteins of the same class. The simplest of these domains is the bZip domain, which forms a continuous α-helix when bound to DNA. The lower basic part of this helix (the b of bZip) forms sequence-specific contacts with DNA, while the upper part can dimerize with an appropriate partner through the formation of a parallel coiled coil, also known as a **leucine zipper** (Section 8.5.A). The simplicity of the coiled-coil interaction means that the two polypeptide chains that interact need not be identical. **The dimer has the appearance of a pair of scissors binding to a palindromic DNA sequence in which each half-site is exposed on opposite faces of DNA.**

bZIP proteins can bend DNA by asymmetric charge neutralization, in which one or two basic side-chains interact with a symmetry-related pair of phosphate groups of the DNA. The nucleotide sequences flanking the target site can affect its ability to bind the bZip peptide specifically, probably by varying the overall DNA conformation.

(A) (B)

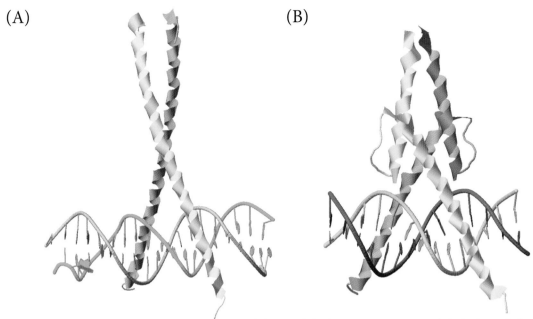

Figure 13-24. The structures of the DNA-bound forms of the bZIP protein GCN4 (A) the basic helix–loop–helix protein MyoD (B). Generated from PDB files *1ysa* and *1mdy* using the program Jmol.

The helix–loop–helix domain binds to DNA in a similar manner, with the difference that the α-helix involved in dimerization is separated from the DNA binding α-helix by a short loop. The four helices in the dimer form a parallel left-handed bundle.

Comprehensive identification of human bZIP interactions with coiled-coil arrays. J. R. Newman & A. E. Keating (2003) *Science* **300**, 2097–2101.

Reflections on apparent DNA bending by charge variants of bZIP proteins. P. R. Hardwidge *et al.* (2003) *Biopolymers* **69**, 110–117.

13.3.E. β-Sheets: Methionine Repressor

The **Met repressor recognizes DNA through β-strands**, rather than the α-helices of most other DNA-recognition proteins. It is a polypeptide chain of 104 amino acid residues that exists as a homodimer (Figure 13-25). Similar to the *trp* repressor (Section 13.3.A.5), **the two subunits of the Met repressor intertwine to form the symmetrical dimer**. The dimer noncooperatively binds two molecules of S-adenosyl-methionine (S-AdoMet):

(13.8)

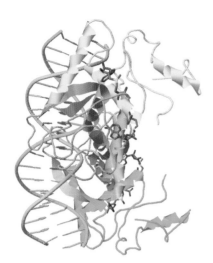

Figure 13-25. The structure of two dimeric methionine repressors bound to and bending dsDNA. The backbone of the protein is depicted, with β-strands depicted as *arrows*; two such strands from each dimer interact with the major groove of the DNA. The S-adenosylmethionine corepressor molecules bound to each monomer are depicted in skeletal form. Interactions between the DNA and the opposite ends of the nearly vertical α-helices (depicted as *coils*) cause bending of the DNA. Generated from PDB file *1mj2* using the program Jmol.

Each monomer of the Met repressor contains one β-strand and three α-helices, which account for about half of the sequence (Figure 13-25). The two β-strands of the dimer generate an antiparallel β-sheet, which packs against the α-helical regions.

This repressor recognizes an eight-nucleotide 'Met box', with the palindromic consensus sequence AGACGTCT, that is present in an altered form two to five times in each DNA region. **Two repressors having bound S-AdoMet bind tightly and cooperatively to DNA fragments containing two consecutive consensus Met boxes, while binding is negligible to nucleotide sequences containing only one Met box**. A DNA fragment containing two adjacent eight-base pair Met boxes binds two dimers. One dimer binds to each half-site, and the two-fold symmetry axis of each binding site coincides with the two-fold axis of the protein dimer. Two neighboring dimers interact through α-helix 1 to generate a tetrameric protein structure,.

The antiparallel protein β-sheet is inserted into the major groove of B-form DNA containing Met box sequences. Amino acid side-chains on the exposed face of the β-sheet form hydrogen bonds directly with specific base pairs. Thr25 from each β-strand contacts an adenine base, and the neighboring residue Lys23 of each strand contacts a guanine. The backbone from the N-terminus of α-helix 2 also makes contacts. The repressor senses the conformation and flexibility of the DNA backbone, so binding can also depend upon the sequences of base pairs that are not in contact with the protein.

S-AdoMet binds to the dimer at two independent and symmetrical sites, one on each monomer. The trivalent S atom of S-AdoMet is positively charged and situated near the partial negative charge at the C-terminus of a helix (Figure 8-11). The adenine moiety of S-AdoMet inserts into a hydrophobic pocket, replacing the side-chain of Phe65, while the methionine moiety lies at the protein surface. Consequently, methionine alone does not bind to the aporepressor, while S-adenosylhomocysteine (which lacks the terminal $-CH_3$ group and the positive charge on the S atom) binds with about half the affinity of S-AdoMet.

How the binding of S-AdoMet to the repressor regulates its binding to DNA is not clear, because it neither causes conformational changes in the protein nor contacts the DNA in the repressor–operator complex. A long-range electrostatic effect involving the positive charge on the S atom seems the most plausible explanation.

Direct and indirect readout in mutant Met repressor–operator complexes. C. W. Garvie & S. E. Phillips (2000) *Structure* **8**, 905–914.

Structural basis for the differential regulation of DNA by the methionine repressor MetJ. A. M. Augustus *et al.* (2006) *J. Biol. Chem.* **281**, 34269–34276.

13.3.F. Histone-Fold

Histones are small basic proteins that are involved in organizing the chromosomal DNA of eukaryotes (Section 3.7). There are normally four distinct histones (H2A, H2B, H3 and H4), but with related sequences, and two of each assemble into an octameric structure. This structure interacts with about 200 base pairs of double-stranded DNA, to form the **nucleosome** (Figure 3-12). It interacts with nearly all the DNA in a chromosome and is not very specific for the nucleotide sequence. **The histone-fold motif is a feature of all four histones**. It consists of a long central α-helix connected by loops to two shorter α-helices (Figure 13-26). The histone-fold is found in other proteins and it might have arisen by gene duplication of a helix–strand–helix motif, with the long central α-helix being produced by fusion of two helices.

Extensive interactions between two histone-folds in a **handshake motif**, with extensive hydrophobic interactions between the nonpolar faces of the two long amphipathic α-helices, produce the heterodimers that exist in **the nucleosome core particle** (Figure 13-26). Each dimer has three independent DNA-binding sites: two contributed by the loops at each end of the dimer, and one formed by the two N-termini of α-helices of each histone-fold at the dimer apex. **Histone–DNA contacts include extensive salt bridges and hydrogen bonds from both main chains and side-chains of the histones to the phosphate backbone, together with nonpolar contacts with DNA sugar groups and electrostatic interactions of the positively charged N-termini of α-helices in the histone-folds with DNA phosphate groups**. Such interactions might be expected for proteins that can bind to a variety of DNA sequences; the number of specific base contacts in the core particle structure is very small (one of them being a nonpolar contact of the 5-methyl group of a thymidine

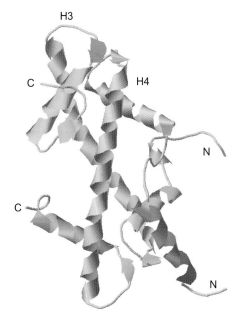

Figure 13-26. The histone-folds in histones H3 and H4 when part of the nucleosome bound to DNA. Histone H3 is *green*, H4 *mauve*. Generated from PDB file *1aoi* using the program Jmol.

in the major groove). Strikingly, Arg side-chains contact the minor groove each of the 14 times it faces inwards, probably to screen electrostatic repulsions between the DNA phosphate groups. **Each histone-fold dimer in the octamer organizes 27–28 base pairs of DNA**, with stretches of four base pairs between contacts.

The histone-fold has also been found to occur in other proteins, with little or no sequence homology with the histones, that are also involved in DNA/protein recognition and protein–protein interactions. For example, nine of the 14 **TAFs** (Section 13.3.B) contain histone-fold domains. The histone-fold motif thus appears to be widespread, and it may primarily be a robust structural motif particularly well-suited to protein dimerization in large complexes.

A database for histone and proteins containing the histone-fold is available at http://research.nhgri.nih.gov/histones/.

Crystal structures of histone S in mutant nucleosomes reveal altered protein–DNA interactions. U. M. Muthurajan *et al.* (2004) *EMBO J.* **23**, 260–271.

Histones and histone modifications. C. L. Peterson & M. A. Laniel (2004) *Curr. Biol.* **14**, R546–551.

Assembly of variant histones into chromatin. S. Henikoff & K. Ahmad (2005) *Ann. Rev. Cell. Dev. Biol.* **21**, 133–153.

Histones in functional diversification. Core histone variants. R. H. Pusarla & P. Bhargava (2005) *FEBS J.* **272**, 5149–5168.

13.3.G. Bacterial Type II DNA-binding Proteins: Heat-unstable (HU) and Integration Host Factor (IHF)

The bacterial type II DNA-binding proteins condense bacterial DNA genomes by binding at multiple sites and inducing coherent bends. They include the abundant **HU** (**heat-unstable**) and **integration host factor** (**IHF**) proteins, plus protein TF1 that is produced by bacteriophages. HU and TF1 bind with little, if any, sequence specificity, but IHF binds at specific sites that have some sequence similarities. All of these proteins bind to DNA as dimers. IHF is normally a heterodimer, while HU can be either a homo- or heterodimer.

These proteins bend the DNA very substantially (Figure 13-7). For example, within the 2.5 turns of double helix that comprise its binding site, IHF causes a bend of at least 160°, perhaps even more than 180°. In the cases of both IHF and HU, the two subunits intertwine to form a compact core, from which are extended two long β-sheet ribbons (Figure 13-7). These arms lie in the minor groove, extending from the inside to the outside of the wrapped DNA; there are two substantial kinks in the DNA where they terminate. IHF also clamps the hairpin by contacts in the minor groove using α-helices from both subunits of the dimer and extensive electrostatic interactions with the backbone phosphate groups.

A Pro side-chain located at the tips of the β-arms is absolutely conserved and produces these two kinks by intercalating partially between adjacent base pairs. This opens the minor groove on the outside of the bend. The narrowed groove on the inside should increase the repulsions between the phosphate groups on opposite strands, but this is diminished by interactions that neutralize their charges.

The conformation of the DNA dictates the sequence specificity of IHF binding, not specific contacts with the bases. Two short sequences separated by approximately half a turn provide the sequence specificity. The conserved sequence CAA is favored over other sequences at the kink site because it can best accommodate the severe distortion induced by the protein. **IHF binds rapidly to straight or partially bent DNA, followed by slower DNA bending** on the 10^{-4} s time scale. The rate of bending is similar to that for opening a single A•T base pair inside duplex DNA. Spontaneous thermal disruption of base pairing at an A•T site may be sufficient to overcome the free energy barrier needed to bend or kink the DNA before forming a tight complex with IHF.

DNA–protein interactions and bacterial chromosome architecture. J. Stavans & A. Oppenheim (2006) *Phys. Biol.* **3**, R1–R10.

Direct observation of DNA bending/unbending kinetics in complex with DNA-bending protein IHF. S. V. Kuznetsov *et al.* (2006) *Proc. Natl. Acad. Sci. USA* **103**, 18515–18520.

Bacterial protein HU dictates the morphology of DNA condensates produced by crowding agents and polyamines. T. Sarkar *et al.* (2007) *Nucleic Acids Res.* **35**, 951–961.

13.3.H. Single-strand DNA-binding Proteins

DNA does not normally exist naturally as a single-strand (ssDNA), except in a few bacterial viruses. It normally occurs only transiently during the replication, repair and recombination of double-stranded DNA molecules, so there are only a few **ssDNA binding proteins (SSBs) but all organisms have at least one**; some of them also bind single-stranded RNA (Section 13.4.C). **These proteins are involved in the replication and repair of DNA and in genetic recombination, when they protect ssDNA from damage by nucleases, prevent hairpin formation and block reannealing of the DNA until the process is completed**. Many ssDNA-binding proteins interact physically and functionally with a variety of other DNA processing proteins. These interactions are thought to order and guide the parade of proteins that 'trade places' on the ssDNA, a model known as **hand-off**, as the processing pathway progresses.

Minimalist protein design: a β-hairpin peptide that binds ssDNA. S. M. Butterfield *et al.* (2005) *J. Am. Chem. Soc.* **127**, 24–25.

---

1. Prokaryotic Single-strand DNA Binding Proteins

SSB proteins from prokaryotes (and mitochondria of eukaryotes) have a tetrameric structure of identical 19-kDa polypeptide chains that binds specifically to ssDNA with high affinity and moderate cooperativity. As a result, **many molecules of SSB coat ssDNA, and they can induce denaturation of double-stranded DNA**. Binding is virtually independent of the nucleotide sequence.

The N-terminal domain of each monomer in the tetrameric structure (Figure 13-27-C) contains a single α-helical segment and two β-pleated sheets that fold to create a barrel from which β-hairpin loops protrude, the oligonucleotide/oligosaccharide binding fold (OB-fold) (Section 13.3.H.3). In contrast, the C-terminal domain, which includes the last third of the polypeptide chain, is disordered in both instances, but it adopts a fixed structure upon interacting with RNA polymerase and other proteins. The monomers form dimers that can then further dimerize, forming the tetramer. The structure has external patches of positive charge that, together with the flexible loops, guide the wrapping of ssDNA around the tetramer. Aromatic residues on the external surface of the assembly are essential for ssDNA binding and interact directly with the nucleotides of the DNA. SSB binds ssDNA with virtually no change in structure. Each subunit of the tetrameric protein contains the same potential ssDNA binding site, so ssDNA can bind to the protein in several ways. Normally, a 65-nucleotide stretch of ssDNA interacts with and wraps around all four subunits of the tetramer; the wrapping occurs on a time scale of tens of microseconds. Binding to ssDNA is accompanied by a substantial negative enthalpy change and negative heat capacity change, due in part to linkage of temperature-dependent protonation and DNA base unstacking equilibria.

An SSB tetramer can be transferred from one ssDNA molecule to another without proceeding through a free protein intermediate, and the rate of transfer is determined by the availability of free DNA binding sites within the initial SSB–ssDNA donor complex. It occurs by a 'direct transfer' mechanism in which an intermediate composed of two DNA molecules bound to one SSB tetramer forms transiently prior to the release of the acceptor DNA.

Microsecond dynamics of protein–DNA interactions: direct observation of the wrapping/unwrapping kinetics of single-stranded DNA around the *E. coli* SSB tetramer. S. V. Kuznetsov *et al.* (2006) *J. Mol. Biol.* **359**, 55–65.

3D structure of *Thermus aquaticus* single-stranded DNA-binding protein gives insight into the functioning of SSB proteins. R. Fedorov *et al.* (2006) *Nucleic Acids Res.* **34**, 6708–6717.

---

2. Eukaryotic Replication Protein A

Replication protein A (RPA) was originally identified as a eukaryotic SSB protein essential for the *in vitro* replication of simian virus 40 DNA, but subsequently it was found to be **indispensable in almost all DNA metabolic pathways**, including DNA replication, repair and recombination, plus cell-cycle and DNA-damage checkpoints. It is a more complex protein than the prokaryotic SSB, consisting of three different polypeptide chains of molecular weights 70, 32 and 14 kDa. The 70-kDa subunit of RPA is composed of multiple domains: an N-terminal domain involved in protein interactions, a central DNA-binding domain, a putative ($C-X_2-C-X_{13}-C-X_2-C$) zinc finger, and a C-terminal domain involved in interacting with the other subunits.

RPA binds tightly to ssDNA, with a dissociation constant of about 10^{-10} M, and it prefers ssDNA over RNA or double-stranded DNA by a factor of about 10^3. The three subunits of human RPA contain six conserved **DNA-binding domains** (DBDs), of which four bind ssDNA (DBD-A, -B and –C on the largest subunit, DBD-D on the 32-kDa subunit) with dissociation constants smaller than 10^{-9} M. Each DBD has an OB-fold (Figure 13-27-D). DBDs A and B are very similar and individually have very low intrinsic affinities for ssDNA, but tandem DBDs (AA, AB, BA and BB) bind ssDNA with about 100-fold greater affinity. DBD-A has the highest intrinsic affinity for ssDNA, while DBD-D has the lowest; it binds ssDNA only after the other three binding sites are occupied. Binding of ssDNA occurs in at least three different stages, with initially eight to 10 nucleotides interacting, then 13–14, and finally 27, when it uses all four of its DBDs. At low salt concentrations, however, RPA tends to bind to only 18 nucleotides of ssDNA, using only three of its DBDs. Binding is not markedly cooperative.

A dynamic model for replication protein A (RPA) function in DNA processing pathways. E. Fanning *et al.* (2006) *Nucleic Acids Res.* **34**, 4126–4137.

Functions of human replication protein A (RPA): from DNA replication to DNA damage and stress responses. Y. Zou *et al.* (2006) *J. Cell. Physiol.* **208**, 267–273.

3. Oligonucleotide/Oligosaccharide Binding Fold

The SSBs share a common core ssDNA-binding domain with a conserved oligonucleotide/ oligosaccharide binding fold (OB-fold). The OB-fold is a small β-barrel formed from five β-strands connected by modulating loops (Figure 13-27); its topology is that of the Greek key (Section 9.1.D.4). Although the OB-fold is most prominent in proteins that bind nucleic acids, it also occurs in many other types of proteins, where it can interact with oligosaccharides or other proteins. In all OB-folds, two or three loops at the same end of the β-barrel act as clamps to bind to their ligands. The loops change in length and sequence to accommodate various ligands. Those that bind nucleic acids usually have little specificity for the nucleotide sequence and bind nonhelical structures, although a few recognize specific single-stranded regions of nucleic acids.

OB-fold domains range between 70 and 150 amino acid residues in length, depending on the lengths of the loops between the β-strands. Strands 3 and 5 can close the β-barrel by hydrogen-bonding in a parallel manner, but in some cases they are not in close contact and the barrel is not completely closed. An α-helix often occurs between strands 3 and 4 and packs against the bottom of the barrel (Figure 13-27). Ligands are usually bound by the loop connecting strands 2 and 3, designated L_{23}, assisted by loops L_{12}, $L_{3\alpha}$, $L_{\alpha 4}$ and L_{45} (the α refers to the α-helix) The loops define a cleft that runs across the surface of the OB-fold perpendicular to the axis of the β-barrel. The majority of nucleic acid ligands bind within this cleft, usually perpendicular to the antiparallel β-strands and with a polarity running 5′ to 3′ in the nucleic acid from strands 4 and 5 to strand 2. The prokaryotic SSB is, however, unusual in binding ssDNA with the opposite polarity. Interactions also occur with the surface of the barrel defined by strands 2 and 3.

It might be expected that SSBs that are not sequence-specific would interact primarily with the backbone of ssDNA, and not with the nucleotides. In fact, **the bases of the ssDNA are usually in close contact with the protein, while the phosphodiester groups are primarily exposed to solvent**; similar binding interactions occur in other proteins that bind single-stranded nucleic acids and loop structures. The bases interact with the protein primarily through stacking interactions with aromatic amino acid side-chains and packing interactions with hydrophobic parts of side-chains, which can also involve the deoxyribose rings. Deleting these side-chains decreases the binding affinity substantially. The nonspecific nature of these nonpolar interactions means that the binding sites recognize single-stranded nucleic acid structures but not the identities of the bases. In addition, these binding sites seem to be intrinsically flexible, so that they can mold themselves to be complementary to nucleic acids with different sequences. Some OB-folds bind their ligands with no change in conformation, whereas others undergo folding from a very disordered structure in the absence of the ligand. The nucleic acid ligand can undergo changes in conformation in order to bind. For example, the local secondary structure can be disrupted in order to bind, with nucleotides bulged out of helices or removed from stacking interactions.

612 CHAPTER 13 Nucleic Acid-Protein Interactions

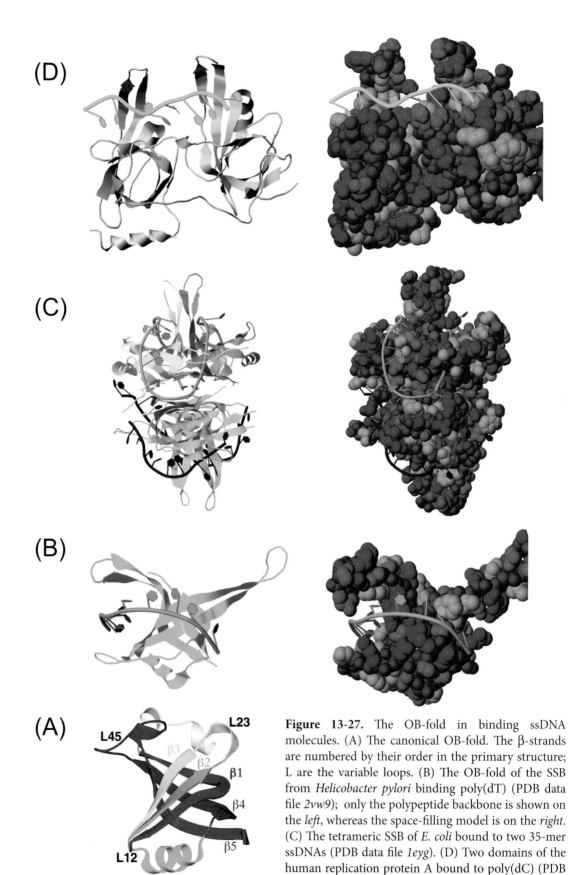

Figure 13-27. The OB-fold in binding ssDNA molecules. (A) The canonical OB-fold. The β-strands are numbered by their order in the primary structure; L are the variable loops. (B) The OB-fold of the SSB from *Helicobacter pylori* binding poly(dT) (PDB data file *2vw9*); only the polypeptide backbone is shown on the *left*, whereas the space-filling model is on the *right*. (C) The tetrameric SSB of *E. coli* bound to two 35-mer ssDNAs (PDB data file *1eyg*). (D) Two domains of the human replication protein A bound to poly(dC) (PDB data file *1jmc*). Figure generated using the program Jmol.

Individual OB domains have low affinity for their ligands, but tethering four of them on the same molecule produces high affinity (Table 12-3). Yet the DNA molecules dissociate at the much faster rates associated with weak binding than the slow rates that would be expected from a single high-affinity binding site. The reason is that a bound ssDNA molecule can dissociate from each single binding site relatively independently of the other sites. **This type of linkage of multiple weak interactions on a dynamic binding scaffold appears to be a highly effective mechanism for multi-step processing in biology.**

The four ssDNA-binding domains of RPA and each of the monomers of the prokaryotic SSB protein have the characteristic OB-fold (Figure 13-27). Each monomer of SSB makes extensive contacts with ssDNA, using a large interaction surface that includes both sides of a protracted two-stranded β-sheet that extends from loop L_{23}. In RPA, binding of ssDNA causes a major reorientation between, and significant conformational changes within, the OB-folds that comprise the DNA-binding domain. Two OB-folds have a tandem orientation in the presence of DNA but adopt multiple orientations in its absence. Within the OB-folds, extended loops implicated in DNA binding change conformation significantly when the DNA dissociates.

Many other proteins use the OB-fold to interact with nucleic acids, including RNA. For example, *E. coli* ribosomal protein S1 is composed of six repeating homologous OB-folds. The second OB-fold in protein S1 is essential for the recognition of transfer messenger RNA, while the four C-terminal OB-folds play a role in stabilizing the complex of S1 and the transfer messenger RNA. Other ribosomal proteins (S17, S12 and IF1) consist of single OB-folds. Telomere end-binding proteins contain four OB-fold domains used for recognition of single-stranded telomeric DNA.

The OB proteins, even those restricted to binding nucleic acids, are renowned for **the absence of any discernible sequence similarities**. Yet the close similarities between their structures and interactions with ligands suggest that they have evolved from a common ancestor. That ancestor must have existed early in evolution, as it is present in all extant life forms.

OB-fold domains: a snapshot of the evolution of sequence, structure and function. V. Arcus (2002) *Curr. Opin. Struct. Biol.* **12**, 794–801.

Insights into ssDNA recognition by the OB-fold from a structural and thermodynamic study of Sulfolobus SSB protein. I. D. Kerr *et al.* (2003) *EMBO J.* **22**, 2561–2570.

Nucleic acid recognition by OB-fold proteins. D. L. Theobald *et al.* (2003) *Ann. Rev. Biophys. Biomol. Structure* **32**, 115–133.

13.4. RNA-BINDING PROTEINS

RNA differs from DNA covalently in only two seemingly minor ways: the 2′-hydroxyl group replaces an H atom and the methyl group of the thymine base is missing (the base is then known as uracil) (Chapter 4). Yet these two small covalent differences have major consequences for the structures and functions of RNA. The structures of RNA are much more diverse than those of DNA, and they even have catalytic activities (Section 14.6). RNA serves as the genetic material in some viruses, but in most cases its main function is, as messenger RNA, to convey the genetic information from the DNA genetic material to the ribosomes for the synthesis of proteins. Other major RNAs are the transfer

RNAs, which serve as adaptors between the genetic information of nucleic acids and the amino acids of proteins (Section 13.4.E). All of these functions of RNA require that they interact with a wide variety of proteins. RNA is also an essential and functional component of many **ribonucleoproteins**, usually abbreviated as RNP. The most prominent example is the **ribosome** (Section 13.4.F).

Not many RNA-binding proteins have been characterized structurally, but sequence homologies detected in a variety of proteins that bind RNA have recognized a number of structural motifs that are believed to participate in binding RNA (Table 13-3). RNA and DNA are recognized in fundamentally similar ways, but **the structural motifs that bind RNA tend to be distinct from those that bind DNA**. Exceptions are the zinc finger (Section 13.3.C.1), OB-fold (Section 13.3.H.3), homeodomain (Section 13.3.A.3) and KH domains (Section 13.4.C), which bind both RNA and DNA.

RNA is most readily distinguished from DNA by the presence of the 2′-hydroxyl group and the A-type double helix that it adopts in place of the B-type helix of DNA. On the other hand, A-form RNA has a deep and narrow major groove and a shallow minor groove, which make the bases relatively inaccessible, so it is more difficult for proteins to recognize specific sequences in double-stranded RNA. Most RNAs are single-stranded, however, and consequently adopt double helices only locally, with hairpins, internal loops, bulges and pseudoknots in the remainder of the RNA molecule. Many proteins recognize irregular RNA structures in which the bases are more exposed for

Table 13-3. Examples of RNA-binding domains and sequence motifs

Sequence motif	Representative proteins	Target RNA
Alu binding module	SRP 14/9 heterodimer	*Alu* domain of SRP RNA
dsRNA binding motif	*E. coli* RNase III	RNA transcripts
	Drosophila Staufen	maternal bicoid mRNA
KH domain	hnRNP protein K	mRNA precursor
OB domain	Class II aminoacyl tRNA synthetase	tRNA
RGG box	hnRNP proteins	mRNA precursor
RNP domain	U1A spliceosomal protein	U1 snRNA
	hnRNP protein C	mRNA precursor
	hnRNP protein A1	mRNA precursor
	Poly(A) binding protein	mRNA poly(A) tail
S1 domain	Ribosomal protein S1	mRNA
	Initiation factor 1	
Sm domain	Spliceosomal core proteins	snRNAs
Zn-finger	TFIIIA	5S rRNA

hnRNP, heterogeneous nuclear RNP; mRNA, messenger RNA; rRNA, ribosomal RNA; RNP, ribonucleoprotein; snRNA, small nuclear RNA; SRP, signal recognition particle.

distinguishing by hydrogen bonds or stacking interactions. Binding to the protein may actually make the RNA structure less regular. Some RNA molecules fold into complex tertiary structures that may be recognized on the basis of their unique shapes.

A **groove binder** class of RNA-binding proteins places an α-helix, 3_{10}-helix, β-ribbon or irregular loop into the groove of an RNA helix, recognizing both the specific sequence of bases and the shape or dimensions of the groove, which are sometimes distorted from the normal A-form structure. Many known RNA-binding proteins are all β- or αβ-proteins; the αβ-folds are comparable to those present in some ribosomal protein subunits and may have evolved from them. These proteins often contain an exposed β-sheet, which is a good RNA-binding surface because aromatic and hydrophobic side-chains emanating from it can interact with the bases of the RNA molecule and create pockets that recognize the bases of single-stranded RNA. Some of these proteins recognize completely unstructured RNA; in others RNA secondary structure indirectly promotes binding by constraining bases in an appropriate orientation. **Binding specificity is generally a result of base-specific hydrogen bonds, nonpolar contacts and mutual accommodation of the protein and RNA surfaces.**

The amino acid residues most involved in contacts with the RNA are Arg, Asn, Ser and Lys; less preferred are Ala, Ile, Leu and Val; Trp and Cys are rarely observed. Of the total number of amino acids located at the interfaces, 22% are hydrophobic, 40% charged (32% positive, 8% negative), 30% polar and 8% Gly. Arg and Lys interact electrostatically with phosphate groups. Pro and Asn prefer to interact with bases more than ribose and phosphate groups; Met, Phe and Tyr prefer ribose over phosphate and bases. **No base predominates at protein–RNA interfaces**, but A bases are preferred by Ile, Pro and Ser residues, C is preferred by Leu, G by Asp and Gly, and U by Asn. Of all the contacts, 72% are van der Waals interactions, 23% involve hydrogen bonds and 5% are very short contacts. Of the potential H bonds, 54% are standard, 33% are of the C–H … O type, and 13% are between ionized groups; the protein main chain is involved in 32% of the hydrogen bonds. Amino acid residues that are not frequently in direct contact with RNA components usually interact with RNA atoms via their polypeptide backbone atoms in indirect hydrogen bonds involving water molecules.

The majority of RNA-binding proteins have modular structures in which an RNA-binding domain is combined with other domains with other functions.

The yeast three-hybrid system has become a useful tool in identifying interactions between RNA and proteins. An RNA sequence is tested in combination with an RNA-binding protein linked to a transcription activation domain for transcription of a reporter gene. A productive RNA–protein interaction activates that reporter gene *in vivo*.

Protein families and RNA recognition. Y. Chen & G. Varani (2005) *FEBS J.* **272**, 2088–2097.

Extending the size of protein RNP RNA complexes studied by nuclear magnetic resonance spectroscopy. C. D. Mackereth *et al.* (2005) *Chembiochem.* **6**, 1578–1584.

RNA RNP protein interactions in the yeast three-hybrid system: affinity, sensitivity and enhanced library screening. B. Hook *et al.* (2005) *RNA* **11**, 227–233.

Crystallization of RNA RNP protein complexes. E. Obayashi *et al.* (2007) *Methods Mol. Biol.* **363**, 259–276.

13.4.A. Ribonucleoprotein (RNP) Domain

The sequence of the ribonucleoprotein (RNP) domain has been found in well over 200 distinct RNA-binding proteins from diverse species, so it probably appeared early in evolution. It is one of the most common protein sequence motifs found in entire genomes. It is also known as **RRM** (RNA recognition motif) or **RNP-CS** (RNP-consensus sequence)-type RNA-binding domain. Some proteins contain multiple copies of the RNP domain, but others contain only one. For example, there are four tandem copies of the RNP domain in the protein that binds to the polyadenylate (poly A) tail of messenger RNA in eukaryotes.

RNP motifs can recognize a disparate range of RNA structures and sequences, with greatly varying affinities and specificities. In some cases they bind RNA very tightly, with dissociation constants of less than 10^{-9} M, and with exquisite sequence specificity, to form stable RNP particles. In other cases, binding to RNA is much weaker, with dissociation constants of roughly 10^{-6} M and poor discrimination between different RNA sequences. Weak binding by a single RNP motif, however, is augmented if there are several such domains in a protein that can bind to the same RNA molecule simultaneously. In many cases, affinity for an RNA is also determined by parts of the protein outside the RNP motif. The diversity of RNA structures recognized by the RNP motif is remarkable, in part because the structures of both the protein and the RNA are malleable and undergo substantial alterations in forming a complex.

An RNP domain generally consists of about 80 amino acid residues. It is characterized by two short sequence motifs called RNP1 (eight amino acid residues) and RNP2 (six residues) that are highly conserved. Structurally, the RNP domain is constructed of a four-stranded antiparallel β-sheet with two α-helices lying on one face (Figure 13-28). The two middle β-strands of the sheet contain

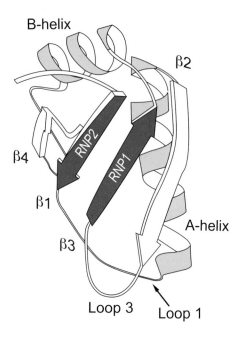

Figure 13-28. The structure of the RNP domain. The amino acid sequences that can identify RNP domains RNP1 and RNP2 comprise two of the β-strands. Adapted from K. Nagai *et al.* (1990) *Nature* **348**, 515–520.

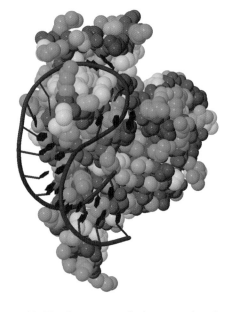

Figure 13-29. Structure of the complex between the U1A spliceosomal protein and its hairpin RNA binding site in U1 snRNA. The RNA is depicted as a skeletal model, whereas the van der Waals surface of the protein is depicted; basic groups are *blue*, acidic groups are *red*. Generated from PDB file *1a9n* using the program Jmol.

the RNP1 and RNP2 sequences, and the side-chains of the three highly conserved aromatic residues within RNP1 and RNP2 project from the surface of the β-sheet. In the case of the U1A protein (Figure 13-29), an RNA hairpin loop of 10 nucleotides binds to the surface of the β-sheet as an open structure; the polypeptide loop 3 between the β2 and β3 strands protrudes through the RNA loop. The first seven nucleotides of the loop fit into a groove formed on the surface of the β-sheet. Their bases are splayed out and make nearly all the contacts between the RNA and protein. These seven bases stack onto adjacent bases and protein side-chains; they are also involved in an intricate hydrogen bond network with the side-chains and polypeptide backbone of protein. **These stacking interactions are comparable to those observed in the binding of ssDNA to proteins** (Section 13.3.H).

RNA recognition by RNP proteins during RNA processing. G. Varani & K. Nagai (1998) *Ann. Rev. Biophys. Biomol. Structure* 27, 407–445.

13.4.B. Double-stranded RNA-binding Domain

The double-stranded RNA-binding domain (dsRBD) is a sequence motif of 65–75 amino acid residues that is found in multiple copies in RNA-binding proteins from diverse origins. These proteins are involved in diverse functions, such as localizing messenger RNA in the cell, editing messenger RNA and processing ribosomal and transfer RNAs. **The dsRBD module consists of a three-stranded antiparallel β-sheet and two α-helices packed on one side** (Figure 13-30-A); their order is α–β–β–β–α in the primary structure. The dsRBD is homologous to the N-terminal domain of ribosomal protein S5, suggesting that dsRBD evolved from a ribosomal protein.

A dsRBD usually recognizes 12 base pairs of double-stranded RNA (dsRNA), interacting with two successive minor grooves and across the intervening major groove on one face of a primarily A-form RNA helix (Figure 13-30-B). The nature of these interactions explains why dsRBD is specific for dsRNA, rather than single-stranded RNA (ssRNA) or double-stranded DNA (dsDNA). A dsRBD of Staufen protein recognizes the shape of the A-form double helix of dsRNA through interactions between conserved residues within loop 2 and the RNA minor groove, and between loop 4 and the phosphodiester backbone across the adjacent major groove. In addition, the A helix interacts with the single-stranded loop that caps the RNA helix. The dsRBD does not bind to dsDNA because it would have the recognizably different B-form and because of contacts with the 2'-OH groups of the ribose moiety in the minor groove. The interface between the protein and RNA is not tightly packed but is relatively flexible, with the protein interacting primarily with the sugar–phosphate backbone, not the bases, so it is not apparent how there can be sequence specificity.

The dsRBD of endonuclease Rnt1p targets this enzyme to its RNA substrates by recognizing hairpins closed by AGNN tetraloops (Section 4.1.B). The tetraloop is retained in the RNA–dsRBD complex (Figure 13-30-B). The dsRBD contacts the RNA at successive minor, major and tetraloop minor grooves on one face of the N-terminal α-helix. Neither the universally conserved G nor the highly conserved A bases of the AGNN tetraloop are recognized by specific hydrogen bonds, but an α-helix fits snugly into the minor groove of the RNA tetraloop and top of the stem, interacting in a nonsequence-specific manner with the sugar–phosphate backbone and the two nonconserved tetraloop bases.

Double-stranded RNA is most common in some animal viruses, and cells often use its presence to detect a viral infection. The dsRBD is usually involved and activates the enzyme dsRNA-dependent

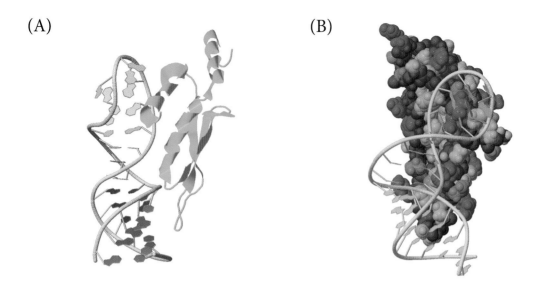

Figure 13-30. Structure of the dsRBD domain from the ribonuclease III Rnt1p complexed with the 5′-terminal hairpin of one of its substrates closed by an AGNN tetraloop. (A) The topology of the protein domain is depicted schematically. (B) The protein domain is shown in space-filling form, with the RNA bound on it surface. This view is rotated from that shown in (A). Generated from PDB data file *1t4l* using the program Jmol.

protein kinase, which can lead to a shutdown in gene translation, thereby inhibiting protein synthesis and replication of the virus.

The double-stranded-RNA-binding motif: interference and much more. B. Tian *et al.* (2004) *Nature Rev. Mol. Cell. Biol.* **5**, 1013–1023.

The double-stranded RNA-binding motif, a versatile macromolecular docking platform. K. Y. Chang & A. Ramos (2005) *FEBS J.* **272**, 2109–2117.

Structure and specific RNA binding of ADAR2 double-stranded RNA binding motifs. R. Stefl *et al.* (2006) *Structure* **14**, 345–355.

13.4.C. KH Domain

Three copies of a sequence motif within hnRNP K, one of the nuclear proteins that bind to precursors of messenger RNA (mRNA) and homologous sequences, are known as **KH domain**s.

The KH domain interacts with ssRNA and ssDNA. It has an αβ-structure containing a three-stranded β-sheet and three α-helices (Figure 13-31). The binding groove is an αβ-structure formed by the juxtaposition of two α-helices, one β-strand and two flanking loops. A groove with a hydrophobic floor accommodates a stretch of three C nucleotides, but specificity for flanking residues is provided by hydrogen-bonding groups in the KH domain.

KH domain: one motif, two folds. N. V. Grishin (2001) *Nucleic Acids Res.* **29**, 638–643.

X-ray crystallographic and NMR studies of the third KH domain of hnRNP K in complex with single-stranded nucleic acids. P. H. Backe *et al.* (2005) *Structure* **13**, 1055–1067.

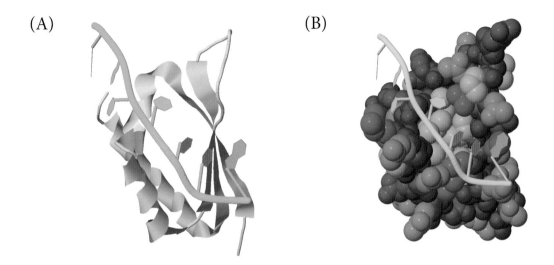

Figure 13-31. The structure of the complex between a KH domain from a human poly(C)-binding protein and a 7-nucleotide segment of ssDNA, with the sequence 5′-AACCCTA-3′. (A) The course of the polypeptide chain of the KH domain is shown schematically. (B) The protein is depicted in space-filling form. Generated from PDB file *2p2r* using the program Jmol.

13.4.D. MS2 Bacteriophage Coat Protein

The coat protein of MS2 bacteriophage is one of the best-characterized RNA-binding proteins. The bacteriophage contains a genomic ssRNA of 3569 nucleotides that is packaged into an icosahedral protein shell made of 180 copies of the coat protein. A dimer of the coat protein binds to a hairpin on the RNA that is near the ribosome-binding site of the virally encoded replicase gene; this inhibits translation of the replicase mRNA and triggers assembly of the coat protein and packaging of the genomic RNA. The RNA hairpin binding site consists of a tetraloop and an adenosine bulge. It binds to the surface of a continuous β-sheet across the dyad axis of the coat protein dimer (Figure 13-32). Consequently, only a single RNA molecule binds asymmetrically to the symmetric protein dimer:

Figure 13-32. Two copies of the RNA hairpin of MS2 bacteriophage binding to 1.5 dimers of the coat protein. Only one subunit of the lower dimer is depicted. The protein is illustrated as *ribbons*, the RNA as a skeletal model. Note that each dimer is symmetrical but can bind only one RNA molecule because it lies on the two-fold axis. Generated from PDB file *2izm* using the program Jmol.

the bulged adenosine and an adenosine in the tetraloop bind to equivalent sets of residues from each subunit. The amino acid residues that are energetically most important for binding the RNA are clustered in the middle of the RNA binding interface and are surrounded by residues that are less important.

RNA–protein interactions in spherical viruses. H. H. Bink & C.W. Pleij (2002) *Arch. Virol.* **147**, 2261–2279.

Investigating the structural basis of purine specificity in the structures of MS2 coat protein RNA translational operator hairpins. C. Halgstrand *et al.* (2002) *Nucleic Acids Res.* **30**, 2678–2685.

Alanine scanning of MS2 coat protein reveals protein–phosphate contacts involved in thermodynamic hot spots. D. Hobson & O. C. Uhlenbeck (2006) *J. Mol. Biol.* **356**, 613–624.

13.4.E. Recognizing Transfer RNAs

Many aspects of the nature of protein–RNA recognition can be illustrated with **transfer RNAs** (tRNA). They are the adaptor molecules that specify the amino acid residues that correspond to each of the triplet codons (Figure 7-21). Their recognition involves two extremes of specificity. Each cell contains 20 aminoacyl-tRNA synthetases, one for each amino acid, that are responsible for attaching the correct amino acid to each of the tRNAs. Each of these enzymes must specifically recognize and aminoacylate *only* their cognate tRNA, using the correct amino acid. In contrast, all the tRNAs interact with the certain proteins. For example, the prokaryotic elongation factor Tu (EF-Tu) introduces aminoacylated tRNAs (aa-tRNA) into the A-site of the ribosome and binds to nearly all the aa-tRNAs, with the exception of the initiator Met tRNA and selenocysteinyl-tRNA. Consequently, it must recognize features that are common to all the elongator aa-tRNAs. Intermediate and varying degrees of specificity are involved in the processing of the tRNA precursors. The various tRNAs are synthesized initially as precursor RNA molecules, which must be processed correctly by specific nucleases and many different enzymes that modify the bases (Figure 4-16). The modification enzymes have varying specificities; some recognize almost all tRNAs, others recognize a subset, while others are specific for a unique tRNA.

Bacteria such as *E. coli* have at least 46 different tRNA molecules, with anticodons that correspond to the various amino acids (Figure 7-21). There are, however, only 20 aminoacyl-tRNA synthetases, one for each of the normal amino acids, so some of the synthetases must recognize more than one tRNA. For example, the seryl-tRNA synthetase recognizes the six tRNAs that accept serine and must ignore all the others, even though all the tRNAs have very similar structures (Figure 4-17). The aminoacyl-tRNA synthetases can be grouped into two classes, I and II, based upon some conserved sequence motifs and the structural architecture and action of their catalytic domains. Yet they are remarkably diverse, with polypeptide chains that vary from 303 to 951 amino acid residues and α, α_2, α_4 and $\alpha_2\beta_2$ quaternary structures. Given that tRNAs have similar secondary and tertiary structures, the molecular basis for the specific recognition between aminoacyl-tRNA synthetases and tRNA resides in the so-called **tRNA identity elements**.

The class I glutaminyl-system and the class II aspartyl- and seryl-systems are best understood. They interact with their tRNA in very different ways but are similar in having a large interface that binds to the tRNA. The large interface probably increases the binding affinity and ensures the correct positioning and orientation of the tRNA. The interactions include nonspecific backbone contacts, often involving basic residues. Those interactions that are specific for the bases and responsible for discriminating between the various tRNAs are largely restricted to the anticodon and the acceptor stem. The second general feature is that protein–RNA contacts involve mutual induced fit via conformational changes in either or both of the protein and the tRNA. In the process, flexible protein loops can become ordered and protein domains can be reoriented and stabilized. In the case of the tRNA, base pairs in the acceptor stem can be broken, the bases in the anticodon loop can be unstacked, and the 3'-end can be distorted.

On the evolution of structure in aminoacyl-tRNA synthetases. P. O'Donoghue & Z. Luthey-Schulten (2003) *Microbiol. Mol. Biol. Rev.* **67**, 550–573.

Two conformations of a crystalline human tRNA synthetase–tRNA complex: implications for protein synthesis. X. L. Yang *et al.* (2006) *EMBO J.* **25**, 2919–2929.

The crystal structure of the ternary complex of phenylalanyl-tRNA synthetase with tRNA^{Phe} and a phenylalanyl-adenylate analogue reveals a conformational switch of the CCA end. N. Moor *et al.* (2006) *Biochemistry* **45**, 10572–10583.

---

1. Class I Glutaminyl-tRNA Synthetase

Glutaminyl-tRNA synthetase (GlnRS) is a monomeric protein and binds a single tRNA^{Gln}. The specificity of GlnRS for tRNA^{Gln} is largely determined by interactions with identity elements in the tRNA acceptor stem and anticodon stem loop (Figure 13-33). Both have very different conformations in the free tRNA. In the complex, the anticodon stem of the tRNA is extended from five to seven base pairs; the two additional base pairs are not the usual Watson–Crick pairs. The interface between the distal two β-barrel domains of the synthetase forms three separate recognition pockets into which the three anticodon bases (CUG) fit by being splayed out. In the active site of the synthetase, the tRNA is oriented in the active site of the synthetase so that base pairs 2 and 3 of the anticodon can be recognized. The 3'-end of the tRNA that accepts the amino acid reaches the catalytic center only by forming an unusual hairpin turn, in which the first base pair is broken but compensated at least to some extent by formation of a new hydrogen bond.

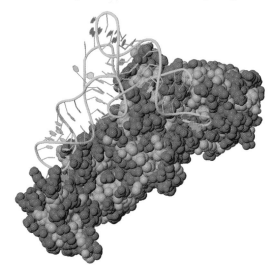

Figure 13-33. The monomeric glutaminyl tRNA synthetase from *E. coli* binding one molecule of tRNA^{Gln}. The tRNA is shown in skeletal form, whereas the protein is illustrated with a space-filling model. Polar amino acid residues are *blue*, nonpolar residues are *gray*. Generated from PDB file *1qrs* using the program Jmol.

Many bacteria lack a glutaminyl-tRNA synthetase, and tRNA^{Gln} is initially mischarged with Glu by a nondiscriminating glutamyl-tRNA synthetase. This enzyme thus charges both tRNA^{Glu} and tRNA^{Gln} with glutamate. The Glu charged to tRNA^{Gln} is subsequently converted to Gln. All eukaryotes and some bacteria contain a discriminating GluRS that only generates Glu–tRNA^{Glu}. The difference between the two types has been tracked down to one amino acid residue: an Arg residue required for discrimination is changed to a Gly that permits both tRNAs to serve as substrate.

A number of archaeal organisms generate Cys–tRNA(Cys) in a two-step pathway, first charging phosphoserine onto tRNA^{Cys} and subsequently converting it to Cys-tRNA^{Cys}.

Crystal structure of a non-discriminating glutamyl-tRNA synthetase. J. O. Schulze *et al.* (2006) *J. Mol. Biol.* **361**, 888–897.

2. Class IIb Aspartyl-tRNA Synthetase

Aspartyl-tRNA synthetase (AspRS) is a dimer that can bind two tRNAs simultaneously and symmetrically, one to each subunit (Figure 13-34). In contrast to the class I case, class II synthetases interact specifically with the major groove side of the acceptor stem of their cognate tRNA, so that the 3′-end of the tRNA can enter the synthetase active site without significantly distorting its structure. Recognition of the anticodon by AspRS recognizes the anticodon using an OB-fold (Section 13.3.H.3) at the N-terminus. The normally compact anticodon loop undergoes a large conformational change upon binding, exposing five bases of the anticodon loop. The three anticodon bases, GUC, lie on the surface of the β-sheet and are recognized by specific hydrogen bonds. The central U stacks with a conserved Phe residue and hydrogen bonds to Gln and Arg residues that are also conserved. The tRNA synthetase recognizes primarily the shape of the tRNA and interacts with the tRNA backbone. Many of the normal bases of the tRNA can be replaced by alternatives, so long as the new nucleotides preserve the original backbone structure of the tRNA.

Human aspartyl-tRNA synthetase contains an extension at its N-terminus that is involved in the transfer of Asp–tRNA to elongation factor 1α. This extension may reduce the rate of dissociation of

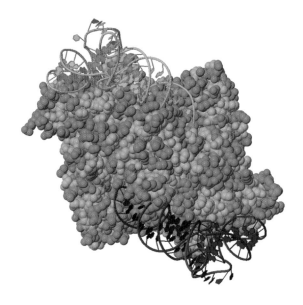

Figure 13-34. Two molecules of tRNA^{Asp} bound to the dimeric aspartyl tRNA synthetase from yeast. The view is along the two-fold symmetry axis. The tRNAs are illustrated as skeletal models, whereas the protein is a space-filling model. Polar amino acid residues are *blue*, nonpolar residues are *mauve*. Generated from PDB file *1asy* using the program Jmol.

Asp–tRNA from human aspartyl-tRNA synthetase and provide enough time for elongation factor 1α to interact with the Asp–tRNA.

Expanding tRNA recognition of a tRNA synthetase by a single amino acid change. L. Feng *et al.* (2003) *Proc. Natl. Acad. Sci. USA* **100**, 5676–5681.

Recognition of acceptor-stem structure of tRNA^{Asp} by *Escherichia coli* aspartyl-tRNA synthetase. H. Choi *et al.* (2003) *RNA* **9**, 386–393.

Structure of the N-terminal extension of human aspartyl-tRNA synthetase: implications for its biological function. H. K. Cheong *et al.* (2003) *Int. J. Biochem. Cell Biol.* **35**, 1548–1557.

---

3. Class II Seryl-tRNA Synthetase

Another dimeric class II synthetase is Seryl-tRNA synthetase (SerRS), but it recognizes tRNA significantly differently from AspRS (Figure 13-35), perhaps because it must recognize six different tRNAs for serine. The six serine anticodons are part of two distinct groups (Figure 7-21), so the serine tRNAs share no common anticodon base and the anticodon is not an identity element for tRNA^{Ser}. One identity element is the long variable arm of tRNA^{Ser}, which is shared only by two other tRNAs. Only one molecule of tRNA^{Ser} can bind to the dimer, because it binds nonsymmetrically across both subunits. SerRS is unique in having a 100-residue N-terminal domain that is a flexible antiparallel coiled coil, 60 Å long and exposed to the solvent, known as the *helical arm*. The helical arm changes its orientation and interacts with the TΨC loop and the long variable arm of the tRNA. The tRNA variable arm has a minimum length in this case because the backbone of six base pairs contacts the protein. The synthetase contacts primarily the tRNA backbone, not the bases, and recognizes primarily the unique shape of tRNA^{Ser}. The shape is largely determined by two bases inserted into the D-loop that make novel interactions in the core of the tRNA. One of these bases is stacked against the first base pair of the long variable arm and determines its spatial orientation.

The synthetase does not contact the anticodon stem loop, so it can recognize any of the six anticodons for serine.

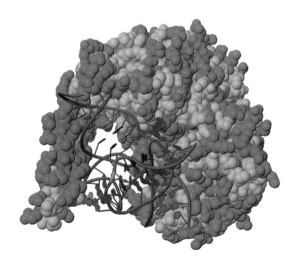

Figure 13-35. Structure of the ternary complex of seryl-tRNA synthetase, tRNA^{Ser} (GGA), and a nonhydrolyzable analog of seryl-adenylate. A space-filling model of the protein is represented, with polar groups *mauve*. The tRNA is depicted as a skeletal model; if it is depicted as a space-filling model, the complex is seen to be closely packed. The tRNA synthetase is a dimer of identical subunits and there are two active sites, each with seryl-adenylate bound, but only a single tRNA molecule binds because it lies on the two-fold axis. Generated from PDB file *1ser* using the program Jmol.

4. Elongation factor EF-Tu

EF-Tu is involved in polypeptide biosynthesis, delivering charged tRNAs to the ribosome. It is a member of the important class of G proteins, which bind and hydrolyze GTP to drive reactions that would otherwise be unfavorable energetically. In the activated GTP-bound state, EF-Tu binds all aminoacylated tRNAs and more tightly than their uncharged forms.

EF-Tu needs to recognize and bind to all tRNAs, so activated EF-Tu makes contacts only to relatively limited and conserved regions of aminoacylated tRNAs: to the aminoacylated CCA end, the 5′-phosphate and the T-stem helix (Figure 13-36). The CCA 3′-end binds in a cleft between domains 1 and 2 of EF-Tu, and conserved residues from domain 2 interact with the base and phosphate of nucleotide A76. The backbone of EF-Tu makes hydrogen bonds to the ester group linking the carboxyl group of the amino acid to the 3′-OH of the ribose and to the free amino group of the amino acid, thereby being specific for charged tRNAs. The amino acid itself projects into a pocket formed between domains 1 and 2 that must be not very specific. The complex is extremely elongated.

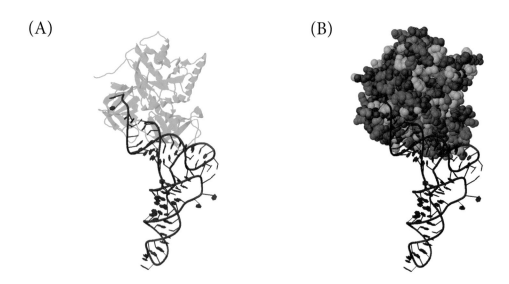

Figure 13-36. Structure of the complex of EF-Tu and a tRNA. The backbone and bases of the tRNA are depicted in skeletal form. (A) The polypeptide backbone of the EF-Tu structure is depicted with *coils* for α-helices and *arrows* for β-strands. (B) The protein is depicted with van der Waals *spheres* for each atom. Generated from PDB file *1b23* using the program Jmol.

Crystal structure of the bovine mitochondrial elongation factor Tu.Ts complex. M. G. Jeppesen *et al.* (2005) *J. Biol. Chem.* **280**, 5071–5081.

A protein extension to shorten RNA: elongated elongation factor-Tu recognizes the D-arm of T-armless tRNAs in nematode mitochondria. M. Sakurai *et al.* (2006) *Biochem. J.* **399**, 249–256.

Directed mutagenesis identifies amino acid residues involved in elongation factor Tu binding to yeast Phe-tRNAPhe. L. E. Sanderson & O. C. Uhlenbeck (2007) *J. Mol. Biol.* **368**, 119–130.

13.4.F. The Ribosome

Ribosomes are the ultimate RNPs, having molecular weights of roughly 2.5 million in prokaryotes and 4 million in eukaryotes. They are so large that they and their component RNA molecules are usually designated by their sedimentation coefficients. A complete 70S prokaryotic ribosome has an irregular structure with a maximum dimension of about 250 Å, and it dissociates into two parts, the 50S and 30S subunits. In *E. coli*, the 50S subunit consists of two RNA molecules, one 23S (2904 nucleotides) and the other 5S (120 nucleotides) plus 33 protein molecules. The 30S subunit consists of one 16S RNA molecule (1542 nucleotides) and 21 different protein molecules. The RNAs make up 66% of the mass. The larger eukaryotic 80S ribosome dissociates into 40S and 60S subunits. The 40S subunit consists of an 18S, 1874-nucleotide RNA and 33 polypeptide chains, while the 60S subunit consists of 28S (4718 nucleotides), 5.8S (160 nucleotides) and 5S (120 nucleotides) RNA molecules, plus 49 polypeptide chains. RNA molecules make up 60% of the total mass.

Ribosomes are the cellular factories where polypeptide chains are assembled using the genetic information in mRNA, so they have very complex functions and interact with a number of different molecules. They also have very complex structures that are known in great detail by X-ray crystallography (Figure 13-37). Much information about the overall architecture had been gained earlier by electron microscopy.

The structures of the individual subunits of the ribosome are dictated primarily by the structures of their large RNA molecules, which have extensive secondary structures (Figure 13-37). The proteins are generally in peripheral positions, but with unstructured loops enmeshed with the RNA. Nearly all of the known types of RNA tertiary structures have been found in ribosomal RNA. Most of the base pairs involved are standard Watson–Crick type (Figure 2-13) and proteins tend to interact most with RNA stems with classical double-helical structures (Figure 4-5). Some of the proteins are homologous to homeodomains, zinc fingers, leucine zippers and other motifs found in proteins that bind nucleic acids. Based on their topology, the protein structures fall into six groups. Many share similar α + β sandwich folds, but the topology of this domain varies considerably, as do the ways in which the proteins interact with RNA.

An extraordinary variety of protein–RNA interactions is observed. The proteins function primarily to stabilize interactions between the RNA domains that maintain the complex's structural integrity. Many electrostatic interactions occur between the phosphate groups of the RNA backbone and numerous Arg and Lys residues, particularly those in flexible extensions at the ends of the polypeptide chain. Bases of the RNA are recognized in both the minor groove and the widened major groove of RNA helices, as well as through hydrophobic binding pockets that capture bulged nucleotides and through insertion of amino acid residues into hydrophobic crevices in the RNA. Many of the proteins have, in addition to globular domains, long extended regions, either in the termini or internal loops, which make extensive contact to the RNA component and are involved in stabilizing the RNA tertiary structure.

Remarkably, the ribosomal RNA is primarily responsible for the major catalytic activities of the ribosome, including the step that forms the peptide bond between the new amino acid and the nascent polypeptide chain, plus the hydrolysis of GTP to fuel the steps that involve moving the nascent chain from one site to another (Section 14.6). The proteins appear to be involved in more peripheral roles and are important for stabilizing the overall structure and interacting with other molecules, such as the tRNAs (Section 13.4.E).

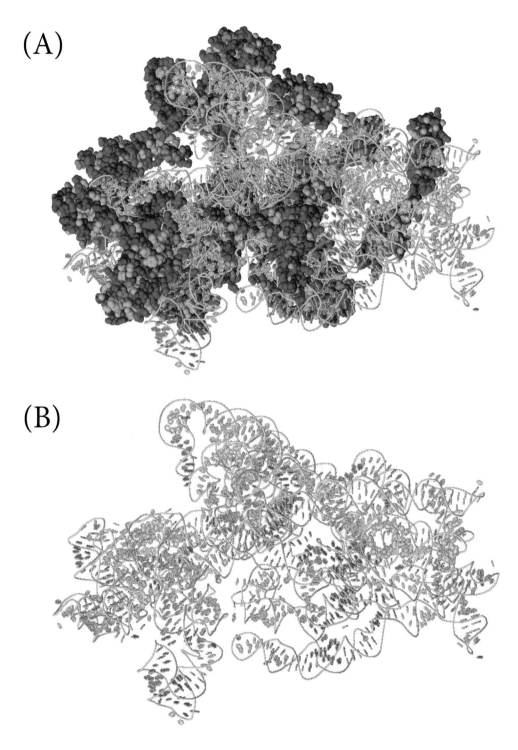

Figure 13-37. Structure of the small (30S) subunit of the ribosome from *Thermus thermophilus*. (A) The entire structure of the ribosomal subunit, with both RNA and proteins. The RNA is shown as a skeletal model, whereas the proteins are shown as *spheres* of the van der Waals radii of the atoms; acidic groups are *red*, basic groups are *blue*, and atoms of the backbone are *mauve*. (B) The structure of the RNA in the ribosome, with the proteins deleted. The secondary structure is described in Figure 4-6-B. Generated from PDB file *2avy* using the program Jmol.

RNA structure: reading the ribosome. H. F. Noller (2005) *Science* **309**, 1508–1514.

Structure of the 70S ribosome complexed with mRNA and tRNA. M. Selmer *et al.* (2006) *Science* **313**, 1935–1942.

Ribosome dynamics: insights from atomic structure modeling into cryo-electron microscopy maps. K. Mitra & J. Frank (2006) *Ann. Rev. Biophys. Biomol. Structure* **35**, 299–317.

~ CHAPTER 14 ~

CATALYSIS

Life is possible only because chemical reactions take place in living organisms at rates that are many orders of magnitude greater than would occur in a solely chemical system. Virtually all biochemical reactions are catalyzed, usually by an **enzyme** but in some cases by RNA catalysts. Enzymes bind ligands in the normal way (Chapter 12) but then produce covalent changes in them; the initial ligands are the enzyme's *substrates* and their final forms are the enzyme's *products*, which together define the chemical reaction catalyzed by the enzyme.

Like all catalysts, enzymes speed up the rates of chemical reactions, but their special importance lies in the fact that they do this at neutral pH and relatively low temperatures, and do so specifically, for only one chemical reaction (Table 14-1). Enzymes may show either absolute or broad specificity for their substrates. For example, catalase acts only on hydrogen peroxide, whereas alcohol dehydrogenase oxidizes a number of aliphatic and aromatic alcohols. The rate increase is due to the lowering of the activation energy barrier by binding and stabilizing the transition state, and can be as much as 10^{17}-fold (Table 14-1). Usually the uncatalyzed reaction is barely detectable, if at all. Enzymes *in vitro* can function at concentrations very much lower than those of the substrates on which they act, although in the cell the enzyme level often exceeds the substrate concentration. Enzymes increase the rate of a reaction in both the forward and reverse directions and, therefore, have no effect on the equilibrium constant (K_{eq}) of the reaction. As catalysts, enzymes remain unchanged at the end of a reaction.

Enzymes have the same general structures as other proteins. The folding of the polypeptide chain gives rise to the final enzyme structure, which contains an **active site** at which the substrates bind and undergo reaction. It is this folding of the polypeptide chain to form the active site that is responsible for the catalytic power and specificity of enzymes. The active form of an enzyme may consist of single or multiple subunits, with molecular weights ranging from about 14,000 to 4,000,000.

Some enzymes bring about energy transductions that include the conversion of light energy to chemical bond energy in photosynthesis, as well as the conversion of chemical bond energy to mechanical energy in muscle contraction and pumping energy for the membrane transport of ions against an unfavorable gradient.

Enzyme catalysis has been studied extensively for over a century, so a thorough description would require much more space than can be provided here. In the classical early studies, much was learned about the kinetics of action of enzymes, but little about the enzymes themselves; emphasis was placed

Table 14-1. Rate enhancements by some enzymes

Enzyme	Uncatalyzed reaction		Catalyzed rate (k_{cat}, s^{-1})	Rate enhancement (k_{cat}/k_{un})
	Half-life	Rate constant (k_{un}, s^{-1})		
OMP decarboxylase	7.8×10^7 years	2.8×10^{-16}	39	1.4×10^{17}
Staphylococcal nuclease	1.3×10^5 years	1.7×10^{-13}	95	5.6×10^{14}
AMP nucleosidase	6.9×10^4 years	1.0×10^{-11}	60	6.0×10^{12}
Carboxypeptidase A	7.3 years	3.0×10^{-9}	578	1.9×10^{11}
Ketosteroid isomerase	7 weeks	1.7×10^{-7}	6.6×10^4	3.9×10^{11}
Triose phosphate isomerase	1.9 days	4.3×10^{-6}	4.3×10^3	1.0×10^9
Chorismate mutase	7.4 h	2.6×10^{-5}	50	1.9×10^6
Carbonic anhydrase	5 s	0.13	1×10^6	7.7×10^6

Data from A. Radzicka & R. Wolfenden (1995) *Science* **267**, 90–93.

on the substrates and products, and the enzymes were present at such small concentrations that they were apparent only by their catalytic activity. All this has changed in the past four decades, when the structures and properties of enzymes and their complexes with ligands (substrates and inhibitors) have been elucidated.

The classical kinetics of enzyme action will be reviewed briefly, as this is still a very important subject. It can give considerable information about the physical properties of an enzyme and about its mode of catalysis, and requires only a quantitative and accurate method for following the course of the reaction catalyzed (which can usually be done spectrophotometrically); the enzyme need not even be pure. Emphasis will then be placed on the structural properties of enzymes and the physical principles by which they catalyze reactions at apparently extraordinary rates. It is now believed that the kinetic properties of enzymes can be at least rationalized on the basis of established physical and chemical principles.

Enzymatic mechanisms of phosphate and sulfate transfer. W. W. Cleland & A. C. Hengge (2006) *Chem. Rev.* **106**, 3252–3278.

Enzymatic catalysis and transfers in solution. I. Theory and computations, a unified view. R. A. Marcus (2006) *J. Chem. Phys.* **125**, 194504.

Evolution of enzyme superfamilies. M. E. Glasner *et al.* (2006) *Curr. Opinion Chem. Biol.* **10**, 492–497.

Enzyme promiscuity: evolutionary and mechanistic aspects. O. Khersonsky *et al.* (2006) *Curr. Opinion Chem. Biol.* **10**, 498–508.

Directed evolution: an approach to engineer enzymes. J. Kaur & R. Sharma (2006) *Crit. Rev. Biotechnol.* **26**, 165–199.

Novel enzymes through design and evolution. K. J. Woycechowsky *et al.* (2007) *Adv. Enzymol.* **75**, 241–294.

14.1. CHEMICAL CATALYSIS

Catalysts increase the rate of a reaction by lowering the free energy barrier that defines the rate of the reaction, without being consumed or changed in the reaction. They do this by lowering the free energy of the transition state, i.e. stabilizing it. Virtually every chemical reaction can be catalyzed in solution, for example by nucleophiles or electrophiles in general acid or base catalysis (Figure 14-1), and the rate of the reaction depends upon the concentrations of the catalysts. Interactions with the solvent are extremely important for determining the rate of a chemical reaction. Charge separations almost invariably occur in any transition state, because electrons are being redistributed

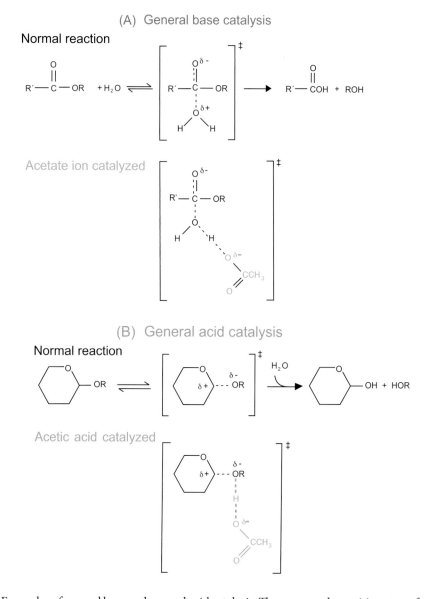

Figure 14-1. Examples of general base and general acid catalysis. The presumed transition states for the catalyzed and uncatalyzed reactions are enclosed in brackets. (A) Acetate ion catalyzes hydrolysis of an ester, presumably by interacting with the transition state to make the charge separation more favorable and lower its free energy. (B) Acetic acid catalyzes hydrolysis of an acetal, presumably by stabilizing the transition state as indicated. Adapted from A. Fersht (1977) *Enzyme Structure and Mechanism*, W. H. Freeman, Reading.

in the process of covalent bond breakage and reformation. Polar solvents can greatly stabilize charges, so reactions involving increased charge separation in the transition state occur much more rapidly in such solvents. Conversely, reactions with less charge separation occur more rapidly in nonpolar solvents. For example, the reaction:

$$\text{(reaction scheme)} \tag{14.1}$$

occurs 10^4–10^5 times faster in ethanol than in water. The intermediate in brackets is believed to approximate the transition state and has the net charge of the original molecule greatly diminished.

Chemical catalysis can be distinguished from that by enzymes because the rate of the chemical reaction is always proportional to the concentrations of the reactants, whereas the rate of the enzyme-catalyzed reaction becomes constant at high reactant concentrations.

14.2. ENZYME KINETICS: MICHAELIS–MENTEN

The first step in catalysis by enzymes is a collision between the enzyme (E) and its substrate (A), to form an enzyme–substrate complex (E•A):

$$E + A \rightleftharpoons E\bullet A \rightleftharpoons E\bullet A^* \rightleftharpoons E\bullet P^* \rightleftharpoons E'\bullet P \rightleftharpoons E' + P$$

[Collision   Clamping]   Conversion   [Unclamping   Dissociation]

Binding   Catalysis   Release

(with transition state ‡ above E•A* ⇌ E•P*) (14.2)

The enzyme at the end of the reaction is depicted as E′, being slightly different from the starting enzyme; this difference is usually believed to be very subtle, a result simply of having bound product rather than substrate. The difference is not significant if E′ quickly reverts to the original form, E, as normally happens. Equation 14.2 involves two reactants, E and A, so it is a second-order reaction. The rate of collisions between E and A will be given by k_d [E] [A], where the brackets indicate the concentration of each species; k_d is the second-order rate constant for diffusion of the two reactants, corrected for any long-range attractions between the two. If the mode of binding is not appropriate for catalysis, the complex will be nonproductive. The rate of binding will be high and the binding will be tight if the structure of a substrate is complementary to that of the active site of the enzyme. On the other hand, a substrate that is not exactly complementary to the binding site will need to use some of its binding energy for conformational changes in either the substrate or the enzyme. This is postulated as the clamping reaction on the enzyme in Equation 14.2. The equilibrium constant for overall binding will be the product of that for the initial interaction and the subsequent conformational change. The enzyme–substrate complex that is present can be E•A, E•A* or both, depending upon their relative free energies and, perhaps, the rates of their interconversion.

The enzyme-bound substrate must undergo activation before the reaction can proceed. Its energy must be raised to that of the **transition state**, which is the most transient species, with the greatest free energy, that occurs during the conversion reaction. The activated molecules can either fall back to the ground state or be converted to E•P* at a rate that, in the simplest case, is postulated to be independent of the structure of the reactants. The activation energy required for the conversion of ground-state molecules to the transition state is very much lower for an enzyme-catalyzed reaction than for either a noncatalyzed reaction or a chemically catalyzed reaction. The reason is that **the activated substrate is stabilized by being bound more tightly to the enzyme than it is in the ground state** (Figure 14-2). Without such tighter binding, there would be no catalysis, and the increase in rate is proportional to the increased binding affinity. After the E•P* complex has been formed, the enzyme undergoes another conformational change, depicted as 'unclamping' in Equation 14.2, before releasing the product and regenerating the free enzyme.

Enzyme kinetics are usually measured by the initial velocity of the reaction before any significant amount of substrate has disappeared (which will decrease the rate) and before any significant amount of product has been generated (products are always inhibitors of the forward reaction, by binding to the free enzyme and contributing a reverse reaction). The entire time–course of an enzyme-catalyzed reaction can be followed until it reaches equilibrium and the entire curve then fit to a complete kinetic scheme that includes both forward and reverse reactions; this requires, however, that the enzyme be totally stable throughout this time and that no other reactions are taking place.

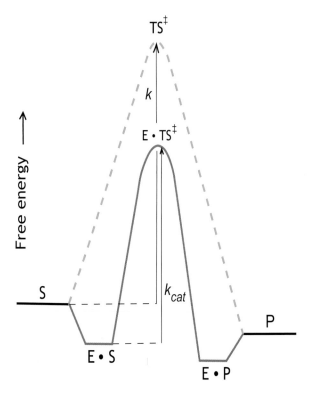

Figure 14-2. Simplified free energy profile of an uncatalyzed reaction (*dashed line*) and one catalyzed by an enzyme (*solid line*). The relative free energies of the substrate (S), transition state (TS‡) and product (P) are shown when free and when bound to the enzyme (E). The reaction is catalyzed by the enzyme when the energy of the transition state is lowered more upon binding to the enzyme than substrate. The relative free energies of S, TS‡ and P when free and bound depend upon their respective affinities for the enzyme and the concentration of the enzyme. The height of the free energy barrier to the catalyzed reaction is given by the free energy of E•TS‡ relative to E•S; this determines the value of k_{cat}/K_m (Section 2.1.D). The free energy of E•S relative to S is determined by the concentrations of E and S and their affinity. The reaction is catalyzed when the enzyme has greater affinity for the transition state than the substrate.

Classical enzyme kinetics are usually observed under steady-state conditions, where all binding steps are rapid and reversible on the time scale of the observations, the concentration of substrate is substantially greater than that of the enzyme, and the enzyme is turning over many times while the concentration of substrate is depleted upon being converted to product. In this case, and in the absence of significant concentrations of the product P, the complete Equation 14.2 can usually be simplified to the **Michaelis–Menten** mechanism:

$$E + A \underset{k_d}{\overset{k_b}{\rightleftharpoons}} E \bullet A \xrightarrow{k_{cat}} E + P \qquad (14.3)$$

The general steady-state rate equation for this reaction is:

$$v = \frac{V_{max}[A]}{K_a + [A]} \qquad (14.4)$$

where v is the observed velocity and V_{max} the maximum velocity, which is given by $[E]k_{cat}$. K_a is known as the **Michaelis constant (K_m)** for substrate A. Reactions that follow this mechanism are described as obeying Michaelis–Menten kinetics. Equations 14.3 and 14.4 will also apply to reactions with multiple substrates so long as the concentration of only a single substrate is varied and those of the other reactants are fixed.

Steady-state conditions require that the concentration of substrate be at least an order of magnitude greater than that of the enzyme. Forming the enzyme–substrate complex will then not reduce the concentration of free substrate significantly. Also, the concentration of the enzyme–substrate complex will be constant while the initial velocity of the reaction is being measured.

Effective experimental design: enzyme kinetics in the bioinformatics era. E. F. Murphy *et al.* (2002) *Drug Discov. Today* **7**, S187–191.

Optimal designs for Michaelis–Menten kinetic studies. J. N. Matthews & G. C. Allcock (2004) *Stat. Med.* **23**, 477–491.

Quasi-steady-state kinetics at enzyme and substrate concentrations in excess of the Michaelis–Menten constant. A. Rami & E. R. Edelman (2007) *J. Theor. Biol.* **245**, 737–748.

14.2.A. The Michaelis–Menten Equation

Plotting the observed velocity as a function of substrate concentration according to the Michaelis–Menten equation (Equation 14.4) produces a rectangular hyperbola (Figure 14-3). The two fundamental parameters are k_{cat} and (k_{cat}/K_a). They are determined by very high and very low concentrations of the substrate. At very high concentrations of A, the velocity is at a maximum, $v = V_{max}$. The reaction is zero-order with respect to A because the enzyme is saturated with substrate and operating at its intrinsic rate. The measured value of V_{max} is directly proportional to the concentration of enzyme $[E]_t$ and the first-order rate constant k_{cat} is equal to $V_{max}/[E]$. This is often referred to as the **turnover number** for the enzyme, as it is a measure of the number of moles of product produced per mole of enzyme (or enzyme active site) per second. The turnover numbers of enzymes vary considerably

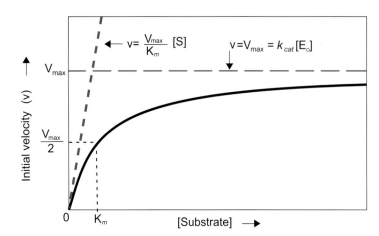

Figure 14-3. Variation of the initial velocity (v) as a function of the concentration of substrate ([S]) for an idealized enzyme exhibiting Michaelis–Menten kinetics.

(Table 14-2). When A is very small, the initial velocity of the reaction is first-order with respect to A. As [A] approaches zero, $v = (V_{max}/K_a)[A]$. The apparent second-order rate constant for the interaction of substrate with enzyme is k_{cat}/K_a, with units of $M^{-1}s^{-1}$. It can be expressed in terms of the microscopic rate constants of Equation 14.3:

$$\frac{k_{cat}}{K_a} = \frac{k_b \, k_{cat}}{k_d + k_{cat}} = \frac{k_b}{1 + \dfrac{k_d}{k_{cat}}} \qquad (14.5)$$

If $k_{cat} \gg k_d$, the rate of the reaction is limited by k_b, the rate of binding of the substrate to the enzyme. The rates of formation of enzyme–substrate complexes vary considerably but they can approach the diffusion rate, which is approximately 10^8–10^9 M^{-1} s^{-1}.

Table 14-2. Examples of kinetic parameters for some enzymes

Enzyme	Substrate	K_a (μM)	k_{cat} (s^{-1})	k_{cat}/K_a (μM^{-1} s^{-1})
Chorismate mutase	Chorismate	100	47	0.5
Chymotrypsin	Acetyl-L-tryptophanamide	5000	100	0.02
Dihydrofolate reductase	Dihydrofolate	1	20	20
Fumarase	Fumarate	5	800	160
β-Lactamase	Benzylpenicillin	50	2000	40
Lysozyme, hen	Hexa-N-acetylglucosamine	6	0.5	0.08
Prephenate dehydrogenase	Prephenate	65	95	1.5

Data from J.F. Morrison (1999) in *Encyclopedia of Molecular Biology* (T. E. Creighton, ed.), Wiley-Interscience, NY, p. 1495.

Single molecule Michaelis–Menten equation beyond quasistatic disorder. X. Xue et al. (2006) *Phys. Rev. E* **74**, 030902.

When does the Michaelis–Menten equation hold for fluctuating enzymes? W. Min et al. (2006) *J. Phys. Chem. B* **110**, 20093–20097.

14.2.B. K_m (Michaelis Constant)

The K_m is the concentration of substrate required to give half-maximum velocity. In the case of substrate A, it is designated as K_a. Its value is independent of enzyme concentration and can be measured using an impure enzyme preparation, so long as the preparation does not contain interfering enzyme activities, inhibitors or activators.

The Michaelis constant is the ratio of the two fundamental parameters V_{max} and (V_{max}/K_a), or k_{cat} and (k_{cat}/K_a), so it is not a fundamental parameter. In general, its value is given by:

$$K_a = \frac{(k_{cat} + k_d)}{k_b} \tag{14.6}$$

The K_m is obviously related to the affinity of the enzyme for its substrate expressed as the dissociation constant, which is given by k_d/k_b, but the K_m is the same as the dissociation constant of the enzyme–substrate complex only when the rate of product formation is slow compared with the rate at which E•S dissociates back into E and S ($k_{cat} \ll k_d$). In this case, E•S is in equilibrium with E and S.

An empirical observation is that the substrate of an enzymatic reaction normally is present *in vivo* at a concentration close to its K_m value.

14.2.C. Turnover Number (k_{cat})

The value of k_{cat} is also known as the **turnover number**, which for an enzyme-catalyzed reaction is defined as **the number of moles of substrate converted to product per mole of enzyme per second under conditions where the concentration of all substrates is saturating**. Turnover numbers are usually expressed in units of s^{-1}. The turnover numbers of enzymes vary drastically (Table 14-2), from 2–5 s^{-1} for the very sluggish enzyme ribulose bisphosphate carboxylase, to 1,000,000 s^{-1} for carbonic anhydrase.

14.2.D. Lineweaver–Burke Plot

Determining the values for V_{max} and K_a from a hyperbolic curve (Figure 14-3) is always difficult, especially using data with experimental errors. A particular problem is defining the asymptote, V_{max}, because this requires concentrations of the substrate much greater than its K_m; for example, a substrate concentration 10 times greater than the K_m will produce a velocity that is only 91% of V_{max}, and a 100 times greater concentration is required to reach 99% V_{max}. Rearranging the Michaelis–Menten equation (Equation 14.3) turns it into a linear form that permits ready extrapolation to infinite substrate concentrations and easy estimation of V_{max} and K_m. Taking reciprocals of each side of Equation 14.3 and rearranging produces:

$$\frac{1}{v} = \frac{K_a}{V_{max}} \frac{1}{[A]} + \frac{1}{V_{max}} \tag{14.7}$$

According to this equation, plotting 1/v versus 1/[A] should yield a straight line. Its slope should be K_a/V_{max} and its intercept of the vertical ordinate should be $1/V_{max}$ (Figure 14-4-A). The vertical ordinate is where the substrate concentration would be infinite. When $1/v = 0$, where the straight line intersects the abscissa, $1/K_a = -1/[A]$.

(A) Lineweaver-Burke

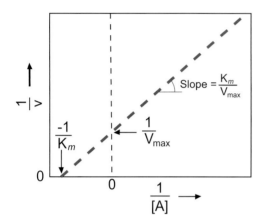

(B) Eadie-Hofstee

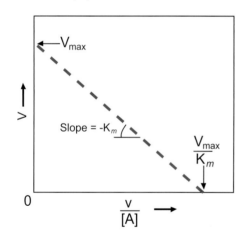

Figure 14-4. The two most popular methods of plotting initial rate data for an enzyme following Michaelis–Menten kinetics. (A) Lineweaver–Burke double-reciprocal plot of the variation of the initial velocity (v) of an enzyme-catalyzed reaction as a function of substrate concentration ([A]). (B) An Eadie–Hofstee plot.

1. Eadie–Hofstee Plot

The Eadie–Hofstee plot uses a different rearrangement of Equation 14.3 to a linear form:

$$v = -K_a \frac{v}{[A]} + V_{max} \qquad (14.8)$$

According to this equation, plotting v versus v/[A] should produce a straight line with a negative slope, a positive intercept on the abscissa, and a positive intercept on the vertical ordinate (Figure 14-4-B). The slope of the line should be equal to $-K_a$, the intercept with the abscissa should be equal to V_{max}/K_a, and the intercept with the ordinate should be equal to V_{max}. **It is equivalent to the Scatchard plot used to analyze binding of ligands to proteins.**

2. Lineweaver–Burke versus Eadie–Hofstee

The Lineweaver–Burke plot has the benefit of showing the straightforward variation of one dependent variable as a function of the concentration of one or two independent variables. It is also ideal for analyzing multi-substrate reactions. It has the disadvantage, however, of being dominated by the measurements at the lowest substrate concentrations, which are usually the least accurate. The Eadie–Hofstee plot, on the other hand, emphasizes the more accurate measurements. But Equation 14.8 becomes very complex with multi-substrate reactions.

There have been many discussions about the relative merits of the Lineweaver–Burke and Eadie–Hofstee plots for obtaining the best estimates of values for kinetic parameters, but computer programs are now available for least-squares fitting of the experimental data, with appropriate weighting factors, to an assumed rate equation. Graphical methods are important, however, for determining the form of the rate equation to which the data are to be fitted and for illustrating the results of kinetic investigations.

The comparison of the estimation of enzyme kinetic parameters by fitting reaction curve to the integrated Michaelis–Menten rate equations of different predictor variables. F. Liao *et al.* (2005) *J. Biochem. Biophys. Methods* **62**, 13–24.

14.2.E. Kinetics of Individual Enzyme Molecules

New insights into enzyme kinetics have been provided by kinetic studies on single enzyme molecules, rather than the ensemble of many molecules studied using the classical techniques. Enzymatic turnovers of individual enzyme molecules can be monitored over substantial time periods by detecting the appearance of one fluorescent product at a time. The time–course of the reaction itself is not observed, only the frequency with which it occurs in a single molecule. Single-molecule enzymatic turnover experiments typically measure the probability density $f(t)$ of the stochastic waiting time t for individual turnovers. While $f(t)$ can be reconciled with ensemble kinetics, it contains more information than the ensemble data; in particular, it can detect dynamic disorder in which the apparent catalytic rate fluctuates, which is usually taken as being due to interconversions of various conformers of the enzyme that have different catalytic rate constants. In this case, $f(t)$ exhibits the expected monoexponential decay at low substrate concentrations but multi-exponential decay at high substrate concentrations. A single-molecule Michaelis–Menten equation for the reciprocal of the first moment of $f(t)$, $1/<t>$, exhibits a hyperbolic dependence on the substrate concentration [S], similar to the ensemble enzymatic velocity. This single-molecule Michaelis–Menten equation holds under many conditions, in particular when the interconversion rates among different enzyme conformers are slower than the catalytic rate. Unlike the conventional interpretation, however, the apparent catalytic rate constant and the apparent Michaelis constant in this single-molecule Michaelis–Menten equation are complicated functions of the catalytic rate constants of individual conformers. The randomness parameter r, defined as $<(t - <t>)^2>/t^2$, can serve as an indicator for dynamic disorder in the catalytic step of the enzymatic reaction, as it becomes greater than unity at high substrate concentrations when there is dynamic disorder. In addition to a fluctuating enzyme conformation, the stochastic nature of substrate concentration fluctuations is another possible source of the complex behavior of single-molecule enzyme kinetics.

The enzyme β-galactosidase exhibits a phenomenon described as 'molecular memory', characterized by clusters of turnover events separated by periods of low activity. Such a 'memory' lasts for time scales ranging from milliseconds to seconds. It is attributed to the presence of interconverting conformers with broadly distributed lifetimes; it is observed only at high substrate concentrations. β-galactosidase is a tetramer, and each molecule has four identical active sites. The clusters of turnovers may then be all four of the active sites of each molecule tending to turn over together; this should be most pronounced at high substrate concentrations, when all the active sites are occupied by substrate.

Kramers' theory of reaction rates is believed to be more applicable to single-molecule kinetics than normal transition state theory, which was derived for homogeneous ensembles of molecules. Direct application of Kramers' flux-over-population method yields analytic expressions for the time-dependent transmission coefficient and the distribution of waiting times for barrier crossing that reproduce the observed trends in simulations and experiments.

Theory of the statistics of kinetic transitions with application to single-molecule enzyme catalysis. I. V. Gopich & A. Szabo (2006) *J. Chem. Phys.* **124**, 154712.

Single molecule studies of enzyme mechanisms. R. D. Smiley & G. G. Hammes (2006) *Chem. Rev.* **106**, 3080–3094.

Michaelis–Menten for single enzyme molecules. A. Doerr (2006) *Nature Methods* **3**, 158.

Michaelis–Menten is dead, long live Michaelis–Menten! N. G. Walter (2006) *Nature Chem. Biol.* **2**, 66–67.

14.3. ENZYME KINETIC MECHANISMS WITH MULTIPLE SUBSTRATES

Enzymes generally use more than one substrate; those that use a single substrate are strictly limited to isomerases. Hydrolases can appear to have a single substrate, but water is a second, although its concentration is usually fixed and it is usually omitted from the equations. **Interacting with two or more substrates increases the number of possible reaction mechanisms**. The kinetic mechanism of an enzyme refers to the order of substrate addition to the enzyme and release of the product. The procedure that is normally used for describing kinetic mechanisms involves the use of the letters A, B and C for substrates and P, Q and R for products, with the Michaelis constants for these reactants being denoted by K_a, K_b, K_p, K_q, etc. E and F are used for stable forms of the enzyme. Inhibition constants for a substrate or product are distinguished from Michaelis constants by having the letter *i* preceding the substrate symbol, for example K_{ia} and K_{ip}. The enzyme is indicated by a horizontal line and the complexes of the enzyme are indicated below the line (Figure 14-5). The reversible addition of substrates and the release of products are denoted by vertical arrows above the line. Rate constants for the forward and reverse steps of each reaction can be written, when necessary, to the left and right of the vertical arrows, respectively.

The expressions for a Michaelis constant in terms of individual rate constants, as in Equation 14.6, increase in complexity with increasing number of substrates and products associated with a reaction. Generally, however, the Michaelis constant for a substrate is considered to be that concentration of the substrate that yields half-maximum velocity when all other substrates are saturating.

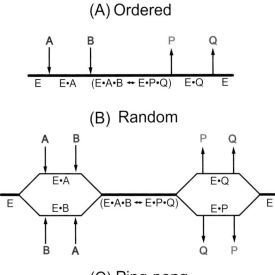

Figure 14-5. Kinetic mechanisms for reactions involving two substrates and two products. (A) and (B) are seqential mechanisms, while (C) is nonsequential.

The types of kinetic mechanism that apply can be illustrated with reactions involving two substrates and two products, known as *Bi–Bi* reactions. A large fraction of all known enzymatic reactions are of this type. Usually, some group or chemical moiety is transferred from one of the substrates to the other. In this case, each product is related more closely to one of the substrates. There are two broad categories of such reactions: sequential and nonsequential (Figure 14-5).

Determining the chemical mechanisms of enzyme-catalyzed reactions by kinetic studies. W. W. Cleland (1977) *Adv. Enzymol.* **45**, 273–387.

Enzyme kinetics and mechanism. A. Initial rate and inhibitor methods. D. L. Purich, ed. (1979) *Methods Enzymol.* **63**.

W. W. Cleland (1986) in *Investigations of Rates and Mechanisms* (C. F. Bernasconi, ed.), John Wiley & Sons, **6**, 791–870.

Mechanistic and kinetic studies of inhibition of enzymes. C. G. Whiteley (2000) *Cell. Biochem. Biophys.* **33**, 217–225.

14.3.A. Sequential Mechanisms

Both substrates must be added to the enzyme before either product is released in this type of mechanism. The addition can be either ordered or random.

1. Ordered Mechanisms

An ordered mechanism is characterized by a compulsory order in which substrates are added or products released (Figure 14-5-A). Such a mechanism can be explained by the free enzyme possessing a binding site for substrate A but not for B; its binding site is created by formation of the E•A complex. The first binary enzyme–substrate complex might undergo a conformational change that enables binding of the second substrate. This is a strictly ordered mechanism. Alternatively, the binding may be synergistic, in which the binding of one simply enhances binding of the second. The substrate that is bound first is simply that with the greater affinity.

Chemical catalysis occurs only after formation of the ternary E•A•B complex, generating an E•P•Q complex. These two complexes are treated as a single entity, because steady-state kinetic studies cannot distinguish between them and because including both does not affect the steady-state rate equations. The product of the first substrate to add is usually the last to dissociate from the enzyme, so the kinetic scheme is symmetrical. This mechanism applies to many dehydrogenases: NAD or NADP is usually the first substrate to add, and NADH or NADPH is the last product to be released.

The mechanism is known as **steady-state ordered** when no assumptions are made about the relative rates of the various reaction steps. When the steady-state concentrations of the ternary complexes are very low, it is known as the **Theorell–Chance** mechanism. It is known as the **equilibrium-ordered** mechanism when E and A bind rapidly and reversibly and their interaction is at thermodynamic equilibrium.

2. Random Mechanisms

Each substrate can add to the enzyme before or after the other in random mechanisms, and the two products can dissociate in any order. The reason is that the enzyme possesses distinct binding sites for each of the two substrate–product pairs (Figure 14-5-B). Binding of one substrate to the enzyme can enhance or hinder the binding of the other, or it can have no effect. One substrate might even be present on the enzyme at the same time as the product of the other substrate. This is a nonproductive dead-end complex that is responsible for inhibition of the enzyme by the presence of the product (Section 14.3.I). It always occurs with the smaller pair of substrate and product, while the larger substrate–product pair may not be possible if there is steric hindrance between them.

A general method for the analysis of random bisubstrate enzyme mechanisms. V. Leskovac *et al.* (2004) *J. Ind. Microbiol. Biotechnol.* **31**, 155–160.

14.3.B. Nonsequential Mechanisms: Ping-Pong

In the single-site ping-pong mechanism (Figure 14-5-C), the moiety that is transferred from one substrate to the other is initially and transiently transferred to the enzyme. The first substrate to bind transfers the donor group to the enzyme (E), to form a new stable form of enzyme (F), and the first product, P, is released. The second substrate, B, reacts at the site that P has vacated, and the group on the enzyme is transferred to B to form the second product, Q. For example, aminotransferases use this mechanism: the amino group of an amino acid is transferred to the cofactor pyridoxal phosphate (Section 14.4.G.1), generating a pyridoxamine phosphate enzyme, and then to an α-keto acid that is the second substrate. Kinases can use the Pong-Pong mechanism, transferring the phosphate group to the enzyme temporarily.

The two substrates of a ping-pong reaction usually have similar structures because they bind to the same site on the enzyme. The F form of the enzyme can usually be prepared in the absence of the second substrate, then isolated and characterized. It should also undergo a stoichiometric reaction with the second substrate, B. A characteristic feature of ping-pong-type enzymes is that **they catalyze partial exchange reactions between the E and F forms of the enzyme**.

Some enzymes catalyze ping-pong mechanisms for which there are two or more sites at which multiple substrates can bind simultaneously. An enzyme-bound carrier such as biotin, lipoic acid or 4-phosphopantetheine usually shuttles the transferred moiety between the separate sites.

14.3.C. Initial Rate Equations

To distinguish between the various types of mechanisms possible with multiple substrates, **the initial velocity of the reaction catalyzed by the enzyme is measured while varying systematically the concentrations of the two substrates and the products**. Analysis of the data is usually carried out with double-reciprocal Lineweaver–Burke plots, noting whether changing each concentration alters the intercept, the slope, or both.

1. Steady-State Ordered and Rapid-Equilibrium Random Mechanisms

Random mechanisms usually behave as though they are of the rapid-equilibrium type, with the interconversions of the ternary E•A•B and E•P•Q complexes being the slowest steps in each direction and all binding steps being rapid and at equilibrium. Alternatively, the reaction might be ordered but the enzyme in a steady-state, in which the concentrations of the various forms of the enzyme remain relatively constant, determined by the rates of the various steps but not at true equilibrium. This makes no difference to the initial velocity equation:

$$v = \frac{V_{max}[A][B]}{K_{ia}K_b + K_a[B] + K_b[A] + [A][B]} \tag{14.9}$$

The four terms in the denominator indicate the distribution of the enzyme among its various forms. They give the relative proportions of total enzyme that are present as free enzyme (E), E•B, E•A and E•A•B, respectively, when the enzyme is at equilibrium. No significant concentrations of E•P and E•Q will be present because the products are released much faster than they are produced by the enzyme (Figure 14-5-B). If the enzyme is in a steady-state rather than at equilibrium, the relative proportion of free enzyme is indicated by the terms $K_{ia}K_b$ and $K_a[B]$, the proportion of E•A complex is represented by the term $K_b[A]$ and the combined concentrations of the E•A•B, E•P•Q and E•Q complexes are represented by the term [A][B] (Figure 14-5-A).

When A is the substrate that is varied, the reciprocal form of Equation 14.9 is:

$$\frac{1}{v} = \frac{K_a}{V_{max}}\left[\frac{K_{ia}K_b}{K_a[B]} + 1\right]\frac{1}{[A]} + \frac{1}{V_{max}}\left[\frac{K_b}{[B]} + 1\right] \tag{14.10}$$

According to this equation, a double-reciprocal plot of 1/v versus 1/[A] at different fixed concentrations of B should produce a family of straight lines that intersect at a common point to the left of the vertical ordinate, and both the slopes and the intercepts of the lines should vary with the concentration of B (Figure 14-6). A double-reciprocal plot with B as the variable substrate should have the same general

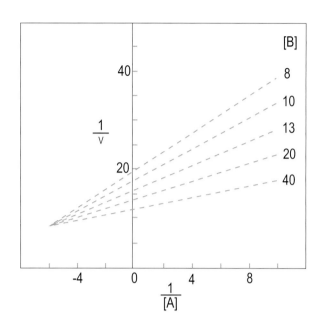

Figure 14-6. Intersecting double-reciprocal plots upon varying the concentration of substrate A in the case of steady-state ordered or rapid-equilibrium random mechanisms, at various fixed concentrations of substrate B. A similar pattern should be observed when B is the varied substrate.

form, because Equation 14.9 is symmetrical. With A as the variable substrate, the lines should intersect at the point $-1/K_{ia}$ and $(1/V_{max})[1 - K_a/K_{ia}]$, and at coordinates $-K_a/(K_{ia}K_b)$ and $(1/V_{max})[1 - K_a/K_{ia}]$ when the concentration of B is varied.

The four parameters K_a, K_{ia}, K_b and K_{ib} are linked functions (Figure 11-14), so $K_{ia}K_b = K_aK_{ib}$ and $K_b/K_{ib} = K_a/K_{ia}$ (Figure 14-5-B). The value of K_a/K_{ia} determines whether the lines of the primary plot of Figure 14-6 cross above, on or below the abscissa, and the abscissa coordinate of the cross-over point does not depend upon which substrate is varied.

Each straight line from the primary Lineweaver–Burke plot will give only the apparent values of V_{max} and K_a for that particular concentration of the nonvaried substrate, say B. The dependence of these parameters on [B] is analyzed by further plots of slopes and the intercepts as a function of [B]. The slopes of the lines should be described by:

$$\text{slope} = \frac{K_{ia}K_b}{V_{max}}\frac{1}{[B]} + \frac{K_a}{V_{max}} \qquad (14.11)$$

Consequently, replotting the slopes of the lines of the primary plot against 1/[B] should produce a straight line that intersects the abscissa at a point equal to $K_{ia}K_b/K_a$. Variation of the intercepts with [B] should be described by:

$$\text{intercept} = \frac{K_b}{V_{max}}\frac{1}{[B]} + \frac{1}{V_{max}} \qquad (14.12)$$

Therefore, replotting the intercepts of the primary plot against 1/[B] should produce a straight line that intersects the vertical ordinate at the true value for V_{max} and the abscissa at the true value for K_b. Values for V_{max}, K_{ia} and K_a would be obtained in a similar manner by starting with plots of 1/v against 1/[B] at different fixed concentrations of A.

Initial velocity measurements cannot distinguish between rapid-equilibrium random mechanisms and steady-state ordered mechanisms. If the enzyme is not at equilibrium, the value of K_{ia} cannot be assigned to a particular substrate. Furthermore, the Lineweaver–Burke plots should not be strictly linear, although the nonlinearity cannot generally be detected with steady-state kinetic measurements.

A generalized numerical approach to rapid-equilibrium enzyme kinetics: application to 17β-HSD. P. Kuzmic (2006) *Mol. Cell. Endocrinol.* **248**, 172–181.

---

2. Equilibrium-ordered Mechanism

The initial velocities expected with an equilibrium-ordered mechanism are described by:

$$v = \frac{V_{max}[A][B]}{K_{ia}K_b + K_b[A] + [A][B]} \tag{14.13}$$

This differs from Equation 14.9 in not being symmetric and not having a denominator term containing only [B]. The reason for the latter is that the value of K_a is essentially zero and the free form of enzyme is represented by the single term $K_{ia}K_b$. Rearrangement of this equation in double-reciprocal form with A and B as the variable substrates, respectively, produces:

$$\frac{1}{v} = \frac{K_{ia}}{V_{max}}\left[\frac{K_b}{[B]}\right]\frac{1}{[A]} + \frac{1}{V_{max}}\left[\frac{K_b}{[B]} + 1\right] \tag{14.14}$$

$$\frac{1}{v} = \frac{K_b}{V_{max}}\left[\frac{K_{ia}}{[A]} + 1\right]\frac{1}{[B]} + \frac{1}{V_{max}} \tag{14.15}$$

Both initial velocity patterns are of the intersecting type, but they are not symmetrical. The straight lines with A as the variable substrate intersect to the left of the vertical ordinate, the intersection point being at coordinates $-1/K_{ia}$ and $1/V_{max}$. In contrast, the lines intersect on the vertical ordinate when B is the variable substrate (Figure 14-7). The intersection point must lie above the abscissa with either variable substrate.

Replotting the slopes of the primary plot with A as the variable substrate versus 1/[B] should produce a straight line that passes through the origin. Replotting the intercepts versus 1/[B] should give the values for V_{max} and K_b. Replotting the slopes of the primary plot with B as the variable substrate versus 1/[A] should give the value of K_{ia}.

3. Ping-Pong Mechanism

The initial velocities of a reaction conforming to a single-site ping-pong mechanism (Figure 14-5-C) are described by:

$$v = \frac{V_{max}[A][B]}{K_a[B] + K_b[A] + [A][B]} \tag{14.16}$$

This equation does not have a constant term in the denominator that lacks [A] and [B] but it is

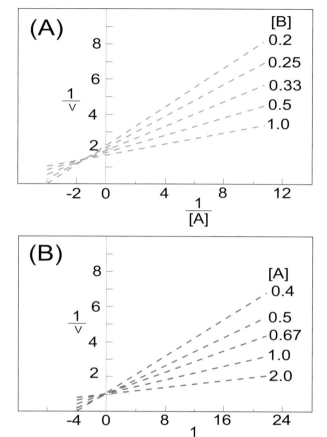

Figure 14-7. Double-reciprocal plots expected for an equilibrium-ordered mechanism. (A) Varying the concentration of substrate A with different fixed concentrations of substrate B. (B) Varying the concentration of substrate B at various fixed concentrations of A. Note that the intersection point in (A) is at negative values of 1/[A], whereas it is at 1/[B] = 0 in (B).

symmetrical. The terms $K_a[B]$ and $K_b[A]$ represent proportions of the E and F forms of enzyme, respectively, and [A][B] represents all the complexes of the enzyme with substrates and products. Rearrangement in double-reciprocal form with A and B as variable substrates, respectively, produces:

$$\frac{1}{v} = \frac{K_a}{V_{max}} \frac{1}{[A]} + \frac{1}{V_{max}}\left[\frac{K_b}{[B]}+1\right] \tag{14.17}$$

$$\frac{1}{v} = \frac{K_b}{V_{max}} \frac{1}{[B]} + \frac{1}{V_{max}}\left[\frac{K_a}{[A]}+1\right] \tag{14.18}$$

These symmetrical equations indicate that the slopes of the lines should not vary with the concentration of the second substrate, either A or B, so both initial velocity patterns should consist of families of parallel straight lines (Figure 14-8). The patterns are symmetrical. Values for V_{max}, K_a and K_b can be obtained by replotting the intercepts as a function of the reciprocals of the concentration of the second substrate that was kept constant in the primary plot.

An initial velocity pattern might only seem to consist of parallel straight lines, whereas the lines might actually intersect far below the abscissa and far to the left of the vertical ordinate. It is advisable to measure the kinetics of the reaction in both directions. A mechanism that is truly ping-pong will produce parallel initial velocity patterns in both directions; any other mechanism will exhibit a parallel pattern in one direction but an intersecting pattern in the other.

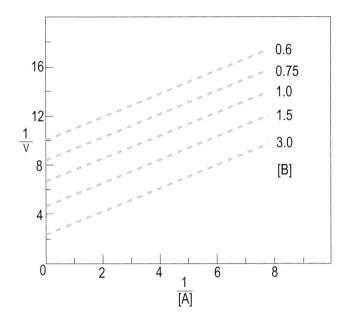

Figure 14-8. Double-reciprocal plots expected for a ping-pong mechanism by varying the concentration of substrate A at different fixed concentrations of substrate B. The straight lines are parallel, which is characteristic of the ping-pong mechanism.

With a ping-pong mechanism, it should be possible to isolate the intermediate form of the enzyme by incubating it with only the first substrate.

Observation of a hybrid random ping-pong mechanism of catalysis for NodST: a mass spectrometry approach. N. Pi *et al.* (2004) *Protein Sci.* **13**, 903–912.

Structures of NADH and CH_3–H_4 folate complexes of *Escherichia coli* methylenetetrahydrofolate reductase reveal a spartan strategy for a ping-pong reaction. R. Pejchal *et al.* (2005) *Biochemistry* **44**, 11447–11457.

14.3.D. Dead-end Inhibitors

A dead-end inhibitor occupies the active site of the enzyme but does not undergo reaction. The enzyme can become active only when the inhibitor dissociates. Dead-end inhibitors are usually structural analogs of a substrate. The patterns of inhibition exhibited can be diagnostic of the kinetic reaction mechanism, so such inhibitors are very useful.

The presence of the inhibitor affects the **slope** of a double-reciprocal Lineweaver–Burke plot when either (1) the inhibitor and variable substrate combine with the same form of the enzyme or (2) the steps between the binding of the inhibitor and the varied substrate are fully reversible, which requires that the varied substrate binds after the inhibitor.

The inhibitor affects the **intercept** of a double-reciprocal plot if the inhibitor and the varied substrate combine with different forms of the enzyme and saturation with the variable substrate does not overcome the inhibition. The intercept will always be affected unless all the steps between binding the varied substrate and the inhibitor are in rapid equilibrium.

The effects expected for the *Bi–Bi* reaction mechanism are summarized in Table 14-3.

Table 14-3. Dead-end inhibition patterns for *Bi–Bi* reaction mechanisms, A + B ⟶ P + Q

Mechanism	Inhibitor combines with	Variable substrate	
		A	**B**
Ordered	E	Competitive	Noncompetitive
	E•A	Uncompetitive	Competitive
	E•Q	Uncompetitive	Uncompetitive
	E•A & E•Q	Uncompetitive	Noncompetitive
Equilibrium-ordered	E	Competitive	Competitive
	E•A	Uncompetitive	Competitive
Ping-pong	E	Competitive	Uncompetitive
	F	Uncompetitive	Competitive
	E & F	Noncompetitive	Noncompetitive
Rapid-equilibrium random	E & E•B	Competitive	Noncompetitive
	E & E•A	Noncompetitive	Competitive

Data from J. F. Morrison (1999) in *Encyclopedia of Molecular Biology* (T. E. Creighton, ed.), Wiley-Interscience, NY, p. 629.

Practical robust fit of enzyme inhibition data. P. Kuzmic *et al.* (2004) *Methods Enzymol.* **383**, 366–381.

Enzyme inhibitors as chemical tools to study enzyme catalysis: rational design, synthesis, and applications. J. Hiratake (2005) *Chem. Rec.* **5**, 209–228.

14.3.E. Competitive Inhibition

Analogs of the substrate of an enzyme can often bind at the same place as the substrate within the enzyme's active site but are not acted upon by the enzyme. They are inhibitors of the enzyme and cause dead-end inhibition. They usually compete with the substrate, which produces **competitive inhibition**, because very high concentrations of the substrate can overwhelm the effect of the inhibitor.

A 2.13 Å structure of *E. coli* dihydrofolate reductase bound to a novel competitive inhibitor reveals a new binding surface involving the M20 loop region. R. L. Summerfield *et al.* (2006) *J. Med. Chem.* **49**, 6977–6986.

A Raman-active competitive inhibitor of OMP decarboxylase. B. P. Callahan *et al.* (2006) *Bioorg. Chem.* **34**, 59–65.

Kinetic characterization of human JNK$^{2\alpha 2}$ reaction mechanism using substrate competitive inhibitors. L. Niu *et al.* (2007) *Biochemistry* **46**, 4775–4784.

1. Linear Competitive Inhibition

The simplest type of competitive inhibition occurs when an inhibitor combines reversibly at the active site of the same form of the enzyme as does the substrate. The inhibitor may be a structural analog of the substrate or a product of that substrate. The inhibitor (I) combines reversibly with the free enzyme to form a dead-end E•I complex that can only dissociate back to the components from which it was formed:

$$E \underset{K_a}{\overset{+A}{\rightleftharpoons}} E \cdot A \xrightarrow{k_{cat}} E + P$$

$$K_i \updownarrow +I$$

$$E \cdot I \qquad (14.19)$$

The general equation that describes this type of inhibition is:

$$v = \frac{V_{max}[A]}{K_a\left(1 + \frac{[I]}{K_{is}}\right) + [A]} \qquad (14.20)$$

[I] is the concentration of inhibitor and K_{is} is the apparent inhibition constant. For a single-substrate reaction, K_{is} would be equal to K_i. The double-reciprocal form of this equation is:

$$\frac{1}{v} = \frac{K_a}{V_{max}}\left(1 + \frac{[I]}{K_{is}}\right)\frac{1}{[A]} + \frac{1}{V_{max}} \qquad (14.21)$$

This equation indicates that plotting 1/v versus 1/[A] at different concentrations of I should produce a family of straight lines that intersect at the same point on the vertical ordinate (Figure 14-9-A), which would indicate the same value of V_{max} in the presence or absence of the inhibitor. In other words, **the substrate can compete with the inhibitor and sufficiently high concentrations will overcome the inhibition completely.**

The slope of each curve should be described by:

$$\text{slope} = \frac{K_a}{V_{max}}\frac{[I]}{K_{is}} + \frac{K_a}{V_{max}} \qquad (14.22)$$

Therefore, plotting the slope versus [I] should produce a straight line that intersects the abscissa where [I] = $-K_{is}$. This secondary plot should be linear, so the inhibition is classified as being **linear competitive**.

When two substrates are involved in a reaction, the value determined for K_{is} by varying one substrate may be only an apparent constant, because its value will depend on the concentration of the other substrate. In this case, the inhibition is said to be competitive with a particular substrate.

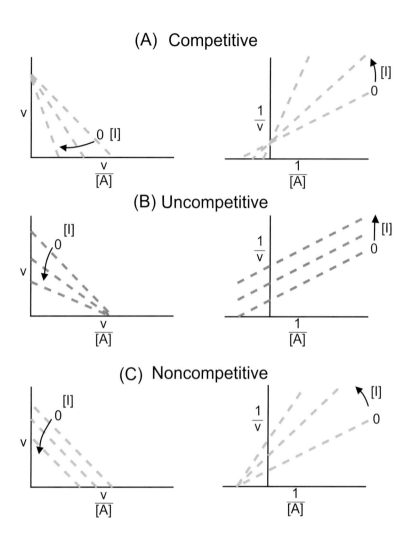

Figure 14-9. Steady-state kinetic analysis of enzyme inhibition. The catalytic activity of a very low concentration of the enzyme, v, is measured at varying concentrations of substrate, A, and inhibitor, I. The relationship between v and [A] is plotted at different values of [I] in Eadie–Hofstee plots (*left*) and Lineweaver–Burke plots (*right*). The *straight line* labeled 0 is that in the absence of added inhibitor, and the *arrow* shows the direction of the changes produced by increasing amounts of inhibitor. (A) Competitive inhibition. (B) Uncompetitive inhibition. (C) Noncompetitive inhibition; the cross-over points may be above, on or below the abscissa.

Linear competitive inhibition will also be observed with a two-substrate equilibrium-ordered reaction when the inhibitor is an analog of the first substrate to bind in a mechanism, A, and the second substrate, B, is varied in concentration:

(14.23)

The inhibition by I will be linear competitive with A because both A and I compete directly for the same form of the enzyme (E). The inhibition will also be competitive with B, even though I and B combine with different forms of the enzyme, because the binding of E and A is at equilibrium. Consequently, increasing concentrations of B will ultimately reduce the concentrations of both E•A and E to zero, so that no free enzyme remains to interact with I.

2. Hyperbolic Competitive Inhibition

This type of inhibition occurs when an inhibitor binds reversibly at a site on the enzyme other than the active site and makes it more difficult for the substrate to combine at the active site. Thus **the inhibitor and substrate can be present on the enzyme at the same time**:

$$\begin{array}{ccccc}
E & \underset{K_a}{\overset{+A}{\rightleftharpoons}} & E\cdot A & \xrightarrow{k_{cat}} & E + P \\
{\scriptstyle K_i}\updownarrow{\scriptstyle +I} & & {\scriptstyle K_{i/a}}\updownarrow{\scriptstyle +I} & & \\
E\cdot I & \underset{K_{a/i}}{\overset{+A}{\rightleftharpoons}} & E\cdot I\cdot A & \xrightarrow{k_{cat}} & E + P + I
\end{array} \quad (14.24)$$

where K_i and $K_{i/A}$ denote dissociation constants of I from E·I and E·I·A, respectively. Binding of I and A are linked functions (Figure 11-14), so it is a thermodynamic requirement that:

$$K_{a/i} = K_a \frac{K_{i/a}}{K_i} \quad (14.25)$$

The inhibition will be competitive only if the inhibitor does not affect k_{cat}. A plot of 1/v versus 1/[A] should be linear but the slopes of the lines will vary as a hyperbolic function of [I]. The intercept of the curves with the vertical ordinate should be independent of [I].

The same kinetic mechanism would apply if I is an activator that enhances the combination of the substrate with the enzyme. In this case, the plot of slope against [I] would yield a concave-up nonrectangular hyperbola.

14.3.F. Noncompetitive Inhibition

The diagnostic feature of **noncompetitive inhibition** is that the **inhibition cannot be overcome completely by increasing the substrate concentration**. Consequently, the vertical intercept of the Lineweaver–Burke plot, the velocity at infinite substrate concentration, is altered by the inhibitor.

1. Linear Noncompetitive Inhibition

Linear noncompetitive inhibition of an enzyme occurs whenever an inhibitory analog of the substrate combines with a form of enzyme other than the one with which the variable substrate combines and a reversible connection exists between the points of addition of inhibitor and the variable substrate. It should be noted that reversible connections are broken either by the release of product at zero concentration or the presence of a nonvaried substrate at an essentially infinite concentration. For a two-substrate ordered kinetic mechanism:

$$\begin{array}{c}
\;\;\;A\;\;\;\;\;\;B\;\;\;\;\;\;\;\;\;\;P\;\;\;\;\;Q\\
\;\;\;\downarrow\;\;\;\;\;\;\downarrow\;\;\;\;\;\;\;\;\;\;\uparrow\;\;\;\;\;\uparrow\\
\overline{E\;\;\;\;E\cdot A\;\;\;(E\cdot A\cdot B \leftrightarrow E\cdot P\cdot Q)\;\;\;E\cdot Q\;\;\;E}\\
{\scriptstyle K_i}\updownarrow{\scriptstyle +I}\\
E\cdot I
\end{array} \quad (14.26)$$

the inhibition by I would be linear competitive with respect to [A]. It would be linear noncompetitive with respect to [B], because B combines with E·A, I combines with E, and binding of A and B are not at equilibrium. The general form of the equation to describe noncompetitive inhibition is:

$$v = \frac{V_{max}[A]}{K_a\left(1+\frac{[I]}{K_{is}}\right) + [A]\left(1+\frac{[I]}{K_{ii}}\right)} \tag{14.27}$$

where K_{is} and K_{ii} are the inhibition constants associated with the slopes and intercepts of the double-reciprocal plot, which has the form:

$$\frac{1}{v} = \frac{K_a}{V_{max}}\left(1+\frac{[I]}{K_{is}}\right)\frac{1}{[A]} + \frac{1}{V_{max}}\left(1+\frac{[I]}{K_{ii}}\right) \tag{14.28}$$

A plot of 1/v against 1/[A], at different concentrations of I, should yield a family of straight lines that intersect at a common point to the left of the vertical ordinate. The 1/v coordinate for the cross-over point is given by $(1/V_{max})[1 - (K_{is}/K_{ii})]$, so the intersection point can occur above, on or below the abscissa, depending on whether K_{is} is less than, equal to or greater than K_{ii}, respectively (Figure 14-9-C).

The term noncompetitive is sometimes used only when the lines of a double-reciprocal plot intersect on the abscissa, with the term **mixed inhibition** being used to describe cases where the lines intersect either above or below the abscissa. Such a distinction is unnecessary, as the position of the cross-over point is simply a function of the relative values of K_{is} and K_{ii}.

The inhibition constants associated with the slopes (K_{is}) and intercepts (K_{ii}) of a double-reciprocal plot are calculated and interpreted as described for competitive inhibition and uncompetitive inhibition, respectively.

2, Hyperbolic Noncompetitive Inhibition

This type of inhibition would be observed for the scheme illustrated under hyperbolic competitive inhibition, when the E·A and E·A·I complexes give rise to product at different rates. The A term of the equation that describes this inhibition would be modified by the factor $(1 + [I]/K_{in})$:

$$v = \frac{V_{max}[A]\left(1+\frac{[I]}{K_{id}}\right)}{K_a\left(1+\frac{[I]}{K_{is}}\right) + [A]\left(1+\frac{[I]}{K_{in}}\right)} \tag{14.29}$$

which, in reciprocal form, is described by:

$$\frac{1}{v} = \frac{K_a}{V_{max}}\left(\frac{1+\frac{[I]}{K_{is}}}{1+\frac{[I]}{K_{id}}}\right)\frac{1}{[A]} + \frac{1}{V_{max}}\left(\frac{1+\frac{[I]}{K_{in}}}{1+\frac{[I]}{K_{id}}}\right) \tag{14.30}$$

Values for K_{is} and K_{id} can be determined from the variation of the slope with varying concentrations of inhibitor, [I], and for K_{in} and K_{id} for the variation of intercept with [I], in the same way as described for hyperbolic competitive inhibition.

Crystallographic identification of a noncompetitive inhibitor binding site on the hepatitis C virus NS5B RNA polymerase enzyme. R. A. Love *et al.* (2003) *J. Virol.* **77**, 7575–7581.

14.3.G. Uncompetitive Inhibition

Linear **uncompetitive inhibition** is produced when the inhibitor I binds to a different form of the enzyme than the substrate that is varied and there is no reversible connection between their binding steps. **In the simplest case, the inhibitor combines only with the enzyme–substrate complex.** A two-substrate reaction with an ordered kinetic mechanism exhibits this type of inhibition when A is the varied substrate and I is an analog of substrate B and combines with E•A (Figure 14-10-A). B and I both bind to E•A, so the inhibition by I would be linear competitive with respect to B. It would be linear uncompetitive with respect to A because I and A combine with different enzyme forms and there is no *reversible* connection between the binding of A to E and I to E•A. Increasing the concentration of A increases the concentration of the E•A complex with which I reacts. An analog of substrate B that binds only to the E•Q complex of an ordered *Bi–Bi* mechanism would produce inhibition that is linear uncompetitive with respect to both substrates (Figure 14-10-B).

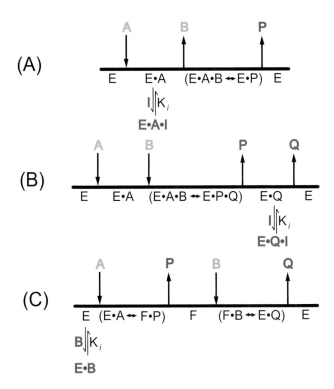

Figure 14-10. Enzyme kinetic mechanisms that can exhibit linear uncompetitive inhibition. The inhibitor is I in (A) and (B) but substrate B in (C).

Structural analogs of either of the two substrates with ping-pong mechanisms commonly produce linear uncompetitive inhibition (Figure 14-10-C). An I that is an analog of A would compete with A for binding to E, which would produce linear competitive inhibition. The inhibition would be linear uncompetitive with respect to B, because B and I combine with different enzyme forms and no reversible connection exists between the binding of I to E and B to F. The connection between the binding of I and B is broken by the release of P or Q, which would not be reversible in the absence of significant concentrations of these products. An inhibitory analog of substrate B would produce similar inhibition patterns.

Linear uncompetitive inhibition is described by:

$$v = \frac{V_{max}[A]}{K_a + [A]\left(1 + \frac{[I]}{K_{ii}}\right)} \tag{14.31}$$

$$\frac{1}{v} = \frac{K_a}{V_{max}}\frac{1}{[A]} + \frac{1}{V_{max}}\left(1 + \frac{[I]}{K_{ii}}\right) \tag{14.32}$$

K_{ii} is the inhibition constant indicated by the intercepts of double-reciprocal Lineweaver–Burke plots. The slope of the curves is independent of [I], so **a plot of 1/v versus 1/[A] should produce a series of parallel straight lines at various [I]** (Figure 14-9-B). Replotting the vertical intercepts versus [I] should yield a straight line:

$$\text{intercept} = \frac{1}{V_{max} K_{ii}}[I] + \frac{1}{V_{max}} \tag{14.33}$$

The value of K_{ii} should be given by the horizontal intercept. K_{ii} would not be a true dissociation constant for the mechanism of Figure 14-10-A but would be equal to $K_i(1 + [B]/K_b)$. This equation can be used to calculate the value of K_i.

A new type of uncompetitive inhibition of tyrosinase induced by Cl⁻ binding. Y. D. Park *et al.* (2005) *Biochimie* **87**, 931–937.

Uncompetitive inhibition of *Xenopus laevis* aldehyde dehydrogenase 1A1 by divalent cations. F. B. Rahman & K. Yamauchi (2006) *Zoolog. Sci.* **23**, 239–244.

14.3.H. Substrate Inhibition

A substrate inhibits its enzyme whenever it can generate a nonproductive dead-end enzyme–substrate complex. This usually occurs only at high substrate concentrations and is due to the substrate binding to a form of enzyme with which the product of the substrate normally binds, as illustrated in Figure 14-11. With an ordered mechanism (Figure 14-11-A) substrate B causes inhibition by behaving like P and binding to E•Q to form a nonproductive E•B•Q complex. The inhibition by B would be uncompetitive with respect to A. Pyruvate behaves in this way for the lactate dehydrogenase reaction:

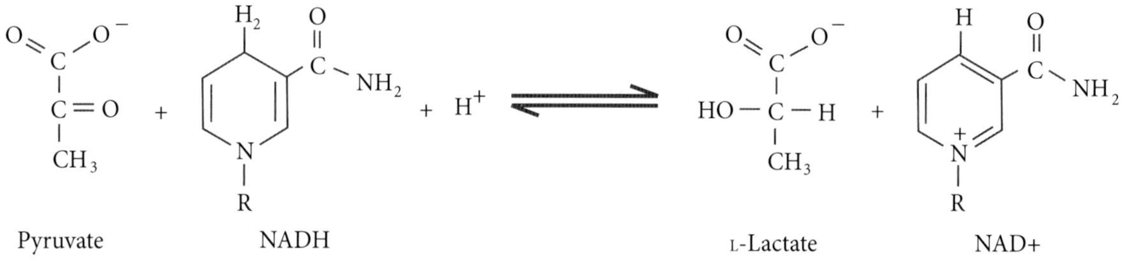

Pyruvate + NADH + H⁺ ⇌ L-Lactate + NAD⁺

(14.34)

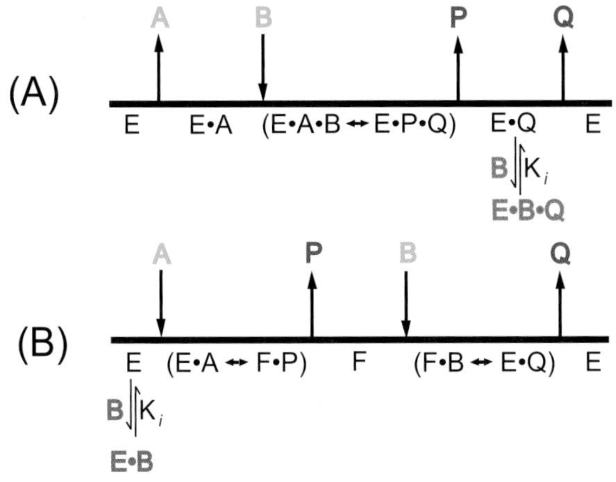

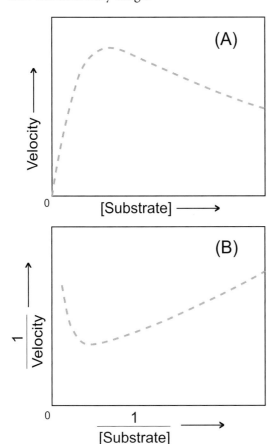

Figure 14-11. Two most common mechanisms of substrate inhibition. (A) Uncompetitive inhibition by substrate B with respect to the variable substrate A for an ordered Bi–Bi kinetic mechanism. (B) Competitive inhibition by substrate B with respect to the variable substrate A for a ping-pong kinetic mechanism

It is an ordered mechanism in which NADH is the first substrate to bind and NAD the last substrate to dissociate. The dead-end complex is E•NAD•pyruvate.

The structural similarities of the free enzyme (E) and the covalently modified enzyme (F) make ping-pong mechanisms susceptible to substrate inhibition. Substrate B in Figure 14-11-B normally combines with F but can also bind to E. Both B and A can bind to the same form of the enzyme, so the inhibition by B would be linear competitive with respect to A. Both E•B and F•A complexes can be formed, so both substrates can inhibit. In this case, the double-reciprocal Lineweaver–Burke plot looks quite complex (Figure 14-12). At high concentrations of the variable substrate, the plot curves upward and the slope of the line increases as the concentration of the other, fixed, substrate moves into the inhibitory range.

Figure 14-12. Typical kinetic consequences of substrate inhibition. (A) Plot of the initial velocity as a function of substrate concentration; the velocity decreases at high substrate concentrations. (B) Double-reciprocal plot.

Substrate inhibition by a single substrate A is described by:

$$v = \frac{V_{max}[A]}{K_a + [A] + \frac{[A]^2}{K_i}} \tag{14.35}$$

$$\frac{1}{v} = \frac{K_a}{V_{max}}\frac{1}{[A]} + \frac{1}{V_{max}} + \frac{[A]}{V_{max} K_i} \tag{14.36}$$

K_i is the inhibition constant of the substrate A. Equation 14.35 demonstrates that the reaction velocity should increase initially as [A] approaches the value of K_a but decrease at higher concentrations, depending upon the value of K_i (Figure 14-12-A). The double-reciprocal plot curves upwards at high concentrations of substrate (Figure 14-12-B). At very high substrate concentrations, Equation 14.36 becomes:

$$\frac{1}{v} = \frac{[A]}{V_{max} K_i} + \frac{1}{V_{max}} \tag{14.37}$$

Therefore, plotting the data from the curved portion of Figure 14-12-B should produce a straight line that intersects the abscissa where $-[A] = K_i$.

The structure of human SULT1A1 crystallized with estradiol. An insight into active site plasticity and substrate inhibition with multi-ring substrates. N. U. Gamage *et al.* (2005) *J. Biol. Chem.* **280**, 41482–41486.

X-ray crystal structures of HMG-CoA synthase from *Enterococcus faecalis* and a complex with its second substrate/inhibitor acetoacetyl-CoA. C. N. Steussy *et al.* (2005) *Biochemistry* **44**, 14256–14267.

14.3.I. Product Inhibition

The product of any enzymatic reaction will inhibit the overall reaction by increasing the rate of the reverse reaction. It should also inhibit the rate of the forward reaction because **a product can bind to the free enzyme in place of the substrate**. The inhibition patterns produced by a product can be very informative in the case of enzyme reactions that produce multiple products.

The **slope** of a Lineweaver–Burke plot is affected by a product inhibitor when it and the varied substrate combine reversibly with the same enzyme form or with different enzyme forms that are connected by a series of reversible steps. Release of a product in the absence of significant concentrations of it is an irreversible step, as is the binding of substrate at a saturating concentration.

The **intercept** of a double-reciprocal plot is affected by a product when it binds reversibly to a different form of the enzyme than the one with which the varied substrate binds and if saturating concentrations of the varied substrate do not overcome the inhibition.

The product inhibition patterns expected with *Bi–Bi* reaction mechanisms involving two substrates and two products are summarized in Table 14-4. The rate equation for a particular mechanism should be used for quantitative analysis of product inhibition data. The concentration of the second product that is not included is set to zero, then the equation is rearranged to the double-reciprocal form with each of the substrates varied. It is then possible to estimate the inhibition constants.

Table 14-4. Product inhibition patterns for *Bi–Bi* reaction mechanisms,
A + B ⟶ P + Q

Mechanism	Inhibitor	Varied substrate	
		A	B
Ordered	P	Noncompetitive	Noncompetitive[a]
	Q	Competitive	Noncompetitive
Equilibrium, ordered	P	Noncompetitive	Noncompetitive
	Q	Competitive	Competitive
Ping-pong	P	Noncompetitive	Competitive
	Q	Competitive	Noncompetitive
Rapid-equilibrium random, plus dead-end E•B•Q	P	Competitive	Competitive
	Q	Competitive	Noncompetitive
Rapid-equilibrium random, plus dead-end E•B•Q & E•A•P	P	Noncompetitive	Competitive
	Q	Competitive	Noncompetitive

[a] Inhibition becomes uncompetitive if B is present at a saturating concentration.

Data from J. F. Morrison (1999) in *Encyclopedia of Molecular Biology* (T. E. Creighton, ed.), Wiley-Interscience, NY, p. 1960.

Deriving the rate equations for product inhibition patterns in bisubstrate enzyme reactions. V. Leskovac *et al.* (2006) *J. Enzyme Inhib. Med. Chem.* **21**, 617–634.

Structural basis for non-competitive product inhibition in human thymidine phosphorylase: implications for drug design. K. El Omari *et al.* (2006) *Biochem. J.* **399**, 199–204.

14.3.J. Haldane Relationship

An enzyme does not alter the equilibrium constant (K_{eq}) for the reaction that it catalyzes, so the kinetic parameters in the forward and reverse directions are not independent. This is expressed by the Haldane relationship, which expresses the relationship between the kinetic and equilibrium properties of an enzymic reaction. In general terms, a Haldane is the ratio of the apparent rate constants for the reactions in the forward and reverse directions when all the substrate concentrations are relatively low.

For a simple unimolecular reaction, A ↔ P, catalyzed by a **mutase** or an **isomerase**, the rates of the forward and reverse reactions at concentrations of reactants well below the values of their Michaelis constant are given by the expressions $(V_f/K_a)[A]$ and $(V_r/K_p)[P]$, respectively. V_f and V_r are the maximum velocities for the forward and reverse reactions, respectively. At equilibrium, the rates of the forward and reverse reactions must be equal, which leads to:

$$K_{eq} = \frac{[P]}{[A]} = \frac{V_f K_p}{V_r K_a} \tag{14.38}$$

With more than one substrate, the apparent rate constant will be given by the value of V_{max}/K_m for the last substrate to add to the enzyme multiplied by the reciprocals of the dissociation constants for the substrates that were added previously. The same definition applies for the products that act as substrates for the reaction in the reverse direction. Thus, for an ordered *Bi–Bi* mechanism, where substrate A binds to the enzyme before B, and product P dissociates before Q:

$$K_{eq} = \frac{\dfrac{V_f}{K_b}\dfrac{1}{K_{ia}}}{\dfrac{V_r}{K_p}\dfrac{1}{K_{iq}}} = \frac{V_f K_p K_{iq}}{V_r K_b K_{ia}} \tag{14.39}$$

In the case of a *Bi–Bi* ping-pong mechanism, there is a Haldane for each half-reaction. The overall equilibrium constant is given by the product of the two:

$$K_{eq} = K_{eq}^1 K_{eq}^2 = \frac{V_f K_p}{K_a V_r} \frac{V_f K_q}{K_b V_r} = \left(\frac{V_f}{V_r}\right)^2 \frac{K_p K_q}{K_a K_b} \tag{14.40}$$

The importance of Haldane relationships is that they provide a means of checking the validity of a proposed kinetic mechanism. The value of K_{eq} specified by the kinetic parameters must agree with that determined experimentally.

Generalized Haldane equation and fluctuation theorem in the steady-state cycle kinetics of single enzymes. H. Qian & X. S. Xie (2006) *Phys. Rev. E* **74**, 010902.

14.3.K. Isotope Exchange at Equilibrium

The rate of exchange of isotopes between substrates and products in the presence of an enzyme can be very useful in kinetic analysis of enzymatic reactions. Radioactive atoms can be transferred from a substrate (or a product) into the product (or the substrate) in the presence of the enzyme under conditions where there is no net reaction. This differs from the kinetic techniques discussed earlier that measure only the net rates of generating a product.

This technique is widely used to measure amino acid activation by amino acyl tRNA synthetases (Section 14.4.M). In the absence of tRNA, to which the correct amino acid becomes attached in the full reaction, the enzyme catalyzes the partial reaction:

$$E + ATP + \text{amino acid} \leftrightarrow \text{amino acyl adenylate} \cdot E + PP_i \tag{14.41}$$

The amino acyl adenylate remains firmly bound to the enzyme. This reaction is readily measured by ^{32}P isotope exchange between ATP and pyrophosphate (PP_i).

1. Ping-Pong Mechanism

A ping-pong mechanism with a single site consists of two half-reactions: substrate A is converted to P, leaving a modified form of the enzyme that then converts substrate B to product Q. The first half-reaction is:

$$E + A \leftrightarrow (E \cdot A \leftrightarrow F \cdot P) \leftrightarrow F + P \tag{14.42}$$

Addition of only substrate A to the enzyme will produce a modified form of the enzyme, F, plus the release of a stoichiometric amount of P. Measurements of the half-reaction by isotope exchange use only low, micromolar concentrations of the enzyme, so the amount of P produced should be small. The half-reaction will reach equilibrium so long as the second substrate, B, is absent. A and P will be being interconverted at identical rates at equilibrium, so isotopically labeled A (or P) that is added will be converted to P (or A) and the isotope will be incorporated into P (or A). Eventually, A and P will have the same specific radioactivity.

This partial exchange reaction in the absence of the other substrate is characteristic of a ping-pong mechanism. Before being taken as evidence for a ping-pong mechanism, however, the exchange reaction should be demonstrated with both halves of the reaction, using both A ↔ P and B ↔ Q. If only one exchange reaction is observed, it could be caused by an impurity in the enzyme.

Starting with labeled A, the initial rate of the exchange reaction observed between A and P, v_{A-P}, should be described by:

$$v_{A-P} = \frac{V_f \frac{K_{ia}}{K_a}[A][P]}{K_{ia}[P] + K_{ip}[A] + [A][P]} \tag{14.43}$$

V_f is the maximum rate at which the enzyme catalyzes the overall reaction. The maximum velocity of the isotope exchange reaction should be given by either $V_f K_{ia}/K_a$ or $V_r K_{ip}/K_p$, where V_r is the maximum rate of the reverse reaction. This equation is similar to that for the initial velocity of a ping-pong mechanism (Equation 14.16) except that the denominator contains the dissociation constants K_{ia} for the E·A complex and K_{ip} for F·P. Plotting $1/v$ versus $1/[A]$ should produce a series of parallel straight lines for different concentrations of P, so long as there is no substrate inhibition, i.e. formation of a dead-end F·A complex.

The initial rate of B ↔ Q exchange should be described by a similar equation:

$$v_{B-Q} = \frac{V_f \frac{K_{ib}}{K_b}[B][Q]}{K_{iq}[B] + K_{ib}[Q] + [B][Q]} \tag{14.44}$$

The maximum velocities of the exchange reactions depend on the values of the dissociation and Michaelis constants. The maximum rates of the chemical (V_f and V_r) should be related to those for the exchange half-reactions:

$$\frac{1}{V_{A\text{-}P}} + \frac{1}{V_{B\text{-}Q}} = \frac{1}{V_f} + \frac{1}{V_r} \tag{14.45}$$

In the case of a single-site ping-pong mechanism:

$$\frac{1}{V_{A\text{-}Q}} = \frac{1}{V_{A\text{-}P}} + \frac{1}{V_{B\text{-}Q}} \tag{14.46}$$

Measuring the rate of the overall exchange reaction, $V_{A\text{-}Q}$, requires adding both substrates or both products. The rates of the partial exchange reactions will be reduced due to formation of the F•B or E•Q complexes, which reduces the concentrations of those forms of enzyme that catalyze the partial exchange reactions.

2. Sequential Reactions

Sequential mechanisms differ from ping-pong mechanisms in that they have no partial exchange reactions, so the overall reaction must be used. In this case, isotope exchange is followed after adding a small amount of a very radioactive form of one of the reactants to a mixture that has come to equilibrium. The initial rate of the exchange of the label from substrate to product is measured as a function of the concentration of a substrate–product pair that is increased in its equilibrium ratio. The equilibrium constant K_{eq} is equal to [P] [Q]/[A] [B] for the reaction:

$$A + B \leftrightarrow P + Q \tag{14.47}$$

Consequently, the concentrations of any of four pairs of reactants ([P]/[A], [P]/[B], [Q]/[A] or [Q]/[B]) can be altered without disturbing the equilibrium. The exchange rate is measured as the concentration of one pair is varied, which is usually a pair of comparable substrate and product. Three exchange reactions can be measured, which can be illustrated using the **creatine kinase** reaction:

$$\text{creatine} + \text{Mg}\cdot\text{ATP}^{2-} \leftrightarrow \text{phosphocreatine} + \text{Mg}\cdot\text{ADP}^- \tag{14.48}$$

The exchanges creatine ↔ phosphocreatine, ATP ↔ phosphocreatine and ATP ↔ ADP can be measured, but not creatine ↔ ADP because no atoms are exchanged between them.

The exchange patterns expected for ordered and rapid-equilibrium random mechanisms are summarized in Table 14-5. The A ↔ Q exchange is inhibited as the concentration of the B/P pair is increased in the case of an ordered mechanism, but not with the random mechanism. All three exchange reactions have equal rates with the rapid-equilibrium random mechanism, as in the case of the creatine kinase reaction (Equation 14.48), but only when all the steps except that for the catalytic conversion are rapid and at equilibrium. In this case, the enzyme has distinct binding sites for each of the two substrates involved in the reaction, so it is possible for one substrate and the product of the other substrate to be present on the enzyme simultaneously; both such dead-end complexes could be formed. The one involving the smaller substrate and smaller product (E•B•Q) will always occur, whereas the other may or may not. Increasing concentrations of the pair of reactants that form a dead-end complex formation will inhibit all three exchanges.

Table 14-5. Isotope exchange patterns for Bi–Bi reaction mechanisms, $A + B \leftrightarrow P + Q$

Mechanism	Varied substrate/ product pair	Exchange measured	Double-reciprocal plot
Ordered	A/Q	$A \leftrightarrow Q$	Linear
		$B \leftrightarrow P$	Linear
	B/P	$A \leftrightarrow Q$	Substrate inhibition[a]
		$B \leftrightarrow P$	Linear
Rapid-equilibrium random	A/Q	$A \leftrightarrow Q$	Linear
		$B \leftrightarrow P$	Linear
	B/P	$A \leftrightarrow Q$	Linear
		$B \leftrightarrow P$	Linear
	B/Q	$A \leftrightarrow Q$	Substrate inhibition[a]
		$B \leftrightarrow P$	Substrate inhibition[a]
		$A \leftrightarrow P$	Substrate inhibition[a]

[a]The rate of isotope exchange decreases as the concentrations of both substrate and product are increased, but at a constant ratio.

Data from J. F. Morrison (1999) in *Encyclopedia of Molecular Biology* (T. E. Creighton, ed.), Wiley-Interscience, NY, p. 1327.

14.3.L. Slow- and Tight-Binding Enzyme Inhibitors

The preceding discussion of enzyme inhibition has assumed that the inhibitor binds rapidly and reversibly to the enzyme. An inhibitor that binds tightly, however, will tend to dissociate slowly and might not fulfill the requirements for rapid equilibration. In this case, the inhibition becomes time-dependent (Figure 14-13) and the inhibitors have been referred to as **slow-binding inhibitors**.

The simplest scheme for slow-binding inhibition of an enzyme-catalyzed reaction is:

$$E \underset{k_2}{\overset{k_1[A]}{\rightleftharpoons}} E \bullet A \xrightarrow{k_7} E + P$$
$$\text{slow } k_4 \Updownarrow k_3[I]$$
$$E \bullet I$$

(14.49)

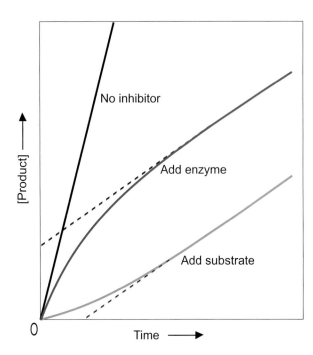

Figure 14-13. Typical progress curves for an enzyme-catalyzed reaction in the absence and presence of a slow-binding inhibitor. The reactions were started at time zero by adding enzyme or adding substrate after pre-incubation of enzyme and inhibitor. The *dashed lines* indicate the steady-state rates that are attained eventually in the presence of the inhibitor.

Formation of an E•I complex might be slow because the concentration of I required to demonstrate inhibition is relatively low, so the rate of forming E•I would also be low. Alternatively, the inhibitor might encounter barriers to its binding so that it would be intrinsically slow.

A second possible mechanism for slow-binding inhibition is:

$$E \underset{k_2}{\overset{k_1[A]}{\rightleftharpoons}} E{\cdot}A \xrightarrow{k_7} E + P$$

$$k_4 \Updownarrow k_3[I]$$

$$E{\cdot}I \underset{k_6}{\overset{k_5}{\rightleftharpoons}} E{\cdot}I^*$$

slow

(14.50)

The collision complex E•I could be formed rapidly but then undergo a slow conformational change or other isomerization reaction to generate a more stable complex, E•I*. The overall dissociation constant for both inhibitor complexes will be:

$$K_i^* = \frac{[E][I]}{[E{\cdot}I]+[E{\cdot}I^*]} = \frac{K_i k_6}{k_5 + k_6} \qquad (14.51)$$

where $K_i = k_4/k_3$. The ratio k_5/k_6 determines the enhancement of the initial binding due to the isomerization reaction. Their absolute values will determine the time scale on which inhibition takes place.

Both mechanisms of Equations 14.49 and 14.50 predict the same equation for the time-dependence at which the product P will be produced when the reaction is initiated by adding the enzyme:

$$[P] = v_s t + (v_0 - v_s)(1 - e^{-kt})/k \qquad (14.52)$$

The terms v_0 and v_s denote initial and final steady-state velocities, respectively; t is time and k is an apparent first-order rate constant that reflects the slow binding of the inhibitor, which depends upon the mechanism. The concentration of the inhibitor should be an order of magnitude greater than the total enzyme concentration, so that its free concentration remains essentially constant. Starting the enzymatic reaction by adding the enzyme to a mixture of substrate and inhibitor produces an initial rapid appearance of the product, which is due to the uninhibited enzyme (Figure 14-13). The rate then decreases to a slower steady-state, due to the slow formation of the enzyme–inhibitor complex.

An alternative protocol is to pre-incubate the enzyme with the inhibitor and then start the reaction by adding the substrate. Equation 14.52 still applies but the rate of the reaction is initially low and slowly increases as the substrate replaces the inhibitor after it dissociates from the enzyme (Figure 14-13).

The progress curves for the mechanisms of Equations 14.49 and 14.50 in the presence of different concentrations of a slow-binding inhibitor, and started by the addition of enzyme, are illustrated in Figure 14-14. They are very similar, except that the initial velocity of the reaction is independent of the inhibitor concentration in the case of the mechanism of Equation 14.49. In contrast, the initial velocity for Equation 14.50 varies as with linear competitive inhibition. As the concentration of the inhibitor increases, so does the rate at which the curves for the two mechanisms reach their steady-state velocities. In the presence of a fixed concentration of substrate A, the rate of turnover is a linear function of the inhibitor concentration for Equation 14.49 but hyperbolic for Equation 14.50. The situation is much more complicated if the initial collision complex E•I in Equation 14.50 is at a steady-state, rather than at thermodynamic equilibrium, and it is difficult to distinguish between the two mechanisms. The same basic approach is used when an enzyme has multiple substrates, but analysis of the data is simplified if all substrates, other than the one for which the inhibitor is an analog, are present at saturating concentrations.

When a slow-binding inhibitor inhibits at very low concentrations, comparable to that of the enzyme, it is of the **slow tight-binding** type. The inhibition can still be described in terms of Equations 14.49 and 14.50 but the changes in the concentration of free inhibitor with time complicate the analysis markedly.

Of the enzymes known to be subject to slow-binding and slow tight-binding inhibition, the predominant mechanism of inhibition is that described by Equation 14.50.

The behavior and significance of slow-binding enzyme inhibitors. J. F. Morrison & C. T. Walsh (1988) *Adv. Enzymol.* **59**, 201–301

Kinetics of slow and tight-binding inhibitors. S. E. Szedlacsek & R. G. Duggleby (1995) *Methods Enzymol.* **249**, 144–180

Analysis of slow-binding enzyme inhibitors at elevated enzyme concentrations. B. Perdicakis *et al.* (2005) *Anal. Biochem.* **337**, 211–223.

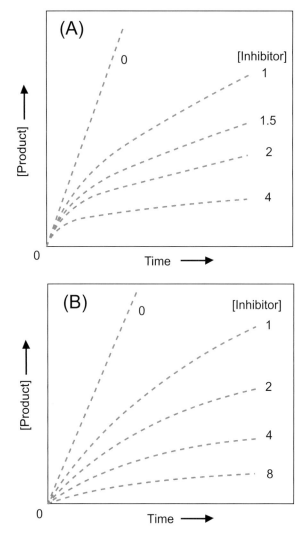

Figure 14-14. Typical progress curves for the inhibition of enzyme-catalyzed reactions by slow-binding inhibitors. The reaction was started in each case by adding enzyme to a mixture of substrate and varying amounts of the inhibitor. (A) The inhibitor binds slowly to the enzyme, as in Equation 14.49. (B) The inhibitor binds rapidly and loosely to the enzyme but the complex isomerizes slowly to a form with greater affinity for the inhibitor (Equation 14.50).

14.4. MECHANISMS OF ENZYME CATALYSIS

The folding of the polypeptide chain that produces the final structure of an enzyme also generates the **active site**, where the substrates bind and the chemical reaction takes place. The active site of an enzyme is a groove, cleft or pocket that comprises only a small part of the total solvent-accessible surface of the protein. The relatively large sizes of enzymes are due at least in part to the need to generate those spatial relationships of the amino acid residues at the active site that are appropriate for the binding of substrates, catalysis and the release of products. As in the binding of other ligands (Chapter 12), the binding of substrates and inhibitors at the active site of an enzyme involves matching the nonpolar groups of the substrate with the nonpolar side-chains of amino acid residues and hydrogen bonding between the polar groups on the substrate with the backbone NH and CO groups within the active site. Chemical catalysis can be produced by reactive groups within the active site, such as acidic, basic and nucleophilic groups of the protein and electrophilic groups of a prosthetic group, coenzyme or cofactor. **Within an enzyme–substrate complex, the effective concentrations of reactive groups can be very high** so their interactions can be energetically much greater than when they act independently in solution.

The most frequently used methods of identifying the roles of individual groups in catalysis involve methodically varying the structure of the substrate or the enzyme and measuring the effects upon k_{cat} and K_m. Varying the substrate can be bedeviled by the ability of substrate analogs to bind to the enzyme active site in remarkably different ways. For example, synthetic substrates for α-chymotrypsin with extraneous hydrophobic residues bind with these groups in the hydrophobic binding site, rather than the Phe or Tyr side-chain that normally occupies this site. Comparing the affinity constants for such a series of substrates, or their values of k_{cat}, would be very misleading in the absence of structural information, as different phenomena would be being compared.

The preferred method is to vary the protein structure, using site-directed mutagenesis (Section 7.6.C). The structure of the protein is less likely to change upon mutation than is a substrate to alter its mode of binding. Site-directed mutagenesis is capable of specifically and effectively removing a single functional group, for example by changing a specific Ser or Cys residue to Ala, an Asp or Glu residue to Asn or Gln, etc. Covalent modification of the enzyme has the disadvantage of usually adding extra groups to the enzyme; simply the extra bulk of these groups can disrupt the enzyme activity, even when the functional group blocked plays no direct role in catalysis.

As an example of the dramatic effects that can be observed using site-directed mutagenesis, the ε-amino group of Lys258 of the pyridoxal phosphate-dependent enzyme aspartate aminotransferase (Section 14.4.G.1) was postulated to be the base responsible for transferring the proton during the transamination of aspartate. Replacing this Lys residue by Ala caused the enzyme activity to decrease by more than six orders of magnitude, even though other partial reactions of the enzyme were unaltered. This mutation demonstrated the major role that Lys258 plays in this enzyme.

Mutating residues that are involved primarily in binding the substrate might be expected to change primarily K_m, whereas changing those involved directly in catalysis should change only k_{cat}. Changing residues outside the active site might not be expected to alter the catalytic activity of the enzyme, unless the mutation alters the structure of the enzyme. Such division of residues, being involved in catalysis, binding or neither, is much too simple, however; in general, **binding is related to catalysis in that the strength of the binding of the transition state, relative to the substrate, determines the rate of the enzymatic reaction**. Residues that are involved in binding are usually involved in catalysis.

Mutagenic data can be misleading if replacements within the active site affect the properties of other residues of the enzyme. Interactions between two residues in a protein may be uncovered by comparing the effects of single mutations of each of the two residues (e.g. X and Y) with the effect of mutating both simultaneously. The effect of the double mutation on the energy of an intermediate or transition state ($\Delta\Delta G°_{(X,Y)}$) should be the sum of the energetic effects of the two individual mutations:

$$\Delta\Delta G°_{(X,Y)} = \Delta\Delta G°_{(X)} + \Delta\Delta G°_{(Y)} \tag{14.53}$$

unless the two residues interact in the enzyme, or if either causes a change in the reaction mechanism or which step in the reaction is rate-limiting.

The effects of mutations are most confidently interpreted if the data exhibit **linear free energy relationships**; in this case, a series of alterations of the enzyme or substrate produces a series of changes in two or more variables of the reaction that are related. Such an approach is well-established for elucidating the structure of the transition state of a reaction. For example, in a **Brønsted plot**, the logarithm of the rate constant for a reaction is plotted versus the pK_a of a reactant or catalyst. If the reaction mechanism and rate-limiting step are unchanged with a series of reactants, a linear plot is generally obtained, with a slope β. When β = 0, the rate is independent of the basicity of the reactant, and its basicity is not evident in the transition state for the reaction. When β = 1, the rate is fully dependent upon the basicity of the reactant, which must be expressed completely in the transition state. **More generally, the rate of a reaction is usually compared with its equilibrium constant.** The two parameters compare the free energies of the transition state and the product (relative to that of the substrate), so the value of the slope of the line gives a measure of how much the transition state is like the substrate or product in terms of the parameter being varied. When β = 0, the transition state resembles the substrate; When β = 1, it resembles the product (Figure 14-15).

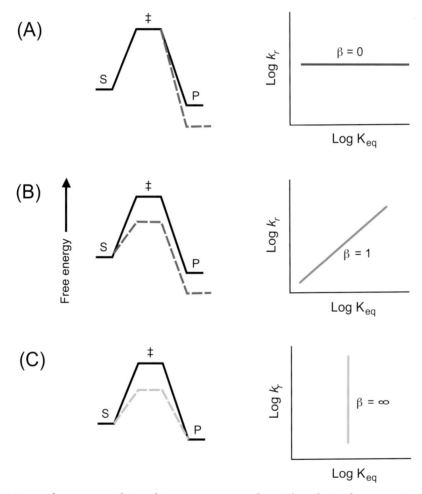

Figure 14-15. Linear free energy relationships in enzyme catalysis. The relative free energies of a reactant (S), product (P) and transition state by which they are interconverted (‡) are depicted on the *left* for two variant forms of the reactant S. (A) The transition state is like the reactant S. (B) The transition state is like the product P. (C) The transition state is unlike both S and P. On the *right* are shown the linear free energy relationships that would be apparent for a series of such variants of S when the logarithm of the rate constant (k_r) is plotted versus that of the equilibrium constant (K_{eq}). The slope of this line is β. In (A), the differences between the variant reactants are apparent only in the relative free energies of the product, so β = 0. In (B), the free energy difference is apparent in both the transition state and the product, so β = 1. In (C), the difference is apparent only in the energy of the transition state, so β = ∞. In general, the free energy difference may be apparent to varying extents in both the transition state and product, so any value of β is possible. From T. E. Creighton (1993) *Proteins: structures and molecular properties*, 2nd edn, W. H. Freeman, NY, p. 412.

With enzyme catalysis, linear free energy relationships are observed in relatively simple cases (Figure 14-16), usually where catalysis is the result primarily of multiple independent binding interactions, as in the case of tyrosyl tRNA synthetase (Section 14.4.M). Where simple relationships are not apparent, the various interactions between enzyme and substrate are not independent or the mutations produce changes in the reaction mechanism, rate-determining step or structure of the enzyme. **Linear free energy relationships suggest that neither the nature of the reaction nor the structure of the enzyme has changed, and they provide valuable evidence about the nature of the transition state**. For example, when the slope of the linear curve, β, is close to zero or ∞, the series of mutations alter either the rate constant or the equilibrium constant. When β =1, the energetic consequences of the mutations are the same in both (Figure 14-15).

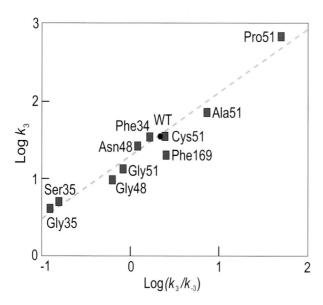

Figure 14-16. Free energy relationship observed in the formation of tyrosyl adenylate from tyrosine and ATP by mutant tyrosyl tRNA synthetases. The variant enzymes are designated by the mutation by which they differ from the normal enzyme (WT) at the indicated residue number. The rate constants k_3 and k_{-3} are for the forward and reverse steps, respectively, in forming tyrosyl adenylate and pyrophosphate from tyrosine and ATP on the enzyme (Equation 14.71). Data from A. R. Fersht *et al.*, *Biochemistry*, **26**, 6030–6038.

If a functional group is removed from an enzyme active site by mutagenesis, it should be possible to replace it by adding a similar small molecule to the solution or, perhaps, by adding it to the substrate. As an example of the first possibility, the Lys258 side-chain of aspartate aminotransferase that had been removed could be replaced by small-molecule amines. The smaller the amine and the greater its pK_a value, the greater its efficiency in assisting catalysis of the reaction on the enzyme. The added amine presumably occupies the void left by the deleted Lys side-chain. In the second case, removing a crucial His residue of subtilisin by mutation to Ala decreased k_{cat}/K_m by a factor of 10^6. This His residue could be replaced to some degree by a His residue within the polypeptide substrate that binds at the corresponding position in the enzyme–substrate complex. Substrates with this His-containing sequence are hydrolyzed by the mutant enzyme much more rapidly than other substrates. Consequently, the mutant enzyme is much more specific for certain substrates than is the normal enzyme.

The active site residues of enzymes occur predominantly in regions of the protein with smaller crystallographic temperature factors than the other residues. This suggests that, in general, the active site residues are less flexible than the other parts of the enzyme molecule. Therefore the vibrational and fast collective motions of the atoms of enzymes would seem not to be functionally important, even though enzyme flexibility and dynamics are often assumed to be crucial for their catalytic functions.

Thermodynamic and extrathermodynamic requirements of enzyme catalysis. R. Wolfenden (2003) *Biophys. Chem.* **105**, 559–572.

Theoretical insights in enzyme catalysis. S. Marti *et al.* (2004) *Chem. Soc. Rev.* **33**, 98–107.

Ab initio quantum chemical and mixed quantum mechanics/molecular mechanics (QM/MM) methods for studying enzymatic catalysis. R. A. Friesner & V. Guallar (2005) *Ann. Rev. Phys. Chem.* **56**, 389–427.

Mapping of the active site of proteases in the 1960s and rational design of inhibitors/drugs in the 1990s. I. Schechter (2005) *Curr. Protein Peptide Sci.* **6**, 501–512.

Mechanisms and free energies of enzymatic reactions. J. Gao *et al.* (2006) *Chem. Rev.* **106**, 3188–3209.

Relating protein motion to catalysis. S. Hammes-Schiffer & S. J. Benkovic (2006) *Ann. Rev. Biochem.* **75**, 519–541.

14.4.A. Reactions on the Enzyme

Steady-state kinetics provide only tantalizing hints of what processes take place on the enzyme to produce catalysis, with postulated complexes of the enzyme with substrates and products, and with intermediate, substituted enzymes in ping-pong-type schemes. To verify the existence of these complexes, and to learn more about what happens between the time when the first substrate is bound and when the last product is released, **the enzyme itself must be studied**. This requires both substrate-like quantities of enzyme and rapid techniques that permit measurements within the turnover time of the enzyme. The turnover time is given by $1/k_{cat}$; its value can be as short as 10^{-6} s, although longer times are more usual (Section 14.2.C).

The existence of enzyme–substrate complexes has been amply demonstrated using primarily spectral techniques like those used to measure ligand binding in general, and with results comparable to those for binding other ligands. Further chemical changes in the bound substrate can often also be detected in this way. A basic difficulty with kinetic studies of multi-step reactions, however, is that only a few intermediates are likely to accumulate to detectable levels. Any correlation between the importance of an intermediate and its probability of accumulating is likely to be an inverse one: **the most important intermediates are those least likely to accumulate**. For example, the transition state is most important for determining the rate of a reaction, but it is the species that accumulates to the least extent.

Intermediates can be detected most readily if widely varying substrates are available for the enzyme, because the various substrates may differ in the rates of individual steps; this can lead to the accumulation of different intermediates along the reaction pathway. Some substrates might undergo partial reaction very rapidly but then not react further at a comparable rate. If a product is released in the first step, that step will not be readily reversed and intermediate E•X will accumulate:

$$E + S \xrightarrow{\text{fast}} E\bullet X \xrightarrow{\text{slow}} E + P_2$$
$$\searrow P_1$$

(14.54)

This kind of kinetic behavior with certain substrates provided some of the evidence for the occurrence of an acyl intermediate of serine proteinases, in which part of the substrate is attached covalently to the enzyme (Section 15.2.D.1 and Equation 14.62). Such substrates have very practical uses as **active site titrants**; upon mixing enzyme with excess substrate, there is an initial 'burst' of formation of product P_1, one mole per mole of enzyme active site (Figure 14-17). This is the best method for measuring the concentration of functional active sites in an enzyme preparation. Subsequent turnovers of the enzyme are limited by the slow second step. In many cases, the intermediate has the second product covalently attached to the enzyme, so the complex can be characterized chemically. With such techniques, intrinsically labile linkages between enzyme and substrate may need to be trapped chemically, for example by irreversibly reducing a Schiff base linkage (Section 14.4.G.1), while others can be stabilized merely by disrupting the enzyme's structure.

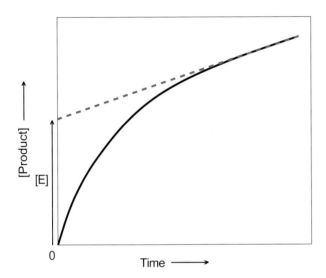

Figure 14-17. Example of an active site titration curve showing rapid formation of an enzyme–intermediate complex that then turns over only slowly. The *dashed line* indicates extrapolation of the linear portion of the curve to zero time, which indicates the molar concentration of enzyme active sites present.

If a substrate normally makes no known covalent intermediate, it may be induced to do so, thus providing the enzymologist with a convenient probe, by adding a reactive group to it. Such **active site-directed irreversible inhibitors** are analogous to affinity labels containing reactive groups that react covalently with ligand binding sites. Further specificity can be introduced by altering the substrate so that a very reactive group is generated during the catalytic cycle, which will then react with an appropriate group nearby in the enzyme active site (Section 14.4.H).

The pH-dependence of each step in an enzyme-catalyzed reaction can measure the pK_a values of crucial acids or bases in both the free enzyme and substrate and in the enzyme–substrate complex. The small, relatively simple enzymes that have been studied most extensively usually demonstrate straightforward pH-dependence of the strength of substrate binding or the rate of catalysis; both binding and catalysis often appear to depend principally upon the ionization of only one or two groups. Care must be exercised, however, in assigning such pK_a values to specific groups. The pK_a values of groups in enzyme active sites are often substantially perturbed from their normal values, and they frequently are even more perturbed in the enzyme–substrate complex. The identities of these and other groups in the active site can be determined by the abolition of substrate binding or catalysis upon their specific covalent modification or mutation.

The most direct and detailed information about enzyme action comes from crystallographic determination of complexes of the enzyme with the substrate at various stages of the reaction, although this is made difficult by the rapid rates at which enzyme-catalyzed reactions occur (Section 14.4.J). It is much more feasible to determine the structures of stable complexes of the enzyme with incomplete sets of the substrates, substrate analogs that react very slowly and inhibitors; many examples of these have been studied crystallographically. The binding of substrates to enzymes is found not to be fundamentally different, at least to a first approximation, to binding of ligands to any specific site on a protein. This type of crystallographic information has provided great insight into how substrates are held in precise orientations at enzyme active sites. Such studies do not observe catalysis directly, however, and extrapolation of the results is necessary to imagine what occurs during actual catalysis, when significant changes in both the enzyme and the substrate could be occurring. Inhibitors can also be misleading because they may be inactive as substrates simply because they bind in a nonproductive mode that is different from the binding of a true substrate.

Plasticity of enzyme active sites. A. E. Todd *et al.* (2002) *Trends Biochem. Sci.* **27**, 419–426.

Hydrogenases: active site puzzles and progress. F. A. Armstrong (2004) *Curr. Opinion Chem. Biol.* **8**, 133–140.

Design of miniproteins by the transfer of active sites onto small-size scaffolds. F. Stricher *et al.* (2006) *Methods Mol. Biol.* **340**, 113–149.

Thymidine phosphorylase from *Escherichia coli*: tight-binding inhibitors as enzyme active-site titrants. A. Gbaj *et al.* (2006) *J. Enzyme Inhib. Med. Chem.* **21**, 69–73.

14.4.B. Stabilizing the Transition State

Like all catalysts, enzymes increase the rate of a chemical reaction by lowering its activation energy (Figure 14-2). The result is that the substrate molecules pass through the transition state at a greater rate than would be the case if the reaction were uncatalyzed. The features of an enzyme-catalyzed reaction and the corresponding nonenzymic reaction can be described in terms of transition-state kinetic theory by the following scheme:

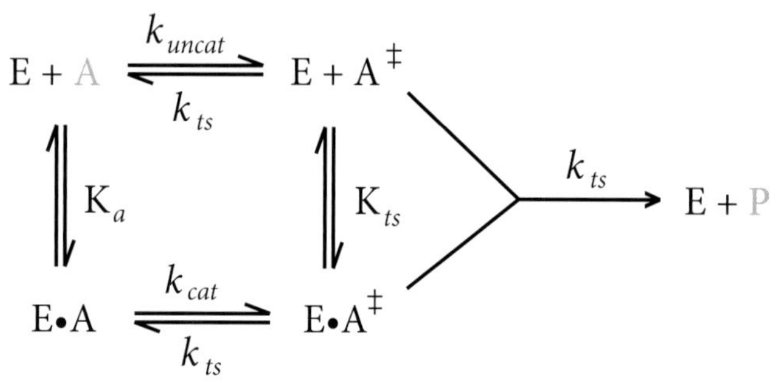

$$\begin{array}{ccc}
E + A & \underset{k_{ts}}{\overset{k_{uncat}}{\rightleftharpoons}} & E + A^{\ddagger} \\
\updownarrow K_a & & \updownarrow K_{ts} \quad \overset{k_{ts}}{\longrightarrow} \ E + P \\
E \cdot A & \underset{k_{ts}}{\overset{k_{cat}}{\rightleftharpoons}} & E \cdot A^{\ddagger}
\end{array} \quad (14.55)$$

In this most simple case, the enzyme E has only a single substrate, A, which is converted to a single product, P, and the release of P from the enzyme is not rate-limiting (the K_m values therefore reflect affinities for the enzyme). The rates of the reaction are k_{cat} and k_{uncat} when bound to the enzyme and when in solution, respectively. Each transition state, designated as $A^{\ddagger}$, is postulated to break down to the original ground state or the product at the same intrinsic rate, k_{ts}, independent of the enzyme. In classical transition-state theory, this rate constant is given by the vibrational frequency of a covalent bond, $k_B T/h$, where k_B is Boltzmann's constant and h is Planck's constant; this rate is 6.2×10^{12} s^{-1} at

25°C. K_a and K_{ts} denote dissociation constants for the E•A and E•A‡ complexes, respectively. From Equation 14.55, it follows that k_{cat}/K_a must be equal to k_{uncat}/K_{ts}, as these are **linked functions** and the free energy cannot change around a cyclic series of reactions (Figure 11-14). The linkage relationship can be expressed as:

$$K_{ts} = (k_{uncat}/k_{cat})K_a \tag{14.56}$$

To the extent that k_{cat} is greater than k_{uncat}, the value of the dissociation constant K_{ts} must be less than K_a, i.e. **the transition state should have a correspondingly greater affinity for the enzyme than the substrate and the active site of the enzyme should be complementary to the transition state rather than the substrate** (Figure 14-2). As enzymes speed up the rates of reactions by as much as 10^{17}-fold (Table 14-1), the binding of the transition state by the enzyme should be very much tighter than the binding of the substrate in the ground state. For an enzyme with a K_a value of 10^{-5} M that increases the rate of reaction of a substrate 10^{17}-fold, the value of K_{ts} should be only 10^{-22} M. Such an extremely low value for K_{ts} would ensure that the transition state is not released from the enzyme. The enhanced binding of the transition state would be due to the additional and favorable interactions between it and the complementary enzyme active site that are not available to the ground state substrate. In effect, **binding becomes catalysis**.

Enzymes may also speed reactions by producing new intermediate states along the reaction pathway. In a single-step reaction, the geometry of the transition state will be somewhere between that of the substrate and product and will have a correspondingly higher free energy than both. With a relatively stable intermediate between reactant and product, each of the two transition states will be closer to either the substrate or the product. Consequently, neither transition state need differ from substrate or product so much in geometry or free energy, and the overall kinetic barrier need not be so high. Therefore, **reactions entailing a number of small changes might occur faster than those involving one large change**. The intermediate need not, and should not, be more stable than either substrate or product for the overall rate to be increased. Being closer to the transition state, however, the intermediate should be bound tightly by the enzyme and consequently be more stable than when free in solution during the normal uncatalyzed reaction.

Do enzymes change the nature of transition states? Mapping the transition state for general acid–base catalysis of a serine protease. R. R. Bott *et al.* (2003) *Biochemistry* **42**, 10545–10553.

Transition state stabilization by general acid catalysis, water expulsion, and enzyme reorganization in *Medicago savita* chalcone isomerase. S. Hur *et al.* (2004) *Proc. Natl. Acad. Sci. USA* **101**, 2730–2735.

Do electrostatic interactions with positively-charged active-site groups tighten the transition state for enzymatic phosphoryl transfer? I. Nokolic-Hughes *et al.* (2004) *J. Am. Chem. Soc.* **126**, 11814–11819.

On the importance of being zwitterionic: enzymatic catalysis of decarboxylation and deprotonation of cationic carbon. J. P. Richard & T. L. Amyes (2004) *Bioorg. Chem.* **32**, 354–366.

---

1. Transition-state analogs

The idea of tight-binding of the transition state prompted the development of **transition-state** analogs with structural features approximating those of the transition state. Such compounds are expected to be highly specific and potent inhibitors of enzymes. On the other hand, **a transition state is extremely unstable and must have an unusual covalent structure, with partial and extended bonds and unusual bond angles**, so it should be difficult, if not impossible, to prepare a good stable analog. Most of the compounds referred to as transition-state analogs are, in effect, **intermediate-state analogs**, as they are analogs of actual intermediates, such as carbanions and tetrahedral adducts

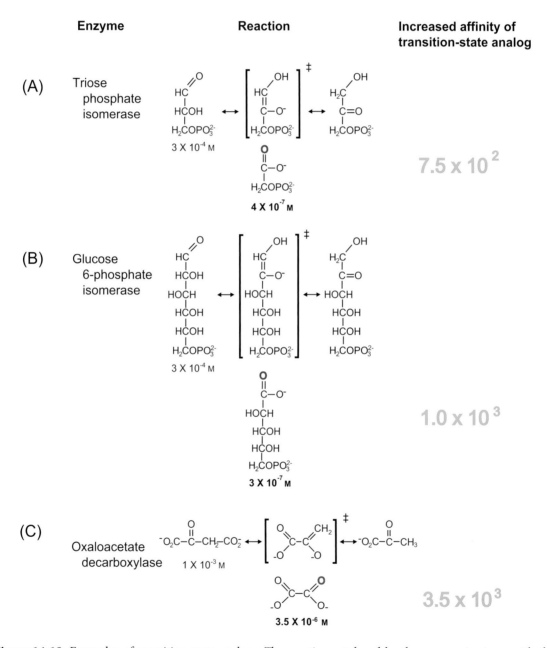

Figure 14-18. Examples of transition-state analogs. The reaction catalyzed by the enzyme is given, with the presumed structure of the transition state in brackets. Below it is the structure of the transition-state analog, along with its observed dissociation constant. The dissociation constant of the normal substrate is given below its structure. The greater affinity of the transition-state analog is the ratio of the two values. From T. E. Creighton (1993) *Proteins: structures and molecular properties*, 2nd edn, W. H. Freeman, NY.

(Figure 14-18). An enzyme would not be expected to bind an intermediate state as tightly as it would a transition state. Nevertheless, molecules designed to approximate the transition state are found to bind much more tightly to enzymes than do substrates.

Many enzymes transfer phosphate groups from one substrate to another, and **vanadate**, VO_3^-, is an excellent mimic of the transition state involved because it is a planar molecule that tends to form pentacovalent complexes exhibiting trigonal bipyramidal geometry:

$$\underset{\substack{\text{Phosphoryl}\\\text{transition}\\\text{state}}}{\begin{array}{c}O\quad\;O^-\\\diagdown\;\diagup\\\cdots\cdots P\cdots\cdots\\|\\O\end{array}}\qquad\underset{\text{Vanadate ion}}{\begin{array}{c}O\quad\;O^-\\\diagdown\;\diagup\\V\\|\\O\end{array}}$$

(14.57)

In many cases, vanadate complexes provide the most accurate transition state that can be reasonably attained. Vanadate readily forms covalent bonds with a variety of ligands and has produced a wide variety of transition state mimics.

Tight binding alone is not evidence that an inhibitor approximates the transition state. Methotrexate, an analog of dihydrofolate, is a potent inhibitor of **dihydrofolate reductase** from *Escherichia coli* and shows other characteristics of the behavior expected of intermediate-state analogs. But the strong inhibition of dihydrofolate reductase by methotrexate is not because it behaves as an intermediate-state analog, for it is bound upside down relative to the binding of dihydrofolate. Instead, there are more interactions between the inhibitor and enzyme bound in that way than between the enzyme and normal substrate. **Enzymes have not been selected by evolution to have very small dissociation constants for their substrates**, which normally approximate the concentration of the substrate that is encountered naturally.

The most convincing evidence of transition state-like behavior is that of the **phosphoroamidate inhibitors of thermolysin**, which mimic the tetrahedral intermediate that is believed to occur transiently in the reaction. A series of related substrates and transition-state analogs demonstrate a reasonable correlation between the K_i values for the transition-state analogs and the ratio K_m/k_{cat} for the substrates, but not with the K_m values alone (Figure 14-19). **The value of K_m/k_{cat} is proportional to the free energy of the enzyme transition state relative to that of the free substrate and enzyme**, so the correlation of this value with K_i for the inhibitors indicates that they are truly transition-state analogs and not just analogs of the substrate. Nevertheless, none of the K_i values of the analogs approach those expected of true transition states.

A notable feature of intermediate-state analogs is that they often bind very slowly (Section 14.3.L). Slow binding often results from the rapid formation of an enzyme–inhibitor complex that subsequently undergoes a slow conformational change, leading to marked enhancement of the binding of the inhibitor (Equation 14.50). This behavior could well be analogous to the conformational change that an enzyme–substrate complex is believed to undergo in the formation of the transition-state complex. The conformational changes in both instances might be responsible for the stronger interactions and tighter binding.

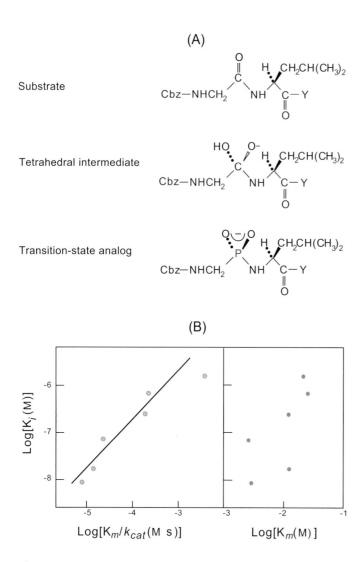

Figure 14-19. Comparison of a series of substrates and transition-state analogs of thermolysin. (A) Structures of the substrate, the tetrahedral intermediate believed to occur during the reaction, and transition-state analogs. The series of substrates and analogs have various Y groups attached to the terminal carboxyl group of the blocked dipeptide Cbz–Gly–Leu–. Cbz is the benzyloxycarbonyl group blocking the terminal amino group. The transition-state analogs have a phosphonamidate replacing the peptide group of the substrate. This was modeled to mimic the tetrahedral intermediate, in which the carbonyl carbon also has tetrahedral geometry. (B) Correlation between the inhibition constant, K_i, for the transition-state analogs of (A) and the catalytic parameters of the corresponding substrates. The measured values of K_i are correlated much more closely with the measured values of K_m/k_{cat} (*left*) than with just their K_m values (*right*). Data from P. A. Bartlett & C. K. Marlowe (1983) *Biochemistry* **22**, 4618–4624.

Enzymatic transition states and transition state analogs. V. L. Schramm (2005) *Curr. Opinion Struct. Biol.* **15**, 604–613,

Enzymatic transition states: thermodynamics, dynamics and analog design. V. L. Schramm (2005) *Arch. Biochem. Biophys.* **433**, 13–26.

A Trojan horse transition state analog generated by MgF_3^- formation in an enzyme active site. N. J. Baxter *et al.* (2006) *Proc. Natl. Acad. Sci. USA* **103**, 14732–14737.

Electrostatic contributions to binding of transition-state analogs can be very different from the corresponding contributions to catalysis: phenolates binding to the oxyanion hole of ketosteroid isomerase. A. Warshel *et al.* (2007) *Biochemistry* **46**, 1466–1476.

14.4.C. Entropic Contributions

The magnitude of enzyme catalysis is given by the ratio of the rate at which conversion of substrate S to product P occurs on the enzyme, k_{cat}, and the uncatalyzed rate in solution in the absence of the enzyme, k_{uncat} (Equation 14.55). With unimolecular reactions this comparison is straightforward, but not with reactions involving two or more substrates. The nonenzymatic rate constants are of second-order, or even higher, whereas the enzymic rate, k_{cat}, is always first-order. A chemical multi-molecular reaction requires the simultaneous encounter of the reactants, so consequently its rate is proportional to each of their concentrations in solution. With enzymes, the reactants are bound to the enzyme, and the reaction occurs in this complex in an essentially unimolecular process.

This points to a major way in which enzymes can catalyze reactions: in unimolecular reactions the effective concentrations of reactants can far exceed those that are possible in bimolecular reactions (Table 14-6). The maximum is believed to be about 10^{10} M, which simply reflects the entropy that must be lost in a bimolecular encounter between reactants in solution that need not occur in a unimolecular reaction. With an enzyme–substrate complex, the substrates have already lost this entropy in binding to the enzyme, and this factor is included in their binding affinities. Consequently, **there is a large entropic advantage in a multi-molecular reaction occurring on an enzyme**. The greater the number of reactants, the greater the advantage. For example, the probability of simultaneous encounters between three or four reactants in solution, when all are present at low concentrations, is essentially nil, whereas ternary and quaternary complexes on an enzyme are not uncommon. Of course, simultaneous binding of the reactants to the enzyme is not sufficient; high effective concentrations occur only if the reactive groups are held in the appropriate positions and orientations for reaction to occur (Table 14-6). Substrates held apart would be prevented from reacting and would have an effective concentration of zero.

This entropic factor might seem not to apply to unimolecular reactions with only a single substrate. Virtually every chemical reaction can be catalyzed in solution, however, for example by nucleophiles or electrophiles in general base or acid catalysis (Figure 14-1), and the rate of the reaction depends upon the concentrations of the catalysts. Enzymes have numerous potential functional groups that could play such a role and could be present at very high effective concentrations to the substrate, if held in correct position within the active site. The structures of enzyme complexes with transition-state analogs indicate a **general tendency for enzyme active sites to close around the substrate** in such a way as to maximize binding contacts, which will increase the effective concentrations of the interacting groups. **Effective concentrations in excess of 10^8 M have been measured**. For example, the reactions of Table 14-6 are catalyzed by acid, hydroxide ion and imidazole, with second-order rate constants of between 10^{-6} and 10^{+6} $M^{-1}s^{-1}$. If just one of these catalysts were present in an enzyme active site in exactly the correct position to present an effective concentration of 10^8 M to the substrate, the reactions catalyzed on the enzyme would have first-order rate constants of between 100 s^{-1} and 10^{14} s^{-1} by this mechanism alone. These hypothetical reactions would occur between 10^8 and 10^{20} times more rapidly than the uncatalyzed reaction in solution. Consequently, **unimolecular reactions may, in theory, also be catalyzed at very high rates on enzymes simply by incorporating a chemical acid or base catalyst into the appropriate position within the active site**.

The presence of multiple substrate molecules or catalytic groups within an enzyme–substrate complex implies that the reactions that occur there may be somewhat different from those that occur in solution. The most favorable mechanism in solution might be limited to bimolecular encounters between reactants or substrates, simply because higher order encounters are so improbable. On the enzyme there is no such entropic restriction on encounters between multiple groups, because of the unimolecular nature of the complex. Concerted reactions between multiple groups can readily occur on an enzyme; for example:

$$\text{B:} \curvearrowright \text{H-N:} \curvearrowright \text{C=O} \rightleftharpoons \text{BH}^+ \quad \text{N-C-O}^- \qquad (14.58)$$

Table 14-6. Relative rates of inter- and intramolecular examples of the esterification reaction

$$R\text{-}\mathbf{OH} + \mathbf{HO}\text{-}\overset{\overset{O}{\|}}{C}\text{-} \longrightarrow R\text{-}O\text{-}\overset{\overset{O}{\|}}{C}\text{-} + H_2O$$

Reactants	Rate constant	Effective concentration of intramolecular groups
phenol + $HO\overset{O}{\overset{\|}{C}}CH_3$	10^{-10} s^{-1} M^{-1}	–
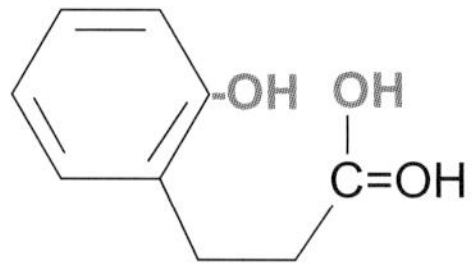 2-(hydroxyphenyl)propanoic acid	3.2×10^{-6} s^{-1}	3.2×10^{4} M
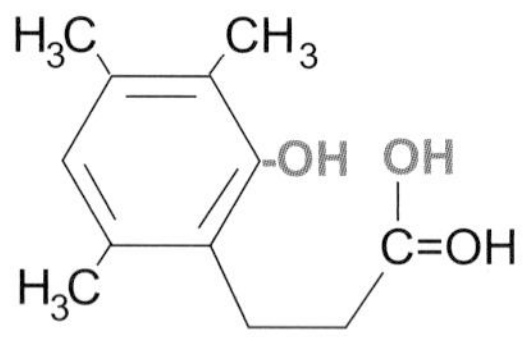 trimethyl-substituted 2-(hydroxyphenyl)propanoic acid	3.3×10^{-6} s^{-1}	3.3×10^{4} M
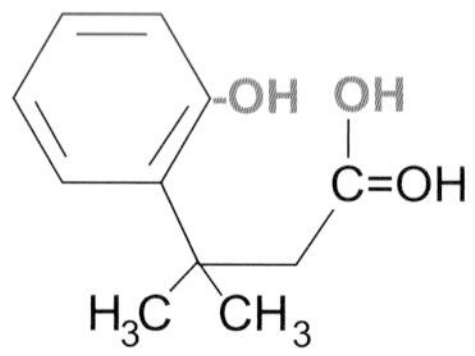 gem-dimethyl 2-(hydroxyphenyl)propanoic acid	3.6×10^{-5} s^{-1}	3.6×10^{5} M
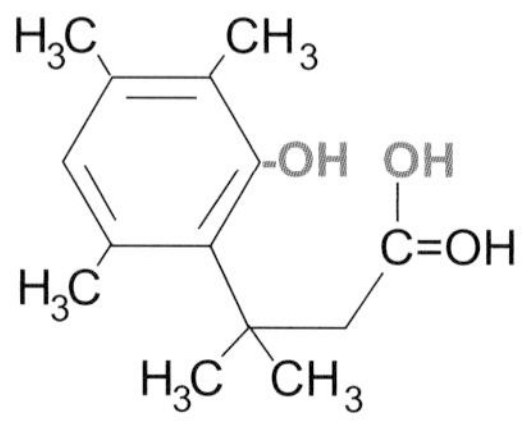 trimethyl + gem-dimethyl 2-(hydroxyphenyl)propanoic acid	8.5×10^{-2} s^{-1}	8.5×10^{8} M

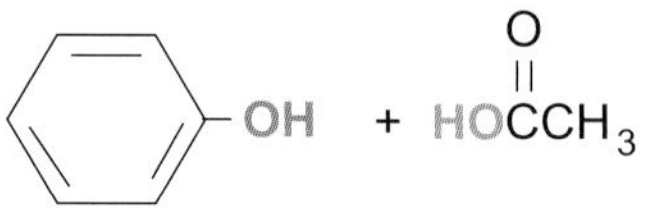

Data from S. Milstien & L. A. Cohen (1970) *Proc. Natl. Acad. Sci. USA* **67**, 1143–1147.

Role of entropy in increased rates of intramolecular reactions. A. A. Armstrong & L. M. Amzel (2003) *J. Am. Chem. Soc.* **125**, 14596–14602.

The effective molarity of the substrate phosphoryl group in the transition state for yeast OMP decarboxylase. A. Sievers & R. Wolfenden (2005) *Bioorg. Chem.* **33**, 45–52.

14.4.D. Bisubstrate Analogs

Bisubstrate analogs were developed originally for mechanistic studies on enzymes that catalyze reactions with two substrates or two products: they embody in a single molecule the structural features of each of the two substrates (or products). Hence it was expected that they would bind simultaneously at the binding sites for the two substrates more tightly than the individual substrate molecules, and act as potent enzyme inhibitors. They have also been referred to as transition-state analogs, but such a classification may not always be appropriate.

Two early, and now classical, examples of bisubstrate enzyme inhibitors are P^1,P^5-di-(adenosine-5′) pentaphosphate (Ap_5A) and *N*-phosphonoacetyl-L-aspartate (PALA).

Bisubstrate inhibition: theory and application to N-acetyltransferases. M. Yu *et al.* (2006) *Biochemistry* **45**, 14788–14794.

Conjugation of adenosine and hexa-(D-arginine) leads to a nanomolar bisubstrate-analog inhibitor of basophilic protein kinases. E. Enkvist *et al.* (2006) *J. Med. Chem.* **49**, 7150–7159.

---

1. Ap_5A

Ap_5A was developed as an inhibitor of **adenylate kinase**, which is a monomeric enzyme that catalyzes the reaction:

$$\text{Mg} \cdot \text{ATP}^{2-} + \text{AMP}^{2-} \leftrightarrow \text{Mg} \cdot \text{ADP}^- + \text{ADP}^{3-} \tag{14.59}$$

The catalyzed reaction conforms to a rapid-equilibrium random kinetic mechanism, which implies that the enzyme possesses two distinct nucleotide-binding sites within its active site. One is for either $\text{Mg} \cdot \text{ATP}^{2-}$ or $\text{Mg} \cdot \text{ADP}^-$ and the other is for ADP^{3-} or AMP^{2-}. Irrespective of whether they are substrates or products, the moieties bound by the enzyme include two adenosine moieties and four phosphate moieties; Ap_5A differs only in having five phosphate groups linked covalently between the adenosine moieties. Ap_5A is a potent inhibitor of the adenylate kinase reaction, and the inhibition is competitive with respect to both Mg·ATP and AMP, as would be expected. The stoichiometry of binding is 1:1, with a dissociation constant of 15 nM. The bisubstrate analog binds to both substrate-binding sites simultaneously and links them (Figure 14-20).

The substrates, transition state and products have only four phosphate groups, so it might be thought that Ap_4A, which is the equivalent of covalently linking ATP to AMP, or ADP to ADP, would be the best inhibitor, but its binding is almost 3000-fold weaker. The reason is probably that the two bonds between the phosphate groups that are being broken and formed in the transition state will be elongated, so an extra phosphate group is the best approximation to the transition state. An increase in the number of phosphoryl groups to six also reduces the binding affinity, by 400-fold.

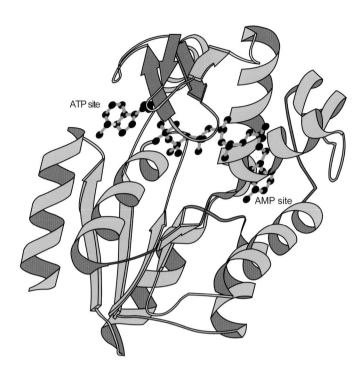

Figure 14-20. The structure of adenylate kinase with the bisubstrate analog Ap$_5$A bound. The course of the polypeptide backbone is indicated schematically, with *arrows* for β-strands and *coils* for α-helices. The ligand is shown as *dark balls* and *sticks*. Adenylate kinase catalyzes the transfer of the terminal phosphate group of ATP to AMP, to produce two molecules of ADP. The binding site for ATP is at the *left*, that for AMP to the *right*. The bisubstrate analog spans and links the two substrate binding sites. Adapted from a drawing by G. E. Schulz.

2. N-Phosphonacetyl-L-Aspartate (PALA)

PALA has been used extensively for kinetic and structural studies on **aspartate transcarbamoylase** (Section 15.1.C), which catalyzes the reaction:

$$\text{carbamoyl phosphate} + \text{aspartate} \leftrightarrow \text{carbamoyl aspartate} + P_i \qquad (14.60)$$

PALA is considered an analog of the two substrates linked covalently (Figure 14-21). It might also be a transition-state analog:

$$\text{PALA} \qquad \text{Putative transition-state complex} \qquad (14.61)$$

One molecule of PALA binds to each catalytic subunit of the enzyme, and the inhibition is linear competitive with respect to carbamoyl phosphate and linear noncompetitive relative to aspartate. The kinetic mechanism for the aspartate transcarbamoylase reaction is essentially ordered, with carbamoyl phosphate being the first substrate to add; this explains the inhibition pattern. The K_i value for the enzyme•PALA complex is 7.2 nM, three orders of magnitude lower than the dissociation constant for the corresponding enzyme•carbamoyl phosphate complex.

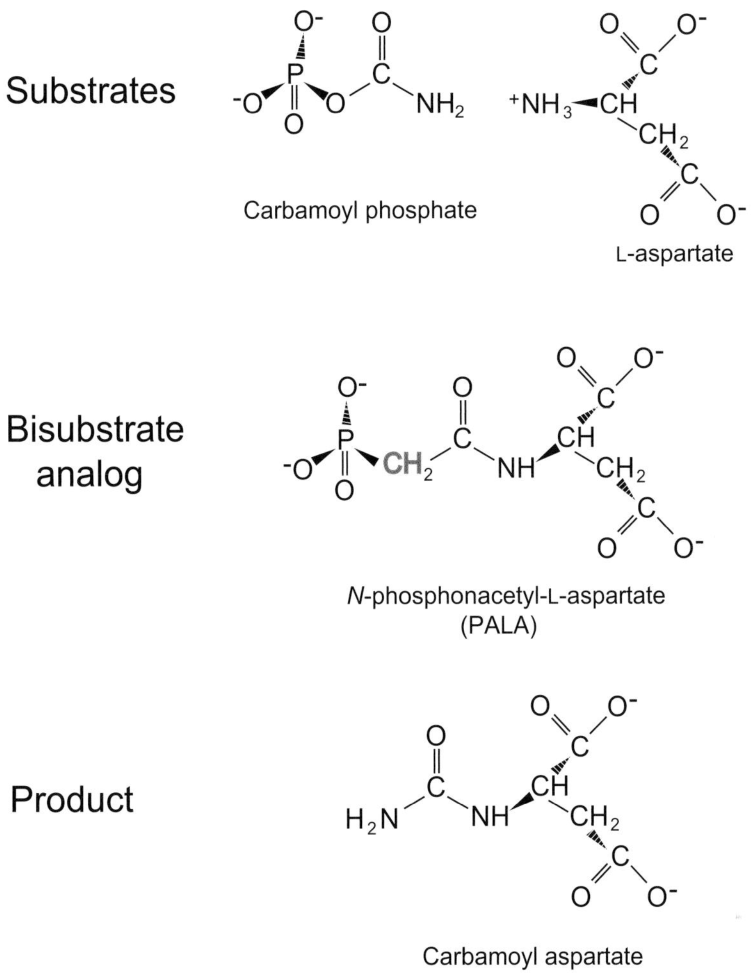

Figure 14-21. Example of a bisubstrate analog. The structures of the substrates of aspartate transcarbamoylase, carbamoyl phosphate and aspartate (*top*), and the product, carbamoyl aspartate (*bottom*), are compared with that of the bisubstrate analog, PALA (*middle*).

Aspartate transcarbamylase from the hyperthermophilic archaeon *Pyrococcus abyssi*: thermostability and 1.8Å resolution crystal structure of the catalytic subunit complexed with the bisubstrate analog N-phosphonacetyl-L-aspartate. S. Van Boxstael *et al.* (2003) *J. Mol. Biol.* **326**, 203–2216.

Design, synthesis and activity of bisubstrate, transition-state analogs and competitive inhibitors of aspartate transcarbamylase. C. Grison *et al.* (2004) *Eur. J. Med. Chem.* **39**, 333–344.

14.4.E. Induced Fit

The specificities of enzymes are determined by how their substrates fit into their active sites. Only the appropriate substrate should fit productively, and it was thought originally that there should be a close steric relationship between an enzyme and its substrate. This suggested the **lock and key** hypothesis of enzyme action. The active site of an enzyme was envisaged to be simply a rigid template for the substrate, which would account for the specificity of an enzyme for only one or a limited number of substrates. Subsequently it was recognized that this mechanism did not account for the very low activities of many enzymes that could use water as a substrate, which is always present at a concentration of 55 M. For example, a kinase that transfers a phosphate group from ATP to a second substrate would be expected to hydrolyze ATP in the absence of the second substrate, simply by substituting water for it. A second puzzle was the failure of enzymes to act on compounds whose structure is very similar to that of the normal substrate. These observations suggested that the active site of an enzyme is not necessarily a template for the substrate, but that binding of the correct substrate to the active site produces a conformational change in the enzyme that induces the correct alignment of the catalytic groups of the enzyme with those of the substrate so that the reaction can occur. This could explain why molecules of similar structure to the normal substrate can combine at the active site but not undergo reaction, because they lack the structural features that are required for the conformational change.

A classic demonstration of substrate-induced conformational change is the binding of glucose to **hexokinase**. This enzyme has two lobes, and the cleft between them contains the active site. Binding of glucose causes rotation of the two lobes and closure of the cleft (Figure 11-5), which facilitates transfer of the phosphoryl group from ATP to glucose, rather than to water, because it is excluded. The apparent affinity of the enzyme for ATP is also increased about 50-fold. ATP binds in the cleft, interacting with both domains in a way that is favored by the conformational change. After rotation of the two lobes, the glucose molecule is almost entirely engulfed by the enzyme, so it cannot enter or leave the active site; dissociation of product from the enzyme is quite slow and rate-limiting and is probably limited by the opening of the cleft.

These induced fit changes upon binding glucose explain why hexokinase does not catalyze efficiently the hydrolysis of ATP in the absence of glucose. If the active site pre-existed, ATP could bind and its phosphate group would be transferred to water at the same rate it is transferred to glucose. Such hydrolysis of ATP would be extremely wasteful physiologically. The conformational changes that occur also explain why the ATPase activity of the enzyme is stimulated by the sugar lyxose, which is not phosphorylated. It interacts at the active site in place of glucose and increases the maximum velocity of the ATPase reaction and lowers the K_m for ATP, presumably by inducing the active enzyme conformation. Lyxose cannot accept the phosphoryl group of ATP but tends instead to be transferred slowly to an adjacent Ser residue of the enzyme.

Further conformational changes probably take place in hexokinase after binding the second substrate, ATP, because the glucose 6-hydroxyl and ATP γ-phosphoryl groups are observed in the individual complexes to be 6 Å apart, too far for the reaction between them to occur in one step.

Numerous changes are observed with the enzyme **adenylate kinase**, which catalyzes the transfer of a phosphoryl group from ATP to AMP, to generate two ADP molecules (Equation 14.59). Upon binding AMP to one domain of this enzyme, that domain closes over the active site, with C^α atoms moving by up to 8.2 Å (Figure 14-22). The binding of ATP alone causes the protein to close over it, but the greatest changes occur when both nucleotides bind. The most dramatic changes are produced by the binding of the Ap_5A bisubstrate analog (Section 14.4.D), which also may be a transition-state analog; the substrates are then removed from contact with water.

Many different kinase enzymes have similar bilobed structures of two or more domains, even though many are not homologous, other than having similar nucleotide-binding motifs (Section 12.4) related to their common use of ATP. All these enzymes are thought to undergo similar movements of the domains relative to each other.

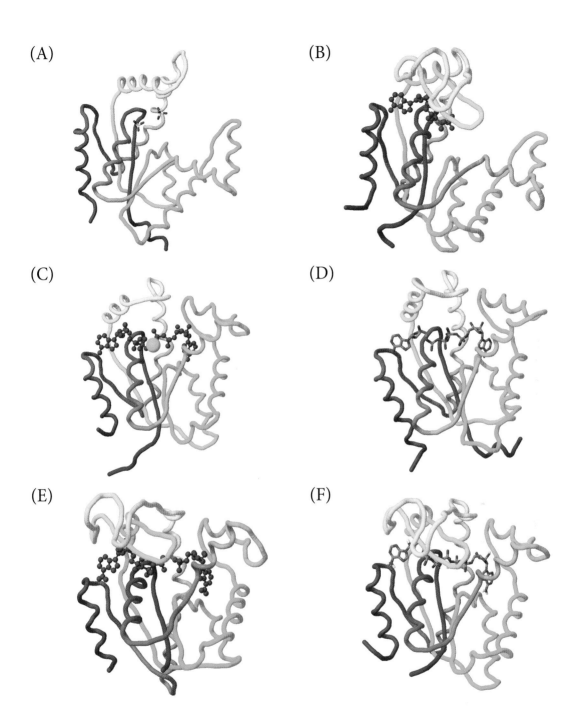

Figure 14-22. Domain movements that occur upon substrate binding to adenylate kinase. Similar views of the protein structure are depicted with various ligands bound. The polypeptide backbone is shown using 'rainbow' mode (Figure 14-2) to facilitate comparison of the structures. The ligands are depicted in various ways. (A) The unliganded enzyme, demonstrating the open nature of the structure. Two sulfate ions are near the active site in the crystalline enzyme. (B) With ATP bound; the *yellow* portion of the backbone closes over it. (C) With two ADP molecules and one Mg^{2+} ion bound, the molecule becomes much more compact. (D) With the ligand P,P-diadenosine tetraphosphate (Ap_4A) bound. (E) With AMP and AMP–PNP bound. (F) With Ap_5A bound, the Ap_5A is presumed to approximate both AMP and ATP bound simultaneously to the active site and may approximate the transition state as well. The proteins originate from various species, so not all the differences need be due to ligand binding. Figure generated from PDB files *3adk, 1dvr, 2cdn, 2c95, 1ank* and *2aky*, respectively, using the program Jmol.

Induced fit changes in the enzyme structure also may indicate that it is necessary for the enzyme to embrace the substrate intimately, bringing reactive groups into its proximity from all directions and preventing the transition state from dissociating. Yet it is also necessary for an enzyme to associate with substrates and release products; the induced fit movements would be a mechanism for providing access to an otherwise inaccessible active site. Indeed, the release of products is often observed to be the rate-determining step in an enzyme-catalyzed reaction. The structural changes in the enzyme during a reaction may also be a reason why transition-state analogs frequently bind so slowly to enzymes (Section 14.3.L).

The phenomenon of induced fit might be expected to be general, but **large conformational changes during the catalytic reaction are observed primarily with kinases**, where it is probably necessary to prevent them from hydrolyzing ATP using water in the absence of the other substrate, the phosphoryl acceptor. The reason that such induced fit seems not to occur more generally may be that it does not contribute to increasing the rate of the reaction; also, structural changes upon binding the substrate increase the K_m for the substrate over what would otherwise be necessary, because some of the binding energy must be used to pull the enzyme into the active conformation. Nevertheless, some small degree of induced fit is apparent in the binding of any ligand to a protein or nucleic acid.

Observing an induced-fit mechanism during sequence-specific DNA methylation. R. A. Estabrook & N. Reich (2006) *J. Biol. Chem.* **281**, 37205–37214.

Structural and kinetic studies of induced fit in xylulose kinase from *Escherichia coli*. E. DiLuccio *et al.* (2007) *J. Mol. Biol.* **365**, 783–798.

Guanylate kinase, induced fit, and the allosteric spring probe. B. Choi & G. Zocchi (2007) *Biophys. J.* **92**, 1651–1658.

14.4.F. Covalent Catalysis

Some reactions catalyzed by an enzyme involve the transient formation of a covalent bond between the substrate and the enzyme. The most obvious examples are the substituted forms of the enzyme that occur in ping-pong mechanisms (Section 14.3.B). **Covalent bonds are formed most commonly as a result of the attack by an enzyme nucleophilic group, such as $-COO^-$, $-NH_2$, $-OH$, $-S^-$ and the N atoms of the imidazole ring of His residues, on an electrophilic moiety of the substrate**. The electrophilic moieties of substrates may be acyl, phosphoryl or glycosyl groups, so the covalent intermediates would be acyl-, phosphoryl- and glycosyl-enzyme complexes, respectively. The reactions involved in the formation of an acyl-enzyme are:

$$\text{Enzyme-X:}^- \quad \overset{R}{\underset{Y}{C}}=O \;\rightleftharpoons\; \left[\text{Enzyme-X-}\overset{R}{\underset{Y}{C}}\text{-O}^- \right]^{\ddagger} \;\rightleftharpoons\; \text{Enzyme-X-}\overset{R}{C}=O + Y^-$$

Acyl-enzyme

(14.62)

The second step in covalent catalysis is the attack by a low molecular weight nucleophile on the covalent intermediate to release the second reaction product. If the attacking nucleophile is water, the enzyme would catalyze a hydrolysis reaction with a kinetic mechanism involving the ordered release of products. If the attacking nucleophile were a second substrate, the reaction would have a ping-pong kinetic mechanism.

Protein molecules have few electrophilic groups, so enzymes tend to use metals or prosthetic groups that act as electron sinks during catalysis (Section 14.4.G). Many enzymes catalyze reactions through the formation of covalent intermediates, so this type of catalysis probably offers some general advantages. Immobilization of a covalent intermediate should provide a significant entropic advantage, and the catalytic efficiency may be increased by having several steps involving bond making and breaking taking place within a single active site.

14.4.G. Cofactors, Coenzymes and Prosthetic Groups

A **cofactor** is an additional compound, besides the enzyme, that is required for an enzymatic reaction but remains unchanged at the end of the reaction. **Coenzymes**, in contrast, are organic molecules that participate in enzymatic reactions. Compounds that are covalently or tightly bound to an enzyme are usually regarded as **prosthetic groups**. They provide groups that the protein component of the enzyme does not possess but are essential for catalysis. Metal ions, in particular, are often incorporated into the active sites of enzymes to provide the electrophilic properties that are not available with any of the normal groups on proteins.

Great metalloclusters in enzymology. D. C. Rees (2002) *Ann. Rev. Biochem.* **71**, 221–246.

---

1. Pyridoxal Phosphate

Pyridoxal 5′-phosphate (pyridoxal-P) is the most spectacular illustration of how a coenzyme can provide chemical functions that are not available with the normal amino acid residues. It is essential for numerous enzymes, many of which function in all living cells. Enzymes that require pyridoxal-P are unrivaled in the diversity of the reactions that they catalyze. Almost 1.5% of all genes in many free-living prokaryotes code for pyridoxal-P-dependent enzymes, but the percentage is substantially lower in higher eukaryotes because these enzymes are involved mainly in basic metabolism.

Pyridoxal-P is a phosphate ester of the aldehyde form of vitamin B_6, a rather simple derivative of 3-hydroxypyridine. There are two enzymatically important forms:

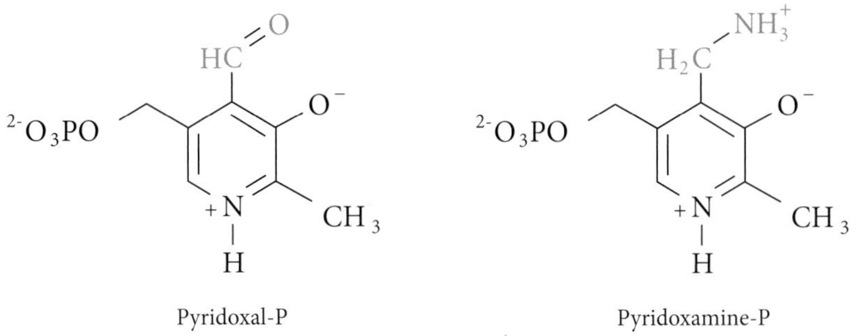

$$\text{Pyridoxal-P} \qquad \text{Pyridoxamine-P} \tag{14.63}$$

The ionic forms illustrated predominate both in solution and in most enzymes.

The carbonyl group of pyridoxal-P usually forms a **Schiff base** (Section 7.2.E.4) with the amino group of a specific Lys side-chain of the enzyme, which is commonly called the **internal aldimine**:

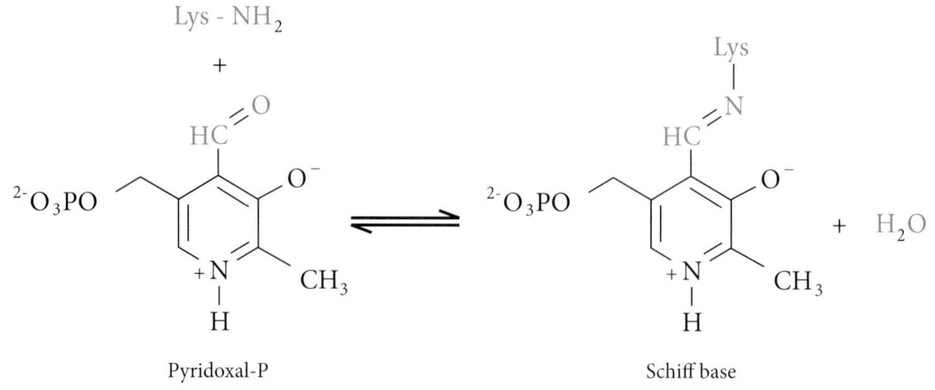

$$\text{Pyridoxal-P} + \text{Lys-NH}_2 \rightleftharpoons \text{Schiff base} + \text{H}_2\text{O} \tag{14.64}$$

When an amino acid substrate binds to the active site, its amino group displaces that of the Lys side-chain, to produce a stable hydrogen-bonded Schiff base with the substrate, the **external aldimine**:

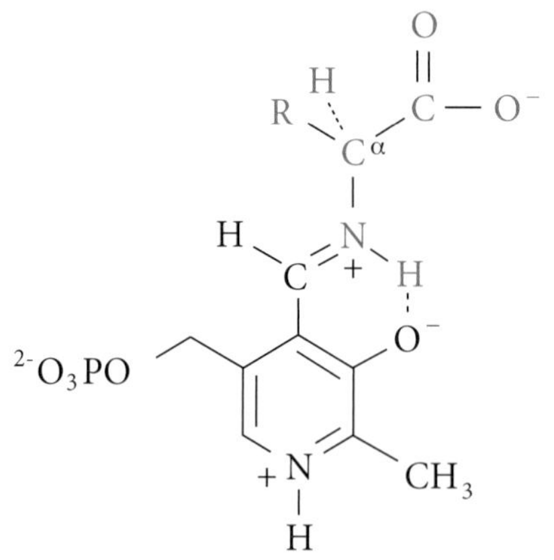

$$\tag{14.65}$$

The atoms of the amino acid substrate are indicated in red.

The protonated ring nitrogen is a powerful electrophilic center that can assist in breaking any one of the three bonds to the C^α atom of the substrate. The result can be removal of the carboxyl group, H atom or side-chain. Removal of the H^α is the first step in a number of reactions, such as those involving elimination of the β- or γ-substituent of the amino acid side-chain. A large number of decarboxylases remove the α-carboxyl group as carbon dioxide. Cleavage of the side-chain is of more limited occurrence but occurs in some important metabolic reactions, such as cleavage of serine by serine hydroxymethylase to produce glycine and formaldehyde. Temporary removal of any of the three groups can produce racemization to the D-isomer. The many resonance forms of the pyridoxal-P cofactor also help to stabilize the electronic transitions that are required for each of these steps. Which type of reaction occurs is determined by the structure of the enzyme active site and the way in which the amino acid part of the Schiff base is held.

Removal of the H atom from the C$^\alpha$ atom produces another Schiff base called the **ketimine**:

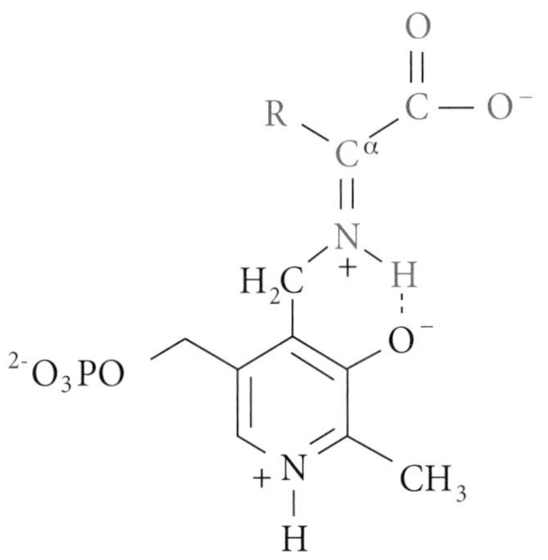

$$\begin{array}{c}\text{[ketimine structure: pyridoxal-P ring with 2-O}_3\text{PO-CH}_2\text{ substituent, protonated ring N-H, 3-O}^-\text{ hydrogen-bonded to iminium N-H, 4-CH=N}^+\text{ linkage to C}^\alpha\text{ which bears R and COO}^-\text{, C}^\alpha\text{=N double bond, 2-CH}_3\text{]}\end{array}$$

(14.66)

Hydrolysis of the double bond to the C$^\alpha$ atom by water produces pyridoxamine-P (Equation 14.63). This is the product of a half-reaction of a transaminase.

Pyridoxal-P is widely used to label amino groups in proteins (Section 7.2.E.4). The Schiff bases formed can be reduced with borohydride, BH_4^-, which may be radiolabeled with deuterium or tritium. The reduced form is stable, in contrast to the original Schiff base.

Pyridoxal phosphate enzymes: mechanistic, structural, and evolutionary considerations. A. C. Eliot & J. F. Kirsch (2004) *Ann. Rev. Biochem.* **73**, 383–415.

Reaction specificity in pyridoxal phosphate enzymes. M. D. Toney (2005) *Arch. Biochem. Biophys.* **433**, 279–287.

Covalent catalysis by pyridoxal: evaluation of the effect of the cofactor on the carbon acidity of glycine. K. Toth & J. P. Richard (2007) *J. Am. Chem. Soc.* **129**, 3013–3021.

14.4.H. Suicide Substrates

A **suicide substrate** is also known as a **mechanism-based inhibitor**. It is a structural analog of an enzyme substrate that contains a latent reactive group that is activated by the enzyme as a result of catalysis of its normal reaction. Consequently, a chemically reactive group is unleashed within the enzyme's active site, which can react covalently with any appropriate active site residue that is nearby. This inactivates the enzyme, which has **catalyzed its own suicide**. The many functional groups that can be catalytically unmasked by their target enzymes include acetylenes, olefins and β-substituted amino acids.

Most suicide substrates are based on the generation of an intermediate that has reactive conjugated double bonds. The classical example is that for the enzyme β-**hydroxy-decanoyl-dehydrase**, which normally catalyzes the reactions:

$$C_6H_{13}-CH=CH-CH_2-CO-NAC$$
$$\downarrow$$
$$C_6H_{13}-CH_2-CH=CH-CO-NAC$$
$$\downarrow +H_2O$$
$$C_6H_{13}-CH_2-CHOH-CH_2-CO-NAC \quad (14.67)$$

where NAC is $-S-CH_2-CH_2-NH-CO-CH_3$. The suicide inhibitor has a triple bond in place of the double bond of the normal substrate, so it is isomerized by the enzyme to a form with two adjacent double bonds. This species is chemically unstable and reacts with a His residue in the enzyme active site:

$$C_6H_{13}-C\equiv C-CH_2-CO-NAC + \text{enzyme}$$
$$\downarrow$$
$$C_6H_{13}-CH=C=CH-CO-NAC\cdot\text{enzyme}$$
$$\downarrow$$
$$C_6H_{13}-CH=\underset{|\atop\text{enzyme}}{C}-CH_2-CO-NAC \quad (14.68)$$

Reactions involving the suicide substrate (SS) can be formulated as:

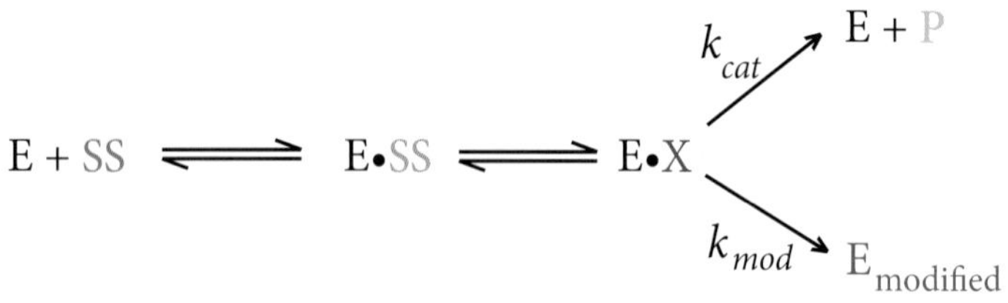

$$E + SS \rightleftharpoons E\cdot SS \rightleftharpoons E\cdot X \begin{array}{c} \overset{k_{cat}}{\nearrow} E + P \\ \underset{k_{mod}}{\searrow} E_{modified} \end{array} \quad (14.69)$$

The enzyme catalyzes conversion of the E•SS complex to an E•X complex in which X is an activated intermediate that can either be released as product, P, or react with the enzyme. The partition ratio that describes the relative rates of the two reactions, k_{cat}/k_{mod}, is a measure of enzyme turnover relative to enzyme inactivation, and it remains constant over the time–course of the reaction.

With a suicide inhibitor as substrate, the rate at which the product appears decreases with time, long before the equilibrium of the reaction is reached. The enzyme would eventually become completely inactivated. These properties are similar to those of a slow-binding enzyme inhibitor (Section 14.3.L) whose action is irreversible. After all the enzyme molecules have been inactivated, the ratio of the total amount of product formed, P_∞, to the total amount of enzyme present, E_t, should equal the partition ratio, k_{cat}/k_{mod}. The smaller the partition ratio, the greater the effectiveness of the suicide inhibitor at modifying the enzyme. Partition ratios have been observed to vary from 1 for the inactivation of GABA aminotransferase by gabaculine and L-aspartate aminotransferase by vinylglycine, when the modification and catalysis steps have the same rate constant, to values in excess of 1000 where the modification reaction is relatively slow.

Structure of the ubiquitin hydrolase UCH-L3 complexed with a suicide substrate. S. Misaghi *et al.* (2005) *J. Biol. Chem.* **280**, 1512–1520.

A fluoro analog of the menadione derivative 6-[2′-(3′-methyl)-1′,4′-naphthoquinolyl]hexanoic acid is a suicide substrate of glutathione reductase. Crystal structure of the alkylated human enzyme. H. Bauer *et al.* (2006) *J. Am. Chem. Soc.* **128**, 10784–10794.

3-nitropropionic acid is a suicide inhibitor of mitochondrial respiration that, upon oxidation by complex II, forms a covalent adduct with a catalytic base arginine in the active site of the enzyme. L. S. Huang *et al.* (2006) *J. Biol. Chem.* **281**, 5965–5972.

14.4.I. Cryoenzymology

Reactions on enzymes normally occur within less than a second, and intermediates are extremely transient and often not populated to significant extents. **Cryoenzymology involves following the kinetics of enzyme action at very low temperatures, down to −70°C, where all reactions become very much slower.** The energy necessary to overcome an activation barrier comes from the kinetic energy of ordinary thermal fluctuations; the rate, therefore, is decreased exponentially as the temperature is lowered to an extent determined by the enthalpy and entropy of the transition state. In this way, the various steps on the enzyme can often be slowed sufficiently to be followed kinetically. Intermediates may accumulate for much longer times, and these productive species may be studied by the most powerful methods. New intermediates may also be detected at low temperatures, because the rates of the various steps are usually slowed to different extents and new steps may become rate-determining. Possible drawbacks of this approach include the need to use antifreeze solvents, such as 50–80% methanol, dimethyl sulfoxide or ethylene glycol; there is always the possibility that the antifreeze, as well as the low temperature, will change the reaction mechanism from that under normal conditions.

A related method is to trap intermediate states by rapid freezing, by suddenly lowering the temperature drastically. This method, known as **flash freezing**, is especially useful with crystalline forms of enzymes, for the structure of the species trapped can be determined directly by X-ray crystallography.

Enzymes from organisms that live naturally in cold environments (psychrophiles) have evolved a range of structural features that give them greater flexibility than their homologs from other organisms. High flexibility, particularly around the active site, is translated into a low activation enthalpy, low substrate affinity and high specific activity at low temperatures. High flexibility is also accompanied by a trade-off in stability, resulting in heat lability and, in the few cases studied, cold lability.

Direct crystallographic observation of an acyl-enzyme intermediate in the elastase-catalyzed hydrolysis of a peptidyl ester substrate: exploiting the 'glass transition' in protein dynamics. X. Ding *et al.* (2006) *Bioorg. Chem.* **34**, 410–423.

Cold-adapted enzymes. K. S. Siddiqui & R. Cavicchioli (2006) *Ann. Rev. Biochem.* **75**, 403–433.

14.4.J. Time-Resolved Crystallography

The ultimate goal of enzymology is to determine the structures and energetics of all the intermediate and transition states along the enzyme reaction coordinate. Of course, an enzyme with only one of its two substrates bound is stable indefinitely and its structure can be determined using classical crystallographic methods, but the species within which catalysis takes place are intrinsically very unstable and short-lived. **It is feasible to determine the structures of very short-lived crystalline species using Laue crystallography**, as crystallographic data may be collected with a pulse of only some 100 picoseconds of multi-wavelength radiation from a synchrotron. The greatest difficulty is to synchronize all the enzyme molecules within the crystal so that all are at the same point in the reaction sequence at the same time. This is generally accomplished by using a 'caged ligand' that can be diffused into the crystal in an inactive form and then activated by a pulse of light to initiate the reaction. This is usually accomplished by releasing a protecting group that masked the reactive part of the substrate.

The caged substrate has been found in some instances to be bound by the enzyme in an unusual mode, but removal of the protecting group permits the liberated substrate to bind rapidly in the correct manner. A technical difficulty encountered is that the protecting group that is released often tends to react with the protein. A more fundamental problem is that only particularly stable intermediate states will be visible by this technique. If such intermediates do not accumulate on a substantial fraction of the enzyme molecules in the crystal, all that will be observed crystallographically will be a shift with time in the proportion of enzyme molecules that have substrate bound to those having product bound. To overcome this, the energies of the various intermediates can often be changed by varying the substrate concentrations. Alternatively, mutant forms of the enzyme lacking a catalytic group can be blocked at a certain step, so the intermediate immediately preceding that step accumulates.

Such studies have elucidated most of the steps that occur during catalysis by a few enzymes. **Somewhat surprisingly, only small movements of no more than 2 Å are observed of the side-chains of the various residues in the active site of the enzyme upon binding of the substrates and during their conversion to products.**

Millisecond Laue structures of an enzyme–product complex using photocaged substrate analogs. B. L. Stoddard et al. (1998) *Nature Struct. Biol.* **5**, 891–897.

Ultrafast time-resolved crystallography. K. Moffat (1998) *Nature Struct. Biol.* **5**, 641–643.

New results using Laue diffraction and time-resolved crystallography. B. L. Stoddard (1998) *Curr. Opinion Struct. Biol.* **8**, 612–618.

14.4.K. Polymeric Substrates: Processivity

Substrates that are extended polymers, such as nucleic acids, polypeptide chains and carbohydrates, differ from those discussed up until now in that they offer an enzyme many sites at which it can attach and act. Enzymes acting on polymers are usually classified as being either **exo** or **endo**, depending upon whether they act at the ends or in the middle of the polymeric substrate. There is also the possibility that the enzyme will continue to work on the next polymeric unit of the substrate without dissociating first, a phenomenon known as being **processive**.

Many examples of processive enzymes are known. For example, β-**amylase** is a starch-hydrolyzing exo-type enzyme that can catalyze the successive liberation of β-maltose units from the ends of α-1,4-

linked glucopyranosyl polymers. In a phenomenon known as multiple or repetitive attack, the enzyme releases several maltose molecules from a single enzyme–substrate complex. The multiple attack action requires that the enzyme slide along the substrate, plus form a stable productive substrate–enzyme complex through a hydrogen bond between the nonreducing end of the substrate and the enzyme.

The **RecBCD** enzyme is a nuclease that hydrolyzes both single- and double-stranded DNA substrates, but it is unusual in that it requires ATP. The degradative reaction is processive, because the degradative reaction is more rapid than dissociation from the DNA; consequently, including excess alternative single-stranded DNA to trap free RecBCD has no effect on the nuclease reaction. The reaction begins near the 3′-end of the DNA substrate, and the major cleavage sites are two to four phosphodiester bonds apart. When the ATP concentration is varied 10-fold, the overall reaction rate varies about 10-fold but the product distribution is unchanged. These observations indicate that **DNA cleavage is tightly coordinated with movement of the enzyme along the DNA**, and ATP hydrolysis is probably required for this movement. The reaction time–courses at low concentrations of ATP exhibit a significant lag period before cleavage products appear, which may reflect movement of the DNA from an initial binding site in the helicase domain of the RecB subunit to the nuclease active site in a separate domain of RecB, which requires ATP hydrolysis.

The enzymes that replicate nucleic acids also tend to process along the template while synthesizing the complementary strand, polymerizing thousands of nucleotides without dissociating (Chapter 6). The protein responsible is usually one part of the polymerase and is known as the **sliding clamp** (Figure 14-23). It is a ring-shaped protein that encircles the DNA and anchors the polymerase to the DNA template. The sliding clamp can thread itself onto (and off) a free end of double-stranded DNA, but otherwise it cannot assemble itself around DNA, so it is loaded onto it by another protein, known as the **clamp loader**. Sliding clamps have circular six-fold symmetry, being composed of six identical subunits that can be two homotrimers or three homodimers (Figure 14-24). The central cavity accommodates double-stranded DNA. The inner surface is positively charged, as would be expected, to interact with the phosphate groups of the double helix, but the outer surfaces are negatively charged, probably to prevent the protein from sticking irreversibly to the DNA template.

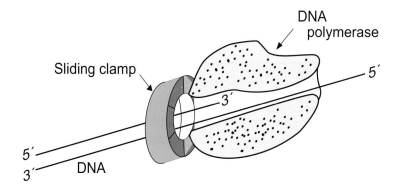

Figure 14-23. Schematic model for a sliding clamp acting as a processivity factor for a DNA polymerse. The sliding clamp encircles the DNA and binds to the polymerase, thereby tethering it to the DNA so that it does not dissociate while synthesizing the complementary strand.

Many other enzymes that act on DNA use sliding clamps.

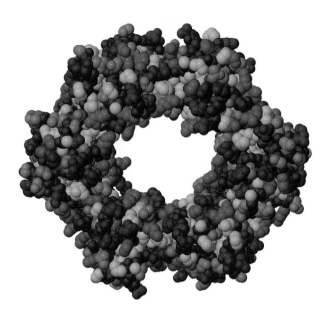

Figure 14-24. Structure of the β-subunit of the DNA polymerase of *E. coli*, which functions as the sliding clamp. The protein is a dimer of homotrimers. Basic groups are depicted in *blue*, acidic groups in *red*. The hole in the center of the ring is sufficient to accommodate double-stranded DNA. Figure generated from PDB file *3d1g* using the program Jmol.

Processive enzyme mimic: kinetics and thermodynamics of the threading and sliding process. R. G. Coumans et al. (2006) *Proc. Natl. Acad. Sci. USA* **103**, 19647–19651.

Kinetics of ATP-stimulated nuclease activity of the *Escherichia coli* RecBCD enzyme. A. Ghatak & D. A. Julin (2006) *J. Mol. Biol.* **361**, 954–968.

Surface density of cellobiohydrolase on crystalline celluloses. A critical parameter to evaluate enzymatic kinetics at a solid–liquid interface. K. Igarashi *et al.* (2006) *FEBS J.* **273**, 2869–2878.

The principles of guiding by RNA: chimeric RNA–protein enzymes. A. Huttenhofer & P. Schattner (2006) *Nature Rev. Genet.* **7**, 475–482.

Kinetic and structural analysis of enzyme sliding on a substrate: multiple attack in β-amylase. K. Ishikawa *et al.* (2007) *Biochemistry* **46**, 792–798.

14.4.L. Enzyme Function *In Vivo*: Towards 'Perfection'

It is relatively straightforward to imagine what properties the most efficient enzyme possible would have, if it is assumed that the ultimate goal would be to have the greatest catalytic capacity so that the smallest amount of enzyme would be needed. The impulsive conclusion would be that such an enzyme should have the greatest conceivable value of k_{cat} and the lowest possible K_m. Such an enzyme, however, would be limited by the rate at which substrate molecules could diffuse to its active site, k_D: in general, **the value of k_{cat}/K_m cannot be greater than k_D**. Increasing the value of k_{cat} beyond that where binding of substrate becomes rate-limiting is unlikely to be beneficial, for such increases in k_{cat} would have to be accompanied by corresponding increases in K_m. In general, **an enzyme is unlikely to be more efficient than is functionally necessary *in vivo***. Selective evolutionary pressures undoubtedly maintain the catalytic properties of enzymes because the majority of mutational events that affect their functional properties are detrimental.

On the basis of what is feasible, a so-called 'perfect' enzyme is considered to be one for which k_{cat}/K_m for the least stable substrate or product has the maximum value of k_D and no intermediate accumulates (Figure 14-25). This requires that the diffusion-limited transition state be that with the highest free

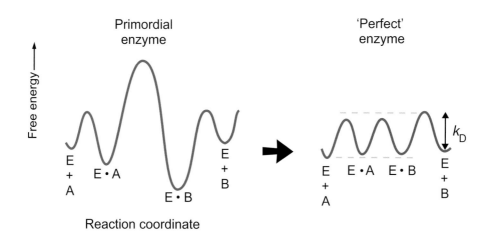

Figure 14-25. Hypothetical illustration of how enzymes would be expected to evolve to approach 'perfection'. The postulated primordial enzyme is inefficient, in that it has a high free energy barrier to catalyzing the reaction. It also has high affinities for the substrate and product, so that the enzyme is usually present as a complex with one or the other, usually B in this case. In the 'perfect' enzyme, the transition state with the highest free energy is limited by the rate of diffusion (k_D) of the less stable substrate, in this case B. the other transition states are of lower free energy. Also, the free energies of the complexes at the physiological concentrations of A and B, and of any intermediates, have higher free energies than does the free enzyme, so that no complexed forms of the enzyme are present at substantial levels. Original kindly provided by J. R. Knowles in T. E. Creighton (1993) *Proteins: structures and molecular properties,* 2nd edn, W. H. Freeman, NY, p. 442.

energy and that the free energies of all the enzyme complexes be somewhat greater than the free energy of the free enzyme and most stable substrate. The relative free energies of the free enzyme, E, and the enzyme–substrate complex, E•A, will depend upon the concentration of the substrate and will be equal when the dissociation constant for the E•A complex is the same as the free substrate concentration, [A]. For E•A and E•B to have comparable free energies, the enzyme will need to have different affinities for the substrate and product to compensate for any difference in their relative free energies in solution, which determine the equilibrium constant for the reaction. The latter parameter is independent of the enzyme. These considerations are frequently found to be obeyed by actual enzymes, in that their K_m values for each substrate are similar to the normal *in vivo* concentration of that substrate, and the most stable forms of the enzyme have similar free energies under such conditions. At least some enzymes are believed to approach optimal efficiency; yeast triose phosphate isomerase is considered to be within 60% of being 'perfect'; evolutionary pressures are considered unlikely to be able to produce a totally perfect enzyme.

Perfection in this sense applies only to one set of conditions, and virtually all enzymes are likely to encounter fluctuating concentrations of their substrates *in vivo*. Moreover, the substrate for one enzyme is usually the product of another enzyme, so the concentration of such a metabolite will depend upon the catalytic activity of both enzymes. A single enzyme should not be considered in isolation of the other components of metabolism.

The wide variation in the maximum velocities for enzyme-catalyzed reactions (Table 14-1) is due to differences in the equilibrium constants for the reactions, as well as to differences in the physiological concentrations of the substrates.

Conditions *in vivo* are often very different from those used normally in studies of enzyme kinetics. For example, the enzymes of glycolysis, a central pathway of metabolism, are often present in cells at high concentrations, with the concentration of each of the enzyme active sites in the range of 10^{-4}–10^{-3} M. **These *in vivo* enzyme concentrations are greater even than the concentrations of some of their substrates**, in particular the metabolic intermediates fructose-1,6-diphosphate, 1,3-diphosphoglycerate and phosphoenolpyruvate, which are present at levels of only about 10^{-4} M. Only the precursors and products of glycolysis, such as glucose-6-phosphate and ATP, occur at concentrations, between 10^{-3} and 10^{-2} M, that are greater than those of the enzymes. That the metabolic intermediates are present at lower concentrations than the enzymes that produce or use them suggests that the intermediates along the metabolic pathway may not usually be free in solution but in enzyme-bound form.

The high *in vivo* concentrations of the glycolysis enzymes make it plausible that they interact physically and that metabolic intermediates may be passed on directly from one enzyme to another, without dissociating into the aqueous milieu. There is some experimental evidence that this can occur *in vitro* between certain pairs of enzymes at high concentrations, and the tendency for different forms of the enzyme to have similar free energies (Figure 14-25) can be explained as the result of evolution to maximize the probability of transfer of a metabolic intermediate between enzymes. On the other hand, the observed properties of individual enzymes *in vitro* have been found sufficient to simulate metabolism *in vivo* if the metabolites are free in solution, so there is no need for such direct transfer between enzymes.

The high concentrations of proteins inside cells and the existence of a molecular architecture within the cellular cytoplasm and organelles make it plausible that many enzymes might exist as multi-molecular aggregates that are not apparent once the cell is disrupted and its contents diluted. A number of multi-molecular enzyme aggregates and polypeptides with multiple enzyme activities are known, and they may simply be the most stable examples. For example, the six enzymes involved in fatty acid synthesis in the mammalian enzyme system are all part of a single large polypeptide chain known as fatty acid synthase. In yeast, they are divided amongst two separate polypeptide chains, while in prokaryotes and plants the six enzymes occur on individual polypeptide chains and can be separated by standard chromatographic procedures. Nevertheless, the individual enzymes are functionally very similar in all of these systems, and the mechanism of fatty acid chain elongation is the same. In each case, it occurs on a large multifunctional complex. During this process, the fatty acid is assembled while attached to the 'acyl carrier protein', which carries it from one active site to another.

Another notable example of variability in the association of various enzyme activities of a biosynthetic pathway with different polypeptide chains is the pathway of tryptophan biosynthesis in microorganisms. Here, only one of the intermediates along the pathway, indole, is kept from dissociating into the aqueous solution. It is an intermediate between two consecutive enzyme activities that are part of the tryptophan synthase bifunctional complex. In some species, the two activities are part of the same polypeptide chain, whereas in others they reside on individual polypeptides, α and β, that aggregate into an $\alpha_2\beta_2$ complex. The most remarkable aspect is that the $\alpha_2\beta_2$ complex has a hydrophobic tunnel through which the indole intermediate is transferred 25–30 Å, from the active site on α where it is produced to that on β where it is used as a substrate.

It is probably necessary for indole to be channeled in this way because it is a very hydrophobic molecule that readily leaks through the cell membrane when released into the cytoplasm. **The other intermediates in tryptophan biosynthesis are ionized or phosphorylated, as are most other metabolic intermediates, so they do not readily pass through membranes.** These intermediates of tryptophan biosynthesis appear to be released into solution, even when produced and used as a substrate by two sequential enzyme activities that are part of the same polypeptide chain. The reasons why several enzyme activities of tryptophan biosynthesis occur on the same polypeptide chain or on different polypeptides that aggregate, and why this varies between species, are not known.

Tunnels are also found in some other, unrelated, enzymes; they transport ammonia, carbon monoxide, acetaldehyde and carbamate.

Tunneling of intermediates in enzyme-catalyzed reactions. A. Weeks *et al.* (2006) *Curr. Opinion Chem. Biol.* **10**, 465–472.

Steady-state kinetic behaviour of two- or n-enzyme systems made of free sequential enzymes involved in a metabolic pathway. G. Legent *et al.* (2006) *C. R. Biol.* **329**, 963–966.

14.4.M. One Example: Tyrosyl tRNA Synthetase

The mechanisms of action of a number of enzymes have been elucidated to significant extents, but the tyrosyl tRNA synthetase from the thermophilic bacterium *Bacillus stearothermophilus* is one of the best-characterized. The structures of some of its complexes have been determined crystallographically, and the energies of all the interactions between it and its substrate, transition state and products have been measured by the effects of deleting each of the groups within the enzyme's active site. The results indicate it to be one of the simplest known enzymes, in that its catalytic powers arise almost entirely from specific binding interactions. There are no polar groups within the active site that are likely to be involved in general acid or base catalysis.

Only the first half of the overall reaction catalyzed by this enzyme has been characterized in detail: formation of the intermediate tyrosyl adenylate (Tyr–AMP) and pyrophosphate (PP_i) from the amino acid tyrosine and ATP:

$$E + \text{tyrosine} + ATP \leftrightarrow E\bullet\text{Tyr–AMP} + PP_i \qquad (14.70)$$

The reaction kinetically is of random order in that either ATP or tyrosine can be bound first:

$$\begin{array}{c}
E \xrightleftharpoons[+\text{Tyrosine}]{K_t} E\bullet\text{Tyrosine} \xrightleftharpoons[+\text{ATP}]{K'_a} E\bullet\text{Tyrosine}\bullet ATP \underset{k_{-3}}{\overset{k_3}{\rightleftharpoons}} E\bullet\text{Tyr-AMP}\bullet PP \overset{K_{pp}}{\rightleftharpoons} E\bullet\text{Tyr-AMP} + PP_i \\
E \xrightleftharpoons[+ATP]{K_a} E\bullet ATP \xrightleftharpoons[+\text{Tyrosine}]{K'_t}
\end{array} \qquad (14.71)$$

The affinity for ATP alone is low, however, and binding tends to be ordered, with tyrosine binding first, ATP second. The transition state relevant to catalysis is the one that occurs between the complex with tyrosine plus ATP and that with the tyrosyl adenylate plus pyrophosphate. Most of the rate and equilibrium constants of the reaction have been measured directly, by equilibrium binding studies, pre-steady-state kinetic methods and steady-state kinetics, so the relative free energies of the relevant forms of the enzyme along the reaction pathway are known.

In the second half of the normal reaction catalyzed by this enzyme, the tyrosyl adenylate is transferred to its specific tRNA, to generate the charged tRNA that is used in protein biosynthesis (Section 13.4.E).

In the absence of the tRNA, the reactive and thermodynamically unstable tyrosyl adenylate is not released from the enzyme, but is sequestered and stabilized in the enzyme active site. Keep in mind that tyrosyl adenylate is not the final product of the enzyme but an intermediate that is stabilized by the enzyme.

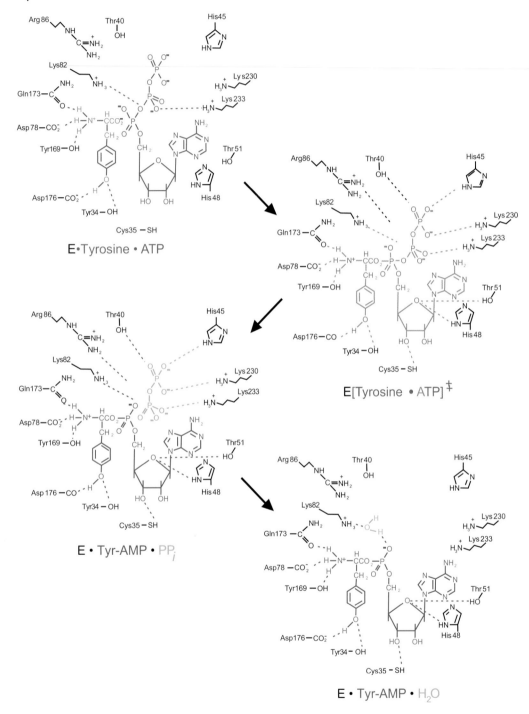

Figure 14-26. Interactions of tyrosyl tRNA synthetase (E) with the substrates tyrosine (Tyr) and ATP, transition state ([Tyr–ATP]‡) and products tyrosyl adenylate (Tyr–AMP) and pyrophosphate (PP_i). The interactions with Tyr and Tyr–AMP have been determined crystallographically; the others are inferred from the results of mutagenesis studies in which the functional groups have been removed by mutation. Adapted from A. R. Fersht et al. (1988) *Biochemistry* **27**, 1581–1587.

The enzyme is dimeric, with two identical 418-residue polypeptides. Only the 320-residue domain responsible for the first half-reaction is observed in crystals; the domain that binds the tRNA is disordered in its absence. The dimeric enzyme exhibits half-of-the-sites reactivity (Section 12.5.C), in that only one molecule of tyrosine is bound to the dimer and only one molecule of tyrosyl adenylate is formed. The dimer in solution is asymmetric, but the structural basis for this is not known, for the crystalline enzyme appears to be symmetrical. Binding of tyrosine induces changes in both active sites.

The interactions that are believed to occur between the enzyme and the substrates, transition state and products, are depicted in Figure 14-26. The roles of the enzyme groups involved in the binding site have been determined by changing the various residues individually and in pairs, using site-directed mutagenesis. The replacements have been to those amino acids that lack the polar groups normally present, but with minimal other changes in overall structure. The measured changes in the relative free energies of each complex indicate the contributions of the group removed to the binding of that reactant (Figure 14-15). For example, residues Asp78, Tyr169 and Glu173 of the enzyme form a binding site for the α-amino group of the bound substrate tyrosine. Mutation of any one of these residues weakens binding of tyrosine by about 3 kcal/mol, and similar effects are observed on other complexes. This indicates that the functions of these mutated residues are only to bind the amino group of the tyrosine substrate, with no substantial change or involvement as the reaction progresses. The specificity for tyrosine rather than, say, phenylalanine, is contributed largely by interactions of its phenolic hydroxyl with the side-chains of Asp176 and Tyr34 of the enzyme. Replacing Tyr34 with Phe decreases the affinity of the enzyme for tyrosine and the stabilities of the other complexes; the Phe34 enzyme still favors tyrosine as substrate, by a factor of 1×10^4 over phenylalanine, but this discrimination has been reduced by a factor of 15. The interactions of Tyr34 with the substrate contribute to binding of tyrosine, but they are of similar magnitude throughout the reaction sequence. Like those involved in binding the α-amino group just described, they do not lower the free energy of the transition state relative to that of the substrates and do not, therefore, contribute directly to catalysis.

Residues Cys35, Thr51 and His48 interact with the ribose ring of the adenylate moiety and contribute substantially to stabilizing the bound adenylate and the transition state, but not the substrate, ATP. Linear free energy relationships observed with the mutations (Figure 14-15) show that, relative to the tyrosyl adenylate complex, 12% of the binding energy of these interactions is realized on the binding of ATP to the E•tyrosine complex, 84% on binding the transition state, and 91% on the complex of tyrosyl adenylate and PP_i. These interactions stabilize the transition state on the enzyme more than the substrates and, therefore, contribute directly to catalysis.

Two residues, Thr40 and His45, do not interact directly with tyrosine, ATP or tyrosyl adenylate in complexes of the enzyme. Changing them to Ala and Gly, respectively, has little effect on the affinities of the enzyme for these species. The mutations do, however, decrease the rate of formation of tyrosyl adenylate by factors of 7000 and 200, respectively; the rate is decreased 3×10^5-fold when both residues are changed. The stability of the E•Tyr–AMP•PP_i complex is also decreased, indicating that these groups interact with the pyrophosphate moiety. Similar observations have been made for the basic side-chains of residues Lys82, Arg86, Lys230 and Lys233. These residues are on flexible loops in the free enzyme and in the complex with tyrosyl adenylate, but they become fixed by interacting with the β- and γ-phosphoryl groups of bound ATP, which become the pyrophosphate product. All of these interactions occur only in the transition state, and so contribute to catalysis, and in the E•Tyr–

AMP•PP$_i$ product, not with the substrate ATP, because of the change in geometry that takes place in the reactants during the reaction: the reaction proceeds by inversion of the configuration of the α-phosphorous atom of ATP, probably through a five-coordinate transition state (Equation 14.57).

A mobile loop of five residues in tyrosyl tRNA synthetase, known as KMSKS for its sequence in one-letter abbreviations (Figure 7-2), is involved in catalysis. It destabilizes the E•Tyr•ATP complex but stabilizes the following E•[Tyr–ATP]‡ transition state for the formation of E•Tyr–AMP. In the absence of this loop, synergistic coupling occurs between the tyrosine and ATP substrates, in that each enhances the binding affinity of the other, which stabilizes the E•Tyr•ATP intermediate preceding the transition-state complex. The mobile loop disrupts this synergism and uses the ATP binding energy to stabilize the transition state for the reaction. The net effect of the mobile loop in the E•Tyr•ATP complex results from several conflicting interactions between side-chains that tend to offset each other, but these conflicting interactions have been minimized in the E•[Tyr–ATP]‡ transition-state complex and replaced by stabilizing pairwise interactions. The mobile loop adopts a highly constrained conformation during formation of the transition-state complex.

Adenosine-5′ tetraphosphate mimics a transition state in which the KMSKS loop develops increasingly tight bonds to the PP$_i$ leaving group, weakening linkage to the α-phosphorous atom of ATP as it is relocated by an energetically favorable domain movement. The KMSKS loop adopts the open form, transiently shifts to the semi-open conformation when the adenosyl moiety binds, and finally assumes the rigid closed form when ATP is bound. This develops high affinity for the adenosine and pyrophosphate moieties of the transition state, which move significantly relative to one another during the catalytic step.

After the amino acid has been activated, the KMSKS loop adopts the semi-open form again, ready to accept the tRNA for the amino acyl transfer reaction. A concerted mechanism for the transfer of tyrosine to tRNA(Tyr) suggests that catalysis of the second step of tRNA(Tyr) amino acylation involves stabilization of a transition state in which the scissile acyl phosphate bond of the tyrosyl adenylate species is strained. Cleavage of the scissile bond on the breakdown of the transition state alleviates this strain.

The tyrosyl tRNA synthetase demonstrates how **simply binding the transition state more tightly than the substrates increases the rate of a reaction, using primarily binding interactions with side-chains of the enzyme. Binding energy can be used to stabilize the transition state, and unfavorable interactions can destabilize the ground state**. What is initially surprising in this case is that the transition state is not stabilized more than the tyrosyl adenylate and pyrophosphate products. Normally, such an enzyme would not be an efficient catalyst, as it will simply bind the products exceedingly tightly. In the case of tyrosyl tRNA synthetase, the tyrosyl adenylate is not the final product of the enzyme but merely an intermediate that is to be transferred to the appropriate tRNA. The intermediate is intrinsically unstable in solution, so tight binding on the enzyme protects it from hydrolysis and stabilizes it so that it is available for transfer to the tRNA. The equilibrium constant for the first half of the reaction, i.e. the ratio of the equilibrium concentrations of tyrosyl adenylate and pyrophosphate relative to those of tyrosine and ATP, is 2.3 on the enzyme but only 3.5×10^{-7} in solution.

Structural snapshots of the KMSKS loop rearrangement for amino acid activation by bacterial tyrosyl-tRNA synthetase. T. Kobayashi *et al.* (2005) *J. Mol. Biol.* **346**, 105–117.

Crystal structures of apo wild-type *M. jannaschii* tyrosyl-tRNA synthetase (TyrRS) and an engineered TyrRS specific for O-methyl-L-tyrosine. Y. Zhang *et al.* (2005) *Protein Sci.* **14**, 1340–1349.

Crystal structure of tryptophanyl-tRNA synthetase complexed with adenosine-5′ tetraphosphate: evidence for distributed use of catalytic binding energy in amino acid activation by class I aminoacyl-tRNA synthetases. P. Retailleau *et al.* (2007) *J. Mol. Biol.* **369**, 108–128.

1. Editing of Amino Acid Activation

The specificity with which tRNA synthetases attach amino acids to tRNA is crucial for protein biosynthesis (Section 13.4.E). The fidelity of assembling the correct sequence of amino acids into a protein depends upon each tRNA synthetase recognizing the correct amino acid, activating it as the amino acyl adenylate, and transferring it to the correct tRNA. Much of the specificity comes from binding interactions; for example, tyrosyl tRNA synthetase binds tyrosine 1.5×10^5 times more tightly than the similar phenylalanine, which differs only in not having the hydroxyl group on the aromatic side-chain. Much of this discrimination comes from the interactions of Asp176 and Tyr34 with the phenolic hydroxyl group of bound tyrosine (Figure 14-26).

There are limits, however, to the degree of discrimination that is possible by just binding interactions. For example, the extra methylene group of isoleucine relative to valine causes the isoleucyl tRNA synthetase to bind valine less tightly by a factor of only 100–200. The *in vivo* concentration of valine is five times greater than that of isoleucine, so an error rate of 2–5% would be expected. Yet the observed error rate is only 0.03%.

The extra specificity is gained by incorporating an editing mechanism involving hydrolysis of incorrect structures, in at least those tRNA synthetases that require it. For example, addition of tRNA–Ile to the incorrect complex of valyl adenylate and isoleucyl tRNA synthetase results in hydrolysis of the valyl adenylate. In some cases, the hydrolysis may be of the adenylate directly but, in those that have been studied, hydrolysis occurs after transfer of the incorrect amino acid to the tRNA. With the correct amino acid, the rate of hydrolysis is much less. As a consequence, the correctly charged tRNA usually predominates.

In the case of the valyl tRNA synthetase, **editing occurs at a separate site** (Figure 14-27). This hydrolytic site is designed to recognize the closely related amino acid threonine and is imagined to have a donor or acceptor for a hydrogen bond to the hydroxyl group of threonine. If tight binding to this site occurs, the amino acid is recognized as being threonine and hydrolyzed from the tRNA.

Such an editing process has been described as a **double sieve** process. In the initial discrimination, upon binding to the tRNA synthetase, larger or markedly different amino acids will not bind with substantial affinity. Smaller or similar amino acids, however, may still bind to a significant extent. These incorrect amino acids are detected because they bind to the second, hydrolytic site more effectively than do the correct amino acids. **Those that are accepted by the second sieve are hydrolyzed**.

Such multiple editing mechanisms are used where extreme selection is required. Another example is **DNA replication**, where error rates must be as low as possible. DNA replication occurs in bacteria with error rates as low as 10^{-8}–10^{-10}. Discriminating between the four nucleotides solely on the basis of binding affinities alone would not be sufficient to attain such accuracy, so the DNA polymerases also use further hydrolytic editing steps (Section 6.1).

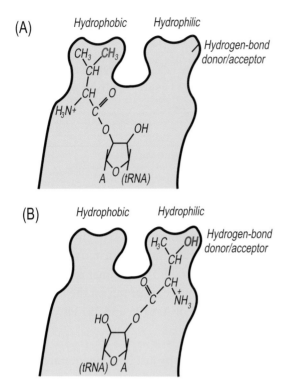

Figure 14-27. Mechanism for the prevention of misacylation of tRNA^{Val} with threonine. The valyl tRNA synthetase is believed to have two distinct sites: an acylation site, where the amino acid is transferred from the adenylate to tRNA; and a hydrolysis site, where amino acids that bind tightly are hydrolyzed from the tRNA. The acylation site is imagined in (A) to be specific for valine by being hydrophobic and complementary in shape to the valine side-chain. The hydrolytic site is depicted in (B) as being specific for threonine by virtue of a hydrogen bond donor or acceptor for its hydroxyl group. Binding to this site produces hydrolysis of the charged tRNA. Adapted from A. R. Fersht & M. Kaethner (1976) *Biochemistry* **15**, 3342–3346.

14.5. CATALYTIC ANTIBODIES

Antibody molecules are produced *in vivo* in response to exposure to an **immunogen**; the molecules that they bind are known as **antigens**. Antibodies are normally complementary to their immunogens and antigens and do not catalyze noticeable reactions on them. On the other hand, **an antibody that happened to be specific for a transition state should function like an enzyme and promote catalysis of the corresponding reaction**. The ability to produce homogeneous monoclonal antibodies has made it possible to exploit the diversity of the immune system to produce antibodies against transition-state analogs. Such antibodies would be anticipated to bind substrates that are related structurally to the transition state and to act on them catalytically according to the nature of the transition state. It should be possible to design appropriate transition-state analogs and produce tailor-made catalysts that will catalyze reactions for which no enzyme is available. Of course, factors other than solely high affinity for the transition state may be required, such as having reactive groups in the correct and precise orientation and being able to release the reaction products from the active site.

Transition-state analogs used as immunogens have been found to produce antibody molecules with catalytic capabilities that are able to catalyze more than 100 distinct chemical reactions. The active sites of the catalytic antibodies that have been produced in response to a given immunogen are found to have very similar amino acid sequences and often to adopt similar structures. Shallow clefts within these sites are complementary to the structural and electronic features of the immunogen. It has subsequently been discovered that natural antibodies also often have enzymatic activities.

Antibody-catalyzed reactions have a number of characteristics expected of an enzyme, in that they exhibit Michaelis–Menten kinetics in which binding of the substrate precedes the chemical transformation step, followed by dissociation of the product; the chemical transformation step is often rate-limiting. **The rate enhancement is generally less than that observed for enzymes catalyzing similar reactions**, probably because transition-state analogs are only approximations of the true transition state. **Enzymes usually exhibit high K_m and k_{cat} values, while antibodies generally have low K_m and k_{cat}.** Antibodies appear to catalyze reactions primarily through tight binding and restrictions on the translational and rotational movement of substrates, and to a lesser extent through the active-site nucleophilic catalysts that occur on enzymes. An enzyme's ability to provide more extensive electrostatic interactions and hydrogen bonding during catalysis may be the basis for its greater catalytic capability. Catalytic antibodies often seem to have a single polar group, for acid or base catalysis, situated within the binding site in reasonable proximity to the bound substrate, and they catalyze primarily simple model reactions, characterized by a high thermodynamic driving force, a low enthalpy of activation and a single rate-determining transition state.

Nevertheless, small differences in the structures of the transition-state analogs used to produce the antibodies can produce large differences in catalytic characteristics, and they can be impressively like enzymes. For example, the antibody pocket can bind the transition state rigidly and provide a highly structured microenvironment for the reaction in which the catalytic group is activated through partial desolvation. A highly polarizable transition state can be stabilized by dispersion interactions with aromatic residues of the antibody, and the leaving group can be solvated by external water.

An antibody generated against 5′-phosphopyridoxyl -N^ε-acetyl-L-lysine (the amino acid lysine linked through its α-amino group to pyridoxal-P; Equation 14.65) catalyzes the reversible transamination of hydrophobic D-amino acids in the presence of added pyridoxal-P (Section 14.4.G.1). The conformation of the adduct of pyridoxal-P and amino acid, and its interactions with the antibody, are similar to those occurring in normal enzymes that use pyridoxal-P, except that the amino acid substrate is only weakly bound and there is no Lys residue that normally binds pyridoxal-P covalently via a Schiff base in the active sites of these enzymes (Equation 14.64). The N-acetyl-L-lysine moiety of the immunogen appears to have selected for two aromatic residues in one antibody hypervariable loop, which accounts for the substrate preference for hydrophobic side-chains. In complexes with pyridoxal-P–alanine aldimine (Equation 14.65), the space vacated by the absent L-lysine side-chain of the immunogen can be filled by a mobile Tyr residue of the antibody. This group stabilizes a hydroxide ion, which abstracts the C^α proton only from D-alanine. No residue capable of deprotonating the C^α of L-alanine is present, which explains the enantiomeric selectivity for D-substrates.

Significant structural rearrangements have been observed along the reaction pathway in an antibody that hydrolyzes cocaine, but they are generally limited to the binding site and substrate. Several side-chains that interact with the substrate either change their rotamers or alter their mobilities to accommodate the various reaction steps. The hypervariable loops of the antibody that contact the substrate change their positions by up to 2.3 Å, and substantial side-chain rearrangements of up to 9 Å alter the shape and size of the antibody active site from 'open' to 'closed' to 'open' for the substrate, transition state and product states, respectively.

Catalytic antibodies elicited against transition-state analogs are the most effective enzyme mimics currently available, but they are not, as yet, capable of efficient catalysis of complex reactions of the type handled so effectively by enzymes.

Complete reaction cycle of a cocaine catalytic antibody at atomic resolution. X. Zhu *et al.* (2006) *Structure* **14**, 205–216.

Structural basis for D-amino acid transamination by the pyridoxal 5′-phosphate-dependent catalytic antibody 15A9. B. Golinelli *et al.* (2006) *J. Biol. Chem.* **281**, 23969–23977.

Evidence that the mechanism of antibody-catalysed hydrolysis of arylcarbamates can be determined by the structure of the immunogen used to elicit the catalytic antibody. G. Boucher et al. (2007) Biochem. J. **401**, 721–726.

Computer-aided rational design of catalytic antibodies: the 1F7 case. S. Marti et al. (2007) Angew. Chem. Int. Ed. Engl. **46**, 286–290.

14.6. CATALYTIC NUCLEIC ACIDS: RIBOZYMES AND DEOXYRIBOZYMES (DNAZYMES)

RNA molecules, in the form of ribozymes, are capable of catalyzing chemical reactions in the absence of proteins; they occur naturally in nature and their identification was the first indication that nucleic acids could have catalytic capabilities. Ribozymes can increase the rates of their reactions by as much as 10^{11}, with specificities of better than 1000:1. **The catalytic RNA folds into a compact three-dimensional structure that is directed towards chemical reaction of a single specific phosphate linkage** (Section 4.2.C). All of the known natural ribozymes participate in RNA maturation or processing and catalyze transesterification reactions of the phosphodiester groups of the RNA backbone. The majority of natural catalytic RNAs catalyze only intramolecular reactions but the ribozyme and 'substrate' can be separated onto individual RNA molecules, when the ribozyme becomes a true catalyst that is capable of multiple turnovers.

The existence of catalytic RNAs supports the idea that life started in a **primeval RNA world** in which RNA molecules alone, without DNA or proteins, were responsible for storing and replicating the genetic information, as well as for catalyzing biochemical reactions. Subsequently, DNA or proteins would have been found to be more efficient at most of these functions and thus replaced the RNA. Present-day catalytic RNAs may be vestiges of the original functions of RNA.

Nucleic acid enzymes. R. Fiammengo & A. Jaschke (2005) Curr. Opinion Biotechnol. **16**, 614–621.

Ribozyme catalysis: not different, just worse. J. A. Doudna & J. R. Lorsch (2005) Nature Struct. Biol. **12**, 395–402.

Structure, folding and mechanisms of ribozymes. D. M. Lilley (2005) Curr. Opinion Struct. Biol. **15**, 313–323.

HDV ribozymes. M. D. Been (2006) Curr. Top. Microbiol. Immunol. **307**, 47–65.

Bacterial RNase P: a new view of an ancient enzyme. A. V. Kazantsev & N. R. Pace (2006) Nature Rev. Microbiol. **4**, 729–740.

14.6.A. Natural Reactions

Catalytic RNAs are involved in diverse functions but they have similar mechanisms of action. Each natural ribozyme uses an activated nucleophile to attack a phosphodiester linkage in-line, which produces inversion of the stereochemistry of the phosphate group (Figure 14-28). The various catalytic RNA reactions differ primarily in the identities of the nucleophile and leaving group.

Four small catalytic RNAs have been identified in **satellite RNAs** from plants and animals: the **hammerhead**, **hairpin**, **hepatitis delta virus** (HDV) and **Varkud satellite** (VS) ribozymes. These

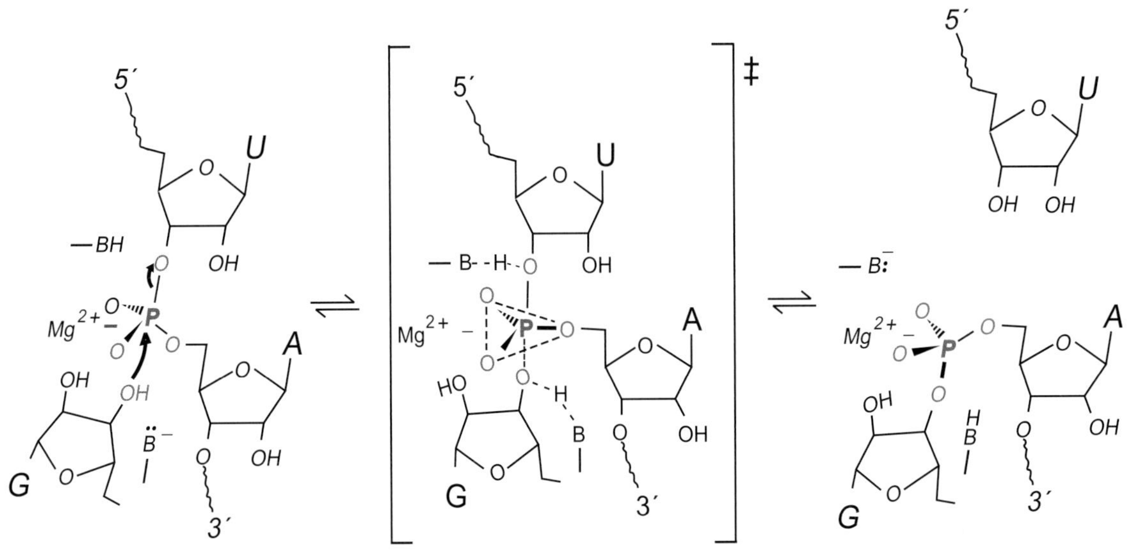

Figure 14-28. The basic transesterification reaction catalyzed by ribozymes. On the *left*, the 3'-hydroxyl group of guanosine acts as the nucleophile in attacking the P atom of the phosphate diester to be cleaved. The reaction is imagined to be assisted by two basic groups, B, and an Mg^{2+} ion complexed to the phosphate. In the transition state (*middle*) the phosphate group is now planar, and it is inverted in the product (*right*), which has the guanosine replacing the chain that has been expelled. Adapted from T. Cech (1987) *Science* **236**, 1532–1539.

catalytic RNAs differ substantially in structure but perform the same cleavage reaction. The four satellite RNAs replicate via a complementary double-stranded RNA intermediate using a rolling circle mechanism, in which replication produces a continuous RNA molecule composed of tandem repeats of the satellite RNA. The function of the ribozymes is to self-cleave the satellite RNA repeats into monomers. All use as internal nucleophile the 2'-OH group that is located immediately adjacent to the phosphate group to be cleaved, (Figure 14-28). In a one-step reaction, the 5'-oxygen of the phosphodiester is displaced and the RNA backbone is cleaved, producing RNAs with 2'–3'-cyclic phosphate and 5'-OH groups at their termini. Chemical alkaline hydrolysis of unstructured RNA produces the same products (Figure 4-2) but the ribozyme reaction is directed towards a single phosphodiester linkage.

A second, more common, type of catalytic RNA comprises the **self-splicing introns**. These RNAs catalyze excision of the intron of which they are a part from a messenger RNA precursor; that includes linking together the flanking exons (Figure 14-29). They can also catalyze the reverse splicing reaction, which permits them to insert into RNA molecules and act as mobile genetic elements and transfer genetic information between organisms. There are two types of self-splicing introns, group I and group II, which differ in their structures and reaction mechanisms (Figure 14-30). **Group I introns** use external nucleophiles to catalyze two consecutive transesterification reactions. In the first, the 3'-OH of an external, bound guanosine attacks the phosphate group at the 5'-end of the intron. This reaction adds the guanosine covalently to the 5'-end of the intron and releases the 5'-end of the exon. The second step is essentially the reverse of the first step, but the 3'-OH of a guanine nucleotide at the 3'-end of the intron is the leaving group. The result is that the intron has been excised, leaving an intact RNA molecule.

700 CHAPTER 14   Catalysis

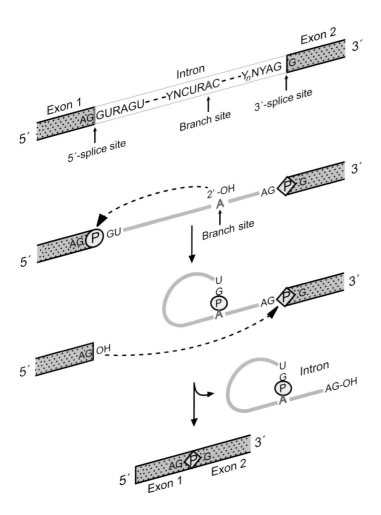

Figure 14-29. The two-step splicing pathway of nuclear messenger RNA precursors. Exon and intron sequences are indicated by *stippled* and *open boxes*, respectively. The sequences given are consensus sequences found at the mammalian 5'- and 3'-splice sites the branch site of U2-dependent introns, where N is any base, Y is a pyrimidine and R is a purine. The *dashed arrows* indicate the nucleophilic attacks on the splice sites by the 2'-OH of the branch site in the first step and the 3'-OH of the cleaved 5'-exon in the second step. The phosphate groups at the 3'- and 5'-splice sites are indicated indicate by P in a *circle* or a *triangle*.

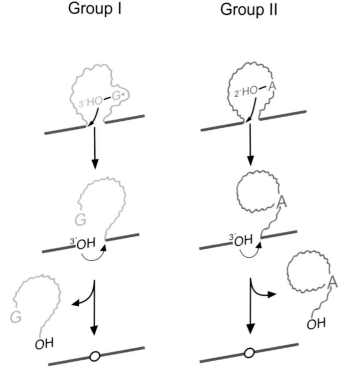

Figure 14-30. Comparison of the mechanisms of action of group I and group II self-splicing introns. The group I ribozymes use as nucleophile a guanosine cofactor that is not part of the polynucleotide chain, which then becomes incorporated covalently into the RNA. The group II ribozymes use a bulged A residue of the RNA molecule (Figure 14-29).

The reaction mechanism of **group II introns** is intermediate between the previous two. They use an internal nucleophile, the 2′-OH of a bulged A nucleotide that it is not adjacent to the phosphate group to be cleaved but much further away, at the 5′-exon–intron boundary (Figure 14-29). As in the case of the group I introns, the 3′-OH group of the 5′-exon is the leaving group for the first step and the nucleophile for the second step. The released intron has a **lariat-like structure**, with the 5′-end of the intron covalently attached to the 2′-OH of the bulged A nucleotide. The same reaction mechanism is employed in the more complex process of messenger RNA precursor splicing that is catalyzed by the **spliceosome**.

A third type of ribozyme is **ribonuclease P**. It participates in the biosynthesis of tRNAs (Section 4.2.B) by catalyzing the removal of extensions from the 5′-ends of tRNA precursors. It consists naturally of both an RNA and a protein component. Both are essential *in vivo* but the RNA molecule alone is sufficient to carry out the tRNA processing reaction, although at a reduced rate. Unlike the other naturally occurring catalytic RNAs that use either the 2′- or 3′-OH groups as nucleophiles, RNaseP uses water, which hydrolyzes the RNA backbone. The original 5′-sequence is removed, producing a tRNA with a terminal 5′-phosphate group. Using an external nucleophile and acting on an external substrate, RNaseP is the only known natural RNA that catalyzes multiple turnovers and acts as a true enzyme *in vivo*.

The **Varkud satellite** ribozyme is the largest nucleolytic ribozyme. It consists of five helical sections, organized by two three-way junctions. The substrate stem–loop binds into a cleft formed between two helices, while making a loop–loop contact with another section of the ribozyme. The phosphate that is to be acted on makes a close contact with an internal loop (the A730 loop), the probable active site of the ribozyme. This loop contains a particularly critical nucleotide, A756, as most changes to this nucleotide lead to three-orders of magnitude slower cleavage. The groups that would participate in Watson–Crick base pairing are especially important.

Many of the functions of the **ribosome** (Section 13.4.F) are catalyzed by the RNA components, rather than the proteins, including the step in which a peptide bond is formed to extend the growing polypeptide chain. The 2×10^7-fold rate enhancement produced by the ribosome is achieved entirely by lowering the entropy of activation. The enthalpy of activation for generating a peptide bond is slightly less favorable on the ribosome than in solution, in contrast with protein enzymes. The ribosome enhances the rate of peptide bond formation mainly by positioning the substrates and/or by excluding water from the active site, rather than by conventional chemical catalysis.

Chemical models for ribozyme action. T. Lonnberg & H. Lonnberg (2005) *Curr. Opinion Chem. Biol.* **9**, 665–673.

The transition state for formation of the peptide bond in the ribosome. A. Gindulyte *et al.* (2006) *Proc. Natl. Acad. Sci. USA* **103**, 13327–13332.

Nucleobase catalysis in ribozyme mechanism. P. C. Becilacqua & R. Yajima (2006) *Curr. Opinion Chem. Biol.* **10**, 455–464.

Alternative roles for metal ions in enzyme catalysis and the implications for ribozyme chemistry. R. K. Sigel & A. M. Pyle (2007) *Chem. Rev.* **107**, 97–113.

14.6.B. Ribozyme Structure and Catalysis

The three-dimensional arrangement of the nucleotides, metal ions, water molecules and cofactors that comprise the RNA structure are vital for catalysis. A binding site on an RNA that can recognize and bind a specific substrate requires that the structure be relatively complex (Section 4.2). All the ribozymes appear to have compact, folded structures reminiscent of proteins and tRNA (Section 4.2.C) in which an active site is pre-organized and located within crevices. Catalytic RNAs generally require divalent metal ions, usually Mg^{2+} or Mn^{2+}. These ions are important for both structure and function. Those involved in catalysis usually serve as an electrophile to activate the nucleophile or stabilize the leaving group.

Understanding how these structures produce catalysis has been hindered by the structural plasticity of RNA molecules, especially in the case of the hammerhead ribozyme. It is relatively simple, and a minimal model consists of only 42 nucleotides. To prevent catalysis and cleavage from taking place, the RNA molecules studied crystallographically consist in one case of a substrate that contains deoxyribose nucleotides; in a second case, a single 2'-O-methyl is substituted for the 2'-OH nucleophile. A third structure is of an all-ribose substrate but in the absence of divalent metal ions. The three structures are similar but they do not have an active site geometry that could explain catalysis; it appears that these structures must change significantly during catalysis.

The situation was resolved by the structure of a full-length hammerhead ribozyme consisting of 63 nucleotides, with a 2'-O-methyl substitution in place of the 2'-OH nucleophile to prevent cleavage of the phosphodiester bond. The additional nucleotides in this structure increase the rate of catalysis 1000-fold and produce very significant changes throughout the folded structure. The structure of the active site now can explain catalysis (Figure 14-31). The active site has the 2'-OH group of the adjacent ribose ring in position to attack the phosphate group to be cleaved, a guanine base positioned to act as the base, and the 2'-OH group of another guanine residue to act as the acid, as shown in Figure 14-28. No ions appear to be involved in catalysis in this case.

Ribozymes are now believed to function using the same principles identified in enzymes, by placing functional groups in the active site in correct proximity to assist the chemical reaction and by binding and stabilizing the transition state.

Structure of ribonuclease P: a universal ribozyme. A. Torres-Larios et al. (2006) *Curr. Opinion Struct. Biol.* **16**, 327–335.

Tertiary contacts distant from the active site prime a ribozyme for catalysis. M. Martick & W. G. Scott (2006) *Cell* **126**, 309–320.

Dissecting the multistep reaction pathway of an RNA enzyme by single-molecule kinetic 'fingerprinting'. S. Liu et al. (2007) *Proc. Natl. Acad. Sci. USA* **104**, 12634–12639.

14.6.C. Selection for Novel Ribozymes and Deoxyribozymes

Large libraries of RNA molecules with up to 10^{16} different sequences have generated novel RNA catalysts. Such libraries are typically used to isolate RNA or DNA molecules with catalytic activity in

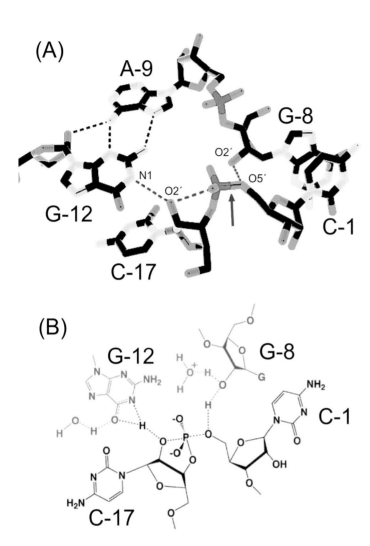

Figure 14-31. The structure of the active site and the inferred transition state of the full-length hammerhead ribozyme. The complete structure of the RNA is shown in Figure 4-18. (A) The structure of the active site determined crystallographically. C atoms are black; O, green; N, yellow; P, mauve. Hydrogen bonds are indicated by *black*, *red* or *blue* dashed lines. The *red* and *blue* ones are postulated to be involved in catalysis. The phosphodiester bond to be cleaved is indicated by the arrow. The 2′-OH of C-17 is in close proximity to the phosphate group, very nearly in-line with the P atom and the 5′-O leaving group, as expected if it is the nucleophile that attacks the P atom (Figure 14-28). The nucleobase of G-12 is in position to act as a base and accept the proton from the 2′-OH group. The 2′-OH of G-8 is in position to act as an electrophile in the bond cleavage. (B) The proposed transition state for the reaction. The general base, G-12, is *red*, the general acid, G--8 *blue* and the substrate *black*. The water molecules are believed to be necessary to accept or supply protons to the basic and acidic catalytic groups of G-12 and G-8, avoiding the necessity of them existing as fully charged ions. Adapted from figures kindly supplied by W. G. Scott.

an *in vitro* selection procedure known as **SELEX** (**S**ystematic **E**volution of **L**igands by **EX**ponential enrichment). SELEX is a combinatorial amplification technique that exerts evolutionary pressure on a population of RNA molecules with random sequences to select and optimize for a particular catalytic function. The population is selected for those RNA molecules that have the desired property, then replicated and subjected to further selection. Only a handful of the original 10^{16} sequences survive many rounds of stringent selection steps. The range of reactions catalyzed by *in vitro*-selected ribozymes is now well beyond those known for natural RNAs. Some of the notable activities identified by this approach include:

(1) an RNA that functions like an RNA polymerase, using nucleoside triphosphates and releasing pyrophosphate as the leaving group (Section 6.2.B);

(2) an RNA that functions like a polynucleotide kinase, being able to transfer the γ-phosphate from ATP to the 5′- or internal 2′-OH of an RNA substrate;

(3) RNAs that use Ca^{2+} or Pb^{2+} instead of Mg^{2+} for catalysis.

RNAs have also been identified that catalyze chemical reactions other than simply transesterification at phosphate groups. One acts like an amino acyl RNA synthetase, in that it can transfer an amino acyl group from Phe–AMP onto the 3′-end of an RNA substrate. Other ribozymes have been selected that catalyze formation of carbon–nitrogen and carbon–sulfur bonds. A transition-state analog used as the bait selected an RNA able to catalyze the isomerization of a bridged biphenyl linkage.

This Darwinian approach to *in vitro* evolution has also demonstrated that nucleic acid-based catalysts are not restricted to RNA, as **DNA can act similarly**. Catalytic DNA molecules are known as **deoxyribozymes** or **DNAzymes**. They appear to act similarly to their RNA counterparts but have not been investigated as thoroughly.

Ribozymes have the potential to use their nucleobases directly in chemical catalysis in a variety of ways reminiscent of enzymes, including hydrogen bonding to the transition state, stabilizing charge development and transferring protons as general acid–base catalysts. Consequently, the chemical reactions found to be catalyzed by RNA and DNA are likely to continue to expand.

Functional gene-discovery systems based on libraries of hammerhead and hairpin ribozymes and short hairpin RNAs. M. Sano *et al.* (2005) *Mol. Biosyst.* **1**, 27–35.

Sensors made of RNA: tailored ribozymes for detection of small organic molecules, metals, nucleic acids and proteins. S. Muller *et al.* (2006) *IEE Proc. Nanobiotechnol.* **153**, 31–40.

Functional DNA nanotechnology: emerging applications of DNAzymes and aptamers. Y. Liu & J. Liu (2006) *Curr. Opinion Biotechnol.* **17**, 580–588.

Revitalization of six abandoned catalytic DNA species reveals a common three-way junction framework and diverse catalytic cores. W. Chiuman & Y. Li (2006) *J. Mol. Biol.* **357**, 748–754

14.6.D. Ligand-Binding Nucleic Acids: Aptamers

Aptamers are artificial small single-stranded RNA or DNA molecules that can be generated *in vitro* to bind with high affinity to a wide range of molecules, ranging from small chemical compounds to large proteins. The name is derived from *aptus*, the Latin word for 'to fit'. Aptamers are isolated from complex libraries of synthetic nucleic acids by a SELEX procedure that involves repetitively reducing the complexity of the library by partitioning on the basis of selective binding to the target molecule, followed by re-amplification. The selected RNA and DNA molecules bind their targets with similar affinities as antibodies and are able to distinguish between isotypes of an enzyme, so aptamers have been also been claimed to be **synthetic antibodies**. Moreover, aptamers can be easily stabilized by chemical modifications for *in vivo* applications. Numerous observations have shown that stabilized aptamers against extracellular targets such as growth factors, receptors, hormones and coagulation factors are very effective inhibitors of the corresponding protein function *in vivo*. RNA aptamers have also been expressed in living cells, where they have inhibited a protein implicated in intracellular signal transduction.

Peptide aptamers are small peptide sequences that have been selected in similar ways to recognize a predetermined target.

Aptamers come of age: at last. D. H. Bunka & P. G. Stockley (2006) *Nature Rev. Microbiol.* **4**, 588–596.

DNA and RNA aptamers: from tools for basic research towards therapeutic applications. H. Ulrich *et al.* (2006) *Comb. Chem. High Throughput Screen.* **9**, 619–632.

Structures of RNA switches: insight into molecular recognition and tertiary structure. H. Schwalbe *et al.* (2007) *Angew. Chem. Int. Ed. Engl.* **46**, 1212–1219.

~ CHAPTER 15 ~

ENZYME REGULATION

The rates of enzyme-catalyzed reactions are regulated in response to extracellular and intracellular signals, maintaining metabolic stability (homeostasis) in response to changes in the environment. The general regulatory mechanisms used include:

(1) the amount of enzyme can be increased or decreased by altering the rates of its biosynthesis or degradation;

(2) the local concentrations of enzymes (or their accessibility to substrates) can be altered by proteins that target them to particular sites, such as the cell membrane;

(3) the catalytic properties of the enzyme (k_{cat} and K_m) can be modified by (a) reversible binding of allosteric substrates, activators and inhibitors, and (b) specific covalent changes, such as phosphorylation, adenylylation and proteolytic cleavage.

Some regulatory enzymes are regulated in several ways. Varying the amount of an enzyme, by controlling its rate of biosynthesis or degradation via mechanism (1), requires significant periods of time and is a genetic phenomenon outside the scope of this volume. Relatively little is known about mechanism (2), whereas altering the catalytic properties of an enzyme, by mechanism (3), can occur very rapidly and is well understood. This type of regulation will be described here.

Enzymes that are subject to regulatory behavior are usually those that play key roles in metabolism. The regulation of those involved in the central metabolic pathway, where their products are used in multiple pathways that must be coordinated, can be very complex and subtle. Metabolic systems tend to be more complex in higher eukaryotes than in more simple organisms. The simplest case is a distinct independent pathway, such as that with a single product, for example an amino acid; in this case, the enzyme catalyzing the first step is usually subject to **feedback inhibition** by the end product of the pathway. Consequently, a surplus of the end product of the pathway decreases the flux of molecules entering that pathway by inhibiting the first enzyme, and the subsequent enzymes are not subject to regulation. The metabolic precursor that would have been used in this pathway is diverted to other pathways.

15.1. ALLOSTERIC ENZYMES

Allosteric interactions have been discussed previously (Section 12.5) in terms of their effects on the affinity for ligands, but in the case of enzymes they can affect both the affinity of the enzyme for substrates and its catalytic potency. **Allosteric regulation is usually apparent from the kinetic behavior of an enzyme deviating from normal Michaelis–Menten behavior** (Section 14.2). The dependence of the initial velocity of the enzyme-catalyzed reaction on substrate concentration is analogous to allosteric effects on ligand binding (Section 12.5), except that both K_m and k_{cat} (or V_{max}) may be altered. The regulation can involve homotropic effects, involving a single type of molecule, and heterotropic effects, where one type of molecule affects the activity of the enzyme towards other molecules serving as substrate. Examples of homotropic effects involving the substrate are **positive cooperativity**, **negative cooperativity** and **substrate inhibition**, which are apparent from the response of the enzyme to variations in the concentration of the substrate (Figure 15-1). In the cases of positive and negative cooperativity, the enzyme velocity data are frequently analyzed as a **Hill plot** (Figure 12-23), and the **Hill coefficient** is used to provide a quantitative measure of the cooperativity. The Hill coefficient is greater than 1 for positive cooperativity, up to a maximum of the number of molecules that can bind simultaneously to the enzyme, and less than 1 for negative cooperativity.

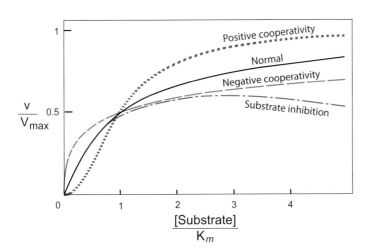

Figure 15-1. Homotropic effects on regulatory enzyme kinetic behavior. The initial velocity relative to the maximum is plotted as a function of the substrate concentration normalized to the Michaelis constant for the substrate, K_m, in the cases of normal nonregulatory hyperbolic Michaelis–Menten kinetics, positive cooperativity, negative cooperativity and substrate inhibition.

Most regulatory enzymes also display heterotropic effects produced by ligands other than the substrate. There are two limiting cases for heterotropic effects (Figure 15-2): the apparent affinity for the substrate (K_m) is altered in a **K-system**, whereas a **V-system** has the maximum reaction velocity (V_{max}) altered. The effects may be positive, with decreased K_m or increased V_{max}, or negative, due to increased K_m or decreased V_{max}. Many enzymes that are involved in the regulation of central metabolic pathways, such as aspartate transcarbamoylase (Section 15.1.C), glycogen phosphorylase (Section 15.2.B) and phosphofructokinase (Section 15.1.D), are K-systems and are best understood. In these cases, the *in vivo* concentrations of the substrates are probably close to their K_m values, so changes in the K_m can cause large changes in the rate of the reaction. V-system behavior is found primarily in metabolic pathways where there is a relatively high fixed level of the substrate. Signal transduction pathways also contain V-system regulatory enzymes, where they appear to function as 'on-off switches' and the heterotropic effector tends to be another protein.

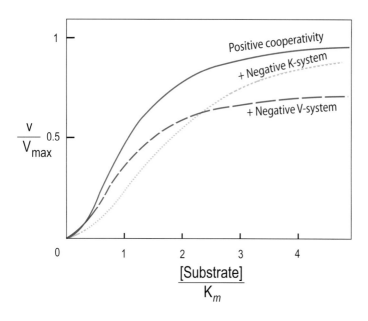

Figure 15-2. Examples of heterotropic effects on the kinetic behavior of regulatory enzymes. The initial velocity relative to the maximum is plotted as a function of the substrate concentration normalized to the Michaelis constant for the substrate, K_m, for the case of positive cooperativity for substrate, with no additional regulatory effector (*solid curve*). A negative V-system heterotropic effector decreases V_{max}, which is apparent here as the downward displacement of the curve; a positive V-system heterotropic effector would increase V_{max}. A negative K-system heterotropic effector increases the K_m, which is apparent here as the shift of the curve to the right; a positive K-system heterotropic effector decreases the K_m.

Searching for new allosteric sites in enzymes. J. A. Hardy & J. A. Wells (2004) *Curr. Opinion Struct. Biol.* **14**, 706–715.

Allosteric mechanisms in ACT domain-containing enzymes involved in amino acid metabolism. J. S. Liberles et al. (2005) *Amino Acids* **28**, 1–12.

Application of a generalized MWC model for the mathematical simulation of metabolic pathways regulated by allosteric enzymes. T. S. Najdi et al. (2006) *J. Bioinform. Comput. Biol.* **4**, 335–355.

15.1.A. Allosteric Models

Allosteric models for enzyme regulation involve the existence of multiple binding sites on the enzyme for various ligands and interactions between the sites (Section 12.5). In the **sequential model**, binding of an effector at one site on an enzyme can have a direct effect on the affinities of other sites on the enzyme or on the catalytic capabilities of its active sites. In the **concerted model**, binding of an effector at one site alters the affinity or catalytic properties at another site only indirectly, as a result of changes in the protein conformation, usually of the quaternary structure. The positive cooperativity seen in initial velocity experiments (Figure 15-1) of classical K-system regulatory enzymes appears to be similar to the cooperative binding of oxygen by vertebrate hemoglobins (Section 12.5.B). Concerted allosteric models for K-systems are most simple if the substrate and enzyme bind rapidly and reversibly and interconvert R- and T-type quaternary conformations of the enzyme comparable to those of hemoglobins. Even if the two conformations are assumed to have identical turnover numbers (V_{max}), which is not strictly necessary, the homotropic and heterotropic effects observed with K-system enzymes can often be explained entirely by interactions between the ligands bound at different sites that affect their K_m value.

V-system enzymes have been much less studied but generally there are no indications of cooperativity in the binding of the substrate. Positive cooperativity is apparent in plots of the initial velocity as a function of the substrate concentration, but solely due to the various enzyme–substrate complexes having different values of V_{max}.

As in the case of binding ligands, deviations from normal kinetic behavior can usually be described with either the sequential or concerted allosteric models (Section 12.5.A), except that the simplest concerted model cannot explain negative cooperativity.

Allosteric mechanisms of signal transduction. J. P. Changeux & S. J. Edelstein (2005) *Science* **308**, 1424–1428.

15.1.B. Structural Aspects

Regulatory enzymes have several structural features in common: **the functional unit is usually an oligomer**, composed of more than one polypeptide chain (subunit), and **multiple structural domains often can be identified in each subunit. The substrate and effector binding sites are often at the interfaces between domains or subunits**. They are separated by great distances, so the effect of one ligand on the binding of another must be indirect and mediated by the protein structure. **The protein conformation is observed to vary in complexes with bound substrates or effector ligands**, providing clues to the mechanism of the regulation. **The conformational differences usually involve rigid body movements of domains and/or subunits relative to one another, and the movements are primarily rotations, with only small translations**. The interacting surfaces between the domains and/or subunits are conserved but the detailed interactions change. A key role for interfaces between subunits or domains in the regulatory behavior of enzymes is not unreasonable because the interactions across the interfaces, although similar to, are weaker than those within the folded subunit and more sensitive to changes resulting from ligand binding.

In a few cases, binding of substrates or allosteric effectors results in changes in the oligomeric structure, i.e. subunit association or dissociation. For example, porphobilinogen synthase has several interconverting quaternary structure isoforms: one monomer conformation directs assembly of a high-activity octamer, whereas an alternative monomer conformation dictates assembly of a low-activity hexamer. Most, if not all, **G protein-coupled receptors** have the propensity to dimerize, which can result in interactions between the protomers and cooperative binding of ligands. The receptors oscillate between two states: the dimer that is cooperative and the monomer that is not.

In classical K-system regulatory enzymes, the substrate and/or allosteric effector binding sites are usually located at the interfaces between the domains or subunits and contain amino acid side-chains from both sides of the interface. Thus **binding of substrates or allosteric effectors affects the interactions between subunits and/or domains directly**. One exception is glycogen phosphorylase (Section 15.2.B.2) but, in this case, the active site is connected directly to the subunit interface by an α-helix that is analogous to the connection between the subunit interface and the oxygen-binding site in hemoglobin (Figure 12-26). The crystal structure of a V-system regulatory enzyme, D-3-phosphoglycerate dehydrogenase, also shows that substrate and/or effector binding sites involve interfaces, but exceptions to this generalization are known.

Several enzymes from signal transduction pathways are monomeric and appear to function as V-systems in which a regulatory protein activates or inhibits the catalytic activity. Normally, but not always, the regulatory protein binds at the active site of the enzyme. In at least one case the regulatory protein activates the enzyme by contributing a catalytic residue to the active site. The known examples of V-system enzymes are generally consistent with the generalization from structural studies of the classical K-system regulatory enzymes, that key aspects of enzyme regulation are oligomeric structures, interfaces between or within the components, and conformational changes as a result of ligand binding. In V-system enzymes, however, the active site and the binding site for the allosteric effector are often very near one another.

Principles: a model for the allosteric interactions between ligand binding sites within a dimeric G-protein coupled receptor. T. Durroux (2005) *Trends Pharmacol. Sci.* **26**, 376–384.

Morpheeins: a new structural paradigm for allosteric regulation. E. K. Jaffe (2005) *Trends Biochem. Sci.* **30**, 490–497.

15.1.C. Aspartate Transcarbamoylase

Aspartate transcarbamoylase (ATCase) from *Escherichia coli* is one of the best-characterized regulatory enzymes. It catalyzes the first unique step in the *de novo* pathway of pyrimidine biosynthesis:

$$\text{Carbamoyl phosphate} + \text{L-aspartate} \longrightarrow N\text{-carbamoyl aspartate} + P_i \tag{15.1}$$

The carbamoyl group is transferred directly from carbamoyl-P to aspartate on the enzyme. Binding of the substrates tends to be ordered, with carbamoyl-P binding first 90% of the time, and the inorganic phosphate being released last. A bisubstrate analog, known as PALA (Section 14.4.D.2), binds at least 10^3 times more tightly than either single substrate.

Carbamoyl aspartate is the first metabolite that is unique to the pyrimidine biosynthetic pathway, so the end products of this pathway, CTP and UTP, repress the biosynthesis of ATCase and inhibit its enzymatic activity. Evolution has developed a variety of regulatory controls for maintaining homeostasis in intracellular nucleotide pools, and various genetic and enzymological regulation mechanisms are critical for balancing the nucleotide precursors of DNA/RNA synthesis in various organisms. The regulatory characteristics of the *E. coli* enzyme are just one of many variations, but it is the best characterized.

The *E. coli* ATCase is inhibited by the end products of the pathway, CTP and UTP, but activated by ATP, presumably to coordinate synthesis of pyrimidine and purine nucleotides. UTP by itself has little effect, but greatly increases the inhibition caused by CTP. These heterotropic allosteric regulators affect primarily K_m, not V_{max} (Figure 15-3-A). CTP, UTP and ATP compete for the same site but produce opposite effects on the cooperativity and apparent K_m values for the substrates (Figure 15-3-A).

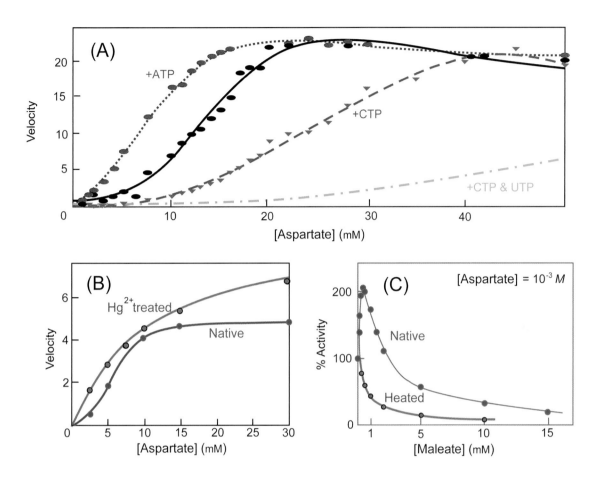

Figure 15-3. Effects of allosteric effectors on the steady-state kinetic behavior of aspartate transcarbamoylase. (A) The sigmoidal dependence of enzyme velocity on the concentration of aspartate, at a fixed concentration of the other substrate, carbamoyl-P (3.6 mM). CTP lowers the apparent affinity for aspartate and increases the cooperativity, whereas ATP has the opposite effect. UTP has little effect by itself, but increases the inhibition in the presence of CTP. (B) The kinetic behavior of native and mercuric ion-treated enzyme. The treated enzyme is dissociated into catalytic trimers and regulatory dimers. The kinetic response to the aspartate concentration of the treated enzyme is hyperbolic, i.e. normal Michaelis–Menten, and the value of V_{max} is increased. (C) The effect of the inhibitor maleate, which competes with aspartate, on the enzymatic activity of normal and heat-dissociated aspartate transcarbamoylase. With the dissociated enzyme, maleate acts as a normal competitive inhibitor, but it activates the native enzyme at low concentrations of both maleate and aspartate. The inhibitory effect of maleate binding at one or a few of the six active sites on the native enzyme is more than compensated for by an allosteric activating effect on the remaining active sites, increasing their affinity for aspartate. Data from J. R. Wild (1999) in *Encyclopedia of Molecular Biology* (T. E. Creighton, ed.), Wiley-Interscience, NY, p. 196; J. C. Gerhart (1970) *Curr. Top. Cell Reg.* **2**, 275–325; J. C. Gerhart & A. B. Pardee (1963) *Cold Spring Harbor Symp. Quant. Biol.* **28**, 491–496.

The enzyme is also subject to homotropic interactions and does not obey normal Michaelis–Menten kinetics, in that the dependence of enzyme velocity on the concentration of either substrate is sigmoidal rather than hyperbolic. A most striking consequence of the homotropic interactions is that **substrate analog inhibitors, such as maleate, that are competitive with the substrate aspartate, and undoubtedly bind at the active site in place of it, activate rather than inhibit the enzyme at low concentrations** (Figure 15-3-C). At high concentrations, such inhibitors display the usual net inhibition by blocking the active sites. The activation by low concentrations is a consequence of the allosteric effect on the other active sites of the same enzyme molecule produced by the inhibitor binding at just one or a few sites. **This activating effect on the other sites more than compensates for the inhibitory effect on the sites to which the inhibitor is bound**; it is the result of converting the quaternary structure from a low-affinity, low-activity T state to a high-affinity, high-activity R state.

The ATCase enzyme consists of six copies of each of two polypeptide chains, designated here as c and r, because one is primarily catalytic, the other regulatory (Figure 15-4). The c_6r_6 complex can be dissociated reversibly by mild treatments with heat, mercurials and urea into two c_3 trimers (often designated C) and three r_2 dimers (or R):

$$c_6r_6 \leftrightarrow 2c_3 + 3r_2$$

or

$$C_2R_3 \leftrightarrow 2C + 3R \tag{15.2}$$

The dissociated enzyme is somewhat more active than the original enzyme and has lost all its allosteric properties (Figure 15-3-B and C). **The catalytic trimer generally exhibits normal enzyme kinetics**, although the Arg105 → Ala mutant form exhibits cooperativity. **The r_2 dimers have no catalytic activity but bind the allosteric effectors**. Both the homotropic and heterotropic allosteric interactions occur only in the presence of both of the two different polypeptide chains, one catalytic, the other regulatory, assembled into the large quaternary structure. The catalytic properties of reconstituted, hybrid complexes are determined largely by the c polypeptide chains, the regulatory properties by the r chains.

ATCase is constructed of two rings of three c subunits associated face-to-face, with three r_2 dimers linking the two rings (Figure 15-4). Each ring of three c subunits is produced by heterologous interactions, whereas the two rings associate by isologous interactions (Section 9.1.H). Both the c and r polypeptide chains are composed of two domains. Each of the domains of the c polypeptide chain binds one of the two substrates, carbamoyl-P or aspartate. **The active site is not contained within one polypeptide chain but is situated between the two domains of one subunit and an adjacent polypeptide chain in the trimer.** One of the two domains of the r polypeptide chain binds Zn^{2+}, the other binds the allosteric effectors ATP and CTP. The Zn^{2+} ion is tetrahedrally chelated by the sulfhydryl groups of four Cys residues and is believed to play primarily a structural role. The Zn^{2+} domain plays a crucial role in linking the allosteric domain to the catalytic polypeptide chain. Mercurials dissociate the c_6r_6 complex because they react with the Cys thiol groups and displace the Zn^{2+} ion. The allosteric effectors bind to the r subunit on the periphery of the enzyme complex, some 60 Å from the active site. The allosteric interaction between these distantly spaced sites consequently would seem most likely to involve changes in quaternary structure in this enzyme.

The homotropic part of the allosteric mechanism of ATCase fits well the concerted allosteric mechanism involving the usual R and T states (Section 12.5.A.2). The unliganded enzyme is designated the low-affinity low-activity T state. It binds carbamoyl-P first, which causes local rearrangements of the structure to provide the binding site for the second substrate, aspartate. Once both substrates are bound, the local structure closes in upon them, forcing them together, optimizing their interactions with the enzyme and contributing to catalysis. At the same time, these changes weaken those intersubunit interactions that stabilize specifically the T quaternary structure. Upon binding both substrates, or the bisubstrate analog PALA, the quaternary structure changes to that of the R state (Figure 15-5). This causes the enzyme to increase its hydrodynamic volume substantially, which is apparent as a decrease in the sedimentation coefficient of the protein. The catalytic dimers re-orient about the three-fold axis by 10° and move apart; the regulatory dimers rotate about each two-fold axis by 15°. As a result, the molecule elongates along its three-fold axis by 11 Å. These conformational changes are almost entirely movements of rigid domains relative to each other. Once the enzyme is converted to the R state by substrate binding, the enzyme remains in the R state until substrates are exhausted.

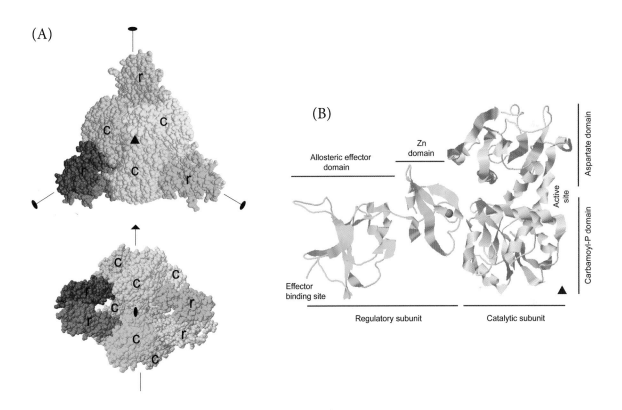

Figure 15-4. The structure of aspartate transcarbamoylase. (A) Two views of the overall quaternary structure in the R state, using a space-filling representation of the atoms. The 12 polypeptide chains are indicated in different colors. *Top*: a view down the three-fold axis, showing a trimer of three catalytic subunits (*c*) lying above a second trimer that is just barely visible below it. The two trimers are linked on the outside by three dimers of regulatory subunits (*r*). Three two-fold axes relating the top and bottom halves are indicated. *Bottom*: a view down one such two-fold axis, orthogonal to the other view. The three-fold axis is indicated. Note the large channels through the interior of the enzyme complex. (B) The tertiary structure of one regulatory, *r*, and one catalytic, *c*, subunit. The view is down the three-fold axis, indicated by the *solid triangle* at the *lower right*. Each polypeptide chain is composed of two domains. The active site is between the two *c* domains, each of which binds predominantly one of the substrates; the adjacent *c* chain of the trimer also contributes residues to the active site. The Zn^{2+} atom bound to the *r* chain is believed to have a solely structural role. The allosteric effectors ATP and CTP bind to the periphery of the *r* chain, some 60 Å from the active site. Figure generated from PDB files *1q95* and *2atc* using the program Jmol.

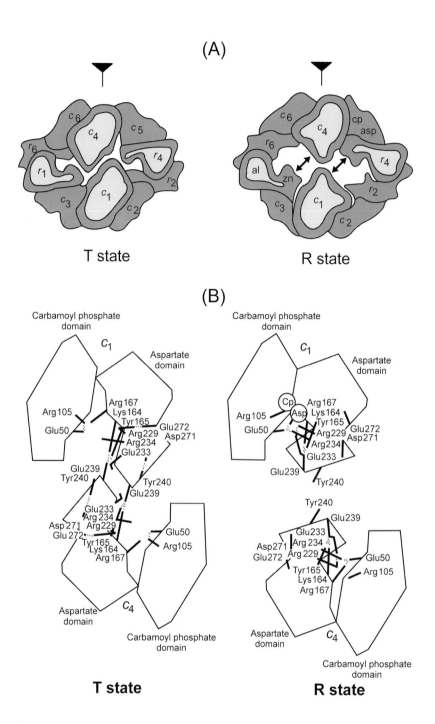

Figure 15-5. Differences in the quaternary structures of the T and R states of aspartate transcarbamoylase and their consequences for the active site. (A) Schematic illustration of the quaternary structure. The view is down one of the two-fold axes, with the three-fold axis vertical. Catalytic subunits c_1 and c_4 are the pair from different trimers that are in direct contact in the enzyme complex (Figure 15-4-A); identical structures are present in the c_2–c_5 and c_3–c_6 pairs. In the other subunits that are visible, al and zn are the allosteric and effector domains of subunit r_1, while cp and asp are the carbamoyl phosphate and aspartate domains of subunit c_5. Adapted from J. E. Gouaux et al. (1989) *Proc. Natl. Acad. Sci. USA* **86**, 8212–8216. (B) Local conformational changes that are believed to link the change in quaternary structure and the catalytic activity of the active site. Adapted from E. R. Kantrowitz & W. N. Lipscomb (1988) *Science* **241**, 669–674.

A single molecule of PALA binding to just one of the six active sites on ATCase causes the quaternary structure to change from that of T to R when ATP is also present. As a result, the five other catalytic sites are converted to the high-affinity high-activity form, explaining the homotropic allosteric effects and the activating effect of PALA and other substrate analogs at low concentrations (Figure 15-3-C). That the transition from T to R is primarily concerted is shown by reconstituting ATCase from its dissociated polypeptide chains in which some of the c chains are inactivated by chemical modification. Hybrid forms of ATCase containing two native and four inactivated c chains can be compared when the two active chains are either part of the same c_3 trimer or of different trimers. The two isomeric hybrids exhibit identical catalytic and regulatory behavior, indicating that **the cooperative unit is the entire ATCase molecule, not some smaller part**. Hybrid molecules also demonstrate that the structural alterations produced by substrate binding to some c subunits are experienced by others. For this purpose, one of the c_3 trimers of the ATCase hybrid has been chemically inactivated so that it can not bind aspartate or its analogs; it has also been nitrated (Section 7.2.L) to introduce a nitrotyrosyl chromophore that is sensitive to the protein conformation. The spectrum of the hybrid ATCase changes when substrate analogs are bound at the other c_3 trimer.

Binding of carbamoyl-P to the active site induces a local conformational change that increases the affinity for aspartate. When both substrates are bound, the two domains of adjacent c subunits move closer together. This change is believed to be facilitated when the quaternary structure change also takes place, because the active site is connected to a loop at the interface between c subunits in different trimers (Figure 15-5). Because each active site is formed by two polypeptide chains, mixing two types of catalytic subunits that have been inactivated in different ways produces hybrid ATCase molecules that possess some catalytic activity as a result of the generation of composite active sites with no modification (Figure 15-6).

The heterotropic interactions caused by CTP, UTP and ATP are complex and not readily explained by the concerted allosteric model and a simple equilibrium between R and T states. The inhibition exhibited by CTP and UTP is synergistic: CTP alone inhibits ATCase activity by up to approximately 60%, but UTP alone has minimal effect, while the two together inhibit ATCase activity >95%. CTP, ATP and UTP bind competitively to the same common allosteric site, although with different affinities. CTP binds most tightly, with a dissociation constant of 5–20 μM, but there are apparently two classes of three sites each, with dissociation constants differing by a factor of 20. The binding of ATP is similar in exhibiting two classes of sites with different affinities, but it is an order of magnitude weaker, with a dissociation constant of 60–100 μM. UTP seems to bind only to three sites, with a dissociation constant of 800 μM, although the affinity of a second class of sites may be too weak to be measured. Remarkably, binding of CTP to three sites seems to enhance almost 100-fold the binding of UTP to the remaining three sites, and the K_d for UTP is only 10 μM in the presence of CTP. These observations indicate that **the ATCase hexamer is not always symmetrical with six identical subunits**. The measured affinities seem to be appropriate for the *in vivo* situation; the intracellular concentrations of CTP (500 μM) and UTP (900 μM) are three- to six-fold lower than that of ATP (3–5 mM) but their greater affinities mean that they can effectively compete with ATP for the allosteric sites of the enzyme.

Even though they have opposite effects, ATP and CTP bind to the effector site of an individual r subunit in very similar ways. Their triphosphate and ribose moieties make similar interactions; only their bases are bound somewhat differently. The larger base of ATP causes its binding site to expand, but how this small difference could be communicated to the remainder of the ATCase molecule is not obvious.

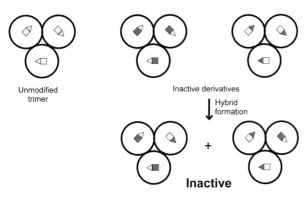
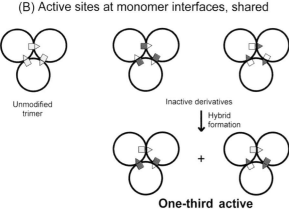

Figure 15-6. Rationale of the use of hybrid oligomers to determine whether pairs of residues contribute to the active site within or between polypeptide chains. *Large circles* designate individual c polypeptide chains that are associated into c_3 trimers. *Squares* and *triangles* indicate the two residues that are being tested; *open symbols* represent native residues, *red* or *blue* symbols are residues that have been altered to inactivate any active site to which they contribute. Functional active sites are *yellow*. Both homogeneous populations in which all of one of the pair of residues has been altered will be inactive. Mixing the dissociated monomers and reconstituting them randomly into trimers will generate active hybrids only if the active site uses the two residues from different polypeptide chains. In this way, Lys84, Ser52 and His134 of ATCase were shown to contribute to an active site that is shared between two polypeptide chains. Adapted from figures by H. K. Schachman.

There are structural differences between the *r* subunits of the R and T states, particularly in the relative orientations of their two domains. The binding of ATP, and probably CTP also, to the *r* subunits affects primarily the R ↔ T equilibrium. Binding of ATP and CTP have opposite effects on the affinities of substrates and analogs that bind to the active site, and they also have opposite structural affects on the *c* subunits, as observed using the nitrated form. Both ATP and CTP bind to the two *r* effector sites with negative cooperativity, which is difficult to explain with the two-state concerted allosteric model. This negative cooperativity is believed to result from direct changes in one *r* subunit upon effector binding to the other *r* subunit of the r_2 dimer unit, independent of any quaternary structure change. The heterotropic interactions in ATCase are not well understood.

Structure of the *E. coli* aspartate transcarbamoylase trapped in the middle of the catalytic cycle. K. A. Stieglitz et al. (2005) *J. Mol. Biol.* **352**, 478–486.

T-state active site of aspartate transcarbamylase: crystal structure of the carbamyl phosphate and L-alanosine ligated enzyme. J. Huang & W. N. Lipscomb (2006) *Biochemistry* **45**, 346–352.

Picosecond dynamics of T and R forms of aspartate transcarbamylase: a neutron scattering study. J. M. Zanotti et al. (2006) *Biochim. Biophys. Acta* **1764**, 1527–1535.

Direct observation in solution of a preexisting structural equilibrium for a mutant of the allosteric aspartate transcarbamoylase. L. Fetler *et al.* (2007) *Proc. Natl. Acad. Sci. USA* **104**, 495–500.

A solution NMR study showing that active site ligands and nucleotides directly perturb the allosteric equilibrium in aspartate transcarbamoylase. A. Velyvis *et al.* (2007) *Proc. Natl. Acad. Sci. USA* **104**, 8815–8820.

15.1.D. Phosphofructokinase

Phosphofructokinase (PFK) catalyzes the step of glycolysis in which fructose-6-phosphate (Fru-6P) is phosphorylated to fructose-1,6-bisphosphate (Fru-1,6-P_2) using either ATP or pyrophosphate (PP_i) as a phosphate donor. PFKs from animals, yeast and most bacteria use ATP as phosphate donor; those found in plants, protozoa and some bacteria use PP_i but ATP-dependent PFKs are often also present. The reaction catalyzed by PFK is virtually irreversible under physiological conditions using either ATP or PP_i. The amino acid sequences of most of the ATP-dependent PFKs are homologous (Section 7.6.A), indicating that they form an evolutionary gene family.

PFK is a highly regulated enzyme with several allosteric effectors. It contributes to the regulation of the overall rate of glycolysis, a key metabolic system, controlling the production of cellular energy and the flux of carbon between glucose and pyruvate. The regulation is often complex, using adenine nucleotides and various metabolites. **Most of these effectors do not appear to modify the maximum velocity but affect only the saturation by Fru-6P, influencing the cooperativity and/ or the half-saturating concentration**. All PFKs are activated by an increase in ADP and/or AMP concentrations, which would indicate a shortage of cell energy. The regulation is especially complex in higher eukaryotes, where AMP, ADP and Fru-2,6-P_2 activate, while ATP and citrate inhibit. Citrate is the first metabolite in the Krebs cycle. The inhibition by ATP is usually severe and allosteric, with nonhyperbolic saturation. It would cause PFK in the liver to be almost inactive with the *in vivo* concentrations of Fru-6P and ATP, and the total PFK activity would be much too low for the glycolytic flux measured *in vivo*. The explanation is due to Fru-2,6-P_2, which is the most potent allosteric activator of eukaryotic PFKs and relieves the allosteric inhibition by ATP. In mammalian cells, the concentration of Fru-2,6-P_2 is controlled by hormones that act on cell metabolism through cyclic AMP (cAMP) and protein kinase A; the concentration of Fru-2,6-P_2 decreases when that of cAMP increases. This hormonal control of PFK activity via Fru-2,6-P_2 regulates the balance between the degradation of glycogen and the synthesis of glucose and/or lipids. Fru-1,6-P_2 can also activate PFKs, but only at higher concentrations than Fru-2,6-P_2.

Regulation is considerably simpler in bacteria, where there is a single activator, ADP. Bacterial PFKs are insensitive to ATP and citrate but are sensitive to feedback inhibition by a single inhibitor, phosphoenolpyruvate (PEP), which is the penultimate end product of glycolysis. PEP is also the substrate of pyruvate kinase, which is activated allosterically by the substrate of PFK, Fru-6P, so the activities of the two enzymes are coupled. Both catalyze irreversible phosphoryl transfer steps, so their coupling is probably crucial for control of the glycolytic flux.

Bacterial and mammalian PFKs are tetramers (Figure 15-7) and are inactivated upon dissociation into dimers. In some cases, the regulatory inhibition involves such a reversible dissociation: the inhibitor binds more tightly to the dissociated form. **The structures of bacterial PFKs explain why an oligomeric structure is required for activity**: the Fru-6P binding site is at the interface between two subunits, with the fructose moiety interacting with one subunit and the 6-phosphate group with the other.

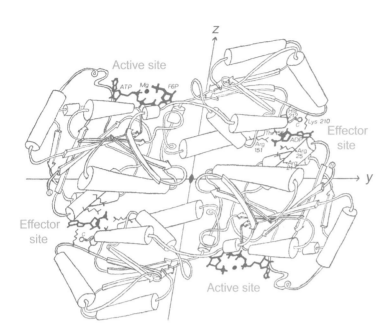

Figure 15-7. Schematic view of two subunits of the PFK from *Bacillus stearothermophilus* that make up one dimer of the tetramer. The other dimer would be below, related by the *y* and *z* two-fold symmetry axes. The dimers are similar in both the R and T states, but differ primarily in their orientations relative to each other, as a result of a rotation about the *x* axis (perpendicular to the page). Two active sites are shown with bound substrates ATP and fructose-6-P (F6P). The regulatory effector sites shown have bound activator ADP. Adapted from P. R. Evans *et al.* (1981) *Philos. Trans. R. Soc. Lond. [Biol]* **293**, 53–62.

Examination of MgATP binding in a tryptophan-shift mutant of phosphofructokinase from *Bacillus stearothermophilus*. M. R. Riley & G. D. Reinhart (2005) *Arch. Biochem. Biophys.* **436**, 178–186.

The structure of the ATP-bound state of *S. cerevisiae* phosphofructokinase determined by cryoelectron microscopy. M. Barcena *et al.* (2007) *J. Struct. Biol.* **159**, 135–143.

Structures of *S. pombe* phosphofructokinase in the F6P-bound and ATP-bound states. S. Benjamin *et al.* (2007) *J. Struct. Biol.* **159**, 498–506.

The first crystal structure of phosphofructokinase from a eukaryote: *Trypanosoma brucei*. J. Martinez *et al.* (2007) *J. Mol. Biol.* **366**, 1185–1198.

---

1. Mechanism of Phosphoryl Transfer

The catalytic activities of all PFKs are sensitive to pH, ionic strength, divalent metal ions and/or their chelators (the real substrate is the ATP•Mg^{2+} complex), phosphate, NH$_4^+$ and K$^+$ ions. **Transfer of the γ-phosphate from ATP to Fru-6P occurs directly within a ternary complex of the enzyme and both substrates, with no phosphoryl enzyme intermediate**. Steady-state kinetics indicate a random order of substrate binding, and the pH dependence of the maximum velocity suggests that Mg^{2+}•ATP^{4-} is the most active ionic form.

The ionized side-chain of Asp127 acts as a base and abstracts a proton from the 1-OH group of Fru-6P to increase its nucleophilicity (Figure 15-8). Replacement of Asp127 by Ser decreases k_{cat} 10^4-fold. The transferred phosphoryl group is stabilized by interactions with the magnesium ion, the positive charge of Arg72 and a hydrogen bond with Thr125. The ionized side-chains of Asp103 and Asp129 hold the magnesium ion that acts as an electrophilic catalyst. The amide NH of Gly104 makes a hydrogen bond with the bridge oxygen to facilitate cleavage of the O–P bond. **The active site of *E. coli* PFK can take two conformations, an 'open' one that binds substrates and releases products, and a 'closed' one in which catalysis occurs**. These conformational changes have been observed to be slow enough to be rate-limiting for the catalytic cycle under some conditions.

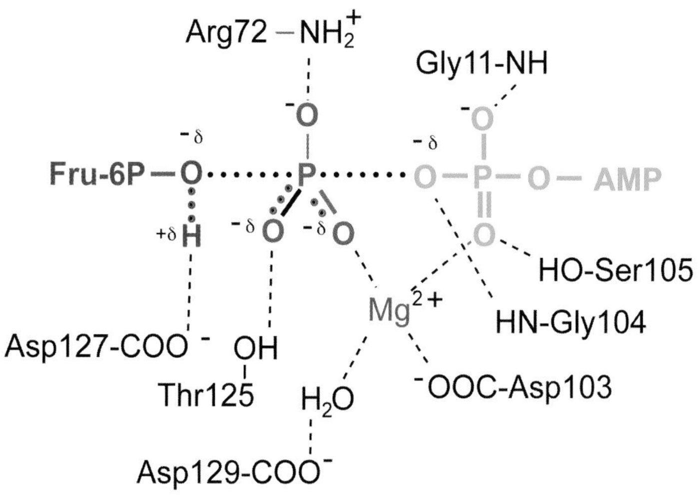

Figure 15-8. The transition state of the reaction catalyzed by PFK, with the main interactions within the active site. The amino acid residues shown are conserved in all the active PFKs except for Thr125, which is replaced by a Ser in eukaryotic PFKs but retains an OH group. The partial covalent bonds in the process of being broken are indicated by *dotted lines*, and the hydrogen and electrostatic bonds by *dashed lines*. Fru6P is fructose-6-P. From data supplied by J. R. Garel.

2. Allosteric Properties of E. coli PFK

The PFK from the bacterium *E. coli* is best understood. It is activated by ADP (or GDP) and inhibited by PEP. **Its steady-state kinetic behavior can be explained remarkably well by the concerted allosteric model** (Section 12.5.A.2). According to this model, the active R state has a much higher affinity for the substrate Fru-6P and the activator ADP (or GDP), while the inactive T state has a much higher affinity for the inhibitor PEP. In the absence of any ligand, the equilibrium between the R and T states of free PFK would be largely in favor of T, with a ratio $T_0/R_0 = 4 \times 10^6$. Both the cooperativity towards Fru-6P and the influence of allosteric effectors can be explained simply by the transition from T to R. Some bacterial PFKs show sigmoidal saturation by Fru-6P only in the presence of the inhibitor PEP, indicating that the transition into an inactive state induced by PEP and the cooperativity towards Fru-6P are related.

The inhibitor PEP binds to a regulatory site that is remote from the active site (Figure 15-7) and causes the protein to changes its quaternary conformation. This change closes the Fru-6P binding site, consistent with the observation that PEP inhibits by decreasing PFK's apparent affinity for Fru-6P. Some residues that bind PEP are also involved in binding the allosteric activator ADP when PFK is in its active conformation (Figure 15-8). The opposite effects of binding activators and inhibitors to a single effector site can be explained by them having opposite effects on the equilibrium between the R and T conformations of the concerted allosteric model. The quaternary structure of the inactive conformation is less stable, which explains why PEP can induce dissociation of some PFKs.

The R and T states of *E. coli* PFK that have been characterized crystallographically do not, however, explain entirely the cooperativity towards Fru-6P. In the absence of ligand, free PFK is found crystallographically to be in the R state rather than the inactive T_0 state, and it binds Fru-6P with high affinity and no cooperativity. The cooperativity is independent to some extent of the transition between the crystallographic R and T states. The binding of substrates and/or effectors to *E. coli* PFK produces conformational changes within the R state, and they are slow enough to be rate-limiting for the catalytic cycle. The cooperativity can be explained by **a change in the rate-limiting step from a conformational change at low Fru-6P concentrations to phosphoryl transfer at high Fru-6P concentrations**. Such a kinetic origin of cooperativity can explain observations with some mutants that are difficult to explain with the concerted model, such as values of the Hill coefficient greater than the number of Fru-6P binding sites and reversal of the influence of effectors, with PEP becoming an activator and GDP an inhibitor.

There are still many unanswered questions about the basis of the cooperativity of this relatively simple regulatory enzyme.

Disentangling the web of allosteric communication in a homotetramer: heterotropic activation in phosphofructokinase from *Escherichia coli*. A. W. Fenton *et al.* (2004) *Biochemistry* **43**, 14104–14110.

Effects of protein–ligand associations on the subunit interactions of phosphofructokinase from *B. stearothermophilus*. R. J. Quinlan & G. D. Reinhart (2006) *Biochemistry* **45**, 11333–11341.

15.1.E. Threonine Synthase

Threonine synthase catalyzes the last step in the synthesis of the amino acid threonine. Its substrate is also used to synthesize another amino acid, methionine, in plants. To keep biosynthesis of the two amino acids in balance, the threonine synthase is activated by the end product of the methionine pathway, S-adenosylmethionine (S-AdoMet, Equation 13-8). The presence of high concentrations of S-AdoMet increases the k_{cat} of threonine synthase eight-fold and decreases its K_m for substrate 25-fold. The effect of S-AdoMet is cooperative, with an apparent Hill constant of 2.9. In this way, the synthesis of threonine is increased when methionine is in ample supply.

Threonine synthase is a dimer of identical subunits, each composed of four domains (Figure 15-9). **The two active sites are at opposite ends of the dimer and between domains**. Each active site contains one molecule of the required cofactor pyridoxal-P (Section 14.4.G.1), which produces only very local changes in the structure. The dimer is symmetrical and the two subunits are identical. The pyridoxal-P is bound in such a way as to block binding of the substrate, which explains the low k_{cat} and high K_m. The active site adopts a more active, closed conformation after S-AdoMet is bound at the allosteric sites, but only in one of the subunits. Surprisingly, **the dimer becomes asymmetric**, for unknown reasons.

The binding sites for the S-AdoMet allosteric effectors are at the interface between the two subunits, as far away from the active sites as possible. Surprisingly, each site binds two S-AdoMet molecules in tandem and in close proximity, each with different conformations and interactions with residues

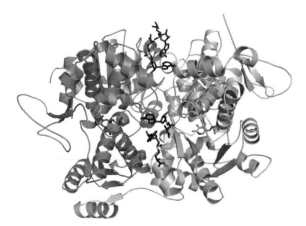

Figure 15-9. The dimeric structure of threonine synthase of *Arabidopsis thaliana* complexed with pyridoxal-P (in *red*) at the two active sites (on the *left* and *right*) and four S-adenosylmethionine (SAM) allosteric effectors (in *blue*) bound between the two subunits. The activated subunit is on the *right*. Kindly provided by R. Dumas.

from both monomers. Binding of the four S-AdoMet molecules is accompanied by a sliding of both monomers along the dimer interface, producing a global movement in which one domain moves towards the interface, another away from it. These movements are of a rigid β-sheet, and only peripheral groups change their conformations significantly.

Threonine synthase is another example of an allosteric enzyme that is regulated by changes to its quaternary structure caused by cooperative binding of allosteric effectors at sites between the subunits. The effects on the active site are not direct but mediated by the movements of relatively rigid domains and subunits relative to each other. Binding of the substrate probably induces further changes in the relative positions of domains at the active site, as occurs in other threonine synthase enzymes from other organisms. Many questions, however, remain to be answered

Allosteric threonine synthase. Reorganization of the pyridoxal phosphate site upon asymmetric activation through S-adenosylmethionine binding to a novel site. C. Mas-Droux *et al.* (2006) *J. Biol. Chem.* **281**, 5188–5196.

15.2. COVALENT REGULATION

Covalent modification of the enzyme is much more important in eukaryotic systems than is allosteric regulation involving reversible binding of substrates and effectors to the enzyme. **In most cases, the covalent modification is introduced by one enzyme and removed by a different one**, and the two reactions are not simply the reverse of each other. The most common reversible covalent modifications are phosphorylation and adenylylation.

15.2.A. Phosphorylation

Protein phosphorylation is a fundamental mechanism of signal transduction used by all cells to regulate the properties of their proteins in response to external or internal stimuli. The list of fundamental biological processes regulated by protein phosphorylation is very long and includes gene expression, energy production and storage, cell division and differentiation, and synaptic plasticity in the nervous system, which is thought to underlie learning and memory. The regulatory phosphorylation reactions that have been best-characterized take place within cells, although secreted proteins and peptides are also known to be phosphorylated.

The phosphorylation of proteins is catalyzed by **protein kinases**, which are a specific class of phosphoryl transferases that generally transfer the phosphoryl group from the γ-position of ATP to the side-chain of a specific amino acid residue of a specific protein (Figure 15-10). The ATP is usually complexed with Mg^{2+} but the two can be replaced *in vitro* in some cases by GTP or Mn^{2+}. Several types of amino acid residues can be phosphorylated:

Ser, Thr	Phosphoester ($-\mathbf{O}-PO_3H_2$)	(15.3)
Tyr	Phosphoester ($-\mathbf{O}-PO_3H_2$)	
His, Arg, Lys	Phosphoramidate ($-\mathbf{N}-PO_3H_2$)	
Cys	Phosphate thioester ($-\mathbf{S}-PO_3H_2$)	
Asp, Glu	Mixed phosphate–carboxylate acid anhydride ($-\mathbf{CO}-\mathbf{O}-PO_3H_2$)	

The atoms in bold originate from the amino acid. The dephosphorylation reaction is hydrolysis of the phosphoester bond, catalyzed by a **protein phosphatase**.

Both the phosphorylation and dephosphorylation reactions are energetically favorable, and the net result of the two steps is hydrolysis of ATP to ADP and inorganic phosphate (P_i). Under artificial conditions of very low levels of ATP and high levels of ADP, protein kinases can catalyze the reverse step, transfer of a phosphoryl group from the phosphor-amino acid of a protein to ADP, generating ATP.

The level of phosphorylation of a protein *in vivo* is determined by a dynamic equilibrium between its rates of phosphorylation and dephosphorylation.

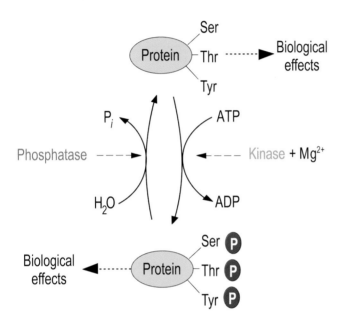

Figure 15-10. The major pathways of regulatory protein phosphorylation/dephosphorylation, observed in all types of living organisms. It involves phosphorylation of Ser, Thr or Tyr residues by protein kinases using ATP•Mg^{2+} as phosphoryl donor. The dephosphorylation is catalyzed by a protein phosphatase, which is usually distinct from the protein kinase. The activity of both the protein kinase and phosphatase can be regulated by intracellular or extracellular stimuli. The phosphorylated or dephosphorylated forms of the protein substrate, or both, may be responsible for the biological response observed. The *circled P* represents a phosphate group.

Charging it up: global analysis of protein phosphorylation. J. Ptacek & M. Snyder (2006) *Trends Genet.* **22**, 545–554.

Analysis of protein phosphorylation: methods and strategies for studying kinases and substrates. S. C. Peck (2006) *Plant J.* **45**, 512–522.

Phosphorylation studies using plant protein microarrays. T. Feilner & B. Kersten (2007) *Methods Mol. Biol.* **355**, 379–390.

Exploring the role of protein phosphorylation in plants: from signalling to metabolism. S. C. Huber (2007) *Biochem. Soc. Trans.* **35**, 28–32.

1. Phosphorylation in Eukaryotes

Regulatory protein phosphorylation in eukaryotes occurs primarily on Ser, Thr and Tyr residues (Figure 15-10). Phosphorylation of protein His residues is now recognized as also being important, but phosphohistidine is relatively unstable and is easily overlooked. Consequently, His phosphorylation has not been as well studied as Ser/Thr or Tyr phosphorylation.

More than a thousand different protein kinases are coded in the genomes of higher eukaryotes. Even the genome of the yeast *Saccharomyces cerevisiae* contains genes for 113 conventional protein kinases. Most have homologous catalytic domains that comprise a small lobe, mostly of β-sheet, a large lobe, mostly of α-helix, and a cleft between the two, in which catalysis takes place. The small lobe is involved in binding ATP, the large lobe in substrate binding. The kinase is active only when a critical loop of the large lobe, known as the **activation loop** or **T loop**, is positioned correctly. In many cases, the loop must be phosphorylated on one or several residues. The phosphorylation can be unregulated or regulated by another activating protein kinase.

Ser/Thr kinases contain additional domains that vary greatly from one kinase to another and play a role in targeting and oligomerization as well as regulation. In several cases, the regulatory domain contains a sequence that corresponds to a substrate or pseudosubstrate, which obstructs the catalytic site and inhibits the Ser/Thr kinase. Activation is usually produced by the binding of a regulatory molecule that removes this intramolecular inhibition and permits substrates to access the active site. Activation can also be produced *in vitro* by limited proteolysis to remove the inhibitory segment and release a free active catalytic domain that is not subject to regulation.

Phosphorylation of Tyr residues occurs much less frequently than Ser and Thr but it is of critical functional importance. Phosphorylation of Tyr residues is catalyzed by either specialized enzymes that phosphorylate only Tyr residues or by protein kinases that have dual specificity and act on both Tyr and Thr residues of specific protein substrates. The catalytic domains of specialized Tyr kinases are very homologous in both sequence and structure and are also closely related to those of Ser/Thr kinases. Tyr kinases are more closely related to each other than to Ser/Thr kinases and have a number of residues in common, which makes it possible to predict the specificity of these enzymes on the basis of just their amino acid sequences.

Tyrosine kinases can be grouped into two categories: **receptor tyrosine kinases** (**RTK**), which have a transmembrane segment and are usually receptors for extracellular ligands, and **nonreceptor tyrosine kinases** (**NRTK**), which are solely intracellular. RTKs have an extracellular domain of varying length that serves as receptor, a single transmembrane segment and an intracellular catalytic domain. In some cases, the polypeptide chain lacks the receptor domain, but it is provided by an associated protein that may be attached to the transmembrane polypeptide by a disulfide bond or to the membrane by a covalently attached membrane anchor, such as a glycosyl-phosphatidyl-inositol (GPI) group.

RTKs exist as monomers or interact with other proteins in the absence of ligand. **Activation requires their dimerization**. Ligand binding in some cases promotes dimerization, either by a conformational change in the extracellular domain or from the divalent nature of the ligand itself. The dimer is pre-formed in other cases, but inactive. Ligand binding causes the intracellular catalytic domains of each of the second subunits to be phosphorylated, which leads to the phosphorylation of Tyr residues located in the crucial activation loop of the catalytic domain and enhances its enzymatic activity. Some receptors are phosphorylated rapidly in the absence of ligand, but this is compensated by a high rate of dephosphorylation; tilting the balance in favor of phosphorylation can produce activation.

NRTKs comprise a heterogeneous group of tyrosine kinases in which the catalytic domain is associated with various other domains. Some of these NRTKs are functionally equivalent to the catalytic domains of RTKs.

Protein phosphatases in eukaryotes have catalytic domains that belong to several gene families: Ser/Thr phosphatases belong to either of two gene families of metal-requiring enzymes, whereas Tyr or dual-specificity phosphatases form a superfamily of genes, grouped into four families, that have a common signature sequence in their active site (Cys–X_5–Arg) and a common catalytic mechanism.

The intricacies of p21 phosphorylation: protein/protein interactions, subcellular localization and stability. E. S. Child & D. J. Mann (2006) *Cell Cycle* **5**, 1313–1319.

Posttranslational phosphorylation of mutant p53 protein in tumor development. M. Matsumoto *et al.* (2006) *Med. Mol. Morphol.* **39**, 79–87.

Tyrosine phosphorylation controls PCNA function through protein stability. S. C. Wang *et al.* (2006) *Nature Cell Biol.* **8**, 1359–1368.

Multisite phosphorylation of the cAMP response element-binding protein (CREB) by a diversity of protein kinases. M. Johannessen & U. Moens (2007) *Front. Biosci.* **12**, 1814–1832.

---

2. Phosphorylation in Prokaryotes

Ser/Thr/Tyr phosphorylation using ATP is as important in bacteria as in eukaryotes. Bacterial phosphorylation systems use protein kinases that have sequences homologous to those of eukaryotes but their protein phosphatases are less-well characterized. The Ser/Thr protein kinases are usually regulated by cellular metabolites, instead of by second messengers. Surprisingly, many bacterial protein kinases do not exhibit any similarity to eukaryotic protein kinases but instead resemble nucleotide-binding proteins and kinases that phosphorylate low-molecular weight substrates. The isocitrate dehydrogenase kinase/phosphatase of enteric bacteria is unusual in being a single polypeptide chain with both phosphatase and kinase activities.

The **sensor kinase/response regulator**, often known as the *two-component system*, is a second system of protein phosphorylation. A sensor kinase that is generally associated with the membrane phosphorylates itself on a His residue in response to external stimuli such as the presence of a nutrient or chemoattractant or changes in osmotic pressure. The phosphoryl group is then transferred to an Asp residue that is located on a different protein or on the same polypeptide chain, which triggers the appropriate biochemical response. This type of signal transduction is present in both eubacteria and archea, so it seems to be very ancient evolutionarily.

In a third phosphorylation system, a phosphoryl group that is donated from PEP is transferred between His and Cys residues down a chain of proteins. Such PEP-dependent systems are implicated in a wide variety of regulations, including carbohydrate metabolism, transcription and adenylate cyclase.

Role of protein phosphorylation on serine/threonine and tyrosine in the virulence of bacterial pathogens. A. J. Cozzone (2005) *J. Mol. Microbiol. Biotechnol.* **9**, 198–213.

Ser/Thr/Tyr protein phosphorylation in bacteria: for long time neglected, now well established. J. Deutscher & M. H. Saier (2005) *J. Mol. Microbiol. Biotechnol.* **9**, 125–131.

Mass spectrometric analysis of protein histidine phosphorylation. X. L. Zu *et al.* (2007) *Amino Acids* **32**, 347–357.

3. Protein Phosphorylation in Signal Transduction Networks

Many signal transduction systems in cells involve various protein systems that greatly amplify and modulate the signal and usually involve protein phosphorylation. **Phosphorylation reactions in cells often occur in cascades** in which activation of a first protein kinase phosphorylates and activates a second protein kinase that, in turn, phosphorylates and activates a third kinase, and so forth. This may account in part for the very large number of protein kinases. **Phosphorylation cascades can amplify the signal** because one molecule of the first activated kinase can phosphorylate several secondary kinases, which can each phosphorylate several tertiary kinases, etc. **These cascades can also exhibit kinetic properties similar to those of allosteric enzymes,** such as positive or negative cooperativity, even if the individual kinases composing the cascade themselves follow classical Michaelis–Menten kinetics. Such protein kinase cascades can convert a signal that varies gradually into one that exhibits virtually on–off behavior.

Protein phosphorylation and signal transduction modulation: chemistry perspectives for small-molecule drug discovery. T. K. Sawyer *et al.* (2005) *Med. Chem.* **1**, 293–319.

Ribosomal protein S6 phosphorylation: from protein synthesis to cell size. I. Ruvinsky & O. Meyuhas (2006) *Trends Biochem. Sci.* **31**, 342–348.

Regulation of Rho proteins by phosphorylation in the cardiovascular system. G. Loirand *et al.* (2006) *Trends Cardiovasc. Med.* **16**, 199–204.

Protein tyrosine phosphorylation and reversible oxidation: two cross-talking posttranslation modifications. P. Chiarugi & F. Buricchi (2007) *Antioxid. Redox Signal.* **9**, 1–24.

---

4. Specificity of Protein Phosphorylation

Protein kinases are very specific for both the residue and the protein that they phosphorylate. Most phosphorylate the side-chains of either Ser/Thr or Tyr residues, and only a few have dual specificity, usually for both Thr and Tyr. Protein kinases also recognize the immediate environment of the residue that is phosphorylated, including its neighboring sequence, and **consensus sequences have been identified for many of them** (Table 15-1). In some cases, the specificity also depends on the more global structure, including sites distant from the phosphorylated residue. The proximity between the protein kinase and its substrate is important in living cells. Specific targeting mechanisms can localize the kinase at particular sites in the cell. A number of scaffolding proteins can bring together several enzymes, including kinases, phosphatases and their substrates.

Protein phosphatases are less specific for the sequence surrounding the phospho-amino acid but more for higher order structural determinants and the targeting processes. Except in the few cases with a single target protein, the substrate specificities of most protein kinases and phosphatases do not match exactly. This suggests that the regulation of groups of these enzymes is complex.

Phospho3D: a database of three-dimensional structures of protein phosphorylation sites. A. Zanzoni *et al.* (2007) *Nucleic Acids Res.* **35**, D229–231.

Activation loop phosphorylation-independent kinase activity of human protein kinase C zeta. S. Ranganathan *et al.* (2007) *Proteins* **67**, 709–719.

Table 15-1. Examples of consensus phosphorylation sites for protein kinases

Protein kinase	Sequence of phosphorylation site
Ca^{2+} calmodulin-dependent kinase II	–Arg–Xaa–Xaa–(**Ser/Thr**)–
cAMP-dependent protein kinase	–(Arg/Lys)–Arg–Xaa–**Ser**–
Casein kinase 1	–Acid-Xaa–Xaa–**Ser**–
Casein kinase 2	–(**Ser/Thr**)–Xaa–Xaa–(Asp/Glu)–
EGF receptor tyrosine kinase	–(Glu/Asp)–**Tyr**–Hyd–
MAP kinases, cyclin-dependent kinases	–(**Ser/Thr**)–Pro–
Protein kinase C	–(Arg/Lys)–Xaa$_{1-2}$–(**Ser/Thr**)–Xaa–(Arg/Lys)–

Amino acid residues are shown with their three-letter code; Xaa is any residue; alternative residues are in parentheses; residues phosphorylated are in bold; Acid corresponds to Asp, Glu or a phosphorylated residue, Hyd to hydrophobic residues Ile, Leu or Val. Data from J. P. Girault (1999) in *Encyclopedia of Molecular Biology* (T. E. Creighton, ed.), Wiley-Interscience, NY, p. 1839.

5. Effects of Phosphorylation on the Properties of Proteins

The physiologically important effects of phosphorylation on the properties of many proteins are numerous and diverse and include the activation and inhibition of enzymes, opening and closing of ion channels, increasing and decreasing activities of transcription factors, aggregation and disassembly of cytoskeletal components, and many others. In only a few instances, however, is the precise molecular mechanism known by which phosphorylation of a specific residue brings about functional changes. Phosphorylation of an amino acid residue can have only local consequences or it can affect the three-dimensional structure of the protein. The interactions of a protein with others can also be affected. A phosphate group is bulky and has two negative charges at physiological pH, so its presence alters dramatically the properties of the amino acid side-chain to which it is bound.

How phosphorylation can modify dramatically the activity of an enzyme by a local effect on the active site is illustrated by **isocitrate dehydrogenase** (IDH), an enzyme that is regulated by phosphorylation in plants and bacteria. It catalyzes the oxidative decarboxylation of isocitrate to α-ketoglutarate, a critical step in the Krebs citric acid cycle, when it is not phosphorylated. In the presence of high levels of ATP, however, IDH is inactivated by phosphorylation of residue Ser113, which is located in the site that binds the substrate. **Simple steric hindrance and electrostatic repulsion by the phosphate group prevent the isocitrate from binding**.

The structures of glycogen phosphorylase in the phosphorylated and nonphosphorylated states (Section 15.2.B) illustrate how phosphorylation of a single residue can have dramatic consequences on the tertiary and quaternary structures of a protein.

The effects of protein phosphorylation on the regulation of protein–protein interactions are illustrated by **SH2** domains. They were recognized first as a 100-residue domain in the protein known as Src, then by homology in many other proteins, so it was named Src-homology 2 (SH2). These domains bind specific peptides, but only if they are phosphorylated on Tyr residues. Specific peptides are recognized using additional residues on the C-terminal side of the phosphorylated Tyr reside. The interactions cause enzymes to be clustered around activated growth-factor receptors and brought into the proximity of their membrane-associated substrates, which triggers a cascade of reactions that produce the effects of the growth factors.

SH2 domains can also maintain a protein in an inactive state, as in the case of Src family tyrosine kinases. A C-terminal phosphorylated Tyr residue interacts with an SH2 domain of the same polypeptide chain, generating a closed conformation in which the enzyme is unable to reach its substrates. Dephosphorylation of the C-terminal Tyr residue results in displacement of the SH2 domain by a competing phosphopeptide and activation of the kinase. Other domains that interact specifically with phosphorylated proteins include the **PTB (phosphotyrosine-binding) domains** and **14-3-3 proteins**, which interact with their partner proteins only when they are phosphorylated on a Ser residue.

Control of isocitrate dehydrogenase catalytic activity by protein phosphorylation in *Escherichia coli*. A. J. Cozzone & M. El-Mansi (2005) *J. Mol. Microbiol. Biotechnol.* **9**, 132–146.

Conformational changes in protein loops and helices induced by post-translational phosphorylation. E. X. Groban *et al.* (2006) *PLoS Comput. Biol.* **2**, e32.

14-3-3 proteins: a historic overview. A. Aitken (2006) *Semin. Cancer Biol.* **16**, 162–172.

Structural determinants of 14-3-3 binding specificities and regulation of subcellular localization of 14-3-3–ligand complexes: a comparison of the X-ray crystal structures of all human 14-3-3 isoforms. A. K. Gardino *et al.* (2006) *Semin. Cancer Biol.* **16**, 173–182.

---

6. Methods to Characterize Protein Phosphorylation

Protein phosphorylation is most frequently detected by radiolabeling cells with inorganic ^{32}P-orthophosphoric acid (P_i). The $^{32}P_i$ is taken up by cells and incorporated into phosphorylated molecules, including ATP. Phosphoproteins can be detected on the basis of their radioactivity and isolated. Phosphorylation can occur at two or more residues, with different physiological consequences, so peptide mapping is required to determine the phosphorylation status of each residue. The amount of ^{32}P incorporated at any residue does not reflect simply the extent of phosphorylation at this site, but is also determined by its rate of dephosphorylation. The level of labeling also depends on the specific activities of the intracellular ATP pools, which may vary and depend on the physiological status of the cell.

The labeling process just described is known as **front-phosphorylation**. In **back-phosphorylation**, unlabeled proteins are isolated while preserving their phosphorylation state. They are then phosphorylated completely *in vitro* using γ-^{32}P-ATP and an appropriate purified kinase. The greater the degree to which the site is phosphorylated *in vivo*, the less the incorporation of radioactive phosphate *in vitro*. This method requires a highly active purified protein kinase and is not very sensitive. An alternative method is to use **antibodies that are specific for the phosphorylation state**. Some such antibodies recognize a phosphorylated amino acid residue irrespective of its environment, such as those specific for phospho-Tyr residues. Antibodies that are specific for a given protein phosphorylated at a particular site can be raised using as immunogen synthetic peptides that include the phosphorylated residue.

Electrophoresis is very useful because **the phosphoryl group introduces a negative charge on the molecule and alters its electrophoretic mobility**. A phosphorylated residue shifts the electrophoretic mobility of the protein towards a more acidic form under nondenaturing conditions. In the presence of the detergent sodium dodecyl sulfate (SDS), phosphorylation should have no effect on the electrophoretic mobility. Yet it has been observed to decrease the mobility and increase the apparent size, probably due to a decreased charge density resulting from decreased binding of SDS to the phosphorylated protein. Phosphorylation at specific sites has also been found to *increase* the mobilities of some proteins.

The type of residues phosphorylated can be determined after acid hydrolysis of the protein to amino acids. Phosphoserine, -threonine and -tyrosine amino acids can be separated and identified by thin-layer electrophoresis. Phosphohistidine and -aspartate are extremely acid-labile and cannot be identified by this approach. Phosphotyrosine is the only amino acid that is resistant to alkaline pH, which can be used to identify and measure it specifically. The phosphorylation status of a protein can be compared qualitatively using two-dimensional peptide maps of ^{32}P-labeled proteins. Phosphopeptides can also be separated by HPLC and the phosphorylated residue and its adjacent protein sequence identified by protein sequencing (Section 7.5). Site-directed mutagenesis can replace a 'phosphorylatable' residue with a 'nonphosphorylatable' one, for example a Ser or Thr residue can be replaced by an Ala, and the effect on phosphorylation of the protein determined. This approach is extremely specific but also indirect, and there is always the possibility that the replacements might have unsuspected consequences on the properties of the protein.

Phosphorylated residues can be readily detected by their increased mass, using mass spectrometry. This is now the method of choice, but identification of protein phosphorylation sites is still challenging.

Mining phosphopeptide signals in liquid chromatography-mass spectrometry data for protein phosphorylation analysis. H. Y. Wu *et al.* (2007) *J. Proteome Res.* **6**, 1812–1821.

Modular mass spectrometric tool for analysis of composition and phosphorylation of protein complexes. J. D. Blethrow *et al.* (2007) *PLoS ONE* **2**, e358.

Electron capture dissociation in the analysis of protein phosphorylation. S. M. Sweet & H. J. Cooper (2007) *Expert Rev. Proteomics* **4**, 149–159.

15.2.B. Glycogen Phosphorylase

Glycogen phosphorylase catalyzes the first step in the intracellular degradation of glycogen, which is a polymeric storage form of sugar. It liberates individual molecules of glucose that can be used for energy, so its activity is regulated in a number of ways. It is one of the best-characterized regulatory enzymes and illustrates the complexity of regulation of metabolic enzymes in higher organisms.

Glycogen consists of glucose molecules joined into long chains by the elimination of water between the first carbon of one glucose and the fourth of another, forming α-1,4 links. Every four to five glucose units, a branch is introduced by an α-1,6 link, resulting in a highly branched molecule that may contain up to a million glucose units but is compact and roughly spherical in shape. Glycogen phosphorylase catalyzes cleavage of the 1,4 links by a phosphate ion:

$$(\alpha\text{-1,4-glucoside})_n + HPO_4^{2-} \leftrightarrow (\alpha\text{-1,4-glucoside})_{n-1} + \alpha\text{-D-glucose-1-}PO_3^{2-} \quad (15.4)$$

The 1,6 branch point links are removed by a separate enzyme, the **glycogen debranching enzyme**. The two enzymes together can degrade glycogen completely.

Glycogen phosphorylase is a dimer that exists in two forms, *a* and *b*, that differ covalently only in that the side-chain of Ser14 has been phosphorylated in form *a* by the specific enzyme **phosphorylase kinase**. Each subunit of glycogen phosphorylase consists of a single polypeptide chain of 842 amino acid residues. The N-terminal Ser1 residue is acetylated, while the cofactor pyridoxal-P is bound to Lys680 via a Schiff base (Section 14.4.G.1). The structure of each subunit is characteristic of αβ-type proteins, with a 50% content of α-helix and 30% β-sheet (Section 9.1.I.3). The dimeric structures of the two forms *a* and *b* are very similar, except that the N-terminal 16 residues are disordered in phosphorylase *b* in the absence of the phosphate group on Ser14. When Ser14 is phosphorylated, these residues adopt a distorted helix that binds to the rest of the protein, displacing four residues from the C-terminus (Figure 15-11). The crucial aspect of phosphorylation appears to be the introduction of the negative charge of the phosphate group. It compensates for the positive charge of basic residues at positions 9, 10, 11 and 16. These four basic residues comprise the specificity sequence of the phosphorylation site, but it is simply the excess of positive charge in this segment that probably is the reason why it is disordered when Ser14 is not phosphorylated.

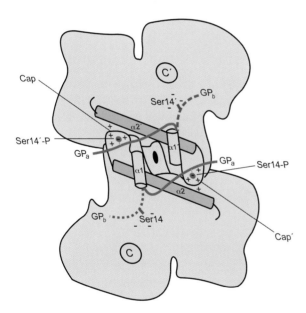

Figure 15-11. Changes in the N-terminal residues upon phosphorylation of Ser14 to convert glycogen phosphorylase from the *b* to the *a* form. The view is down the two-fold axis of the dimer. The N-terminal residues 10–19 shown as the *dashed line* are mobile in the *b* form and Ser14 is located close to negatively charged groups. From residue 19, the chain folds into α-helix 1, followed by the cap and α-helix 2. Upon phosphorylation, the N-terminal residues swing through about 150° and become more ordered (*solid line*). The Ser14-P group is located at the subunit interface and interacts with positively charged basic residues, from its own subunit and the cap region of the other subunit. These charged groups are distant from the catalytic site (C). Original figure kindly provided by L. Johnson. Adapted from N. J. Darby & T. E. Creighton (1993) *Protein Structure: in focus*, IRL Press, Oxford, p. 87.

Each monomer comprises two large domains (Figure 15-12). The N-terminal domain consists of 490 residues and a core β-sheet of nine β-strands. The sheet is twisted by 180° and flanked by helical segments, loops and some additional β-strands. This domain also contains the site where the allosteric activator AMP and the allosteric inhibitors **ATP** and **glucose-6-phosphate** bind. It is only 10 Å from the Ser14–phosphate group. Some 30 Å away, in a helical part of the domain, is the **glycogen storage site**. It binds oligosaccharides in a helical form that is like that of amylose but slightly distorted. Amylose binds here some 20-fold tighter than at the active site, so phosphorylase acting on a glycogen particle is probably anchored to it by these sites. The C-terminal domain consists of a core of six parallel β-strands surrounded by α-helices, similar to the nucleotide-binding motif of the dehydrogenases (Section 12.4), although more buried by additional helices and loops. **The deep cleft between the two domains contains the active site**, which is buried deeply, next to the phosphate

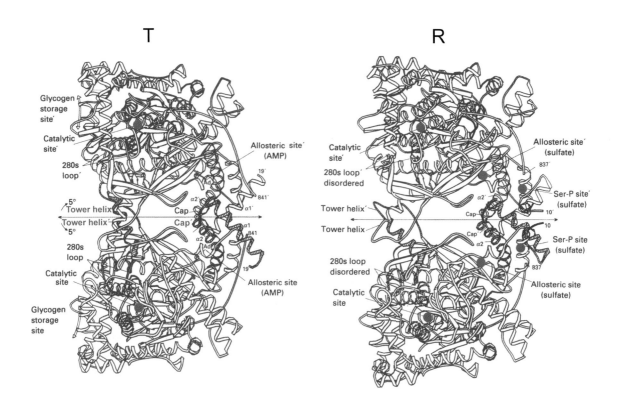

Figure 15-12. Ribbon representations of the T state (*left*) and R state (*right*) structures of glycogen phosphorylase *b*. The catalytic, glycogen storage and allosteric effector sites are indicated, with any ligands present in the structure given in brackets. Sulfate ions used to crystallize the R state occupy sites that normally are occupied by phosphate groups. The catalytic site is close to the '280s' loop at the end of the tower helix. The allosteric effector site and the Ser14-P site are situated between the 'cap' and the α2 helix. Adapted from a figure by L. N. Johnson.

group of pyridoxal-P and an inhibitor site for caffeine. The entrance for substrates is between Phe285 and Tyr613, which is 'locked' when caffeine binds.

When two subunits associate to form the dimer, they generate shared binding sites for AMP, the phosphate of Ser14 and a pair of α-helices (known as the **tower helices**) that extend reciprocally towards the symmetry-related active sites. The catalytic face of the dimer is concave and contains the active sites and glycogen storage sites. The convex control face may be exposed to the action of phosphorylase kinase or phosphatase even while bound to glycogen (Figure 15-12).

The coenzyme **pyridoxal-P plays a structural role** and is sandwiched between the two domains; removing it produces inactive monomers. It also has a role in catalysis, but very different from its usual role (Section 14.4.G.1). Its phosphate group is the only part required for catalysis, and it must be capable of forming a dianion. Phosphorylase *b* is inactive when reconstituted with pyridoxal, which lacks the phosphate group, but it can regain appreciable activity in the presence of added P_i, phosphite or fluorophosphate. Only one equivalent of pyrophosphate binds to each monomer, but it inhibits competitively with both glucose-1-P and the activating phosphite. When the enzyme is reconstituted with pyridoxal pyrophosphate glucose, which has the substrate and coenzyme covalently linked through the pyrophosphate group, the enzyme transfers the glucose moiety to an added oligosaccharide. Phosphorylase can phosphorylate the unusual sugar heptenitol to heptulose 2-P, which is a potent inhibitor. It binds with a hydrogen bond between its phosphate group and that of the pyridoxal-P.

All these observations suggest that the normal degradation of glycogen involves the proton of the substrate HPO_4^{2-} group attacking and cleaving the glycosidic bond of glycogen while immediately gaining a proton from the phosphate group of pyridoxal-P.

Structural relationships among regulated and unregulated phosphorylases. J. L. Buchbinder *et al.* (2001) *Ann. Rev. Biophys. Biomol. Struct.* **30**, 191–209.

Glycogen phosphorylase inhibitors. B. R. Henke & S. M. Sparks (2006) *Mini Rev. Med. Chem.* **6**, 845–857.

---

1. Regulation by Phosphorylation

The nervous signal for contraction of muscle is release of Ca^{2+}, which binds to calmodulin (Section 12.3) and thus activates phosphorylase kinase, converting the inactive phosphorylase *b* of resting muscle to the more active *a* form. The phosphorylase in the liver of a well-fed animal is inactive but can be activated by a signal from the pancreatic hormone **glucagon** that blood sugar levels are low, again via the mediation of phosphorylase kinase (Figure 15-13). The hormones **adrenaline** and **glucagon** act by stimulating production of **cAMP**, the common second messenger that is formed from ATP by **adenylate cyclase**, a membrane-bound complex of at least three different proteins. One is the receptor for the specific hormone and on the outer surface of the membrane. Another is the catalytic protein, on the inner side of the membrane, which produces cAMP from ATP within the cell, but only when hormone is bound to the receptor. Communication between the two components is mediated by the third component, a **G protein** or **GTPase**, but only when it has GTP bound. The interactions between these components are intricate and only partly understood.

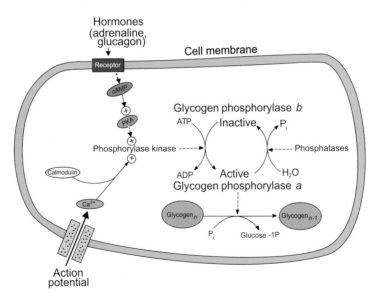

Figure 15-13. The main pathways regulating glycogen degradation. Glycogen is degraded stepwise by glycogen phosphorylase, to liberate glucose-1-P. Glycogen phosphorylase exists in either a phosphorylated active form (*a*) or a dephosphorylated inactive form (*b*). Phosphorylase kinase catalyzes the conversion of *b*-form to *a*-form and is activated by two independent mechanisms. In the first, it is activated by phosphorylation by cAMP-dependent protein kinase (PKA). PKA is stimulated directly by cAMP, which is generated in response to the stimulation of adenylyl cyclase by various G protein-coupled receptors, such as those for the hormones adrenaline and glucagon. In the second pathway, Ca^{2+} enters the skeletal muscle cell during nerve stimulation and binds to calmodulin. The complex activates phosphorylase kinase directly. Glycogen phosphorylase *a* is dephosphorylated and inactivated by several phosphatases. P_i is inorganic phosphate.

The increased level of cAMP caused by hormone binding to the cell surface activates two different **cAMP-dependent kinases**. In the absence of cAMP, both kinases are catalytically inactive. Both are composed of two regulatory and two catalytic subunits, R_2C_2; they differ only in the R chains. cAMP dissociates this complex by binding more tightly to the free regulatory dimers, R_2, thereby releasing active catalytic subunits:

$$R_2C_2 + 4\,\text{cAMP} \rightleftharpoons (R\cdot\text{cAMP}_2)_2 + 2\,C \tag{15.5}$$
$\quad$ Inactive $\qquad\qquad\qquad\qquad\qquad$ Active

The R subunit has a substrate-like sequence that is essential for interacting with the C subunit. It has two cAMP-binding domains that are not equivalent and to which cAMP binds cooperatively.

Activation of the cAMP-dependent protein kinase is also regulated by covalent modification and allosteric regulation. One type of R subunit can be phosphorylated by the catalytic subunit at its substrate-like site, thereby increasing its tendency to dissociate. With the other type of regulatory subunit, ATP binds tightly and cooperatively to the R_2C_2 form, thereby inhibiting its tendency to be dissociated and activated by cAMP.

The active cAMP-dependent kinases phosphorylate a number of different enzymes, each on accessible Ser (or Thr) residues occurring in the sequences –Lys–Arg–X–X–Ser or Arg–X–Arg–X–Ser. X can be almost any amino acid, and Thr can replace the Ser residue. As a consequence of these phosphorylations, **degradative enzymes such as glycogen phosphorylase are activated, whereas biosynthetic enzymes such as glycogen synthetase are inhibited**. These regulatory effects often are not direct but occur via intermediary enzymes in cascades. For example, cAMP-dependent kinases directly phosphorylate phosphorylase kinases, not glycogen phosphorylase. The phosphorylated phosphorylase kinase then phosphorylates phosphorylase *b*.

Phosphorylase kinase is a complex protein, consisting of four copies of each of four different polypeptide chains. Two of these types of chains can each be phosphorylated on a specific Ser side-chain. Phosphorylation of one chain increases the rate of phosphorylation of the other. The complex is active as a kinase only when the first subunit is phosphorylated and if another of the subunits, **calmodulin**, has bound Ca^{2+} ion (Section 12.3.A). This makes activation of glycogen phosphorylase sensitive to muscle contraction, which is triggered by Ca^{2+} ion release (Figure 15-13). The activated phosphorylase kinase then specifically phosphorylates Ser14 of phosphorylase *b*. The phosphorylation of Ser14 of each glycogen phosphorylase subunit by phosphorylase kinase occurs by a rapid-equilibrium random *Bi–Bi* kinetic mechanism, meaning that either substrate (ATP or glycogen phosphorylase) can bind first, the measured K_m values approximate binding constants, and the rate-limiting steps occur in reactions on the ternary complex of the two substrates and the enzyme (Section 14.3.C.2).

All of these activation steps are balanced by reverse steps, catalyzed by phosphatases. Under normal conditions there is probably a steady-state condition of phosphorylation balanced by dephosphorylation, even though the net result is hydrolysis of ATP (Figure 15-10). The adenylate cyclase is regulated by its middle component hydrolyzing its bound GTP and thereby temporarily stopping the hormone from causing formation of cAMP. Existing cAMP is hydrolyzed to AMP by a specific phosphodiesterase, thereby favoring inhibition of the cAMP-dependent kinase by aggregation with its regulatory subunits (the reverse of Equation 15.5). The two phosphoryl groups on phosphorylase kinase are removed by two different phosphatases; the phosphoryl group that was added first is also removed first. The responsible phosphatase removes the phosphoryl group on Ser14 of phosphorylase *a* and glycogen synthetase. The rate at which it acts on Ser14 probably depends upon the conformational state of phosphorylase *a*. The phosphatase appears to bind tightly to phosphorylase *a* but to cleave the phosphoryl group only when it becomes accessible. Therefore this covalent modification is also susceptible to allosteric control within the enzyme.

The activities of these two phosphatases are also regulated directly, one by calmodulin and Ca^{2+} binding, the other by two protein inhibitors. One of these inhibitors is active only if phosphorylated by cAMP-dependent protein kinase. This amplifies the effect of the hormone, because the inhibitor is activated and then inhibits the phosphatase; consequently, the rate of inactivation of phosphorylase *a* is decreased, which tends to increase the rate of glycogen breakdown.

The amino-terminal tail of glycogen phosphorylase is a switch for controlling phosphorylase conformation, activation, and response to ligands. A. C. Bjorn & D. J. Graves (2001) *Biochemistry* **40**, 5181–5189.

Three-dimensional structure of phosphorylase kinase at 22-Å resolution and its complex with glycogen phosphorylase *b*. C. Venien-Bryan *et al.* (2002) *Structure* **10**, 33–41.

---

2. Allosteric Properties of Glycogen Phosphorylase

In addition to regulation by phosphorylation, **phosphorylase is regulated by a number of low-molecular weight effectors that bind to four different sites on the enzyme**. The endogenous inhibitors are glucose, ADP, ATP, fructose 1-P, glucose 6-P and UDP-glucose, while AMP is an activator. The effects of these molecules binding to the various binding sites on the *a* and *b* forms on the properties of the enzyme and the interactions between them are complex and not readily summarized. In general, the *a* and *b* forms of glycogen phosphorylase differ markedly in their allosteric regulatory properties (Figure 15-14). The *b* form is activated by AMP or IMP and is inhibited by ATP, ADP, glucose and glucose-6-P. ATP and glucose-6-P compete for the AMP site; the effect of ATP is believed to reflect the energy status of the cell, whereas that of glucose-6-P is a type of end product allosteric control. The *a* form is not subject to these controls, except for still being inhibited by glucose, which binds at the active site and is a competitive inhibitor of glucose-1-P; this can be considered a primitive type of end product feedback inhibition. Phosphorylase *a* may be said to have 'escaped' the allosteric control of the *b* form, which is superseded by regulation using covalent modification under extracellular hormonal and neuronal regulation.

The allosteric control of phosphorylase is usually analyzed in terms of whether the allosteric effectors stabilize the T or R form (Figure 15-12). The two forms interconvert by rotating one subunit relative to the other by ~ 10° about an axis perpendicular to the two-fold symmetry axis of the dimer. In the process, the 'tower' helices pull apart, tilt, slide past each other and adopt an alternative mode of packing. The geometry of one helix is constrained by that of the symmetry-related helix, preserving the molecular symmetry; the conformational change appears to be concerted. There are, however, many variations on both of these quaternary structures. When the *b* form changes from the T to the R form, dramatic conformational changes occur in the N-termini of both subunits, in a loop between residues 282 and 286 (the '280s loop') that controls access of the substrate to the active site (Figure 15-12).

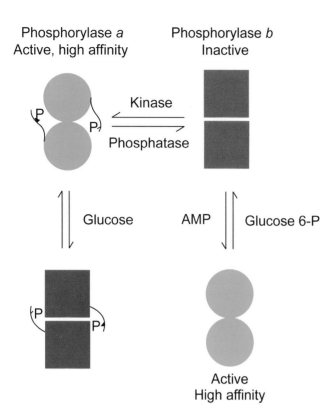

Figure 15-14. Simplified diagram of the allosteric and covalent regulation of glycogen phosphorylase. *Circles* depict the R state, *squares* the T state. The *a* form has a phosphate group (P) on Ser14, while the *b* form does not. Glucose binds more tightly to the T state of the *a* form, while glucose 6-P binds more tightly to the T state of the *b* form. AMP binds more tightly to the R state. Adapted from a figure by L. N. Johnson.

The allosteric ratio T:R for unliganded *b* is believed to be of the order of 3000, but that for the *a* form is only 10. This permits the latter to be pulled into the R state by substrates alone, and the slight activation of the *a* form by AMP is inhibited by ATP. Since AMP binds 200 times more tightly to the *a* than the *b* form, a large portion of its binding energy to the *b* form is used in the allosteric conformational change, and the interactions of the Ser14-P group must be responsible. The *b* form requires AMP for activity, but full conversion to the active R state requires substrate as well. The effect of binding ligands on the quaternary structure is believed to be transmitted to the interface between the two subunits by the two 'tower' α-helices (Figure 15-12).

The effect on the structure of phosphorylating Ser14 has been determined by comparing the structures of the two forms in the T state: crystals of phosphorylase *a* grown in the presence of glucose are in the allosterically inhibited T state, while those of phosphorylase *b* grown with IMP are also in the T state (Figure 15-11). This reveals that the disordered N-terminal segment of the *b* form, residues 5–16 including Ser14, becomes an ordered, but irregular, helix that lies across the dimer interface in the *a* form, with eight new intersubunit hydrogen bonds between polar groups plus additional hydrophobic interactions (Figure 15-11). A series of tertiary and quaternary conformational changes also take place that open access to the active site. The AMP binding site is more fully formed in the *a* form and capable of greater interactions with this activating ligand, while the affinities for the inhibitors are decreased. The tighter subunit associations of the dimer in the *a* form were originally thought to account for it obeying the concerted allosteric model, whereas phosphorylase *b* tends to follow the sequential model (with intermediate forms occurring). Phosphorylase *a* has, however, also been found to adopt intermediate structures.

Glycogen phosphorylase is an example of an extremely complex protein molecule that resists description with simple allosteric models.

Crystallographic studies on acyl ureas, a new class of glycogen phosphorylase inhibitors, as potential antidiabetic drugs. N. G. Oikonomakos *et al.* (2005) *Protein Sci.* **14**, 1760–1771.

The crystal structure of human muscle glycogen phosphorylase *a* with bound glucose and AMP: an intermediate conformation with T-state and R-state features. C. M. Lukas *et al.* (2006) *Proteins* **63**, 1123–1126.

Bioactivity of glycogen phosphorylase inhibitors that bind to the purine nucleoside site. L. J. Hampson *et al.* (2006) *Bioorg. Med. Chem.* **14**, 7835–7845.

15.2.C. Adenylylation

Adenylylation is the attachment of adenosine-5′-monophosphate (AMP) covalently to another molecule via a phosphodiester or phosphoramidate linkage. The AMP moiety is most often derived from ATP (Figure 15-15), but the coenzyme $NADP^+$ (Equation 12.5) is the source in some bacteria. Similarly, **deadenylylation is the removal of AMP from the adenylylated molecule**. The adenylylation and deadenylylation reactions are not the reverse of each other. Deadenylylation requires P_i instead of PP_i, and the reaction product is ADP instead of ATP. The adenylylation/deadenylylation reaction is involved in the regulatory control of enzyme activity and is analogous to phosphorylation.

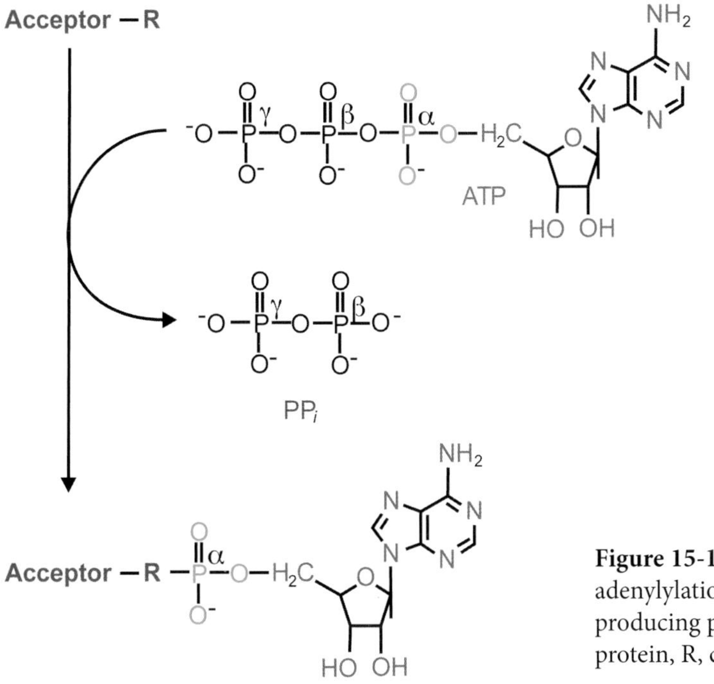

Figure 15-15. Chemical reaction of ATP-dependent adenylylation. ATP is used to adenylylate the protein, producing pyrophosphate (PP_i). The acceptor on the protein, R, can be an O or N atom.

Adenylylation must be distinguished from **adenylation**, which is the covalent attachment of ADP via the β-phosphoryl group. Adenylation is not very common, perhaps limited to adenylation of carbohydrates to yield, for example, glucose-1-ADP. Adenylation and adenylylation may be distinguished using the appropriate radiolabeled substrates: adenylylation will radiolabel an acceptor using α-^{32}P-ATP, but not with β- or γ-^{32}P-ATP. ATP that is radiolabeled with ^{3}H or ^{13}C in the adenine moiety is also used to decide whether ^{32}P label incorporated into the acceptor is a result of phosphorylation or adenylylation. They may also be distinguished by their sensitivity to **phosphodiesterases**, which will hydrolyze the AMP from an adenylylated substrate but will not remove a phosphate group.

1. Glutamine Synthetase

Glutamine synthetase (GS) catalyzes the condensation of ammonia with glutamate, in the presence of ATP, to yield glutamine:

$$^-O_2C-CH_2-CH_2-\underset{CO_2^-}{\underset{|}{\overset{NH_3^+}{\overset{|}{CH}}}} \underset{ADP}{\overset{ATP}{\rightleftharpoons}} {}^{2-}O_3PO-\overset{O}{\overset{\|}{C}}-CH_2-CH_2-\underset{CO_2^-}{\underset{|}{\overset{NH_3^+}{\overset{|}{CH}}}} \underset{P_i}{\overset{NH_3}{\rightleftharpoons}} H_2N-\overset{O}{\overset{\|}{C}}-CH_2-CH_2-\underset{CO_2^-}{\underset{|}{\overset{NH_3^+}{\overset{|}{CH}}}}$$

Glutamate   γ-Glutamyl phosphate   Glutamine

(15.6)

This is a crucial step in intermediary metabolism, especially that of nitrogen compounds. The glutamine synthetase from *E. coli* isolated from cultures grown in a medium that is rich in nitrogen exhibits dramatically different kinetic properties from that isolated from cultures grown under nitrogen starvation. The difference in the two forms of the enzyme is that residue Tyr397 is adenylylated in an environment rich in nitrogen, when the catalytic activity of GS is not required. In contrast, mammalian and plant GS and those of some microorganisms are not regulated by adenylylation.

Bacterial GS is a dodecameric oligomer formed by isologous face-to-face association of two hexameric rings (Figure 15-16). Residue Tyr397 is on the surface of the enzyme, and the adenylylated and unadenylylated forms of GS do not appear to differ greatly in their physical properties. The adenylylation state of GS *in vivo* may vary from zero to 12 AMP groups/dodecamer. The enzymatic activity decreases with increasing extent of adenylylation, which increases the K_m for substrates and the Mg^{2+} or Mn^{2+} ions that are required. The structural basis for this decrease in activity is not certain. Tyr397 is involved in some way in catalysis, because replacing it with other amino acid residues produces substantial effects on the activity.

Both the adenylylation (AT) and adenylyl-removing (AR) reactions are catalyzed by the enzyme **adenylyl transferase** (ATase). It consists of two conserved nucleotidyltransferase domains linked by a central region of approximately 200 amino acids. The N-terminal nucleotidyltransferase domain

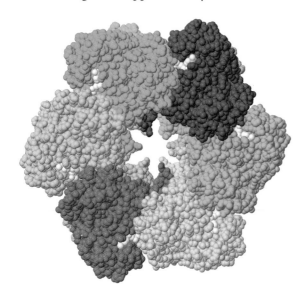

Figure 15-16. The dodecameric structure of the glutamine synthetase from *Salmonella typhiumurium*. The protein is depicted with van der Waals spheres for the atoms. Each of the 12 polypeptide chains is shown in a different color. The view is down the six-fold symmetry axis, so the second hexameric ring is below and just barely visible. The enzyme was not adenylated but bound AMP (the *white* atoms) indicates the active site. Figure generated from PDB file *1lgr* using the program Jmol.

contains the AR active site, and the C-terminal domain contains the AT active site. The two functional domains of ATase exhibit significant homology in amino acid sequences but they have very distinct functional properties.

A minimal consensus sequence required for efficient adenylylation of GS by ATase includes the Tyr397 to be adenylylated and residue Pro400 in the sequence –Met–Asp–Lys–Asn–Leu–**Tyr**–Asp–Leu–**Pro**–Pro–Glu–Glu–Ser–Lys–. Changing several other residues in this sequence has negligible or modest effects on the rate of adenylylation by ATase, but replacement of Pro400 nearly abolishes the activity. Introducing a Pro400 residue in a distantly related GS by replacing a Ser residue is sufficient to make that GS a substrate for the ATase from *E. coli*.

Cycling of the adenylylation and deadenylylation reactions would convert ATP and P_i to ADP and PP_i, but such futile and wasteful cycling is prevented by regulating the adenylylation of GS using a complex bicyclic cascade (Figure 15-17). The cascade includes ATase, a signal transduction enzyme (P_{II}) and a uridylyl transferase (UTase). **It is UTase that responds directly to nitrogen levels, and the adenylylation cycle is coupled to the uridylylation cycle of P_{II}.** When the nitrogen levels are low, UTase uridylylates P_{II} at Tyr51 to form P_{II}–UMP. When nitrogen levels are high, the UTase cleaves the UMP from P_{II}–UMP to generate P_{II}. The AT reaction is activated by glutamine and the unmodified form of P_{II} and is inhibited by P_{II}–UMP. Conversely, the AR reaction is activated by P_{II}–UMP and inhibited by glutamine and P_{II}. Both AT and AR reactions are regulated by α-ketoglutarate, which binds to P_{II} and P_{II}–UMP. ATase contains a binding site for glutamine, and its binding there increases the affinity for P_{II}.

The adenylylation and uridylylation cycles are similar, differing mainly in the use of ATP in one and UTP in the other. The two cycles are coupled because unmodified P_{II} protein stimulates the adenylylation of glutamine synthetase, whereas the uridylylated form of P_{II} is required for removal of the adenylyl groups. Ultimately, these two modification cycles are regulated by a variety of metabolites that influence the activities of the two modifying enzymes.

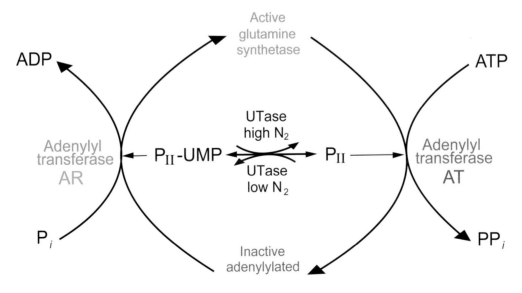

Figure 15-17. General reaction scheme for regulation by ATP-dependent adenylylation. ATase catalyzes the adenylylation/deadenylylation of glutamine synthetase; P_{II} stimulates the adenylylation activity (AT) and uridylylated P_{II} (P_{II}–UMP) stimulates the deadenylylation activity (AR). Uridylylation of P_{II} is catalyzed by UTase. UTase uridylylates P_{II} when nitrogen levels are low, and it hydrolyzes the uridylyl group from P_{II}–UMP when nitrogen levels are high.

The P_{II} protein is also involved in the regulation of transcription of the gene for GS. Its synthesis is induced when the source of nitrogen is limited, which is when protein P_{II} is uridylylated. Synthesis of the enzyme is repressed when P_{II} is unmodified. Consequently the nitrogen status of the cell affects both the biosynthesis and the activity of glutamine synthetase.

Mutation of the adenylylated tyrosine of glutamine synthetase alters its catalytic properties. S. Luo *et al.* (2005) *Biochemistry* **44**, 9441–9446.

Feedback-resistant mutations in *Bacillus subtilis* glutamine synthetase are clustered in the active site. S. H. Fisher & L. V. Wray (2006) *J. Bacteriol.* **188**, 5966–5974.

Escherichia coli glutamine synthetase adenylyltransferase: kinetic characterization of regulation by PII, PII-UMP, glutamine, and α-ketoglutarate. P. Jiang *et al.* (2007) *Biochemistry* **46**, 4133–4146.

15.2.D. Proteolysis: Turning Zymogens into Proteinases

Proteinases hydrolyze the peptide bonds of proteins and most are destructive; consequently it is vital that their catalytic activities are not unleashed before they are required. Probably for that reason, proteinases are generally synthesized as larger precursors, known as **zymogens**, that are catalytically inactive. All are activated by proteolysis, but the mechanism of the activation and the basis for the inactivity of the precursor are different in each proteinase family.

Proteolysis differs from the other types of covalent enzyme regulation in that it is irreversible in most cases under physiological conditions. Only when the new terminal amino and carboxyl groups are held in close and appropriate proximity is resynthesis of a peptide bond significant. This occurs in protein inhibitors of proteinases, which bind tightly to the active site of the proteinase and do not dissociate once cleaved. The proteinase can then resynthesize the peptide bond and there is an equilibrium between cleaved and uncleaved forms bound to the proteinase.

The peptidase zymogen proregions: nature's way of preventing undesired activation and proteolysis. C. Lazure (2002) *Curr. Pharm. Des.* **8**, 511–531.

Creation of a zymogen. P. Plainkum *et al.* (2003) *Nature Struct. Biol.* **10**, 115–119.

---

1. Trypsin Family of Serine Proteinases

Serine proteinases are so-called because of the reactive residue Ser195 that forms an acyl-enzyme intermediate in the catalytic mechanism (Equation 14.62). The trypsin family in mammals includes **trypsin, chymotrypsin** and **elastase**. They are very similar in their three-dimensional structures and catalytic activities. All are synthesized as the corresponding zymogens **trypsinogen, chymotrypsinogen, proelastase**, etc., with polypeptide extensions at their N-termini. **The natural zymogens are folded but inactive as proteinases**; their low activities towards small substrates are no greater than that of a comparable solution of imidazole. Activation of these folded zymogens occurs by proteolytic cleavage of the peptide bond between residues 15 and 16 (using the numbering system of chymotrypsin). Further cleavages may then take place on the surface of the protein, such as after residues 13, 146 and 148 in chymotrypsinogen A, to release the dipeptides 14–15 and 147–148 and give the final disulfide cross-linked, three-polypeptide chain α-chymotrypsin (Figure 15-18). Only the initial cleavage, however, is necessary for the generation of catalytic activity. In the mammalian gut, activating the serine proteinases that are found there and are involved in digestion is initiated by a proteolytic enzyme, **enterokinase**, that is attached to the external side of the brush border membrane. It is extremely specific for activating trypsinogen; the product, active trypsin, then activates other molecules of trypsinogen and also of proelastase and chymotrypsinogens A, B and C. Consequently active proteinases are generated only when they are useful.

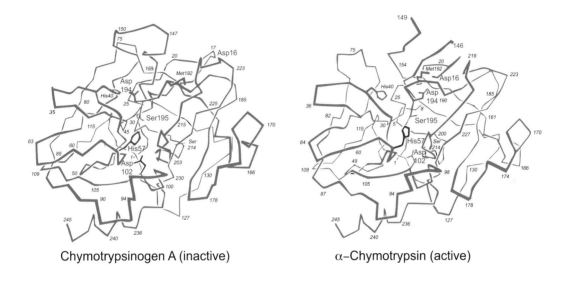

Figure 15-18. Comparison of the polypeptide backbone conformations of chymotrypsinogen A (*left*) and α-chymotrypsin (*right*). The latter has had residues 14, 15, 147 and 148 removed proteolytically upon activation. The side-chains of some residues important for catalysis and activation are indicated in red. Adapted from S. T. Freer *et al.* (1970) *Biochemistry* **9**, 1997–2009.

The structural basis of this activation became apparent by comparing the structures of the initial zymogen and the activated proteinase. **That the structures are so similar was initially surprising** (Figure 15-18). In particular, their active sites are largely intact, with a stereochemically acceptable **catalytic triad** of Asp102, His57 and Ser195 that is believed to be important for activating the side-chain hydroxyl group of Ser195 (Figure 15-19). In trypsinogen, four segments of residues 16–19, 142–152, 184–193 and 216–223 that border the active site are very flexible in the zymogen but become fixed upon activation. They also become fixed in a trypsin-like conformation if the protein inhibitor BPTI (Figure 9-2) is bound to the active site; the affinity of BPTI for trypsinogen is, however, only 10^{-7} that for trypsin, because some of the binding energy must be used to pull the enzyme into its specific conformation (Figure 12-5). Only the loop 142–152 is flexible in chymotrypsinogen, where it is generally cleaved proteolytically to release the dipeptide of residues 147–148. The other three segments of chymotrypsin adopt only slightly different conformations in the zymogen and the active enzyme. These small differences affect the binding pocket for the substrate in the zymogen and disrupt the **oxyanion binding site** of the active site, which consists of the backbone –NH– groups of residues 193 and 195 and serves to orient the substrate appropriately for catalysis. The –NH– of residue 193 points in the wrong direction in the zymogen, where it forms a hydrogen bond with the carbonyl oxygen of residue 180 in a turn.

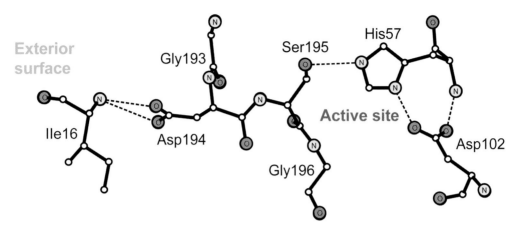

Figure 15-19. The position of the Ile16 α-amino salt bridge to Asp194 within the interior of the α-chymotrypsin molecule, relative to the catalytic triad of Asp102, His57 and Ser195 at the active site. Presumed hydrogen bonds are shown as *dashed lines*. These residues virtually span the interior of the globular folded structure of the protein, with the entrance to the active site at the *right*, front; the rear surface of the protein is at the *left* (Figure 15-18). Original figure kindly provided by D. M. Blow. From T. E. Creighton (1993) *Proteins: structures and molecular properties*, 2nd edn, W. H. Freeman, NY, p. 436.

The most important change upon activation of the zymogens is the generation of the new α-amino group of residue 16. This amino group forms a salt bridge with the side-chain of Asp194 in the active enzyme (Figure 15-19), which involves a substantial rearrangement of residues 189–194. The α-amino group of residue 16 does not exist in the zymogen, and Asp194 is in a polar environment, with its side-chain interacting with His40. The Ile16 side-chain is accessible to solvent in the zymogen, but upon activation it displaces the buried side-chain of Met192, which moves into the solvent (Figure 15-18). As a result of these conformational rearrangements of residues 189–194, the –NH– group of residue 193 adopts its position appropriate for participating in the oxyanion hole. The changes extend through the middle of the protein molecule, for the activating cleavage of the peptide bond preceding residue 16 occurs on the surface of the molecule opposite the active site (Figure 15-18).

Prevention of β-strand movement into a zymogen-like position does not confer higher activity to coagulation factor VIIa. O. H. Olsen *et al.* (2004) *Biochemistry* **43**, 14096–14103.

Kinetic analysis of zymogen autoactivation in the presence of a reversible inhibitor. W. N. Wang *et al.* (2004) *Eur. J. Biochem.* **271**, 4638–4645.

Developments in the characterisation of the catalytic triad of α-chymotrypsin: effect of the protonation state of Asp102 on the ^{1}H NMR signals of His57. G. Bruylants *et al.* (2007) *Chembiochem.* **8**, 51–54.

---

2. Carboxyl Proteinases

The acidic proteinase **pepsin** is generated from the zymogen **pepsinogen** by the proteolytic removal of 44 residues from its N-terminus. The zymogen is stable at neutral pH but spontaneously activates when the pH is decreased to less than 5, as in the stomach where the proteinase functions. The proteolytic cleavage can be catalyzed by another molecule of pepsin, or it can be a solely intramolecular process in which the active site of the zymogen cleaves its own polypeptide chain. The intramolecular reaction is characterized by the rate constant being independent of the zymogen concentration and not affected by the presence of a substrate for pepsin. Activation also occurs even if the pepsinogen is immobilized by covalent attachment to a resin, where different molecules cannot come into contact.

Upon placing pepsinogen into acidic solution, a conformational change is observed spectrally to occur very rapidly, before formation of active pepsin, indicating an intermediate in the activation process. The first peptide bond is hydrolyzed more slowly and is that between residues 16–17 of the pro segment; this produces enzymatically active **pseudopepsin**. Further processing to remove the remainder of the pro segment and generate mature pepsin occurs intermolecularly.

The structural basis of this phenomenon is suggested by the crystal structure of pepsinogen at pH 6 (Figure 15-20). Residues 11–44 of the pro enzyme segment bind in the substrate-binding cleft of the pepsin portion of the zymogen and block access to the two catalytically active Asp residues. The pepsin portion of the zymogen is not detectably different from the mature, active enzyme. The proenzyme segment is held in place by a number of electrostatic interactions between basic side-chains of the pro enzyme segment and acidic residues of the pepsin portion. At least some of these interactions are believed to be weakened upon lowering the pH and protonating the carboxyl groups, and electrostatic repulsions between the resulting unpaired basic residues are imagined to alter the conformation. This permits the 16–17 peptide bond to enter the active site and be cleaved by the pepsin portion of the molecule. In pepsinogen, therefore, **the pro enzyme segment acts as an inhibitor of an otherwise active enzyme**. Like some other proteinases (Section 11.4.B), unfolded pepsin can fold to its native conformation only in the presence of the pro segment.

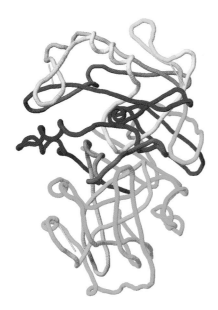

Figure 15-20. The structure of porcine pepsinogen. Only the path of the polypeptide chain is depicted, with the N-terminus colored *blue*, through the colors of the rainbow to *red* at the C-terminus. The pro peptide is at the N-terminus and thus is the *bluest* part. It can be seen to lie in the active site cleft between the two domains. Figure generated from PDB file *3psg* using the program Jmol.

Structure and mechanism of the pepsin-like family of aspartic peptidases. B. M. Dunn (2002) *Chem. Rev.* **102**, 4431–4458.

Kinetics of autocatalytic zymogen activation measured by a coupled reaction: pepsinogen autoactivation. M. E. Fuentes *et al.* (2005) *Biol. Chem.* **386**, 689–698.

Comparison of solution structures and stabilities of native, partially unfolded and partially refolded pepsin. D. Dee *et al.* (2006) *Biochemistry* **45**, 13982–13992.

3. Metalloproteinases

The zymogens of this class of proteinase appear to be similar to those of the carboxyl proteinases, in that the zymogen contains an intrinsically active proteinase that is inhibited by the proenzyme segment (Figure 15-21). **Procarboxypeptidase** has a 94-residue extension at its N-terminus but displays substantial amounts of enzyme activity. The pro enzyme segment of 94 residues in isolation has a folded conformation typical of a globular protein. It binds to the active site and is a strong inhibitor of carboxypeptidase A.

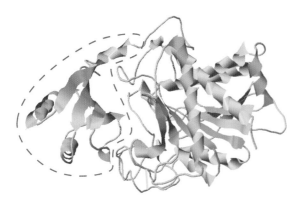

Procarboxypeptidase B

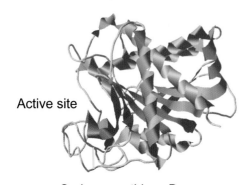

Active site

Carboxypeptidase B

Figure 15-21. Comparison of the structures of the procarboxypeptidase B zymogen and active carboxypeptidase B. The pro segment is *circled* in the structure of the zymogen. The polypeptide backbone of the protein is depicted schematically; the *yellow* bars are disulfide bonds. Figure generated from PDB files *1nsa* and *1z5r* using the program Jmol.

In some metalloproteinases, the pro enzyme segment is believed to have a Cys residue that interacts with the Zn^{2+} atom at the active site.

The propeptide in the precursor form of carboxypeptidase Y ensures cooperative unfolding and the carbohydrate moiety exerts a protective effect against heat and pressure. M. Kato *et al.* (2003) *Eur. J. Biochem.* **279**, 4587–4593.

Detailed molecular comparison between the inhibition mode of A/B-type carboxypeptidases in the zymogen state and by the endogenous inhibitor latexin. R. Garcia *et al.* (2005) *Cell. Mol. Life Sci.* **62**, 1996–2014.

INDEX

Page numbers for major entries, illustrations and tables are in **bold**.
Page numbers for items that are primarily a reference are in *italics*.

14-3-3 proteins 727
2-D electrophoresis **102**, *239*, **289-91**
280s loop **730**, 733
3_{10} helix **339-42**
3D-1D compatibility method 430-1
434 repressor 582-4, **586**
5′-cap 205
5S ribosomal RNA **115**, 186, 614, 625
16S ribosomal RNA **115-7**, 134, 625
23S ribosomal RNA *118*, 625
30S ribosome **625-6**
50S ribosome *118*, 625
70S ribosome *129*, 625, *627*
7TM superfamily 412

A-form DNA, RNA 30, **49-50**, 53-4, 72, 78, **111-2**, *115*, 116, 126, 148, 165, 175-6, 615, 617
A-minor motif **122-3**
ab initio prediction *xxv*, *46*, **425-6**, *666*
absorbance 33, 79, **140**, 144, 157, 168, 246, 254, **256-8**, 267-8, 270, 281-2, 287, 370, 459-60, 463, 566
acceptor arm, stem, tRNA **127-9**, *153*, 621
acceptor, fluorescence 21, 168, 202, 335, 337
acceptor, hydrogen bond 13, **33-4**, 61, 113, 232, 240, 243, 257, 261, 339, 342, 370, 389, 394, 492, **565**, 572, 695
accessible surface 37, **52**, 68, 76, 80, 84, 112-3, 118, 136, 138, *191*, 245, 252, **260**, *263*, **264**, 266, **388-9**, 392, **402-3**, 440-2, 447, **470**, **473**, 480, 484, 515, 523, 532, 540, **548-9**, 564, 582, 614, 662, 680, 740
acetaldehyde 81, 690
acetals 220-1, **631**
acetate 99, 235, 281, 476, 527, **631**
acetic acid **631**
acetic anhydride 216, 224, **235-6**, 257, *299*, **443**
acetoacetyl-CoA 654

acetonitrile 212, 215-6, 224, 281
acetyl chloride 232
acetyl-Ala 333
acetylamides **262**, **392**
acetylation, amino groups 22, 85, **236**, **304-5**
acetylation, hydroxyl groups 232, **256-7**, *305*
acetylcytidine (Ac⁴C) **110**
acetylenes 683
acetylgalactosamine 305
acetylglucosamine **364**
acetylimidazole **235**
acetyllysine 697
acetylsuccinimide **235**
acetyltransferases **363**, **396**, *675*
acetyltryptophan amide **634**
acetyltyrosine **256**
achaete **588**
acid catalysis 76, **631**, *669*, 673
acid chlorides, fluorides 273
acid hydrolysis 218, **281-2**, **289**, 294, 728
acid proteinases **740-1**
actin 397, 405, **439**
activated ester 273
activated intermediate 684
activated ligand 686, 694
activated state 213, 272-3, 632, 683, 697-8
activation energy *xxv*, 158, **455**, 628, 632, 668, 685, 701
activation loop 723, *725*
active site titrants **667**
active site-directed irreversible inhibitors 667
active site, enzyme *xxv*, 74, 178, **181**, **186**, *246*, 259, 286, 323, 400-1, 408, 417, 422, 441-2, *445*, *453*, 511, **520**, *535*, **561**, 621-3, 628, 631, 633, 638, **645-701**, **708-42**
active site, nucleic acid 131, *133*, **701-5**
activity, chemical *xxv*, 580
acycloguanosine **39**
acyclovir **39**
acyl carrier protein 690
acyl enzyme **680**, *685*
acyl group 211, 222, 225, 277, 703
acyl intermediate 667
acyl ureas *735*
acylation 84-5, **234-5**, **694-5**
adaptive evolution 323-4
adenine, adenosine **26-9**, 211, 225, 521, 541-3, 571, 577, 583, 605-6, 727, 735
adenosine platform **123**, 127

adenylate cyclase 724, 731-2
adenylate kinase 543, **675-6**, **678-9**
adenylation 205, 735
adenylyl transferase (ATase) **732-4**
adenylylation **181-2**, **731-4**
adiabatic compressibility *xxv*, **71**, **439**, **457-8**
ADP **542**, 658, **675-6**, **678-9**, **717-9**, 722, 735-7
adrenaline 731
adrenocorticotropic hormone **303**
AF-2 602-3
affinity chromatography 194, 327, 566
affinity, ligand 486-7, 502, 523-4, 527-9, **535-63**, 568, 572-3, 578-620, 632-40, **670**, 685, 691-6, 706-19, 737-9
affinity label 562, 667
agar 280
agarose 79, 95, **100**, **568-9**
aggregation 44, 99, 158, 285, 322, 343, *351*, 353-8, **360-3**, 394, 437, **464**, 482, 504, **516-8**, 690, 726
agonist 602-3
AIDS 189
alanine (Ala) residue **4**, *22*, **229-32**, **260-2**, 265-6, 281, **286**, 302, 311, **317-8**, 325, **332-6**, 347, *348*, 351-2, **355**, 358, **390-2**, **419**, **440**, 443, **470**, *484*, 496, 543, 615, 663, **665**, 693, 697, 728
alanine scanning *619*
alanosine *716*
albumin **320**, 326, **439**, 457
alcohol dehydrogenase **403**, **522**, 541, 628
alcohols 50, 53, 485, *497*, 628
aldehydes 9, 12, 108, 238, 681
aldehyde dehydrogenase *652*
aldimine **682-3**, 697
aldolase **396-7**
aldose **13**
aldose reductase 325
alfalfa mosaic virus *185*
aligning sequences 311, **313**
alkaline Bohr effect 556
alkaline hydrolysis 29, 109, **209**, 282, 571, 699
alkaline phosphatase 168, 205
alkylation **82**, **234-5**, **245-6**, 249, **442**
allopurinol **39**
allostery *xxv*, 523, **543-63**
alpha-fibrous proteins **354-8**
alpha-lytic proteinase 418, **511**
alpha/beta $(\alpha/\beta)_8$ barrel **373**, **405-8**, 422, 432

alpha1-antitrypsin **512**
Alu **614**
Alzheimer's disease 343, 484
amidation 241, **304**
amides 9, *22*, 84, 171, 176, 241, **242-4**, 259, **262**, *275*, 281, 333, 337, 339, 346, 376-7, 392, 447, 543, 718
amidination **234-6**
amines 51, 54, 84, 108, 126, 133, 171, 228, **234-9**, 241, 269, 273, 496, **528**, 609, 665
amino acid analysis 274, **280-5**, 295
amino acid detection **267-70**
amino acid hydantoins **294-9**
amino acid identification **280-5**
amino acid residues **227-328**, **329-61**
amino acid sequence 161, 190-8, 227, 243, **280-328**, 338, 341, 346, 353-61, 362-3, 384-7, 393, 400-1, **415-33**, 463, **498-502**, 504, 530-2, 581, 588, 617, 696, 717, 723, 737
amino acyl adenylate **656-7**, **691-6**
amino acyl tRNA synthetases **619-23**, **691-6**, 703
amino butyric acid 281
amino ethylglycine backbone **171**
amino groups 31-4, 56, 76-8, 81, 84, 126, 141, 144, 171, 182, 195, **227-30**, **234-9**, 246, 256, 268-9, **271-9**, **283-5**, 287, 292, **294-7**, 305, 393, **443**, 527, **555-6**, 624, 640, 663, 672, 682-3, 693, 697, 740
amino terminus 173, *269*, 286, **294-7**, 511, *733*
amino Tyr residue **256**
amino-adenine 56
aminoisobutyric acid 273, *377*
aminopeptidase 286, 294-5
aminopurine **39**
aminotransferase 640, 663-5, 684
ammonia 211-8, 221, 223, 225, 267, 269, 281, 292, 690, 736
ammonium sulfate 496
AMP nucleosidase **629**
amphipathic **341**, *372*, 383, 414, 603, 607
amphiphilic *xxv*, 538
amyloid proteins 343, *380*
β-amylase **686-7**
amylose 729
ancestral proteins 308, *317*, 324, 411
anchoring 305, 393, 409, 414, 538, 687, 723, 729

angiotensinogen 325
anglecin 80
anhydrides and amino group 235-6, **283-4**, 292, **443**
anhydrides and carboxyl group *299*
anhydrides and hydroxyl group 216, 224
anhydrides and thiol group 246
anhydrides and Tyr residue 257
anhydrides, mixed 273, **298**, 721
aniline **206**
anilinonaphthalene-1-sulfonate (ANS) 484-5
anions *xxv*, 28, 145, 173, 245-6, 250-3, 275, 472, 497, 519, **523**, 533, 552, 557, 573, 578, 730
anisotropy *xxv*, 74
ankyrin *516*
annealing *xxvi*, 146, **157-70**, *179*, 183, 609
anode 293
antagonist *xxvi*, 603
anti conformation **14-5**, **27**, 31, *40*
antibiotic 57, 77-8, 188, 193
antibody 185, 192, 290, 327, 378, 395, 402, 435, **486-7**, **520**, **696-7**, 704, 727
anticodon **127-9**, 620-3
anticooperativity 500
antifreeze 685
antigen 185, 378, 395, 402, 696
*anti*periplanar conformation **14-5**
antisense RNA *xxvi*, 57, 133, **190-1**
antisigma factor 187
antitrypsin **512**
Ap$_5$A **675-6**, **678-9**
ApA **44**
apoferritin **399**, 532
apomyoglobin **475**, **483**, 485
apoprotein **520**, 525-7
appropriator 187
aptamers *xxvi*, **704-5**
ApU **44**
aqueous solutions 4, 13, 35, 44, 68, 70, *71*, 79, 203, 254, 263, 266, 275, 296, 344, 348-9, *377*, 409, 438, 456, 462, 468, 474, 502, 529, 690
arabinose-binding protein **366-8**
arabinosyladenine **39-40**
arc repressor **588**
archea 723
arginine (Arg) residues *190*, **229**, **231**, **233-4**, 260-6, **286**, *287*, **292**, **302-3**, 307, 317-8, 322, **347**, 348, **355**, 370, 376, **390**, 392, 395, 410, **419**, 428, *527*, 557, 572, 577, 600, 602, 607, 615, 621, 622, 625, *675*, *685*, 721, 726
argininosuccinate lyase 325
argon (Ar) 299
aromatic rings 33-5, 78, **255-6**, **264**, 282, 287, 311, 337, 393, 410-4, **454-5**, **459-60**, 473, 483, 531, 610-1, 615-6, 695, 697
aromatic stack 380
Arrhenius plot **506-8**
artificial molecules *52*, 191-2, 601, 704
Asp-N proteinase **286-7**
asparagine (Asn) residues **229**, **242-4**, 380, **390**, 411, **419**, 457, 465, **539-40**, 587, 615, 663
asparagine ladder 380
aspartate aminotransferase 663, 665, 684
aspartate transcarbamoylase (ATCase) 394, **399**, **522**, 561, **676-7**, 707, **710-7**
aspartate-β-semialdehyde dehydrogenase *562*
aspartic acid (Asp) residues **229**, **231**, **241-4**, **260-6**, 281, **286-7**, **289**, **302**, **317-8**, 340, **347**, 348, **355**, 376, **390**, 392, **419**, 436-8, **446**, 477, 523, 525, 530, 533, **539-40**, **542**, **572**, 587, 615, **622-3**, 663, 721, 724, 728, **739-40**
aspartyl proteinases *287*, *741*
aspartyl-tRNA synthetase 620, **622-3**
asymmetric unit **398**
asymmetry 2, 6, 7, 29, 114, 118, 136, **228-32**, **400-2**, *460*, *548*, 560-2, 584, 591, 596, 604, 619, 692, 720-1
ATP **27**, 95, 181-2, 199, 202-9, 219, 409, **541-3**, 554, 561, **656-8**, 665, **675-6**, **678-80**, 687, 689, **691-4**, 703, **710-37**
ATP synthase 400, 409
ATP-binding motif 540, **543**
ATPase 678
autoactivation 739-40
autoradiography *xxvi*, 202-3, 206, 208, **569-70**
avian pancreatic peptide **403**
azacytidine **39**
azides 273
azidoadenine *567*
azidophenyl acetimidate 85
azidothymidine **39**
azurin **522**, **533-5**

B-form DNA **24-85, 86-107**, 113-5, 148, 175-6, **564-627**
B-value 394, 452, 458
back-phosphorylation 727
backbone **1-2, 25, 29-30, 171-2, 329-61, 366-86**, 662, 698, 701, 739
bacteria 39, 40, 79, 86, 89, 116, 128, *164*, 187-8, 191-5, 280, 325, 399, 417, **511**, 575, **608-9**, 620-1, 696, 717, **724**
bacterial artificial chromosome 191-2
bacteriochlorophyll protein 397, 521
bacteriophages *xxvi*, 38, **188**, 191, 195, 199, 420, 591-2, **619**
bacteriorhodopsin 399, **412**, 428
baculovirus *194*, 327
*Bam*H1 **196-7**
bandshift electrophoresis **568-9**
barium (Ba) 525
barrels **373-4**, 381-4, **405-10**, 413, 422, 432, 503, 534, 521, 609, 611, 621
β-barrels **373-4**, 381-4, **405-10**, 413, 422, 432, 503, 534, 521, 609, 611, 621
base catalysis 76, 225, **630-1**, *669*, 673, 691, 697, *701*, 704
base pairs *13*, 22, *24*, 25, **40-50**, 52-68, 73-8, 80-1, 84, 93, 96, **112-3**, **118-37**, **139-96**, **565-6**, 572, 583, 609, 701
base stacking **42-6**, *75*, 112, 116, 119-20, 123, 127, 134, *143*, 175
base triplet 119-20, 175
Bayesian approach *134*
beads 279-80
bending DNA 31, **72-5**, 103, 107, 118, *196*, 569, 572, **582-5**, 590, 597-8, 605-6, 609
benzene 255, 457
benzoyl protecting group 225
benzyl-dimethyl-alkyl-ammonium chloride 99
betaine **144**, 146, 496, *582*
bifunctional reagent 141
bilayer **409-5, 460, 502-3**
bimolecular reactions *63*, 143, 158-60, 515, 673
binding energies 77, 117, 573, *581*, 632, 680, 693-4, 734, 739
binding, preferential 144, 414, 496, 546, 558
bioluminescence 168
biotin 168, 641
birds 321, 323
bisnaphthalimide **74**

bisphosphoglycerate (DPG) **554-5**, 557, 689
bisubstrate analog **675-9**, 710, 713
bisulfite **81-2**
bitopic 409, **414**
biuret reaction **267-8**
BLAST 313
blender 71
blind mole rat 320
blood clotting 305, 322, 354, 385, 535, 539-40
blood clotting factor X **309**
blotting 166, 267, 271, 296, 566
blue-copper proteins **534-5**
blunt ends 196
BNPA-skatole 258, 289
Boc (*t*-butyloxycarbonyl) group 172, **272-4**, 277, **279**
Bohr effect *xxvi*, **552-7**
bond angles, lengths, rotations **14, 19, 29-31, 329-33, 336-8, 455**, 670
bone 359, 361
borate 234
borohydride **206, 235, 238**, 249, 683
bovine serum albumin (BSA) 457
BPTI (bovine pancreatic trypsin inhibitor) **284, 293, 364-5**, 367, **386-7, 391, 447-8**, 452, **454-5**, 483, 493-4, 499, *508*, **513-5**, 739
Bradford assay 270
bromelain **286**
bromine 40
bromo-acetate, -propionate, -pyruvate **442**
bromoacetaldehyde **81**
bromoalanine 275
N-bromosuccinimide 257
bromouracil **39-40**
Brønsted plot 443, **664**
buckle, base pair **43**
buffers *xxvi*, 75, 142, 246, 254, 475, 569, 578
bulged nucleotides 84-5, **113-4, 118**, 132, **151**, 153, 155, **158-9, 165-7**, 611, 614, 619, 625, 699-700
β-bulge 372, **374-5**, 406
buoyant densities 40, **70**, 437
bushy stunt virus *185*, 382, 401
butanedione **233**
butyl group **35**, 274
bZip domains 575, 588, **604-5**

C-terminus *190*, 273, **275-9**, 288-9, 294-5, **298-9**, 305, 327, 339, 348, 360
C_0t curves **163-5**
cadmium (Cd) 247, **526**, **528**, **531-2**
caffeine 730
caged ligands 686
calcium-binding proteins 242, 309, 326, 384, *531*, **535-40**
calmodulin 371, **537-9**, **731-3**
calories 264
calorimetry **467-70**, 471, **475**, 482, *503*, *529*, *537*, *581*
Cambridge Structural Database (CSD) 24
cAMP-dependent kinase **726**, **731-3**
capillary electrophoresis 7, *95*, 203, *236*, *242*, *267*, *270*, *283*, 285, 290, 568, *570*
carbamate **555-6**, 690
carbamylation 235, **239**
carbobenzoxy (Z) group **272**
carbodiimide **241-2**, 273, 279
carbon monoxide (CO) **552**, 690
carbon dioxide *557*, 682
carbon-13 NMR **336**
carbonate 533
carbonic anhydrase **439**, 485, **552**, **629**, 635
carboxamidomethyl group 246, **284**
carboxyglutamic acid (Gla) residue 309, **539-40**
carboxyl groups **227-8**, **241-2**, 244, **271-5**, 289, 291, **298-9**, 305, *348*, **444**, **525-30**, 537, **539-40**, 577, 624, 682, 721, 738
carboxyl proteinases 286, **740-1**
carboxylase **396**, 399, 402, 635
carboxylation **302-4**, **539-40**
carboxymethyl group 246, 281, **284**, **291-3**, 513
carboxypeptidase 285, 295, *299*, 368, 390, 422, **520**, **522**, **629**, **742**
carcinogen 79
cartilage 361
cascade 168, 540, 725, 727, 732, 737
casein **320**, 726
CASP 426
cassette mutagenesis **218-9**
catabolite gene activator protein (CAP) **596-7**
catabolite repression 596
catalase 401, 624
catalysis *xxvi*, 244, 400-1, *422*, **628-704**, **705-41**

catalytic antibody **696-8**
catalytic nucleic acids **130-3**, 191, **698-704**
catalytic triad 286, **739-40**
catenane *23*, **87-9**, 95
CATH domain structure database *24*, 409
cavitein *383*
cavities 389, 440, 446, **456-9**, 479, 500, *525*, 548, 555-6, 572, 687
CD spectra *68*, *345*, 460, 463, 477, **481-2**, 484
ccDNA (closed circular DNA) **86-106**
cDNA (complementary DNA) 189-90, **194-5**, 205, *303*
cell adhesion 423
cellobiose 38
cellulose *687*
centrifugation 40, 70, 269, 458, 566
cesium (Cs) 61, **70**, 95
chain-cleavage DNA sequencing 84, **204**
chain-termination DNA sequencing **199-204**, 207, 571
chalcone isomerase 669
channels, ion 409, 726
chaperone 187, **399**, **518**
characteristic ratio **21-2**, **334**
charge-transfer interaction 44
chelate effect **527-9**
chelation *xxvi*, 84, 126, **527-9**, 532, 536, 599, 712, 718
chemical exchange 451
chemical ligation **275-7**, *383*
chemical modification 56, **80-5**, 136, **230-60**, 292, **441-4**, 464, 715
chemical potential 497
chemical shift *13*, *150*, *238*, 468, **480-1**, 484, *497*
chemiluminescence *xxvi*, 168, 170
chiral center *xxvii*, **4-13**, 540
chiral persistence length 175
chirality **3-13**, *97*, 103, 136, 172, 175-6, *220*, 228, 232, 272, 333, 381, 398, 460
chloramphenicol acetyltransferase **363**, **396**
chloride ion 472, 554-6, 578
chlorine 282
chloro-acetaldehyde **81**
chlorophenyl protection **211-2**, 216, **222-3**
chloroplasts 304, 400
chorismate mutase **629**, **634**
Chou & Fasman **426-7**
chromatin *xxvii*, *96*, 187, 603, *608*

chromatography *5*, *7*, 31, 142, 164, 169, 172, **205-9**, 214-5, 218, 251, *253*, **281-3**, 289-91, 294, 296, 299, **513**, 517, 566-7, 690, *728*
chromophores *xxvii*, 78, 238, 269, 459-60, 531, 715
chromosomes *xxvii*, 26, 60, 86, *96*, 106-7, 170, 191-2, 198-9, 564, 575, 603, 607, *609*
chymotrypsin, chymotrypsinogen **286-7**, 293, **387-8**, 417-8, **439**, 441, **443**, **475**, **490**, 511, **634**, 663, **738-40**
circular dichroism (CD) 4-5, 33, 140, *345*, 463, *474*, **481-2**, 484, 567
circular molecules 23, 72, **86-106**, **166**, **184**, **192-3**, *515*, **687-8**
cis conformation 3, 7, **9-10**, 14, 259, **330-4**, 344, 368-9, 376-7
cis-trans isomerization **337-8**, 345, 360, **505-6**, 518
citraconic anhydride 235-6
citrate **11**, 716
citrate synthase **403**
citric acid cycle 716, 721
clamp loader **687-8**
clamping **577**, 598, 608, 611, **630-2**
cleavage of nucleic acids 75, 84, *90*, 95, **109**, 154, 163, 192, **196-8**, **204-7**, **220-1**, 225, **570-1**, 687, **698-703**
cleavage of proteins 234, *242*, 273, **285-93**, 303-4, 322, 327, 464, 706, **738-42**
Cleland's reagent **252-3**
clinal conformation **14-5**
clock, evolutionary 320
cloning *xxvii*, 184, **191-9**, 203, 218
closed-circular duplex DNA (ccDNA) **86-106**
clostripain **286-7**
clotting of blood 305, 322, 354, 385, 535, 539-40
CNBr (cyanogen bromide) 255, **286-9**, 292
co-solvents *xxvii*, 353, 470, 496, 519
coactivator 564, 582, 594, 602-3
coaxial helices 66, **121-2**, **124**, 131
cobalt (Co) 247, **528**
cobalt hexamine 54, 66, 130
cocaine 697
codons *xxvii*, 127, 193, **301-3**, 317, 322, 327, 620
coenzymes *xxvii*, 111, 181, 368-9, **521**, 662, **681-3**, 729, 734

cofactors *xxvii*, 37, *188*, 238, 383, **411**, *505*, 521, *535*, 545, 552, 560, 640, 662, **681-3**, 700-1, 720, 729
cohesive end **195**
coiled-coils **354-8**, **604-5**
cold unfolding **475**, 494
collagen 304-5, 307, **319**, 353, **359-61**, 370, 462
compaction, DNA 86, 107
competitive inhibitors **646-51**, 653, **655**, 661, 675-7, 711, 715, 730, 733
competitive labeling **442-4**
complement component nine **309**
complementary DNA (cDNA) 189-90, **194-5**, 205, **303**
compressibility **70-1**, **439**, **457-9**, 478-9, 484
computer simulations 53, *97*, *98*, *455*, *457*, 459, *484*, *497*, *510*, *512*, 638, *708*
concanavalin A 397
concerted allosteric model **544-7**, 557-60, 708-9, 712, 715-6, 719, 733-4
condensed ions **69**, 145, **581-2**
condensing agent 211, 213-4
configuration *xxvii*, **3-13**, 29, 38, *98*, 368, 693
conformation *xxvii*, 3, **13-24**
conformational ensembles 22, 154-5, *476*, 482, *505*, *554*, 637-8
conformational entropy **17-8**, **23-4**, 332, 334, 338, 346-8, 480, *484*, **494-6**, 501, 509
conformational switch 118, *424*, *621*
conformer **13-4**, **66-7**, 637-8
contact map **386-8**, 417
controlled-pore glass 216, 225
convergent evolution **309**, **408**, **421-2**
coomassie blue 267, **270-1**
cooperativity *xxvii*, 14, 49, 119, 150-4, 163, *175*, 345, 350-2, 424, 447, 459, **465-9**, 472-3, 476-80, 485, 492, **494-5**, 500, **507-11**, 513-6, 533, 537, **543-63**, 576, 591-2, 596-7, 605-6, 609-10, **707-21**, 725, 732, *742*
coordination number *247*, **525-6**, 530-1, 534-5, 540, 548
copolymer 1
copper (Cu) 240, 247, **267**, **525-6**, **528**, **533-5**
copy number 133, 191
corepressor *xxvii*, 564, 582, 595, 603, 606
CORN rule 228
coronavirus *125*
correlation time 73, 453

corticotropin-like intermediate lobe peptide (CLIP) 303
cosmid **191-2**
counterions 28, **47-8**, 50, *53*, 57, **68-71**, 99, 144-5, 173, 445, 485, 519, **581-2**
counting residues *245*, **283-5**
coupled reactions *740*
coupling methods **211-26, 271-80**
covalent bond, structure 1-2
covalent catalysis **680-3**
covalent regulation **721-42**
CpG **44**, 54, 74-5, 583
crambin 394, 436
creatine kinase **658**
critical micelle concentration (cmc) 409
Cro repressor 581-3, **589, 591-2**
crocodiles 323
cross-links **23-4, 79-83**, 85, 136, 144, 203, 274, 277, 322, 361, **444**, 480, 483, 516, 567
cross-strand stack 118
crowding, macromolecular *140, 472*, 499, *518, 609*
CRP (cyclic AMP receptor protein) 574, 581-2, **584, 587-9, 596-7**
cruciform **64-6**, 84, 99
cryo-electron microscopy *xxvii, 627, 718*
cryo-enzymology *xxvii*, **685**
cryoprotectant *xxvii*, 497
cryosolvent *xxviii*, 456, 552
crystal lattice 13, 53, 66, 130, 364, **393-5**, 409, 435-6, 448, 450, **451-2**, 456
crystallin **319-20**
crystallization *8*, 47, 53, *120*, 364, 394, 409, 436-7, 451-2, 560, *567*, 579, 585, *615, 654*, 730
crystallography *7*, 24, **40**, 47, 53, *55*, 69, 74, 130, 364, 370, 393-4, 409, 415, **451-2**, 585, *618*, 625, *651*, 665, 667-8, **685-6, 691-2, 702-3**, 719, 735
Ctmp protecting group **220-1**
cubic symmetry **398-9**
cucumber necrosis virus *400*
cupredoxins 534
cyanate **235, 239, 294-5**
cyanide 249
cyanoethyl protecting group *167*, **212-3, 217-8**, *222*, 225
cyanogen bromide (CNBr) 255, **286-9**, 292
cyanylation **286, 288-9**

cycle sequencing **201-2**
cyclic AMP (cAMP) **27-8**, 596-8, 717, **731-3**
cyclic AMP receptor protein (CRP) / catabolite gene activator protein (CAP) 581, **596-8**
cyclic AMP response element binding protein (CREB) 720, *724*
cyclic AMP-dependent protein kinase **726**
cyclic GMP (cGMP) 28
cyclic symmetry **398-400**
cyclo-oxygenase 414
cyclohexane 35, 261-3, 266, 389
cyclohexyl group 171, *173*
cyclophilin 518
cycloviolacin **381**
cysteic acid 248, 282, 292
cysteine (Cys) residue **6**, **229**, **231**, 237, **245-54**, **258-62**, **264-5**, *269*, **275-7**, **281-9**, **291-3**, 302, 305, 318, 335, *341*, **347**, 348, **355**, 362, **384-6**, **390**, 393, 418-9, 464, **480**, **493, 513-5**, 522, 525, **530-2**, 534, 599, **602-4**, 615, 621, 663, 712, 721, 724, 742
cystine 282
cystine knot **381**
cytochrome *b* **522**
cytochrome *c* *24*, 99, 141, **308**, **314-5**, **318-9**, 322, 415, **417-8**, **439**, 441, **449**, 473, **475**, 485, **490**, **522**
cytochrome *c* oxidase **410-1**, 428
cytochrome c_3 521
cytoplasm **304-5**, 533, 690
cytosine, cytidine **26-9**, 31, **33-42**, 54, 56-60, 63, 76, **81-2**, 84, **109**, 139, 144-7, 173-4, 195, 205, **210-1**, 225, **577**

D-arm, -loop, -stem, tRNA 119, **127-9**, 623, *624*
D_2O (^2H_2O) **447-9**
dansyl *236*, **294-6**, 335
databases 24, *53*, *115*, *120*, 301, **306**, 313, *315*, **408-9**, *425*, *460*, 586, 607, *725*
daunomycin 78
Dbmb protecting group **222-3**
ddNTP (dideoxynucleoside triphosphate) **199-202**, 207
de novo design 343, 380, **432-3**
dead-end inhibitors 640, **645-53**, 655, 657-9
deamidation *242*, **243-4**, 289, 465
deamination 29, 37, **81-2**

deazapurine 68
Debye 331
decarboxylase 399, **629**, *646*, **670**, *674*, 682
decarboxylation *669*, 726
defensin *375*
deformed homeodomain 581
dehydrogenases **541-2**, 640, 729
deletion of bases, residues 78, **162**, 165, 198, 200, 218, 310-5, 357, 417-20, 428-9, 458, 500, 521, 591, 611, 665, 691
delta virus **131**, 698
denaturants *xxviii*, 139, **144-5**, 149-50, 167-8, 353, **449-50**, **463-5**, **468-75**, *478*, 480-5, **497-9**, 503, 505, 507-8, 513, 517
denaturation mapping **141**
denaturation, nucleic acids *59*, 99, 103, **139-76**, 183, 201, 609
denaturation, proteins 364, 413, 450, 458, **462-518**
density-gradient centrifugation 40, 70
deoxyribonucleotides **26**, 31, 39
deoxyribose 26, **29-31**, 53-4, 61, 75, 84, 108, 571, 611, 702
deoxyribozymes (DNAzymes) *220*, **698-704**
depurination 145
design of proteins *343*, *380*, **432-3**
desmin 354
destabilizers **497**
detection of nucleic acids **79**, 168-70, 198, 205, *208*
detection of proteins *245*, **267-71**, 282, 285
detergent 409, 413-4, 497, 503, 728
dextrorotatory 5
di-isopropylphosphoramidite **212-3**, **217**
diagonal dot-plot **312**, 325
diagonal map **291-3**
dialdehyde 108
diastereomer *xxviii*, 3, **7-13**, 172, 190
diazoacetyls, amide 241
dichloromethane 224-5
dicyclohexylcarbodiimide 273
dideoxycytidine **39**
dideoxynucleosides **199-204**
dielectric constant **446**
diethyl pyrocarbonate (DEPC) **81**, **206**
diethylsulfoxide *145*
diffuse ions 69, *121*, 126
diffusion *xxviii*, 21, *73*, 75, 89, *117*, 158, 170, 337, 413, 434, 441, 460, 521, 552, 556, 566, 579, *593*, 631, 634, 686, 688-9
dihedral angles **14-7**, *173*, 329, *333*, 338, 362, **368-70**, 375-8, 430
dihedral symmetry **399**
dihydrofolate reductase **634**, *646*, 671
dihydrouracil **38**
dihydrouridine (D) **110**, **208**
dihydroxyacetone phosphate **12**
diisopropyl fluorophosphate (DFP) **441-2**
dimerization 37, 61-2, 79, 106-7, 347, **400-4**, 411-2, **432-3**, 517, 521, 531, 548, 553, 558-60, 575-6, 589-92, **594-614**, **619-23**, 687-8, 691-2, **709-21**, 723, 729-34
dimethylether **2-3**
dimethylguanosine (m$^2_2$G) **110**
dimethylsulfate (DMS) 37, **81-2**, 204-6, 571
dimethylsulfide (Me$_2$S) 255
dinucleotide platform **123**
dinucleotide-binding motif **541-43**
dioxane 261-2, 266
diphenylphosphoroisothiocyanatidate **298**
diphosphoglycerate (DPG) **554-5**, 557, 689
dipoles *xxviii*, 44, **331**, 339, 341, **348**, 350-1, 383, 393-4, 432, 445, 501, 521
directed evolution *423*, *629*
dispersion forces **697**
dissociation constant (K_d) 78, 527, 538, 564-9, **573-82**, 591-3, 598, 610, 616, 635, 649, 652, 656-60, 669-71, 675-6, 688, 715
distamycin A 77
distance distribution **20-1**
distance geometry 136
distance map **386-8**, 417
disulfide bonds *xxviii*, *245*, **248-54**, 282-3, 285, **292-3**, 304-5, 327, 335, 362, 367, 381, 384-6, 393, 418, 454, 464-5, 467, **480**, 483, **485**, **492-4**, 501, 504, **513-5**, 518, 723
dithiobis-(2-nitrobenzoic acid) (DTNB, Ellmans' reagent) **253-4**
dithiothreitol (DTT), dithioerythritol (DTE) **251-2**
divergent evolution **306-26**, **415-29**, *422*, 423, 590
DMTr ((di-*p*-anisyl)phenylmethyl) protection **210-18**, 224-5
DNA **25-85**, **86-107**, **139-176**, **177-218**, **564-627**, **698-705**
DNA Data Bank of Japan **306**
DNA gyrase 95

DNA ligase **86-8**, 93, 104, 166, **180-2**, 184-5, 192, 219
DNA polymerase 40, 95, **178-81**, 183-4, 187-90, **199-201**, 219, 314, 571, **688**, 696
DNA-dependent RNA polymerases **185-8**, 193-4, 204, 564, 596-8, 610, *651*
DNA•protein complexes **564-611**
DNA•RNA hybrids 48, 54, **149**, 154, 160, 164, 168, 189
DNase I 571
DNAzymes *220*, **698-704**
dNTP (deoxynucleoside triphosphate) 178-80, **199-207**, 219
domain, nucleic acid **86-92**, 103, 107, **115**, 119, 128-32, 151-4, 157
domain, protein *24*, *179*, *196*, **308-9**, 326-7, 358, **362-3**, **366-9**, 371, 378, 383-6, 388, **400-8**, 414, 420-2, *429*, 434-5, 447, *449*, 452, 456, 458, 464-9, 476-8, 483, 488, 505-10, **516-8**, **521-3**, 531-3, 538-43, *562*, *582*, *586*, 590-605, 609-25, **678-9**, 694, **709**, **712-16**, 720-21, **723**, 726-7, 729-30, 732, 736-7, 741
domain swapping **403-4**, 464
domain, topological **86-92**, 103, 107
donor, electron 534
donor, fluorescence 21, 168-9, 202, 335-7
donor, hydrogen bond 13, 33, 61, 113, 232, 240, 243, 257, 259, 261, 339, 342, 370, 389, 394, 492, **565**, 572, **695-6**
dot blot 166
dot plot **312**, 325
double helix **25**, **46-55**
double reciprocal plot **635-59**
double sieve editing **695-6**
double-stranded RNA-binding domain (dsRBD) **616-8**
DPG (bisphosphoglycerate) **554-5**, 557, 689
drag-tag **203**
dsRBD (double-stranded RNA binding domain) **616-8**
dsRNA-dependent protein kinase 617-8
DTNB **253-4**
DTT (dithiothreitol) **251-2**
dyad axis **41-2**, 106-7, 195, 596-7
dynamic programming 135

E-box **577**
E12 transcription factor **574**

E47 protein **577**
Eadie-Hofstee plot **636-7**, **648**
echinomycin 57
eclipsed conformation **14-5**
*Eco*R1 75, 192, **195-7**, **574**
*Eco*RV *196*, **574**, 579, 582
editing amino acid activation **695-6**
Edman sequencing 243, 289, **295-7**, 300
EDTA 84, 116, 168, **527-9**, 571
EF-hand **536-9**
effective concentration **23**, 72, 253, 335, 338, 350, 404, 434, **491-4**, 511, 662, **673-4**
EGF motif **309**, **384-5**, 406, 540
EGF receptor **726**
EGTA **527-9**
elastase **286**, **387-8**, 417-8, 441, *685*, 738-9
elastic stretch modulus 72
electron capture dissociation *728*
electron crystallography 364, 409
electron density 272, *351*, 364, 393-4, 451-2
electron microscopy 66, **99-100**, 141, **169**, 360, 567, 625, *627*, *718*
electron paramagnetic resonance 534
electron tunneling *691*
electrophoresis *xxviii*, 31, 73, *79*, **94-5**, **99-104**, 140, 166, **196-209**, *236*, *239*, *242*, 267, 270, **283-4**, **289-93**, 296, 503, **567-70**, 728
electrospray ionization MS 207, 497, **527**
electrostatic interactions 44, 69, 72-3, 77, 103, 119, 126, 154-5, 173, 176, 252-3, 263, **348**, 354-7, 383, 393, 432, 437, 441, **445-7**, 464, 469, 476-8, 485, 494, 497-9, 501, 521, 543, 555, 560, 569, 573, 577-84, 606-8, 615, 625, 669, 672, 696, 719, 726, 741
electrostriction 438
ELISA assay 166
Elk-1 **588**
Ellman's reagent (DTNB) **253-4**
elongation by duplication **325-6**
elongation factors *188*, 620, **624**
EMBL sequence database **306**
enantiomer *xxviii*, **3-11**, *242*, 697
enantiotopic 10,
encounter complex 160
end-label 198, **203-6**, **290-1**
end-to-end distance **19-22**, 72, 334
endo conformation 14, **29-30**, **49**, 53-4, 78, 116, 165, 176
endonucleases *xxviii*, 163, 195, 206, **574**, 582, 617, 686

endoplasmic reticulum **303-5**
endoproteinases 286-7, 686
enediol **12**
energy barrier 6, 10, 152, 248, 336, **509-10**, 514-5, 609, 628, **630-2**, 689
energy distribution **104-5**
energy fluctuations **435**
energy minimization *135*, 138
energy transduction 628
enhancer *xxviii*, *594*
enol tautomer 13, **34**
enolase 325
enterokinase 738
enthalpy (H) 76, 104, *135*, 145-6, 253, **264-5**, 346, **455**, 468, 472-5, **478**, 482, **488-94**, 507-8, 579-80, 610, 685, 697, 701
enthalpy-entropy compensation 488
entropy (S) 17-8, 23, 77, 104, 117, *135*, 146, 173, 175, **264-5**, 332, 334, 337-8, 346-50, **455**, 457, 462, 472-6, 480, *484*, **488-96**, 501, 507-11, 577-81, **672-5**, 681, 685, 701
enzyme-linked immunosorbent assay (ELISA) 166
enzymes *xxviii*, **628-695, 706-742**
epidermal growth factor (EGF) **309**, **384-5**, 406, 540
epimers, epimerization **8**, 13, 40
EPR (electron paramagnetic resonance) 534
equilibrium dialysis 567
equilibrium-ordered mechanism 640, **643-8**
error rate 180, 185, 189, 695-6
erythrose **7**
ESI (electrospray ionization) 207, 497, *527*
ester hydrolysis **631**
estrogen receptor **600-3**
ethanol 2-3, *46*, 99, **144**, 225, 232, 261-2, 266, 630
etheno derivatives **81**
ethidium bromide 68, **78-9**, 99, **101-3**, 106
ethyl-urea **144**
ethylene glycol **14**, 144-5, 685
ethylenimine 241, **247**, 287
N-ethylmaleimide **246-7**, 251, 292
evolution 3, *24*, **135-6**, *138*, 164, 185, 198, **306-26**, *374*, *381*, *384*, *386*, *388*, 394, *404*, 408, **414-33**, *453*, 498, 500, 503, 511, *531*, *537*, 590, 613, 616, *621*, *629*, 671, *683*, 688-90, 702-3, 710, 717, 724
EX1, EX2 mechanisms **450-1**

excluded volume 18-9, **22-3**, *150*, 168, 337, 460, 491, *496*
exo conformation **29-30**, 53
exons *xxix*, 131, *324*, 699-700
exonuclease *xxix*, 180, 195, 200, 204, 207
exopeptidase *xxix*, 295
exoproteinase *xxix*, 286
expressed sequence tag (EST) 194
expression vectors **193-4**, 327
extinction coefficient 267
extremophile *499*
extruded helix 118, *120*
eye lens 320, 325

FAD (flavin adenine dinucleotide) **521**, 541, 543
farnesyl group 305
FASTA 313
fatty acid synthase 690
fatty acids *xxix*, 305, 406, 409
feedback inhibition 706, 717, 733, *738*
Fenton reaction **571**
Fermi-Dirac probability distribution *393*
ferredoxin **319**, **325-6**, **400-1**, *522*
ferritin **383**, **399**, 532-4
ferrous ion 84, 532, **547-50**
fiber diffraction **47-8**, 149
fibrillin 384
fibrin 354, 385
fibrinogen 322, 354
fibrinopeptides 314, **318-20**, 322-3
fibronectin **309**
fibrous proteins 307, **319**, **353-61**, **405**, **439**, 457
β-fibrous proteins 353, **358**
filamentous arrays 353-5, 358, **405**
filter, membrane 166
filter-binding assays **567-9**
filtration 215, 269, 277, 567
fingerprint **205**, **208**, 289-90, 386, 542-3, *702*
FIS protein **586**
Fischer projection **5**
FISH (fluorescence *in situ* hybridization) 170
fish 323
flash freezing 685
flavin adenine dinucleotide (FAD) **521**, 541, 543
flavodoxin **363**, **389**
flexibility, conformational 14, 21, 165, *579*, *582*

flexibility, DNA 53, **71-7**, 84-5, **86-107**, 171, 572, 583, *585*, 598, 606
flexibility, proteins *259*, 314, **329-38**, 394, 418-20, 423, 434-5, 441, **447-58**, 484, 511, 515, 521, **523-4**, 552, 593, *635*, 637, 665, 685
flexibility, RNA 108, 113-5, 126, 130
Fli-1 **588**
flipping of protein aromatic side-chains **454-5**
flipping of bases 55, 74-5, 103, 118,
fluorescamine 267, **269-70**, 282
fluorescein 202
fluorescence *xxix*, 21, 35, *68*, *75*, **79**, 81, 168-70, 184, 202-3, *236*, 246, 258, 267, 269-70, 285, 290, 294, 335, 458-60, 463, 484, 566-7, 637
fluorescence correlation spectroscopy *163*, 337
fluorescence *in situ* hybridization (FISH) 170
fluorescence resonance energy transfer (FRET) 21, 168-70, **335-7**, 459
fluorinated bases 39-40
fluoro-2,4-dinitrobenzene **235**, 292
fluorenylmethoxy carbonyl (Fmoc) 172, **272-7**, *383*
fluorophosphate 729
fluorouracil **39-40**
Fmoc (fluorenylmethoxy carbonyl) 172, **272-7**, *383*
fold recognition 430
folding unit 366
folding, protein 293, 322, *351*, 360, *375*, *377*, *381*, *383*, 404, *407*, *410*, *413*, *423*, 424-5, **430**, 435, *437*, *449*, **454**, *459*, **462-518**, *525*, 527, 534-5, 611
folding, RNA **111**, *117*, 119-21, *127*, 130, 134, 136, **150-7**, *163*, *698*
Folin phenol assay **268**
footprinting **570-1**
formaldehyde 81, **83**, 141, 144, 235, 682
formamide **144**, 146-9, 159, 168
formate dehydrogenase 259
N-formylkynurenine **258**
Fos-Jun **588**
four-helix bundle **383**, 385, 405-6, 412, 414, **433**, 576, 590, 593-4
four-stranded DNA **60-3**
four-stranded proteins 356
four-way helical junction **66-7**, 119, 131
Fpmp protecting group **220-1**, 225
fragmentation, DNA 71, 195, **570**

fragmentation, proteins **285-9**, **299-300**
frame shift 78, 120
free energy 6-7, 14, 18, **45**, 76, 90, 95, 99, **103-7**, 119-20, 134-6, 146-7, 151-3, 156, 160, 248, **264-6**, 333, 336, **390**, 424-5, 435-7, **449-52**, **465**, **469**, **472-9**, **488-504**, **508-11**, 580-4, 591, **630-2**, 669-71, **688-93**
free energy of transfer **35**, 261-3, **392**, 427, **470**
free energy relationships **663-6**
freely rotating chain 19
FRET (fluorescence resonance energy transfer) 21, 168-70, **335-7**, 459
front-phosphorylation 727
fructose phosphates 689, **717-9**, 733
FTIR (fourier-transform infrared) spectroscopy *68*, *236*, *351*
fumarase 634
fumarate 3, 9, 634
furanose **9**, 14
furocoumarins *80*
fushi tarazu **588**

G proteins 624, 709-10, 731
G quartet **60-4**, 133
GABA aminotransferase 684
gabaculine 684
GAGA **588**
GAL4 motif **530**, **588**, **603-4**
galactose **8**
β-galactosidase 590, 638
ganciclovir **39**
GATA-1 **579**, **588**, **600**
gauche conformation **14**
Gaussian chain theory 337
Gaussian distribution **20**, **93-4**, 105
GCN4 leucine zipper **354**, **574**, 578, **588**, **605**
gel electrophoresis 73, *79*, **94**, 99, **101-4**, 196, 207, **209**, 290, 503, **569-70**
gel retardation assay **568-70**
gelatin 360, **439**, 462
GenBank database **306**
gene duplication 313, **324-6**, 421, 607
gene expression *89*, 133, 186-7, **190-1**, 198, 564, 575, 594, 598, 721
gene fusion **326**
gene rearrangements **324-6**
gene silencing 187, 191
genetic algorithm 426
genetic code *xxix*, 194, 198, 280, **301-2**, 310, 317, 362

genomes *xxix*, 25, 37, 60, *79*, 86, *89*, 108, 113, 129, *136*, 138, 141, 160, **163-5**, 175, 177, 185-6, 189, 191, 195, 198-9, 203, *205*, *207*, 280, *309*, 315, 317, *320*, *324*, 409, *415*, 420, 504, 564, **566**, **575**, 591, 599, 603, 608, 616, 619, 723
geranylgeranyl group 305
Gibbs free energy (G) 6-7, 14, 18, **45**, 76, 90, 95, 99, **103-7**, 119-20, 134-6, 146-7, 151-3, 156, 160, 248, **264-6**, 333, 336, **390**, 424-5, 435-7, **449-52**, **465**, **469**, **472-9**, **488-504**, **508-11**, 580-4, 591, **630-2**, 669-71, **688-93**
Gla residues 540
glass support 216, 225
glass transition **456**, *685*
globins *234*, **312**, **314-6**, **318-9**, **321**, **323-5**, 369, 394, 399-400, 403, 405-6, **415-21**, 439, 445, 456-9, 473, **475-6**, **479**, **483**, **485**, **490**, **552**, **545-60**, 708-9
globular proteins *xxix*, 353, 356-7, 360, **363-409**, 423, **434-60**, **462-518**, **519-562**, 590, 595, 625, 742
glucagon *242*, 319, 435, **731**
glucocorticoid receptor **578-9**, **588**, 602-3
glucose 8, **12-3**, 38, 496, **524**, **678**, 717, 728-35
glucose-6-phosphate isomerase **670**
glucosyl-5-hydroxymethylcytosine 38
glutamate 578, 621, 735
glutamic acid (Glu) residues **229-31**, 241-3, **260-6**, 281, **286-7**, **302-7**, **317-8**, **347**, **348**, **355**, 357, **390**, 392, **419**, 432, 436-7, **444-6**, 457, 525, 530, **539-40**, 542-3, 577, 587, 663, 721
glutamine (Gln) residues **229-31**, **242-4**, **260-6**, 281, 294, **302**, **318**, **347**, **355**, 358, 372, 376, **390**, 411, **419**, 465, 550, 587, 663
glutamine synthetase **736-8**
glutaminyl tRNA synthetase **621-2**
glutathione (GSH) 253, 327, 492
glutathione disulfide (GSSG) 492, 513, **521**
glutathione peroxidase **137**, 259, **403**
glutathione reductase **521**, *685*
glutathione S-transferase 323, 325-7
glyceraldehyde-3-phosphate **12**
glyceraldehyde-3-phosphate dehydrogenase (GPD) 314, 560
glycerol 438, 456, 496
glycine 4, **10**, 144, 496, *582*, 682-3

glycine betaine **144**, 146
glycine (Gly) residues 171, 228, **230-2**, 244, **262**, 273, 275-6, **286**, 289, **302**, 305, 307, **317-8**, **332-3**, *348*, 446, 457, *484*, 501, 542-3, **587**, 615, 621
glycogen debranching enzyme 729
glycogen phosphorylase **728-35**
glycogen synthetase 732-3
glycolysis 689-90, 717
glycosidic (glycosyl) bond 26, **29**, **31**, 41-2, **49**, 54, 67, 82, 211, 731
glycosyl phosphatidylinositol (GPI) groups **304-5**, 723
glycosylation 38, **304-6**, 325, 327, 360, 680
glyoxal **141**
gold *59*
Golgi apparatus **304**, 325
GpC **44**, **46**, 54
Greek key topology **379**, **381-2**, 406, 534, 611
GroEL chaperonin 399
groove binder 176, 576, 590, 615
group I self-splicing introns 120, 127, **129-32**, 151, **699-702**
group II self-splicing introns 129, **131-2**, **699-702**
growth factor 381, 704, 727
growth hormone **287**, **319**
GRP motif **588**
GSSG (glutathione disulfide) 492, 513, **521**
GTP 412, 543, 554, 624-5, 721, 731-2
GTP cyclohydrolase 399
GTPase 624-5, 730
guanidination **235**, **237-8**
guanidinium chloride (GdmCl) 144, 254, 449, 465, **469**, **471-2**
guanidinium ion **469**, **471-2**
guanido group 233, 292, 395, **572**
guanine **26-9**, 31, **33-36**, 40, 42, 53-7, **60-4**, 69, 75-7, 81-2, 139, 144, 146-7, 161-2, 210-1, 223, 225, 571-2, 602, 606, 699, 702
guanine quartet **60-4**, 127
guanyl-3,5,-dimethylpyrazole **235-7**
guanylate kinase *680*
guide RNA 133
gyrases 94-5

H-DNA **60**, 93, 99
β-hairpins, protein *343*, *351*, **375-7**, **379-80**, *508*, 584, 599-600, *609*

hairpins, nucleic acid 61, **64-5**, **113-7**, **119-25**, 131-2, 149-51, **154-63**, 167-8, 608-9, **614-21**
hairpin loops **116, 123-4**
hairpin ribozyme **131**, *133*, 698, *704*
Haldane effect 556-7
Haldane relationship **655-6**
half-life, -time *xxix*, **244**, 337, 511, 534, 567, **629**
half-of-sites reactivity *562*
halophile 498-9
hammerhead ribozyme **130**, 133, *150*, 698, **702-4**
hand-off 609
handshake motif 607
haptoglobin 325
hard-soft acids, bases *xxix*, 525
HDV (hepatitis delta virus) ribozyme *125*, **131**, 698
heat capacity *xxx*, 37, *150*, **263-6**, 435-7, 458, **467-8**, **470-5**, 478, 482, 488-90, **506-8**, 511, 579-81, 610
helical arm 623
helical junction **66**, 113, 119, 131
helical pitch **17, 49**, 54
helical repeat **17**, 73, 93, 103, 107, 342
helical wheel **341**, **357**
helicase 177, 185, 687
helices, nucleic acid **46-55**, **57-60**, 66-8, 70-2, 78, **90-1**, 112-24, 131-5, **139-55**, 173-6, 183, 565
helices, protein **339-42**, **344-61**, **367-72**, **375-8**, **383**, 387, **405-14**, **418-9**, **423**, **426-8**, 432-3, 456-8, 482, 491, 497-8, 500, 503, **523**, **536-43**, 547, **576-86**, **603-11**, **730**
helix **17, 25, 46-55, 58-9**
β-helix **380, 423**
helix-loop-helix domains **536-9, 575-6, 588, 604-5**
helix-turn-helix motif **576, 586-97**
heme 369, **415-8, 521-2, 547-59**
hemerythrin 369, **399, 403, 522**
hemoglobin *234*, **312**, **314-25**, 369, 394, 399-400, 403, 405, **415-8**, **545-60**, 708-9
hemolysin **399**, 413
hepatitis delta virus (HDV) ribozyme *125*, **131**, 698
heptad repeat **354-8**, 590
heptenitol 730

heptulose 730
heteroduplex *xxx*, **147-9**, 154, 161, **168-9**, 190, **219**
heterologous association **395-405**, 561, 580, 712
heterotropic **543**, 546, **554-5**, 707-12, 715-6, *720*
hexamer *11*, 399, **403**, 560, 575, 709, **712-5**, **736**
hexamine 54, 130
hexokinase **524**, 678
HF (hydrogen fluoride) 172, 277, 279
hidden Markov method 426-7
high-mobility-group (HMG) protein *585*
high-potential iron protein **522**
Hill coefficient, plot **551-3**, 560-2, 706, 719-20
Hin recombinase **588**
histidine (His) residues **229**, **239-40**, 246, **260-2**, **264-6**, 268, 292, **302**, 307, **318**, 327, **347**, **355**, **390**, **418-9**, **442**, 530, 722-4, 728, **739-40**
histones *xxx*, **106-7**, 187, 314, **319**, 568, **588**, **607-8**
HIV 531
HIV proteinase **399-401**
HIV reverse transcriptase **189-90**, 401-2
HMG-CoA synthase *654*
HNF3 **588**
hnRNP (heterogeneous nuclear RNP) **614**, 618
Hofmeister series 145, **471-2**, 496-7, 578
Holliday junction **66**
holo protein **520**, 560, 595, 597
homeodomains **588-9**, **592-4**, **614**, 625
homeostasis *xxx*, 707, 710
homoarginine 237
homocitrulline 239
homogenizer 71
homology *xxx*, 63, 66, 134-6, 166, **169**, **307-26**, **387-9**, 399, **400-1**, 408-9, 413, **415-24**, **427-31**, 467, 511, 533, **549**, 593, 607, 613-4, 616, 618, 625, 678, 685, 717, 723-4, 726, 737
homopolymers 1, 160-1, *172*, 352
homoserine residues **288**
homotrimer 363, 398-9, **687-8**
homotropic *xxx*, 543, 560, 707-8, 711-2, 715
Hoogsteen base pairing **42**, **56-60**, 81, **173-6**
hormones *xxx*, 285-7, *295*, **303**, 305, **319**, 322, 325, *381*, 435, 520, 704, 717, 731-3

hormone receptors **530-1**, **602-3**
horseradish peroxidase 168
HPLC *xxii*, 251-2, *270*, *283*, 285, 287, 290, 727
HU protein **608-9**
hybridization *xxx*, 149, 154, **160-75**, 191, *195*, 293
hydantoin **294-5**
hydration *xxx*, 40, 46, 53-5, **68-71**, 77, 103, *126*, *145*, **263-5**, *394*, *403*, **437-40**, 456-9, 491, 496-7, **579-82**, *591*
hydrazine **204-6**, **234**
hydrodynamic volume **101**, 140, 482, 713
hydrodynamics 157-8, 469, 476
hydrogen bonds *xxx*, 13, *21*, **25**, 31, **33-5**, 37, **41-4**, **49**, **53-6**, 61, *75*, 112-3, 118-20, **124**, 144, 232, 261, **338-53**, 359-60, **370-8**, **387-96**, 406-11, 414, 418, 445, 462, 469, 472, **491-5**, 497, 503, **521-3**, 542, 547-8, 550, 559, **565**, **572**, **577**, 579, 582, 587, 591, 596, **602**, 606-7, 611, 615-8, 621-4, 662, 687, **695-6**, **703-4**, **719**, 734, **739-40**
hydrogen exchange *3*, **75-7**, **447-51**, 458, 468, 484
hydrogen ions 557
hydrogen peroxide **571**, 628
hydrogenases 668
hydropathy 263, **427-8**
hydrophilicity *xxx*, **36**, 175, 232, *257*, **261-2**, **311**, 317, 395, 405, 413-4, 427, 431, 436-7, 542, 695
hydrophobic moment 341
hydrophobicity *xxx*, 21, **35-6**, 44, 54, 144, 257-8, **260-7**, **311**, 317, **341**, 346, 353, **392-4**, 406, 409-14, 427-8, 457, 462, **469-70**, 475, 484, **490-1**, **494-503**, 549, 690, 695
hydrosulfite 257
hydroxamate group 289
hydroxide ion 12, 51, 75, 239, **249**, 444, 673, 697
hydroxy-decanoyl-dehydrase **683-4**
hydroxy-Lys residue 307, **359**
hydroxy-Pro (Hyp) residue 269, 307, **359-60**
hydroxyapatite 142, 164, 169, 361, 535
hydroxyaspartic acid residues **540**
hydroxyl radical 84, 532, **571**
hydroxylamine 37, **81-2**, 244, 286, 289
hydroxylation **304-5**, **359-60**, 540

hydroxymethylcytosine **38**
hydroxymethyluracil **38**
hydroxysuccinimide **235**
hyperchromism **140**, 143
hyperfine splitting 534
hyperreactivity 441-2, 570
hypersensitivity 60
hyperthermophile **583-4**
hypervariability *xxx*, 323, 378, 697
hypochromism **140**
hypodermic syringe 71
hypoxanthine **35-6**, **39**, 82
hysteresis *xxx*, 157

i-motif 63-4
icosahedral symmetry **398-401**, 619
iFRET (induced FRET) 168
IgG 452
isoleucyl tRNA synthetase 695
imidazole 237, **239-40**, 286, 446, 525-6, 534, 547, 556, 673, 680, 738
imino acid **228**, **258-9**, 269
imino group **75**
iminol group **75**
iminolactone 288
iminothiolane 85
immobile junction 66
immobilized macromolecules 166, 170, 741
immobilized NMR spectrum 454
immunoassay 185, **486-7**
immunogen *xxx*, 190, 696-7, 728
immunoglobulins **320**, 398, **520**
immunoPCR 185
improper angle 16
in silico xxx, 430
inclination, base pair **43**, 150
inclusion bodies 327, 517-8
indel 310, 429
indole **257-8**, **443-4**, 455, 690
induced fit 582, 621, **677-80**
inducer *xxxi*, 194, 590
infrared (IR) spectroscopy 68, *236*, *342*, *351*, 435
inhibition constant 638, **672**
inhibitors, enzyme 639-40, **645-55**, 657, **659-62**, 666-8, **670-1**, **675-7**, **683-5**, 706-7, **710-1**, 715, 717-20, 723, 726, 729-35, 738-42
initial velocity 632-6, **641-5**, 653, 657, 661, 707-8

inosine (I) **110**, 161, 194, *599*
inositol 496
inositol phosphates 554, 558
insulin **319**, 322, **403**, **447-8**, 464, **522**
integration host factor (IHF) **574**, **582-4**, **588**, **608-9**
intercalation 56-7, **63**, **74-5**, **78-80**, 84, 93, **101-3**, 106, 119, 174, 176, 583-5, 608
interference analysis *123*
interferon 583
interhelical stacking **121-2**
interleukin-1 motif **384-5**, 406
intermediate filament **354-7**
intermediate-state analog **670-2**
intermediates, enzyme 645, 666-7, **669-70**, 680-1, **683-6**, **689-91**, 694, 718, 738
intermediates, folding **150-1**, **156-8**, 163, **293**, 424, **454**, 465-6, **483-5**, **507-15**
intermediates, kinetic **151**, 484, **666**
intermediates, metastable **154-5**, 503, **512**
intermediates, reaction **12**, **76**, **146**, 141-2, 174, **182**, 213-6, 248, 273, 275-8, **444**, 455, 610, **630**, 689-90, 741
intermediates, tetrahedral **671-2**
intermolecular interactions 62, 133, 251, **253**, *292*, 342, 364, **404**, **462**, **464**, **491-2**, 515, **674**, 741
internal friction 337
internal loop 84, **113-4**, **118-20**, 123, 131, 137, 142, **158-9**, 614, 625, 701
interwound superhelix **90-1**, 95, 99
intramolecular interactions 21, 23, 60, 113-4, 139, 143, 146, 149, 158, **160-3**, 165, **252-3**, 263, 288, 293, 322, 342, 346, 364, **404**, **443-4**, **462-4**, **491-2**, **514-5**, **674**, 698, 723, 740
introns *xxxi*, 64, **127-33**, 151, 193-4, 301, **699-700**
inverse folding problem **430**
inverse PCR **184-5**
inverted repeats **64-6**, 195, 578, **589**
iodination **256**, 441
iodine 213-4, 224, **257-8**, **443-4**
iodotyrosine 256
iodoacetamide, iodocetate **235**, **245-6**, 251, **254**, **283-5**, **291-3**, **442**, 513
iodosobenzoic acid 258, 286, 289
iodouracil **39**
ion channel 409, 725

ion-exchange chromatography 31, 282-3, 513, 517
ionic strength 55, 72, 95, 103, 144-5, 173-4, 436, 446, 464, 469, 475, 499, 569, 578, 718
ionic zipper 355
ionization 28, **32-3**, 39, *46*, 76, 145, *246*, 251, 253, 348, 443, **445-6**, **476-7**, 482, 667
IR (infrared) spectroscopy *68*, *236*, *342*, *351*, 435
iron (Fe) 84, 240, 247, 369, 418, 525, 529, **547-52**, 558
iron-storage proteins **532-4**
iron-sulfur proteins **401**
iron-transport proteins **532-4**
irregular conformations 426
isoAsp residues **244**
isocitrate dehydrogenase **724-7**
isocytosine (iC) 68
isodeoxyadenosine (iA) 56
isoelectric focusing (IEF) 283
isoelectric point (pI) *xxxi*, **436-7**, 476
isoguanosine (iG) 68
isoleucine (Ile) residues **229**, **232**, **260-2**, **265-6**, 273, 281, 287, **302**, 306, **317-8**, 340, **347**, **354-5**, **390-2**, 411, **419**, 438, 615, 695
isoleucyl tRNA synthetase 695
isologous association **395-9**, 402-3, 561, 712, 735
isomers *xxxi*, **2-13**, 90, 171, 220, **228-9**, 232, 244, 272, **337-8**, 360, 368, 505-6, **540**, 682, 715
isomerase 319, **369**, 373, 397, **403**, 405, 407, 422, 518, 629, 638, 656, 669-70, *672*, 689
isomerization 242, **337-8**, 345, 360, *455*, 504-6, 509, 517-8, 660, 662, 684, 703
isopentenyladenosine (i⁶A) **110**, **207**
isopropanol 50
isoschizomer **196**
isothermal compressibility *xxxi*, **439**, **457-8**
isothiocyanate **298**
isotopes *xxxi*, 6-7, 10-13, 168, 202, 443, 447, 451
isotope effect *238*, *438*
isotope exchange **447-51**, **656-9**

jelly roll motif **382**, 406

K-system, allosteric **707-10**
K11 bacteriophage 188

keratin 358
ketimine **682-3**
keto tautomer 13, **34**
ketoglutarate 726, 737-8
ketosteroid isomerase **629**, *672*
KH domain **614**, **618-9**
kinases *xxxi*, 187, *194*, 204-7, 366, 399, 402-3, **524**, 539, 543, 618, 640, 658, **675-6**, **678-80**, 707, **717-33**
kissing hairpin loops **123-4**, 133
Kjeldahl protein assay 267
Klenow fragment **179-81**, 188
K_m (Michaelis constant) **632-6**, 656, 663, 665, 668, 671-2, 678, 680, 688-9, 696, 706-10, 720, 732, 736
knot 136, 368, 381, 87-9, 95, *98*, *105*
knotted DNA **87-9**, 95
Kramers' theory 638
Krebs citric acid cycle 716, 726
kringle domains **309**, **384-6**, 467

lac repressor 398, **576**, 578, 582, 586, 589, **590-1**, 597
lactalbumin **320**, 325, **439**, 458, **481**, **485**, 515
lactam tautomer **34**
β-lactamase 485, **634**
lactate **11**
lactate dehydrogenase (LDH) **319**, 325, **369**, 541, 545, **652-3**
lactim tautomer **34**
lactoferrin *534*
β-lactoglobulin **439**
lactose 590, 597
lactose synthesis 325
ladder sequencing 204-7, *295*, **299-301**
lag period 505-6, 509, 687
lambda (λ) repressors 576, 581, 588, **591-2**
laminin 354, 405
latexin *741*
lattice, crystal 13, 53, 66, 130, 364, **393-5**, 409, 435-6, 448-50, **451-2**, 456
lattice models 425
Laue crystallography 686
lead ions (Pb^{2+}) **526-8**, 703
lectins *382*, 520
left-handed helices **17**, 175, **340-1**, 344, 354, 359, 370-2
Leu-rich repeats **422-4**
leucine (Leu) residues **229-32**, 244, **260-2**, **265-6**, 281, **286-7**, **302**, 306, **317-8**, **347**, 354, **355**, *383*, **390**, 392, 395, 411, **419**, 422-4, 432, 438, 457-8, 470, 550, 583, 615
leucine aminopeptidase 295
leucine zipper **354**, *358*, 578, 590, **604-5**, 625
leucine-rich variant protein 423
leucyl tRNA 128-9
levorotatory 5
LiBr **471**, 497
libraries **194-5**, **278-80**, *379*, 429, 431, *615*, 702-4
lifetimes 35, 79, 511, 569, 638
Lifson-Roig model **349-53**
ligands *12*, 470, 686
ligand binding, nucleic acids 50, 56-7, **63**, 68-9, **74-80**, 84, 93, **101-3**, **105-6**, **111**, 113, 118-9, *140*, 174, 176, **564-627**, 704
ligand binding, proteins 327, 368, *393*, 401, 406-7, 412, 436, 484, 496, 502, **519-63**, **564-628**, 667, **675-80**, 686, **706-20**, 723
ligand binding and macromolecule flexibility 448, **524**
ligand binding and refolding 508, 516
ligation, peptide **273-7**, *383*
ligation and ligands 485, 534, 553
light-activated ligands 686
light-harvesting complex 397
linear free energy relationship **663-6**, 693
linear programing *432*
Lineweaver-Burke plot **635-58**
linkage relationships **493**, 502, 669
linked functions 477, **493**, **501**, 546, 554-6, 642, 649, 669
linker DNA 107
linking difference **92-3**
linking number **87-9**, **91-105**
lipids *xxxi*, **409-15**, **460-1**, 502-3, 519, 716
lipoic acid 248, 251, 641
lipoproteins *xxxi*, 308, 385
liquid chromatography 251, *253*, *728*
liquid packing density 389, 479, 494-5, 500
liquid compressibilities 457
LNA probe *167*
local unfolding **448-52**, 511
locked nucleic acids 165
loops, DNA 60-4, 68, **86-90**, *143*, *196*, 576, *579*, *585*, 590
loops, hairpin **116-7**, **123-4**, 149-51, **158-163**, 600, 609

loops, internal 84, **113-4**, **118-20**, 123, 131, 137, 142, **158-9**, 162-8, 614, 625, 701
loops, omega 375, 378
loops, protein *23*, 364, 375, 383, 395, 405, 407-8, 423, 428, 432, 434, 452-4, 504, 523, 535-7, 542-5, 547, 589, 592, 597, 602, 605, 607, **609-18**, 621, 693-4, 697, 715, 723, *727*, 730, 739
loops, RNA **117-39**, **150-7**, 621-3, 701
low-density lipoprotein (LDL) receptor **308-9**
Lowry protein assay **267-8**
LU-79553 antitumor drug **74**
luminescence 168
lutropin 319
lysidine (L) **110**
lysine (Lys) residues 176, **182**, **229-31**, **234-9**, 246, **260-6**, **283-7**, 292, **302-7**, 317-8, **336**, **347**, 348, *351*, **355**, **359-61**, 385, **390**, **392**, 410, **419**, 428, 432, **445-6**, 615, 625, 663-5, **682**, 693, 697, **721**, 729
lysozymes **325**, 353, 420, **437-9**, **443-4**, *452*, **456-8**, **472-8**, 483, **488-90**, **494-7**, 506, *515*, **520**, **634**
lysylendopeptidase 287
α-lytic protease **511**
lyxose 678

M13 bacteriophage 199, *280*
macro-conformation 14-5, 18
macrodipole 348
macromolecular crowding *140*, *472*, *499*, *518*, *609*
magnesium ions (Mg^{2+}) 51, 66, 69, 95, 119-21, 126-7, 130-2, 150-4, 157, 196, **526-9**, 535, 538, 540, 679, 699, 701, 703, 718, 721-2, 736
maize mosaic virus *207*
major groove, double helix *13*, **41**, **49-50**, 52-7, 61, 68-9, 75, 84, 112, 119, 126, 176, **564-5**, 571-2, 576, 578-9, **583-607**, 614, 617, 622, 625
malate dehydrogenase 541
MALDI (matrix-assisted laser desorption/ionization) mass spectrometry *207*, *237*, 300
maleate **3**, 9, **711**
maleic anhydride **235-6**, 246, 283
maleimide **247**, 251, 292
manganese ions (Mn^{2+}) 130, 196, 247, **526-8**, 701, 721, 736

manganese superoxide dismutase 397
mannitol 496
mannose **8**
MASH-1 **574-5**, 580
mass mapping *237*, *645*
mass spectrometry 203, 207, *234*, *240*, *249*, 290, *295*, **299-301**, *305*, 397, *408*, *410*, *449*, 497, *525*, *527*, *566*, 569, *571*, *645*, *724*, 728
Maxam-Gilbert sequencing **204**
maximum parsimony 313
β-meander **379-80**, 406
mechanism-based inhibitors **683-5**
MEF-2C transcription factor **574**, 583
melanocyte-stimulating hormone (MSH) 303
mellitin 403
melting DNA 14, 66, 95, **142-8**, 163
melting proteins **472-4**, 498
melting RNA 119, 127, 135, **148-150**
melting temperature *xxxii*, 135, 143, **145-9**, 160-1, 190, 472, 498
membrane, blotting 166, 267, 271, 296, 567-8
membrane, cell 28, 39, 171, 304-5, 341, 354, 362, 399, **414-5**, 434, 540, 690, 706, 723, 727, 739
membrane proteins 306, 325, *383*, 399, **409-15**, **427-8**, *433*, 434, 436, **460-1**, **502-3**, 519-20, 706, 723, 731
mercaptoethanol 248, **252-3**, 269, 292
mercury (Hg) 247, **526**, 711-2
meso compounds **4**
mesophile *xxxii*, 498-9
messenger RNA (mRNA) *xxxii*, 86, 108, 120, 127-9, 133, 137, 164, 170, 186, 189-90, 193-4, 205, **301-3**, 327, 564, 613-6, 618-9, 625-7, 699-700
met repressor **396**, 588, **605-7**
metal ions and proteins 232, 240, **247**, 253, 368, 383, 485, 508, **521-2**, **525-35**, 540, 681, see also metalloproteins
metal ions and nucleic acids 51, 54, 61, 66, **126-7**, **130-3**, 153-5, 701-2
metal shadow **99-100**
metalloproteins *436*, *505*, 508, **525-35**
metalloproteinases **742**
metallothioneins **531-2**
methanol **144**, 685
methemoglobin 549
methionine biosynthesis 720

methionine (Met) residues **229-31**, 246, **254-5**, 258, **260**, **262**, **264-6**, **286-9**, 292, **302-4**, **318**, 347, *352*, **355**, 376, **390-2**, 411, **418-9**, **470**, 522, 534, *535*, 538, 615
methotrexate 671
methoxymethylene protecting group 222-3
methyl acetamidate **235**
methyl amines 496
methyl groups 29, 31, **35**, 40, **68**, 74, 108, 453, 572, 577, 586, 607, 613
methyl iodide 254
methyl isourea **235-7**
 N-methylacetamide **261**
methyladenine **37-8**, 42
methyladenosine (m¹A) **110**, **208**
methylation 37, 127, 195, 204, *680*
methylcytidine (m³C) **110**
methylcytosine **37-8**, 195, **208**
methylglyoxal *234*
methylguanine **38**, **82**
methylguanosine (m⁷G) **110**, **208**
methylimidazole 224
methylisourea **234-6**
methylribose 110
methylthymine **42**
methyltransferase 37, 195
methyluridine 29, 40, **208**
metmyoglobin **475-6**, **490**
micelle 337
Michael addition *233*
Michaelis-Menten **630-62**, 696, 707-11, 725
micro-chip array 166
micro-conformation 14-8, 22, 436
microarrays *166*, *168*, *214*, *722*
microcalorimetry *472*, *537*
microdipole 348
microfibrils 361
microRNA 187, *195*
microscopy 66, **99-100**, 141, **169**, 360-1, *460*, *516*, 567, *625*, *627*, *718*
microtubule 484
mimicry *403*, *573*
minicircles *92*, *104*
minimal mutations procedure 313
miniprotein 668
minor groove of double helix **25**, **41**, **49-55**, 61-3, 68-9, 73, **77-8**, 113, 119-20, 122-3, 127, 150, 176, **571**, 576, 583-4, **588-92**, 598, 600, 607-8, 614, 617, 625

mirror plane, image **3-4**, 7, 344, 378, 395, 398
mitochondria 128-9, **304**, 400, 411, 417, 609, *624*, 685
mixed anhydrides 273, **298**, 721
mixed disulfide 252-4
mixed inhibition 650
Mn-porphyrin 571
mobile genetic element 699
mobility shift assay **568-70**
modules 112, 119, 123, 326, 384, 385, 423, 516, *531*, 599-601, 614-6
molecular beacons 168
molecular dynamics *16*, *53*, *97*, *118*, 138, *140*, 333, *404*, *454*, *457*
molecular mechanics *22*, 138, *666*
molecular memory 638
molecular recognition 378, 385, 423, *437*, *525*, 594, *704*
molecular volume 52, 479
molecular weight 1, 71, 282-3, 299-301, 306, 334, 388-9, 457, 503, 625, 628
molten globule **481**, **484-5**, 507, 511, 515
molybdate 247, 268
monotopic **409-10**, **413-4**
monoclonal antibody *xxxii*, 696
Monod-Wyman-Changeux model **545-6**
mononucleotide-binding motif 422, **541-3**
mosaic proteins **308-9**, 326
MS2 bacteriophage **619-20**
MSNT coupling **211-2**, 223
Mthp protecting group **220-3**
muconolactone isomerase 397
multiplicity, quaternary structure 398
muramidase 406
murine leukemia virus *190*
muscle 324, 354, 357, *405*, *539*, 547, *549*, 628, 731-2
mutagens 78-9, 85
mutagenesis, site-directed **218-9**, 327, **498-500**, 503, *624*, 663-5, 692-3, 728
mutant proteins *358*, 361, *452*, 458, **499-500**, *548*, *607*, **665-6**, 686, 712, *717*, *718*, 719, *724*
mutase 656
mutation data (MD) matrix **311**
myc **588**
MyoD transcription factor **574-5**, **605**
myoglobin **314-6**, **319**, **321**, 324, 403, 405, **415-6**, **439**, 445, 456-9, **473**, **475-6**, **479**, **483**, 485, **490**, 522, **547-60**

myosin 354-7, **439**
myosin light chain kinase (MLCK) **539**
myristoylation **304-5**

N-terminus 180, 200, 234, **243**, 272, **275-8**, 283, **288-9**, **294-7**, 300, 302, **304-5**, 330, **339-41**, 343, 346, **348**, 361, *377*, 384-5, 405, 412, 518, 523, **590-8**, 606-7, 609-10, 617, 622-3, 729, 733-4, 736, 737, 740-42
NAD, NADH 181-2, 369, 504, **540-3**, 545, 560, 640, *645*, **652-3**
NADP, NADPH **521**, 640, **735**
nanoparticle *59*, *191*
nanotechnology *209*, *704*
naphthalene 335, 484-5
native conformation 14-5, 47, 76, 94, 119,129, 140, 154-5, 353, 360, 424-5, 431, 434, 441, **444**, 447, 449, 451, 462-5, 469, 472-3, 476-88, 491-2, 496, 501, 503-4, 507-17, *539*, 540, 545, 711, 715-6, 741
native chemical ligation 275, **277**, *383*
natively unfolded proteins 484
natural antisense transcript 190-1
natural selection 307-9, 317, **321-4**, 421, *497*, 501, 511
nearest-neighbor approximation 134, 147, 352
Needleman-Wunsch algorithm 313
negative cooperativity 544, 546, 553, **560-3**, **707-9**, 716, 725
negative selection **322-3**
neighbor exclusion 78
neomycin 54
neoschizomer **196**
nested primers 184
neural networks 426
neuraminidase 397, 399, 407
neurotoxins **320**
neutral mutations **321-3**
neutron diffraction *55*, 370, 451
neutron scattering *436*, *438*, 451, *715*
nicked circular duplex DNA 87, 92-3, **98-106**
nickel ions (Ni^{2+}) 327, **526-8**
nicotinamide 521, **541**
ninhydrin **267-9**, 282, 293
nitration **256-7**, 715-6
nitric oxide 257
nitro-Tyr residues **256-7**, 715
nitrobenzaldoxime **212-3**, 215, 223
nitrobenzyl protecting group **221**

nitrocellulose 166, *271*, **567-8**
nitrogen mustard **83**, 85
nitrophenyl group 273
nitrophenylsulfenyl chloride 292
nitropropionic acid *685*
nitrosoureas **84**
nitrothiocyanobenzoic acid **288**
nitrothiosulfobenzoate (NTSB) **249**
nitrous acid 37, **81-3**
nitrous oxide 558
NMR (nuclear magnetic resonance) 7, *12-3*, 17, 24, 25, *40*, 47, *57*, 59, *73*, 77, *112*, *118*, *120*, *126*, 130, 136, 163, *238*, 240, *275*, 364, 370, 429, 447, **453-5**, *459*, 468, *476*, **480-4**, *507*, *518*, *537*, 585, *591*, *593*, *618*, *717*, *740*
NOE (nuclear Overhauser effect) 47
noncompetitive inhibition **646-51**, 655, 676
noncrystallographic symmetry 452
nonideality 580
nonreceptor Tyr kinases (NRTK) 723
norleucine (Nle) 274, 281
normal mode *400*, 459
Northern blot 166
nuclear hormone receptors 530-1, 602
nuclear magnetic resonance, *see* NMR
nuclear Overhauser effect (NOE) 47
nucleases *xxxii*, 60, 165, *167*, 171, 190, 570-1, 599, 609, 620, 687
nucleation 158-60, **349-53**, 360, *377*, 509, 532
nucleic acid **25-85**, **86-107**, **108-138**, **139-176**, **177-226**, **564-627**, **698-704**
nucleolus 186
nucleophiles *xxxii*, 37, 51, 84, 131-2, 178, 182, 232, 234, 239-40, 245-6, 249-51, 254, 259, 273, 288, *295*, 442, 630, 662, 673, 680-1, 696, 698-703, 718
nucleoporin 484
nucleosides **26-9**
nucleosome, *xxxiii*, *92*, *97*, **106-7**, 607-8
nucleotides *xxxiii*, **26-7**
nucleotide-binding proteins 378, 407, 422, **540-3**, 675, 678, 724, 729
nucleus, cell 86, 190
nylon filters 166

OB fold **609-14**, 622
OBF-1 594
Oct transcription factors **585**, 588, 594

octahedral geometry **398-9**, 529-30, 535-6
octamers *53*, 106, 396, **402-3**, 591, 607, 709
octanol 261-3
octaplex DNA *64*
octarellin 432
oil drop 479
olefins 683
oligoamide **171**
oligonucleotides **25-85, 86-107, 108-138, 139-176, 177-226, 564-627, 698-704**
omega loop 375, 378
OMP decarboxylase **629**, *646*, *674*
OmpF porin **410**
open duplex DNA 87
opiomelanocortin **303**
optical activity 5
optical rotation 4-5, 13, 463
optical trap 155, 204
optical tweezers *157*
ORD (optical rotatory dispersion) 5, **463**
order parameter (S) **453-4**
ordered enzyme mechanism **640**, 642-4, 652-3, 658
organelles 127-8, 302, 690
origin of replication **193-4**
ornithine (Orn) residue **234**, 281
orthogonal protection 273
orthologous **314-7**, 323-4
osmium tetroxide **81**
osmolytes *xxxiii*, *145*, 497, 580-1
ovalbumin 325, **439**
overwound superhelix 93, 585
oxaloacetate decarboxylase **670**
oxidation *xxxiii*, 37-8, 84, 213, 217, 224, **247-8**, 255, **257-8**, 282, **285-7**, 293, **443-4**, 480, 532, *725*
oximes 212, 215
oxindolealanine **257-8**
oxyanion binding site *672*, 739-40
oxygen (O_2) 84, 247, 281, **319**, 323-4, 415, 521, 525, *534*, **549-59**
oxygen radicals 37

^{32}P 202-9, 657, 727-8, 735
P-loop 378, 543
p160 coactivator 602
p21 *723*
P22 Arc repressor **574**
p53 **600**, *724*

packing density, protein **389-91**, 435, 440, 457-8, 484, 490, 495, 500-3
padlock probe **166-7**
PAGE (polyacrylamide gel electrophoresis) 201, 207, 209, 218
pair-distance distribution function **21**
PALA (*N*-phosphonacetyl aspartate) **675-7**, 710, **713-5**
palindromes *xxxiii*, 64-5, 154-5, 576, 582, 589-91, 595-7, 602-6
palladium 100
palmitoyl groups **304-5**
parahelices *372*
parallel DNA duplexes **66-8**
paralogous **314-7**, 324, 415, 417
partial volumes **70**, 231, 260-1, **438-40**, 456-8
partition coefficient 35, 104, **261-2**
partition function 104-5, 351-2
partition ratio 684
parvalbumin *320*, *324*, 520, **536-8**
pBR322 93-5
PCNA *724*
pectate lyase **380**, 423
pentraxin 399
pepsin, pepsinogen **286**, **739-40**
peptide bond *xxxiii*, **9**, 14, 227-8, 232, 242, **244**, 258-9, 261, 267, **272-82**, 285-9, 299, **330-4, 337-9, 343-9**, 360, **368-9**, 372, 375-7, 410, 481, 505-6, 518, 625, 701, 740-41
peptide library **278-80**
peptide mapping 249, **285-7, 289-91**, 293, 727-8
peptide nucleic acids (PNA) **170-6**, 190
peptide sequencing 280, **293-306**
peptide synthesis 172, **271-80**, *383*
peptides **227-328, 329-61**
perfect enzyme **688-91**
performic acid 248, 282, 291-3, 336
periodate 108
periplanar conformation **14-5**
permanganate 81
peroxidases 137, 168, 259, 403, **439**
peroxide 255, 532, 571, 628
persistence length *21*, **22, 71-2**, 113, 175
perturbation theory *150*, 502
phage, *see* bacteriophage
phage display *280*, *423*, *520*
phaseolin **399**

phenol 268, 282
phenol reagent 268
phenolic groups 256-7, 525, 693-5
phenylalanine (Phe) residues **231**, **255**, **260-2**, **265-6**, 282, **286-7**, **302**, 306, **317-8**, 347, *352*, **355**, **390**, 392, 411, **419**, **454-5**, **459-60**, **470**, 584, 598, 615, *621*, 663, 693-5
phenylisothiocyanate (PTC) 281-2, **295**
phenylthiocarbamyl (PTC) **295-6**
phenylthiohydantoin (PTH) **295**
phenylxanthenyl (Px) protection **210-1**
phi value **508**
phosphatases, protein **721-4**, **729-32**
phosphate groups 27-8, 31, 50-2, **68-70**, 75, 103, 145, 190, 202, 204, 578, 596, 607-8, 615, 625, 640, **671**, 675-6, 678, **698-703**, 726, 729
phosphines 249
phosphite 730
phosphite triester **212**, 216
phosphodiesters 27-9, 51, 84, 86, 108, 130, 166, 172, 181-2, 204, 211, 216, 220, 584, 611, 698-9, 735
phosphodiester synthesis approach 211
phosphodiesterase 207, 731, 735
phosphoenolpyruvate 689, 717
phosphofructokinase (PFK) 366, **399**, **402-3**, 707, **717-20**
phosphoglycerate dehydrogenase 709
phosphohistidine 722, 728
phospholipase 418
N-phosphonacetyl aspartate (PALA) **675-7**, **710-7**
phosphonamidates **672**
H-phosphonate coupling **213-6**, 220
phosphopantetheine 641
phosphopantetheine adenylyltransferase **561-2**
phosphoramidates 721, 735
phosphoramidite coupling **212-7**, 220, 225-6
phosphoribosylaminoimidazole carboxylase **396**, 399, 402
phosphoric acid 213, 727
phosphorothioate 190-1, 207, 216, *220*
phosphorous acid 213
phosphorylase kinase 728-32, 728
phosphorylation 205, 209, 305, 706, **721-33**, 735
phosphotriester coupling **211-3**, 215-6, 220, 222

photo-cross-linking **79-80**
photosynthetic reaction center 409, 411-2, 428
photochromism *238*
o-phthalaldehyde **269-70**
phthalimide 74, 216
phylogenetics 116, 131, **134-6**, **307-9**, 314-5, *422*
Pi (Π) DNA 66
pi (π) electrons 33, 44
pi turn **125**
ping-pong mechanism **639-46**, **651-8**, 666, 680-1
piperidine 84, 274
Pit-1 594
pitch, helix **17**, **49**, 52, 54, 92, 99, 176, 357, 359,
pivaloyl chloride 214
plasmid *xxxiv*, 60, 86, *89*, 93, 102, 133, 141, 188, 191, **193**, 199, **219**
plasmin, plasminogen 385-6, **467**
plasminogen activator inhibitor 512
plastocyanin **319**, **522**, 534-5
platinum 99
pleated sheets, *see* β-sheets
plectonemic superhelix *xxxiv*, **90-1**, 95, 99
point accepted mutation (PAM) matrix **311**
point group **398-400**
polarizability 44, *53*, *127*, 245, 445, 535, 538, 697
polarized light *xxxiv*, 4-5, *79*
pol I, II, III, IV **186-7**
poly-(Ala) 22
poly-His tag 327
poly(A) 57, 188,
poly(A) tail 303, 614
poly(A)•poly(dT) 52
poly(A)•poly(U) **48**, 52
poly(dA-dC)•poly(dG-dT) 48
poly(dA-dG)•poly(dC-dT) 48
poly(dA)•poly(dT) **48**, 52
poly(dA-dT) 47, 52
poly(dA)•poly(U) **48**, 52
poly(dG-dC) 48
poly(dG)•poly(dC) **48**
poly(dGdC) **52**, 70
poly(dI)•poly(C) 48
poly(dIdC) **52**, 70
poly(dT) 188
poly(dT)•poly(dA)•poly(dT) **48**
poly(I)•poly(C) **48**

poly(I)•poly(dC) **48**
poly(T) 194
poly(U) 57
poly(U)•poly(A)•poly(U) **48**
polyacrylamide 199, 203, 207, 267, 270-1, 290, *299*, **568-70**
polyamines 51, **54**, 69, 126, 133, 270, *609*
polyamino acids 227, 334, 338, 342, 344-5, 349, 351
polydimethylacrylamide (PDMA) 274
polyelectrolytes 28, 51, **68-71**, 76, 126, 145-6, 577-8, 581
polyethylene glycol *104*, *438*
polyglycine **339**, **344**
polymerase chain reaction (PCR) 170, **182-5**, *190*
polynucleotide kinase 204-5, 703
polynucleotides **26-40**
polyproline **339-40**, **344-5**, *372*
polyprotein **304**
polystyrene 172, 216, 225, 277
polytopic *503*
ponceau S 267, **271**
porin **399**, **410**, 413
porphobilinogen synthase 708
positive cooperativity 544-6, 551, 560, 707-8, 725
positive selection **321-4**
positive-inside rule 410, 428
post-transcriptional processing *xxxiv*, **128**, 205
post-translational modifications *xxxiv*, 106, 227, 232, 280, 289, 293, 299, **302-6**, 327, 540, *727*
potassium channel 412
potential energy functions *430*
POU domains **593-4**
PRDI 583
prealbumin **396-7**
predicting structures *98*, **132-8**, 152, 341, 352, *393*, *403*, *405*, *406*, *407*, 412, *413*, *414*, **424-33**, 504, *539*, 542
preferential binding, hydration 69, 144, 414, *438*, 458, *472*, 496-7, 546, 558, 580
prephenate dehydrogenase **634**
pressure 139-40, 363, 436, 438, 448, 450, 455, **457-9**, 463, 465, **478-9**, 485, *491*, 498, *742*
primary structure *xxxiv*, **24**, **111**, 113, 127, 205, 227, 300, **306-28**, **362-3**, 400-1, 413, **415-22**, 427, 431, 458, 483, **498-504**, 534, 542-3, 587

primers **178-9**, **183-5**, 189, 194, **199-203**, 208, 218, 587
primeval world 698
primordial enzyme **689**
pro-proteins **304**, 511
probes, DNA *xxxiv*, 161, **165-70**, 189, 194, 293, 575
processivity **686-8**
prochiral **10-2**,
procollagen **304**, 360-1
product inhibition, enzyme **654-5**
profile table **431-2**
proinsulin **322**, 464
prokaryotes *xxxiv*, 37, 133, 193-4, 205, 306, 327, 566, 575, 587-9, **609-13**, 620, 625, 681, 690, **724**
prolactin 319
proline (Pro) residues **228-9**, **231**, **258-8**, **260-2**, **265**, 269, 282, **286-7**, 302, 318, *333*, **337-8**, 339, 344-5, **347**, **359-61**, *372*, **390-3**, *413*, **419**
prolylendopeptidase **286-7**
promiscuity, enzyme *629*
promoter *xxxiv*, 186-7, 193, 574-6, 581, 583, 593, 597-8, *604*
proofreading 180, 185, 187, 189
propeller twist **43**, 53, 73, 92, 584
propionamide **144**
prostaglandin synthase **410**
prosthetic groups *xxxiv*, 368, 411-2, 485, 525, 547, 662, **681-3**
protecting groups **210-8**, **220-5**, *269*, **271-8**, 686
protection factor 447, **451**
proteins **227-328**, **353-61**, **362-433**, **434-61**, **462-518**, **519-563**, **564-627**, **628-97**, **706-42**
Protein Data Bank (PDB) **24**, **47**, **112**, 362
Protein Identification Resource (PIR) **306**
protein kinases *194*, 618, 675, 717, **721-7**, 731-3
protein phosphatases **722-5**
Protein Research Foundation **306**
proteinase inhibitors **323**, 395, **512-3**
proteinases *xxxiv*, 170, 247, 273, 282, **285-7**, 300-1, 325-7, **385-7**, 395, 401-2, **417-8**, 422, 434, **441-2**, 467, 504, **511-2**, 535, 540, 667, **738-42**

proteoglycans 361
proteolysis **303-4**, 322-5, 361, 401, 464, 467, 511, 706, 723, **738-42**
proteome *xxxv*, *238*, *270*, *280*, *289*, *292*, *300-1*
prothrombin **309**, *386*, 540
protomer **398**, 404, 709
pseudoatom **136**
pseudoknot *xxxv*, **120-1**, 131, 135, 150-1, **189**, 614
pseudolinking number 92
pseudopeptide 171
pseudouridine (Ψ) **110**, **208**
psoralen **79-80**
psychrophile 498, 685
PTB (phosphotyrosine binding) domains 727
purines **26-31**, **35-6**
purine repressor (PurR) 583
putrescine **54**, 126
pX **95**,
pyranose **13**
pyridine 81, 211-4, 223-4, 298
pyridone *40*
pyridoxal phosphate **238**, 640, 663, **681-3**, 697, 720-1, 729-31
pyrimidin-2-one **35-6**
pyrimidine **26-9**, **36**
pyroglutamyl aminopeptidase 294
pyrophosphate (PP_i) 75, 178-82, 542, 657, 665, 691-4, 702, 717, 730, 735
pyrrolidine amide backbone 171-2
pyrrolidone carboxylic acid **243**, 294
pyruvate **11**, **652-3**, 717

quadruplex DNA **60-4**, *143*
quantum yield 35, 79
quarter-staggered array 361
quasi-native 154, 513-5
quasiequivalence 400-1
quaternary structure, RNA *xxxv*, 111, 133
quaternary structure, protein *xxxv*, **362-3**, **394-404**, 434, 474, 504, **516-7**, **545-59**, 561, 582, 594, 620, 706-9, 712, 716, 719-21, 726, 733-4
quenching, fluorescence *68*, *79*, 168, 170, 458-9

R state, allosteric **544-6**, **550-8**, **712-4**, 719, 730, 734-5
$R_0 t$ curves **163**

racemate, racemic mixture **6-8**, 175, 682
racemization 242-4, 273, 276
radial distribution function **20**, **335**
radius of gyration (R_g) **21-3**, **334-5**
Ramachandran plot 29, **329-34**, **339-40**, 342, **370**
Raman spectroscopy *79*, 435, *646*
random coil *xxxv*, **18-23**, 72, 149-50, **333-8**, *345*, **346-50**, 353, 362, 364, 435, **480-4**, 486-7, 504, 514
random enzyme mechanism **639-43**, 646, 655, 658-9, 675, 691, 716, 732
random polymer, *see* random coil
random-flight chain **20-2**
randomness parameter 637
rapid-equilibrium mechanism **641-3**, 645-6, 655, 658-9, 675, 732
reading frame 193, 301-2, 327
real-time PCR 184-5, *190*
RecBCD 687
receptors *xxxv*, **126**, 308-9, 412, 519-20, 530-1, 533, *562*, 578, 588, 600, **602-3**, 704, 709-10, 723, 726-7, 731
receptor Tyr kinases (RTK) 722, 725
recombination, DNA 66, 87, 91, *321*, 564, 609-10
redox (reduction/oxidation) *xxxv*, 395, **533-5**, 541, 549
relaxation time *xxxv*, **336**, 453
relaxed circular DNA **87**, **90-5**, 98-105
renaturation, nucleic acids **139-76**, 183
renaturation, proteins **462-518**
replicase **185**, 619
replication, nucleic acid *xxxv*, 37-40, 75, 78-9, 84, 86, 90-1, 96, 108, 120, *125*, 129, **177-85**, 191, 193, 198, 208, 219, 564, 609, 618, 687, 696, 698, 702
replication protein A **610-2**
reporter genes 615
reporter groups 165, 246, 257
repressors *xxxv*, 64, 395-6, 400, 517, 520, 564, **574**, **576-97**, **605-7**
residence time 78, 438
residual dipolar couplings (RDC) 3, 120, 47, *484*
resonance, chemical *xxxv*, 233, 240, 259, 330-1, 682
restriction enzymes *xxxv*, 64, 75, 184, 192, **195-6**, 218-9, **574-5**, 579

restriction fragments 218
restriction maps **196-8**
retinal 10, 238
retrovirus 189, 531
reverse gyrase 94-5
reverse Sanger sequencing **204**, **207**
reverse transcriptase *xxxv*, **188-90**, 194, 200, 205, 207, **401-2**
reverse transcription *xxxv*, **189-90**, 194
reverse turns 343, **375-8**, 418-9, 432
reverse Watson-Crick base pairs **42**
reverse-phase chromatography 218, 281-3
RGG box **614**
rho protein *725*
rhodamine 202
ribonuclease III **618**
ribonuclease A 206, **284**, **320**, **336**, **404**, *423*, **439**, **442**, **463**, **471**, **473**, **475**, **483-4**, **490**, 515
ribonuclease H 154, 189-90
ribonuclease inhibitor **423**
ribonuclease P 129, **701-2**
ribonuclease Sa *240*
ribonuclease T1 206, **463**
ribonuclease T2 207-9
ribonuclease V$_1$ 142
ribonucleases 205-6, 225
ribonucleoprotein (RNP) 133, **614-7**, **625-7**
ribonucleotides **26**, 205, **208**
ribose 2, **5**, 9, 14, 26, 29-30, 35, 40, 42, 112-3, 165
ribose zipper **124-5**
ribosome 108, 111, 115-6, 119, 125, *129*, 193, 613-4, 619-20, **624-7**, 701
riboswitch **111**
ribothymidine **110**
ribozyme *xxxv*, 108, 117, *125*, **129-33**, 149-50, **153-7**, **698-704**
ribulose bisphosphate carboxylase 635
rifampicin 188
right-handed helices **17**
RNA **108-38**, **139-70**, **185-98**, **205-8**, **220-6**, **613-27**, **698-704**
RNA interference 108, 191
RNA ligase 204-5
RNA polymerase **185-8**, 193-4, 204, 564, 596-8, 610, *651*, 703
RNA replicase **185-6**
RNA world 111-2, 698

RNA•DNA hybrids **48**, 54, **149**, 154, 160, 164, 168, 189
RNP (ribonucleoprotein) domain 133, **614-7**, **625-7**
roll, base-pair **43**, 73, 92
β-roll **380**
rolling circle DNA replication 699
Root effect 323
Rossman fold 541
rotamer library 429
rotamers **16**, 339, **371**, 697
rotary shadow **99-100**
rotational diffusion 10, **71-3**, 336-7, 453-5
rotational isomeric state 19
Rouse-Zimm theory *23*
RRM (RNA recognition motif) 616
rubredoxin **522**
Ruhemann's purple **268**

^{35}S (sulfur-35) 202
S-adenosyl-methionine 37, **605-6**, 718-21
S1 domain **614**
S1 nuclease 84, 142, 164
S1 ribosomal protein 613
salt bridges 338, 392, 395, 443, 445, 477, 498, 500-1, 556-8, 607, **740**
Sanger DNA sequencing method 199-204, 207
sarcosine **144**, 496
satellite RNA 129, **698-9**, 701
scaffolding proteins *343*, *668*, 725
scanning calorimetry **467**, 472
scanning force microscopy *100*
scanning transmission electron microscopyy *567*
Scatchard plot 636
Schiff base *13*, **238**, *667*, **682-3**, 697, 729
SCOP database 408
SCOR (structural classification of RNA) database *120*
screw rotation 17, 398
*Sda*I restriction enzyme *195*
SDS-PAGE **290**, 569
second messengers 535, 724, **731**
secondary structure, nucleic acid *xxxv*, **111-21**, **127-38**, 150-4, 160, 165, 327, 625-6
secondary structure, protein *xxxv*, 338-9, 344-5, 353-4, **362-3**, **369-78**, 380, 386-9, 392, 395, **404-11**, 418, 421-2, **425-8**, 458, 460, 484, 503, 514, 542, 611, 615

secretory vesicles, granules 304
sedimentation *xxxvi*, **99-101**, 567, 625, 712
selection, natural 307-9, 317, **321-4**, 421, *497*, 501, 511
selection marker gene **193**
selenocysteine 128, **136**, **259-60**
SELEX 703-4
self-splicing introns 120, 127, **129-32**, 151, **699-700**
seminal ribonuclease **404**
sensor kinase/response regulator 723
sequence alignment **136-7**, 158-9, 312-3, *317*, 429
sequencing DNA **198-205**
sequencing peptides **280-306**
sequencing RNA **205-8**
sequential allosteric model **544-6**, 559, 708-9, 734
sequential enzyme mechanism **639-40**, **658-9**
Ser/Thr protein kinases *194*, **722-6**
serine (Ser) residues **232-3**, **260-2**, **264-6**, 282, 288-9, **302**, 305, **317-8**, 340, **347**, **355**, 358, 376, **390**, 417, **419**, 426, 436-7, **441-2**, 446, 615, 663, 678, 718, **721-6, 729-35**, 737, 738-40
serine carboxypeptidase 422
serine hydroxymethylase 682
serine proteinases 286, *305*, 417-8, **441-2**, 467, 511, 535, 540, 667-9, **738-40**
serine stack 380
serpins 325, **512**
serum albumin 326, *439*, 457
serum response factor (SRF) **584-5**
seryl-tRNA synthetase 620, **623**
seven-helix fold *358*, 412
SH2 domains **725-6**
shear number 373, **406-7**
sheared base pair 162
shearing DNA 71, 163
β-sheets *338*, *339*, 342-4, *351*, 358, 369, 372-87, **396**, **405-8**, 418, 423-7, 447, 456, 458, 491, 503, **512**, **514**, 541, 543, **605-8**, 615-9, 721, 720
Shine-Dalgarno sequence **192-3**
shotgun DNA sequencing 203, *207*, *300*
side-chain 1-2, 23, 175, 227
side chains, amino acid **227-324**, **329-61**, **364-96**, **406-18**, **427-38**, **446**, **452-9**, **469-70**, 477, 483-4, 500-1, 522, 525, 530, 533, 535-42, 547-50, 556-9, 577-8, 582-91, 599-617, 662-5, 682-6, 693-7, 709, 718, 721, 725-6, 732, 739-41
sigma factor **187**
signal peptide **303-4**
signal recognition particle (SRP) *118*, **614**
signal transduction 28, 423, *537*, **544**, 602, 704, 705, 709, 721, **724-5**, 737
silica gel 215
silk **357-9**
silver 247
silver stain 267, *270*
similarity index 310
simulated annealing 429
single-strand DNA-binding proteins **609-13**
single-stranded oligonucleotides 22, **28-30**, 40, 60, **72**, *75*, 77, 79, 81, 84, 108, 111, 113-35, **141-6**, **149-56**, 175, 178-9, 182-3, 190-1, 195-9, 204-5, 208, 218-9, 567, **609-18**, 687, 704
site-directed mutagenesis **218-9**, 327, 498-9, *548*, 577, **663-6**, 693, 728
skatole 258, 289
skin 79, 359-61
sliding clamps **687-8**
slot blot 166
slow-binding enzyme inhibitors **659-62**, 671, 684
Sm domain **614**
Smad 588
small nuclear RNP (snRNP) 133, 186-7, 614
small-angle neutron scattering *438*
small-angle X-ray scattering *484*
snake venom phosphodiesterase 207
snake venom toxins 320
snRNA (small nuclear RNA) 186-7, 614
sodium dodecyl sulfate (SDS) 497, 503, 728
β-solenoids 380
solid-phase folding **517-8**
solid-phase sequencing **296**
solid-phase synthesis 172, 212-4, **216-8**, 220, **222-6**, 272, **274-80**, *383*
solubility 22, **144-5**, 261-2, 269, 353, **436-40**, 464, **469-70**, 498-9, 532
solution-phase synthesis 211, **215-6**, 220, **222-3**, **271-7**
solvation *13*, 17, **22-3**, 31, **69-71**, *127*, 245, 263, 333, 338, 341, 344-5, 353, 375, 388, **392-4**, 432, 436, 440, 445, 456-8, 473, 467, 475, 482, 492, **494-9**, 521-3, 577, 580, **630-1**, 697

solvent 12, 22, *54*, 491
solvent, cryogenic *456, 502*, 685
solvent, crystalline 394, 452
solvent, denaturation *145*, 474, 496-7
solvent, theta 22-3, 481
solvent exchange 76, 240, 438, **447-51**, 484
solvent exposure 44, **52-4**, 68, 79, 130, 138, 181, 245, 347, *348*, 372, 388, 392-4, 402, 412, 432, 436, 440, 447, 459, 469, **473**, 484, 492, 494, 521, 523, 532, 540, 549, 552, 564, 579-80, 611, 623, 662, 740
solvent penetration 438, **446-7**, **449**, 458
solvent perturbation 459
sonication 71
sorbitol 496
southern bean mosaic virus 401
Southern blot 166
soybean trypsin inhibitor **439**, 457
SP6 phage 188
specific linking difference 94-5
spectral properties, nucleic acid 33, 140
spectral properties, protein **255-8**, **459-60**, **481**, 666, 742
spermidine **54-5**, 126, 130
spermine **54**, 69, 126
spine, hydration 53, 68, 77
spliceosome **614**, 701
splicing RNA *xxxvi*, 108, 120, 127, **129-33**, 151, 190, 205, **699-701**
Src 725-6
SRP (signal recognition particle) *118*, **614**
SRY **588**
stabilizers **496-7**
stacked X structure **66-7**
stacking, base **42-6**, 49, 52-3, 55, 57, **66-7**, 72, **74-5**, 78, 112, 116, **118-23**, 125, 127-8, 131, 134, 140, *143*, 150, 154, 163, 165, 175-6, 583, 610-1, 615, 617, 621, 623
staggered end **195**
stain 99, 141, 169, **267-71**, 293, 361
stammer 357
staphylococcal nuclease 445, **476-7**, 483, *487*, 535, **629**
star activity **196**
starch 686-7
start codon **193**, 302
statistical segment **22**
staufen **614**, 617
steady-state enzyme kinetics **633**, **640-3**, 648, *656*, **660-1**, 666, *691*, 711, 718-9, 732
stellacyanin *535*
stem-loop **113-7**, *125*, 133, 168, 621, 623, 701
stereochemistry **3-13**
stereoisomers *xxxvi*, 2-3, 7, 13, 171
steroid hormone receptors **602-3**
sticky ends *xxxvi*, **87-8**, **184**, **192**, **195**
stop codons **192-3**
strain, conformational 29, 55, 90-1, 103, 248, **251-3**, *404*, 492, 495, 511, 694
β-strand **342-4**, **358-387**, **396**, **404-8**, **413-4**, **418-9**, **422-6**, 500, **512**, **514**, 521, 534, **541-3**, 588, 600, **605-7**, **611-4**, **616-8**, 729, *740*
strand invasion **173-5**
streptavidin 168
stretching 31, **71-8**, **150**, **155-7**
stringency 161, **167-8**
structural proteins 305, 325, 363, 397, 423, 436, 520
structure prediction, protein **424-33**
structure prediction, RNA **134-8**
stutter 357
submaxillary proteinase 287
substrate enzyme inhibition **652-4**, 657, **659**, 675, **707**
subtilisin 422, **445-6**, 535, 665
subtractive hybridization *195*
succinic anhydride **235-6**, **283-4**
succinimides **235**, **244**, 257-8, 289, 292
suicide substrates **683-5**
sulfate-binding protein **523**
sulfates 496
sulfation **304**
sulfenate 248
sulfhydryl (thiol) group 6, 235, 237, 240, **245-55**, 259, 275, 277, **282-4**, 286-7, **291-3**, 305, 348, 442, 464, 480, 492-3, **513-5**, **525-6**, 531, 534, 712
sulfinate 248
sulfite **249**
sulfones **255**, **291-3**
sulfonic acid 172, 235, 248, 282
sulfonium salts 246, **254**, **258**, 292
sulfoxides *145*, **255**, 685
sulfur atoms 207, 393, 530, 704
SULT1A1 *654*,
supercoiling 55, 60, *64*, 66, **87-106**, 357, 423
supercooled water *476*

superhelix **90-107**, 354, 406, 423
superoxide dismutase 397, **403**, **522**
superoxide ion 532
supersecondary structures 362, **378-86**, 422
supershift 569
surface plasmon resonance *567*
surface areas 37, 44, **52**, **260**, **263-6**, 342, 366, **388-9**, 392, 395, **402-3**, 440, 457, **470**, **473**, 475, 491, 494, 561, 579-80, 583
surfaces, DNA 37, 44, **50**, 53, **68-71**, *97*, 103, 144, **564-66**
surfaces, proteins 322, **341-3**, **354-6**, 360-1, 364, 366-8, 372, 375, 378, **388-97**, **402-6**, **409-23**, 429, **434-48**, 452-3, 456-7, 460, **469-77**, 480, 484, 486, **490-504**, 508, 515, 521, *525*, 534-5, 538, 572, 575, 582, 586-7, 597, 599, 602, 606, 610-1, 613, 615-9, 622, 662, 687, 709, 736, 738, 740
surfaces, RNA 127, 138
SV-11 RNA **154-5**
SV40 virus 101, 610
SWI5 600
SWISS-PROT **306**
symmetry 2, 4, 7, 20, 29, 41-2, **47-8**, 61-2, 64, 66, 106, 114, 118, 137, 195, 228-9, 232, 256, 332-4, 344, 384, **395-403**, 407, 411, 421, 435, 452, 455, 460, **545-55**, 560-2, 575-6, 584, 589-91, 596-7, 603-6, 619, **622**, 640-4, 687, 693, 715, **718**, **720-1**, 730, 733, **736**
syn conformation **14-5**, 27, **30**, 55
*syn*clinal conformation **14**
synergy 500, **527-9**, 640, 694, 715
*syn*periplanar conformation **14**
synthesizers 216, 272

T arm, loop, tRNA 119, **127-9**, *624*
T loop, kinase 723
T state, allosteric **544-59**, **712-9**, 730, **734-5**
T-even bacteriophages 38
T4 bacteriophage 38, 182, 204-7, 420
T4 lysozyme *452*, 477
T7 phage 186, 188, 194, 201, *280*
TAF (TBP-associated factor) **588**, 598, 608
Taq DNA polymerase **200-1**
TAR RNA *157*
tartaric acid **4**
TATA box, binding protein (TBP) 193, **574**, 579, 581, **584**, **588**, **598-99**

tau protein 484
taurinomethyluridine *220*
tautomers *xxxvi*, 3, **12-3**, 29, 33-4, 42, **75-6**, *238*, 240
TBDMS protecting group 220, **222**, **224-5**
TBP (TATA binding protein) **574**, 579, 581, **584**, **588**, **598-99**
TCF-1 **588**
telomerase, telomere **60-3**, *220*, 613
temperature factor 74, 393, **452-3**, 456, 471, 665
tendon 359, 361
tertiary structures *xxxvi*, 47, 90, **111**, **118-34**, **136-8**, 151-5, *236*, 353, **362-3**, **366-9**, 385-6, 400, **405-9**, **415-25**, 427, 431, 434, 458, 484, 503, 511, 516, 558, 615, 620, 625
tetrahedral atoms **4-13**
tetrahedral intermediates **670-2**
tetrahedral symmetry *xxxvi*, **397-9**, **526**, **530-1**, 534, 712
tetrahydrofuran 214
tetrahydrofuryl group **35**
tetrahydrophthaloyl anhydride 236
Tetrahymena ribozyme **130-2**, **154**, **156-7**
tetraloops 117-8, *120*, **126**, 153, **617-20**
tetramethyl guanidine **211-2**, 215, 223
tetranitromethane 257, 441
tetrazole 212-3, 216, 224
TF1 **588**, 608
TGGE (transverse gradient gel electrophoresis) 95
Theorell-Chance mechanism 640
thermal expansion **456**, 458
thermochromism *238*
thermodynamic approach, RNA structure **134-5**
thermolysin **286-7**, 422, **671-2**,
thermophiles 182, *190*, 200, **498-9**, 501, 583-4, *677*, 691
theta solvent **22-3**, 481
thiacytidine **39**
thiamine pyrophosphate **111**
thin-layer chromatography **207-8**, 291
thin-layer electrophoresis 291, 728
thio-dATP **202**
thioguanine **39**
thiohydantoins 296, **298-9**

thiol (sulfhydryl) group 6, 235, 237, 240, **245-55**, 259, 275, 277, **282-4**, 286-7, **291-3**, 305, 348, 442, 464, 480, 492-3, **513-5**, **525-6**, 531, 534, 712
thiol-disulfide exchange 246, **248-54**, **492-3**, 513
thiolate ion **245-54**, 275, 534
thiophosphate **204**, 207
thiouracil 38
thiouridine (s⁴U) **110**, 208
Thp protecting group **220-3**
threading protein sequences 425, **430-2**
three-hybrid system 615
threonine (Thr) residues **229**, **231-3**, **260**, **262**, **264-6**, 282, 288-9, **302**, 305, **317-8**, **336**, 340, **347**, **355**, 376, **390**, **419**, 446, 457, **470**, 542, **695-6**, **721-8**, 732
threonine synthase 401, **720-21**
threose 7
thrombin, prothrombin **309**, 322, 385-6, 417, 540
thymine phosphorylase *651*
thymine, thymidine **26-42**, **56-8**, 60
tilt, base-pair **43**, **47-9**, 53, 92
TIM barrel **373**, **405-8**, *422*
tipping, base-pair **43**
tissue plasminogen activator **309**
titin 405
titration curve, pH 445, 476
T_m (melting temperature) **116**, **142-4**, **146-9**, **158-61**, 168, 173, 175, 184, **471-2**, 489
TNBS (trinitrobenzene sulfonate) **235**
tobacco mosaic virus (TMV) 397
toluene 255, 261
tomato bushy stunt virus *185*, **382**, 401
topoisomers **86-107**
topoisomerases 93, **95-6**, 100-4
topological bond 90-1,
topological domain **86-92**, 103, 107
toroidal superhelix **90-1**, 99
torsion angle **14-19**, 29-30, 74, **248**, 253, **329-30**, 334, 337, **339**, **364-5**, **371**, 418
torsional persistence length 72
tower helices **730-3**
tramtrack **588**
trans conformation 3, 7, **9-10**, 14, **330-4**, **337-8**, **344-5**, 361, 368, 481, **505-6**, 518
transcription *xxxvi*, 55, 57, 60, 86, 90, 96, 108, *120*, 128, 133, *153*, **186-91**, 193-4, 204-5, 564, *573*, *590*, 592-4, 596-8, **602-4**, 615, 724, 738
transcription factors *xxxvi*, 186-7, 193, 529, 531, 564, **574-5**, 579-80, 586, 593-4, 596, **598**, **600**, 604, 726
transfer RNA (tRNA) 29, 57, **110**, 113, 116, 119, **127-9**, 138, 149-50, **153**, 186-7, 189, 205, *208*, 614, 617, **620-7**, 656, 691-6, 701-2
transferrin 325, **533-4**
transition metals *xxxvi*, 84, 532
transition state *xxxvi*, 76, 84, 152, 175, 246, 250, *377*, 455, **506-11**, 515, 628, **630-2**, 638, **663-6**, **668-74**, **691-4**, **679**, 685-6, **688-9**, 696, **699**, *701*, **702-4**, *719*
transition state analogs **669-73**, **675-8**, 680, **696-7**
translation *xxxvi*, 108, 120, 127, 186, **189-93**, **301-5**, 327, 564, 618-20
translational diffusion *xxviii*, *21*, *73*, 75, *89*, *117*, 158, 170, 337, 413, 434, 441, 460, 521, 552, 556, 566, 579, *593*, 631, 634, 686, 688-9, 697, 709
transmembrane polypeptides *383*, **410-4**, **427-8**, 503, 723
transposable element 64
transverse relaxation time (T_2) 453
trefoil **89**
β-trefoil **384-5**
trehalose *456*, 496
trialkyloxonium salts 241
tributyl phosphine 249
trichloroacetic acid (TCA) 217, 224-5, 497
triethylamine 225, 279, 281
triethyloxonium tetrafluoroborate 241
trifluoroacetic acid (TFA) 232, 274, 279, 281, 297
trifluoroethanol **353**
trifluoromethanesulfonic acid 172
trihydrofluoride 225
trimethylamine-*N*-oxide 496
trimethylsilanolate **298-9**
(trimethylsilyl)ethanol 275
trinitrobenzene sulfonate (TNBS) **235**
triose phosphate isomerase (TIM) **373**, **405-8**, *422*, **629**, **670**, 689
triostine 57
triple helices (triplexes) **57-60**, *80*, 145, **154**, **172-6**, 305, **359-61**

tRNA synthetases *129*, 560, 614, **620-3**, 656, **665**, **691-6**
tropomyosin **355-8**, 405, **439**, *539*
troponin 371, 405, **537-9**
trp repressor **395-6**, 517, **574**, 579, 582, **595-6**, 605
trypsin inhibitors **284**, **293**, **364-5**, **367**, **387**, **391**, **439**, **448**, 457
trypsin, trypsinogen 247, **286-7**, 293, 300, **319**, 325, **382**, 417-8, 422, **439**, **441**, 452, 457, 511, 535, **738-40**
tryptophan **595-6**, *695*
tryptophan biosynthesis *326*, 690
tryptophan (Trp) residues **229**, **231**, 255, **257-8**, **260**, **262**, **264-8**, 282, **286-7**, 289, 292, **302**, **311**, **318**, **347**, **355**, 376, **390**, 410-1, **419**, **440**, **443-4**, 455, 458-60, **470**, 531, 615, *718*
tumbling, DNA **73**
tungstate 268
tunneling, intermediates 690-1
turnover, enzyme **633**, **635**, **637-8**, 661, 666-7, 684, 698, 701, 708
β-turns *343*, 358, **375-80**, 385
twist (Tw), DNA **43**, 53-4, **71-7**, **88**, **90-107**, 119, 584-5
twist, β-sheet 342, 358, 372, 374-5, 378, 406-8, 729
two-component system 724
tyrosinase *652*,
tyrosine kinases **723-8**
tyrosine (Tyr) residues 95, **229**, **256-7**, **260**, **262**, **264-8**, 282, **286-7**, **302**, **304-6**, **311**, **317-8**, **347**, **355**, **390-1**, 410-1, **419**, **440-2**, **454-5**, **459-60**, **470**, 476, 525, 531, 533, 615, 663, 697, 715, **721-30**, **736-8**
tyrosyl-tRNA synthetase 560, **665**, **691-5**

U1 spliceosomal protein **614**
UBF **588**
ubiquitin 7, *375*, *377*, *459*, 685
UDP-glucose 733
ultrabithorax 581
uncompetitive enzyme inhibition **646**, **648**, **650-5**
underwound superhelix 90, 93-5, 101, 585
unfolded proteins *19*, *21*, 227, 282-3, 332, **334-8**, 345, 353, 360-1, 389, 434, 453, 531, 534, 582, 741

unfolding, nucleic acids 108, 127, **139-57**, 203
unfolding, proteins 14, 140, 254, 345-6, 349, 353, 360, 424, 435, 437, 440, 445, **447-51**, **462-518**, 741
unperturbed random coil, state **19-23**
upstream 188, 193, 598
UpU 44
uracil, uridine **26-9**, **31-2**, **35-40**, **42**, 75, 81-2, **108-10**, 112, 139, **208**, 223, 225, 613
uranyl acetate 99
urea **144-5**, 150, 203, 239, 284, 448, 463-4, **468-71**, 473-4, 481-2, 488, 497, 499, 503-4, 517, **582**, 712
urethane **272**, 274, 276
uridine turn **125**, **130**
uridylyl transferase **737-9**
uridylylation **737-9**
urokinase **309**
uteroglobin **403**
UV absorbance **33**, 37, **80**, **140**, 144, *153*, 157, 168, 245, 253, 255-6, 258, 267, 282, **459-60**, **463**, 481, 566

V-8 proteinase **286-7**
V-system **707-10**
valine (Val) residues **229**, **231-2**, **260**, **262**, **265-6**, 273, 281, 287, **302**, 306, **311**, **317-8**, *333*, 340, 347, **354-5**, **371**, **390**, **392**, 411, **419**, 615
valyl tRNA synthetase **695-6**
Van der Waals interactions 44, 270, 338-9, 389, 462, **491-5**, 501, 538, 547, 572, 577, 582, 587, 591, 615
Van der Waals radius 25, 69, 106, 179, 189, 366, 370, 389, 412, 438, 520, 601
Van der Waals surface 103, 367-8, 416, 616
Van der Waals volume **231**, **260**, 389, 423, 438, 440, 548-9
van't Hoff plot 468, **506**
vanadate **671**
Varkud satellite ribozyme 698, 701
vectors, cloning 188, **191-4**, 199, 203, **218-9**
vectors, expression **192-4**, 327
vicinal hydroxyl groups *xxxvii*, 108
vinyl pyridine 285
vinylglycine 684
virtual bond 19-20, 334, 367

viruses *xxxvii*, 24, 39-40, 101, 108, 113, *125*, 131, 141, 185, 189, 195, 199, *205*, *207*, 280, 304, 382, 397, **399-401**, 407, 424, 531, 609-10, 613, 617-8, *620*, *651*, 698
viscosity 140, 337, 456, **463**, **477**, **482-3**
vitamin K 540
volume fluctuations **439**, 458

water 12-3, **35**, 44, 49-53, 56, *64*, **68-71**, 75-7, 79, 113, 117, 126, 144-6, 244-5, **261-6**, 269, 273, 286-9, 333, 336, 338, 343, 346, 352-3, 358, 372, **388-96**, 407, **409-15**, 427, **436-41**, 444-5, **447-51**, 456-9, 462, 464, **469-76**, 479, 482, **490-503**, 519, 521, *525*, 537, 548, 552, 572, **578-81**, 587, 595-6, *599*, 615, 631, 638, *669*, 677-83, 697, 701-3
Watson-Crick 25, **41-2**, 46, 52, 56-8, 61, 66-8, 88, 112, 115-20, 123-5, 133-6, **147-8**, 171-5, 178, 572, 621, 625, 701
winged helix fold **588-9**
worm-like chain 72
writhe (Wr), topological **96-103**, **107**
wyosine (Wyo) **110**

X-ray crystallography 7, 8, 24, **46-8**, 53, *57*, 74, 106, 364, 370, 409, 428-9, 451, *453*, 471, 585, *618*, 625, *654*, 685, 727
X-ray scattering *484*
X-ray absorption spectroscopy 525
xanthine **82**
xylulose kinase *680*

yeast 327, 615, 690, 716, 723
yeast artificial chromosomes 191-2
Young modulus 72

Z-form DNA 30, **46-50**, **54-6**, 69, 72, 93, 99
Zif268 **600-1**
Zimm-Bragg model 150, **349-52**
zinc fingers **572**, **588**, **599-602**, 610, 614, 625
zinc ions 240, 247, 286, 520, **526-32**, 534, 588, **599-604**, **712-3**, 742
zinc proteinases 286, **742**
zipper, leucine **354**, *358*, 578, 590, **604-5**, 625
zipper, ribose **124-5**
zippering 158, 160, 345, 361
zwitterion *xxxvi*, 228, 498, *669*
zymogens 540, **738-42**